58th International Conference on Experimental Stress Analysis (EAN 2020)

AA005059

Online, Czech Republic
19-22 October 2020

Editors:

Martin Fusek
Jakub Cienciala
Michal Korinek
Vojtech Machalla
Jiri Smach

ISBN: 978-1-7138-2783-2

Printed from e-media with permission by:

Curran Associates, Inc.
57 Morehouse Lane
Red Hook, NY 12571

Some format issues inherent in the e-media version may also appear in this print version.

Copyright© (2020) by Czech Society for Mechanics (CSM)
All rights reserved.

Printed with permission by Curran Associates, Inc. (2021)

For permission requests, please contact Czech Society for Mechanics (CSM)
at the address below.

Czech Society for Mechanics (CSM)
Dolejskova 5
Praha 8
182 00
Czech Republic

Phone: +420 266 053 045
Fax: +420 286 587 784

csm@csm.cz

Additional copies of this publication are available from:

Curran Associates, Inc.
57 Morehouse Lane
Red Hook, NY 12571 USA
Phone: 845-758-0400
Fax: 845-758-2633
Email: curran@proceedings.com
Web: www.proceedings.com

EXPERIMENTAL STRESS ANALYSIS

58th International Scientific Conference

October 19 – October 22, 2020, online, Czech Republic

EAN 2020

58th conference on experimental stress analysis

BOOK OF FULL PAPERS

Editors: Martin Fusek, Jakub Cienciala, Michal Kořínek, Vojtěch Machalla, Jiří Šmach.

Czech Society for Mechanics
VŠB – Technical University of Ostrava, Faculty of Mechanical Engineering

All rights reserved. No parts of this publication may be reproduced, stored in retrieval system or transmitted in any form or by means of: electronic, magnetic tape mechanical, photocopying, recording or otherwise without permission in writing from the publication

Title:	Experimental Stress Analysis 2020 Book of full Papers

Editors: Martin Fusek
Jakub Cienciala
Michal Kořínek
Vojtěch Machalla
Jiří Šmach

Published by: VŠB – Technical University of Ostrava

The complete set of EAN 2020 Conference Proceedings consists of two parts:

- Book of Extended Abstracts
- USB Flash Drive (Full Papers)

Ostrava, October 2020, first edition

Conference organized by:

Department of Applied Mechanics
Faculty of Mechanical Engineering
VŠB-TU Ostrava

Conference Partners:

Contents

EAN 2020
58th conference on experimental stress analysis

58th International Conference on Experimental Stress Analysis - EAN2020, October 19–22, 2020, Czech Republic

HALAMA R.[1,a], FUSEK M.[1,b]

[1]Deparment of Applied Mechanics, Faculty of Mechanical Engineering, VŠB – Technical University of Ostrava, 17. listopadu 2172/15, 708 00 Ostrava, Czech Republic

[a]radim.halama@vsb.cz, [b]martin.fusek@vsb.cz

Introduction

The Experimental Stress Analysis 2020 ("Experimentální Analýza Napětí 2020" in Czech, EAN2020) is organized with the support of Czech Society for Mechanics, Expert Group of Experimental Mechanics. The 58th continuation of the series of international conferences Experimental Stress Analysis, taking place usually in Czech Republic and Slovakia, held in online, Czech Republic, 19–22 October 2020, under the patronage of the dean of Faculty of Mechanical Engineering VŠB-Technical University of Ostrava, prof. Ing. Robert Čep, Ph.D., on the occasion of the 70th anniversary of FME VSB-TUO.

Basic information

The *goal of EAN2020* is the meeting and presentation of scientific works, sharing professional experience and discussing new theoretical and practical findings. The objective of the conference is to identify the current situation, exchange experiences, and establish and strengthen relationships between universities, companies, and scientists from the field of experimental mechanics in mechanical and civil engineering.

Topics of the conference include experimental research on materials and structures subjected to mechanical, thermal–mechanical, or dynamic loading, including damage, fatigue, creep and fracture analysis. Special attention shall be paid to advances in optical methods and other nondestructive testing techniques (DIC, DIV, ESPI, photoelasticity, X-ray and neutron diffraction, etc.) and sensor technology (strain gages, optical fiber applications, accelerometers, etc.). The conference hosted a competition of young scientists (up to 35 years) on the best presented paper and poster. The winners were financially awarded by the Czech Society for Mechanics and by the University.

Invited Speakers:
Prof. Tasnim Hassan (North Carolina State University, USA)
Prof. Xu Chen (Tianjin University, China)
Dr. Kyriakos Kourousis (University of Limerick, Ireland)

The book of full papers contains 85 scientific contributions. Results of thirty scientific works were presented in the poster session. All accepted contributions will be send for indexing in Scopus database. Thirty selected papers will be sent for peer-review of the special issue dedicated to EAN2020 in Materials journal (MDPI, Open Access) and the successful ones should be published in 2020.

Webpages:
http://ean.vsb.cz/

Webpages of SI Materials:
https://www.mdpi.com/journal/materials/special_issues/EAN2020

Scientific Program:

- Development of experimental methods in mechanics
- Development of experimental methods in biomechanics
- New methods and applications of deformation and stress analysis in mechanical structures
- New methods and applications of deformation and stress analysis in civil and other structures
- Experiment as a tool for the verification of analytical and numerical methods
- Experimental research and prediction of strength, life and operational reliability of structures and devices
- Monitoring of operational loadings and operational states of structures and devices
- Mechanics of advanced materials, composites and 3D print materials
- Methods and means of teaching experimental methods

Scientific Committee of EAN2020
prof. Ing. Nikolaj Ganev, CSc.
prof. Ing. Radim Halama, Ph.D.
prof. RNDr. Miroslav Hrabovský, DrSc.
prof. Dr. Ing. František Holešovský
prof. Ing Stanislav Holý, CSc.
prof. Ing. Josef Jíra, CSc.
prof. Ing. František Plánička, CSc.
prof. Ing. Milan Růžička, CSc.
doc. Ing. Peter Frankovský, PhD.
doc. Ing. Róbert Huňady, PhD.
doc. Ing. Tomáš Návrat, Ph.D.
doc. Ing. Iva Petríková, Ph.D.
doc. Ing. Pavel Padevět, Ph.D.
doc. Ing. Miloš Vlk, CSc.
doc. Ing. Robert Zemčík, Ph.D.
Ing. Karel Doubrava, Ph.D.
Ing. Lubomír Houfek, Ph.D.
Ing. Kamil Kolařík, Ph.D.
Ing. Aleš Lufinka, Ph.D.
Dr. Ing. Roman Růžek
Ing. Jaroslav Václavík

Organizing Committee of EAN2020
doc. Ing. Martin Fusek, Ph.D.
prof. Ing. Radim Halama, Ph.D.
doc. Ing. František Fojtík, Ph.D.
doc. Ing. Zdeněk Poruba, Ph.D.
doc. Ing. Michal Šofer, Ph.D.
Ing. Jakub Cienciala
Ing. Michal Kořínek
Ing. Mgr. Dagmar Ličková
Ing. Vojtěch Machalla
Ing. Zbyněk Paška
Ing. Kateřina Šimečková
Ing. Jiří Šmach

Sponsors of EAN2020 conference

Dewetron - Praha spol s.r.o.	LabControl s.r.o.	TA Instruments
HBP měřicí technika s.r.o.	LABOR machine s.r.o.	X SIGHT s.r.o.
INOVA Praha s.r.o.	LABORTECH s.r.o.	

Proceedings of EAN conferences
http://www.experimentalni-mechanika.cz/cs/konference/konference.html

LCF Properties Investigation of Sn-3.5Ag-0.75Cu Solder Paste

ANTOŠ D.[1,a], HALAMA R.[1,b], MARKOPOULOS A.[1,c]

[1]Deparment of Applied Mechanics, Faculty of Mechanical Engineering, VŠB – Technical University of Ostrava, 17. listopadu 2172/15, 708 00 Ostrava, Czech Republic

[a]daniel.antos.st@vsb.cz, [b]radim.halama@vsb.cz, [c]alexandros.markopoulos@vsb.cz

Keywords: Sn-3.5Ag-0.75Cu, solder paste, low-cycle fatigue, viscoplasticity

Abstract.
This study is focused on research in tension-compression fatigue and uniaxial ratcheting. Strain controlled fatigue tests were done under constant strain amplitude loading considering a constant strain rate on a specially designed specimen. The influence of the mean stress and the stress amplitude on uniaxial ratcheting has been studied for three load frequencies. Results are consistent with the stress-strain behaviour observed under monotonic loading. Aim of this and previous investigations is to obtain experimental data of Sn-3.5Ag-0.75Cu solder paste for calibration of a viscoplastic material model for future finite element analysis and to propose appropriate fatigue damage model for subsequent computational fatigue analysis. First results gained by a non-unified constitutive model are also presented.

Introduction

Tin based materials show strongly rate dependent stress-strain behaviour. Most of the studies which investigate the lead-free solder paste mechanical properties are mainly focused on rate-dependent and thermal-dependent plasticity in monotonic loading [1] or secondary creep properties [2]. Uniaxial low-cycle fatigue was investigated for example by Ohguchi et al [3], but only stabilized hysteresis loops were published. The transient behavior seems to be out of scope for most of scientific studies. There are only few published works emphasized on accumulation of plastic strains due to non-zero mean stress (ratcheting), for example [4], [5].

This study follows the contributions, which were focused on measurement of mechanical properties in monotonic tension, monotonic torsion and creep [6] and low cycle fatigue in torsion of a lead-free solder paste [7].

Experiment description

First, LCF tension-compression fatigue tests were performed for the constant strain rate of 0.015 $mm \cdot s^{-1}$. Special solderjoint specimens were manufactured [6]. Dog bone round cross-section specimens consist of two copper pieces between which 0.5mm of solder paste Sn-3.5Ag-0.75Cu was applied [6]. The number of cycles to failure was evaluated according to the criterion of the percentage drop in tensile force in relation to the level determined during the measurement. The percentage drop value is typically in range from 2% to 30% [8]. In our case, the crack initiation was determined as the number of cycles corresponding to a 20% force reaction drop of extrapolated force over the upper curve of tensile force vs. number of cycles. In the Fig. 1, the main results of the test evaluation are shown, whereas the number of cycles to failure was determined to 85. There is also presented the uniaxial hysteresis loop in half-life on the right side of the Fig.1.

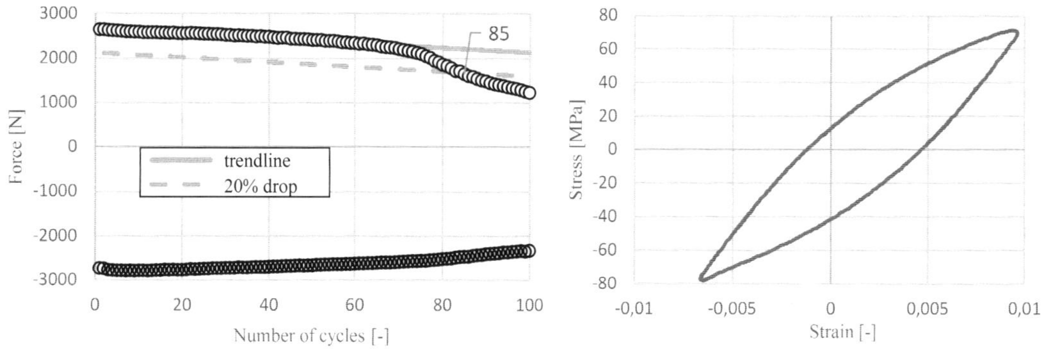

Fig. 1: Evaluation of the crack initialization (left), half-life hysteresis loop (right)

These tension-compression fatigue tests were performed for four strain amplitude levels. Strain amplitudes from hysteresis loops in the half-life were used to construct the e-N curve (Fig. 2). The stiffness of cooper parts was taking into account in evaluations considering Hooke's law with elastic properties of copper (Young modulus E=137.6 GPa, Poisson ratio ν=0.34).

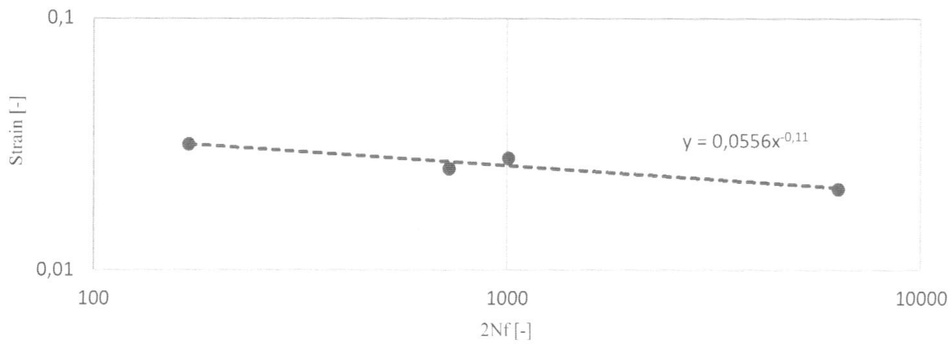

Fig. 2: Lifetime curve for Sn-3.5Ag-0.75Cu of tension-compression fatigue

The second study will presents the influence of mean stress and stress amplitude on the accumulation of axial strain in specimens. Uniaxial ratcheting was investigated for three different loading frequencies 0.25Hz, 0.35Hz and 0.5Hz. The test consisted of seven loading blocks. In the first three loading blocks with duration of 50 cycles per block, the force amplitude was incrementally increased while mean force value remained the same. In the following four loading blocks with duration of 25 cycles per block, the mean force value was step-wisely increased while force amplitude remained constant, see Table 1.

Table 1: Uniaxial ratcheting test conditions

Number of cycles per set	50	50	50	25	25	25	25
Force amplitude (N)	1500	1650	1800	1800	1800	1800	1800
Mean force (N)	200	200	200	250	300	350	400

The influence of stress amplitude on ratcheting in solderjoint was obtained from first three blocks in each test (Fig.3 left). The ratcheting strain is evaluated from measured elongation on 25 mm gauge length after subtracting elastic deformation of copper parts and considering the solderjoint thickness. From following blocks, the dependency of the mean stress on the ratcheting strain rate in solderjoint has been evaluated (Fig.3 right). The impact of mean stress as well as stress amplitude on ratcheting rate is almost linear in particular lower stress intervals.

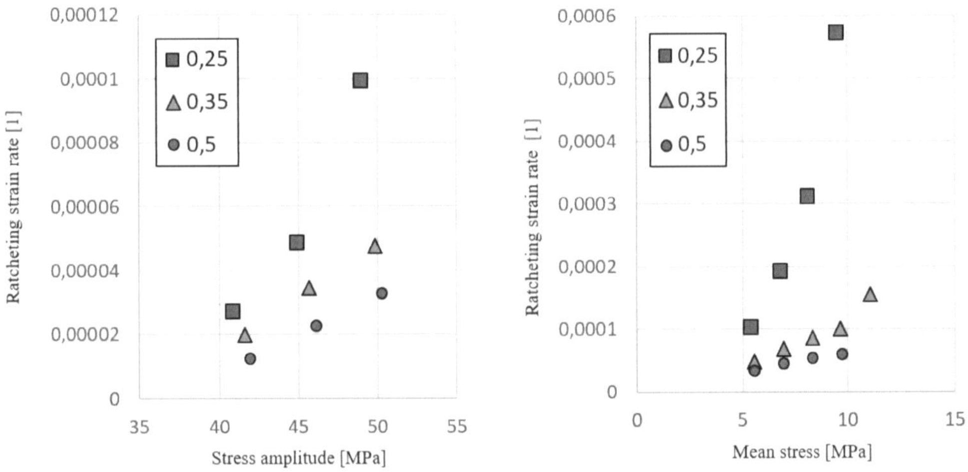

Fig. 3: Dependency of the ratcheting strain rate on stress amplitude (left) and dependency of the ratcheting strain rate on mean stress (right)

Viscoplastic model calibration

Data from described fatigue tests will be used to validate a viscoplastic material model using FE analysis. In this paper, we focused on the calibration of the non-unified material model [9] implemented into the ANSYS 2020R1 by a user-subroutine written in Fortran. AbdelKarim-Ohno kinematic hardening rule was also introduced in order to describe ratcheting behaviour properly. Basic constitutive model equations and material parameters optimized by analytical formulas based on experimental data from literature [3] are stated in Table 2 and Table 3 respectively. The results of predictions are shown in the Fig. 4. The advanced material model for capturing transient cyclic behaviour will be used also in a future study for simulations of fatigue tests presented here.

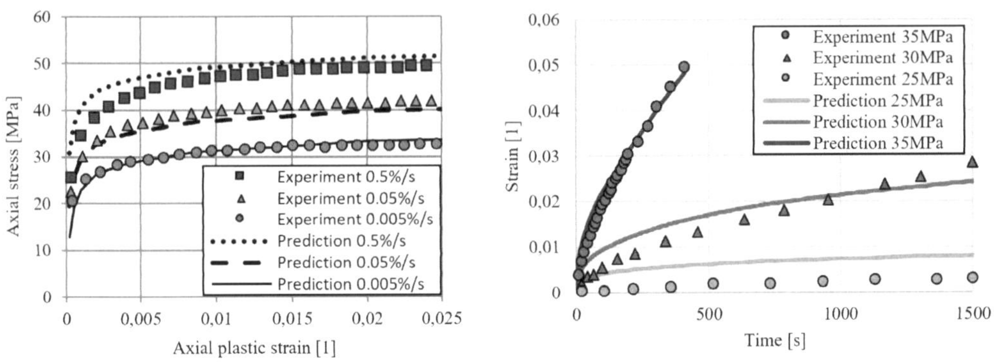

Fig. 4: Calibrated advanced material model considering monotonic tensile curves (left) and creep curves for Sn-3.5Ag-0.75Cu (experiment taken from [3])

Table 2: Viscoplastic non-unified model description

$$\boldsymbol{\varepsilon} = \boldsymbol{\varepsilon}^{e} + \boldsymbol{\varepsilon}^{tr} + \boldsymbol{\varepsilon}^{ss},$$

$$\boldsymbol{\sigma} = \boldsymbol{D}^{e} : \boldsymbol{\varepsilon}^{e},$$

$$\dot{\boldsymbol{\varepsilon}}^{tr} = \frac{2}{3} \cdot g^{tr}(\bar{y}, p) \cdot \frac{\boldsymbol{s}-\boldsymbol{a}}{\bar{y}},\ g^{tr} = A^{tr}\bar{y}^{m^{tr}},$$

$$\dot{\boldsymbol{\varepsilon}}^{ss} = \frac{2}{3} \cdot g^{ss}(\bar{s}) \cdot \frac{\boldsymbol{s}}{\bar{s}},\ g^{ss} = A^{ss}\bar{s}^{m^{ss}},$$

$$\boldsymbol{a} = \sum_{i=1}^{M} \boldsymbol{a}^{i},$$

$$\dot{\boldsymbol{a}}^{i} = \frac{2}{3} h^{i} \cdot \dot{\boldsymbol{\varepsilon}}^{tr} - \mu \cdot \xi^{i} \cdot \boldsymbol{a}^{i} \cdot \dot{p} - H(f^{i}) \cdot \xi^{i} \cdot \langle \dot{\lambda}^{i} \rangle \cdot \boldsymbol{a}^{i},$$

$$\dot{\lambda}^{i} = \dot{\boldsymbol{\varepsilon}}^{tr} : \frac{\boldsymbol{a}^{i}}{h^{i}/\xi^{i}} - \mu \dot{p},\ f^{i} = \frac{3}{2} \boldsymbol{a}^{i} : \boldsymbol{a}^{i} - \left(\frac{h^{i}}{\xi^{i}}\right)^{2},$$

$$\bar{y} = \sqrt{\frac{3}{2}} \|\boldsymbol{s} - \boldsymbol{a}\|,\ \bar{s} = \sqrt{\frac{3}{2}} \|\boldsymbol{s}\|,\ \dot{p} = \sqrt{\frac{2}{3}} \|\dot{\boldsymbol{\varepsilon}}^{tr}\|,$$

where $\langle x \rangle$ means the Macaulay bracket ($\langle x \rangle = (x + |x|)/2$), $H(f^{i})$ is the Heaviside step function and

$\boldsymbol{\varepsilon}^{e}, \boldsymbol{\varepsilon}^{tr}, \boldsymbol{\varepsilon}^{ss}$ - elastic, transient and steady state strain tensors

$\boldsymbol{\sigma}, \boldsymbol{s}$ - stress tensor and its deviatoric part

$\boldsymbol{a}, \boldsymbol{a}^{i}$ - backstress and its i-th part

$\boldsymbol{D}^{e}$ - material elastic stiffness 4th order tensor

$\bar{y}, \bar{s}$ - effective and equivalent stress

p - accumulated plastic strain

M – number of backstress parts

Table 3: Material parameters obtained for Sn-3.5Ag-0.75Cu

E=29500MPa, ν=0.35, $A^{tr} = 7.2e-9$, m^{tr} =4.1, $A^{ss} = 5e-25$, m^{ss} =12, $\mu = 0.1$
$h^1 = 14000MPa, \xi^1 = 2000, h^2 = 4000MPa, \xi^2 = 1000, h^3 = 1800MPa, \xi^3 = 600,$
$h^4 = 675MPa, \xi^4 = 450, h^5 = 500MPa, \xi^5 = 250, h^6 = 375MPa, \xi^6 = 125,$
$h^7 = 200MPa, \xi^7 = 50, h^8 = 20MPa, \xi^8 = 20$

Conclusions

The experimental program of Sn-3.5Ag-0.75Cu solder paste has been briefly described. The low-cycle fatigue tests realized under strain control will serve as a basis for identification of appropriate fatigue criterion. The response of considered lead-free solder paste reveals ratcheting with steady state. The ratcheting tests show important relationship between ratcheting rate and mean stress as well as stress amplitude.

With regard to stress-strain behaviour observed in the experimental campaign the most powerful nonlinear kinematic hardening rule has been stated according to AbdelKarim-Ohno [10]. Separate flow rules have been introduced leading to a non-unified theory. Material parameters of the advanced viscoplastic model were identified using data from literature based on analytical relations. The calibrated material model will be used in future simulation of fatigue tests.

Based on stress-strain history from described fatigue tests the viscoplastic material model will be calibrated for elevated temperatures. The optimization task will be done using the optiSlang software [11] similarly as presented elsewhere for Anand's viscoplastic model [12].

Acknowledgement

This work was supported by The Ministry of Education, Youth and Sports from the Specific Research Project (SP2020/23) by The Technology Agency of the Czech Republic in the frame of the project TN01000024 National Competence Center-Cybernetics and Artificial Intelligence and the Grant Agency of the Czech Republic (GACR) project No. 19-03282S. The support of Continental Automotive Czech Republic company is also acknowledged.

References

[1] M. Motalab, Z. Cai, J. C. Suhling, P. Lall, Determination of Anand constants for SAC solders using stress-strain or creep data, Proceedings of the 13th IEEE ITherm Conference, San Diego, CA, May 30 - June 1, 2012, pp. 910-922.

[2] M. Amagai, Characterization of chip scale packaging materials, Microelectronics Reliability 39 (1999) 1365-1377.

[3] K. Ohguchi, K. Sasaki, M. Ishibashi, T. Hoshino, Plasticity-creep separation method for viscoplastic deformation of lead-free solders, JSME Int. J. Series A 47/3 (2004) 371-379.

[4] K. Sasaki, K. Ohguchi, Uniaxial ratchetting behavior of solder alloys and its simulation by an elasto-plastic-creep constitutive model. Journal of Elec. Materi. 40 (2011) 2403.

[5] T. Kobayashi, K. Sasaki, Experiments and simulations of uniaxial ratchetting deformation of Sn-3Ag-0.5Cu and Sn-37Pb solder alloys, J. Mater. Sci.: Mater. Electron. 20 (2009) 343-353.

[6] D. Antoš, R. Halama, M. Bartecký, Measurement of mechanical properties of lead free alloy, In: Proceedings of 56th International conference on Experimental Stress Analysis 2018 (EAN2018), Hotel Sklář, Harrachov, 2018, Czech Republic.

[7] D. Antoš, R. Halama, M. Bartecký, Cyclic Plastic Properties of a Lead Free Solder, Key Eng. Mater. 810 (2019) 46-51.

[8] ČSN ISO 12106, Kovové materiály - Zkoušení únavy - Metoda řízení osové deformace.

[9] M. Kobayashi, M. Mukai, H. Takahashi, N. Ohno, T. Kawakami and T. Ishikawa, Implicit integration and consistent tangent modulus of time-dependent non-unified constitutive model, Int. J. Numer. Meth. Eng. 58 (2003) 1523-1543.

[10] M. Abdel-Karim, N. Ohno, Kinematic hardening model suitable for ratchetting with steady-state. Int. J. Plast. 16 (2000) 225–240.

[11] ANSYS optiSLang, Dynardo, 2020, https://www.dynardo.de/en/software/optislang.html

[12] D. Antoš, R. Halama, Viscoplastic modeling of stress-strain behavior of a lead free solder, Proceedings of Applied Mechanics 2019 conference, Ostravice, April 15-17, 2019, Czech Republic.

Dynamic Properties of 316L Identified by Split Hopkinson Pressure Bar

BĚLÍK R.[1,a], FUSEK M.[1,b]

[1]Deparment of Applied Mechanics, Faculty of Mechanical Engineering, VŠB – Technical University of Ostrava, 17. listopadu 2172/15, 708 00 Ostrava, Czech Republic

[a]radomir.belik@vsb.cz, [b]martin.fusek@vsb.cz

Keywords: Stainless steel, A316L, SHPB, Johnson-Cook, Dynamic testing, Regression

Abstract. This work presents a dynamic testing of conventional rolled stainless steel A316L on self-constructed Split Hopkinson Pressure Bar (SHPB) apparatus with three different impact speed of striker. The ten cylindrical samples with slenderness ratio equalled to one were tested. Acquired signal of pressure wave propagation were separated into three different waves and stress-strain dependence were evaluated. The coefficients of simplified Johnson-Cook material model were estimated by Power law regression.

Introduction

In modern age of computer aided engineering and finite element analysis solutions is mandatory to know proper mechanical properties of used materials include dynamic response. Dynamic response of material can be tested by lot of different techniques. One of them is Split Hopkinson Pressure Bar (SHPB) invented by Bertram Hopkinson [1] and modified by Herbert Kolsky [2]. In this article self-constructed SHPB is used for identifying of dynamic properties of rolled stainless steel A316L at various strain rate.

Fig. 1: Self-constructed Split Hopkinson Pressure Bar apparatus

SHPB testing. Cylindrical samples of A316L with diameter d = 10 mm and slenderness ratio d/l = 1 was tested at three different speed of 200mm long striker. Sample is compressed between two prismatic bars equipped by resistance strain gauge. Main goal of test is record pressure wave propagation in prismatic bars caused by accelerated striker to incident bar. The dynamic response of material is evaluated from strain gauge signal. Acquired signals from incident and transmitted prismatic bar are depicted on Fig. 2.

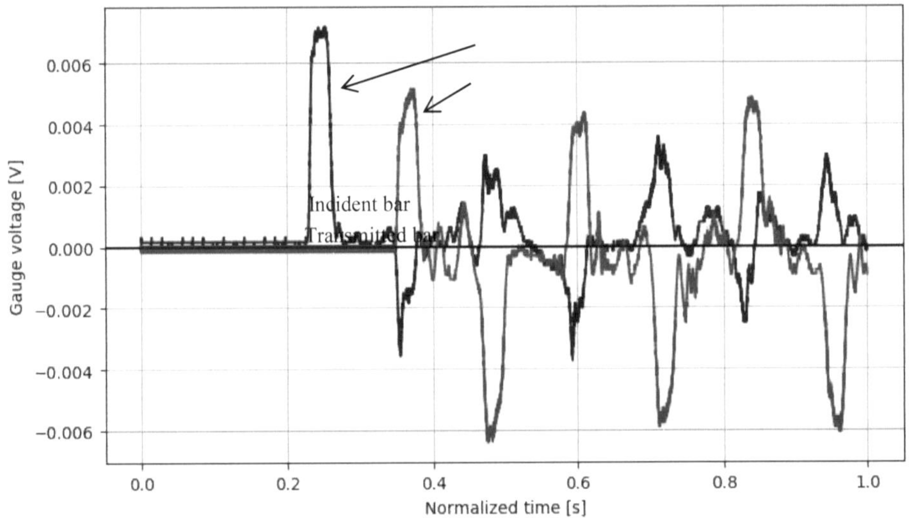

Fig. 2: SHPB test record of sample

Three types of wave are identified in signals - incident wave, reflected wave from first prismatic bar and transmitted wave through sample to second prismatic bar - Fig. 3.

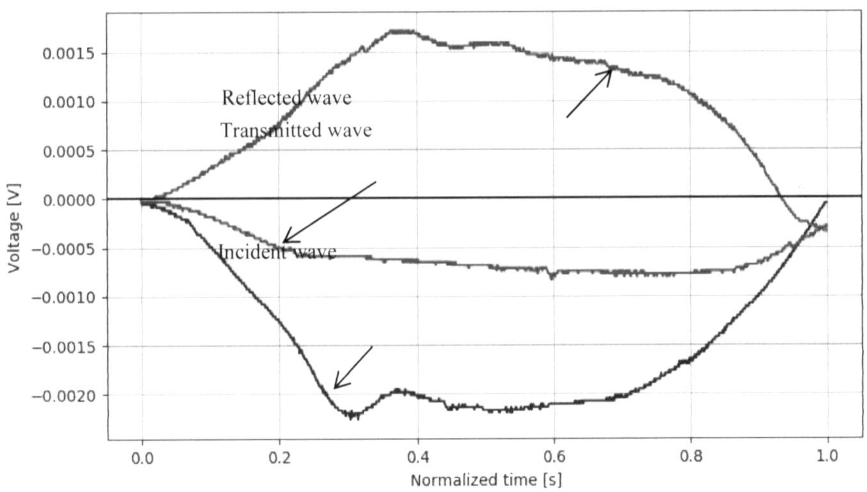

Fig. 2: SHPB test record of sample

Separated waves are used for evaluation of stress, strain and strain rate by Eq. (1 - 3).

$$\sigma(t) = \frac{A_{bar} \cdot E_{bar}}{A_{sample}} \cdot \varepsilon_{trans}(t) \tag{1}$$

$$\dot{\varepsilon}(t) = \frac{2 \cdot C_{bar}}{L_{sample}} \cdot \varepsilon_{ref}(t) \tag{2}$$

$$\varepsilon(t) = \frac{2 \cdot C_{bar}}{L_{sample}} \cdot \int_0^t \varepsilon_{ref}(t)\, dt \tag{3}$$

where A_{bar} and A_{sample} is the cross-section area, E_{bar} is Young modulus of prismatic bars, L_{sample} is length of sample, C_{bar} is speed of sound in prismatic bars and ε_{trans}, ε_{ref} are identified waves in signal.

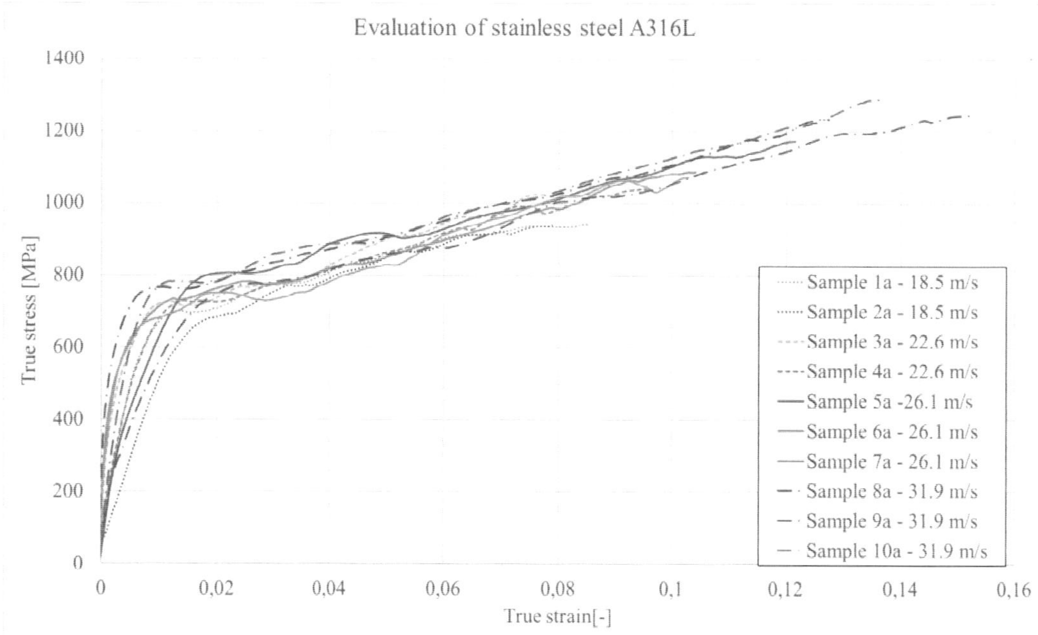

Fig. 3: Stress-strain dependence of tested samples

Stress–strain dependences of tested samples are depicted on Figure 3. The curves of dependences lie close to each other, which shows the stable material behaviour thought different strain rate.

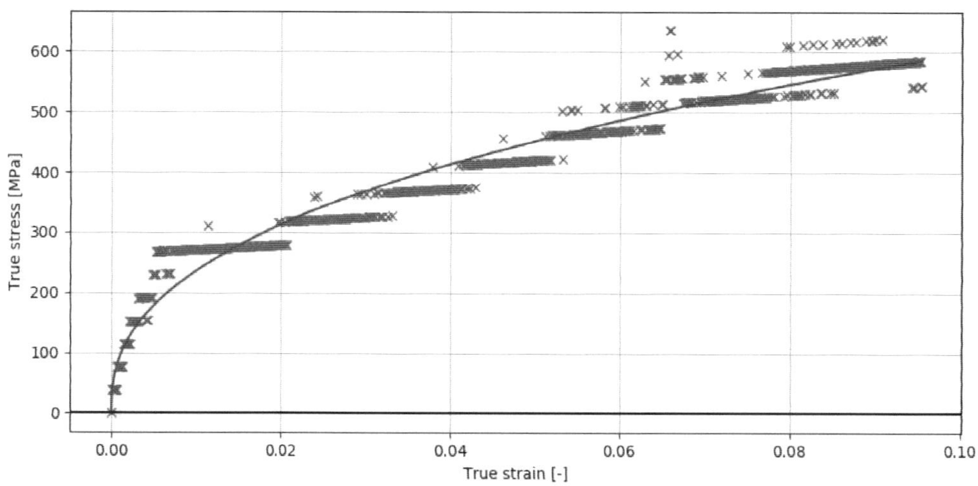

Fig. 4: Power law regression of sample data

Evaluated stress-strain curves are fitted by power law regression (Fig. 4) on the simplified Johnson-Cook material model [3], described by Eq. (4).

$$\sigma = (A + B \cdot \varepsilon^n)(1 + C \cdot \ln \dot{\varepsilon}) \tag{4}$$

where A, B, C, ε, n are material constants. Estimated coefficients of Johnson-Cook material model are summarized in Table 1.

Table 1: Estimated coefficient of Johnson-Cook material model

	A [MPa]	B [MPa]	n [-]
Stainless steel A316L	453	1536	0,593

Conclusions

The aim of the work was to determine the dynamic compressive response of conventional stainless steel of A316L on self-constructed Split Hopkinson Pressure Bar apparatus. Tests with three different strain rates were performed and evaluated to stress-strain curves. Stainless steel A316L shows insensitivity to different load speeds. Finally evaluated stress-strain curves were fitted to the simplified Johnson-Cook material model for purposes of finite element analysis.

Acknowledgement

This work was supported by The Ministry of Education, Youth and Sports from the Specific Research Project (SP2020/23) and has been done in connection with the DMS project reg. no. CZ.02.1.01/0.0/17_049/0008407 financed by Structural Funds of Europe Union.

References

[1] B. Hopkinson, (1914). A method of measuring the pressure produced in the detonation of high explosives or by the impact of bullets. Phil. Trans. Roy. Soc. London A. (213): 437-452.

[2] H. Kolsky, (1949) An investigation of the mechanical properties at very high strain rates of loading. Proc. Phys. Soc. B, (62): 676-701.

[3] G. R. Johnson, W. H. Cook, A constitutive model and data for metals subjected to large strains, high strain rates and high temperatures. Proceedings of the 7th International Symposium on Ballistics. 1983. p. 541-547.

The Effect of Imperfections on the Prediction of the Stability Loss of Composite Shear Panels

CABRNOCH B.[1,a], KRUML J.[2,b], KRÁL M.[2,c], VÍZI I.[2,d]

[1]Czech Aerospace Research Centre, Materials and Technologies Department, Space Division, Beranových 130,199 05 Praha - Letňany, Czech Republic

[2]Czech Aerospace Research Centre, Design and Development Department, Space Division, Beranových 130,199 05 Praha - Letňany, Czech Republic

[a]cabrnoch@vzlu.cz, [b]kruml@vzlu.cz, [c]kral@vzlu.cz, [d]vizi@vzlu.cz

Keywords: post-buckling, finite element method, composite materials, experimental testing

Abstract. The aim of this work was to gain a general overview of the behaviour of composite materials when their stability is lost. For this purpose, one monolithic laminate and one sandwich plate specimens were chosen that were both loaded with a diagonal force. The models of these tasks were created in FEM software. The problem could not be solved by standard linear buckling analysis tools as they are unable to model deformation shapes after loss of stability. The load bearing capacity could thus be investigated only until the critical force of the loss of stability. However, in some cases, it can be assumed that the loss of stability could be very close to the real bearing capacity of the panel, therefore the task was modelled by both linear buckling and nonlinear buckling (with application of imperfections). After the FE analysis an experimental measurement was performed in order to show the precision of simulation and the quality of model. The results of the FE analysis were compared with experimental results.

Introduction

Composite materials have been used in various areas of industry for a long time. Their field of application is still growing and new materials with more beneficial properties are still being developed. However, unlike metals and other materials, composite materials do not tend to create permanent strain before failure, therefore careful consideration should be given to their design and to experimental verification of their properties. Composite structures can be stressed by all kinds of loads such as tension, compression, torsion, bending etc. The limit state of a composite structure loaded in compression can occur when the compressive strength is exceeded or the stability of the structure is lost (buckling). Therefore, it is necessary to understand this phenomenon that occurs when the limit values are reached [1].

The concept of buckling is closely related to the concept of equilibrium. The graph in Fig. 1 shows the so-called equilibrium paths. As the load increases, the loaded structure initially follows its primary equilibrium path, also called the pre-buckling path. In linear buckling, the pre-buckling path ends at the bifurcation point. This point represents the critical load value, also known as the buckling load, at which a sudden change in deformations occurs. Large deformations can result in the immediate failure of the loaded structure. The value of the point is predicted by eigenvalue analysis. From the bifurcation point, the equilibrium path continues along the secondary equilibrium path, also called the post-buckling path. The increasing slope of the post-buckling path indicates that the structure has a load carrying capacity to resist the load after buckling. This equilibrium path is called stable. The decreasing slope of the post-buckling path indicates that the structure is not able to resist the load after buckling. The buckling load in this case is the maximum load of the structure. This equilibrium path is called unstable. In a nonlinear buckling, where structure imperfections are considered, the equilibrium path has a significantly smoother course. The primary equilibrium path ends here at the failure point, which represents the load at which the structure collapses. The value of the failure point can be predicted by nonlinear analysis [2].

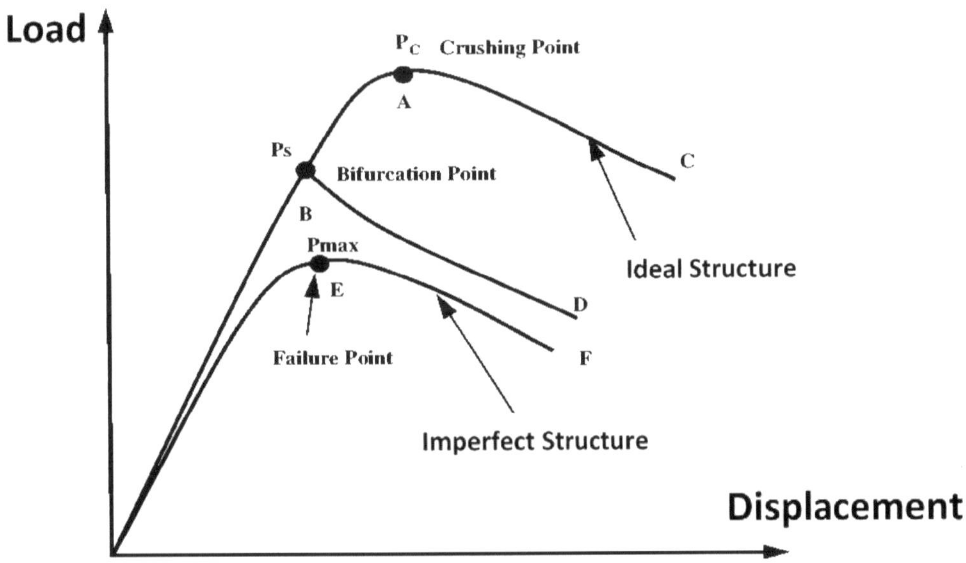

Fig. 1: Equilibrium paths for linear and nonlinear buckling [3]

The loss of stability of the composite panels results in a two-dimensional sine deformation out-of-plane of the panel. As the load increases, the panel shortens in the load direction until stability is lost. After loss of stability, the panel is able to carry the load further, anyway only with reduced stiffness. It is the difference from the loss of stability of the rod, where the load at the loss of stability is the ultimate load.

Special attention should be paid to the loss of stability of sandwich panels. A sandwich panel is made of two thin skins bonded to a core of a lightweight material of greater thickness. The skins transfer load (tension/compression, shear and bending) in-plane of sandwich panel while the core transfers load in out-of-plane direction (shear). Failures of sandwich panels can be divided into three basic categories: insufficient strength, loss of local stability and loss of overall stability. The insufficient strength is caused by exceeding the skin, the core or the interface strength limit. The loss of local stability is caused by too low skins thickness, however too large honeycomb cells size must be also considered. A typical mode of the loss of local stability is wrinkling. The loss of overall stability does not result in skins or core failure, which occurs only under continuing loading. A typical mode of the loss of overall stability is crimping. It usually appears suddenly and it seems as the loss of local stability, but in fact it is sort of the loss of overall stability [1, 4].

The stability of sandwich materials is influenced by a wide range of parameters. The most important ones include boundary conditions and imperfections. Imperfections can be caused by damages and inaccuracies that are formed both in the manufacturing process and in the service. They may occur in form of cavities, delaminations, debondings, cracks in matrix, fractures of fibres etc. These damages are often not visible and easy detectable, but they can cause considerable stiffness degradation and affect strength and stability of a structure. Theoretical predictions and models always require a comprehensive knowledge of geometric and material characteristics, boundary conditions and considerable amount of other influences that are not easy to determine. Including imperfections, the models become much more complicated to solve [1].

The paper tries to quantify the influence of shape imperfections of typical composite structures on the loss of its stability. For this purpose, monolithic laminate and sandwich panels loaded by in-plane shear were chosen.

Description of test specimens

Two composite panels were tested. The first one was a sandwich panel. The sandwich panel consisted of Nomex honeycomb core (cell size 3.2 mm, density 29 kg/m^3) at thickness 5 mm and 2 composite skins. The skins were made of one ply of EP121-C20-45 prepreg with carbon fabric reinforcement (Twill 2/2, 204 g/m^2) and epoxy resin oriented at 45 deg. according to the coordinate system shown in Fig. 2.

The second composite panel was a monolithic laminate panel composed of 4 plies with stacking sequence [+45°/-45°/-45°/+45°] according to the coordinate system shown in Fig. 3. The panel was made of EHG250-68-50 prepreg with glass fabric reinforcement (8H satin, 296 g/m^2) and epoxy resin. Material properties are given in Tab. 1 and Tab. 2 [4].

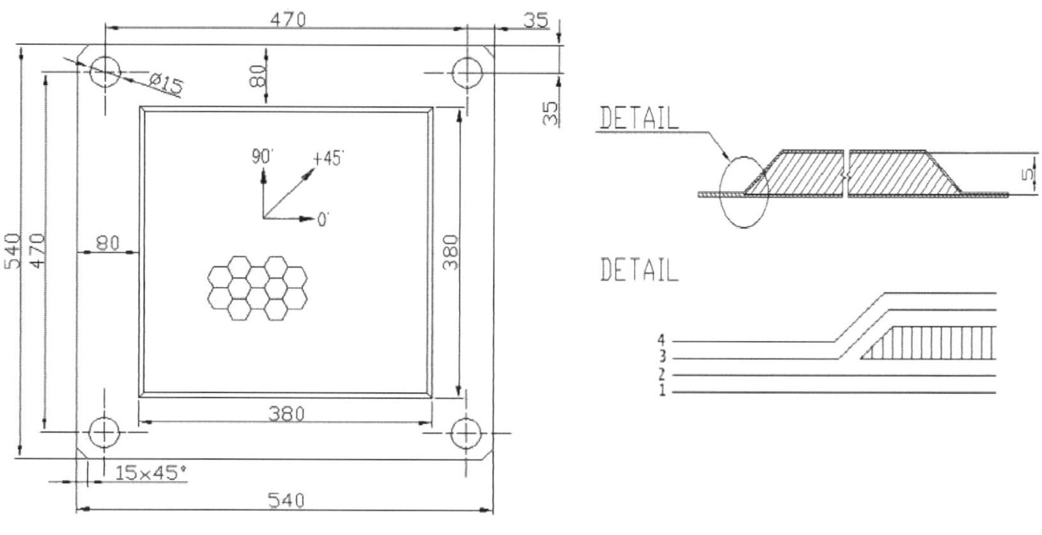

Fig. 2: Tested sandwich panel [4]

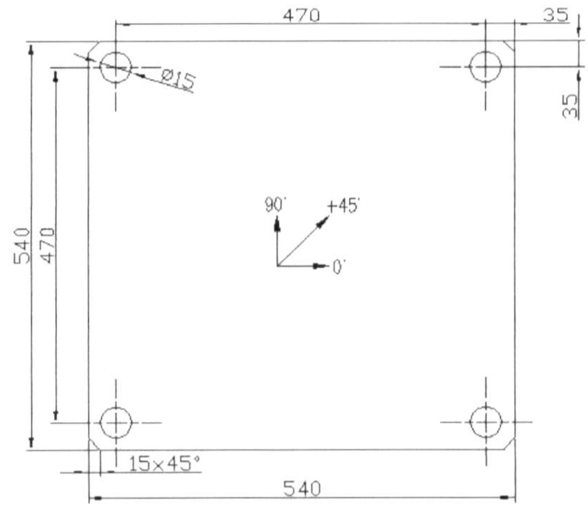

Fig. 3: Tested monolithic laminate panel [4]

	Units	EHG250-68-50	EP121-C20-45
Longitudinal Tensile Modulus **E_L**	[GPa]	23.5	41.8
Transverse Tensile Modulus **E_T**	[GPa]	20.5	41.7
Poisson´s Ratio **ν_{LT}**	[-]	0.15	0.05
In-plane Shear Modulus **G_{LT}**	[GPa]	4.2	4.8

Tab. 1: Material properties of prepreg EHG250-68-50 and EP121-C20-45 [4]

	Units	Nomex ECA-3.2-29
Compression Modulus	[MPa]	138
Compression Strength	[MPa]	2.1
Shear Modulus, L-direction **G_L**	[MPa]	48
Shear Strength **τ_L**	[MPa]	1.32
Shear Modulus, W-direction **G_W**	[MPa]	30
Shear Strength **τ_W**	[MPa]	0.73

Tab. 2: Material properties of honeycomb Nomex ECA-3.2-29 [4]

FEM analyses

Models were prepared and evaluated in Femap software. Analyses were performed using Nastran solver. The sandwich panel was modelled by 2 different approaches. In the first case (Fig. 4), the panel was meshed only by 2D laminate type elements (skins and core, both using 2D orthotropic material models), while in the second case (Fig. 4) the panel was meshed by a combination of 2D laminate type elements (skins, using 2D orthotropic material model) and 3D solid elements (core, using 3D isotropic continuum model). The monolithic laminate plate (Fig. 5) was modelled by 2D laminate type elements only (using 2D orthotropic material model). The effect of the fixture frame was represented by 1D beam elements at the edges of the panels in all three cases. Fig. 6 shows a detailed representation of 1D elements. Fig. 7 shows a detailed representation of the arrangement of the sandwich models.

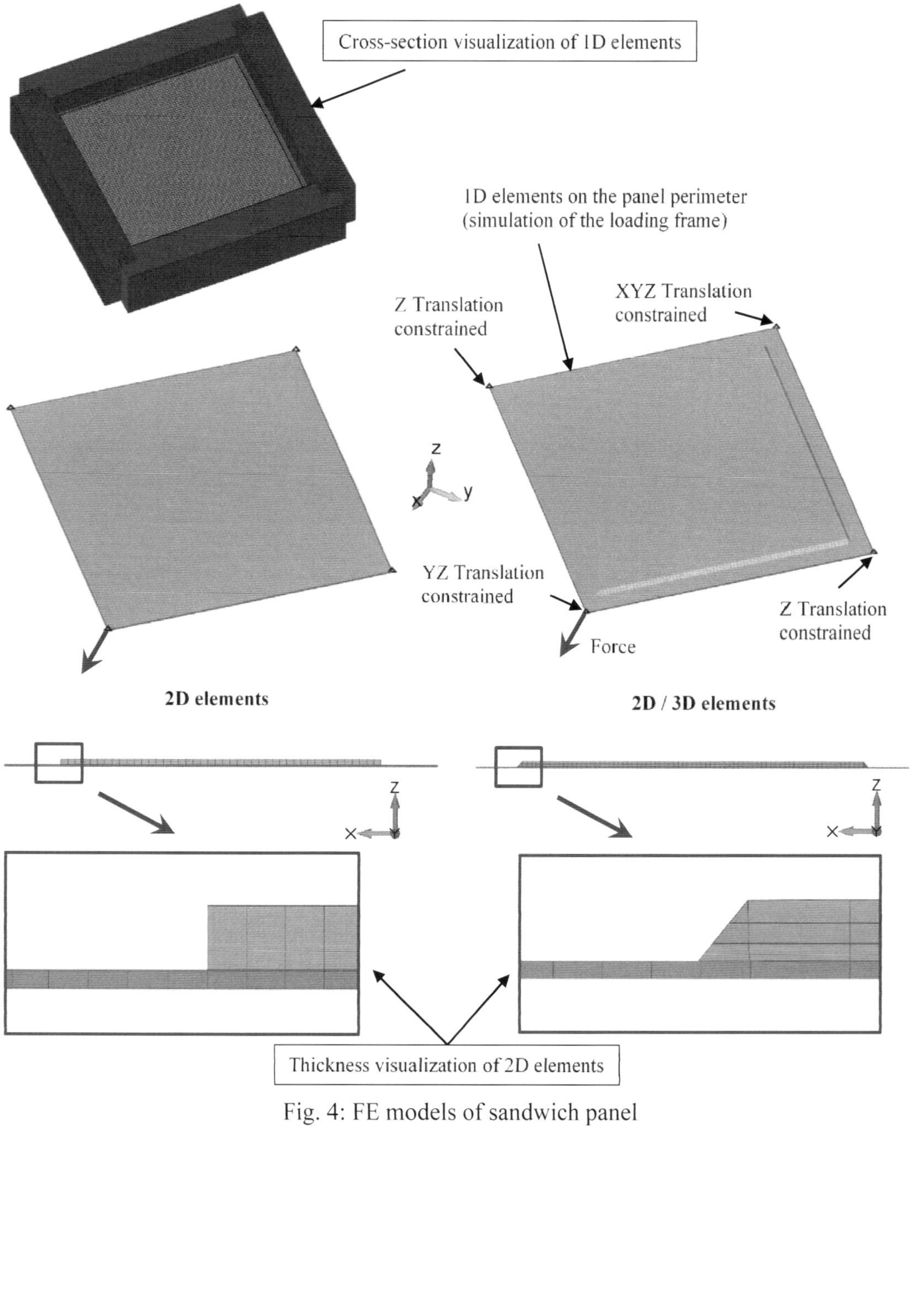

Fig. 4: FE models of sandwich panel

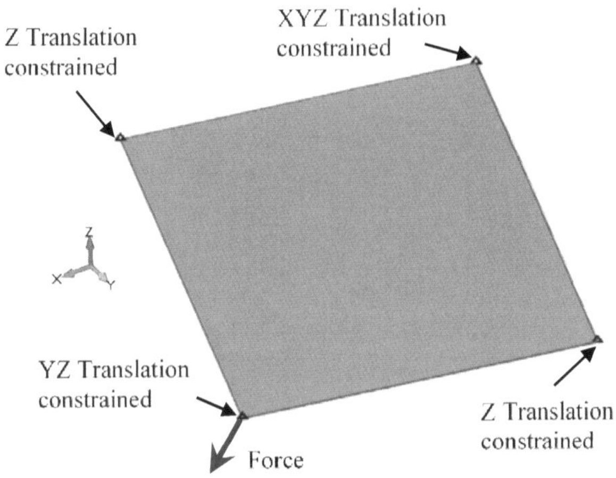

Fig. 5: FE model of monolithic laminate panel

The 2D model of sandwich panel was made of 7 396 2D layered shell elements and 344 1D beam elements. The 3D model of sandwich panel contained 17 328 3D solid elements, 13 776 2D laminate type elements and 344 1D beam elements. The model of monolithic laminate panel consisted of 7 396 2D laminate type elements and 344 1D beam elements.

The all three FE model configurations were validated and verified by standard geometry and numerical checks. In addition, the necessary FEM mesh density was verified to obtain reliable results. The models were verified by adjusting the element size when solving the linear buckling. In all three model configurations, 15 elements were selected as the initial number of elements on the edge of the panel. The number of elements on the edge of the panels was gradually increased to 60 for the monolithic and the 2D sandwich panel and 90 for the 3D sandwich panel. With a further increasing in the number of elements, the value of the critical force did not change. Therefore, FE mesh density with 60 elements on the edge of the panel for the monolithic and the 2D sandwich panel and 90 elements on the edge of the panel for the 3D sandwich panel were used in the FEM models.

Results of analyses

First, a linear solution was performed for all the 3 cases mentioned in the previous paragraphs. The critical force for the sandwich panel created only by 2D elements was 38 102 N, while for the panel created by a combination of 2D and 3D elements it was 33 085 N. The critical force of the monolithic laminate panel was 1 662 N. The analyses also included a determination of eigenshapes gained by the linear buckling solution (Fig. 8 and Fig. 9) to validate analysis by experimental data.

The next step was to modify computational models by geometrical imperfections. The initial geometrical imperfections of the original planar panels were modified by induced deflection in the middle of the panel. Four different deflection sizes (0.25 mm, 0.50 mm, 0.75 mm and 1.00 mm) were used to determine their effect on the critical force value. The critical force for all 3 model types were obtained by a nonlinear solution. The critical force for the sandwich panel created only by 2D elements was on average (from imperfection deflection sizes) 35 719 N, while for the panel created by a combination of 2D and 3D elements it was 21 780 N. The critical force of the monolithic laminate panel dropped on average to 1 605 N, where it loses stability, but does not come to collapse.

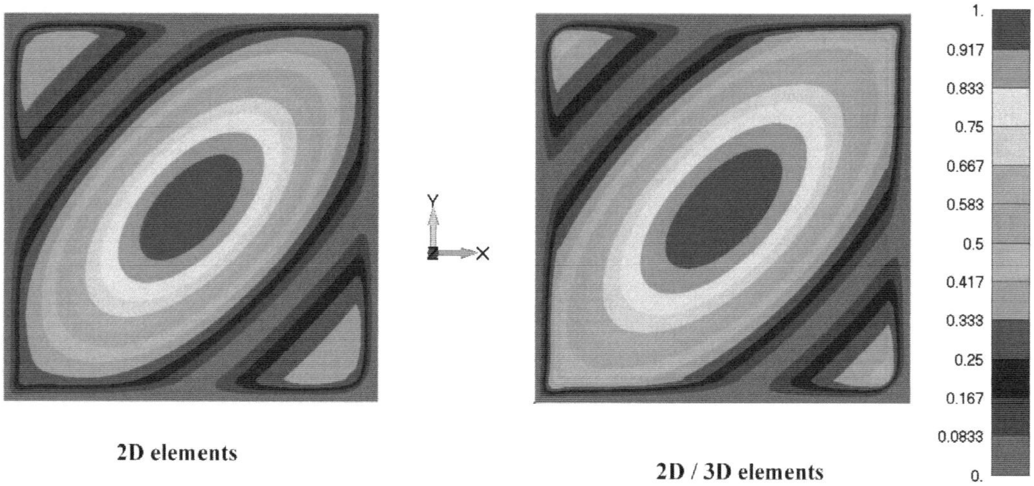

Fig. 8: The first eigenshape of the sandwich panel

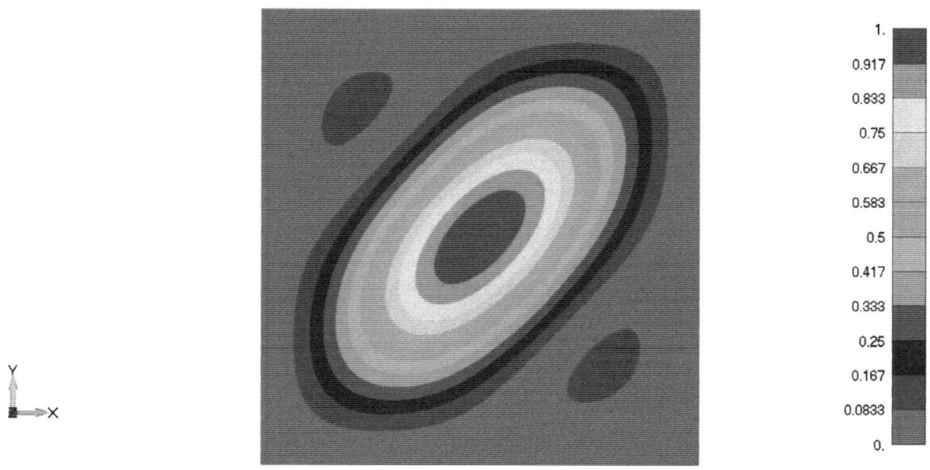

Fig. 9: The first eigenshape of the monolithic laminate panel

Experimental measurements

An example of test set-up is shown in Fig. 10 (left). The tests were performed in VZLU. A tested panel was clamped in the test fixture (frame) by bolts on its perimeter. The test fixture consists of four pinned arms. The frame was loaded by tensile force acting in the diagonal direction. The loading was performed by the Hydropuls servohydraulic system with a maximum loading force of 100 kN.

The considered tested area of the panel has dimensions 400 × 400 mm section of the panel. The remainder area of the panel was designed to transfer loads between the fixture and the tested area. The loss of panel stability was evaluated based on load – displacement curve and also by using the Aramis HS system that enable optical measurement of 3D displacements. The measurement section of the panel (panel area with foil) is shown in Fig. 10 (right). The output of this measuring device was a displacement map [4].

Fig. 10: The test set-up (left) and the measurement section (right) [4]

Discussion of results

An overview of the sandwich panels results for both approaches is shown in Fig. 11. The values of the critical forces show a considerable difference even before the imperfections were applied. The critical force for the sandwich panel created only by 2D elements was 38 102 N, while for the panel created by a combination of 2D and 3D elements 33 085 N. In the experimental part, the value of the sandwich panel critical force was set at 22 000 N. The difference in both computational approaches increases even more with the application of imperfections. The critical force for the sandwich panel created only by 2D elements was on average (from all imperfection sizes) 35 719 N, while for the panel created by a combination of 2D and 3D elements 21 780 N. As can be expected, increasing size of imperfection decreasing the value of the critical force.

Fig. 12 shows an overview of the monolithic laminate panels results. It shows a rather good agreement in computational and experimental values. However, regarding the experimental results, it is important to mention that the accurate determination of the monolithic laminate panel critical force was more less estimation. The monolithic exhibits loss of stability practically from the beginning of loading. The value of the critical force was determined on approximately 1 500 N based on more significant change in slope of load-displacement curve. Fig. 12 further shows that for small sizes of imperfections (from 0.25 mm to 1.00 mm) the critical force values remain almost constant compare to sandwich panel. This fact evokes that the size of the imperfection has a greater effect on the critical force of the sandwich panels than on the monolithic panels, but it is not in line with common response of this type of structures. Usually stiffer panels are less sensitive to initial imperfections. The reasons of this behaviour will be the ability of the models to simulate real response of the panels and also uncertainty in behaviour of thin low stiffness structures where any irregularity in the structure (geometrical, internal arrangement, ...) can cause large deviations.

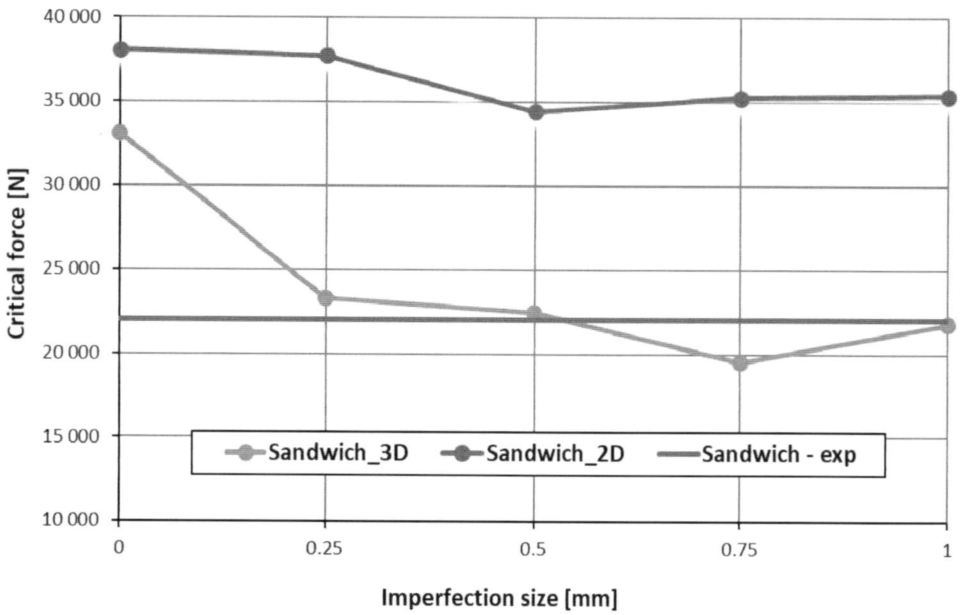

Fig. 11: Dependency of critical force on imperfections size – sandwich panel

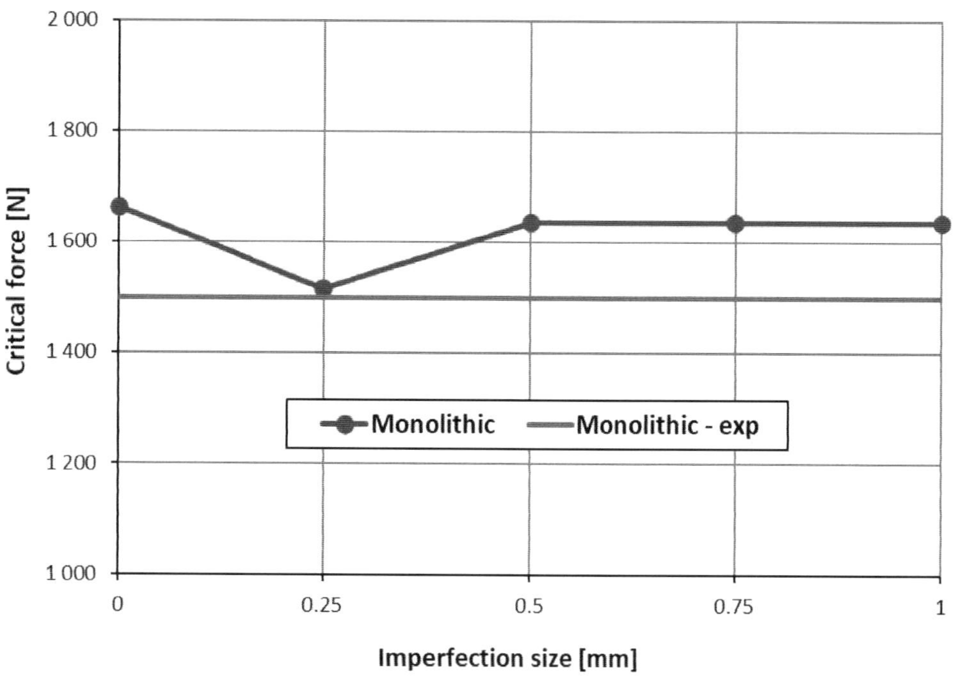

Fig. 12: Dependency of critical force on imperfections size – monolithic panel

Conclusions

From the obtained results it is clear that neglecting the influence of even relatively small geometric imperfections can lead to relatively large errors in simulations of their behaviour. This behaviour is even important for relatively stiff structures like sandwich panels (decreasing about 30% between planar and 0.25 mm deflected sandwich panel). Another interesting result is also a comparison of 2D and 3D models of sandwich structures. The 2D model of sandwich structure is sufficient for linear static analysis but for nonlinear simulations exhibits significantly higher critical force (up to 60%) than 3D model what could be very danger in real structures.

Thus, it is obvious that experimental testing still plays a very important role in the development and the validation of numerical models as well as consideration of possible structural irregularities including geometrical imperfections.

Acknowledgments

This article was created under the project No. LTARF18040 with the financial support of the Ministry of Education, Youth and Sports of Czech Republic within the INTER-EXCELLENCE program.

MŠMT
MINISTRY OF EDUCATION, YOUTH AND SPORTS

References

[1] M. Baumruk (2007). Mechanics of composite materials, analytical methods for calculation of stability problems of composite plates and sandwich panels and their comparison with FEM, *Report*, Czech technical university in Prague, Faculty of Mechanical Engineering, Aerospace Research Centre, Prague.

[2] B.H. Sun, K.Y. Yeh, F.P.J. Rimrott (1995). On the Buckling of Structures, *Technische Mechanik*, Vol. 15, No. 2, Pp. 129-140.
Retrieved September 18, 2020 from:
http://www15.ovgu.de/ifme/zeitschrift_tm/1995_Heft2/Sun_Yeh.pdf

[3] P. Bala, A.V. Raviprakash, V. Ananthapathmanaban (2010). Parametric study on buckling behaviour of thin stainless steel cylindrical shells for circular dent dimensional variations under uniform axial compression, *International Journal of Engineering, Science and Technology*, Vol. 2, No. 4, Pp. 134-149.
Retrieved September 18, 2020 from:
https://pdfs.semanticscholar.org/3f54/1846ba61f0e70048254c6c667663902fa346.pdf?_ga=2.180015012.1825113141.1600435050-136078452.1599031507

[4] P. Průcha (2013). Properties of composite elements used in aircraft primary structures, *Doctoral dissertation*, Czech technical university in Prague, Faculty of Mechanical Engineering, Department of Aerospace Engineering, Prague.

[5] T. Abbey (2015). Linear and Nonlinear Buckling in FEA, *Digital Engineering 247*.
Retrieved April 20, 2020 from:
https://www.digitalengineering247.com/article/linear-and-nonlinear-buckling-in-fea/

EAN 2020
58th conference on experimental stress analysis

Comparison of Residual Stresses and Mechanical Properties of Unconventionally Welded Steel Plates

ČAPEK J.[1,a], TROJAN K.[1,b], KEC J.[2,c], ČERNÝ I.[2,d], GANEV N.[1,e], KOLAŘÍK K. [1,f], NĚMEČEK S.[3,g]

[1]Department of Solid State Engineering, Faculty of Nuclear Sciences and Physical Engineering, Czech Technical University in Prague, Trojanova 13, 120 00 Prague, Czech Republic

[2]SVÚM a.s., Tovární 2053, 250 88 Čelákovice, Czech Republic

[3]RAPTECH s.r.o., U Vodárny 473, 330 08 Zruč-Senec, Czech Republic

[a]jiri.capek@fjfi.cvut.cz, [b]karel.trojan@fjfi.cvut.cz, [c]kec@svum.cz, [d]ivo.cerny@seznam.cz, [e]nikolaj.ganev@fjfi.cvut.cz, [f]kamil.kolarik@fjfi.cvut.cz, [g]nemecek@raptech.cz

Keywords: Unconventional welding, X-ray diffraction, Residual stresses, Real structure, High-cycle fatigue

Abstract. The paper contains results of macroscopic residual stresses and high-cycle fatigue life of unconventionally welded steel plates. Generally, the manufacturing processes of a machine component introduce residual stresses that have an important and may have even essential influence on their behaviour during service life. Therefore, the goal of this contribution was to compare laser and electron weld concerning the distribution of surface macroscopic residual stresses and high-cycle fatigue life. The results, important for further applications of the used technology, are discussed particularly from the viewpoint of possible critical areas near the welds and their impact on fatigue loading.

Introduction

Welding is a widely used method for joining materials in production of ships, trains, steel bridges, pressure vessels, etc. Therefore, high demands are laid on mechanical properties and durability of welds used to connect two or more components. However, because of the heterogeneous application of energy and localized fusion, which occur during the welding process, high undesirable residual stresses (RS) can be present in the region near weld and also in the weld itself. These RS occur as a result of the superposition of thermal (tensile effect) and transformation (compressive effect) stresses, see Fig. 1, and could reach high values and subsequently could cause fatigue life reduction or promote of brittle fractures [1]. Therefore, it is necessary to reduce RS values and size of the area affected by the welding. These RS, microstructure and mechanical properties of the weld can be affected by the used welding method, parameters, and the treatment of the material before and after welding.

The state of residual stress across the laser and electron welds was examined using X-ray diffraction. Unconventional welding methods that bring less heat into the weld and surrounding material and therefore less affect the mechanical properties of the material have been compared. They also show high productivity and automation capability compared to manual welding [2]. In this paper, butt welds made of P355 and P460 steel plates for high-pressure vessels were analysed.

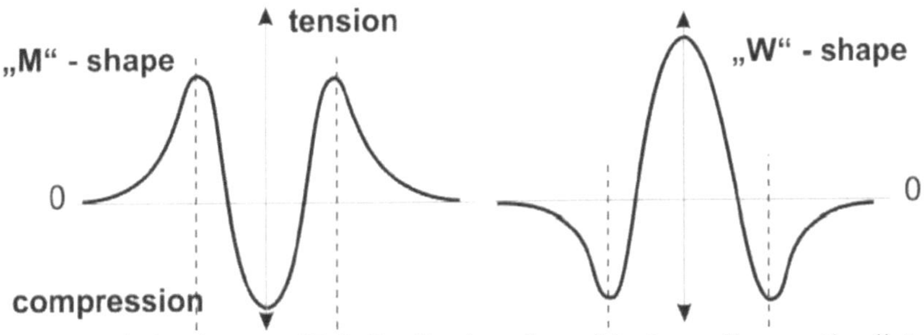

Fig. 1: Characteristic shapes of RS distributions in welds depending on the distance from the weld axis in the perpendicular (left) and parallel (right) direction to the weld [2]

Experiment

The material studied in this work was P355NL1 and P460NL1 low carbon alloy steel formed into the plates 300×300×20 mm in size. These fine-grained ferritic-perlitic steels are widely used for fabrication of high-pressure vessels, steam boiler parts, pressure piping, compressors, etc. The laser and electron beam welded samples were prepared using a high-power diode laser by the company RAPTECH s.r.o. and at Faculty of Materials Science and Technology in Trnava, respectively. For welding parameters, see Tab. 1. Laser welded plates were machined into the double V-groove shape, see Fig. 2. In the middle part of thickness, the plates were double-side welded, where top sides "U" were welded as the first and bottom sides "D" as the second one. Further, the V-shape groove with a depth of 6 mm was filled with filling wire. Plates welded by the electron beam were joined (single-side weld, where top sides "U" were welded) without filling wire.

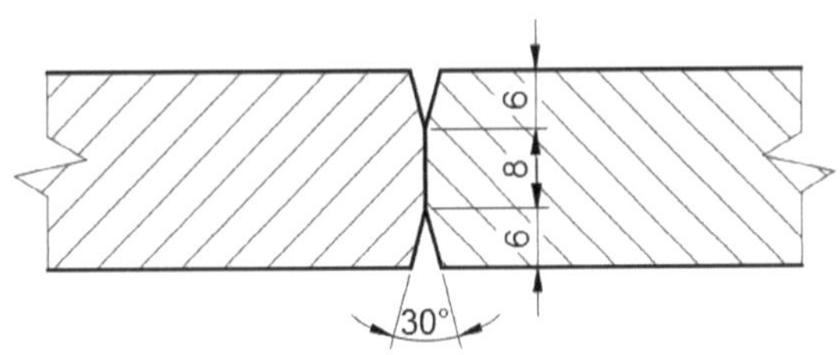

Fig. 2: The double V-groove shape of laser welded plates

Table 1: Welding parameters; P – power of laser beam, v – speed of welding, λ – wavelength of laser beam, U – voltage, I – current

	P [W]	v [mm·s^{-1}]	mode	λ [nm]	U [kV]	I [mA]
laser	3000	5.5	Continuous	900–1080	-	-
electron	-	30	-	-	55	220

The chemical composition of the used steels, measured on SPECTROMAXx OES analyser, is given in Tab. 2. The results show that both materials are Mn alloyed low carbon steels. The Si content was reduced to the absolute minimum due to its negative effect on Charpy impact test properties at low temperatures. Both sheets of steel revealed high metallurgical purity, which resulted in a low content of inclusions. Nevertheless, the measured chemical composition meets the requirements for P355NL1 and P460NL1 steels. The tensile test was carried out at room temperature using the EUS 40 testing machine. For mechanical properties, see Tab. 3.

Table 2: Chemical composition of P355NL1 and P460NL1 steels, in wt%

	C	Si	Mn	P	S	N	Cu	Mo	Cr	Ni	V
P355NL1	0.11	-	0.81	0.016	0.007	-	0.021	-	0.025	0.019	-
P460NL1	0.12	-	1.22	0.015	0.008	-	0.019	-	0.017	0.013	0.041

Table 3: Mechanical properties of laser welded plates of the P355 and P460 steels

Designation [-]	Material [-]	*Rm* [MPa]	KV_2 *0°C* [J]
P355	P355NL1	562	66
P460	P460NL1	624	51

The microstructure of both steels consists of polygonal ferrite and perlite. P355NL1 steel contains a higher volume fraction of perlite in the structure and also the size of the ferritic grains is coarser, which translates into yield strength and brittle fracture properties, as finer grains result in an increase in yield strength and lower DBTT (Ductile-to-Brittle Transition Temperature). The mean grain size of P355NL1 and P460NL1 is 5.7 μm and 3.8 μm, respectively. As can be seen in Fig. 7a, along the fusion boundary, the effect of grain coarsening in HAZ was observed. The weld metal microstructure consists of acicular ferrite (AF) and grain boundary ferrite (GBF), see Fig. 3b.

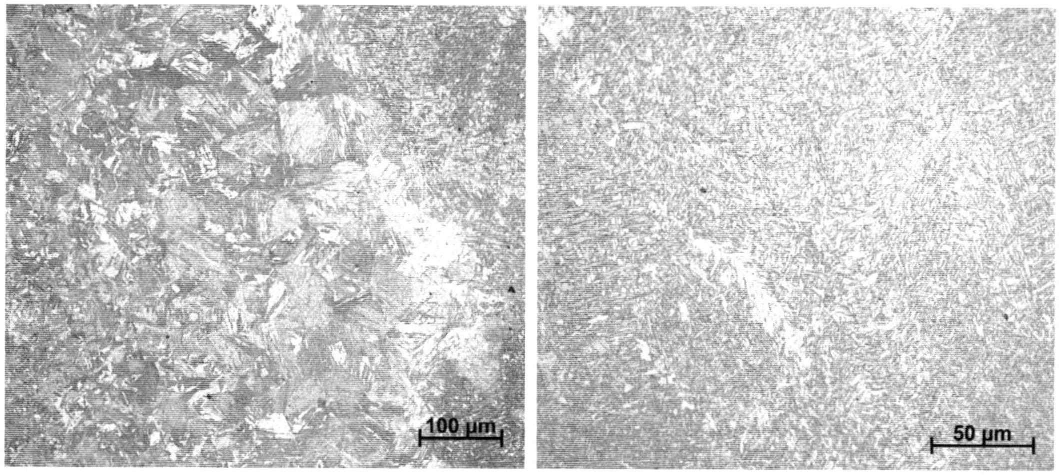

Fig. 3: Microstructure in a) HAZ and b) weld metal

High-cycle fatigue tests were performed at different stress ranges to obtain the whole S-N curve including endurance limit. Load asymmetry was R = 0, frequency 33 Hz.

The *PROTO iXRD COMBO* diffractometer in ω-goniometer set-up with chromium radiation was used for determination of lattice deformations of ferrite phase. Diffraction angles $2\theta^{hkl}$ were determined using *Gaussian function* and *Absolute peak* method from the peaks of the diffraction lines $K\alpha_1$ of the *{211}* planes. To describe the state of RS, the *Winholtz & Cohen* method [3] and X-ray elastic constants $\frac{1}{2}s_2 = 5.75$ TPa^{-1}, $s_1 = -1.25$ TPa^{-1} were used. The samples were analysed in the direction perpendicular "T" and parallel "L" to the welds (welding direction) on both sides of the samples. Welded plates were analysed in the middle part of plates in 18 points (x = 0–30 mm from the axis of the weld). The irradiated area of the primary beam was approx. 11×1 mm with the average effective penetration depth of the X-ray radiation approx. 4 μm [4].

Results and discussions

A comparison of RS for both unconventional technologies is presented in Figs. 4 and 5, where L, E, U, and D indicate laser welded specimen, electron beam welded specimen (with thickness of 20 mm), the side which was welded first, and the other side, respectively.

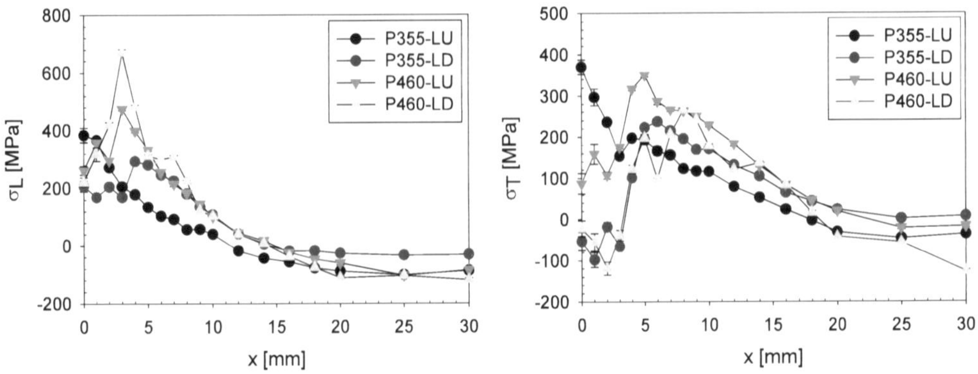

Fig. 4a: Dependence of RS on the distance from the weld axis for the laser weld

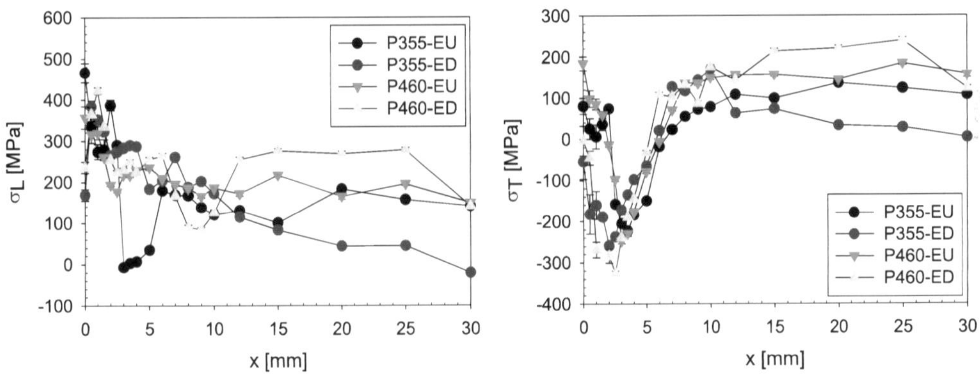

Fig. 4b: Dependence of RS on the distance from the weld axis for the electron beam weld

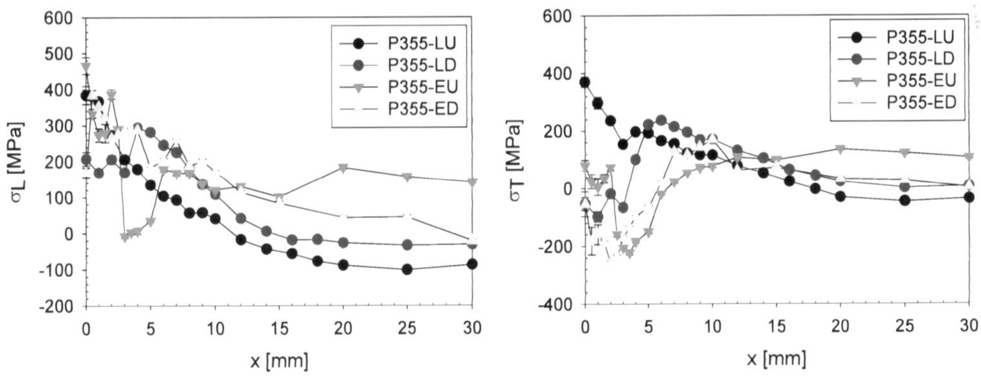

Fig. 5a: Dependence of RS on the distance from the weld axis for P355 steel

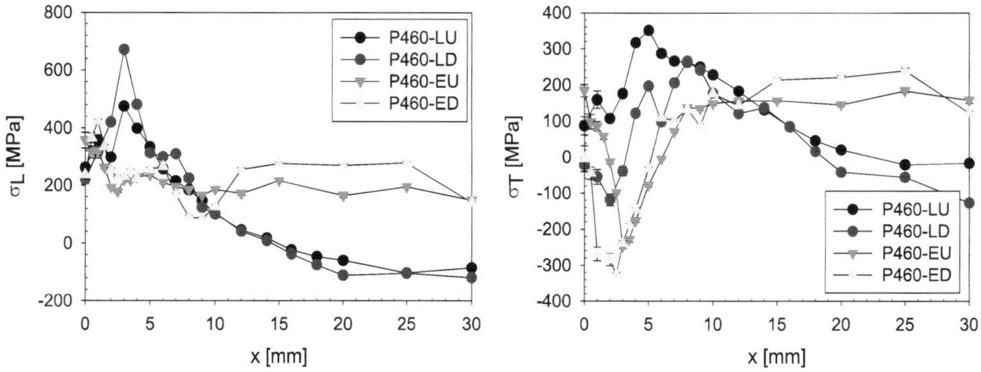

Fig. 5b: Dependence of RS on the distance from the weld axis for P460 steel

From the obtained surface RS profiles, it could be said that there are typical trends of RS on the surface, i.e. tensile RS in the L direction, and both compressive and tensile RS in the T direction, compare Figs. 1 and 4. Nevertheless, higher tensile RS were found for the laser weld on the top side, while lower compressive stresses are on the bottom side, see Fig. 4a. These stresses are probably generated due to the filling of the groove during laser welding and the associated deformation during cooling. RS for electron weld reach higher compressive values in the same area in the T direction. The decreases of RS for electron on the top side at a distance of 3–10 mm in the L direction weld are caused by mechanical removing of the thick oxide layer, see Fig. 4b. The differences in RS values of the bulk material are caused by the steel plates′ fabrication, i.e. rolling and/or removing of the thick oxide layer. Analysed RS near the weld are higher than yield strength (approx. 355 MPa and 460 MPa for P355 and P460 steel, respectively), and even than tensile strength *Rm* for the laser welded plate of the P460 steel, see Tab. 3 and Figs. 4. This effect could be caused by higher hardness of weld, which would result in different X-ray elastic constants. Heat affected zone (HAZ) for both electron welds is narrower for the bottom side which could result from single-side welding and different amount of heat on each side.

From fatigue life point of view, the surface tensile RS are one of indicator of critical area of possible surface crack initialization. However, the transversal direction T is more important from the technological point of view, in this direction, the main stresses of pressure vessels occur. Unfortunately, unlike electron beam weld, the top side RS in and near laser weld areas in the T direction can be considered as significant from the engineering viewpoint, see Fig. 5.

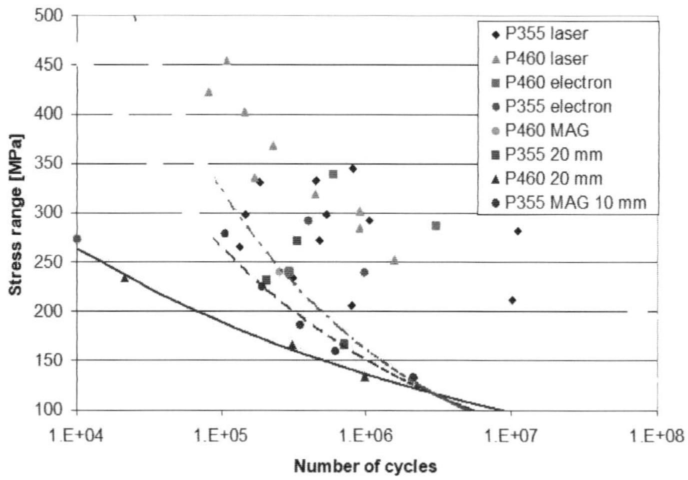

Fig. 6: Results of selected fatigue tests

Results of high-cycle fatigue tests are in Fig. 6, where previous results of investigations performed on welded plates of lower thickness up to 10 mm (single points) are plotted together with S-N curves of 20 mm thick laser welded plates of the P355 and P460 steels and 10 mm thick MAG welded plate of the P355 steel, respectively (points with power regression lines) for a comparison. In general, there are two main groups of points in the diagram. The first group represents tests of plates of the thickness up to 10 mm, mostly between 5–8 mm, welded using either laser or electron beam. If scatter is not considered, particularly in case of the P355 steel caused by internal lack of fusion of some specimens [5], it can be concluded that mean values of fatigue strength are similar. The second group is represented by laser welded plates of 20 mm thickness with fatigue strength incomparably lower indicating that laser welding of thick plates may be problematic and needs special approaches, particularly in case of high strength steels like P460. On the other hand, the results are not so bad in comparison with conventional MAG welding particularly considering that only 10 mm thick plates were evaluated.

What is very important is failure mode of the 20 mm thick laser welded specimens. Fatigue cracking was initiated either in the centre of the welds on internal welding defects or on the specimen surface. In the latter case, all cracking occurred exactly in fusion zone, i.e. at the distance approx. 2 mm from the weld centre. Looking at Fig. 4a, it is clear that cracking did not initiated at the area of maximum tensile RS, the RS were contrarily low or even negative there. As a result of the mentioned internal welding defects, the real cross-sections of investigated specimens were lower in the welds. If the fatigue resistance were re-calculated by considering only the real cross-section of welds without lack of fusion and pores, the correction would result in a significant increase in fatigue resistance. Therefore, the initialization of the surface crack in the fusion zone was caused by a combination of the internal defects (i.e. reduction of the real cross-section of the sample) and small surface compressive or tensile RS.

Conclusions

From surface RS point of view, the critical area for a potential surface crack initialization was on the top side of both laser welded steels in the weld zone and HAZ. The interface of fusion zone could be most critical because of higher hardness, different microstructure, and the presence of notches.

During laser welding, there was no perfect penetration, especially in the middle part of the weld, which was produced without the filling wire. It is also possible to see that there was a lack of inter-run fusion in the weld beads. The positive finding was that despite the weld discontinuities, the final fracture was localized to the base material during the tensile tests.

Investigations of fatigue properties of laser welded plates of 20 mm thickness showed that fatigue strength was incomparably lower in comparison with previous results of laser welded plates of the thickness up to 10 mm indicating that laser welding of thick plates may be an issue and needs special approaches, particularly in case of high strength steels like P460. Failure mode on the specimen surface exactly in fusion zone was caused by the combination of several effects, mainly low compressive or tensile surface RS and internal weld defects (reduction of the real cross-section).

Acknowledgement

This work was supported by the project TH02010664 of the Technology Agency of the Czech Republic and by the Grant Agency of the Czech Technical University in Prague, grant No. SGS19/190/OHK4/3T/14.

References

[1] D. Radaj, Heat effects of welding: temperature field, residual stress, distortion, Springer Science & Business Media, 2012.

[2] T. Nitschke-Pagel, K. Dilger, Sources and consequences of residual stresses due to welding, Mater. Sci. Forum 783 (2014) 2777–2785.

[3] U. Welzel, J. Ligot, P. Lamparter, A.C. Vermeulen, E.J. Mittemeijer, Stress analysis of polycrystalline thin films and surface regions by X-ray diffraction, J. Appl. Crystallogr. 38 (2005) 1–29.

[4] J. Čapek, Z. Pala, Programs for Visualizing Results of Diffraction Experiments, Mater. Struct. 22 (2015) 78–81.

[5] I. Černý, J. Kec, Evaluation of Fatigue Resistance of Laser Welded High Pressure Vessels Steel P355 Considering Fracture Mechanics Approach, Key Eng. Materi. 827 (2019) 428–433.

EAN 2020
58th conference on experimental stress analysis

Experimental and Numerical Determination of Strength of Portable Firefighter Ladder

DOUBRAVA K.[1,a], ŠPANIEL M.[1,b], KUŽELKA J.[1,c], VYSTRČIL V.[2,d], KUBÁSEK J.[3,e]

[1]Faculty of Mechanical Engineering, Czech Technical University in Prague, Technická 4, 166 07 Prague, Czech Republic

[2]Technical Institute of Fire Protection in Prague, Písková 42, 143 00 Prague, Czech Republic

[3]Departments of Metal and Corrosion Engineering, University of Chemistry and Technology Prague, Technická 5 166 28 Prague, Czech Republic

[a]Karel.Doubrava@fs.cvut.cz, [b]Miroslav.Spaniel@fs.cvut.cz, [c]Jiri.Kuzelka@fs.cvut.cz, [d]vaclav.vystrcil@tupo.izscr.cz, [e]jiri.kubasek@vscht.cz

Keywords: Portable firefighter ladder, Three-point bending test, Johnson – Cook plasticity model, Aluminium alloy thermal exposition

Abstract. Portable fire and rescue service ladder are widely used during fire and rescue works. Ladders are mostly made of aluminium alloy at present and this material changes its mechanical properties upon thermal exposure. Determination of the load capacity of ladders using experimental and numerical simulation is described in this paper.

Introduction

Portable ladders are universally used equipment of fire brigade units, where they belong to the equipment of most fire engines. Fire Rescue Service of the Czech Republic uses portable four-piece aluminium ladders from different manufacturers. Ladders are complexly loaded both mechanically and thermally. At present, the ladder's operational suitability is verified according to the standard [1]. The ladder is loaded with a three-point bend, see Fig 1 and any residual deflection determines the further usability.

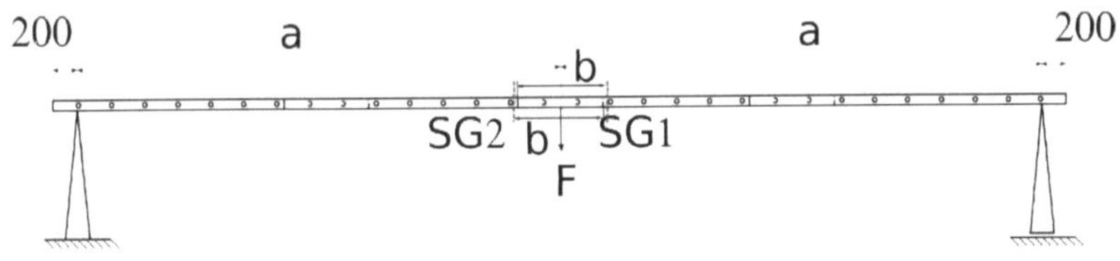

Fig. 1: Scheme of the test portable ladder

The vast majority of ladders used are from only two manufacturers and therefore only these two types were examined in the following. Numerical models for both types were created and verified experimentally.

Experiments

Experimental verification of ladder strength was based on the loading setup according to the mentioned standard [1]. With regard to the available resources, the water tank was chosen to load the ladder, see Fig. 2. The load was monitored using a load cell and a potentiometric displacement sensor was used for deflection measuring of assembled ladder. Strains around the connection ladder detail were monitored using resistance strain gauges. Material properties of the aluminium alloy used were determined by tensile testing, when the tests were carried out at normal temperature at elevated temperatures and on material exposed to different temperature levels.

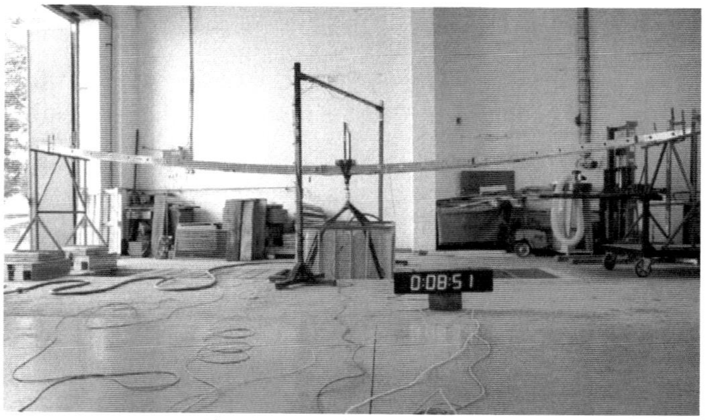

Fig. 2: Portable ladder test

The tests were performed on temperature-unaffected ladders, where three specimens from both main manufacturers were tested. The ladder assemblies were loaded to total destruction, failure of one specimen can be seen in Fig. 3. Results of all destructive bending tests can be seen in Fig. 4.

Fig. 3: Detail of ladder brake

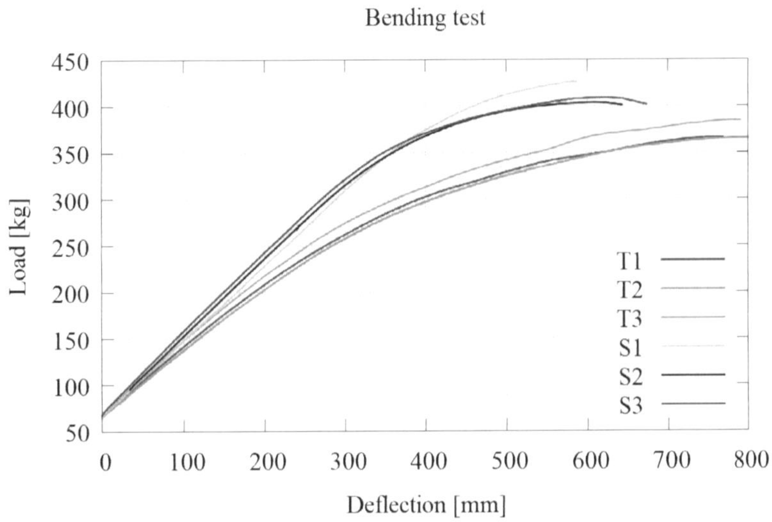

Fig. 4: Experiment results

Numerical simulation

Within numerical solutions, material model parameters were identified for unexposed and exposed state. The FE ladder model was created on the basis of geometry measured on real pieces. A numerical simulation of experiments on an unexposed ladder was performed and the results of the calculations were compared with the measured values, see Fig. 5.

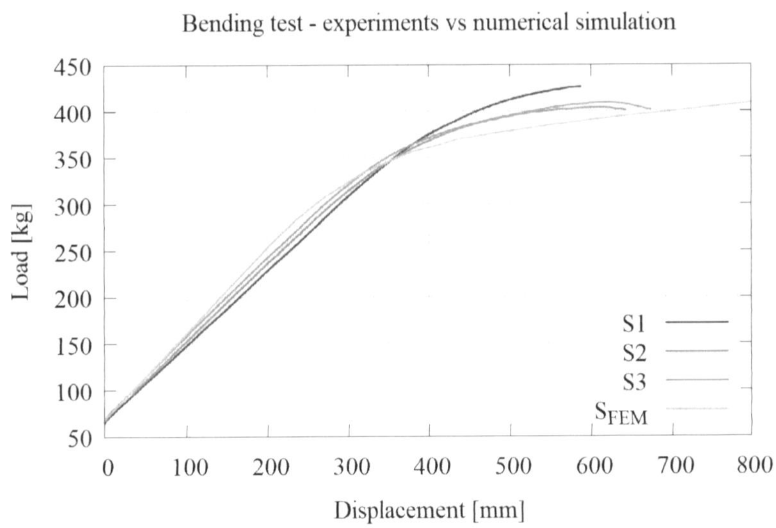

Fig. 5: Comparison of experimental results and numerical simulation

Furthermore, numerical simulations of hypothetical cases of ladders subjected as a whole to the same temperature exposures as samples for material tests were performed. Finally, the case of partially temperature-affected ladder was solved. A simple model of plasticity and ductile failure according to Johnson – Cook [2] was chosen for phenomenological description of mechanical properties of ladder material, when yield stress can be expressed as

$$\sigma_y = A + B(\varepsilon_{pl})^n \qquad (1)$$

Material parameters A, B and n were calibrated using the Levenberg-Marquardt nonlinear least square algorithm. The material model also includes a ductile damage cumulation [1] in the form:

$$\varepsilon^f = d_1 + d_2 e^{-d_3 \eta} \tag{2}$$

where ε^f is fracture locus, η is stress triaxiality and $d_{1\text{-}3}$ are material parameters. Due to lack of experimental data only parameter d_1 was calibrated considering tensile tests and d_2 and d_3 were adopted from [3].

All parts of ladder are modelled as continuum solids. FE model contains about 230 000 linear elements with full integration. In the first step, the ladder is loaded by own weight and by weight of empty container. In the second step, only weight of container increases. The simulations were done for all cases when whole ladder had degraded mechanical properties. Since whole ladder is not exposed to higher temperatures in reality the simulation of this case was made. There was focus on two cases. In the first the exposed part is on the end of ladder and in second the exposed part is in the middle.

Conclusions

Strength tests were carried out on portable ladders that are most commonly used. Numerical models were created on the basis of material and structural tests. Numerical simulations of partially exposed temperature ladders show the limits of existing standards. Tests of thermally loaded ladders are being prepared to confirm obtained numerical results.

Acknowledgments

This work has been supported by project No. VI20162020021 of the Security Research Program of the Czech Republic, Minister of the Interior of the Czech Republic

References

[1] ČSN EN 1147:2010 Portable ladders for fire service use, CEN 2010-04-29.

[2] G. R. Johnson, W. H. Cook, Fracture Characteristics of Three Metals Subjected to Various Strains, Strain rates, Temperatures and Pressures, Engineering Fracture Mechanic, vol. 21, no.1, p. 31–48, 1985.

[3] W. Zhenyu, Z. Yang, L. Xu, H. Zhiguo, Analysis of the Dynamic Response in Blast-Loaded CFRP-Strengthened Metallic Beams, Advances in Materials Science and Engineering, Volume 2013, p. 13, 2013.

[4] J. Kuželka, M. Španiel, K. Doubrava, Fire ladder test simulation, Computational mechanics 2019, 2019.

EAN 2020
58th conference on experimental stress analysis

Design and Optimisation of Thermoplastic Composite Aircraft Passenger Door

DOUBRAVA R.[1,a], RŮŽEK R.[1,b], KŘENA J.[2,c]

[1] VZLÚ (Czech Aerospace Research Centre), Beranových 130, 199 05, Prague – Letňany, Czech Republic

[2]Latecoere, Beranových 65, 199 02, Prague – Letňany, Czech Republic

[a]doubrava@vzlu.cz, [b]ruzek@vzlu.cz, [c]Josef.Krena@latecoere.aero

Keywords: Finite element,

Abstract. The papers deals with optimization of selected parts of an aircraft passenger door manufactured from high performance thermoplastic material. Work associated with the design and optimization of a novel shape, dimensioning of structural parts and developing of optimized manufacturing process for those parts is presented. The paper is focuses on horizontal beams, vertical frames and a lock system structural part. The results tend to achieve more cost and strength effective design of selected parts of the door structure. The novel design of the beam and frame significantly reduces manufacturing process time in comparison to current thermoset technology.

Introduction

Composite materials are increasingly finding use in new structures. With ever increasing demands on structures, especially in the area of its testing needed to approve a structure and obtain certification, both design and composite material applications cannot do without numerical simulations. Experimental testing of each design, or only partial changes in the design of the composite structure, is currently unrealistic, both in terms of time and money. Numerical simulations thus play an irreplaceable role in the application of composite materials. In the design of pressurised aircraft fuselage, the main role in weight reduction. The passenger door was selected for application of thermoplastic composites from point of view of its complex design integrated. The main focus is concentrated on the design of composite profiles and load transfer between door and fuselage.

Global FE model

The preliminary design of pressurised transport aircraft passenger door from composites materials was used for analysis of boundary condition and load distribution (Figure 1 left). The finite element (FE) model of passenger door was analysed in NX/Nastran FE code from point of view of displacements and reaction forces in the attachments points for defined pressurise loading [1].

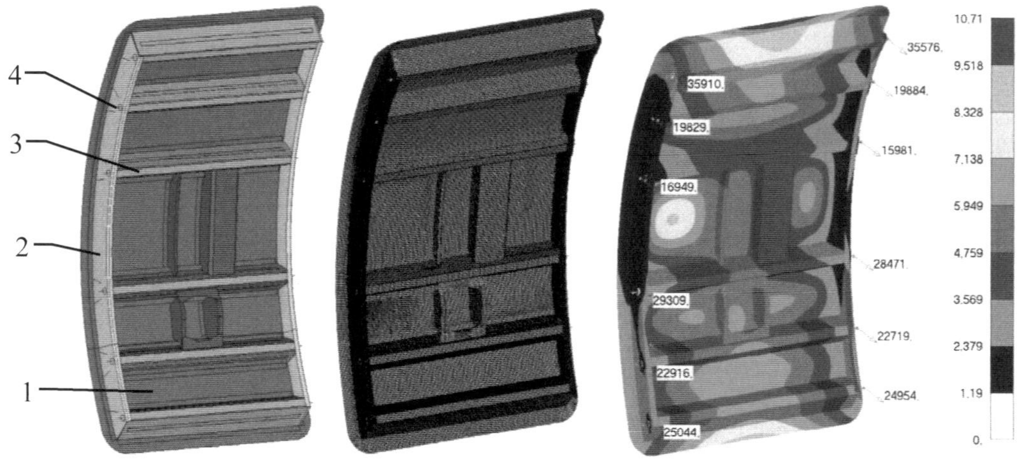

Fig. 1: CAD, FE model and analyses of preliminary design of pressurised transport aircraft passenger door from composites materials (1-skin, 2-vertical frame, 3-horizontal beam, 4-attachment points)

Local FE model

The cut-out section from global FE model was used for detail design and optimisation of connection between main parts of attachment point (skin, vertical frame, horizontal beam) [1].

The different shape o horizontal beam and metal bracket were design and numerically verified on the cut-out section of door (see Fig. 2).

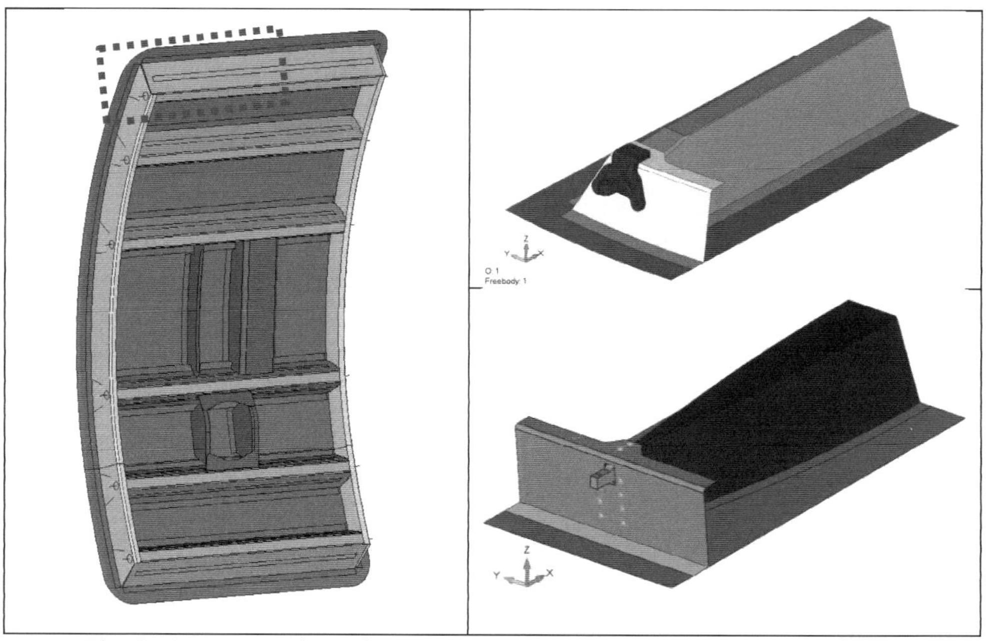

Fig. 2: Cut-out section of passenger door (left) and design proposals (left).

The optimal design was selected on the basis of these criteria:

- Displacement of skin part (requirements from point of view of aerodinamics)
- Allowable strain in composite parts
- Stess and bolt load distibution on metal fitting
- manufacturability

Optimisation

The topology optimisation was used for weight optimisation of metal fittings [2]. Optimization according to the maximum maxiumum Von Misses stress parameter Eq. (1) [5] was performed using the optimization algorithm of the TOSCA [4] code with the MSC/Nastran solver.

$$sig\_topo\_mises = \max \left| \frac{(\sigma_{vMises})^2}{(f(\rho_i)\sigma_y)^2} \cdot \sigma_y \right| \tag{1}$$

where σ_{vMises} is the von Mises stress in the center of the element, σ_y is the reference stress, $f(\rho_i)\sigma_y$ is the stress interpolation factor on elements that have a reduced density during topological optimization.

The total number of iterations of the optimization algorithm was 55. The boundary condfition for load and attachment was used from cut-out section analsyis. Figure 3 shows steps for topology weight optimisation.

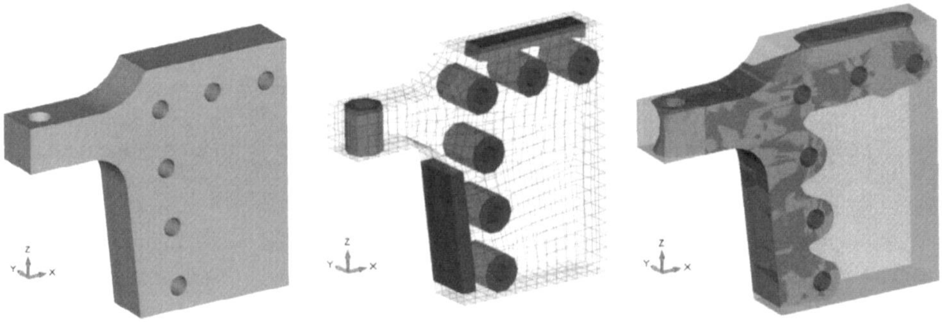

Fig. 3: Initial volume (left), restriction area (centre) and optimised design (right) of metal fitting.

The weight of metal fitting was reduced about 2.3 times for initial volume. In comparison with the previous design, the weight reduction is about 50 g. The total weight reduction for 12 fittings on the one aircraft door is about 0.6 kg.

Verification

The design of optimised metal bracket and attachment in composite horizontal beam was verified by experiments on simplified specimens (see Fig. 4) [3]. The test specimen design match boundary condition requirements of FE model of cut-out section and global FE model of the door. The aim of the test was verified optimised design of metal fitting, fasteners and lay-up of horizontal beam representing by composite plates.

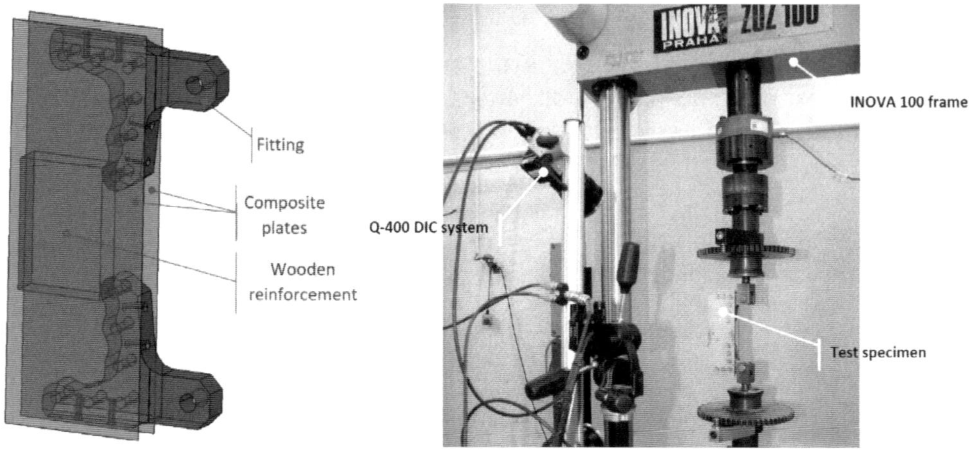

Fig. 4: Design of test specimen and arrangement of the test

Figure 5 shows proposed test specimen representative for the door fitting and comparison between experimental and simulation data for static load of 30 kN. The contactless strain measurement system Dantec Q-400 was used for the field of strain on the composite plate during loading.

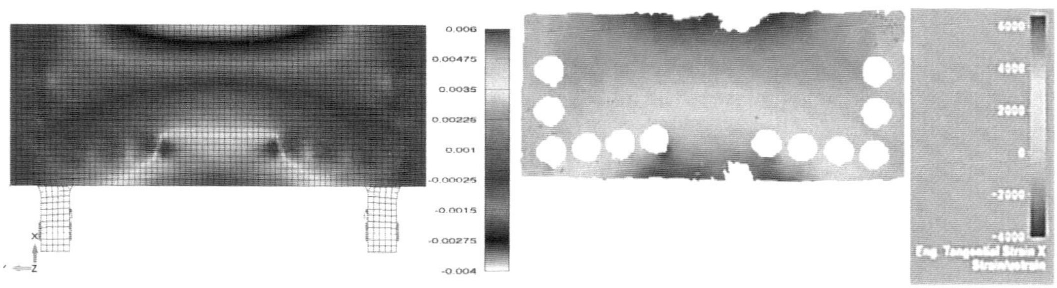

Fig. 5: Comparison of strain field between simulation (mm/mm) and experimental (μm/m) data on composite plates.

Comparison between test and simulation shows good agreement in the maximum and minimum value of strain and load distribution in the field between bolts for optimised fitting attachment.

Conclusions

The novel structural elements were designed. Numerical results were verified by experiments. Good agreement between numerical and experimental data was achieved. The results are to be used for design, manufacturing and experimental verification of a demonstrator.

Acknowledgment

This work was funded by the Ministry of Industry and Trade of the Czech Republic in the framework of TRIO platform, project No. FV30033 – Design and manufacturing process development of primary aircraft parts of advanced shapes of reinforced thermoplastics.

References

[1] R. Doubrava, Pevnostní analýzy vývoje demenstrátoru dveří dopravního letadla z uhlíkového kompozitu s termoplastickou matricí- souhrn prací, report VZLÚ R-7155 (2019).

[2] R. Doubrava, Topologická optimalizace návrhu kování dveří trupu dopravního letadla, report VZLÚ R-7143, (2019).

[3] R. Růžek, M. Kadlec, P. Kucharský, I. Mlch, R. Doubrava, Experimentální verifikace mechanických vlastností vybraných uzlů konstrukce dveří letounu, report VZLÚ R-7162, (2019).

[4] SIMULIA User Assistance 2017. Tosca Structure User Guide. Dassault Systemes, 2017.

[5] Bendsoe, M. P., Sigmund, O. (2004). Topology Optimization: Theory, Methods and Applications. Springer. ISBN: 9783540429920

Sensor for Deep Core Drilling Operational Parameters – Determination of the Drilling Capacity

DUB M.[1,a], HAVLÍČEK M.[1,b], ŠTOČEK O.[1,c]

[1]Czech Technical University in Prague, Faculty of Mechanical Engineering, Department of Designing and Machine Components, Technická 4, 166 07, Prague 6, Czech Republic

[a]martin.dub@fs.cvut.cz, [b]martin.havlicek@fs.cvut.cz, [c]ondrej.stocek@fs.cvut.cz

Keywords: Deep Core Drilling, Sensor, Drilling Capacity, Strain Gauge, Operational Parameters Identification

Abstract. The deep core drilling has currently wide application - construction drills, geological exploration, drills for measuring pendulums in dams etc. The most important parameters for the optimal drilling process are torque, pressure force and rotation speed. These parameters depend on the strength of the drilled material (drilling capacity) and of course on the drilled diameter and applied pressure force. The contribution describes the design of the special sensor for measurement of drilling torque, force and rotational speed. These data are further used for determination of the drilling capacity of the drilled material. There is no methodology to calculate this parameter. Two approaches are used and compared to determine desired parameter of the drilling capacity.

Introduction

The deep core drilling has currently wide application - construction drills, geological exploration, drills for measuring pendulums in dams etc. The most important parameters for the optimal drilling process are torque, pressure force and rotation speed. These parameters depend on the strength of the drilled material (drilling capacity) and of course on the drilled diameter and applied pressure force. In-time identification of the operational parameters enables the optimization of the drilling process directly during the drilling process, [1, 2]. This contribution builds on EAN 2019 contribution, where the design and sensor calibration were introduced. This contribution describes the final design of the sensor for monitoring above mentioned parameters. Data from new terrain measurement are further presented and analysed in detail to determine the drilling capacity of the drilled material. The usage of the sensor is described in the scheme in Fig. 1.

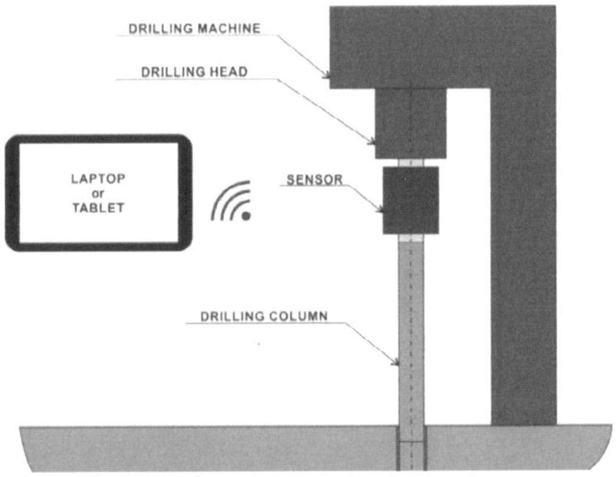

Fig. 1: Scheme of the sensor usage

Sensor final design, data acquisition and telemetry

The sensor design (see Fig. 2) and the principle of measurement and data acquisition were described in detail in [2]. Sensor design is based on the common drilling rod with hexagonal couplings, which is used in short version as a deformation member. Thus, it can be easily connected to the various drilling machines.

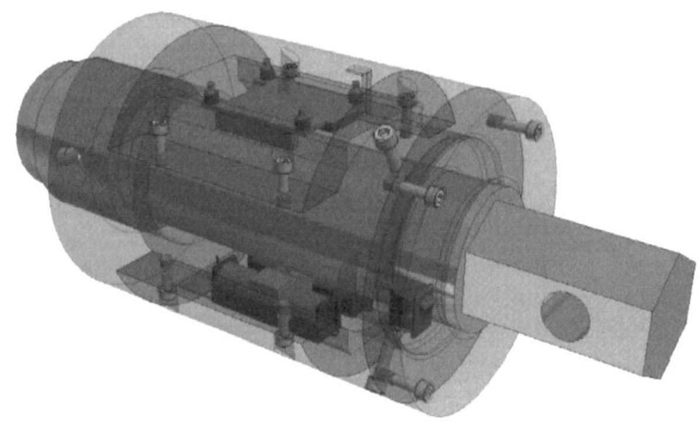

Fig. 2: Sensor design - CAD model and final product with Digi XBee receiver

Above mentioned operational parameters are identified by means of foil strain gauge rosettes (torque – HBM 1-XY21-6/120, axial force – HBM 1-XY91-6/120) and Hall probe in combination with the neodymium magnet (rotational speed), [3, 4]. Data are acquired by means of the special datalogger LPDR with 24bit A/D converter and with sampling frequency 500 Hz per channel. The telemetry is realized by the Digi XBee 2.4 GHz transmission. The power supply is realized by means of two Li-ion batteries of type 18650.

After the problem with insufficient strength of steel S355, the deformation member material was changed to 42CrMo4 with heat treatment. Whole sensor is encapsulated by the FDM printed cover made of PETG material. The design was optimised to achieve better ergonomics.

Terrain measurement

The terrain measurement is carried out with the final sensor and the RDBS mini drilling machine, see Fig 3. The location of the drill is in Pátek near Poděbrady. From the regional geological point of view, the location belongs to the Czech Cretaceous Basin. Turonian sedimentation is represented by calcareous siltstones (marlstone) with positions of limestone. Sediments of the lower and middle turon (marlstone) in the total thickness of about 90 m, are the part of the artesian collector in Poděbrady hot-spring structure. Quaternary cover consists mainly of sandy to clayey-sandy loam. The thickness of the coverings does not exceed steady ground water level, usually 2 m below ground level. The drilled core can be seen in Fig. 4.

Fig. 3: Drilling in Pátek, Poděbrady region

Fig. 4: Drilled core – 1.4 m of backfill then marlstone (one column represents 1 m depth)

Data evaluation

Acquired date are post-processed and evaluated to get the time-behaviour of the drilling torque, pressure and rotational speed. These data are correlated with the depth of the drill and the geological description of the drilled core. The time behaviour of the torque and pressure force can be seen in Fig. 5. The first 1.4 m through the backfill are very quick to drill. The rest of the drill goes through the marlstone. There are several depth measurements during the drilling. These measurements are done by the tape measure. The graph of rotation speed is not presented here, because the rotation speed remains constant 30 rpm.

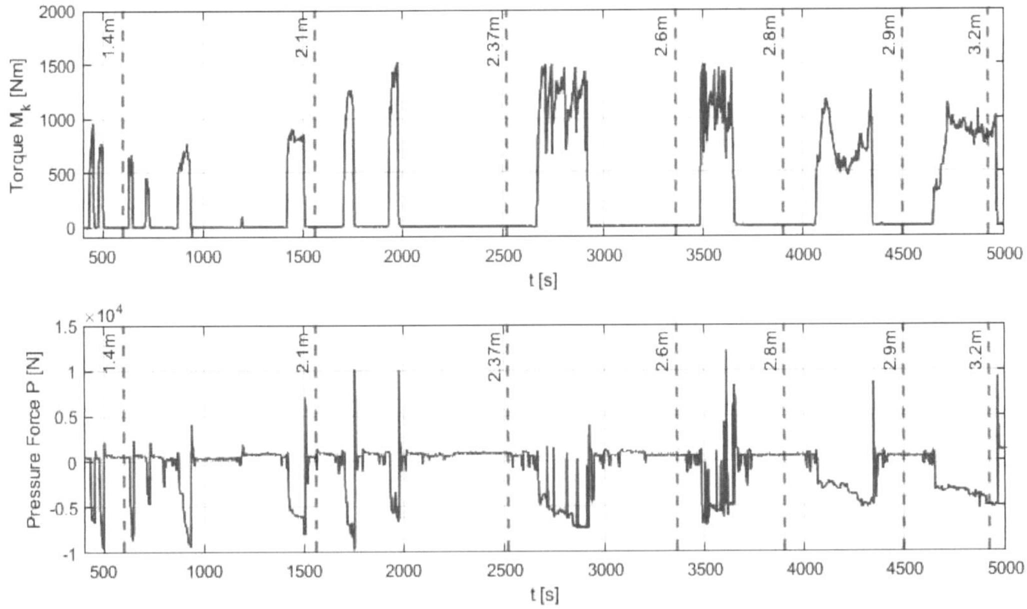

Fig. 5: Time behaviour of the drilling torque and pressure force during terrain measurement

The drilling torque is dependent on the input pressure force and also on the strength of the drilled material. To get the information about the strength of the material – drilling capacity, the torque is divided by the pressure force – drilling coefficient, see Fig. 6 top. The higher the drilling coefficient, the higher the material strength. There is evident increase of the drilling coefficient during the depth of the drill. Especially, there is significant increase in 2 m depth of the drill. Approximate depth of the interface of the backfill and the marlstone is about 1.4 m.

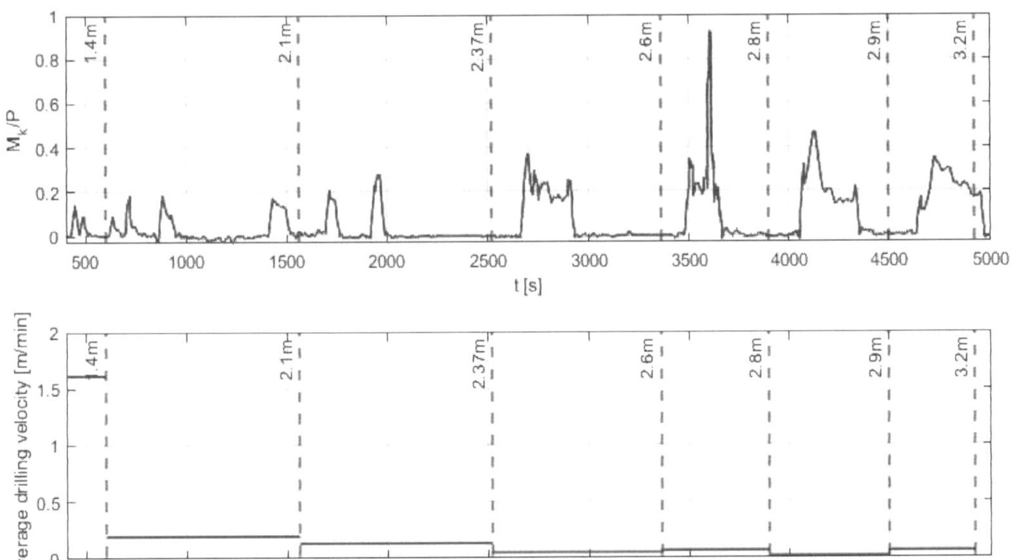

Fig. 6: Time behaviour of the drilling coefficient and the average drilling velocity

As a second indicator of the drilled material strength, the average drilling velocity is calculated. That is calculated between every depth measurement from the time duration of the drilling sequences. The time behaviour is described in Fig. 6, bottom. The difference between the drilling capacity of the backfill and the marlstone is distinct. The average drilling velocity in the marlstone is approximately 10x – 20x lower than in the backfill. The average drilling velocity decreases slowly till the depth 2.37 m and then remains almost constant.

Conclusions

The contribution summarizes the development and testing of the sensor for monitoring deep core drilling parameters. Measured data from the drill in Pátek location are presented and analysed. The main goal of deep core drilling parameters monitoring is the drilling capacity assessment. For this purpose, two quantities were calculated form measured data. The first one is the drilling coefficient – the ratio of the drilling torque and the pressure force, the second one is the average drilling velocity. Both parameters indicate the significant difference of the strength – drilling capacity of the backfill and marlstone in Pátek drilling location. During next improvement of the sensor there should be developed the online drill depth measurement and also online drilling velocity measurement. The main goal is to obtain the detail information of the drilling to determine the continuous information about the strength of the drilled material.

Acknowledgement

This work was supported by The Technology Agency of the Czech Republic project No: TH02010958 "Sensor for measuring of operational parameters of deep core drilling".

References

[1] M. Dub, V. Dynybyl, Sensor for In-time Identification of Deep Core Drilling Parameters, in: Lecture Notes in Mechanical Engineering (2020), 243-249.

[2] M. Dub, O. Štoček, F. Haas, T. Veselý, L. Kučera, V. Dynybyl, Monitoring of deep core drilling operational parameters, in: Experimental Stress Analysis - 57th International Scientific Conference, EAN 2019 - Conference Proceedings (2019), 58-65.

[3] P. Bauer, J. Kalivoda, System of axle-box force measurement for experimental railway bogie. In: Experimental Stress Analysis - 56th International Scientific Conference, EAN 2018 - Conference Proceedings (2018), 9-16.

[4] J. Kalivoda, P. Bauer, Measurement of wheel-rail contact forces at the experimental roller rig, in: Experimental Stress Analysis - 57th International Scientific Conference, EAN 2019 - Conference Proceedings (2019), 194-201.

EAN 2020
58th conference on experimental stress analysis

Embedded Fibre Optic Sensors for Structural Analysis of MTB Bicycle Frame

DVOŘÁK M.[1,a], PONÍŽIL T.[2,b], SCHMIDOVÁ N.[1,c], DOUBRAVA K.[1,d], KROPÍK B. [1,e], RŮŽIČKA M.[1,f]

[1]Department of Mechanics, Biomechanics and Mechatronics, Faculty of Mechanical Engineering, Czech Technical University in Prague, Technická 4, 160 00 Praha 6, Czech Republic

[2]Compo Tech PLUS, spol. s r.o., Nová 1316, 342 01 Sušice, Czech Republic

[a]Milan.Dvorak@fs.cvut.cz, [b]tomas.ponizil@compotech.com, [c]Nikola.Schmidova@fs.cvut.cz, [d]Karel.Doubrava@fs.cvut.cz, [e]Bohumil.Kropik@fs.cvut.cz, [f]Milan.Ruzicka@fs.cvut.cz

Keywords: Composite bicycle frame, FBG sensor, Embedded sensor, Structural analysis, Carbon composite wound tube

Abstract. The article is focused on structural analysis of the composite bicycle frame. Head tube joints were equipped with embedded optical FBG sensors. Frame was tested according to ISO test cases. Additional structural strength laboratory tests were done to evaluate range of limit case strain ranges. Digital image correlation method was used for the strain distribution evaluation in the head tube area. It was found that bicycle frame was designed with sufficient safety reserve.

Introduction

Bicycle frames made from Carbon Fibre Reinforced Polymers (CFRP) are widely commercially available nowadays. Besides prepreg moulding technology the composite wound tube technology is the most common bicycle frame manufacturing method. It allows to design the frame to customer's needs. In general, the most problematic places of the bicycle frame construction based on wound tubes are joints of individual tubes. Common way of joining the tubes using the glued sleeve joints has its performance and visual limits. So research was focused on improvement of joints made using the hand lay-up lamination technology. Structural strain analysis is an integral part of such research. However, commonly used electric strain gauges method has crucial disadvantage. Strain gauge cannot be embedded inside the composite lay-up or adhesive joint, so only a surface strain can be measured. This could be insufficient in a case of need of strain analysis of the place with complicated geometry. Structural analysis of the specific layers of composite lay-up or joints can be performed using the Fibre Bragg Grating (FBG) sensors measurement technology, because FBGs are small enough to be placed between the particular filaments.

Experimental Methods

Two strain analysis methods were used. Embedded FBG sensors were used for the local strain analysis. Digital Image Correlation (DIC) method was used to examine complex mechanical strain distribution in the head tube area.

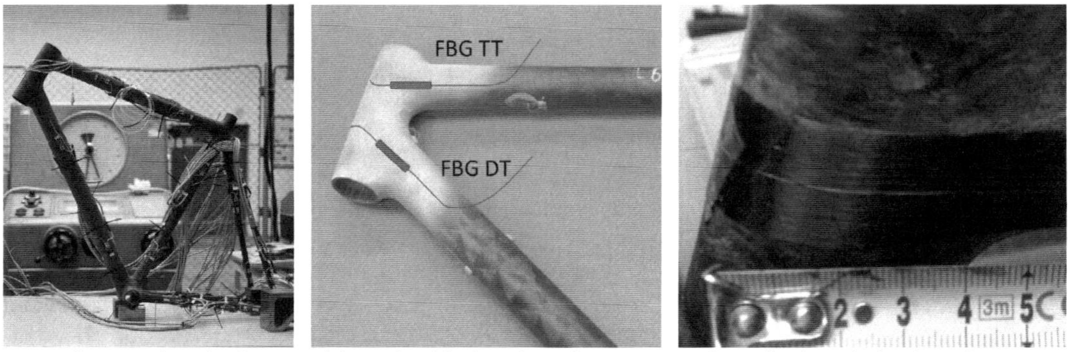

Fig. 1: Bicycle frame with embedded and surface sensors installation (left); locations of FBG sensors (blue marking, centre); detail of FBG sensor placement during the hand lamination (right)

Structural FBG sensors. Bicycle frame was instrumented with surface-mounted strain gauges and structural FBG sensors (Fig. 1 left). Optical fibres were embedded into the joint of the top tube (FBG TT) with head tube and into the joint of down tube (FBG DT) with head tube (Fig. 1 centre), directly between the unidirectional carbon filament layers (Fig. 1 right). These places are critical for transfer of horizontal forces from the safety point of view. Two ORMOCER® coated optical fibres with outer diameter of 0.2 mm, instrumented with one FBG sensor per fibre (with central wavelength of 1550 nm) were used. Surface routing of the fibres was protected using the plastic tubes. Mechanical strain indicated by the FBG sensors was measured and evaluated using the Safibra FBGuard optical interrogation system.

Digital Image Correlation. 3D DIC method was used to measure surface strain field in the head tube area. Dantec Dynamics Q400 commercial DIC system was equipped with two 16 Mpix digital cameras. Istra 4D software was used for data evaluation. To compare measurement methods, average longitudinal strain was evaluated along the line representing FBG sensor routing in its sensing area.

Experimental Procedure

In the first phase of the experiments, the bicycle frame was tested in cooperation with the EFBE Prüftechnik GmbH. The aim was to get a comprehensive idea of the behaviour of the bike frame in the ISO defined load cases. This was followed by the ergometer test and the laboratory testing up to structural failure to examine the structural strength of the frame [1].

Standardized Load Cases. Set of frame loading procedures was based on ISO 4210-6 [2] bicycle safety standard supplemented by specific load cases: 1. pedal forces; 2. horizontal forces (front/back load through the fork); 3. vertical forces (loading through the seat tube); 4. bottom bracket stiffness; 5. head tube torsion stiffness (measured between the front and rear wheel plane, see Fig. 2 left). Mechanical strain indicated along the FBG sensor axis was measured (see Fig. 2 right for results of test no. 5) and was then used to evaluate local stresses for each load case.

Ergometer test. The frame was fitted with components and loaded by pedal forces on an ergometer. The ergometer is an exercise machine that loads the rear wheel and creates pedalling resistance. The rider performed a series of sprints in order to obtain the maximum load from pedalling.

Structural Strength Tests. The work was focused on the most critical load case in terms of a rider safety, the horizontal forces test. Forces acting on the fork during jump, riding through a pothole or a frontal impact were simulated using the hydraulic actuator. Experimental setup and an example of DIC method evaluation is in the Fig. 3. Pushing load case was examined because of the highest expected deformations.

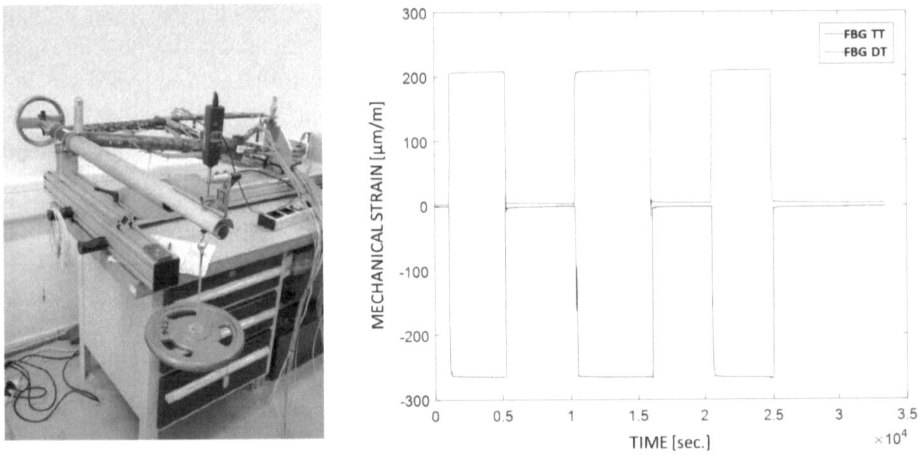

Fig. 2: Head tube torsion test set-up (left), evolution of mechanical strain measured by FBG sensors over time (right)

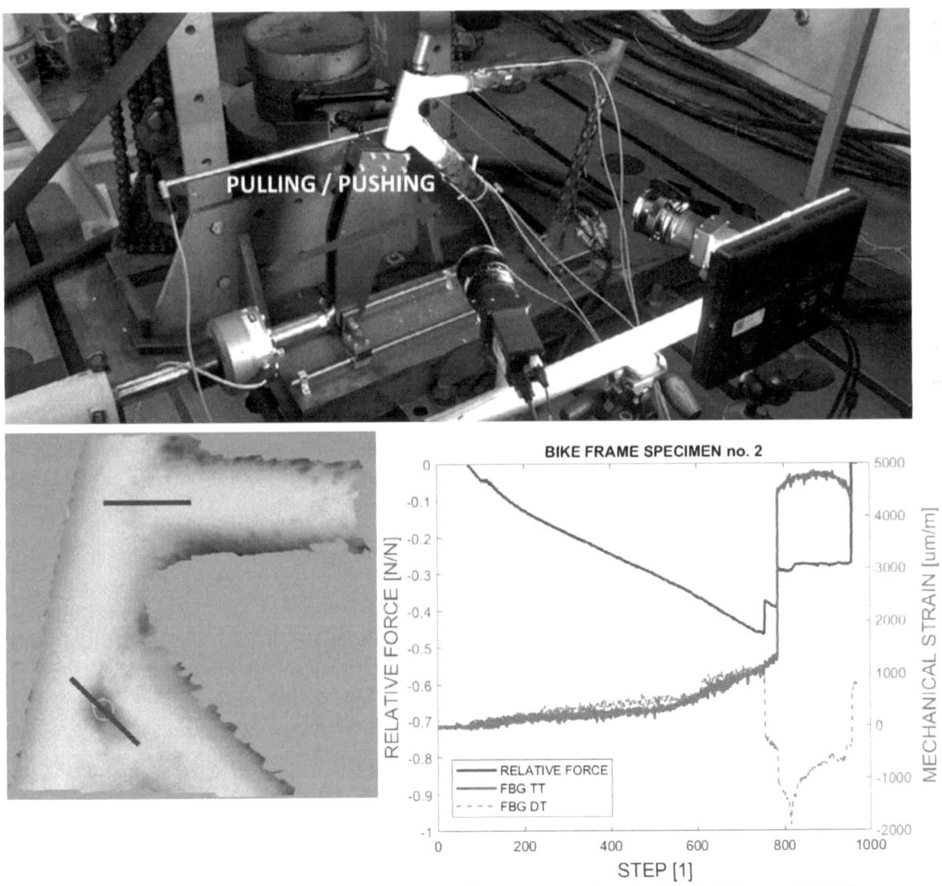

Fig. 3: Experimental setup for horizontal forces test (top), definition of FBG sensors locations for DIC method evaluation (bottom left), mechanical strain in the FBG TT and FBG DT sensors locations (bottom right)

Results and Discussion

Experimental results are presented in the Table 1. Maximum strain values for individual load cases were evaluated separately for the FBG DT and FBG TT areas. It can be seen that the highest strain values are resulting from pedal forces. Regarding the tests feasible in the laboratory and usable for the verification of finite element models, the highest strain values were found in the case of horizontal forces. Down tube is apparently the main load bearing bike frame member. The first five rows represent ISO defined test cases. Compared to the maximum strain values evaluated from strength test it is obvious that the tested bike frame design has sufficient safety reserve.

Table 1: Summary of experimental results

	FBG DT [μm/m]	FBG TT [μm/m]	comments
pedal forces	735 / -595	575 / 614	loading from left side / right side
horizontal forces	349 / -357	133 / -177	pulling / pushing
vertical forces	12	-14	
bottom bracket stiffness	-38	41	
head tube torsion stiffness	-267	209	
ergometer test	963 / -689	627 / -766	loading from left side / right side
structural strength test (data evaluated using DIC method)	-1 800 ÷ 940	-10 ÷ 1 700	range of maximum strain values in the FBG sensor areas (pushing load case)
	2 000 ÷ 4 000	6 200 ÷ 13 000	range of maximum strain values in the head tube area (pushing load case)

Conclusions

FBG sensors were used for evaluation of critical stresses inside the structural joints of bicycle frame tubes during the ISO defined load cases. Strain value ranges evaluated for this mandatory load cases using the sensors embedded inside the joints were compared to limit strain values evaluated during the strength tests. It was confirmed that critical, from the point of view of rider safety, strain values are well above the values resulting from the required ISO defined standard tests. Moreover, results were used for improving the numerical model of frame and for the further improvement of joint design. Operational tests using the FBG sensors are planned and currently postponed because of the need of miniature optoelectronic measurement device, which is not commonly available.

Acknowledgments

The authors would like to thank the Technology Agency of the Czech Republic for supporting this research with project no. TJ02000252.

References

[1] T. Ponížil, Analýza deformací a pevnosti kompozitového rámu silničního bicyklu, Praha, 2019, diplomová práce, FS ČVUT v Praze

[2] International Organization for Standardization. ISO 4210-6:2015(E) Cycles — Safety requirements for bicycles —Part 6: Frame and fork test methods.

Modal Analysis of Alkali Activated Slab Degraded by Ultraviolet Light

DVOŘAK R.[1,a], TOPOLÁŘ I.[1,b], PAZDERA L.[1,c], BÍLEK V.[2,d], HRUBÝ P.[2,e], PLSKOVÁ I.[1,f]

[1]Institute of Physics, Faculty of Civil Engineering, BUT – Brno University of Technology, Brno University of Technology, Veveří 331/95 602 00 Brno, Czech Republic

[2]Institute of material science, BUT – Brno University of Technology, Purkyňova 464/118, Královo Pole, 61200 Brno 12

[a]dvorak.r1@fce.vutbr.cz, [b]libor.topolar@vutbr.cz, [c]pazdera.l@fce.vutbr.cz, [d]bilek@fch.vut.cz, [e]xchrubyp@fch.vut.cz, [f]plskova.i@fce.vutbr.cz

Keywords: Modal analysis, frequency response function, materials, alkali activated systems, non-destructive testing

Abstract. In the field of vibroacoustic testing methods an experimental modal analysis is vastly used, which allows inspecting natural frequencies and damping ratios of test specimens. In the literature a numerous approach is being used for laboratory testing setups which mostly consist of an elastic suspension of test specimen and usage of different types of accelerometers, shakers, and modal hammers. The presented paper focuses on the comparison of different test setups of used sensors and impact exciting of a test specimen of slab made of alkali-activated slag mortar, which was degraded by ultraviolet light. The paper aims to find optimal setup combination for testing of such material by experimental modal analysis and decide if ultraviolet light alter the vibroacoustic properties of test specimens.

Introduction

Based on statistics across the globe, the main construction material used in civil engineering is concrete [1-4], where main used binder component is cement. The manufacturers are pushed to reduce the amount of released CO_2 per ton of cement. either partial substitution of clinker by secondary raw latent hydraulic additives (such as blast granulated slag, fly ash etc.) or usage of completely different binders.

Alkali-activated systems represent applicable alternative binder to conventional cement. Such material has advantage in the utilization of secondary raw materials and much lower input energy and carbon dioxide emissions compared to the manufacturing of Portland cement. Although it is relatively well-known material [5], its usage in the industry is still limited to prefabricated elements and monolithic structures mostly in Ukraine [6]. Alkali-activated materials are formed by the reaction between suitable alumina-siliate precursors with an alkaline solution. The whole process of alkaline activation starts with the dissolution of the reactive precursor particles in highly alkaline environment and continues with the nucleation and poly-condensation reactions to form main hydration products like C-A-S-H in case of alkali-activated blast furnace slag or N-A-S-H for the alkali-activation of fly ash. The hydration products formed are amorphous or semi-crystalline. Some minor crystalline phases can also be found with dependence of alumino-silicate precursor composition [7]. Spatially crosslinked structure of C-A-S-H gels is produced by this reaction [8], which leads to more homogenous and partially amorphous structure apart to crystalline structure of cement stone.

Testing of such material consists mainly of physical-mechanical and physical-chemical procedures. One of the yet unknow property of this composite is its resistance to ultra-violet radiation. The presented paper consist of an experiment of preparation of 4 test specimens of alkali-activated mortar of dimension 320×82×19 mm, which were substituted to cycling for 200 and 268 days of UV radiation, change of temperature and humidity in the QUV-Sun [9].

Test specimens were stored in QUV chamber from 7.8.2019 to 19.02.2020 and the specimens were subdued to UV radiation for 6432 hours. The chamber individual cycle consists of 4 hours of UV radiation of intensity 0.89 W/m^2 at temperature 60 °C and 4 hours of water condensation at temperature 50 °C.

Degradation in QUV chamber is mostly used on polymers, exterior facade elements etc. for fast aging procedures. Similar measurement on alkali-activated specimens is fairly narrow or even non-existing among the scientific papers. As example can be presented study which used UV chambers for accelerated aging of resin treated concrete surface [10]. The conclusion of cited paper sums, that concrete itself is quite durable in terms of UV degradation, and the only part which suffers are polymer coatings. It is therefore expected that alkali-activated composite will behave similar to standard concrete.

Designed mixture

The mixture composition is based on alkali-activation of granulated blast furnace slag (GBFS). The GBFS was obtained by Kotouč Štramberk Ltd., from city Štramberk in Czech Republic. The GBFS composition determined by XRF is shown in Table 1 . It had a specific surface area according to Blaine about 400 m^2·kg^{-1}. A fine aggregate of siliceous sand of three different fraction with maximal grain size 2 mm is used according to the EN 196-1. The 50% solution of sodium hydroxide in dosage of 6% Na_2O by the slag weight was used as an alkaline activator. Water to slag ratio (including extra added water and water from activator) was 0.45. To achieve the desired workability, the commercial superplasticizer based on lignosulfonates (ChrysoPlast 461) was used in 1% dose with respect to the slag weight.

Table 1. The BFS chemical composition (XRF)

Chemical composition in wt. %									
CaO	SiO_2	MgO	Al_2O_3	SO_3	TiO_2	K_2O	MnO	Na_2O	Fe_2O_3
41.1	34.7	10.5	9.1	1.4	1.0	0.9	0.6	0.4	0.3

Firstly a liquid components with slag were mixed for 3 minutes, and afterwards the sand was added to mixture. Whole mixing took 9 minutes. De-molding of test specimens stored in autogenous conditions (sealed with stretch foil) took place 24 hours after the mixing under the autogenous conditions at temperature circa approximately 25 °C.

Used methods

The specimens were tested before and after the degradation by experimental modal analysis. In the evaluation a natural frequencies Nf and damping ratios ζ_i are compared. Testing setup is illustrated in Figure 1. Test slabs of diameters 0.320×0.082×0.019 were placed on two rubber pads in the transverse direction. There were 4 test slabs test in total, which were divided into two groups. First group with specimens names A and B was placed in the QUV testing chamber facing by bottom surface to the source of ultraviolet light and second group with specimens names C and D was facing by upper surface to the source of ultraviolet light.

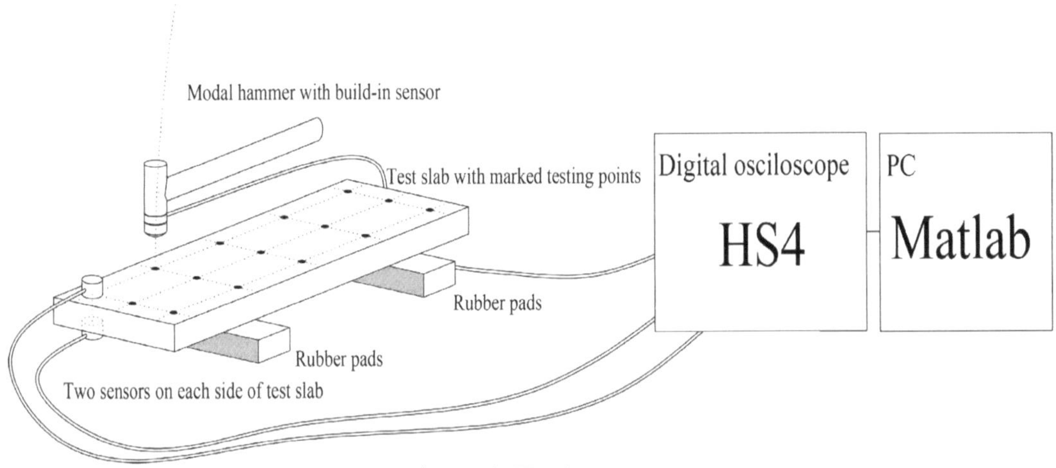

Figure 1: Testing setup

During testing two piezoelectric accelerometers were used, one attached to upper side, and second attached to bottom side of the test slab. For impacting the test specimen, a modal hammer with build-in force sensor was used. Both surfaces of testing slab were divided to 15 testing points, where first point was occupied by receiving sensor on both sides. Each testing point was then impacted three times. After all testing points were tested, the test specimen was turned upside down, and the same measurement was done on the other side. For each testing point an input signal from hammer sensor and response signals from both upper and bottom receiving senors attached to test specimen were recorded.

For the data acquisition a digital oscilloscope HS4 was used. The sampling frequency of board was set to 195 KHz in 16-bit resolution. Each measured signal consists of $65 \cdot 10^3$ samples and is 0.3 s long with 5% pretrigger. Sensors were attached to the surface of test specimens by beeswax, which is most common approach in vibroacoustic testing in civil engineering materials.

During computing of frequency response function [11], a signal recorded by build-in hammer sensor and signal of receiving sensor attached to surface of test specimens are used for estimation of natural frequencies and damping ratios. In this specific measurement we approach to the whole process as a Single Input-Single Output (SISO) state, despite the possibility of usage of Single Input-Multiple Outputs (SIMO) approach. This is done mainly to compare the resulting estimated natural frequencies in all combination. For processing of measured data, a Matlab software was used, precisely Signal Processing Toolbox and its frequency response function [12] and for extraction of natural frequencies and damping ratios a stabilization diagram function was used [13].

By the used setup we receive in total 8 combination of used signals from hammer and sensor for each degradation state of test specimen. The resulting interpretation of results then focus on the analysis of natural frequencies before and after the degradation of test specimens by ultraviolet light, and comparison of used testing setup and its influence on success rate of natural frequencies retrieving.

Results of testing

The procedure of retrieving of natural frequencies and damping ratios of each testing specimen consist of several steps, which will be described on the example of one test specimens. The final table of retrieved natural frequencies sums up all conducted measurements along with the table comparing all used testing setups.

Example of exciting signal from model hammer and signal measured by both sensors is illustrated at Figure 2. This example shows situation where impacts were aimed at the upper surface of the test specimen. This side is uneven, which is mainly caused by the way the test specimens are manufactured in the mould. By this standard procedure specimens has even surface on sides and bottom surface, but upper surface might suffer on the presence of pores, cracks and other defects, which can influence the testing. As expected, the signal from upper sensor has higher amplitude, then sensor from lower surface. [14]

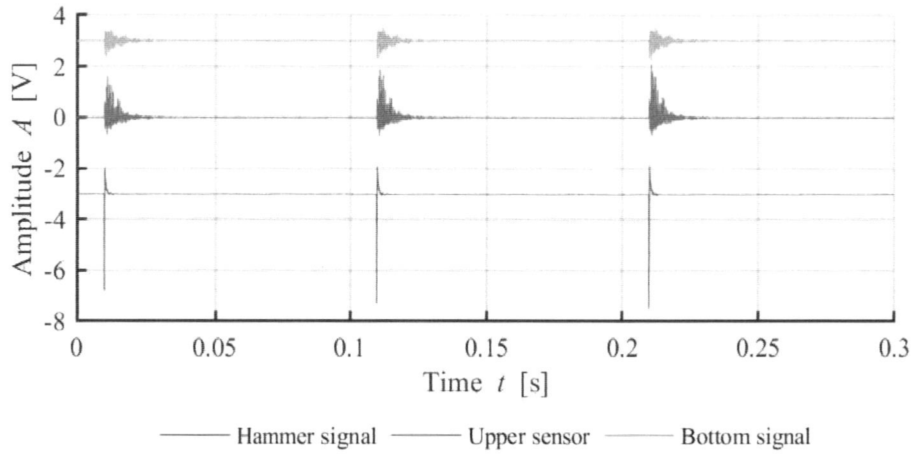

Figure 2: Example of hammer signal and signal response from upper and bottom sensor

From the acquired signals a frequency response function is obtained using Welch's method using least-squares complex exponential (LSCE) algorithm [15]. An example of magnitude and phase of measurement from first row and first column of testing grid is shown on Figure 3. Subsequent localization of present normal frequency modes and their damping ratios are retrieved using subspace method.

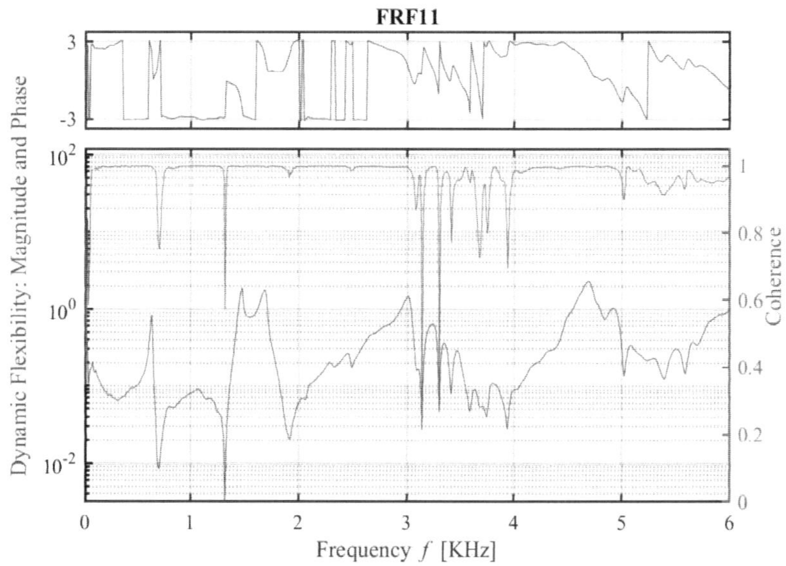

Figure 3: Frequency response function of acquired signals (first column, first row of testing grid on upper side)

The subspace aliased method is fairly used for estimation of modal shapes [16-18]. Example of stability diagram with localized natural frequency modes is shown on Figure 4. In modal analysis, a stabilization diagram is an important tool that is often used to assist the user in separating the physical system poles from mathematical ones [19]. Each given pole is classified as stable in frequency if its natural frequency changes by less than 0.01% as the model order increases. Classify a given pole as stable in damping if the damping ratio estimate changes by less than 0.2% as the model order increases [20]. In the present study the number of localized mode order was upon the initial testing set to 30.

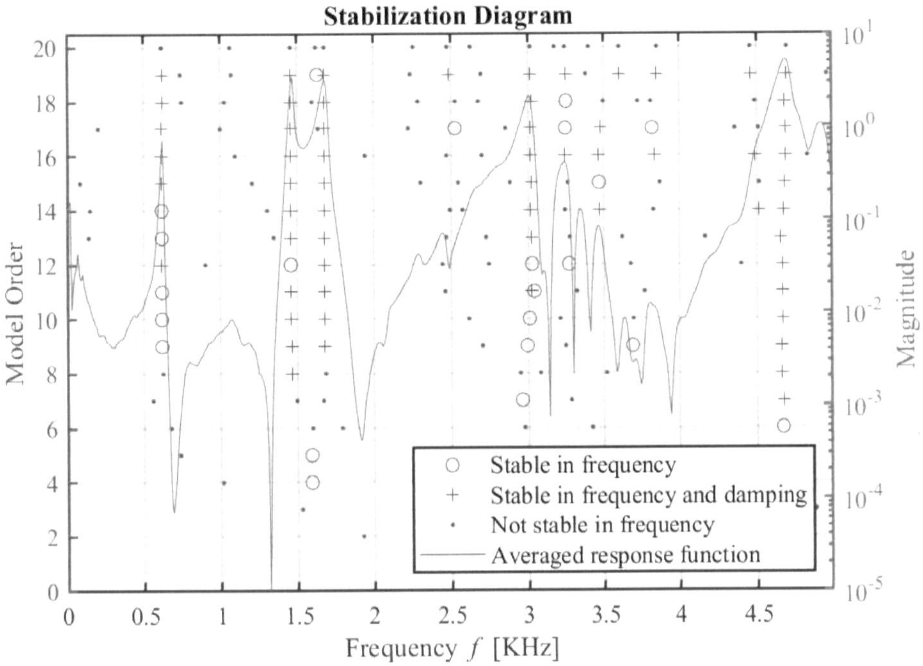

Figure 4: Estimation of stable frequencies of single testing position

The stabilization diagram clearly shows automatically localized natural stable frequencies. Histogram of damping ratios of all conducted tests shown at Figure 5 illustrates damping ratios of all localized frequency modes from all testing points of upper and bottom side of test specimens.

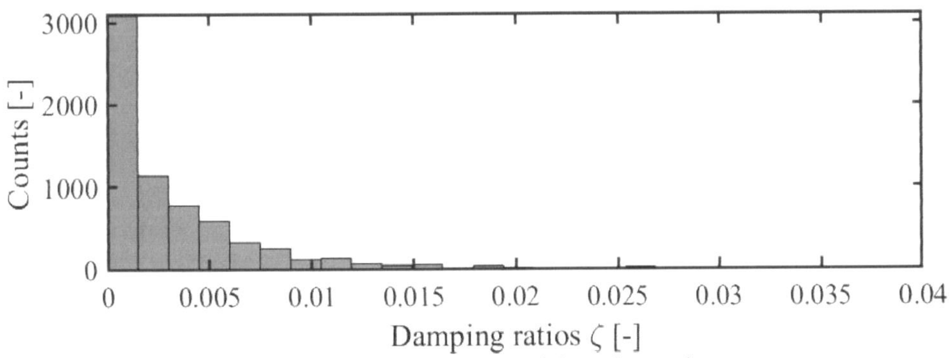

Figure 5: Histogram of damping ratios

The measured natural frequency modes exhibit relatively low damping ratio with exponential distribution across all conducted measurements. Most of the localized natural frequencies are very stable, and for next interpretation only natural frequencies with damping ratios bellow 0.02 are compared. By this step, we compare only stable modal shapes.

Frequencies retrieved from stabilization diagram and filtered by damping ratios threshold are then stacked up from all different setups for each testing specimen. Given by subspace method processing, each mode shape is represented by different amount of the close values of specific frequency [21-22]. Different measuring setups can result in different amount of detected stable frequency values; thus, each setup can be compared by number of individual frequency values that are parts of each stable natural frequency mode. The Figure 6 compares all frequencies from testing of specimen C, where dots represent each setup combination consisting of side of test specimen, which was degraded, surface which was impacted by modal hammer and receiving sensor, whom signal was used for frequency estimation by subspace method.

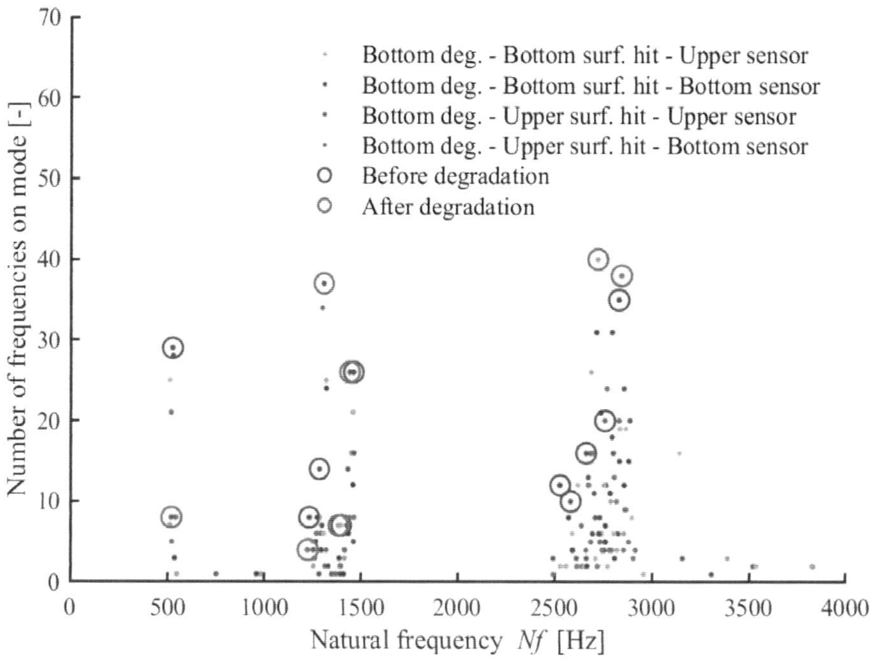

Figure 6: Example of natural frequency histograms of all testing combinations for test specimen C; marked frequencies are dominant in reference and degraded state of sample

The rings then illustrate which frequencies consist of highest amount of values of each state before and after the degradation by UV light. This process was used on all measurements which results in first 10 frequency modes of each specimen before and after the degradation.

For modal analysis computed in Ansys was used poison ratio equal to 0.2, density of material to 2320 $kg \cdot m^{-3}$ and dynamic modulus of elasticity to 1100 GPa. These material characteristics were measured in past publication [23]. Subsequently first ten frequency modes are compared with first ten modes obtained from experimental modal testing, in type of setup hit-sensors which had highest frequency counts based on Figure 6.

Table 2: Exported natural frequencies before and after the degradation with comparison of natural frequencies obtained from ANSYS model

n	Mode frequencies *Nf* [Hz] of test specimens								ANSYS
	A		B		C		D		
	before	after	before	after	before	after	before	after	
1	636	653	645	672	525	593	539	607	697
2	1482	1518	1367	1485	1262	1304	1280	1472	1581
3	1665	1764	1493	1540	1314	1377	1383	1582	1884
4	2933	1853	1678	1790	1446	1453	1436	1664	2479
5	3013	2282	1776	1919	2682	2657	1961	2980	3226
6	3069	3094	2142	3178	2735	2711	2383	3034	3589
7	3262	3163	2422	3445	2808	2816	2495	3139	4992
8	3331	3354	3069	4889	2893	2922	2646	3200	5536
9	4548	3437	3193	4940	4059	4141	2706	4586	5620
10	4671	4817	3264	4995	4165	4221	2804	4666	5723

At the table 2, for each test specimen is present state before and after the degradation by UV light. At all frequency modes we can observe a slight increase in first three frequency modes, in range 2-12 %. Natural frequencies obtained from Ansys model are higher than measured natural frequency modes in real specimens, but if we compare first frequency mode with average first frequency modes of all specimens, the difference is 0.33%, which is almost negligible difference. Higher frequency modes are much different from Ansys model with average difference in range 6-18%. The further optimization of Ansys model could lead to estimation of body stiffness of tested specimens and other material mechanical properties.

Table 3: Comparison of detected frequency modes expressed as a percentage of all detected frequency modes in test specimens

N [%]	Bottom side degraded				Upper side degraded			
	Bottom surface hit		Upper surface hit		Bottom surface hit		Upper surface hit	
	Upper sensor	Bottom sensor	Upper sensor	Bottom sensor	Upper sensor	Bottom sensor	Upper sensor	Bottom sensor
	25	100	56	69	56	88	44	63

For comparison of all conducted measurements setup we can interpreted the effectivity of each setup by expressing the number of successfully localized frequency modes in each setup, with reference to setup with highest stable frequency count and lowest damping ratios. At the Table 3 each setup is present with relative percentage of successfully detected frequencies.

It seems that, best position for testing was setup, where receiving sensor was placed on the bottom flat side of the test specimen, and the hits took place on the same flat side. Other variants of testing resulted in less obtainable information in terms of stable frequency modes.

Conclusion

Designed experiment enabled application of experimental modal analysis as effective method for evaluation of structural integrity. The change of frequency modes is mainly caused by ongoing hydration process in the test specimens, which caused a slight 10-15% increase in all frequencies in both long term cycling. This leads to main conclusions:

- hydration process continued even in QUV chamber, and strong UV radiation did not significantly affect the mechanical properties such as frequency modes or damping ratios,
- there was observed more chaotic behaviour in frequency modes shift after the degradation, where specific frequency modes could not be detected at some of the testing points of test grids,
- during the interpretation phase the most effective position of receiving accelerometer was at the flat bottom surface of the test specimens, without difference if top rough surface or bottom flat surface of specimens was exposed to UV radiation,
- the change in frequency modes indicates a change in inner integrity, where first three frequency modes are not constant at all test points of test specimens, as was in the initial state before degradation.

The conducted experiment also helped to set the basic conditions for testing of small-scale mortar test specimens by experimental modal analysis.

Acknowledgment

This paper has been worked out under the project Czech Science Foundation GACR No. GA 19-04703S supported by Czech Science Foundation and the project No. LO1408 "AdMaS UP - Advanced Materials, Structures and Technologies", supported by Ministry of Education, Youth and Sports under the „National Sustainability Programme I".

Reference

[1] K. Scrivener and R. Kirkpatrick, "Innovation in use and research on cementitious material", Cement and Concrete Research, vol. 38, no. 2, pp. 128-136, 2008.

[2] J. Damtoft, J. Lukasik, D. Herfort, D. Sorrentino and E. Gartner, "Sustainable development and climate change initiatives", *Cement and Concrete Research*, vol. 38, no. 2, pp. 115-127, 2008.

[3] J. Crow, "The concrete conundrum", *Chemistry World*, pp. 62-66, 2008.

[4] "Major countries in worldwide cement production from 2014 to 2018", *Statista*. [Online]. Available: https://www.statista.com/statistics/267364/world-cement-production-by-country/. [Accessed: 2020-01-22].

[5] V. Glukhovsky, "Soil silicates.", vol. 1959.

[6] A. Palomo, P. Krivenko, I. Garcia-Lodeiro, E. Kavalerova, O. Maltseva and A. Fernández-Jiménez, "A review on alkaline activation: new analytical perspectives", *Materiales de Construcción*, vol. 64, no. 315, 2014.

[7] *Alkali Activated Materials*. Dordrecht, 2014.

[8] T. Luukkonen, Z. Abdollahnejad, J. Yliniemi, P. Kinnunen and M. Illikainen, "One-part alkali-activated materials: A review", *Cement and Concrete Research*, vol. 103, pp. 21-34, 2018.

[9] "QUV ACCELERATED WEATHERING TESTER". [Online]. Available: https://www.q-lab.com/products/quv-weathering-tester/quv. [Accessed: 2020-03-08].

[10] T. Guo and X. Weng, "Evaluation of the freeze-thaw durability of surface-treated airport pavement concrete under adverse conditions", *Construction and Building Materials*, vol. 206, pp. 519-530, 2019.

[11] D. Ewins, *Modal testing: theory, practice, and application*, 2nd ed. Philadelphia, PA: Research Studies Press, 2000.

[12] A. Brandt, *Noise and vibration analysis: signal analysis and experimental procedures*. Chichester: Wiley, 2011.

[13] A. Arda Ozdemir and S. Gumussoy, "Transfer Function Estimation in System Identification Toolbox via Vector Fitting", *IFAC-PapersOnLine*, vol. 50, no. 1, pp. 6232-6237, 2017.

[14] X. Bao, H. Sun, G. Iglesias, T. Wang and C. Shi, "Signal denoising method for modal analysis of an offshore platform", *Journal of Loss Prevention in the Process Industries*, vol. 63, 2020.

[15] P. Welch, „The use of fast Fourier transform for the estimation of power spectra: A method based on time averaging over short, modified periodograms." IEEE *Transactions on Audio and Electroacoustics*. 1967, 15(2), 70-73. ISSN 0018-9278. doi:10.1109/TAU.1967.1161901

[16] G. Tondreau and A. Deraemaeker, "Numerical and experimental analysis of uncertainty on modal parameters estimated with the stochastic subspace method", *Journal of Sound and Vibration*, vol. 333, no. 18, pp. 4376-4401, 2014.

[17] M. W. M. a. R. J. SZCZEPAŃSKI, „MODAL ANALYSIS OF REAL TIMBER FRAME HOUSES WITH DIFFERENT INSULATION MATERIALS," *Advances in Science and Technology Research Journal,* pp. 215-221, 2016.

[18] G. Ivan and B. Moor, "Subspace identification combined with new mode selection techniques for modal analysis of an airplane", *IFAC Proceedings Volumes*, vol. 36, no. 16, pp. 675-680, 2003.

[19] Heylen W., S. Lammens and P. Sas. „Modal Analysis Theory and Testing", *KULeuven* (ISBN 90-73802-61-X), 1998.

[20] A. OZDEMIR, A. and S. GUMUSSOY, „Transfer Function Estimation in System Identification Toolbox via Vector Fitting", *IFAC-PapersOnLine. 2017*, 50(1), 6232-6237. ISSN 24058963. doi:10.1016/j.ifacol.2017.08.1026

[21] H. Vold, K. Napolitano and D. Hensley. „Aliasing in Modal Parameter Estimation." *SAE 2007 Transactions Journal of Passenger Cars: Mechanical Systems*. 2007, 2007-05-15, V116(6), 11. ISSN 0148-7191.

[22] X. Li, J. He, M. Li, H. Jiang and Y. Huang, "Modal analysis method for tensegrity structures via stiffness transformation from node space to task space", *Engineering Structures*, vol. 203, 2020.

[23] H. Šimonová, B. Kucharczyková, V. Bílek, L. Topolář and D. Kocáb, „Influence of curing conditions on mechanical and fracture properties of alkali activated slag concrete", *IOP Conference Series: Materials Science and Engineering.* 2019, 660. ISSN 1757-899X. doi:10.1088/1757-899X/660/1/012004

EAN 2020
58th conference on experimental stress analysis

Creep Behaviour of 316L Stainless Steel Srepared by 3D Printing

DYMÁČEK P.[1,a], KLOC L.[1,b], GABRIEL D.[2,c], MASÁK J.[2,d], PAGÁČ M.[3,e], HALAMA R.[3,f]

[1]Institute of Physics of Materials, Czech Academy of Sciences, Žižkova 22, 616 62 Brno, Czech Republic

[2]Institute of Thermomechanics, Czech Academy of Sciences, Dolejškova 1402/5 182 00 Praha, Czech Republic

[3]Faculty of Mechanical Engineering, VŠB – Technical University of Ostrava, 17. Listopadu 2172/15, 708 00 Ostrava, Czech Republic

[a]pdymacek@ipm.cz, [b]kloc@ipm.cz, [c]gabriel@it.cas.cz, [d]masak@it.cas.cz
[e]marek.pagac@vsb.cz, [f]radim.halama@vsb.cz

Keywords: 3D print, Selective laser melting, 316L Stainless steel, Creep

Abstract. The 3D printed 316L stainless steel produced by selective laser melting (SLM) was subjected to short term creep testing at 700 °C. Two directions of printing: i) horizontal and ii) vertical, were selected to test the creep performance of the steel. Comparison with the open literature data shows very good short-term creep properties of 3D printed steel that are superior to conventional steel. Both studied directions of printing show similar results so the steel can be considered from creep point of view as isotropic. The solution annealing prior to the creep testing slightly, but not substantially, lowers the creep performance of the steel.

Introduction

The 3D printing by selective laser melting (SLM) is a prospective method of metal powder consolidation and offers enormous possibilities for parts production. The mechanical behaviour of SLM prepared samples such as static, fatigue and creep properties has been intensively studied recently [1,2]. The necessity of experiments for a phenomenological description of fatigue, creep or creep-fatigue interaction behaviour in technical practice is evident [3,4] to perform reliable numerical analysis in various industries [5].

The short term creep behaviour at 700 °C of 316L cylindrical samples produced in horizontal and vertical direction by SLM technology is the subject of this study.

Material and testing

The technology of additive manufacturing usually named as Selective Laser Melting was used for the production of 3D printed 316L samples. The principle lies in the application of thin layers of a powdery material, which, according to the STL (an abbreviation of "stereolithography") model, are sintered with one laser beam layer after layer. The most important parameters of 3D print process set in Renishaw AM400 were: laser power 200 W, scanning speed 650 mm/s, exposure time 80 μs, focus size (laser beam diameter) 80 μm, layer thickness 50 μm, strategy Meander.

With respect to the orientation of the parts in the construction chamber, support material was designed to eliminate residual stresses and thermal distortion. When finishing 3D printing, the 316L steel has a Ra surface roughness of 5 to 10 μm [6]. Material for finishing operations by machining was added to the wall of the sample to achieve the desired surface roughness and manufacturing tolerances. The samples were scanned by the Meander strategy, which is characterized by a higher melting rate, and after each created layer, the scan direction is rotated by 67°. This controlled rotation ensures that the scanned laser trajectories of each successive layer are not identical. The chemical composition of the powder is shown in Table 1.

Table 1: Chemical composition of 316L powder in wt. %

Chem. Comp.	C	Cr	Ni	Mo	Mn	Si	N	O	P	S	Fe
%	<0.02	16–18	10–14	2–3	≤1.0	≤1.0	≤0.1	≤0.1	≤0.04	≤0.03	ball.

The steel was 3D printed in the form of cylinders (d = 12 mm, L = 75 mm) as shown in Fig. 1. Eight specimens were produced for each direction. After 3D print one group of cylinders for each direction was left in an as received while the second one was solution annealed at 1150 °C for 10 min. Subsequently cylindrical specimens with M10 threaded heads (d_{gauge} 6 mm, L_{gauge} 35 mm) were produced from 3D printed cylinders.

Optical micrographs of microstructure taken by Olympus GX51 Inverted Metallographic Microscope in bright field at 100× magnification are shown in Fig. 2. The arrows indicate the cylindrical specimen axial direction. The grinding and polishing with OP-S colloidal silica followed by electro-etching (10 V, 45 s) in 10% oxalic acid was applied for the metallographic surface preparation. Orientation of the layers for the two different building directions is clearly visible in as received specimens. The weld pool boundaries have fully dissolved after solution annealing. Some grains resemble twins but there is need of more detailed analysis of the microstructure by scanning microscope techniques. Some cavities of size between 20–200 μm are visible and some of them contain powder particles.

The specimens were crept in lever arm creep testing machines in Ar protective atmosphere at 700 °C and 120, 160 and 200 MPa stress in constant load conditions.

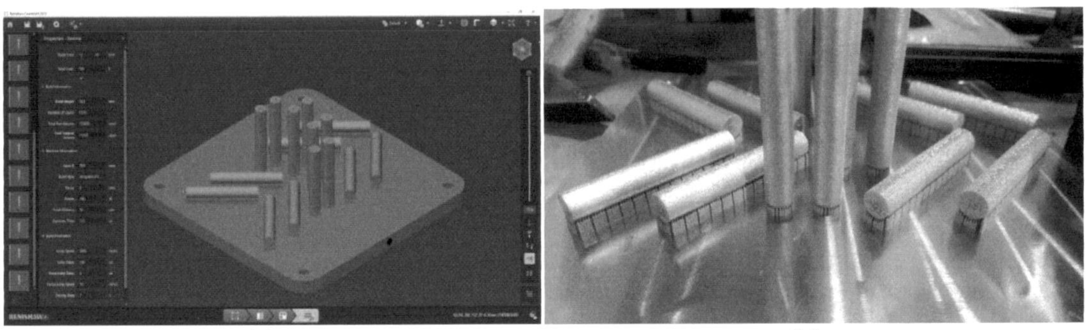

(a) (b)

Fig. 1: Specimen layout (a) in the printing software (b) after printing on the supporting bed

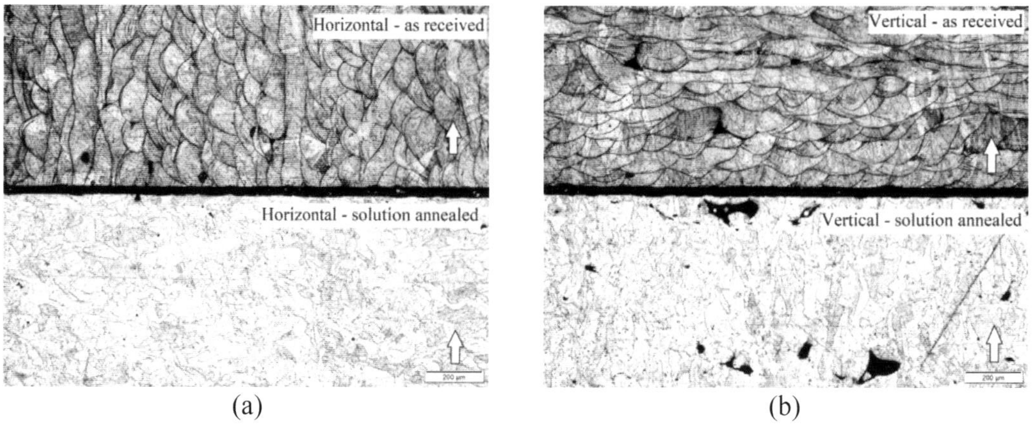

Fig. 2: Microstructure of (a) horizontally (b) vertically built specimens (arrows show specimen axial direction)

Results and discussion

The particular creep curves for all three selected stress levels are shown in Fig. 3. There is an obvious difference between as received and annealed specimens creep curves, the annealed steel performs slightly worse. The overall creep test results in Fig. 4 show excellent creep properties of the SLM. The orientation of the specimens plays minor role in as received state. In the solution annealed state the horizontal direction seems better than vertical. However, most 3D printed specimens perform better than the conventional steel tested in [7,8].

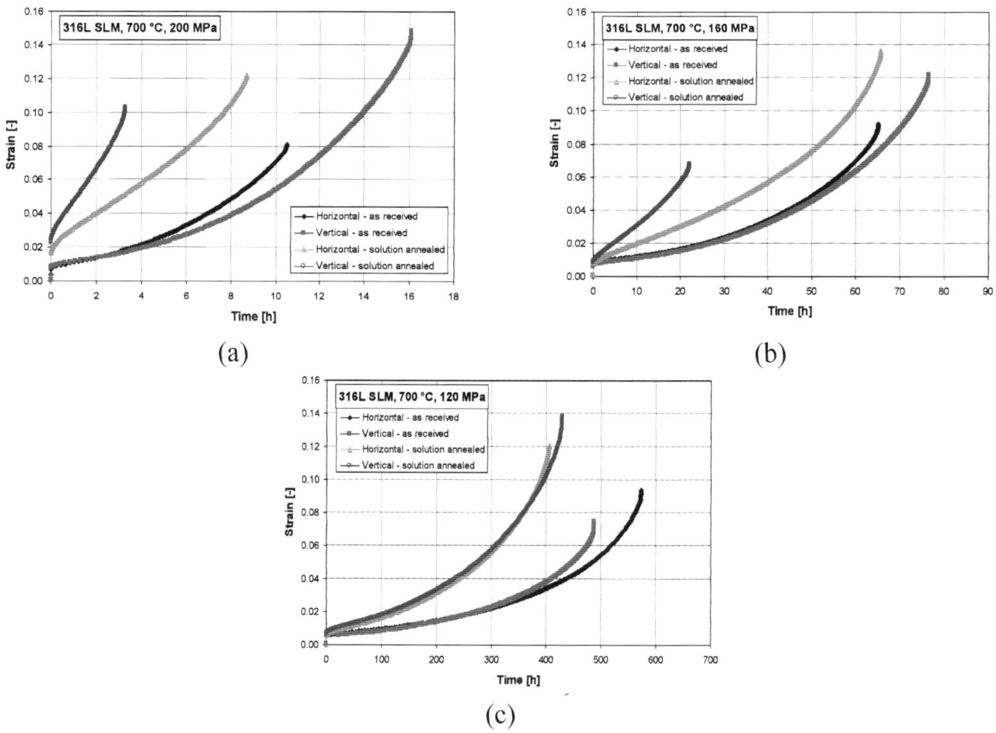

Fig. 3: Creep curves for all tested conditions

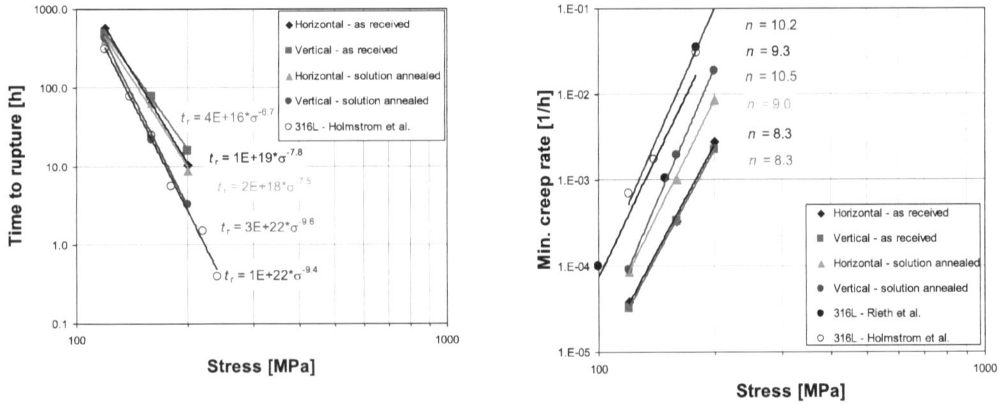

(a) (b)

Fig. 4: Creep rupture time (a) and minimum creep rate (b) dependence on applied stress at 700 °C

With lower stress σ and higher time to rupture t_r there is a tendency to receive comparable results with the conventionally prepared steel, as shown in Fig. 4a, but the minimum creep rates still remain notably lower for 3D printed steel which is obvious from Fig 4b. The stress exponent n is comparable or slightly lower for SLM steel compared to conventional steel.

Fractographs after creep in Fig. 5 show ductile fracture with quite limited necking of the as received specimens. There are marked areas of porosity (by the yellow line) that are more frequent and have larger areas on the horizontal specimen in Fig. 5a compared to vertical in Fig. 5b. Vertical specimen has more necking, but at lower stresses, there is no difference. The rupture strains are comparable for all four groups of specimens, which is obvious from the creep curves in Fig. 3. Some scatter in creep ductility is natural. The details of the fracture surfaces near the pores with remaining powder are shown in Fig. 6.

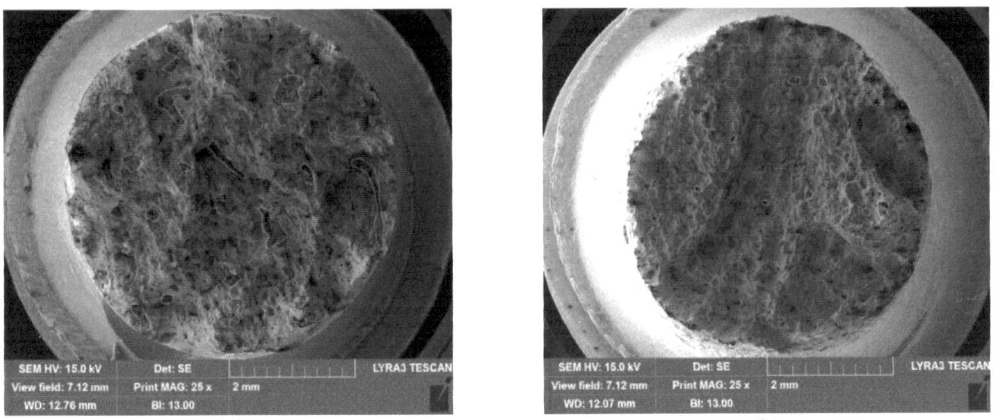

(a) (b)

Fig. 5: Fracture surfaces of tests at 200 MPa, (a) horizontal, t_r = 10.5 h, (b) vertical, t_r = 16.1 h

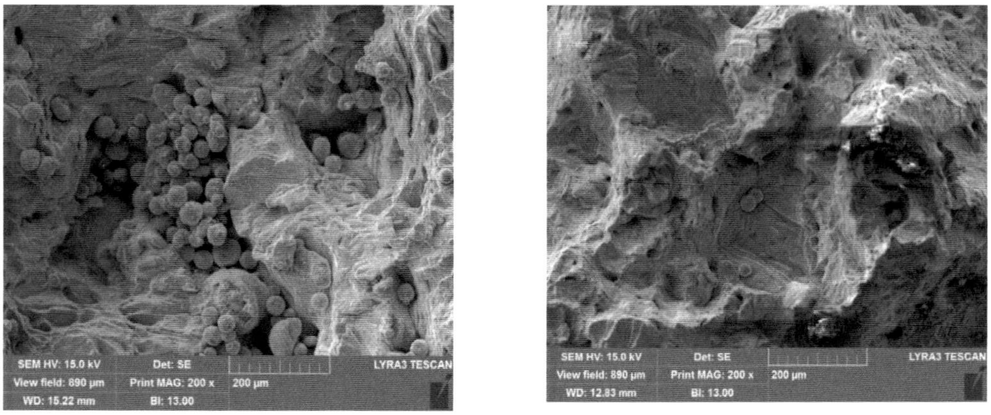

(a) (b)

Fig. 6: Detail of fracture near pores with unmelted powder of tests at 200 MPa, (a) horizontal, $t_r = 10.5$ h, (b) vertical, $t_r = 16.1$ h

Despite higher porosity, the better creep performance of 3D printed steel can be attributed to its advantageous microstructure. The microstructural defects and inhomogeneities introduced by SLM can in reality act as hardening objects (similar as precipitates or dislocations) and at the same time as the nucleation sites for the creep damage. One can expect that the rather homogeneously distributed SLM defects cause a significant hardening against dislocation motion and the dominant creep strain is mainly due to cavitation and fracture processes. This idea is supported by the fact of very small primary creep strain. It has to be still verified by detailed microstructure analysis of the crept specimens.

However, longer-term (up to 10 kh) creep behaviour of standard and sub-sized specimens should be further studied in order to assess the high-temperature performance of thin walled structures that might be more sensitive to manufacturing porosity than standard specimens. The microstructure analyses after creep are still under investigation.

Conclusions

In overall the 3D printed 316L stainless steel shows very good short-term creep properties superior to conventional steel. There is not any significant difference in creep performance between the horizontal or vertical direction of printing. Since both studied directions show similar results, the steel can be considered from a creep point of view as isotropic. The solution annealing leads to softening of the steel, higher minimum creep rates and lower times to rupture, which are still comparable or better than of conventionally produced steel.

Acknowledgments

This manuscript has been completed in association with project reg. no. CZ.02.1.01/0.0/0.0/17_049/0008407 financed by Structural Funds of the European Union and within the project TN01000024 financed by The Technology Agency of the Czech Republic.

References

[1] M. Kraus, V. Mareš, Properties of AISI 316 steel prepared by SLM method, in: Proceedings of New Methods of Damage and Failure Analysis of Structural Parts, 2018, Ostrava, Czech Republic http://konference.fmt.vsb.cz/work2018/abstracts/Kraus.pdf.

[2] R. Halama, M. Pagáč, Z. Paška, P. Pavlíček, X. Chen, Ratcheting Behaviour of 3D Printed and Conventionally Produced SS316L Material, in: Proceedings of the ASME 2019 Pressure Vessels & Piping Conference PVP2019, , San Antonio, TX, USA, 2019, paper number PVP2019-93384.

[3] P.R. Barrett, T. Hassan, A unified constitutive model in simulating creep strains in addition to fatigue responses of Haynes 230, Int. J. Solid. Struct. 185-186 (2020) 394–409.

[4] K.I. Kourousis, D. Agius, C. Wang, A. Subic, Constitutive modeling of additive manufactured Ti-6Al-4V cyclic elastoplastic behaviour, Tech. Mech. 36 (2016) 57–72.

[5] H.P. Mahajan, T. Hassan, Finite element analysis of printed circuit heat exchanger core for creep and creep-fatigue responses, in: Proceedings of the ASME 2019 Pressure Vessels & Piping Conference PVP2019, San Antonio, TX, USA, 2019 paper number PVP2019-93416.

[6] J. Hajnys, M. Pagac, O. Kotera, J. Petru, S. Scholz, Influence of Basic Process Parameters on Mechanical and Internal Properties of 316L Steel in SLM Process for Renishaw AM400. MM Sci. J. 2019 (1) 2790–2794.

[7] S. Holmström, Y. Li, P. Dymacek, E. Vacchieri, S.P. Jeffs. R.J. Lancaster, D. Omacht, Z. Kubon, E. Anelli, J. Rantalla, A. Tonti, S. Komazaki, Naveena, M. Bruchausen, R.C. Hurst, P. Hähner, M. Ruchardson, D. Andres, Creep strength and minimum strain rate estimation from Small Punch Creep tests, Mater. Sci. Eng. A 731 (2018) 161–172.

[8] M. Rieth, A. Falkenstein, P. Graf, S. Heger, U. Jäntsch, M. Klimiankou, E. Materna-Moris, H. Zimmermann, Creep of the Austenitic Steel 316L(N) Experiments and models, report FZKA 7065, Nov. 2004, Forschungszentrum Karlsruhe.

EAN 2020
58th conference on experimental stress analysis

Influence of Plasma Treatment on Polypropylene Microfibers in Foamed Cement Paste Containing Recycled Concrete

ĎUREJE J.[1,a], TREJBAL J.[1,b], TESÁREK P.[1,c]

[1]Faculty of Civil Engineering, Czech Technical University in Prague, Thákurova 7, 166 29 Prague 6, Czech Republic

[a]jakub.dureje@fsv.cvut.cz, [b]jan.trejbal@fsv.cvut.cz, [c]pavel.tesarek@fsv.cvut.cz

Keywords: Foamed cement paste, Oxygen plasma modification, Polypropylene microfibres modification, Recycled concrete

Abstract. This article deals the effect of oxygen plasma modification on mechanical properties of lightweight cement matrix. There is also described the process of plasma modification. Test specimens contain polypropylene microfibers, cement, recycled concrete, foaming agent and mixing water. Two sets of lightweight test samples with different density and a set of non-lightweight samples were made. To determine the effect of the plasma modification on the mechanical properties of the resulting cementitious composite, the samples were divided into two groups. Half of the samples contain oxygen plasma treated microfibers; the other half of the samples have the same composition but contain microfibers without plasma treatment. The mechanical properties of the test specimens were measured and the compressive strength of samples containing plasma modified fibers were compared with samples containing fibers without plasma modification.

Introduction

The cohesion of the cement matrix with the fibers depends on the interfacial transition zone (ITZ). The properties of the ITZ depend on the processing and properties of the cement matrix and the fiber surface. Appropriate processing of the matrix with fine fillers and the use of microfibers can cause a significant reduction of the transition layer, resulting in a much higher bonding of the fiber with the matrix [1,2]. The surface of the polypropylene fibers can be mechanically or chemically modified to increase the cohesion with the cement matrix [3,4]. Plasma treatment can modify the surface both mechanically and chemically. The effect of plasma modification on the fiber surface depends on many parameters - for example, on the pressure during modification (low or atmospheric) or on the gas used. The resulting effect also depends on the type of used device and its settings [5,6]. Polypropylene fibers are poorly wettable in water, oxygen plasma modification can be used to increase their wettability. During plasma modification, free electrons and ions impinge on the surface of the material, disrupt the surface bonds and form various functional chemical groups improving adhesion with the matrix. The mechanical effect of the plasma modification on fibers surface is caused by ions which, due to their mass about 2000 times greater than electrons, can disrupt the surface of the modified material. The energy that the ion strikes the fibre, can cause releasing particles from the fibre surface or disrupt its surface structure. [7,8].

Materials and Samples

Polypropylene microfibers are poorly wettable in water, to increase coherence with cement matrix it is advisable to increase their wettability in water. In order to increase the wettability, the fiber surface was modified by oxygen plasma. Based on previous measurements, the modification by low-pressure oxygen plasma for 60 seconds was chosen [9]. Plasma modifications were performed by Tesla VT 214. First, the fibers were placed in the chamber of the Tesla VT 214 and then almost vacuum was created. The air pressure in the chamber was about 22 Pa. Next, the chamber was filled by oxygen, the pressure in the chamber increased to 60 Pa. After that, the plasma modification process was initiated. The plasma was generated by a radio frequency source with 100 W input power. Finally, after plasma modification process, the chamber was filled by air to atmospheric pressure and the fibers were removed.

Pordtland cement CEM 42.5R (Radotin) was used as a binder. Micronized concrete recyclate was used as a filler. The foaming agent SIKA Lightcrete 400 was used to light-weight the cement matrix. The water ratio was determined to be 0.32 based on previous measurements. To determine the effect of plasma modification, specimens containing plasma modified fibers and fibers without plasma modification were made. The composition of the samples is given in Table 1.

Table 1: Composition of the samples

Set	Cement [g]	Recyclate [g]	Water [ml]	Foaming agent [ml]	W/C+R	Fibers [g]	Plasma treatment
X	750	750	480	0	0.32	0.75	NO
P	750	750	480	0	0.32	0.75	YES
VX	750	750	480	1.5	0.32	1.43	NO
VP	750	750	480	1.5	0.32	1.43	YES
VVX	750	750	480	3	0.32	1.875	NO
VVP	750	750	480	3	0.32	1.875	YES

Experimental Methods and Results

A total of 6 sets of test samples were made, each set containing 6 samples. Half of the samples contain oxygen plasma treated microfibers (*P*), the other half of samples have the same composition, but contain microfibers without plasma treatment (*X*). There are three groups of samples with different density - very lightweight samples with a density of about 1250 $kg \cdot m^{-3}$ (*VV*), lightweight samples with a density of about 1400 $kg \cdot m^{-3}$ (*V*) and non-lightweight samples with a density of about 1800 $kg \cdot m^{-3}$.

The dimensions of the test samples were 40 × 40 × 160 mm. The samples were stored for 28 days under water in a laboratory environment at 22 ± 1 ° C. Subsequently, the specimens were halved during bending tests. After that, compression-testing were performed by hydraulic press. Samples were loaded by a controlled displacement in a hydraulic press Heckert FP100. The loading speed during compression-testing was set to 0.8 mm per second. The measured compressive strength of samples is given in Fig. 1. The percent comparison of compressive strength between samples containing plasma treated fibers and samples containing fibers without plasma treatment is shown in Fig. 2.

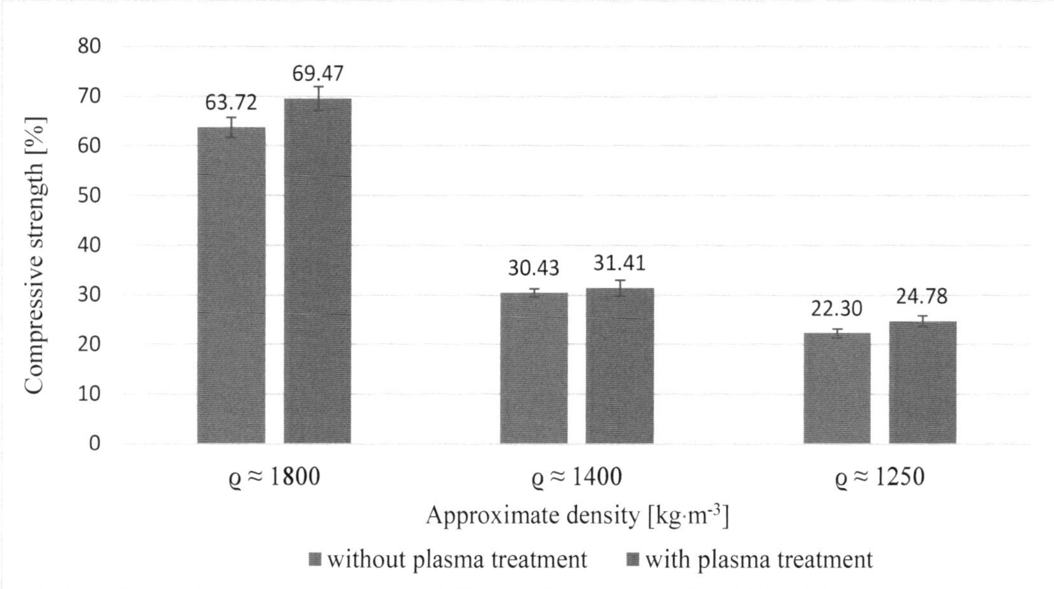

Fig. 1: Compressive strength of samples

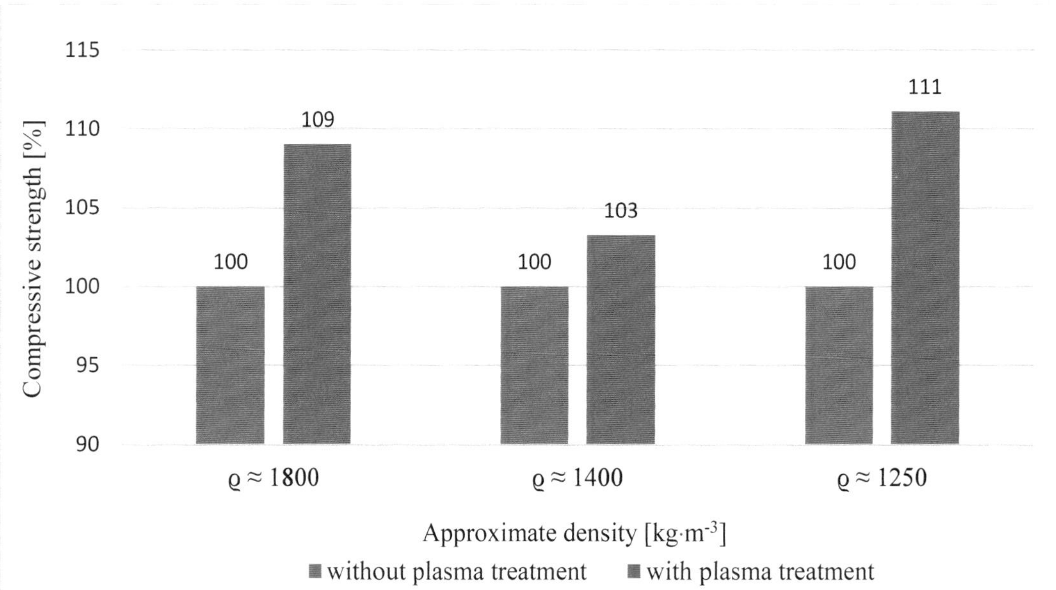

Fig. 2: The percent comparison of compressive strength between samples containing plasma treated fibers and samples containing fibers without plasma treatment

Conclusions

Based on the performed experiment, the positive effect of oxygen plasma on the surface of PP fibers, which were applied to a sufficiently lightweight cement matrix containing recycled concrete, was proved. Compressive strength of samples with a density of about 1250 kg · m^{-3} increased for samples containing plasma modified fibers by approximately 10 % compared to samples containing fibers without plasma modification. Furthermore, the positive effect of plasma treatment of PP fibers applied to non-lightweight cement composite materials was confirmed.

Acknowledgements

This work was financially supported by the Czech Technical University in Prague – the project SGS19/148/OHK1/3T/11. The authors also thank to the Center for Nanotechnology in Civil Engineering at the Faculty of Civil Engineering, Czech Technical University in Prague.

References

[1] P. Mondal, S. P. Shah, L. D. Marks, Nanomechanical Properties of Interfacial Transition Zone in Concrete. Nanotechnology in Construction 3. Berlin, Heidelberg: Springer Berlin Heidelberg, 2009, 315-320. DOI: 10.1007/978-3-642-00980-8_42.

[2] A. Bentur, S. Mindess, Fibre reinforced cementitious composites. 2nd ed. New York. ISBN 978-020-3088-722, 2007.

[3] Z. Dai, F. Shi, B. Zhang, M. Li, Z. Zhang. Effect of sizing on carbon fiber surface properties and fibers/epoxy interfacial adhesion. Applied Surface Science. 257(15), 2011, 6980-6985. DOI: 10.1016/j.apsusc.2011.03.047.

[4] M. M. Kabir, H. Wang, K. T. Lau, F. Cardona, Effects of chemical treatments on hemp fibre structure. Applied Surface Science. 276, 2013, 13-23. DOI: 10.1016/j.apsusc.2013.02.086. ISSN 01694332.

[5] M. Noeske, J. Degenhardt, S. Strudthoff, U. Lommatzsch, Plasma jet treatment of five polymers at atmospheric pressure: surface modifications and the relevance for adhesion. International Journal of Adhesion and Adhesives. 24(2), 2004, 171-177. DOI: 10.1016/j.ijadhadh.2003.09.006.

[6] J. L. Hodgkinson, H. M. Yates, A. Walter, D. Sacchetto, S. J. Moon, S. Nicolay, Roll to roll atmospheric pressure plasma enhanced CVD of titania as a step towards the realisation of large area perovskite solar cell technology. Journal of Materials Chemistry C. 6(8), 2018, 1988-1995. DOI: 10.1039/C8TC00110C.

[7] S R. Sabreen, Cold Gas Plasma Surface Modification – Optimize Plastics Bonding Adhesion [online], Adhesion bonding industry Resources and Connections adhesion bonding, 2012.

[8] D. Hegemann, H. Brunner, Ch. Oehr, Plasma treatment of polymers for surface and adhesion improvement. Nuclear Instruments and Methods in Physics Research Section B: Beam Interactions with Materials and Atoms. 2003, 281-286. DOI: 10.1016/S0168-583X(03)00644-X

[9] J. Ďureje, Z. Prošek, J. Trejbal, P. Tesárek, Plasma Modification of Microfiber Reinforcement to Improve Adhesion with Cement Matrix Containing Recycled Concrete, In: EAN 2018 56th conference on experimental stress analysis, Conference Proceedings. Praha: Česká společnost pro mechaniku, 2018, 91-95. ISBN 978-80-270-4062-9.

Verification of Multiaxial Stress Methods in the Field of Lifetimes Fatigue of Specimens From Steel ČSN 41 1523

FOJTÍK F.[1,a], PAPUGA J.[2,b], PAŠKA Z.[1,c], FUSEK M.[1,d], HALAMA R.[1,e]

[1]Deparment of Applied Mechanics, Faculty of Mechanical Engineering, VŠB – Technical University of Ostrava; 17. listopadu 2172/15, 708 00 Ostrava, Czech Republic

[2]FME Czech Technical University in Prague, Technická 4, 166 07 Prague 6, Czech Republic

[a]frantisek.fojtik@vsb.cz, [b]jan.papuga@fs.cvut.cz, [c]zbynek.paska@vsb.cz, [d]martin.fusek@vsb.cz, [e]radim.halama@vsb.cz

Keywords: high-cycle fatigue, multiaxial fatigue, experiments

Abstract. The contribution describes predictions of fatigue criteria in the area of high cycle fatigue and analysis of fatigue limit. The influence of multiaxial loading is taken into account. The study is performed on the pipe specimens, which were made from steel ČSN 41 1523. The specimens were subjected to different types of combined loading: tension-compression, bending, torsion and inner-outer pressure. Methods for prediction are rated in the interval from 100 000 cycles to 1 000 000 cycles. No significant deviation in prediction abilities of these methods was observed. The three methods - PCRN, QCP and Liu-Zenner (LZ) give the best predictions. Despite the slightly less accurate results of the MMP method, it remains also at stake due to its fastness.

Introduction

The contribution extents results presented by authors in [1,2]. In publications mentioned, authors informed about extensive fatigue research on hollow specimens made of construction steel ČSN 41 1523 (often used for welded constructions). The new method suitable for multiaxial fatigue solution denoted as MMP is also presented in publication [1]. It is extension of Manson–McKnight (MMK) method. The advantage of this method is its simplicity whereas any tabular processor (MS Excel for example) is suitable for the calculation. Here, the method uses Bergman´s approximation is used and therefore it is marked as MMPB. This improved method leads to better prediction of fatigue in comparison with MMK method. Results of undertaken analyses show that results obtained by MMP method can reach the similar quality as results produced by more complex criterions of multiaxial fatigue.

The analysis of fatigue strength stated in [1] was carried out for the lifetime of 750 000 cycles and for all 24 S–N curves acquired from experiments. In publication [2] are presented results of prediction of chosen criterions, each for set of four cross sections in the area of oblique branch of S–N curves acquired from Kohout–Věchet approximation in the range from 100 000 to 750 000 cycles. The aim of this publication is to extend the area tested by the data deducted from S–N curves at the value of 1 000 000 cycles using the new approximation and further to compare the prediction quality using different multiaxial calculation methods and then to compare mutual results of individual methods

Experiments

Three types of specimens were used for experiments. The specimens´ dimensions are given in [1]. Three different types of hollow specimens with following diameters (measured in critical cross section) were used: D=20 mm a d=18 mm or D=11 mm a d=8 mm. Specimens were made of construction steel ČSN 41 1523 melt T31052. The static material properties are summarized in Table 1.

Table 1: Static material parameters of the material investigated

Designation	Ultimate tensile strength [MPa]	Tensile yield stress [MPa]	Elongation at fracture [%]	Reduction of area at fracture [%]	True fracture strength in torsion [MPa]
ČSN 41 1523	560	400	31.1	74.0	516.6

Description and the realization of experiments can be found in [1]. Ways of specimens loading denoted as FFXX are schematically depicted in Fig. 1.. Each experiment of 24 different loading cases was performed in the force control mode.

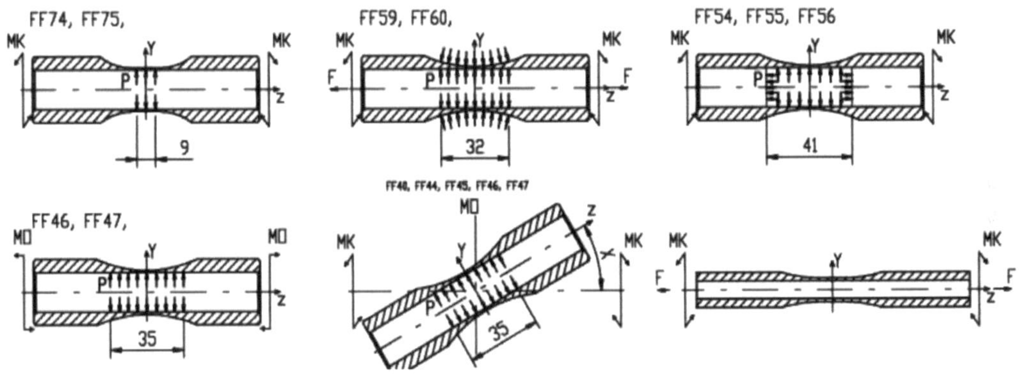

Fig. 1: Overview of various setups of experiments. F – push-pull, M0 - bending moment, MK – torque and P – pressure [1]

Each S–N curve in [1] was acquired from at least 5 experiments by linear regression (Basquine model) or by nonlinear regression (Kohout–Věchet model, see [3]). In some experimental sets (over 750 000 cycles of lifetime), the insufficient agreement between data calculated and experimental results was observed. Therefore, the new regression method (first mentioned in [4]) was used and also the regression of the influence of the mean stress. The regression function proposed is of the form:

$$\sigma = \sigma_0 - (\sigma_0 - \sigma_C) \cdot \sin\left\{\frac{\pi}{2} \cdot [log(4 \cdot N)/log(4 \cdot N_C)]^{a_2}\right\}. \quad (1)$$

This regression can be used as one parametric, where a_2 is a parameter of regression determined by the regression, σ_0 is a static strength and σ_C is a fatigue limit corresponding to the number of cycles N_C. Further, this method can be used as multi parametric where the fatigue limit or static strength can be determined by the regression. The third possible way of use is to estimate the corresponding value of the stress for chosen number of cycles (in the range of 0.25 cycles and the first crack on the specimen). In order to obtain desired loading combination, it was necessary to use different experimental machines and appropriate gripping jigs – reconstructed machine Schenck PWXN, biaxial servo hydraulics machine LABCONTROL 10kN/1000Nm and servo hydraulics machine INSTRON 8802.

Due to unusual load combinations, in some cases, it is not immediately obvious which surface of the hollow specimen is loaded more. For this reason, it was necessary to analyse the critical points on both surfaces during the fatigue prediction phase. The stress components for all load cases on individual specimens were obtained by FE analyses.

Calculations

A total of 21 computational methods were analyzed in the framework of the predictive evaluation. With the exception of the MMP method (see [1]) and the adjusted PCRN method [6], all prediction methods are described in [5]. Stress states in each loading channel corresponding to number of cycles *Nx* were calculated from regression curves. These regression curves were derived using the new regression method. In this contribution, the evaluation for five lifetime levels is carried out: 100 000, 200 000, 500 000, 750 000 and 1 000 000 cycles acquired from regression curves using the new regression.

The relative error between the damage parameter DP calculated (equivalent stress amplitude) and e.g. the fully reversed fatigue tensile strength p_{-1} (tensile loading) or b_{-1} (bending) is determined using fatigue index error ΔFI:

$$\Delta FI = \left(\frac{D_P - p_{-1}}{p_{-1}}\right) \cdot 100\% \,. \qquad (2)$$

The contribution evaluates the quality of the prediction based on this parameter. The calculations were performed in the software PragTic (www.pragtic.com). Results obtained for 24 different load cases and for five different lifetimes (read from curves using the new regression) were processed statistically here. Here, the mean value ΔFI and the sum of ΔFI squares are evaluated.

Discussion of results

The results obtained for all prediction methods are statistically processed and are depicted in Table 2.

Table 2: Results of the ΔFI statistics for 21 methods at different N_x – sum of squares and mean values

Computational method	Sum of ΔFI squares	Mean value of ΔFI for N_x					
		All	100 000	200 000	500 000	750 000	1 000 000
PCRN	46%	2.67%	2.45%	3.54%	2.93%	2.46%	1.95%
DV	121%	2.64%	2.95%	2.82%	2.73%	2.49%	2.21%
SINES	451%	7.88%	9.18%	8.71%	7.89%	7.09%	6.56%
CROSSLAND	135%	-3.23%	-2.84%	-2.92%	-3.15%	-3.46%	-3.76%
PDS	46%	-0.25%	0.97%	0.51%	-0.28%	-0.73%	-1.69%
KK	161%	5.45%	6.84%	6.36%	5.45%	4.62%	3.99%
MMPB	59%	0.28%	1.33%	0.96%	0.32%	-0.29%	-0.92%
MCDMD	175%	1.06%	3.07%	2.07%	0.76%	0.18%	-0.76%
ROGERT	87%	9.44%	11.00%	10.44%	9.44%	8.54%	7.78%
PAPADO	142%	-2.15%	-1.89%	-1.91%	-2.05%	-2.31%	-2.58%
FINDLEY	418%	11.63%	11.22%	11.61%	11.95%	11.78%	11.57%
LZ	57%	0.62%	1.75%	1.38%	0.64%	-0.02%	-0.67%
PIR	87%	1.84%	2.38%	2.25%	1.88%	1.46%	1.22%
PCR	50%	1.64%	2.08%	1.96%	1.69%	1.40%	1.05%
MATAKE	435%	10.59%	10.17%	10.54%	10.89%	10.76%	10.58%
GAM	149%	2.60%	2.54%	2.72%	2.81%	2.57%	2.37%
CS_MD	61%	1.50%	2.22%	1.95%	1.53%	1.22%	0.56%

LM	61%	-2.77%	-2.01%	-2.17%	-2.58%	-3.21%	-3.86%
QCP	40%	-1.06%	-0.04%	-0.40%	-1.06%	-1.57%	-2.24%
SUSMEL	139%	3.30%	4.99%	3.89%	2.99%	2.53%	2.10%
PAPADO_CPA	105%	3.88%	3.96%	4.04%	4.03%	3.83%	3.55%

Table 2 summarizes the sum of squares *ΔFI*, which describe well the remoteness of the general trend from the correct one. To distinguish the general shift of all results to a conservative or non–conservative prediction from the variance around the mean value, Table 2 also contains mean relative error values for individual lifetimes and for all lifetimes. The best results show QCP, PCRN and PDS methods, the worst results then SINES, MATAKE and FINDLEY methods.

An important question in the overall assessment is whether individual methods will show a substantial difference in different lifetimes. Table 2 shows that the mean values of the fatigue index error change slightly. The most methods show the difference of mean value up to 2%.

Conclusions

The analysis presented in this contribution is focused on comparison 21 different damage prediction methods, which were tested on the set of 24 S–N curves for different loading cases. S–N curves were determined by new regression. The experimental results described earlier [1] were obtained on hollow specimens made of structural steel ČSN 41 1523 of melting T31052.

There are presented and compared results of calculations of 21 selected multiaxial methods and the statistical processing of results in the form of an mean value *ΔFI* and the sum of their squares is performed. It can be stated that the QCP and PCRN methods lead to the results with the lower variance unlike the Dang Van method (DV), which is the most commonly used method in the industry. In the range of 100 000 to 1 000 000 cycles, the quality of prediction of individual methods remains relatively stable and, for example, the change in the average error index value is negligible compared to the overall variance of the prediction results.

Acknowledgement

The paper has been done in connection with project: National Competence Centre of Mechatronics and Smart Technologies for Mechanical Engineering, reg. no. TN01000071. This work was also supported by Grant Agency of the Czech Republic (GACR) project No. 19-03282S, and by the specific research SP2020/23 project, supported by the Ministry of Education, Youth and Sports of the Czech Republic.

References

[1] J. Papuga, F. Fojtík, Multiaxial fatigue strength of common structural steel and the response of some estimation methods, International Journal of Fatigue 104 (2017) 27-42.

[2] J. Papuga, F. Fojtík, M. Fusek, Efficient Lifetime Estimation Techniques for General Multiaxial Loading, in: Proceedings of the 56th International Scientific Conference on Experimental Stress Analysis 2018, 2018 pp. 96-101.

[3] J. Kohout, S. Věchet, A new function for fatigue curves characterization and its multiple merits, International Journal of Fatigue 23 (2001) 175-183.

[4] F. Fojtík, J. Fuxa, New modification of conjugated strength criterion, Transactions of the VŠB - Technical University of Ostrava – Mechanical Series, Vol. LVI, No. 1, (2010), 53–60.

[5] J. Papuga, A survey on evaluating the fatigue limit under multiaxial loading, International Journal of Fatigue 33 (2011) 153-165.

[6] J. Papuga, R. Halama, Mean stress effect in multiaxial fatigue limit criteria, Archive of Applied Mechanics, (2018), 1–12.

EAN 2020
58th conference on experimental stress analysis

Stress Fields Optimization of Rotating Saw Blades

FRANKOVSKÝ P.[1,a], KICKO M.[1,b], DELYOVÁ I.[1,c], PÁSTOR M.[1,d], TREBUŇOVÁ M.[2,e] NEUMANN V.[3,f]

[1]Deparment of Applied Mechanics and Mechanical Engineering, Faculty of Mechanical Engineering, Technical University of Košice, Letná 9, 042 00 Košice, Slovakia

[2]Department of Biomedical Engineering and Measurement, Faculty of Mechanical Engineering, Technical University of Košice, Letná 9, 042 00 Košice, Slovakia

[3]Department of Mechatronics, Faculty of Mechanical Engineering, Technical University of Košice, Letná 9, 042 00 Košice, Slovakia

[a]peter.frankovsky@tuke.sk, [b]michal.kicko@tuke.sk, [c]Ingrid.delyova@tuke.sk, [d]miroslav.pastor@tuke.sk, [e]marianna.trebunova@tuke.sk, [f]vojtech.neumann@grob.de

Keywords: Reflective photoelasticity (Photostress), Centrifugal forces, Rotational saw blades, Anti – vibration grooves

Abstract. In practice, conventional saw blades are at the present optimized with different types of grooves. It has been found that by cutting various symmetrical shapes in the body of the rotational saw blades, vibration and noise are significantly reduced. On the other hand, by making such openings, the integrity of the discs will be weakened, causing an increase in the stresses induced by the centrifugal forces. These advantages can also be demonstrated by the experimental mechanics of rigid bodies by following certain procedures and selecting the appropriate experimental method. For this purpose, the work will compare 3 alternative designs of anti-vibration discs, which will be analyzed using the optical method of experimental mechanics, namely the PhotoStress method. The expected output will be an experimental analysis of the stresses arising from the centrifugal forces in the aforementioned alternative proposals.

Introduction

Based on the primary analysis of the time response of vibrations and the measurement of noisiness of the authors Kopecký and Rousek, an apparent relationship was found between the excessive vibration of the disc body and noisiness. In the area of n = 3200 and 3400 rpm, an apparent peak increase was noted of the disc noise signal by 5 to 7 dB above the standard noise level. An absolute increase reached a level of Lp = 99 dB. These conditions were gone with by unpleasant whistling. [1]

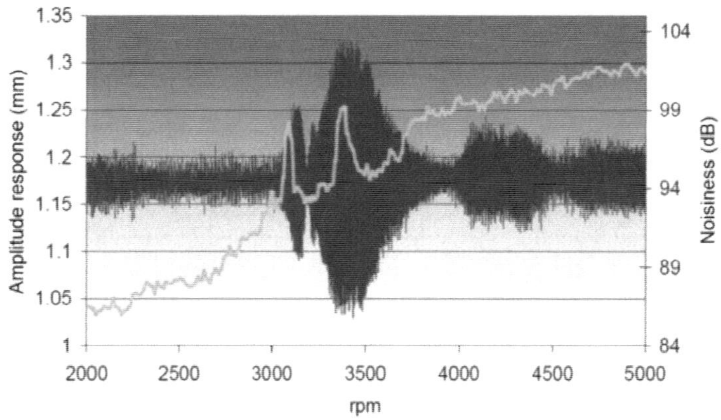

Fig. 1: Analysed saw blades [1]

Description of cutting forces in x and y directions are shown in Fig. 2. The forces are greater in both directions when the maximum number of teeth is in the workpiece just before time t2 and before cutting through the workpiece. As the blade passes the time t2 and is fed downward, the force reduces while the number of teeth that are in touch with the workpiece decreases. It is also shown that a dull blade requires greater force in both directions to cut a slot which shorten the tool life, increase the machining cost, and need to be replaced or re-sharpened. [2]

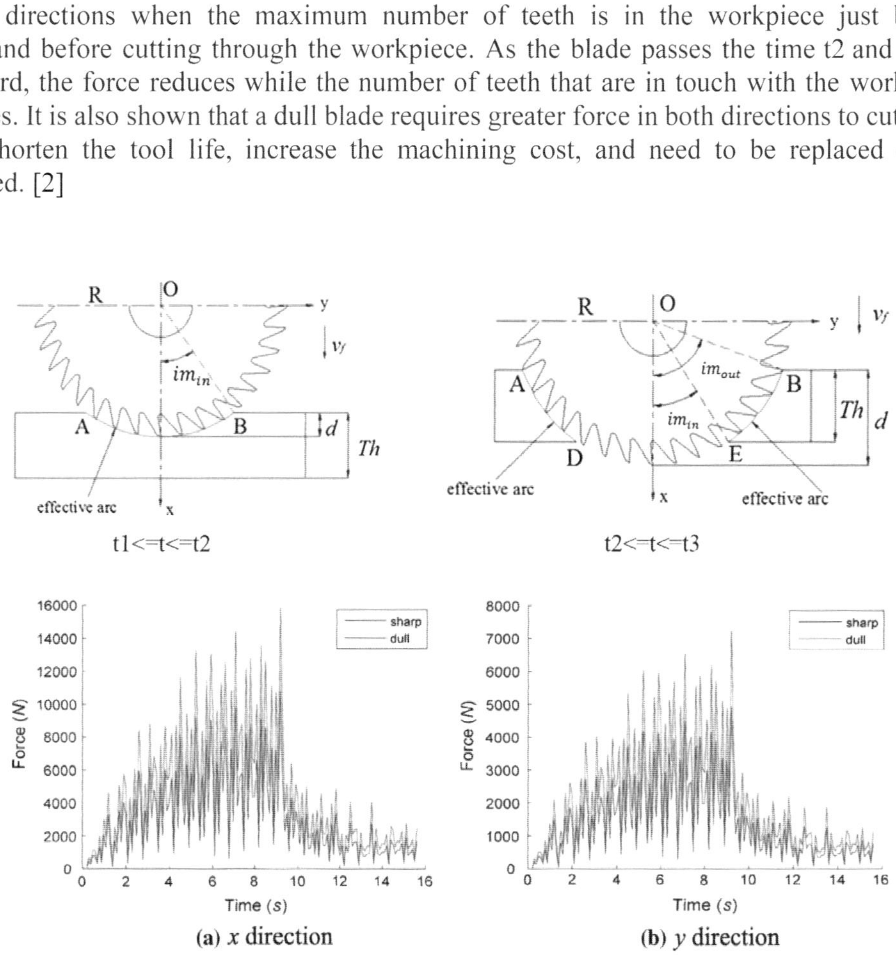

Fig. 2: Analysed saw blades [2]

In most cases, the wood is cut at high speed to achieve better surface quality and efficiency. For that reason, it is necessary to reduce blade vibrations that cause manufacturing inaccuracies, faster tool degradation, noise, higher energy costs. As a result, certain anti-vibration openings have been made, which are cut out in shapes of rounded grooves, which reduce these adverse effects. On the other hand, by cutting such optimization openings, the integrity of the disc will be weakened. Here comes the question of how the maximum stress amplitude caused by the centrifugal forces of such anti - vibration discs will increase. 0-4Chyba! Nenalezen zdroj odkazů.

Submitted paper is focused on measuring the stresses caused by the centrifugal forces on rotating discs using reflective photoelasticimetry, (PhotoStress), which is one of the experimental methods of mechanics. This issue has already been addressed by the authors in articles [5-8], the most important contribution to this work was the analysis of the rotor model, by which the experimental part will be based on. The experimental samples will represent the components of the real use i.e. saw blades, which will be made of PS- 1A photo elastic material, cut by water jet method. The experimental analysis of these samples will be comparing in terms of increase in maximum stress response at the anti-vibration discs caused by centrifugal forces [9-12].

Production of experimental samples. The experiment, which is realized, should correspond as much as possible to the real operating conditions. On this basis, samples respectively models were produced that match the exact dimensions of real saw blades.

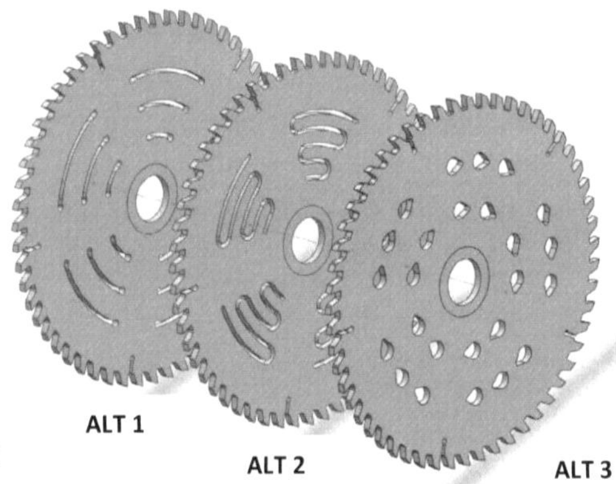

Fig. 3: Analysed saw blades

For the production of samples was used photosensitive material PS - 1A 3 mm thick from Vishay company. The proposed discs shapes were made using the water jet cutting method of Watting, s.r.o. - Prešov. Fig. 1 shows the design of all saw blades alternatives with optimization grooves that will be analysed. It is possible to notice five symmetrically spaced grooves around the periphery in every case, representing the so-called expansion grooves designed to protect teeth, resp. cutting segments against degradation. Mounting hole has a diameter of 20 mm which is commonly used for small circular saws.

The most important part of design are anti-vibration grooves. They are used to reduce the amplitude of the vibration frequency at the desired speed on the rotating discs. Their principle is that by adding a certain number of holes in the body of disc, its integrity is reduced as well as stiffness, so the oscillation will be reduced. These optimization openings, on the other hand, also have the influence of adjacent notches, when appropriately positioned that cause optimal distribution of stress response during cutting process.

Modification of the construction part for mounting rotating discs. The threaded end of the servomotor shaft with an M5 thread and the following 7,5 mm step Fig. 2 had to be adjusted for the possibility of mounting a disc with a 20 mm hole in order to achieve the most realistic results compared to practice, accordingly the mounting device shown in Fig. 2. The protruding shaft of the motor was relatively short and therefore it was necessary to provide the additional fixation - by nut M5 at the threaded end of the shaft. The M20 thread on the flange has been cut in the sense of counter-rotation of the tool for obvious reasons.

Fig. 4: Modification of the mounting part of structure

On the servomotor shaft was clamped and locked reducing flange to which a disc made of photosensitive material was attached. The speed was adjusted using a phase converter in a sequence of 3000, 3500, 4000, 4500, 5000 RPM, which corresponds to the most commonly used velocities in practice.

Analysis of centrifugal forces on rotating discs using by LF/Z-2 polariscope. To measure the stress response, it was necessary to assemble a photoelasticimetric assembly that can be seen on Fig. 3. A LF / Z-2 polariscope (5) was attached to the stand plate (tripod) and a strobe (4) Strobex Model 135 M-11 was then mounted. Another part of the measuring chain consisted of a frame (1) made of aluminum MB profiles, on which was mounted a servomotor (2) controlled by a phase converter (3) for speed setting.

Fig. 5: LF/Z-2 polariscope measuring chain

Further there was a camera (6) to document the results of the experiment and a speedometer to set the desired speed.

The distribution of the stress fields due to centrifugal forces was measured for 3000, 3500, 4000, 4500, 5000 RPM on 3 altenative saw blades designs.

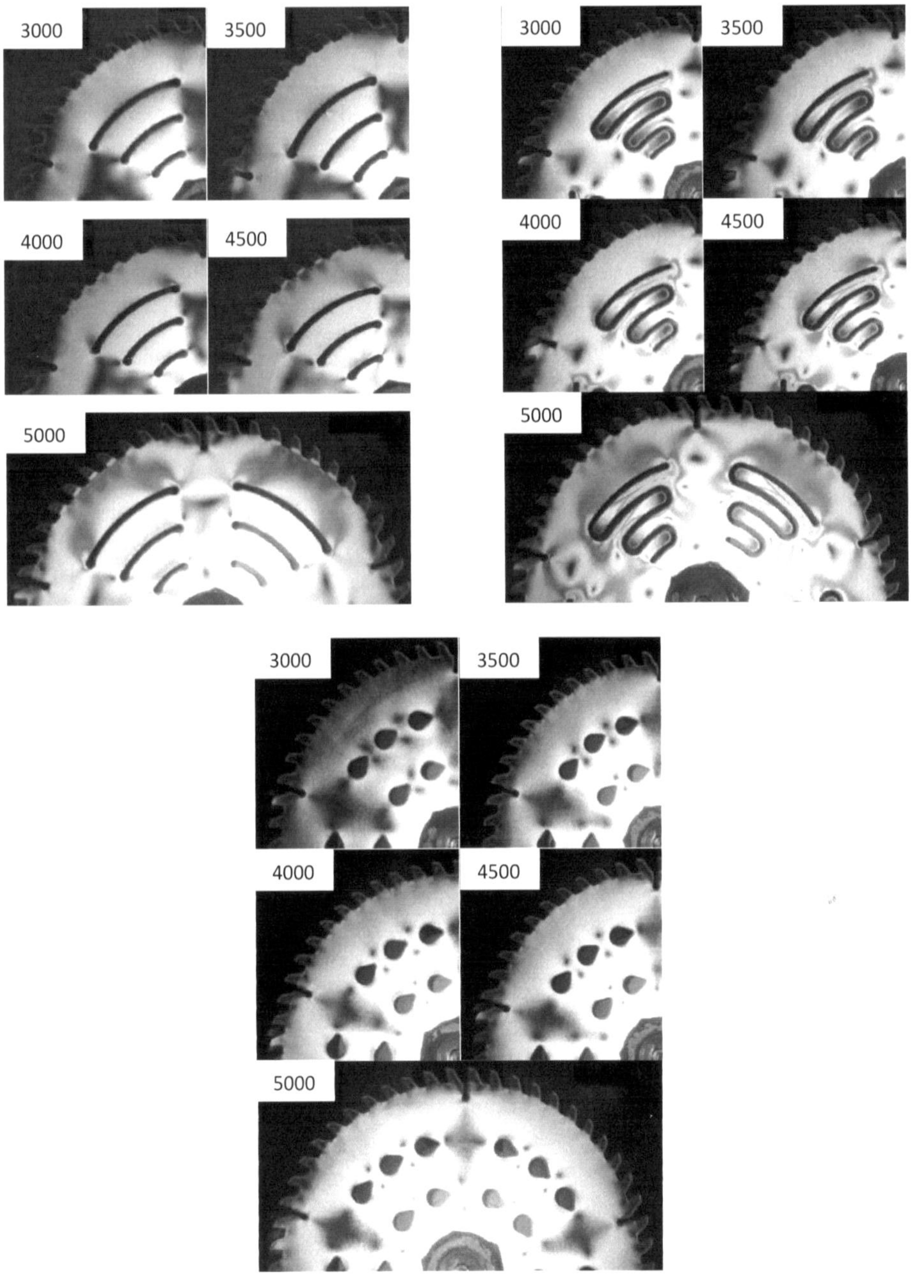

Fig 6: Series of stress responses from CF

By the using of compensator the following main stress responses were found Table 1. Data up to 5000 RPM were measured by experiment, and further values up to 10000 RPM from the safety point of view were determined numerically.

The stress values in the table are given in units of MPa and represent the intensity of the stress, i.e. difference of principal normal stresses.

Table 1: Results of experimental measurements

RPM	Alt 1 [MPa] (PS1-A)	Alt 2 [MPa] (PS1-A)	Alt 3 [MPa] (PS1-A)
3000	1,85	4,89	1,46
3500	2,52	6,65	1,99
4000	3,29	8,69	2,6
4500	4,16	11	3,29
5000	5,14	13,58	4,06
6000	7,4	19,55	5,85
7000	10,01	26,61	7,96
8000	13,15	34,75	10,4
9000	16,65	43,98	13,16
10000	20,55	54,3	16,25

The graph on Fig. 5 represents the dependence of the number of revolutions on the stress response on the measured samples. According to the graph, it is clear that at the same speed, the alt 3 disc produces the lowest stress response and the alt 2 disc produces the greatest stress response due to centrifugal forces.

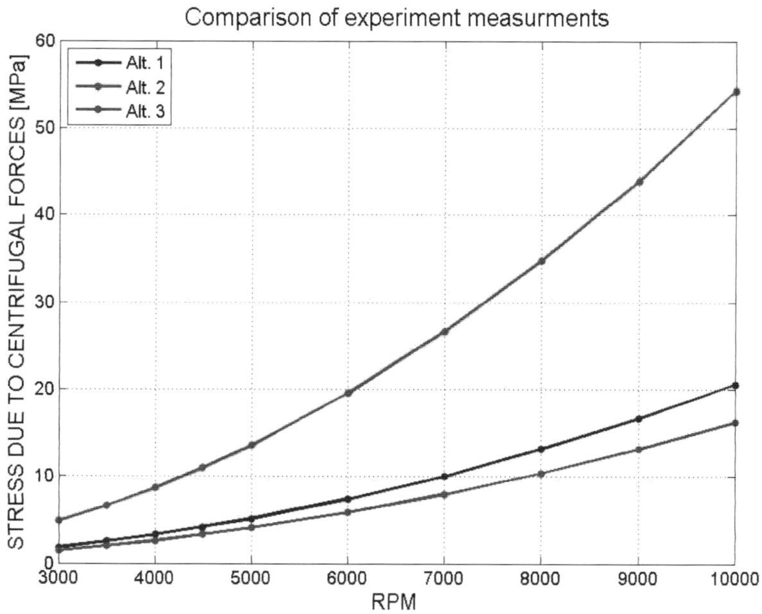

Fig 7: Interpretation of measured data

However, since these samples do not correspond to real material, it is necessary to recalculate the given experimental results for a real steel disc based on equilibrium of the finite element of rotating disk.

In accordance with the similarity criterion that takes into account the ratio of the actual part to the model, we can write relationship

$$\frac{F_{os}}{F_{om}} = \frac{m_s \omega_s^2 r_s}{m_m \omega_m^2 r_m}, \tag{1}$$

where index s denotes the real part and index m denotes the model.
Obviously, the angular velocity ω and the dimension of the disc radius r will be the same when comparing the photo elastic and steel components and therefore it is possible to write

$$\frac{F_{os}}{F_{om}} = \frac{m_s \omega^2 r}{m_m \omega^2 r} = \frac{m_s}{m_m}, \tag{2}$$

The difference between the magnitude of the centrifugal forces, respectively the stresses exerted by these forces will be linearly dependent. On this basis, it is sufficient to derive the ratio between individual materials for only one sample, in this case for the alternative 1. Mass of the photo elastic model of conventional disc m_m = 77,05 g and the real steel disc m_s = 469,54 g were measured. Consequently, it is possible to derive

$$\frac{F_{os}}{F_{om}} = \frac{m_s}{m_m} = \frac{469{,}54g}{77{,}05g} = 6{,}1, \tag{3}$$

The results of the sample experiments can be easily transformed to the maximum stress values for a real steel discs multiplying by this constant.
Thus, the results of the maximum stress differences for 10000 RPM scenario will be 20,55 MPa for the 1st alternative, 54,3 MPa for second one and 16,25 MPa for the last variation of the samples. By multiplying the results of the experiments on photosensitive samples by the ratio according to (10), one obtain the stress values for real steel samples, i.e. from material 75Cr1 (DIN 1.2003), (STN 19418) for which yield strength is 415 MPa. Thus, the maximum stress from the centripetal forces on each steel disc at 10000 RPM will be as follows: Alt.1 = 125,36 MPa, Alt.2 = 331,23 MPa and Alt.3 = 99,16 MPa. Since none of these values exceeded the yield strength, all 3 alternatives are suitable.

Conclusion

Experimental comparison of 3 alternating saw blades with anti-vibration grooves applied showed an increase in the stress response caused by centrifugal forces compared to commonly used blades. Since none of these discs was exceeded by the yield strength, it is possible to proceed with another experimental measurement, namely the measurement of oscillation during rotation by means of digital image correlation and subsequently the measurement of temperature field propagation.

Funding

This research was funded by Ministry of Education of Slovakia Foundation, Grant project VEGA No. 1/0290/18, VEGA No. 1/0500/20 and project KEGA No. 027TUKE-4/2020.

References

[1] Z. Kopecký, M. Rousek, (2012). Impact of dominant vibrations on noise level of dimension circular sawblades. WOOD RESEARCH, 57 (1): 2012 151-160.

[2] Y. Kordestany, Y. Ma, (2017). Force calculation using analytical and CAE methods for thin-blade slotting process.

[3] S. Nishio, E. Marui, (1996). Effects of slots on the lateral vibration of a circular saw blade. International Journal of Machine Tools and Manufacture, 36(7), 771-787.

[4] L. Jakubovičová, M. Sága, (2014). Computational analysis of contact stress distribution in the case of mutual slewing of roller bearing rings. In Applied Mechanics and Materials (Vol. 474, pp. 363-368). Trans Tech Publications Ltd

[5] A. Kaczmarek, K. A. Orlowski, L. Javorek, (2016). A brief review and comparison of selected experimental methods for measuring natural frequencies of circular saw blades. Drewno. Prace Naukowe. Doniesienia. Komunikaty, 59(197).

[6] A. Droba, L. Javorek, J. Svoren, D. Pauliny, (2015). New design of circular saw blade body and its influence on critical rotational speed. Drewno. Prace Naukowe. Doniesienia. Komunikaty, 58(194).

[7] P. Frankovský, F. Trebuňa, (2014). Application of photostress method in stress analysis of a rotating disc. Metalurgija, 53(4), 541-544.

[8] P. Frankovský, O. Ostertag, E. Ostertagová, F. Trebuňa, J. Kostka, M. Výrostek, (2017). Experimental analysis of stress fields of rotating structural elements by means of reflection photoelasticity. Applied optics, 56(11), 3064-3070.

[9] J. Kostka, P. Frankovský, F. Trebuňa, M. Pástor, F. Šimčák, (2014). Stress analysis of a rotating body by means of photostress method and using solidworks programme. American journal of mechanical engineering, 2, 226-230.

[10] J. Kostka, P. Frankovský, M. Pástor, F. Trebuňa, F. Šimčák, (2014). Application of PhotoStress method in stress analysis of structural elements under consideration of centrifugal force effect. Procedia Engineering, 96, 235-241.

[11] M. Kicko a kol. (2018). Návrh rezného kotúča využitím numerických metód mechaniky. Novus Scientia 2018. Košice: TU, S. 81-83.

[12] K. Doubrava, J. Kuželka, (2014). Turbine Blade Loading and Experimental Verification. In: Applied Mechanics and Materials (Vol. 486, pp. 273-276). Trans Tech Publications Ltd.

EAN 2020
58th conference on experimental stress analysis

Identification of Material Parameters and Material Model for 3D Printed Structure

FUSEK M.[1,a], PAŠKA Z.[1,b], ROJÍČEK J. [1,c]

[1]Deparment of Applied Mechanics, Faculty of Mechanical Engineering, VŠB – Technical University of Ostrava, 17. listopadu 2172/15, 708 00 Ostrava, Czech Republic

[a]martin.fusek@vsb.cz, [b]zbynek.paska@vsb.cz, [c]jaroslav.rojicek@vsb.cz

Keywords: Anand material model, Material parameters, ABS-M30, Indentation test, Geneetic algorithm

Abstract. The paper focuses on the determination of material parameters of the Anand material model for ABS-M30 material. The material parameters were determined by the method of Geneetic algorithm (GA) using FEM calculations. Several experimental tensile and indentation tests were performed. The configuration of the individual tests was chosen to capture relaxation and creep, which are important for this type of material. The FEM simulations were then adjusted to the experimental tests carried out to achieve the greatest possible agreement.

Introduction

One of the most advanced methods of component manufacture is the 3D printing process. For the needs of the correct design of components it is necessary, among other things, to know the correct material parameters. The parameters of printed 3D structures are highly dependent on the technology of 3D printing, the laying of the filament and the setting of the printing process (e.g. temperature). This article describes the procedure for determining the material parameters of a printed structure produced by the 3D printing process using printer Fortus 450mc [1]. The material used for printing is ABS-M30 [2].

Experiments

The experiments were performed on a Testometric M500 50CT testing machine at room temperature. The mentioned test machine will soon be supplemented by a furnace, so that it will be possible to perform further experiments at different temperatures. The following tests were performed: Indentation tests, tensile tests, and relaxation tests. The load was applied in different ways so that the specimen was tested in all directions to detect the anisotropy of the test material.

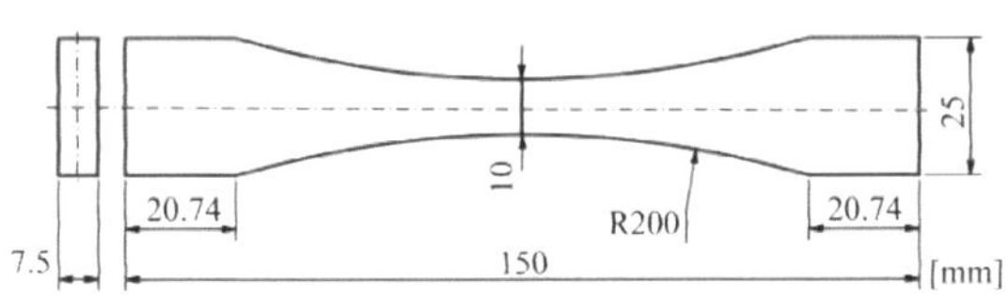

Fig. 1: Specimen scheme for tensile and relaxation tests

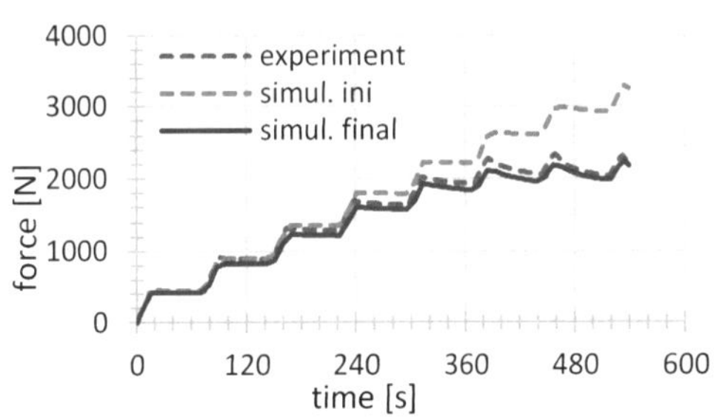

Fig. 2: Graded tensile test and FEM simulation (identification)

The specimen shown in Fig. 1 was used for the tensile test and the relaxation test. All tensile tests were deformation-controlled. Fig. 2 shows a graded tensile test with the following configuration: The specimen was elongated by 0.25 mm at an elongation rate of 0.017 mm s^{-1}. Time delay at the given strain value was always 60 s. This was done until the specimen failed. During the graded tensile test, the phenomenon of creep was partially detected. Up to the value of the tensile force of approx. 800 N, there is no significant relaxation of the tension, see Fig. 2. Fig. 3 is a simple tensile test in which the crossbeam is moved at rate of 0.017 mm s^{-1} until the specimen failed. Fig. 4 is a tensile test with this configuration: The crossbeam rate was 0.017 mm s^{-1}. At a specimen elongation value equal to 1 mm, the time delay was 60 s. This was followed by the displacement of the crossbar at the same rate until the specimen failed.

A cube-shaped specimen with a side length of s = 50 mm was used for the indentation test. The indentation test was carried out with an indenter in the form of a sphere with a radius of 2.5 mm. The indenter rate, the load values and the loading modes were chosen differently to obtain more data that better characterize the behaviour of the printed material. With the help of FEM calculations, it was determined how precisely determined material parameters of the material model corresponds to the results of experimental tests - the validation was done by a graphical comparison of the results of FEM solution and indentation (experiment).

All indentation tests were deformation-controlled. In Fig. 5 is an indentation test with the following settings: The indenter was gradually pressed into the surface of the specimen with rate 0.017 mm s^{-1} to a depth of about 0.8 mm and then the indenter was gradually pulled out of the specimen without delay. In Fig. 6, is a test with the following settings: The indenter rate is 0.017 mm s^{-1} to a depth of 0.8 mm. This is followed by a time delay of 600 s. The indenter is then pulled out at the same speed.

At the same time, specimens with different orientations in the chamber of the 3D printer were examined in this way. The different orientation of the specimen in the chamber of the 3D printer results in different laying of the individual material fibres and leads to anisotropy.

Material model

One of the material models used for the visco-plastic materials is the Anand material model [3]. The Anand model was proposed for use in the analysis of the rate-dependent deformation of metals at high temperatures. The Anand visco-plastic model is pre-built in commercial finite-element software system Ansys. The Anand model is typically used for solder alloys [4] but previously mentioned benefits led us to test it for materials used in 3D printing (ABS-M30). Anand material model [3] does not allow anisotropy, so only 4 selected tests with the same fibre orientation were used for identification. (3 tensile tests and 1 indentation test) and in this phase of material model testing with a one temperature 20°C. The indentation test was then used to validate the numerical results with the experiments. The material model is given by 11 material parameters: Poisson ratio (μ), Young's modulus (E), Initial value of deformation resistance (s_0), Activation energy/Universal gas constant (Q/R), Pre-exponential factor (A), Stress multiplier (x_i), Strain rate sensitivity of stress (m), Hardening/softening constant (h_0), Coefficient for deformation resistance saturation value ($\hat{S}$), Strain rate sensitivity of saturation (deformation resistance) value (n) and Strain rate sensitivity of hardening or softening (a).

Algorithm for material parameters identification

The material parameter identification can be described mathematically as:

$$f(X) = minimum, \tag{1}$$

where $f(X)$ is an objective function and X represents a vector of material parameters. For minimization, the Finite Element Model Updating (FEMU) approach was used, which is described in [5]. In short, the material parameters were determined by repeated calculations - optimization in our own program written in Python. The solution procedure is as follows: First, FEM models are created (with APDL) to simulate experiments. These models/simulations were created in a commercial program (Ansys) as blocks, where the input is the values of material parameters and the output is the value corresponding to the data set obtained from the measurement (load, displacement and time). The difference between the results of the simulations and the measured data is represented by the value of the objective function. These simulations are solved in cycles in which the input values change in relation to finding the minimum value of the objective function. All experiments were deformation-controlled, so that the values of the forces were chosen for the design of the objective function. The difference between the measured data and the data from the simulations for an experiment was solved as follows (an individual objective function $f_i(X)$ for i-th experiment):

$$f_i(X) = \sum_{j=1}^{N_i} \left| \frac{F_j^{EXP} - F_j^{FEM}(X)}{F_j^{EXP}} \right|, \tag{2}$$

where N_i is number of measurement points for i-th experiment, F_j^{EXP} is an experimental force, $F_j^{FEM}(X)$ is a force obtained from simulation.

The value of objective function for all experiments ($f(X)$) was solved as:

$$f(X) = \sqrt{\frac{\sum_{i=1}^{N} f_i(X)^2}{N}}, \tag{3}$$

where N is a number of experiments.

One of the algorithms used to identify material parameters is GA, see e.g. [6]. A detailed description of GA can be found e.g. in [7]. This GA is relatively simple and available in a variety of variants. The advantage lies in its easy programmability and can be used for parallel solutions (supercomputers). For these reasons, this GA is chosen. The chromosome is given by a set of material parameters and a genee is a one material parameter:

$$X = \left\{E, \mu, s_0, \frac{Q}{R}, A, x_i, m, h_0, \hat{S}, n, a\right\}. \tag{4}$$

A population was given by 10 solutions and it was geneerated randomly by Hill climbing algorithm [8]. For example, for the first parameter was used:

$$E_{New} = E_{Best}[1 + \mathrm{Rand}(-0.01, 0.01)], \tag{5}$$

where E_{New} is a new value of the parameter, E_{Best} is a value of parameter for the solution with the minimum value of the objective function for actual population, and $Rand$ generates random values (uniform distribution) from the interval. The population is modified in each cycle. With a probability of 40%, a randomly generated incomer is added, which is generated with the Hill climbing algorithm" (previous equation). With a probability of 60% a child (crossover) is added. The child chromosome was created from three chromosomes of parents, where the parents were randomly selected from the population (uniform distribution). The calculation a value of a gene is presented only for the first parameter (E), the other parameters are calculated by the same approach. The values are calculated from ($\Delta E = k(E_1 - E_2)$) and ($\Delta E = k(E_1 - E_3)$), where equations are selected randomly (with 70% and 30% probability, respectively), E_1 is the gene value of the first parent, E_2 is the gene value of the second parent, E_3 is the gene value of the third parent, ΔE is gene change proposal and k is an increase coefficient ($0.1 - 1.0$). The parents are ordered by its value of the objective function (the first parent has a minimum value and parent three has a maximum value). The value of ΔE must place to interval $(0.0001|E_{Best}|, 0.05|E_{Best}|)$ or it is geneerated randomly by:

$$\Delta E = \mathrm{Rand}(-0.001,0.001)E_{Best}. \tag{6}$$

The child value of the gene (E_{Child}) is finally calculated as:

$$E_{Child} = E_{Best} + \Delta E\big(1 + \mathrm{Rand}(-0.01,0.01)\big). \tag{7}$$

The initial value of material parameters follow: The value of the Poisson ratio $\mu = 0.3$ and the value of Young's modulus $E = 1950$ MPa were taken from [9] for material ABS-M30. The Anand material parameters were taken from [10] for material SAC309 (Sn-3.0Ag-0.5Cu), these parameters were primarily identified by the algorithm. The initial parameters are shown in Table 1.

Table 1: Initial parameters of Anand material model

$s_0[MPa]$	Q/R	$A[1/s]$	x_i	m	$h_0[MPa]$	$\hat{S}[MPa]$	n	a
21	9320	3501	4	0.25	180000	30.2	0.01	1.9

Due to the data set used to identification, the Poisson ratio was not identified. Due to the use of a limited data set, the effect of individual parameters on the value of the objective function was tested. The procedure is presented for the first parameter again. Three values of the objective function were calculated with an analysed value of parameter:

$$\begin{aligned} f &= f\left(\left\{E, \mu, s_0, \frac{Q}{R}, A, x_i, m, h_0, \hat{S}, n, a\right\}\right), \\ f^{+\Delta E} &= f\left(\left\{E(1+0.05), \mu, s_0, \frac{Q}{R}, A, x_i, m, h_0, \hat{S}, n, a\right\}\right), \\ f^{-\Delta E} &= f\left(\left\{E(1-0.05), \mu, s_0, \frac{Q}{R}, A, x_i, m, h_0, \hat{S}, n, a\right\}\right). \end{aligned} \tag{8}$$

The first criterion for excluding a parameter is ($f \approx f^{+\Delta E} \approx f^{-\Delta E}$) with respect to the behaviour of the criterion for other parameters. This criterion show that the parameter has a negligible effect on the result with respect to the other parameters. The second criterion for excluding a parameter is ($f^{+\Delta E} \gg f \ll f^{-\Delta E}$) with respect to the behaviour of the criterion for other parameters. The second criterion shows that the parameter has in the value the local minimum. On the other side, the optimum parameter for identification meets conditions ($f^{+\Delta E} < f < f^{-\Delta E}$) and ($f^{+\Delta E} > f > f^{-\Delta E}$), respectively.

Four experiments was used for identification (Fig. 2, 3, 4 and 5), the fifth experiment was used to material parameter validation (Fig. 6). The identification process required 200 cycles, the values of the objective function are in Table 2 and $f_V(X)$ represents the value of objective function for the verification experiment.

Table 2: Values of the objective functions

	$f_1(X)$ [%]	$f_2(X)$ [%]	$f_3(X)$ [%]	$f_4(X)$ [%]	$f(X)$ [%]	$f_V(X)$ [%]
final	2.2	5.8	3.2	12	6.94	10.6

The resulting parameters of the Anand material model are summarized in a Table 3 and the value of the Poisson ratio is $\mu = 0.33$ and the Young's modulus is $E = 1780$ MPa.

Table 3: Result parameters of Anand material model

$s_0[MPa]$	Q/R	$A[1/s]$	x_i	m	$h_0[MPa]$	$\hat{S}[MPa]$	n	a
19.54	8350	3134	5.18	0.2466	183245	31.0	0.0098	1.524

In Fig. 3 the simple tensile test is compared with the FEM simulation. Fig. 4 again shows the tensile test, but at a certain value of strain of the specimen the loading was interrupted for 60 s to see if relaxation occurred, then the specimen was loaded until failure.

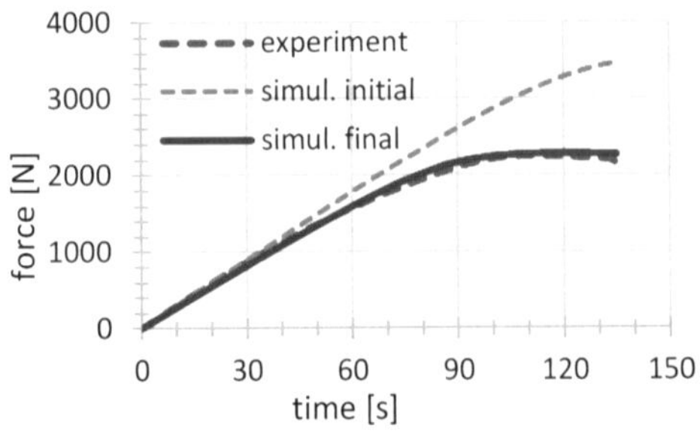

Fig. 3: Tensile test (identification)

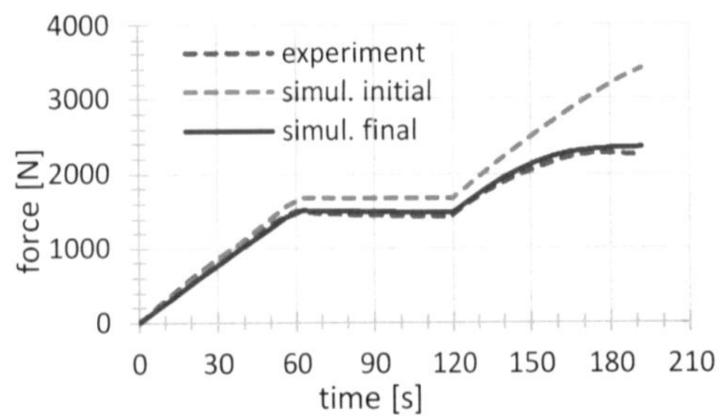

Fig. 4: Tensile test with time delay (identification)

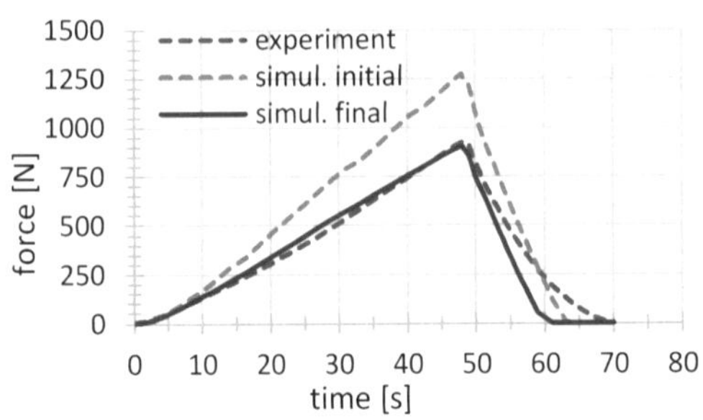

Fig. 5: Indentation with relief of load (identification)

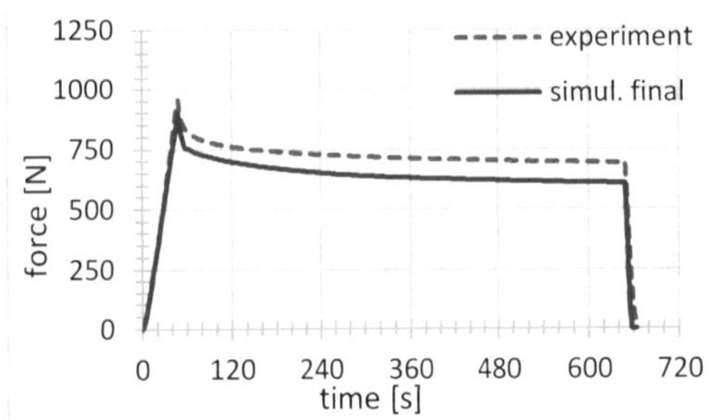

Fig. 6: Indentation with relaxation effect (validation)

Fig. 5 is a recording of the indentation test with FEM simulation. Fig. 6 compare the indentation test with the FEM simulation. This is an indentation test with a time delay to monitor the effect of material relaxation on the magnitude of the loading force.

Conclusions

The results obtained show the possibility of finding the material model parameters with help of an indentation or tensile tests. The results of the FEM calculations are in good agreement with the experimentally obtained data. Our department has recently acquired a temperature chamber and experiments under different temperatures will be possible soon. Anand material model can be used for specimens loaded at different temperatures. However, Anand model is not able to sufficiently capture the anisotropic behaviour of the investigated material. We are currently working on the use of a more advanced material model that includes the effect of material anisotropy and the use of a complete data set from experiments.

Acknowledgment

This article has been elaborated under support of the project Research Centre of Advanced Mechatronic Systems, reg. no. CZ.02.1.01/0.0/0.0/16_019/0000867 in the frame of the Operational Program Research, Development and Education.

This work was supported by The Ministry of Education, Youth and Sports from the Specific Research Project (SP2020/23) and has been done in connection with the DMS project reg. no. CZ.02.1.01/0.0/17_049/0008407 financed by Structural Funds of Europe Union.

References

[1] Fortus 380mc and Fortus 450mc. Stratasys [online]. 2020 [cit. 2020-02-28]. Available from: https://www.stratasys.com/3d-printers/fortus-380mc-450mc.

[2] J.R.C. Dizon, et al. Mechanical characterization of 3D-printed polymers. Additive Manufacturing, 2018, 20: 44-67.

[3] L. Anand. "Constitutive Equations for Hot-Working of Metals". International Journal of Plasticity. Vol. 1. 213-231. 1985.

[4] Z.N. Cheng, et al. Viscoplastic Anand model for solder alloys and its application. Soldering & Surface Mount Technology, 2000.

[5] J. Rojicek. Identification of material parameters by FEM. Mod. Mach. Sci. J, 2010, 2: 185-188.

[6] N. Dusunceli, O.U. Colak, C. Filiz. Determination of material parameters of a viscoplastic model by geneetic algorithm. Materials & Design, 2010, 31.3: 1250-1255.

[7] M.D. Vose. The simple geneetic algorithm: foundations and theory. MIT press, 1999.

[8] D. Simon. Evolutionary optimization algorithms: biologically-Inspired and population-based approaches to computer intelligenece. Hoboken, New Jersey, John Wiley & Sons, 2013, ISBN: 978-0-470-93741-9

[9] Z. Paska, J. Rojicek, P. Ferfecki, M. Fusek, D. Heczko, V. Krys. Methodology of arm design for mobile robot manipulator using topological optimization. MM Science Journal, 2020(June), 3918-3925. doi:10.17973/MMSJ.2020_06_2020008

[10] M. Motalab, et al. Determination of Anand constants for SAC solders using stress-strain or creep data. In: 13th InterSociety Conference on Thermal and Thermomechanical Phenomena in Electronic Systems. IEEE, 2012. p. 910-922.

EAN 2020
58th conference on experimental stress analysis

Analysis of Mechanical Properties of Printed Polyamid PA11

FUSEK M.[1,a], PORUBA Z.[1,b], PAVLÍČEK P. [1,c], FOJTÍK F.[1,d], CIENCIALA J.[1,e] KREJČÍ L.[2,f], NEZDAŘILOVÁ A.[1,g]

[1]Deparment of Applied Mechanics, Faculty of Mechanical Engineering, VŠB – Technical University of Ostrava; 17. listopadu 2172/15, 708 00 Ostrava, Czech Republic

[2]Deparment of Mechanical Technology, Faculty of Mechanical Engineering, VŠB – Technical University of Ostrava; 17. listopadu 2172/15, 708 00 Ostrava, Czech Republic

[a]martin.fusek@vsb.cz, [b]zdenek.poruba@vsb.cz, [c]pavel.pavlicek@vsb.cz, [d]frantisek.fojtik@vsb.cz, [e]jakub.cienciala@vsb.cz, [f]lucie.krejci@vsb.cz. [g]anna.nezdarilova.st@vsb.cz

Keywords: Polyamid, PA11, tension test, mechanical properties, fracture surface, 3D printing

Abstract. The contribution deals with the analysis of mechanical properties of printed polyamide PA11. More specifically, the mechanical properties are evaluated for three different orientations of specimens when printed and for different positions of specimens in printing chamber. Values of yield strength, ultimate stress and ductility are stated. Significant differences of results were found out. In order to explain these differences, the microscopic analysis of fracture surfaces was performed. Uneven structure of fracture surface can be observed.

Introduction

For needs of the contemporary design which reaches limits of material properties the correct values of material properties are required. The point of interest of this contribution are mechanical properties of PA11 material printed by HP Multi Jet Fusion Technology on the printer HP Jet Fusion 4200 Series. Since the further strength analysis of printed products by finite element method will be required, the tension test was chosen as the most common analysis providing necessary parameters such as Young´s modulus, tangent modulus, yield strength, ultimate strength, etc.

HP Multi Jet Fusion is a new technology placed on the market in 2014. HP Multi Jet Fusion technology uses the powder material without the use of lasers and can be used even for series production. [1] The process of the 3D print starts with laying down the thin layer of material in the printing area. Further, the carriage containing an HP Thermal Inkjet array passes from left to right side, printing chemical agents across the full printing area. The layering and energy processes are combined in a continuous pass of the second carriage from top to bottom side. The process continues, layer by layer until a complete part is formed. At each layer, the carriages change its direction. [2]

The material PA11 is the polyamide of natural origin obtained from renewable resources – ricin oil. It is the versatile material used for applications requiring properties such as chemical, thermal or mechanical endurance.[3] One of typical products made of material PA11 are orthopedic aids, such as prostheses and braces. Since medical product are subjected to the strict safety requirements, designers require as much as possible accurate mechanical parameters of materials used.

Experiments

Tension tests on flat specimens, see Fig. 1, printed from the material PA11 on the printer HP Jet Fusion 4200 Series were performed using the testing machine TESTOMETRIC M500-50CT equipped by the extensometer which measures the elongation of the specimen in its active part. The loading speed was set to 10 millimeters per minute. The output of tension tests performed is the tension diagram – graph of engineering stress versus engineering strain values of particular specimens whereas the directional orientation and position of specimens inside the printing chamber were taken in the account.

In order to evaluate the influence of the printing direction on mechanical properties of the structure, three kinds of specimens printed in different orientation were used. Layers of specimen A were laying down in the direction of its thickness, layers of specimen B in the direction of its width and layers of specimen C in the direction of its height. The orientation of specimens when printed is obvious from Fig. 2.

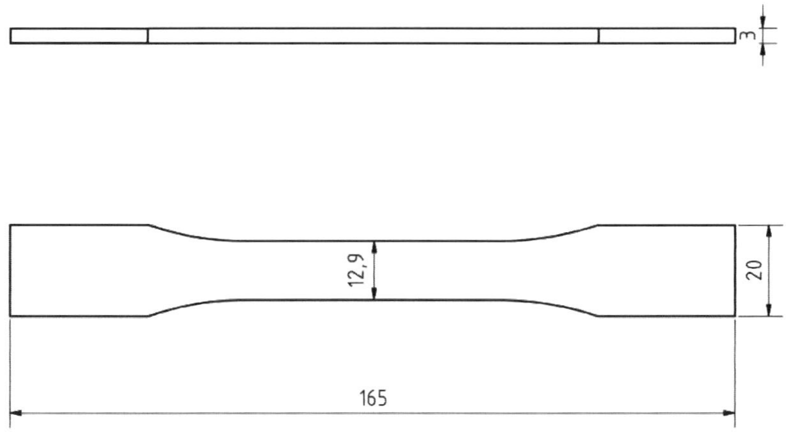

Fig. 1: Specimen used for tension tests

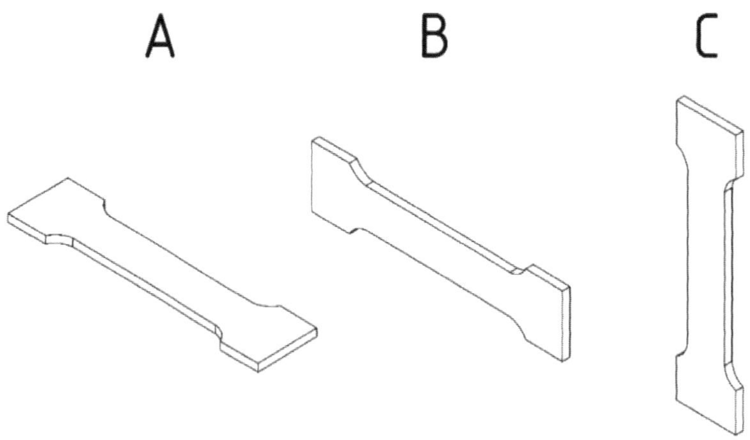

Fig. 2: Orientation of specimens when printed

In order to discover the influence of the position of the printed structure in the chamber on its mechanical properties, the location of each specimen in the printing chamber was recorded. Specimens A10-A50, B10-B50 and C10-C53 were distributed in each corner and in the middle part of the printing chamber.

The printing chamber was filled by two series of eighteen specimens whereas results of tension tests of the first series is presented in this contribution, the other series will be used later in order to evaluate the influence of material aging on its mechanical properties. Except two series of specimens mentioned above, the printing chamber did not contain any other objects to be printed.

Results

Results of individual tension tests are presented in Fig. 3 in the form of tension diagrams – graph of engineering stress versus engineering strain values of each specimen.

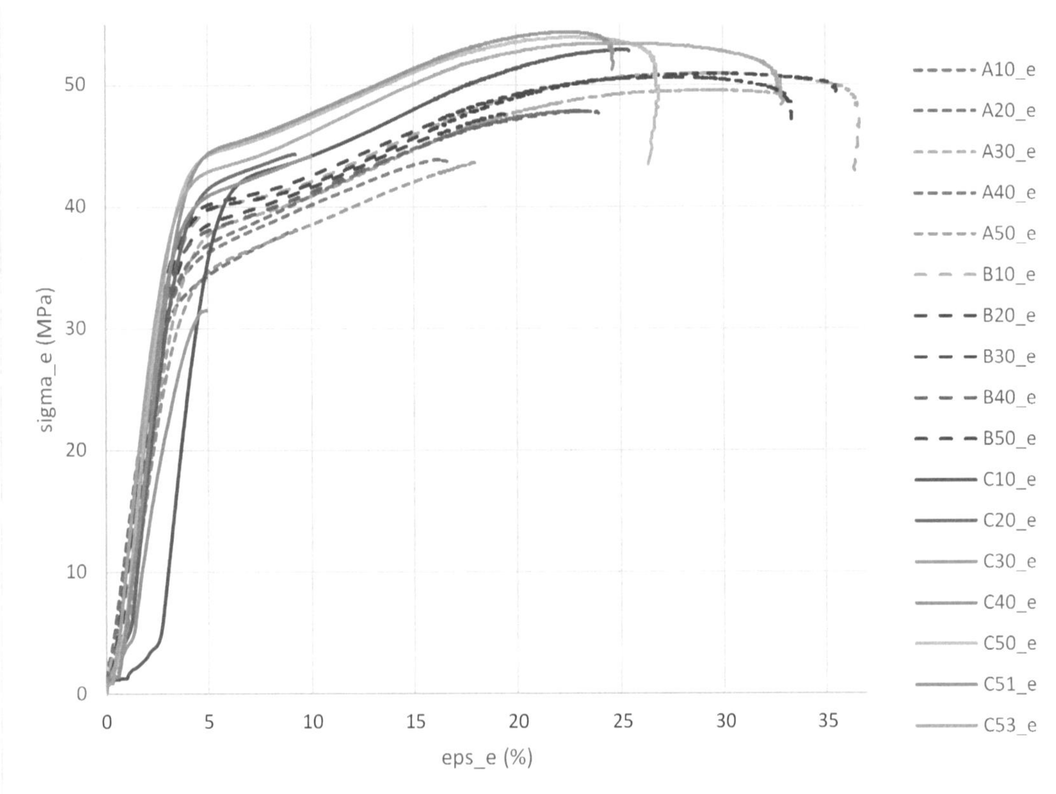

Fig. 3: Tension diagrams of tests performed

Stress-strain curves of specimens printed according to the orientation A (see Fig. 2) are plotted in Fig. 3 by solid lines, orientation B by dashed line and orientation C by dotted line.

Significant differences in stress-strain curves can observed not only for different orientations (A, B, C) of specimens but also for their positions in the printing chamber.

Concerning the yield strength and ultimate stress, the highest values were reached largely for specimens with the printing orientation C – (25-40) MPa and (32-56) MPa respectively, lowest values then for specimens with orientation A – (28-30) MPa, (38-43) MPa, respectively.

Observing the ductility of specimens, its highest values were reached largely for printing orientation B – (18-36)%, lowest values then for the orientation A – (8-16)%.

In order to explain differences among specimens printed in one direction, the microscopic analysis of selected specimens was performed. Fracture surfaces of specimens A20, A30, B20, B30, C30 and C51 are depicted in Fig. 4 to Fig. 9. Fracture surfaces were chosen so that specimens with better and worse material properties is presented whereas specimens with better mechanical properties (A30, B30, C51) are depicted in right columns of table, specimen with worse properties in left columns (A20, B20, C30).

From presented figures it is obvious that even individual fracture surfaces contain differences in the surface structures. Areas with smooth surface, flat furrows and deep cavities leading in the direction which is perpendicular to the fracture surface can be observed. It is also obvious from Fig. 3 and Fig. 4-9 that deeper structure located in the fracture surface improves material properties significantly.

Fig. 4: Fracture surface – specimen A20

Fig. 5: Fracture surface – specimen A30

Fig. 6: Fracture surface – specimen B20

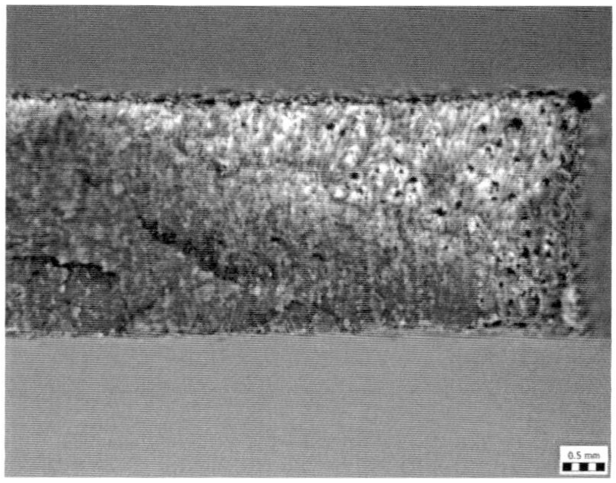

Fig. 7: Fracture surface – specimen B30

Fig. 8: Fracture surface – specimen C30

Fig. 9: Fracture surface – specimen C51

Conclusion

The mechanical analyses of the set of specimens printed from polyamide PA11 was performed. Three kinds of specimens with different printing directions were evaluated. In order to incorporate the influence of specimen´s position in the printing chamber, the location of each specimen was recorded. Except specimens printed, the printing chamber did not contain any other objects to be printed. In order to explain differences in results, the microscopic analysis of fracture surfaces was done.

Based on results obtained, the significant differences of mechanical properties for three printing directions and also for different position in printing chamber were found out. Concerning the yield strength and ultimate stress, best properties showed largely specimens with printing orientation C, worse parameters were found out for specimens with printing orientation A. Observing the ductility, the highest values show mostly specimens with orientation B, lowest values then specimens oriented according to A.

Significant differences in mechanical behavior of specimens printed in the same direction were found out as well. However, no strong connection between mechanical properties and position of the specimen in the printing chamber was found out in the set of experiments performed.

The microscopy analysis of fracture surfaces of chosen specimen showed different kind of material structure. Smooth surfaces, flat furrows and deep cavities leading in the direction which is perpendicular to the fracture surface can be observed. Authors of this contribution explain these structure inconsistences by the uneven heat distribution during the printing process and/or by the use of the combination of new and used PA11 powder. The use of this powder mixture is the common process in the printing practice.

Finally, it can be stated that printed specimens do not show isotropic material properties. Better prediction of mechanical behavior of printed parts can be done under the assumption of orthotropic properties. However, different mechanical properties can be also observed for specimens with the same printing direction but in different location in the printing chamber. According to figures presented in this contribution, it can be observed that the presence of deep cavities leading perpendicular to the fracture surface improve significantly mechanical properties of the structure printed. On the other hand, the prevailing smooth fracture surface determines poorer mechanical properties.

Acknowledgement

This work was supported by The Ministry of Education, Youth and Sports from the Specific Research Project (SP2020/23) and has been done in connection with the DMS project reg. no. CZ.02.1.01/0.0/17_049/0008407 financed by Structural Funds of Europe Union.

References

[1] MJF (Multi Jet Fusion) | cotu.cz. *3D Print | cotu.cz* [online]. Copyright © 2020 [cit. 20.04.2020]. Available from: https://www.cotu.cz/blog/138/mjf-multi-jet-fusion.

[2] HP 3D Jet Fusion 4200. Laptop Computers, Desktops, Printers, Ink & Toner | HP® Official Site [online]. HP, 2020 [cit. 2020-07-13]. Dostupné z: https://www8.hp.com/us/en/printers/3d-printers/products/multi-jet-fusion-4200.html.

[3] PA 11, Rilsan, Arkema - Resinex. *Resinex* [online]. Copyright © 2020 RESINEX Group, Czech Republic [cit. 08.03.2020]. Available from: https://www.resinex.cz/produkty/rilsan.html.

On Cyclic Hardening/Softening Behaviour of Conventional and 3D Printed SS316L

HALAMA R.[1,a], GÁL P.[1,b], PAGÁČ M.[1,c], GOVINDARAJ B.[1,d]
KOCICH R.[2,e], KUNČICKÁ L.[2,f]

[1]Faculty of Mechanical Engineering, VŠB – Technical University of Ostrava, 17. listopadu 2172/15, 708 00 Ostrava, Czech Republic

[2]Faculty of Materials Science and Technology, VŠB – Technical University of Ostrava, 17. listopadu 2172/15, 708 00 Ostrava, Czech Republic

[a]radim.halama@vsb.cz, [b]petr.gal.st@vsb.cz, [c]marek.pagac@vsb.cz, [d]bhuvanesh.govindaraj.st@vsb.cz, [e]radim.kocich@vsb.cz, [f]lenka.kuncicka@vsb.cz

Keywords: non-proportional hardening, SS316L, additive manufacturing, low cycle fatigue

Abstract.
This study is focused mainly on the stress-strain behaviour investigation under proportional as well as non-proportional loading on SS316L made by casting and prepared by Selective Laser Melting. Strain controlled fatigue tests have been performed under constant strain amplitude loading considering a constant strain rate. Influence of non-proportionality degree on the additional cyclic hardening is also investigated showing that the square loading path lead to the maximal level of non-proportional hardening. Both variants of SS316L are compared from the mechanical as well as microstructural point of view.

Introduction

The increasing demands on modern materials go hand in hand with the requirement to develop innovative alloys to widen the possibilities of application of metallic materials, as well as to enhance the utility properties and increase the longevity of products manufactured based on the known material systems. Among the favourable ways how to enhance the properties of metallic materials is to alter their structure, preferably having in mind the Hall-Petch relation [1] which stating that decreasing the grain size within the structure introduces increased mechanical properties, especially strength. Various non-conventional methods the grain size by which can be decreased have been introduced, such as the severe plastic deformation (SPD) technologies [2-5]. Nevertheless, the absolute values of the average grain size after processing also depends on the initial grain size of the original material. By this reason, fabrication of modern materials from initial powders has gained a lot of attention recently [6].

Selective Laser Melting (SLM) belongs to the group of additive manufacturing, or rapid prototyping, technologies using powders to build solid bulk materials [7]. Among its main advantages is the rapid cooling rate (as high as 10^6 K/s), which introduces the possibility to manufacture complex fine-grained structure geometries having microstructures composed of a wide range of crystallographic phases in a relatively easy manner. SLM has successfully been used to produce various materials, from Al-based alloys and composites [8], through Ti-based [9] and Zr-based [10] alloys and high entropy alloy [11], to steels [12].

This work is a continuation of the contribution presented at the ASME PVP2019 conference in Texas [13], where the ratcheting behaviour of 3D printed and conventionally prepared 316L Stainless Steel was presented, including the effects of the applied strain rate on both the materials. Complex microstructures, as the one presented, lead to a higher yield limit [14,15] of the SS316L material, resulting in higher resistance in load-controlled fatigue tests [13]. The SLM technology shows a significant anisotropic behaviour, studied on SS316L for instance in [16]. The high cycle fatigue behaviour of the material under investigation were reported e.g. by Blinn et al. in ref. [17]. The stress-strain behaviour research presented in [13] shows similar softening behaviour during ratcheting tests and uniaxial low-cycle tests as observed on the Ti-6A-4V [18]. The conventional stainless steel has been investigated in special loading modes. There are published works showing the influence of strain rate on the behaviour of the material under room temperature [19,20].

This study is focused on the comparison of stress-strain behaviours of the 316L steel prepared via conventional casting and SLM technology, with the main focus on strain control in fatigue testing. The 3D printing process used for preparation of the herein presented specimens was identical to the process reported in the previous study [13]. Cyclic hardening/softening material behaviour has been evaluated for proportional, as well as non-proportional loading cases. Results of the technique using DIC for cyclic stress-strain curve estimation from a single low-cycle fatigue test are presented too.

Uniaxial fatigue tests

All tests in this study have been realised under constant strain amplitude and zero mean strain on the LabControl 100kN/1000Nm hydraulic testing machine in VSB-Technical University of Ostrava. Tension-compression fatigue tests were performed using a solid specimen with a diameter of 5mm. The comparison of standard cyclic stress-strain curve for uniaxial loading for additively and conventionally produced SS316L is shown in Fig. 1a. The curves and lifetimes are very similar for both production technologies, see Fig. 1b.

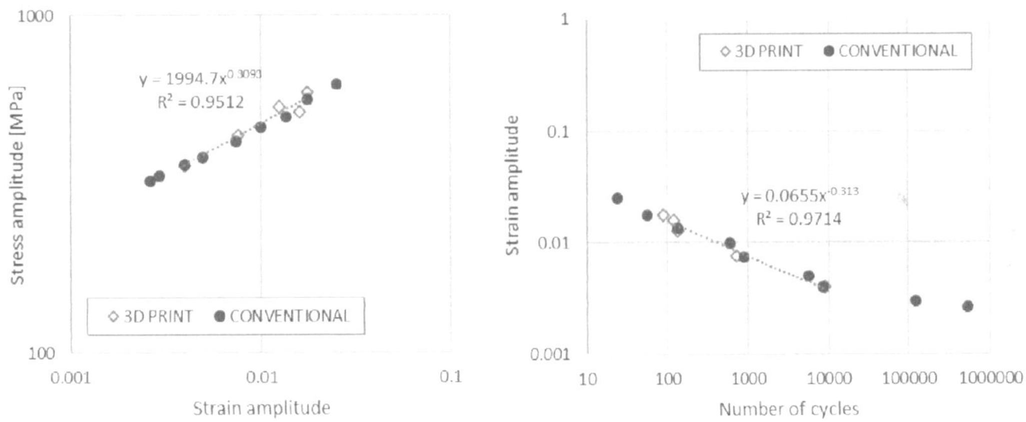

a) b)

Fig. 1: Comparison of uniaxial fatigue test results on conventional and 3D printed SS316L: a) Cyclic stress-strain curves; b) E-N curves

Multiaxial fatigue tests

Multiaxial fatigue testing has been done on a thin-walled specimen with an inner diameter of 10mm and an outer diameter of 12.5 mm [14]. The EPSILON 3550 biaxial extensometer was used to control shear and axial strain (25 mm gauge length) in that case. First, Low-Cycle Fatigue (LCF) tests for different non-proportional load paths and different levels of loading were realised on the conventional SS316L. This steel has low stacking fault energy (SFE) [23] which is sensitive to the additional hardening dependent on the type of material and also on loading path.

First, the stress-strain behaviour of conventional SS316L will be described. The influence of strain paths on transient behaviour during the first several load cycles has been evaluated. In the case of a rhombic path, see Fig. 2a, it is obvious that there is characteristic cyclic hardening following by softening till fracture for a higher level of loading (solid curve). In case of lower loading levels cyclic hardening is followed by the almost steady state after several cycles. In Fig. 2b the axial hysteresis loops are shown for rhombic loading path under 1% of strain amplitude. From this picture is obvious the strong non-proportional hardening during the first several cycles followed by cyclic softening.

A significant change in the shape of the hysteresis loop between the first and tenth cycle is evident from Fig. 2b. A small change on shape is visible between the tenth and twentieth cycle. Other interesting results are reported in [21].

In the case of circle path, see Fig. 3, it is obvious that cyclic hardening, during the first several cycles, is followed almost by the steady state (Fig. 4a). This is in contrast with 3D printed SS316L, where the cyclic softening begins immediately and occurs during the whole test. An interesting observation is that the hysteresis loops in the half-life are very similar in shape and size, see Fig. 4b.

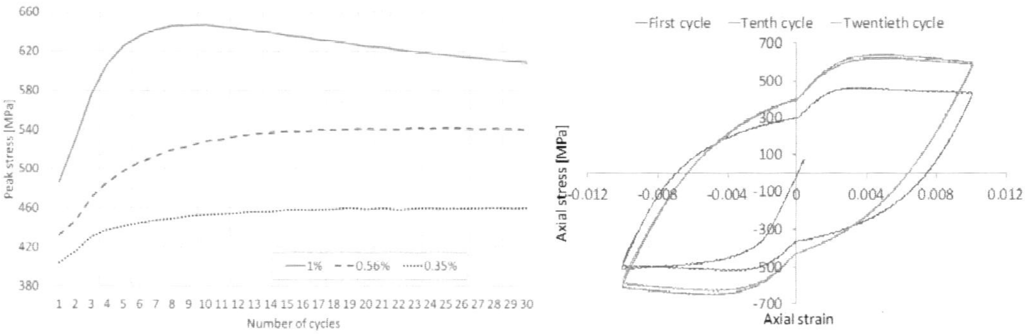

a) b)

Fig. 2: Results of additional hardening due to non-proportional loading: a) Cyclic hardening for rhombic path - conventional SS316L; b) Axial hysteresis loops for the rhombic path under strain amplitude of 1%

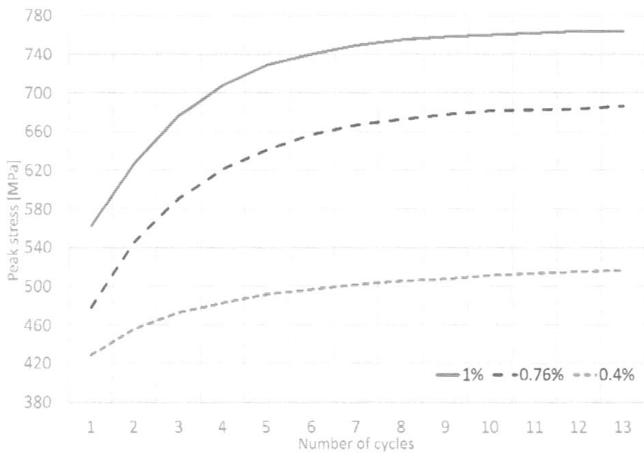

Fig. 3: Results of additional hardening due to non-proportional loading for circle path - conventional SS316L

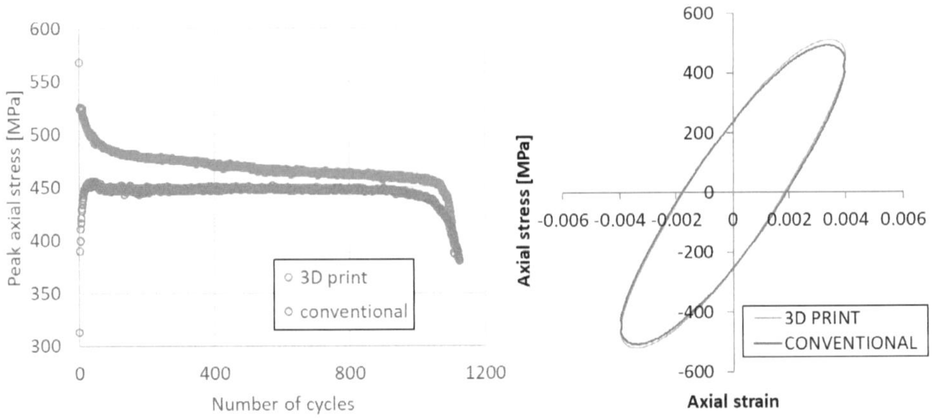

a) b)

Fig. 4: Experimental results for both variants of SS316L: a) Cyclic hardening-softening curves in 0.4% strain amplitude test with circle path b) Axial hysteresis loops in the half-life

The biggest influence on additional hardening during the first several cycles has the square loading path. Lower influence on non-proportional hardening has the circle and rhombic loading paths. This is evident from Fig. 5, where the experimental results from non-proportional experiments are compared with proportional experimental results (tension/compression tests). Square loading path reveals almost two times bigger stress amplitude than proportional loading path, when comparing higher strain amplitudes. In the case of smaller amplitudes, the differences are quite smaller.

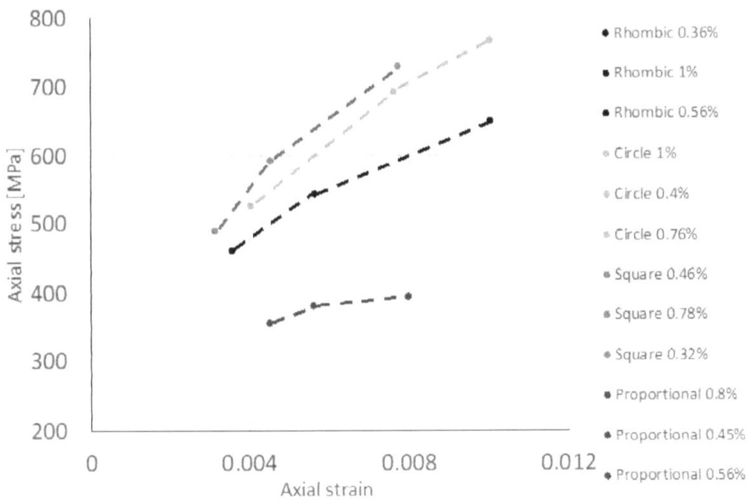

Fig. 5: Peak axial stress vs peak axial strain for different loading paths and strain amplitudes

In cyclic plasticity is the additional hardening usually described by the non-proportional parameter Φ (or different designation). This parameter is usually used at construction of cyclic plasticity or life-time models, like [22]. Some inspiration could be found in [23], where the authors suggest the non-proportional parameter, where they incorporate the additional strain hardening parameter in this form

$$\alpha = \frac{\sigma^n - \sigma^p}{\sigma^p} = \frac{\sigma^n}{\sigma^p} - 1. \qquad (1)$$

The quantities σ^p and σ^n are the basic maximum values of von Mises equivalent stresses under cyclic proportional (tension, compression or torsion) and non-proportional loading paths deformation, respectively. Using equation (1) the value of non-proportionality for different loading paths and levels of loading can be determined as stated in Tab.1.

Tab. 1 Additional hardening for different loading paths

Path	**Square 0.32 %**	**Square 0.46 %**	**Square 0.78 %**
σ_a [MPa]	525.4	618.9	749.8
α [-]	55.36 %	65.77 %	73.82 %
Path	**Circle 0.4 %**	**Circle 0.76 %**	**Circle 1 %**
σ_a [MPa]	527.2	702.4	753.2
α [-]	45.78 %	63.47 %	62.79 %
Path	**Rhombic 0.36 %**	**Rhombic 0.56 %**	**Rhombic 1 %**
σ_a [MPa]	477.7	574.4	661.3
α [-]	36.61 %	45.12 %	42.88 %

As apparent from Tab. 1, the highest level of additional hardening evokes the Square paths under 0.78 % of strain amplitude of loading. The additional hardening observed is almost 74%. On the other side, the Rhombic loading path is present, which evokes the hardening of 37% under 0.36 % of strain amplitude of loading. It is evident that additional hardening depends not only on the strain path but also on the level of loading. The higher amplitude of loading the more significant non-proportional hardening. The results are consistent with imagination, that under high load level more grains are affected by plastic deformation.

Evaluation of cyclic stress-strain curve by Digital Image Correlation

Digital Image Correlation (DIC) measurement has been done on 3D printed uniaxial specimens to verify the accelerated experimental technique published in [24]. The DIC technique was applied to get the cyclic stress-strain curve of conventional SS316L material from a single low-cycle fatigue test. A detailed description of the experiment and its evaluation is reported elsewhere [25].

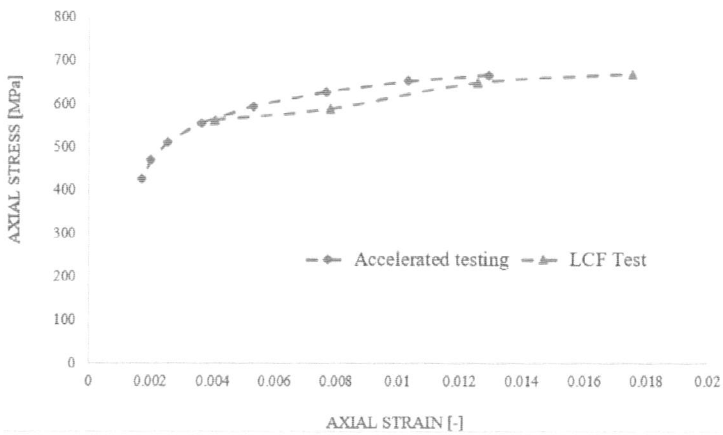

Fig. 6: Peak axial stress vs peak axial strain for different loading paths and strain amplitudes

Structure observations

Analyses of phase composition of both the materials were performed using Scanning Electron Microscopy (SEM-EDX), see Fig. 7. The results of the phase composition analyses of the examined samples show that both the cast and powder structures consisted mostly of austenite. Ferrite occurred mainly in the cast structure, however, only locally, primarily at the boundaries of austenitic grains.

Fig. 8 shows SEM-BSE (back-scattered electrons) images of both the examined materials the low angle (misorientation angle lower than 15°) and high angle (misorientation angle greater than 15°) boundaries (LAGBs and HAGBs) in which are highlighted in red (<15°) and black (> 15°) colours, respectively. Analysis of the grain boundary misorientations showed a fully recovered structure in the cast sample, as the structure exhibited the majority of HAGBs and a negligible presence of LAGBs. This sample also showed the presence of twins.

In contrast, the sample of the material originally printed from powders exhibited a more significant presence of LAGBS, indicating that the grains were not fully recovered and pointing to substructure development. Nevertheless, the content of HAGBs was still prevailing for this sample.

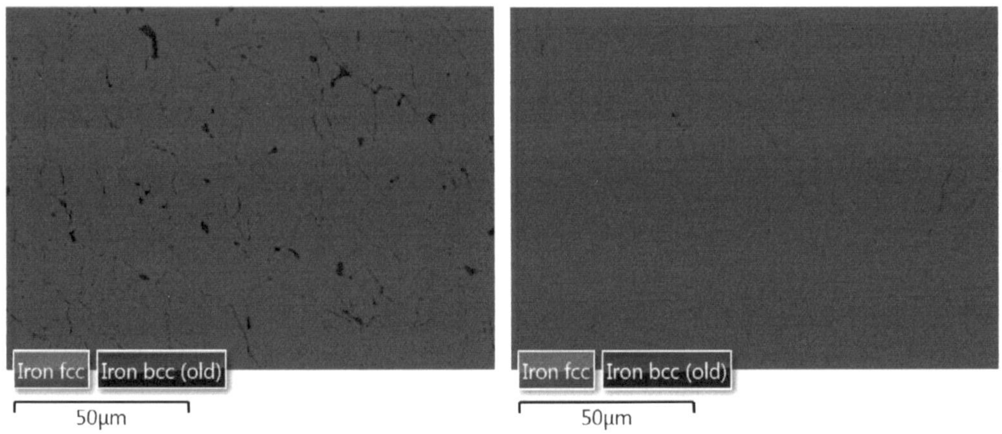

a) b)

Fig. 7: Results of phase composition analysis for both variants of SS316L: a) cast structure b) powder structure

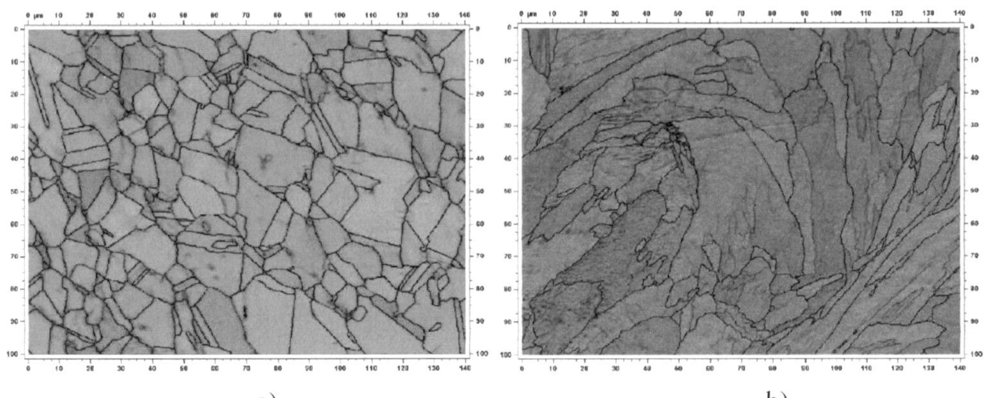

a) b)

Fig. 8: Analyses of grain boundaries for both variants of SS316L: a) cast structure b) powder structure

Conclusion

More extensive study has been done for conventional and 3D printed SS316L as follow up to preliminary ratcheting investigation [13]. Cyclic hardening and subsequent cyclic softening were observed for high loading levels. The square strain path has the strongest influence on additional hardening on the same level of loading. Lower additional hardening due to non-proportional loading has been observed even for circle strain path, i.e. in the 90 degree out of phase test. Cyclic hardening/softening behaviour of conventionally produced specimens differs to 3D printed ones significantly, but the stress-strain behaviour in the half of lifetimes is very similar as visible in Fig. 4b. Fatigue resistance is also comparable for both production technologies considering strain-controlled testing. An accelerated technique based on DIC method application for cyclic stress-strain curve estimation was shown to be comparable with the classical cyclic stress-strain curve. This study has been also done to prepare data for finite element modelling in the area of cyclic plasticity.

Acknowledgement: This research was funded by the Czech Science Foundation (GACR), grant No. 19-03282S, by The Technology Agency of the Czech Republic in the frame of the project TN01000024 and has been done in connection with the DMS project reg. no. CZ.02.1.01/0.0/17_049/0008407 financed by Structural Funds of Europe Union.

References

[1] S. Chen, Z. H. Aitken, Z. Wu, Z. Yu, R. Banerjee, Y-W. Zhang, Hall-Petch and inverse Hall-Petch relations in high-entropy CoNiFeAlxCu1-x alloys. Materials Science and Engineering: A, 773 (2020) 138873.

[2] L. Kunčická, R. Kocich, J. Drápala, V. A. Andreyachshenko, FEM simulations and comparison of the ecap and ECAP-PBP influence on Ti6Al4V alloy's deformation behaviour. In: Proceedings of METAL 2013 - 22nd International Conference on Metallurgy and Materials, 15-17 May 2013, Brno, Czech Republic (2013) 391–396.

[3] R. Kocich, L. Kunčická, A. Macháčková, Twist Channel Multi-Angular Pressing (TCMAP) as a method for increasing the efficiency of SPD. IOP Conf. Ser. Mater. Sci. Eng. 63 (2014) 012006. https://doi.org/10.1088/1757-899X/63/1/012006.

[4] R. Kocich, L. Kunčická, P. Král, A. Macháčková, Sub-structure and mechanical properties of twist channel angular pressed aluminium. Materials Characterization 119 (2016) 75–83. doi:10.1016/j.matchar.2016.07.020.

[5] L. Kunčická, R. Kocich, V. Ryukhtin, J. C. T. Cullen, N. P. Lavery, Study of structure of naturally aged aluminium after twist channel angular pressing, Materials Characterization 152 (2019) 94–100. doi: 10.1016/j.matchar.2019.03.045.

[6] P. C. Angelo, R. Subramanian, Powder metallurgy: Science, technology and applications, 2nd ed., PHI Learning Private Limited, New Delhi, 2008.

[7] R. Zhao, J. Gao, H. Liao, N. Fenineche, C. Coddet, Selective laser melting of elemental powder blends for fabrication of homogeneous bulk material of near-eutectic Ni–Sn composition, Additive Manufacturing 34 (2020) 101261

[8] M. P. Behera, T. Dougherty, S. Singamneni, K. De Silva, Selective laser melting of aluminium metal-matrix composites and the challenges. Materials Today: Proceedings In press, 2020.

[9] S. Pal, N. Gubeljak, R. Hudák, G. Lojen, V. Rajťúková, T. Brajlih, I. Drstvenšek, Evolution of the metallurgical properties of Ti-6Al-4V, produced with different laser processing parameters, at constant energy density in selective laser melting. Results in Physics 17 (2020) 103186

[10] J. J. Marattukalam, V. Pacheco, D. Karlsson, L. Riekehr, J. Lindwall, F. Forsberg U. Jansson, M. Sahlberg, B. Hjörvarsson, Development of process parameters for selective laser melting of a Zr-based bulk metallic glass. Additive Manufacturing 33 (2020) 101124.

[11] N. Li, S. Wu, D. Ouyang, J. Zhang, L. Liu, Fe-based metallic glass reinforced FeCoCrNiMn high entropy alloy through selective laser melting. Journal of Alloys and Compounds 822 (2020) 153695.

[12] J. Ghorbani, J. Li, A. K. Srivastava, Application of optimized laser surface re-melting process on selective laser melted 316L stainless steel inclined parts. Journal of Manufacturing Processes 56(A) (2020) 726-734.

[13] R. Halama, M. Pagáč, Z. Paška, P. Pavlíček, X. Chen, Ratcheting Behaviour of 3D Printed and Conventionally Produced SS316L Material. In: Proceedings of the ASME 2019 Pressure Vessels & Piping Conference PVP2019, July 14-19, 2019, San Antonio, TX, USA, paper number PVP2019- 93384.

[14] J. Hajnyš, M. Pagáč, J. Měsíček, J. Petrů, M. Krol, Influence of scanning strategies parameters on residual stress in SLM process according to bridge curvature method for stainless steel AISI 316L. Materials 13 (2020) 1527. doi: 10.3390/ma13071659.

[15] J. Hajnyš, M. Pagáč, O. Kotera, J. Petrů, S. Scholz, Influence of Basic Process Parameters on Mechanical and Internal Properties of 316L Steel in SLM Process for Renishaw AM400. MM Science Journal, 2019. Pgs: 2790-2794. DOI: 10.17973/MMSJ.2019_03_2018127

[16] L. Hitzler, J. Hirsch, B. Heine, M. Merkel, W. Hall, A. Öchsner, On the anisotropic mechanical properties of selective laser-melted stainless steel. Materials 10 (2017), 1136. DOI: 10.3390/ma10101136

[17] B. Blinn, F. Krebs, M. Ley, C. Gläßner, M. Smaga, J. C. Aurich, R. Teutsch, T. Beck, Influence of the Chemical Composition of the Used Powder on the Fatigue Behavior of Additively Manufactured Materials. Metals 9 (2019), 1285. DOI: 10.3390/met9121285

[18] D. Agius, K. I. Kourousis, C. Wallbrink, A Review of the As-Built SLM Ti-6Al-4V Mechanical Properties towards Achieving Fatigue Resistant Designs. Metals 8 (2018) 75. DOI: 10.3390/ma10101136

[19] M. Sapieta, A. Sapietova, V. Dekys, Comparison of the Thermoelastic Phenomenon Expressions in Stainless Steels During Cyclic Loading. Metalurgija 56 (2017) 203-206. ISSN 0543-5846

[20] P. Kubik, F. Šebek, J. Petruška, J. Hulka, N. Park, H. Huh, Comparative Investigation of Ductile Fracture with 316L Austenitic Stainless Steel in Small Punch Tests: Experiments and Simulations. Theoretical and applied fracture mechanics 98 (2018) 186-198. DOI: 10.1016/j.tafmec.2018.10.005

[21] M. Fusek, D. Ličková, R. Halama, Z. Poruba, F. Fojtík, Evaluation of multiaxial fatigue criteria on LCF data of SS316L. In: Proceedings of Experimental Stress Analysis 2018, Czech Republic: Czech Society for Mechanics, 2018, p. 108-113.

[22] M. Fusek, R. Halama, D. Ličková, Two modification of Jiang criterion for constant amplitude loading of AA2124-T851 AND SS316L. Special issue "Computational Mechanics of Materials" of the Continuum Mechanics and Thermodynamics journal, submitted in 2020.

[23] M. Borodii, S. Shukaev, Additional cyclic strain hardening and its relation to material structure, mechanical characteristics, and lifetime. International Journal of Fatigue 29 (2007) 1184-1191.

[24] R. Halama, P. Gál, Z. Paška, J. Sedlák, A new accelerated technique for validation of cyclic plasticity models. In: Proceedings of MMS 2017 - 22nd Conference on Machine Modeling and Simulation 2017, Sklené Teplice, Slovakia, September 5 – 8 September 2017, Vol. 157, p. 05008. DOI 10.1051/matecconf/201815705008

[25] B. Govindaraj, Digital Image Correlation and its Application for Accelerated Testing of Specimens Manufactured by 3D Printing. Ostrava: VSB - Technical University of Ostrava, Faculty of Mechanical Engineering, Department of Applied Mechanics, 2020, 68 p. Supervisor: prof. Ing. Radim Halama, Ph.D.

EAN 2020
58th conference on experimental stress analysis

Algorithms for Automatic Billet Straightening Machine

HALAMA R.[1,a], SIKORA J.[2,b], FUSEK M.[1,c], MAREK M.[1,d], BARTECKÁ J.[1,e], CZEBE J.[2,f], GURÁŠ R.[2,g], WAGNEROVÁ R.[2,h], MAHDAL M.[2,i]

[1]Department of Applied Mechanics, Faculty of Mechanical Engineering, VŠB – Technical University of Ostrava, 17. listopadu 2172/15, 708 00 Ostrava, Czech Republic

[2] Department of Control Systems and Instrumentation, Faculty of Mechanical Engineering, VŠB – Technical University of Ostrava, 17. Listopadu 2172/15, 708 00 Ostrava, Czech Republic

[a]radim.halama@vsb.cz, [b]jan.sikora@vsb.cz, [c]martin.fusek@vsb.cz, [d]martin.marek@vsb.cz, [e]jana.bartecka@vsb.cz, [f]jiri.czebe@vsb.cz, [g]radek.guras@vsb.cz, [h]renata.wagnerova@vsb.cz, [i]miroslav.mahdal@vsb.cz

Keywords: Straightening process, 3-point bending, Control strategy, Billet straightening

Abstract. This scientific contribution presents current results of cooperation focused on the automatic billet straightening machine development. An experimental study of three-point bending is shortly presented to explain basic ideas of the straightening. Main variants of straightening are briefly described. Some material parameters have to be properly determined for a new material in the straightening process, what is solved by creation of software for their estimation from a mapping regime of straightening.

Introduction

There are many applications requiring straightening of a long metallic material, for example billets [1], elevator guide rails [2] or more general long linear guideways [3].

In technical practice, two methods of straightening profile bars are usually used. The first possibility is continuous straighteners [4, 5], which is however problematic for straightening of large cross-sections, mainly due to the need to use large bearings.

The billet straightening, which is described in this contribution, is necessary operation done in ironworks before grinding of billets. Straightening by means of three-point bending is more accurate and flexible in terms of straightened cross-sections than continuous straightening. The billet straightening by 3-point bending is usually done manually by operators in manual regime based on human vision and joystick control [1]. In the automatic straightening machine an algorithm must be adapted with respect to the input curvature of the profile bar (e.g. single-arc shape, "S" shape or shape with multiple vertices [6]). This is usually the so-called multi-step straightening mechanism [3, 6]. For a multi-step straightening mechanism, it is then necessary to correctly determine, for example, the number of straightening steps, the distance of supports, the size of the straightening force / straightening stroke, with different settings, differently aligned bars can be obtained [2]. The crucial thing is an accurate prediction of springback [7].

The aim of this contribution is to present results currently gained in the frame of a long term project focused on the development of automatic billet straightening machine. The machine has been constructed by KOMA Industry company for Třinecké železárny. The camera vision and visualisation of straightening was developed by experts from Elcom company. The main aim is to show basic ideas of the newly proposed algorithm and to explain the necessary optimisation procedure to get some process parameters for reliable straightening.

Three-point Bending

For better understanding of the basic idea of algorithm the terminology should be introduced. The irreversible deflection w_{pl} remaining after unloading will be called the plastic deflection, see Fig. 1. This quantity is also important input for the algorithm and is measured by a camera system in the developed automatic straightening machine.

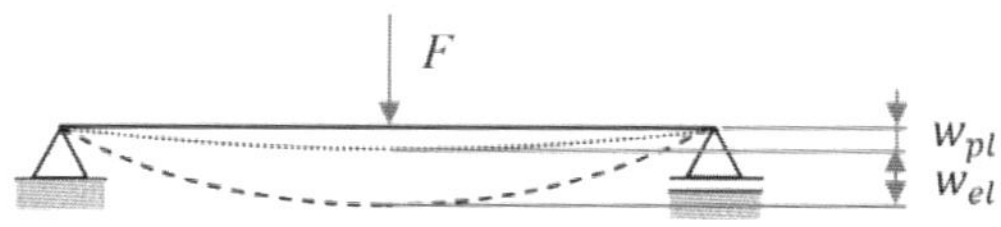

Fig. 1: A scheme of 3-point bending case

Then, the total deflection w is composed of the plastic deflection w_{pl} and the elastic deflection w_{el}, thus

$$w = w_{el} + w_{pl} \,. \tag{1}$$

The plastic deflection w_{pl} can be calculated considering an elastic stiffness k_{el} and applied bending force F according to the analogy to Hooke's law

$$w_{pl} = w - w_{el} = w - \frac{F}{k_{el}} \,. \tag{2}$$

Experiments

First of all, an experimental study including tensile tests and 3-point bending tests performed on the 51CrV4 material will be presented. The bending tests were realised on specimens with the rectangular cross-section of variety of dimensions D and distances of supports L. An example of obtained force vs deflection diagrams is shown in Fig. 2. In order to visualise the difference of cross-section dimensions a photo of some specimens is presented in Fig. 3.

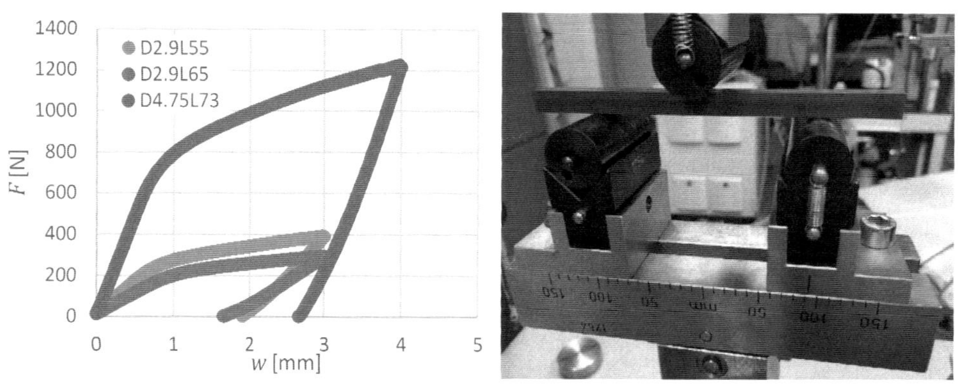

Fig. 2: Dependency of force on deflection from a bending test (left), a photo of measurement

Fig. 3: A photo of selected deformed specimens

For straightening of billets with different dimensions of a square cross-section it is important to investigate how the dependences of the total deflection on the plastic deflection differ for individual cross-sectional sizes. The main result of the experimental study is the graph in Fig. 4, where all the obtained dependences of the total deflection on the plastic deflection for individual dimensions of the cross-section are shown (e.g. the designation 2.9x2.9 corresponds to the dimension of the side of the cross-section 2.9 mm).

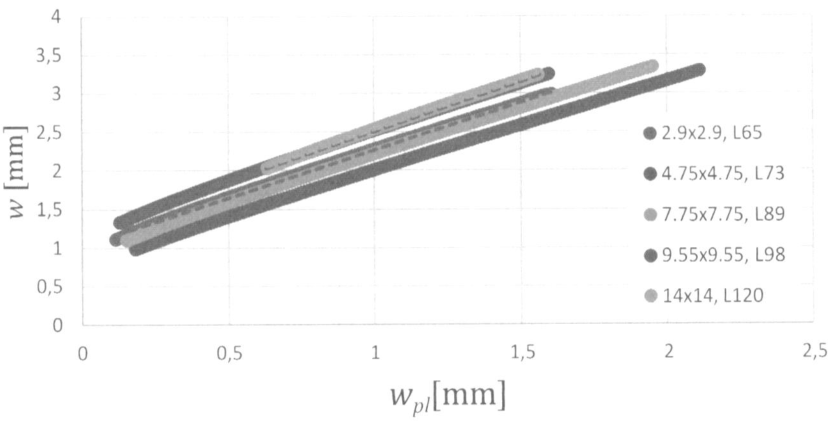

Fig. 4: Dependence of total deflection on plastic deflection from experiments performed in laboratory conditions at room temperature

An important finding from the performed experimental study is the fact that although the cross-sections are significantly different in cross-sectional dimensions, the slope of the dependence $\boldsymbol{w(w_{pl})}$ remains approximately the same. The curves therefore differ only in the vertical offset. However, it is necessary to mention, that the distances of supports *L* were considered differently for each case based on a Finite Element simulations (for instance L65 in the legend of the graph in Fig.4 means *L*=65mm).

Material parameters

The straightening run is designed by dividing the billet into sections corresponding to the distance of the supports (e.g. L = 1000 mm) and for the measured deviation from the straightness w_{pl} the required stroke w is calculated from the relation

$$w = w_Y + k_w w_{pl} , \tag{3}$$

where w_Y, k_w are material parameters.

There are also two other material parameters w_{ignor} and w_{max}, which should be appropriately determined for given material. After estimation of stroke w, it is checked whether $w > w_{ignor}$ is valid in the individual sections. If the condition does not apply, the intervention is omitted in the given section. If $w > w_{max}$, the release is performed by three (or two only) strokes to prevent billet breaking. Estimation of the size of such particular strokes can be done by means of numerical analyses [8]. An application has been coded in Python, which is used for estimation of material parameters mentioned above, see Fig. 5. A mapping regime of straightening is necessary, when a new material should be straightened, to be able to apply equations (2) and (3).

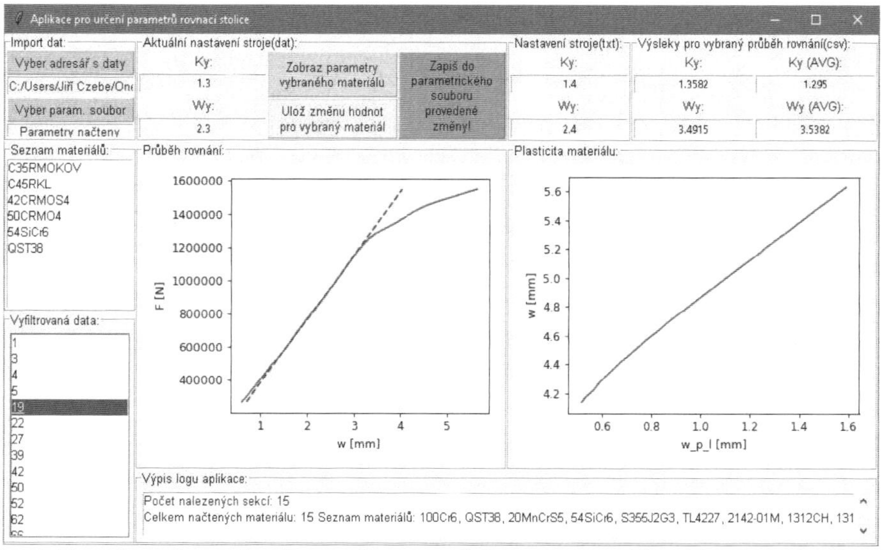

Fig. 5: User interface of software for material parameters identification (In Czech)

Straightening algorithm

Two criteria are used to assess the straightness of the billet. The first direct access to the entire length of the billet, which is the sum of the heights and the minimum deviation from the linear regression equation. This value is denoted as p_1 in the algorithm. The second characteristic is the straightness per meter of billet length. This value is available as the sum of the payments and the minimum deviation from the regression line equation determined by the meter. This value is denoted as p_2 in the algorithm. The critical value of parameter p_1 will be marked as $p_{1\text{crit}}$ and is generally dependent on the billet length. The critical value of parameter p_2 will be marked as $p_{2\text{crit}}$ and influences also the output accuracy of straightening process. According to current state of experimental validation, the following variants of straightening are possible. The border between strongly crooked and slightly curved billets is p_0.

Variant 1 ($p_1 > p_0$ and $p_2 > p_{2\text{crit}}$): This variant is usually the most effective for "snake-like" billets. The billet is divided into meter sections. In each section, a regression line is determined and the value of w is calculated. If $w > w_{ignor}$ then an intervention is performed in the given section. This variant is usually quite time consuming for a large number of interventions.

Variant 2 ($p_1 < p_0$ and $p_2 < p_{2\text{crit}}$): It is assumed that the billet can have two possible shapes (single-arc shape or "S" shape). Then it is straightened by either two or one stroke. The same parameter p_0 is considered for all materials.

Variant 3 ($p_1 > p_{max}$): This is a regime useful for very crooked billets (occurs usually for single-arc shape). The value of p_{max} is a limit of p_1 for the billet to be considered for this variant of straightening. The straightening is boosted based on experience with the given material. The same parameter p_{max} is considered for all materials.

Variant 4 ($p_1 < p_{1crit}$ and $p_2 > p_{2crit}$): This variant is very important in the case, when the billet is curved just in one place. The largest deflection is found and the stroke is proposed analogously as in the Variant 1.

The main program for determination of the position and stroke size proposal was developed in NI LabView. Selection of variants is done from the parent system.

Conclusions

The main finding of this scientific study is the importance of plastic hardening parameter k_w to be able to perform the automatic straightening for different cross-section size in technical practice. There is a possibility to increase the support distance with increasing cross-section size to keep the plastic hardening parameter k_w exactly the same. The second important parameter w_Y depends on the yield strength of the material and the cross-section dimensions of the billet. Both material parameters, k_w and w_Y must be identified for the considered material of billet at least on one selected cross-section size data.

Many numerical simulations were done based on Chaboche material model with two backstress parts [9] to propose the nature of algorithm described in this contribution [8]. The basic variants of straightening considering different shapes of the billet have been described. The algorithm was verified on chosen steels in The New Long Billet Treatment Plant of Třinecké železárny. The process and material parameters are optimised using a Python code. An example of one pass of the billet using variant 1 of the algorithm is shown in Fig. 6.

The next step of research is the application of rigid body movement calculations [3, 6] to speed up the straightening process (to minimize necessity of scanning).

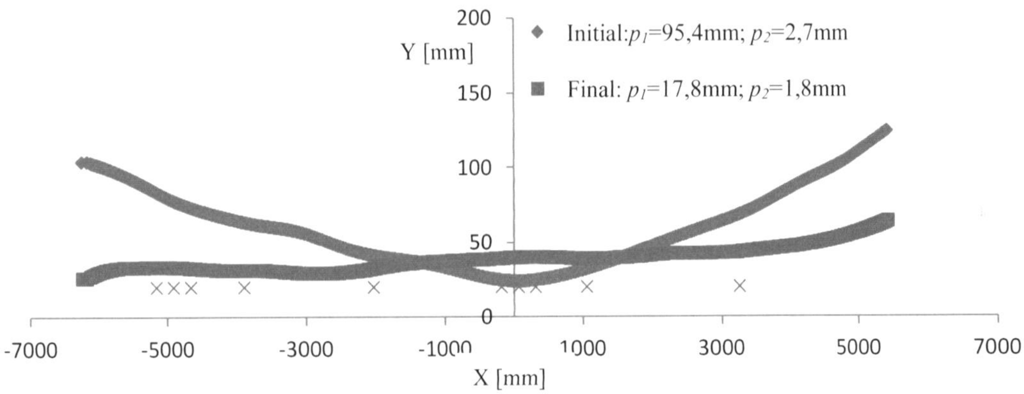

Fig. 6: Shape of the billet before (blue) and after (red) straightening by variant 1

Acknowledgement: This work was supported by The Technology Agency of the Czech Republic in the frame of the project TN01000024 National Competence Center-Cybernetics and Artificial Intelligence and by the European Regional Development Fund in the Research Centre of Advanced Mechatronic Systems project, CZ.02.1.01/0.0/0.0/16_019/0000867 and the project SP2020/57 Research and Development of Advanced Methods in the Area of Machines and Process Control supported by the Ministry of Education, Youth and Sports.

References

[1] M. Fusek, R. Halama, Z. Poruba, Calibration of Material Parameters during Billet Straightening, in: Proceedings of the Fifty Fifth International Conference on Experimental Stress Analysis, Nový Smokovec, Slovakia, 2017, pp. 611–615.

[2] K. Wang, B. Wang, Y. Chuancheng, Research on the multi-step straightening for the elevator guide rail. Procedia Engineering 16 (2011) 459-466.

[3] Y. Zhang, H. Lu, X. Zhang, H. Ling, A novel control strategy for the multi-step straightening process of long/extra-long linear guideways. Proceedings of the Institution of Mechanical Engineers: Journal of Mechanical Engineering Science 233 (2019) 2959-2975.

[4] J. Petruška, T. Návrat, F. Šebek, Novel approach to computational simulation of cross roll straightening of bars. Journal of Materials Processing Technology 233 (2016) 53-67.

[5] B.J. Wu, L.C. Chan, T.C. Lee, L.W. Ao, A study on the precision modeling of the bars produced in two cross-roll straightening. Journal of Materials Processing Technology [online]. Elsevier, 2000, 99, 202-206.

[6] Y. Zhang, H. Lu, Y. Wang, X. Zhang, J. Zhang, H. Ling, Variable Span Multistep Straightening Process for Long/Extra-Long Linear Guideways. IEEE Access 7 (2019) 107491-107505.

[7] P.A. Eggertsen, K. Mattiasson, On the identification of kinematic hardening material parameters for accurate springback predictions. Int J Mater Form 4 (2011) 103–120.

[8] R. Halama, M. Marek, M. Fusek, Optimization of the billet straightening process. Technical report for KOMA Industry. VŠB-Technical University of Ostrava, 2018, 20p. (In Czech).

[9] J.L. Chaboche, J. Lemaitre, Mechanics of Solid Materials, First ed., Cambridge University Press, Cambridge, 1990.

EAN 2020
58th conference on experimental stress analysis

Tensile Tests of 3D Printed PLA and Calibration of an Orthotropic Material Model

HECZKO J.[1,a], KRYSTEK J.[1,b], KOTTNER R.[1,c]

[1]NTIS - New Technologies for the Information Society, Faculty of Applied Sciences, University of West Bohemia, Technicka 8, Pilsen, Czech Republic

[a]jheczko@ntis.zcu.cz, [b]krystek@ntis.zcu.cz, [c]kottner@ntis.zcu.cz

Keywords: 3D printing, PLA, Tensile test, Anisotropic material, Orthotropic material model

Abstract. Tensile tests of a desktop 3D printed polylactic acid (PLA) material are presented. Four different angles of infill orientation were measured, all with the same infill density of 100 %. Poisson's ratios were determined in another set of tests using strain gauges and larger specimens. Calibration of a linear orthotropic material model was performed.

Introduction

Additive manufacturing and 3D printing are nowadays a widely used tool for not only rapid prototyping, but manufacturing of both functional prototypes and final products. Many technologies emerged over the past decades, ranging from stereolithography and fused deposition modelling (FDM) to sintering and jetting methods. Further development of these technologies includes printing from a vast amount of materials (plastic composites, concrete, food [1], biological tissues [2]), continuous fibre reinforcement, non-planar printing or 4D printing [3].

The study of mechanical properties of objects made by additive manufacturing is somewhat lacking this rapid development and the results may even be a little contradictory [4]. These are, however, crucial for ensuring functionality of the printed parts and should be taken into account in the design process. Technical data sheets often provide strength and stiffness of the raw material, but the resulting properties may depend heavily on the technology and setup of the printing process, e.g. (in the case of fused deposition modelling (FDM) technology) nozzle temperature, volume flow ratio, printing speed, part orientation or layer thickness [4-6].

This work focuses on stiffness and strength of PLA printed on a desktop FDM machine, since this is usually a starting point for most 3D printing projects. Moreover, the results, especially testing methods and modelling, may be applicable to the more sophisticated printing technologies and products of industrial machines. Previous works relevant to this particular choice of objectives include the following.

Andrzejewska et al. [7] performed monotonic and fatigue tests of both injection molded and 3D printed PLA. They used two different orientations (±45° and 0/90°) and two different densities (30 % and 90 %) of a rectilinear infill pattern and one type of honeycomb pattern. They found that the usual dogbone shape of specimens (ISO 527) is suitable for both injection molded and FDM/FFF 3D printed material.

Fernandez-Vicente et al. [8] evaluated the effects of infill pattern and its density on ultimate stress, ultimate strength, and Young's modulus ABS for a single orientation of specimens showing that 100 % infill may reach similar properties as the raw material.

Nomani et al. [9] compared different values of layer thickness between 0.2 mm and 0.8 mm, showing that strength and stiffness of ABS increase with decreasing layer thickness.

Pandzic et al. [10] studied the effect of PLA colour on various mechanical properties. The Young's modulus ranges between 2.7 and 3.2 GPa and tensile strength ranges between 30 and 41 MPa.

Verbeeten et al. [11] tested the anisotropy of rate-dependent behaviour of PLA showing that both elastic and viscous properties are anisotropic.

Doungkom and Jiamjiroch [12] compared the impact of specimen orientation with respect to the build plate in the case of PLA and different densities of the honeycomb infill pattern.

Cooling of the printed object has been shown to impact both dimensional quality and mechanical properties. Higher speeds of cooling air led to better accuracy but lower tensile strength, especially if the loading direction is perpendicular to the direction of printing [13]. Lower strength is explained by higher void ratio and lower crystallinity.

The influence of many parameters still remains to be investigated, e.g. perimeters, variable temperature and printing speed or extrusion width. There also seems to be lack of experiments in deformation modes other than uniaxial tension.

Material

The raw material is grey PLA made by the company Filament-PM. The printing properties are 200-230 °C nozzle temperature and 20-60 °C bed temperature. The mechanical properties given by the technical datasheet are: impact strength 16 kJ/m^2 (ISO 179), and flexural modulus 3500 MPa (ISO 178).

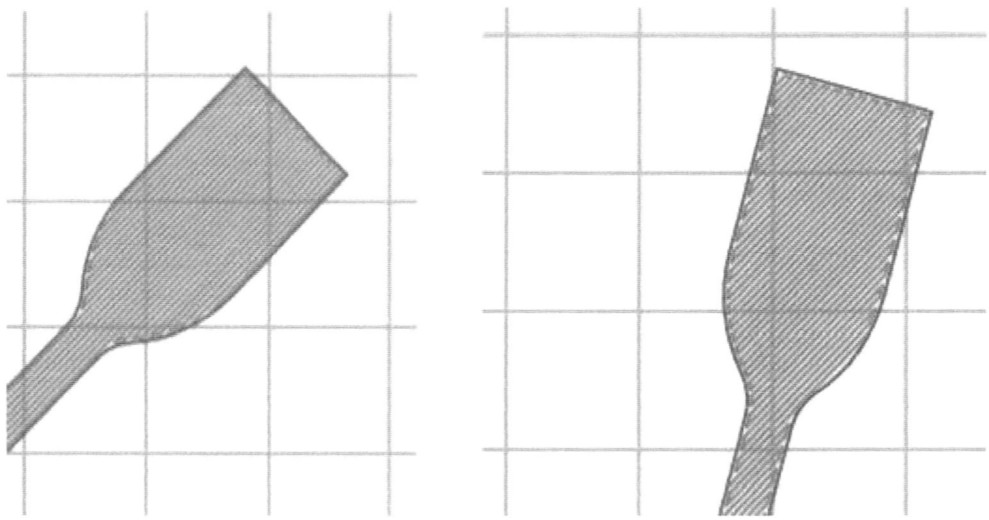

Figure 1. Infills with different orientation angles; 0° left, 30° right

Tensile tests

The outer shape of test specimens was dogbone, type 5A by the standard ČSN EN ISO 527-2 [14]. The instructions for the printer (g-code) were automatically generated using the Slic3r Prusa Edition 1.40.0 with the default settings for the particular material and the Prusa i3 MK3 printer that was used. The most important parameters were:

- layer height 0.15 mm (first layer 0.20 mm),
- nozzle temperature 210 °C (first layer 215 °C),
- bed temperature 60 °C,

- infill density 100 %,
- nozzle diameter 0.4 mm,
- default extrusion width 0.45 mm.

The specimens were labeled according to infill orientation: 0, 30, 60, 90 (angle between the specimen axis and extrusion path - see Figure 1). Specimens of the same infill orientation were distinguished by sequential numbers.

The tests were performed on an electromechanical universal testing machine Zwick/Roell Z050 using a 50 kN load cell. Mechanical extensometer was used to measure strain in the thin middle region. The initial distance of its arms was 15 mm. Monotonic displacement was prescribed with the speed of the clamp of the machine being 2 mm/min in the case of specimen 0/90_01 and 1 mm/min otherwise. All tests were performed at the room temperature, 23±2 °C. Tests 30_02 and 60_01 have been removed from the data due to apparent extensometer slip.

The stress-strain curves in Figure 2 show that the lowest ultimate strength was measured with the 90° infill orientation (43 MPa), the 0° infill exhibits higher ultimate strength (48.0 MPa), and 30° and 60° orientations seem to be even a little stronger (48.8 MPa and 51.3 MPa, respectively). Because the fracture occurred out of the central area in most specimens, see Figure 3, no conclusions were made regarding ultimate strain.

Stiffness was evaluated according to ČSN EN ISO 527-2 [11], Table 1 shows the mean values. The fact that 0° orientation exhibits lower stiffness than 90° is slightly surprising. The reason might lie in the low number of test specimens used for each orientation.

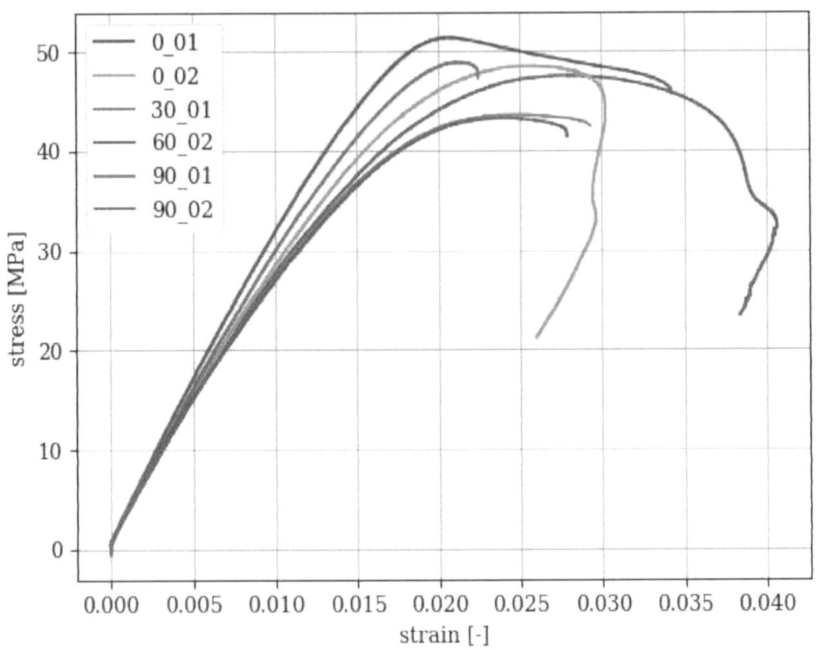

Figure 2. Stress-strain curves.

Table 1: Values of Young's moduli at different infill orientations

Infill angle [°]	0	30	60	90
Young's modulus [GPa]	2.967	3.127	3.287	3.052

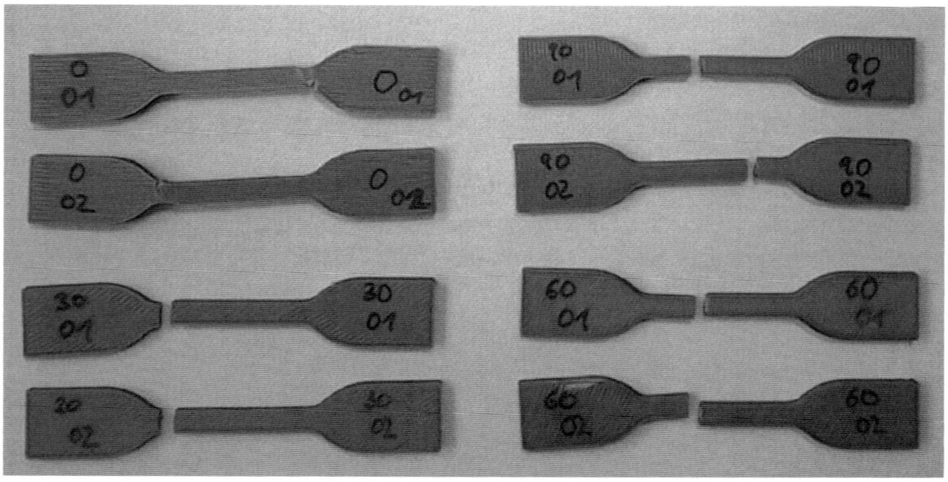

Figure 3. Tensile specimens after the tests.

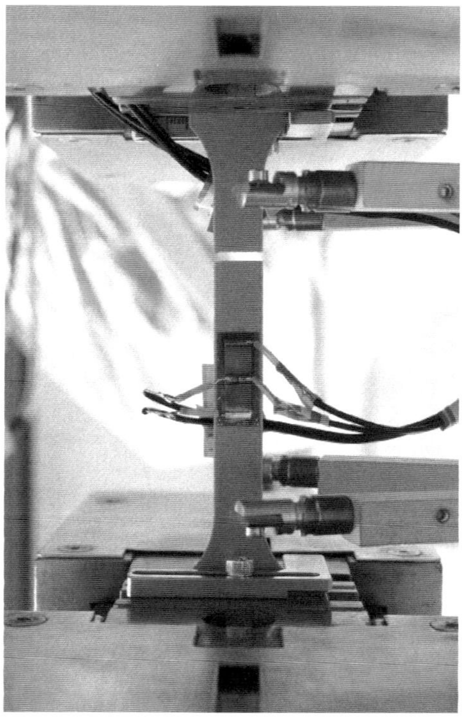

Figure 4. Tensile test with a strain gauge to measure the Poisson's ratio.

Poisson's ratio

Specimens for the evaluation of Poisson's ratio were type 1A by the standard ČSN EN ISO 527-2 [11]. They were printed with the same settings as the tensile specimens. Strain gauges were used to measure strain in the axial (0°) and transverse (90°) directions, see Figure 4. The following values of Poisson's ratios were determined: $\nu_{xy} = 0.3289$, $\nu_{yx} = 0.3267$. These seem to lie within the margin of error.

Material model

Linear material model is considered here, i.e. it obeys the Hooke's law

$$\sigma_{ij} = D_{ijkl}\varepsilon_{kl}. \tag{1}$$

An orthotropic model was considered in plane stress conditions. This leads to the following stress-strain relationship (using the more compact Voigt notation)

$$\begin{bmatrix}\sigma_{xx}\\ \sigma_{yy}\\ \sigma_{xy}\end{bmatrix} = \frac{1}{1-\nu_{xy}\nu_{yx}}\begin{bmatrix}E_x & \nu_{yx}E_x & 0\\ \nu_{xy}E_y & E_y & 0\\ 0 & 0 & G_{xy}(1-\nu_{xy}\nu_{yx})\end{bmatrix}\begin{bmatrix}\varepsilon_{xx}\\ \varepsilon_{yy}\\ 2\varepsilon_{xy}\end{bmatrix}, \tag{2}$$

where the components of the stiffness tensor D^0_{mnpq} transform in the usual way as

$$D^1_{ijkl} = Q_{im}Q_{jn}Q_{kp}Q_{lq}D^0_{mnpq}, \tag{3}$$

Q being the rotation matrix that has the following form

$$Q = \begin{bmatrix}\cos(\alpha) & -\sin(\alpha) & 0\\ \sin(\alpha) & \cos(\alpha) & 0\\ 0 & 0 & 1\end{bmatrix}, \tag{4}$$

upon rotating the specimen infill by the angle α about the z-axis. The Young's modulus in the direction of loading E is computed using the uniaxial relation

$$F = A_0E\varepsilon_{11} = A_0\sigma_{11} = A_0D^1_{11kl}\varepsilon_{kl}, \tag{5}$$

where the transversal strain is computed from the condition

$$\sigma_{22} = D^1_{22kl}\varepsilon_{kl} = D^1_{2211}\varepsilon_{11} + D^1_{2222}\varepsilon_{22} = 0 \tag{6}$$

as

$$\varepsilon_{22} = \frac{-D^1_{2211}\varepsilon_{11}}{D^1_{2222}}. \tag{7}$$

Thus, the Young's modulus is

$$E = \frac{F}{A_0\varepsilon_{11}} = D^1_{1111} + D^1_{1122}\frac{\varepsilon_{22}}{\varepsilon_{11}} = D^1_{1111} - D^1_{1122}\frac{D^1_{2211}}{D^1_{2222}}. \tag{8}$$

This relation is, in general, not linear with respect to the material parameters E_x, E_y, ν_{xy}, ν_{yx}, and G_{xy}.

Simple shear test was suggested to overcome non-uniqueness of the calibration process. In this case, the components of strain are $\varepsilon = [0, 0, 2\varepsilon_{12}]^{\mathrm{T}}$ and the stress-strain relationship reduces to $\sigma_{12} = 2D_{1212}\varepsilon_{12}$ with the other stress components being zero.

Calibration

The calibration problem is to find the parameters $\bar{x}$ of the model that minimize a suitable objective function:

$$\bar{x} = \underset{x}{\operatorname{argmin}} f(x). \tag{9}$$

The objective function was chosen in the form of sum of squares of residuals

$$f(x) = \sum_i (E_i(x) - \bar{E}_i)^2, \tag{10}$$

where $E_i(x)$ is the Young's modulus as computed by the material model (8), $\bar{E}_i$ is the corresponding measured value (see Table 1), and different infill angles are denoted by the index $i \in \{0, 30°, 60°, 90°\}$. The Levenberg-Marquardt method (see e.g. [15]) was used to solve problem (9)-(10).

As expected, the solution to the calibration problem is not unique with respect to the Poisson's ratios ν_{xy} and ν_{yx}. Therefore, their values were measured using the strain gauges and fixed. Thus, only three optimization parameters remained, $x = [E_x, E_y, G_{xy}]$. The resulting values are shown in Table 2. History of the objective function and optimization parameters is shown in Figure 5 and 6, respectively. The data (Young's moduli) are compared to the prediction of the calibrated model in Figure 7.

Table 2: Calibration results

E_x [GPa]	E_y [GPa]	G_{xy} [GPa]
2.943	3.075	1.288

The following procedure was employed in order to quantify the limits of use of the calibrated material (see Figure 8). Upper and lower bounds of stress were chosen for each infill angle as:

$$\sigma_{\text{upper}}(\varepsilon) = 1.05 \max_i \{\sigma_i(\varepsilon)\}, \tag{11}$$

$$\sigma_{\text{lower}}(\varepsilon) = 0.95 \min_i \{\sigma_i(\varepsilon)\}, \tag{12}$$

where $\sigma_i(\varepsilon)$ denotes the stress-strain curves determined experimentally. The model is considered usable at strain ε if the stress predicted by the model, $\sigma(\varepsilon)$, lies between the upper and lower bound

$$\sigma_{\text{lower}}(\varepsilon) \leq \sigma(\varepsilon) \leq \sigma_{\text{upper}}(\varepsilon). \tag{13}$$

Table 3 shows the limit strains beyond which the above relation no longer holds.

Table 3: Limits of use of the calibrated model in terms of tensile strain

Infill angle [°]	0	30	60	90
Strain limit of use [%]	0.91	0.79	1.06	0.47

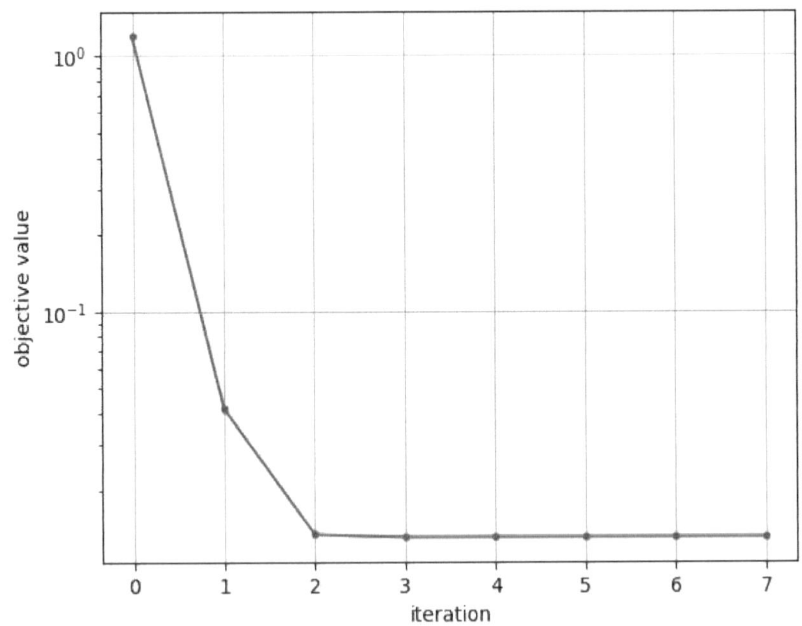

Figure 5. History of the objective function.

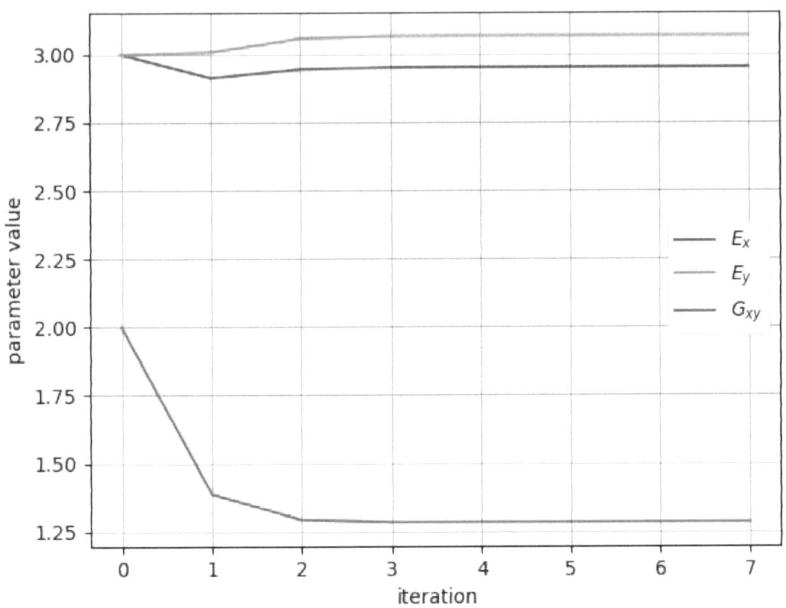

Figure 6. History of the optimization parameters.

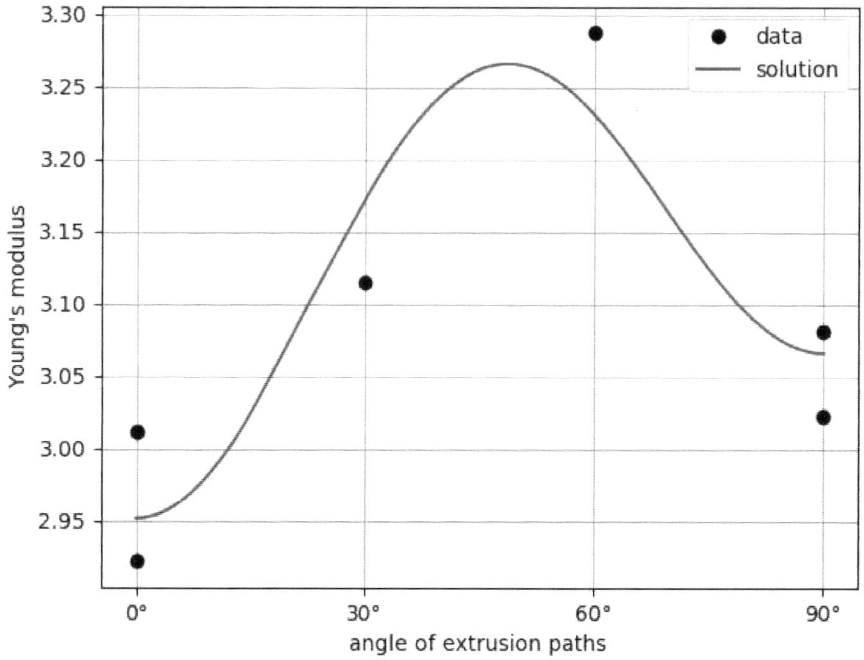

Figure 7. Data (Young's moduli for different infill angles) and the calibrated model.

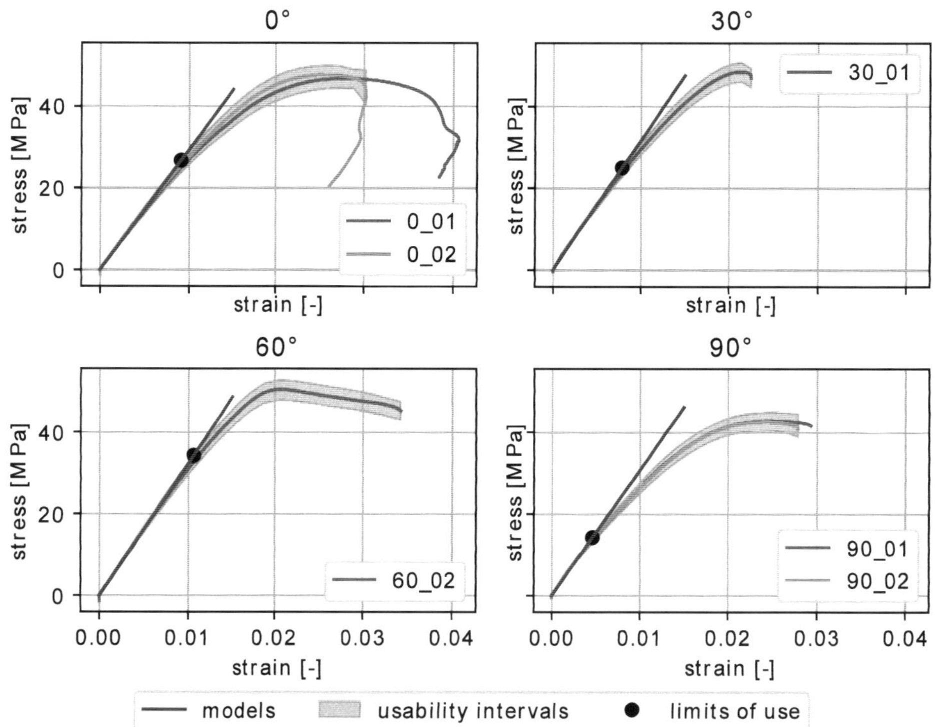

Figure 8. Limits of use of the calibrated model. The grey areas represent the intervals inside which the model is considered to describe the data accurately enough. The black dots mark the points at which the model exits these intervals.

Conclusions

Contrary to the expectation, stiffness in the direction of the extruded paths (0° angle) is lower than in the perpendicular direction (90°), while the opposite holds for ultimate strength. The directions in between (30° and 60°) exhibit higher stiffness.

The linear orthotropic model is capable of modeling the material for reasonably small strains. Successful calibration of this model requires knowledge of Poisson's ratios and tensile tests in other than the principal directions or a shear test, since the shear modulus, G_{xy}, strongly affect the stiffness in rotated coordinates.

Acknowledgement

This publication was supported by the project LO1506 of the Czech Ministry of Education, Youth and Sports under the program NPU I.

References

[1] A. Derossi, M. Paolillo, R. Caporizzi, C. Severini, Extending the 3D food printing tests at high speed. Material deposition and effect of non-printing movements on the final quality of printed structures, Journal of Food Engineering 275 (2020) Art. No. 109865.

[2] E. Saygilia, A. A. Dogan-Gurbuzb, O. Yesil-Celiktasa, M.S. Draz, Review article3D bioprinting: A powerful tool to leverage tissue engineering andmicrobial systems, Bioprinting 18 (2020) Art. No. e00071.

[3] X. Kuang, D. J. Roach, J. Wu, C. M. Hamel, Z. Ding, T. Wang, M. L. Dunn, H. J. Qi, Advances in 4D Printing: Materials and Applications, Adv. Funct. Mater. 29 (2019), Art. No. 1805290.

[4] A.C. Abbott, G.P. Tandona, R.L. Bradford, H. Koerner, J.W. Baur, Process-structure-property effects on ABS bond strength in fused filament fabrication, Additive Manufacturing 19 (2018) 29-38.

[5] B.M. Tymrak, M. Kreiger, J.M. Pearce, Mechanical properties of components fabricated with open-source 3-D printers under realistic environmental conditions, Materials and Design 58 (2014) 242-246.

[6] P. Ramkumar, Investigation on the Effect of Process Parameters on Impact Strength of Fused Deposition Modelling Specimens, in: IOP Conf. Series: Materials Science and Engineering 491 (2019) Art. No. 012026.

[7] A. Andrzejewska, Ł. Pejkowski, T. Topoliński, Tensile and Fatigue Behavior of Additive Manufactured Polylactide, 3D Printing and Additive Manufacturing, Vol. 6, No. 5 (2019) 272-280.

[8] M. Fernandez-Vicente, W. Calle, S. Ferrandiz, A. Conejero, Effect of Infill Parameters on Tensile Mechanical Behavior in Desktop 3D Printing, 3D Printing and Additive Manufacturing, Vol. 3, No. 3 (2016), 183-192.

[9] J. Nomania, D. Wilsonb, M. Paulinob, M. I. Mohammed, Effect of layer thickness and cross-section geometry on the tensile and compression properties of 3D printed ABS, Materials Today Communications 22 (2020) Art. No. 100626.

[10] A. Pandzic, D. Hodzic, A. Milovanovic, Influence of Material Colour on Mechanical Properties of PLA Material in FDM Technology, in: Proceedings of the 30th DAAAM International Symposium, Vienna, Austria, 2019, pp. 0555-0561.

[11] W. M. H. Verbeetena, M. Lorenzo-Bañuelosa, P. J. Arribas-Subiñas, Anisotropic rate-dependent mechanical behavior of Poly(Lactic Acid) processed by Material Extrusion Additive Manufacturing, Additive Manufacturing 31 (2020) Art. No. 100968.

[12] P. Doungkom, K. Jiamjiroch, Analysis of Printing Pattern and Infiltration Percent over the Tensile Properties of PLA Printed Parts by a Fuse Deposition Modelling Printer, IOP Conf. Series: Materials Science and Engineering 501 (2019) 012028.

[13] C.-Y. Lee, C.-Y. Liu, The influence of forced-air cooling on a 3D printed PLA part manufactured by fused filament fabrication, Additive Manufacturing 25 (2019) 196-203.

[14] ČSN EN ISO 527-2 (2012) Plastics - Determination of tensile properties - Part 2: Test conditions for moulding and extrusion plastics.

[15] W. H. Press, S. A. Teukolsky, W. T. Vetterling, B. P. Flannery, Numerical Recipes in C, second ed., Cambridge University Press, Cambridge, 1992.

EAN 2020
58th conference on experimental stress analysis

Performance of Selected Thermally Sprayed Coatings in Comparison with a Competitive Surface Treatment

HOUDKOVÁ Š.[1,a], ČESÁNEK Z.[1,b], POLACH P.[1,c]

[1]Research and Testing Institute Plzeň, Tylova 1581/46, 301 00 Plzeň, Czech Republic

[a]houdkova@vzuplzen.cz, [b]cesanek@vzuplzen.cz, [c]polach@vzuplzen.cz

Keywords: Thermally sprayed coatings, Wear resistance, High temperature, Corrosive environment

Abstract. The selected High Pressure/High Velocity Oxygen Fuel (HP/HVOF) sprayed coatings were tested in terms of mechanical, tribological and corrosion behavior and their performance was compared to the performance of competitive types of surface treatment, relevant to the aimed applications. The main purpose of the intended surface treatments is the protection of components in the power industry. The HP/HVOF sprayed coatings were chosen on the basis of their potential to resist wear, high temperature oxidation and corrosion in an aggressive environment. The completive types of surface treatment were chosen according to their relevance to a specific type of loading.

Introduction

In power industry, the surfaces suffer from various kinds of degradation caused by external loading as well as by the influence of increased temperature and corrosion aggressive environment. To protect the critical components surface from such a degradation, surface treatments are applied. With respect to the type of the component, its shape complexity and requirements given by the type of the loading, different types of surface treatment are chosen: surface hardening and nitriding for increased hardness in the case of low-temperature sliding wear, suitable for small complex-shaped components, hard surfacing for protection from oxidation, corrosion or wear, applicable on large components, where the temperature influence of a substrate material is not limitation, Physical Vapour Deposition (PVD) layers protecting the surface from wear, etc.

One of the applicable surface treatment technologies is a thermal spraying. This technology enables to create coatings from a wide range of materials including ceramics, metals and their alloys or hardmetals. Based on the used coating materials, different functionalities of the surface can be addressed. Thickness of the thermally sprayed coatings usually reached 0.2–0.3 mm, which is comparable to hard surfacing. The advantage of thermal spraying lies in the absence of the thermally affected zone in the underlying substrate. On the other hand, the mechanical locking of the coating on the surface asperities does not provide the adhesive strength comparable to e.g. the laser or the electron beam cladding [1].

In the group of thermally sprayed technologies, High Velocity Oxygen Fuel (HVOF) offers the possibility to deposit coatings of metal alloys and hardmetals in a superior quality [1]. The kinetic energy of the coating materials' particles impacting against the substrate is responsible for creating the coating of a low porosity and a high cohesive strength. On the other hand, the relative low flame temperature (in comparison with e.g. atmospheric plasma spraying) keeps the undesirable phase changes in the coating material at a low level [2].

The hardmetal coatings are characterized by the combination of hard carbide particles with a tough metal matrix. While the carbides ensure the wear resistance, the matrix is mainly responsible for the corrosion and the oxidation resistance. Two types of hardmetal systems are widely spread: WC-based hardmetals, suitable for low temperature applications up to 350 °C and Cr_3C_2-based hardmetal for the application up to 900 °C [3]. As the intended coatings application requires the oxidation resistance of the coating material in the hot steam environment (up to 610°C), attention is paid to the $C_{r3}C_2$-based hardmetals with 3 different types of matrix: NiCr; NiCrMoNb and CoNiCrAlY. From this group, the Cr3C2-25%NiCr composition is the most investigated one, both the HVOF and the Atmospheric Plasma Sprayed (APS) [4–7]. Up to now, less attention was paid to the coating of a variable matrix composition. The effect of thermal treatment on the tribological properties of Cr_3C_2-50%CoNiCrAlY was evaluated in [8]. Recently, the HVOF and the High Velocity Air Fuel (HVAF) sprayed Cr_3C_2-50%NiCrMoNb was analyzed in detail in [9].

For high temperature applications, the superalloys based on Co and Ni are used. A typical representative is a Co-based alloy known as Stellite©, the Hastelloy© Ni-based alloy or the NiCrBSi alloy. These materials often serve as coatings deposited using a laser cladding technology [10, 11]. Alternatively, they can also be thermally sprayed, with the benefit of maintaining the original chemical composition without diluted zones, typical for cladding.

The aim of the paper is to test the response of the above-mentioned coatings to the relevant loading conditions and to compare it to the alternative surface treatments to select the best surface treatments for specific applications. The coating materials were chosen on the basis of their potential to resist wear at a high temperature and a corrosive environment [12,13]. The completive types of surface treatment were chosen according to their relevance to the specific type of loading [14].

Experiments

The study involved the materials based on chromium carbide hardmetals and on Co and Ni-based alloys. The types of competitive surface treatment were laser cladding of Co-based alloy (Stellite 6), PVD thin film TiAlN and gas nitrided X22CrMoV12-1 steel. Laser clad Stellite 6 coating was chosen for the comparison of microhardness HV 0.3, abrasive and sliding wear resistance and corrosion resistance in the aggressive 18%$Na_2SO_4$82%$Fe_2(SO_4)_3$ environment. The PVD layer was tested and compared with the HP/HVOF sprayed coatings for hard particle erosion resistance and water droplet erosion resistance.

The microstructures were evaluated on the coatings' cross sections prepared by the standard metallographic procedure. Both the optical (OM) and the scanning electron (SEM) microscopy were used for the microstructure evaluation. The microhardness was measured on the coatings' polished cross-sections using the HV0.3 method. For each coating, at least 7 indents were done and the average value is reported.

Wear resistance to three-body abrasion and solid particle erosion was measured using the Dry Sand/Rubber Wheel test in accordance to ASTM G65 and the centrifugal erosion test [15]. The impact angle of erosive media (Al_2O_3; F70) varied from 15° to 90°. The coatings volume loss was determined from the mass loss using the values of density, previously determined using the Archimedes method. For each coating at least 3 independent measurements were done and the average value is reported. The worn surfaces were subsequently observed by SEM.

The sliding wear resistance was evaluated using the Ball-on-Flat test, according to ASTM G133. The test parameters were as follows: 25 N load; AISI 440C steel; 6 mm diameter ball counterpart; 5 Hz oscillating frequency; 10 mm stroke length; 1000 s testing time. Three different measurements were performed for each coating. The wear tracks' profiles were measured by the KLA-Tencor P-6 Profiler profilometer, at three different places, and the wear volume was calculated. Prior to sliding-wear tests, the coating surface was ground and polished to the 0.04 ± 0.02 μm Ra value.

The water droplet erosion test was performed in Doosan Škoda Power Ltd. Company in accordance with the company internal test prescription. The testing apparatus consists of a cylindrical vessel, in which a fixed strength disk is rotating, with the test pieces attached to the periphery. The nozzle creates a stream of droplets, through which the samples pass, causing the droplets to strike the samples at a prescribed velocity. Standard testing parameters are as follows: absolute pressure in the chamber: 3 kPa; shaft speed: 12000 RPM; impact velocity: 524 m/s; droplet size: 0.41 mm; total test duration: 3 hours 30 minutes. The samples were weighted in the defined period. The tested samples surface was smoothed using a metallographic procedure prior to the testing. Four samples of each type of coating were tested and the average value is reported.

The protective properties of the coatings in the high temperature aggressive environment of 18% Na_2SO_4 82% $Fe_2(SO_4)_3$ was tested. The thermal cycling from the room temperature up to 600 °C was repeated 50 times, the weight gain was recorded and evaluated. The high temperature test is described in detail in [12].

Performance of selected thermally sprayed coatings in comparison with a competitive surface treatment

The coatings' cross section microstructures can be seen in Fig. 1. No macro-defects, such as cracks or delamination, are visible in the observed coatings. The porosity is also negligible, except the Hastelloy C-276 coating (Fig. 1f), where pores are recognizable between the individual splats. The laser clad Stellite coating (Fig. 1f) has a dendritic structure consisting of the Co-based solid solution. Solitary small pores are concentrated in the interdimeric areas. As they do not connect each other, they do not represent a thread regarding the protection of the substrate from the corrosive environment.

The cross section microhardness is presented in the graph in Fig. 2. While the Cr_3C_2-based hardmetal coatings and the NiCrBSi coating, whose microstructure is strengthened by fine hard boride and carbide particles, are comparable, the alloy HVOF sprayed Stellite and Hastelloy coatings are softer. The laser clad Stellite coating showed a lower hardness then the HVOF sprayed coating of a similar composition.

Generally, the coatings microhardness determines the resistance of the coatings to the abrasive wear (Fig. 3) and the sliding wear (Fig. 4). The carbide-based hardmetal HP/HVOF sprayed coatings were the hardest and also the most abrasive and sliding wear resistant. On the contrary, the HVOF alloy coatings, namely Ni-based Hastelloy, were less wear resistant. Comparing the HVOF sprayed and the laser clad Stellite coatings, the HVOF sprayed coating is harder, and more wear resistant under both type of loading – abrasive and sliding. The mechanism of wear is clearly different (Fig. 5). While in the case of the HVOF coating the delamination of whole splats appeared, a significant adhesive wear can be observed in laser welding.

On the other hand, the solid particle erosion test (Fig. 5) showed the poorer erosion resistance of brittle hardmetal coatings while ductile alloy coatings suffered from lower erosion. Although the erosion tests of the laser clad Stellite coating were not provided, based on the previous experience, a high erosion resistance could be expected.

Comparing to the base material and the PVD TiAlN layer, the HVOF sprayed coatings are less erosion resistant. It is caused probably by a lower cohesive strength of the coating, where delamination of individual splats can appear as a result of repeated impacting of hard erosive particles. Similarly, the water droplet erosion revealed poorer resistance of the HVOF sprayed coating compared to the basic material or the PVD layer under such a type of loading (Fig. 6).

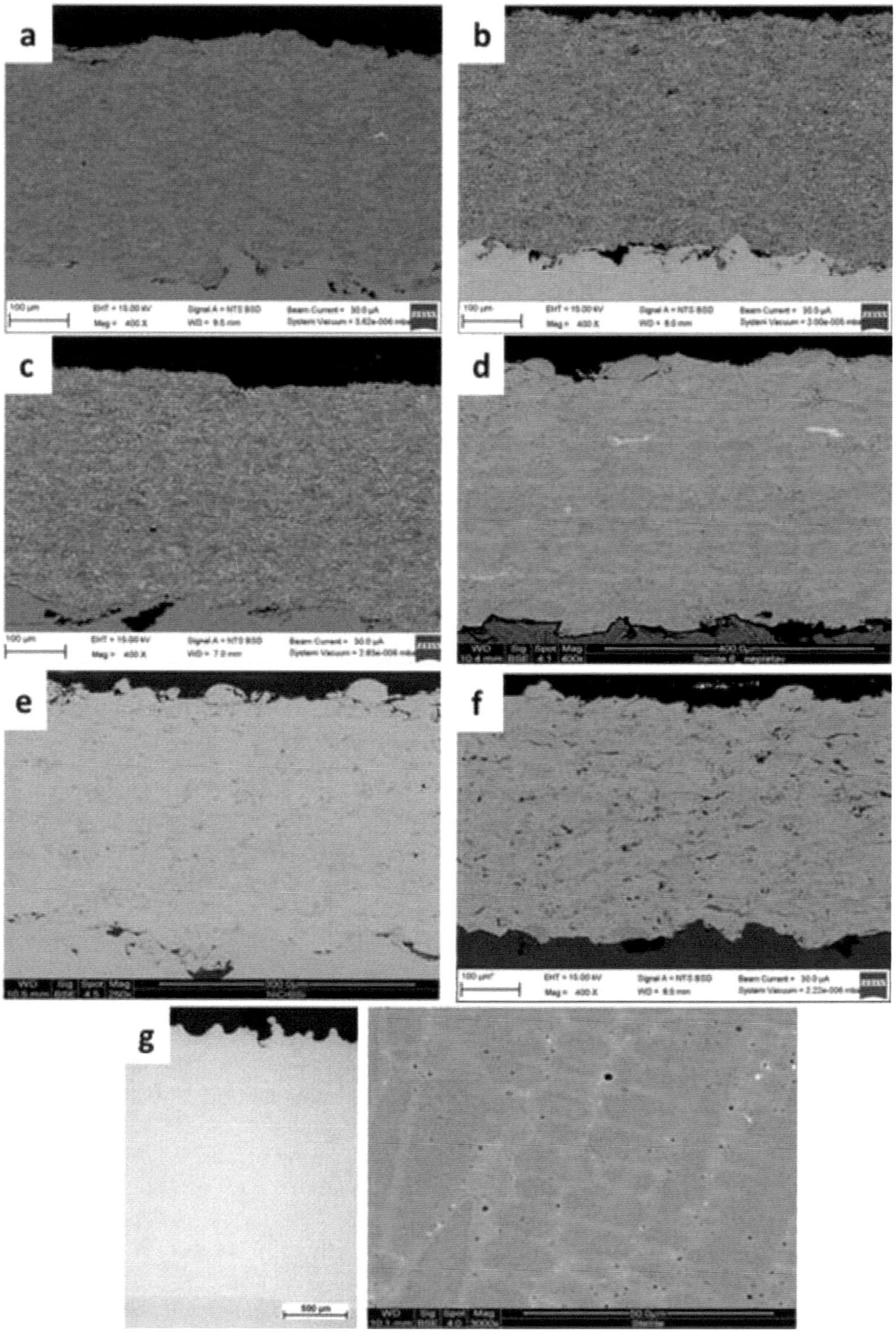

Fig. 1: Microstructure of coatings' cross sections: (a) Cr_3C_2-25%NiCr; (b) Cr_3C_2-5%CoNiCrAlY; (c) Cr_3C_2-50%NoCrMoNb; (d) Stellite HVOF sprayed coating; (e) NiCrBSi HVOF sprayed coating; (f) Hasteloy C-276 HVOF sprayed coating; (g) Stellite laser clad coating

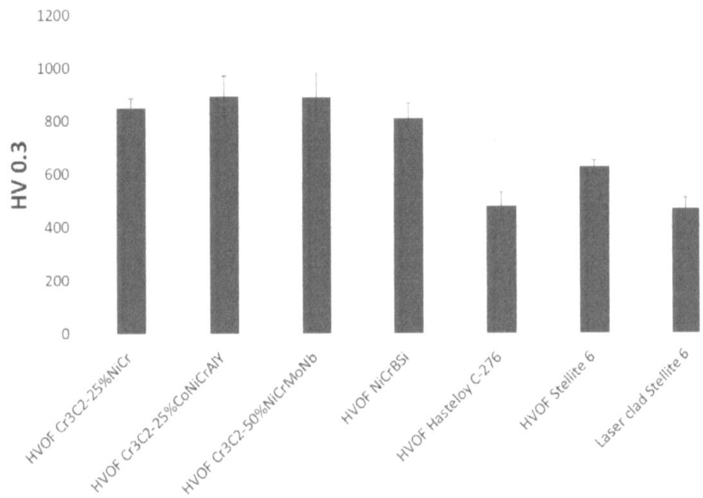

Fig. 2: Microhardness of selected HVOF sprayed coatings and the laser clad Stellite 6 coating

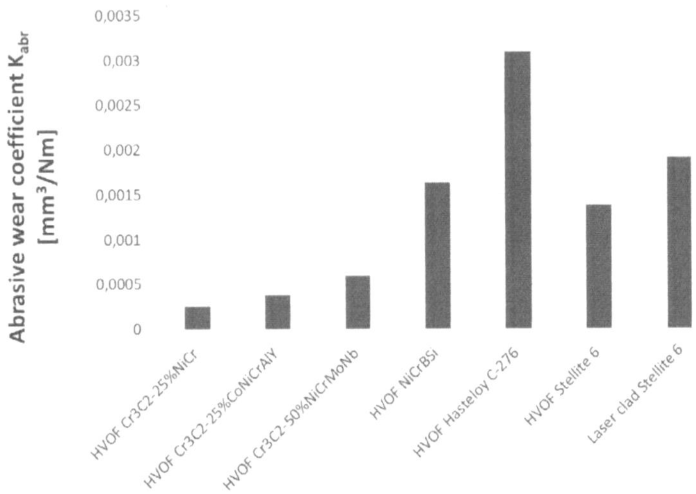

Fig. 3: Abrasive wear of selected HVOF sprayed coatings and the laser clad Stellite 6 coating

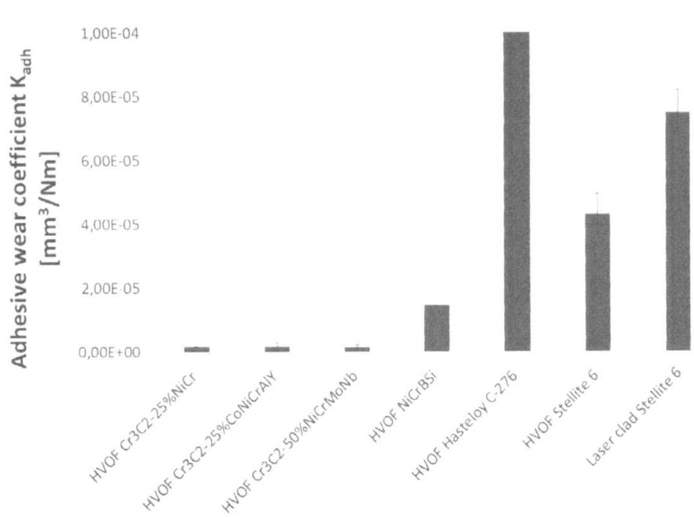

Fig. 4: Sliding wear of selected HVOF sprayed coatings and the laser clad Stellite 6 coating

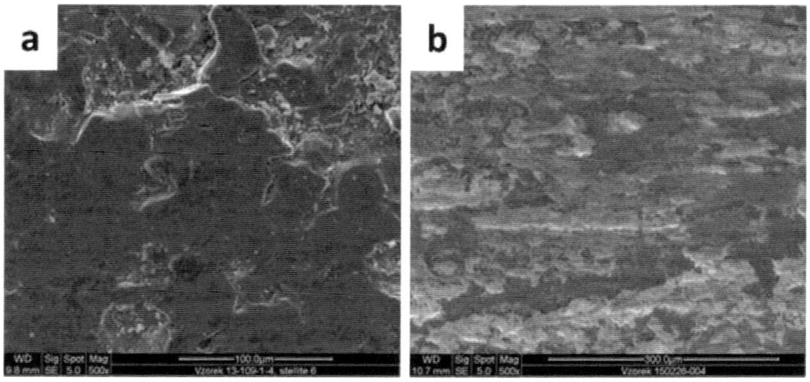

Fig. 5: SEM of wear scar after the sliding wear ASTM G-133 test of (a) the HVOF sprayed Stellite coating; (b) the laser clad Stellite coating

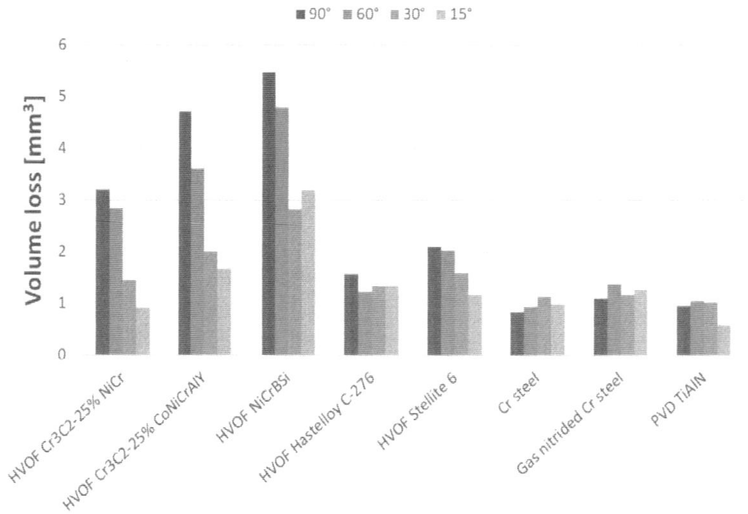

Fig. 6: Solid particle selected HP/HVOF sprayed coatings, erosion of the base material, the gas nitrided surface and the PVD TiAlN layer

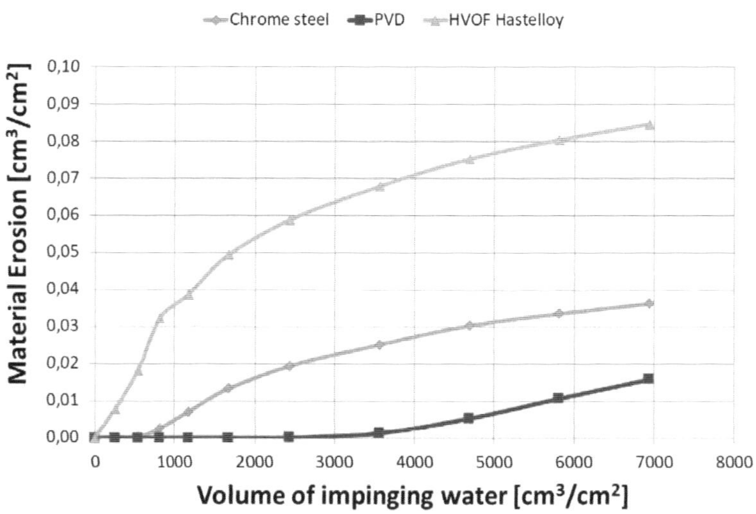

Fig. 7: Water droplet erosion of the base material, the HP/HVOF sprayed Hastelloy coatings and the PVD TiAlN layer

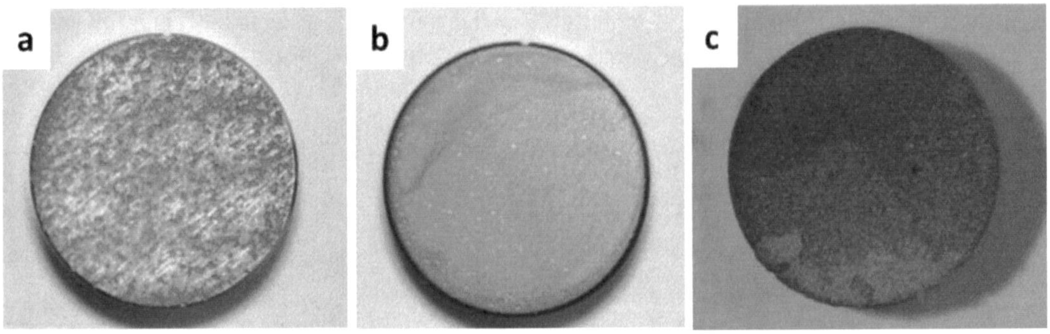

Fig. 8: The surface of eroded surfaces after 130 min of the Water droplet erosion testing: (a) base material; (b) PVD TiAlN layer; (c) HVOF sprayed Hastelloy C-276 coating

Results of the high temperature corrosion testing in the aggressive environment of 18% Na_2SO_4 82% $Fe_2(SO_4)_3$ are shown in Fig. 9. All the tested HVOF sprayed coatings proved their ability to protect the underlying substrate and improve their lifetime. Their protective ability was comparable to that of the laser clad Stellite 6 coating.

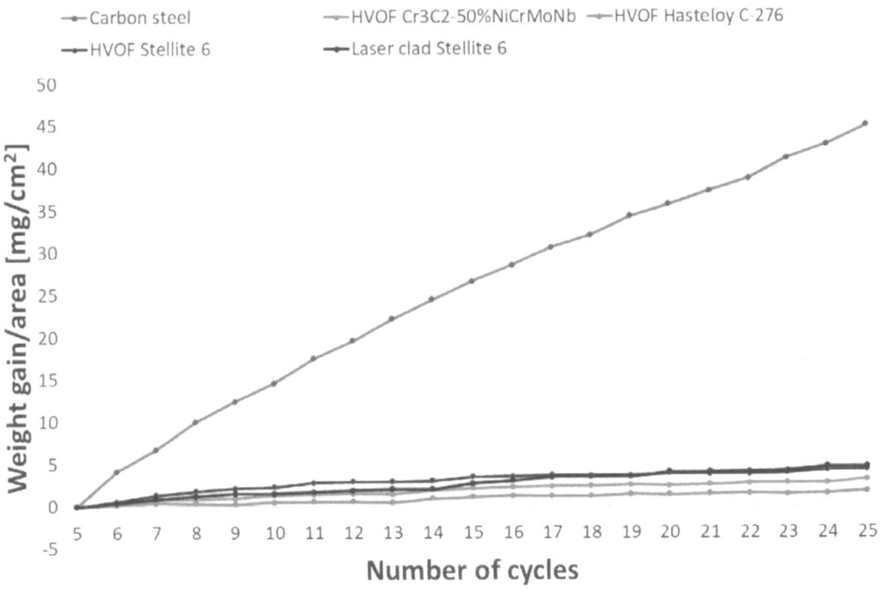

Fig. 9: Cumulative mass gain in dependence on the number of cycles at the high temperature corrosive aggressive environment

Conclusions

Based on the above reported results, the superior behaviour of the HVOF sprayed coatings, namely the Cr_3C_2-based hardmetals, was proved under the abrasive and the sliding wear loadings. On the other hand, the lamellar structure of the thermally sprayed coatings, consisting of individual splats, weakens the resistance of the HVOF coating under repeated loading, such as the solid hard particle erosion or the water droplet erosion. In this type of loading, the alloy coatings provide better results than the hardmetal coatings, but lower that alternative PVD layer. Regarding the protection of the substrate from the corrosive environment, they are comparable to the laser clad Stellite coatings. Based on these findings, the application on the components suffering from oxidation and corrosion can be recommended if the part is loaded mostly with sliding or abrasive wear.

Acknowledgement: The presented results take their origin thanks to solving the project TE01020068 "Centre of Research and Experimental Development of Reliable Energy Production" in the framework the "Competence Centers" program of the Technology Agency of the Czech Republic, work package "Development of advanced surface treatment of components used in parts of turbines working under the condition of operational temperatures of steam using the HP/HVOF technology of thermal spraying".

References

[1] L. Pawlowski, The Science and Engineering of Thermal Spray Coatings, Second Edition, John Wiley & Sons, New York (USA), 2008.

[2] A. Vaidaya, T. Streibl, L. Li, O. Kovarik, R. Greenlaw, An integrated study of thermal spray process structure – property correlations: A case study for plasma sprayed molybdenum coatings, Mat. Sci. Eng. A 403 (2005) 191–204.

[3] L. M. Berger, Application of hardmetals as thermal spray coatings, Int. J. Refract. Hard M. 49 (2015) 350–364.

[4] G. Bolelli, L. M. Berger, T. Börner, H. Koivuluoto, V. Matikainen, L. Lusvarghi, C. Lyphout, N. Markocsan, P. Nylén, P. Sassatelli, R. Trache, P. Vuoristo, Sliding and abrasive wear behaviour of HVOF- and HVAF-sprayed Cr3C2-NiCr hardmetal coatings, Wear 358–359 (2016) 32–50.

[5] N. Espallargas, J. Berget, J. M. Guilemany, A. V. Benedetti, P. H. Suegama, Cr3C2-NiCr and WC-Ni thermal spray coatings as alternatives to hard chromium for erosion-corrosion resistance, Surf. Coat. Tech. 202 (2008) 1405–1417.

[6] I. Hussainova, J. Pirso, M. Antonov, K. Juhani, S. Letunovitš, Erosion and abrasion of chromium carbide based cermets produced by different methods, Wear 263 (2007) 905–911.

[7] S. Matthews, Development of high carbide dissolution/low carbon loss Cr3C2-NiCr coatings by shrouded plasma spraying, Surf. Coat. Tech. 258 (2014) 886–900.

[8] J. A. Picas, M. Punset, S. Menargues, E. Martín, M. T. Baile, Microstructural and tribological studies of as-sprayed and heat-treated HVOF Cr3C2-CoNiCrAlY coatings with a CoNiCrAlY bond coat, Surf. Coat. Tech. 268 (2015) 317–324.

[9] V. Matikainen, G. Bolelli, H. Koivuluoto, L. Lusvarghi, P. Vuoristo, Sliding wear behaviour of thermally sprayed Cr3C2-based coatings, Wear 388–389 (2017) 57–71.

[10] A. Frenk, M. Vandyoussefi, J.-D. Wagnière, W. Kurz, A. Zryd, Analysis of the laser-cladding process for Stellite on steel, Metall. Mater. Trans. B 28 (1997) 501–508.

[11] N. Serres, F. Hlawka, S. Costil, C. Langlade, F. Machi, A. Cornet. Dry coatings and ecodesign part. 1 – Environmental performances and chemical properties, Surf. Coat. Tech. 204 (2009) 187–196.

[12] T. S. Sidhu, S. Prakash, R. D. Agrawal, Characterization and hot corrosion resistance of Cr_3C_2-NiCr coating on Ni-base superalloys in an aggressive environment, J. Therm. Spray Technol. 15 (2006) 811–816.

[12] Z. Česánek, Š. Houdková, F. Lukáč, High-temperature corrosion behavior of selected thermally sprayed coatings in corrosive aggressive environment, Mater. Res. Express 6 (2019) 016426.

[14] Š. Houdková, Z. Pala, E. Smazalová, M. Vostřák, Z. Česánek, Microstructure and sliding wear properties of HVOF sprayed, laser remelted and laser clad Stellite 6 coatings, Surf. Coat. Tech. 318 (2017) 129–141.

[15] T. Deng, A comparison of the gas-blast and centrifugal-accelerator erosion testers: The influence of particle dynamics, Wear 265 (2008) 945–955.

Effect of Temperature on the Interlaminar Strength of Carbon Fibre Reinforced Thermoplastic

HRON R.[1,a], KADLEC M.[1,b], RŮŽEK R.[1,c]

[1]VZLU – Czech Aerospace Research Centre, Beranových 130, Prague, Czech Republic

[a]hron@vzlu.cz, [b]kadlec@vzlu.cz, [c]ruzek@vzlu.cz

Keywords: interlaminar strength, composite, thermoplastic, polyfenylensulfid, temperature

Abstract. Application of thermoplastics composites (TPCs) in aircraft constructions is growing. Thermoplastics have the greatest potential in the interior and in the secondary construction, where they can fully replace aluminum parts and glass laminate parts. In the aircraft interior, they are used mainly for excellent fire, smoke and toxicity properties and on exterior mainly for their excellent impact resistance. Thanks to thermoplastics, today we can talk about real recycling of composites. The most used TPCs in aircraft constructions is polyfenylensulfid (PPS). This material was also chosen for our experiment where we compared two most used test method (ASTM and AITM) which are used for measuring of the interlaminar shear strength by the loading of the curved beam. The aim was to find out whether the results measured according to different standards are comparable (whether it is possible to neglect the influence of the method when comparing interlaminar shear strength, measured using these two standards). For more complex results, the tests were performed at three different temperatures. At room temperature, at 70°C and at –50°C. Both the selection of the test method and the test environment had a significant influence on the measured values of the interlaminar strength.

Introduction

More than 95 percent of composites used in aerospace industry are thermosets [1]. However, the share of high-performance thermoplastic composites (TPCs) in aeronautical industry is rising year after year even at the expense of the thermosets. It is given by attractive properties such as fracture resistance [2, 3, 4] formability[5, 6], welding [7, 8], self-healing possibilities [9, 10] and recyclability [11]. With regard to modern trends and requirements, we can say that the recyclability of composites belongs and will belong (compared to metals) among their weakest aspects. Thanks to thermoplastics, today we can talk about real recycling of composites. Thermoplastics soften when heated and become more fluid as additional heat is applied. The curing process is completely reversible as no chemical bonding takes place. This characteristic allows thermoplastics to be repeatedly cured and recycled without negatively affecting the material's physical properties. Softening by heating further enables the welding of subcomponents. This leads to the eliminating of fasteners and adhesives as is showed in [12]. The requested performance of structural TPCs parts can be easily achieved by using stacking of tailored blanks with combination of thermoforming process – this is, for example, demonstrated on thermoplastic rib in Ref. [13].

This paper present results of curved beam strength of PPS samples at different temperatures, concretely at room temperature (RT), at –55 °C (CT) and at 80°C (HTA). Before these tests started, a comparison of two most used test methods was performed (ASTM D6415 - Standard test method for Measuring the Curved Beam Strength of Fiber-reinforced Polymer-Matrix Composite [14] and AITM 1-0069 - Determination of curved-beam failure load [15]).

Experiment

PPS thermoplastics resin samples (see Fig. 1) with T300 3K, 5HS, 280 gm^{-2} FAW, 43% RC (50% BY VOLUME) carbon fabric 280 gm^{-2} were tested. Used lay-up was: $[[(0,90)/(\pm 45)]_4]$. The coupons were manufactured by thermoforming. Thermoforming is used to convert a flat consolidated continuous fibre reinforced laminate into a complex shape with no change in original laminate thickness. The laminates were heated to the required temperature and then quickly formed by pressure with a few minutes dwell time. Average width of the tested samples was $w = 25.14$ mm, average thickness $t = 4.95$ mm and average angle $\alpha = 89.5$ deg.

The tests were performed on electro-mechanical loading machine Instron 55R1185 with installed load cell with the capacity of ±10 kN. The test setup is shown in Fig. 2. Test specimen was placed on the bottom cylindrical bars. Then, extensometer Instron 2620-604 with a base of 50 mm was installed. Extensometer recorded axial displacement between the upper and lower parts of the fixture. The specimen was loaded by constant crosshead speed of 2 mm/min and test was ended when the loading went rapidly down.

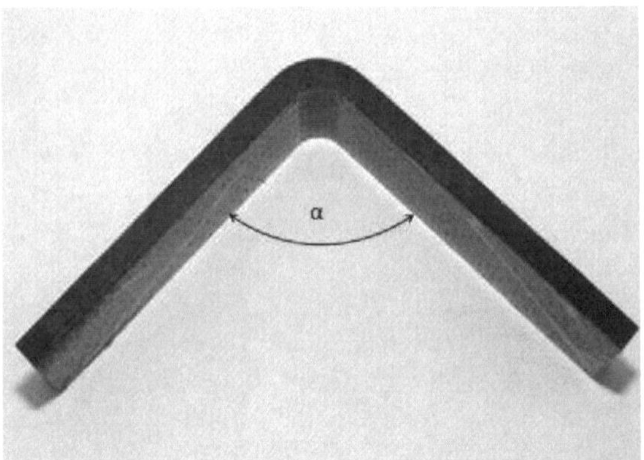

Fig. 1: Curved beam strength sample

Fig. 2: Curved beam strength test

At first, test methods were compared. Three test set-ups were used, see Table 1. The main difference between the test methods is, that the ASTM method used fixed distances between the lower/upper rollers. In contrast, AITM method defines span length of the fixture based on sample geometry – Equations (1) to (4):

$$l_t > 2 \cdot \left(\left((R_i + t + \tfrac{D}{2} \right) \cdot \sin(\varphi) + \left(\tfrac{t}{4} + 1 \right) \cdot \cos(\varphi) \right) \quad \pm 0{,}5 \tag{1}$$

$$l_t < 2 \cdot \left(\left((R_i + t + \tfrac{D}{2} \right) \cdot \sin(\varphi) + \left(\tfrac{t}{2} + 1 \right) \cdot \cos(\varphi) \right) \quad \pm 0{,}5 \tag{2}$$

$$l_b > l_t + t + 10 \quad \pm 0{,}5 \tag{3}$$

$$l_b < l_t + t + 20 \quad \pm 0{,}5 \tag{4}$$

where: l_t is span of top fixture, l_b is span of the bottom fixture, R_i is inner radius, t is thickness of sample, D is roller diameter, and ϕ is angle from horizontal of the sample legs.

Table 1: Tests set-up, overview

Test method	Distance of lower rollers [mm]	Distance of upper rollers [mm]
ASTM	100	75
ASTM mod	75	50
AITM	51.3	26.35

Results

All measured data (see Fig. 3) were tested by Dixon's test for outliers and on the basis of this evaluation, the lowest measured value of 63.5 MPa from the AITM respectively RT set was excluded.

	PPS		
	ASTM	ASTM mod	AITM
	69.3	70.1	83.6
	70.8	85.6	84.7
	72.9	71.4	80.0
	72.0	68.9	63.5*
	78.8	78.7	77.4
	65.4	75.9	-
	70.0	-	-
Mean	**71.3**	**75.1**	**81.4**
S.D.	4.09	6.36	3.33
C.V.	5.73	8.47	4.09
Min.	65.4	68.9	84.7
Max.	78.8	85.6	77.4

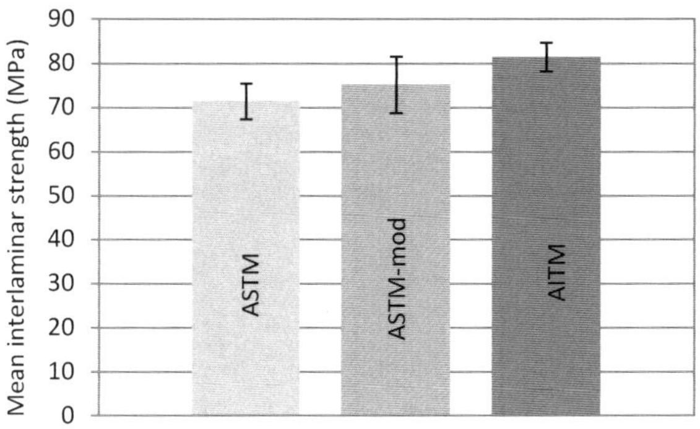

* outlier, not included into the statistical evaluation

Fig. 3: Measured mean interlaminar strength σ (MPa) - test methods comparison.

The statistical analysis of data showed that the test method have statistically significant effect (p-value = 0.05), see Table 2. Specifically, AITM data were statistically significantly different from the other two sets. For ASTM and ASTM-mod sets was the difference evaluated as statistically insignificant.

Table 2: Statistical comparison of test methods. D files are different, ND files are not different, p-value = 0.05.

	RT		
	ASTM	ASTM-mod	AITM
ASTM	-	ND	D
ASTM-mod	ND	-	D
AITM	D	D	-

For further testing (influence of temperature), the AITM test method was chosen. The main reason was to take sample geometry into account for test fixture setting (span length). All measured data are shown in Figure 4.

	PPS		
	RT	CT	HTA
	83.6	89.8	76.4
	84.7	89.5	71.4
	0.0	95.3	68.9
	63.5*	73.1	74.2
	77.4	77.8	73.8
Mean	81.4	85.1	72.9
S.D.	3.33	9.26	2.86
C.V.	4.09	10.89	3.93
Min.	84.7	73.1	68.9
Max	77.4	95.3	76.4

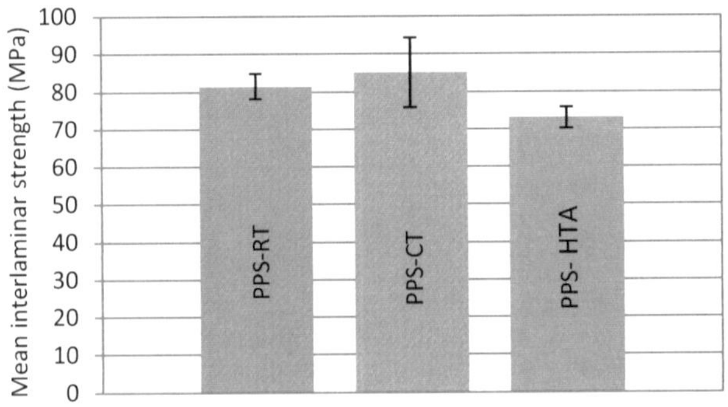

* outlier, not included into the statistical evaluation

Fig. 4: Measured mean interlaminar strength by AITM method, σ (MPa) – temperature effect.

The statistical analysis of data showed that the temperature have statistically significant effect (p-value = 0.05), see Table 3. Specifically HTA (80 °C) data were statistically significantly different from the others two sets. For RT and CT (–55 °C) sets was the difference evaluated as statistically insignificant.

Table 3: Statistical comparison of temperature effect. D- files are different, ND files are not different, p-value = 0.05.

	RT	CT	HTA
RT	-	ND	D
CT	ND	-	D
HTA	D	D	-

Conclusions

The measured values clearly demonstrated the influence of the temperature on the interlaminar strength. Furthermore, it has been shown that the choice of test method has a significant effect on the measured values.

Comparison of the test methods showed that the selection of the test method has a significant influence on the measured values of interlaminar strength. The highest values were measured in tests where the test set up was set according to the AITM 1-0069 method. This method was chosen to determine the effect of temperature on interlaminar strength values. The main reason was to consider the geometry of the test specimen for setting up of the test assembly.

The highest interlaminar strength values were measured on specimens tested at cold temperature. Mean value was approximately 5% higher than the interlaminar strength measured at room temperature. The difference between these two sets was evaluated as statistically insignificant. The samples tested at hot temperature showed a 12 % decrease in strength compared to RT set. The HTA set was evaluated as statistically significant different from the other two sets.

This paper showed that it is not possible to compare values measured on the same material tested by different methods, although the differences in the methods may seem to be insignificant.

Acknowledgement

This work was funded by the Ministry of Industry and Trade of the Czech Republic in the frame of "TRIO" call. Project No. FV 3003 – Design and technological development of primary aircraft parts of advanced shapes made of composite with thermoplastic matrix.

References

[1] B. Dale, Thermosets vs. Thermoplastics: Is the battle over?, www.compositesword.com, 2015.

[2] L. Ye, A. Beehag, K. Friedrich, Mesostructural Aspects of Interlaminar Fracture in Thermoplastic Composites: Is Crystallinity a Key?, Composites Science and Technology, 53, (1995), 167–173.

[3] K. A. Brown, R. Brooks, N. Warrior, Characterizing the Strain Rate Sensitivity of the Tensile Mechanical Properties of a Thermoplastic Composite, JOM, 61, (2009), 43–46.

[4] O. Čoban, M. Ö. Bora, T. Sinmazçelik, İ. Cürgül, V. Günay, Fracture Morphology and Deformation Characteristics of Repeatedly Impacted Thermoplastic Matrix Composites, Mater. Des., 30, (2009), 628–634.

[5] J. H. Lee and J. H. Vogel, An Investigation of the Formability of Long Fiber Thermoplastic Composite Sheets, J. Eng. Mater. Technol., 117, (1995), 127–132.

[6] T. A. Martin, D. Bhattacharyya, R. B. Pipes, Deformation Characteristics and Formability of Fibre-reinforced Thermoplastic Sheets, Compos. Manuf., 3, (1992), 165–172.

[7] C. Ageorges, L. Ye, Resistance Welding of Thermosetting Composite/Thermoplastic Composite Joints, Composites, Part A, 32, (2001), 1603–1612.

[8] A. Yousefpour, M. Hojjati, J. P. Immarigeon, Fusion Bonding/Welding of Thermoplastic Composites, Journal of Thermoplastic Composite Materials, 17, (2004), 303–341.

[9] Y. C. Yuan, T. Yin, M. Z. Rong, and M. Q. Zhang, Self Healing in Polymers and Polymer Composites. Concepts, Realization and Outlook: A Review, eXPRESS Polym. Lett., 2, (2008), 238 250.

[10] L. Yao, M. Z. Rong, M. Q. Zhang, Y. C. Yuan, Selfhealing of Thermoplastics via Reversible Addition–- Fragmentation Chain Transfer Polymerization, J. Mater. Chem., 21, (2011), 9060–9065.

[11] G. Schinner, J. Brandt, H. Richter, Recycling Carbon-fiber-reinforced Thermoplastic Composites, Journal of Thermoplastic Composite Materials, 9, (1996), 239–245.

[12] J. Šedek, R. Hron, M. Kadlec, Bond Joint Analysis of Thermoplastic Composite Made from Stacked Tailored Blanks, Applied Mechanics and Materials, 827, (2016), 161-168. DOI: 10.4028/www.scientific.net/AMM.827.161.

[13] R. Růžek, J. Šedek, M. Kadlec, P. Kucharský: Mechanical behavior of thermoplastic rib under loading representing real structure conditions. EAN 2016 - 54th International Conference on Experimental Stress Analysis. Srní (CZ), 30 May – 2 June 2016.

[14] ASTM D6415- Standard test method for Measuring the Curved Beam Strength of Fiber-reinforced Polymer-Matrix Composite

[15] AITM 1-0069, Issue 2, Determination of curved-beam failure. Airbus, S.A.S, Engineering Directorate, Blagnac, France, 2011.

EAN 2020
58th conference on experimental stress analysis

Equivalence of Boundary Conditions of FEM Model

HUŇADY R.[1,a], LENGVARSKÝ P.[1,b], KAĽAVSKÝ A.[1,c]

[1]Deparment of Applied Mechanics and Mechanical Engineering, Faculty of Mechanical Engineering, Technical University of Košice, Letná 9/B, 042 00 Košice, Slovak Republic

[a]robert.hunady@tuke.sk, [b]pavol.lengvarsky@tuke.sk, [c]adam.kalavsky@tuke.sk

Keywords: equivalent model, boundary conditions, FEM, modal analysis, stiffness

Abstract. The paper deals with two methods of equivalence of boundary conditions in the finite element model. The proposed methods are based on the determination of the stiffness parameters in the section plane, in which a removed part of the model is replaced by a constraint with elastic elements or bushing connector. In the first case, the stiffness parameters are determined by a series of static finite element (FE) analyzes that are used to obtain the response of the removed part to the six basic types of loading. The second method is a combination of experimental and numerical approach. The natural frequencies obtained by the measurement are used in FE optimization, in which the response of the model is tuned by changing the stiffness parameters of the bushing. Both methods provide a good estimate of the stiffness at the point where the model is replaced by an equivalent boundary condition. This increases the accuracy of the numerical model and at the same time saves computational capacity and the required time due to the reduction of elements.

Introduction

The correctness of the numerical solution results in the finite element method is largely dependent on the accuracy of the FEM model itself. The basis is the correct definition of material properties, initial conditions, loads and boundary conditions such as constraints, connections and contacts in accordance with the real ones. Inaccuracies in the model definition lead to deviations or errors in the load response. Many dynamic problems, such as impact tests [1], [2], fatigue analyzes [3], [4], are quite sensitive to boundary conditions and material parameters. The most common causes of differences in model behavior may be constraint stiffness and structural damping [5], [6]. Computational models are usually verified by correlating the results of numerical calculation and experimental measurement. A suitable approach is to compare modal parameters, i.e. natural frequencies, mode shapes, e.g. using the MAC criterion [7], [8], [9]. In order to refine the model, it is then possible to perform a sensitivity analysis, where the response is adjusted by changing the input parameters.

The paper deals with procedures aimed at improving boundary conditions. Conventional boundary conditions are replaced by equivalent ones, which provide a significant simplification of the model while maintaining their accuracy. Depending on the application, their accuracy may even be increased in some cases. Examples are constraints that are considered to be perfectly rigid in the FEM model, while in point of fact they are flexible. Perfectly rigid constraints distort both the static and dynamic response of the model to external loads. This can lead to a considerable influence on the results and their inconsistency with the results of experimental tests such as vibration analysis, fatigue analysis, etc. For this reason, an equivalence method has been proposed to replace the conventional boundary condition or some part of the model. The equivalent boundary condition is represented by a bushing connector whose stiffness can be determined either by a series of numerical static analyses or experimentally by modal analysis in combination with FE optimization. The similar approach has been used to determine the material constants of a homogenized material [10], [11].

The first equivalence method

The principle of the first method is explained on the example of a beam whose one end is fixed to the frame by means of two preloaded bolts (Fig. 1a). This part of the model will be replaced by bushing whose stiffness parameters are determined in FE static analysis, based on acting of three forces and three moments (Fig. 1b), respectively. Fig. 1c shows the equivalent model of the beam where the part of the model representing the beam attachment is replaced by bushing connector connected to ground. The stiffness parameters were calculated using the following formulas and their values are given in Table 1

$$k_{x,y,z} = \frac{F_{x,y,z}}{\delta_{x,y,z}} \quad \text{(N/mm)}, \tag{1}$$

where k_x, k_y, k_z are equivalent translational stiffness in x-, y-, z- direction, respectively,
δ_x, δ_y, δ_z are maximum translations in x-, y-, z- direction, respectively,
F_x, F_y, F_z are loading forces acting in x-, y-, z- direction, respectively,

and

$$k_{\varphi_x,\varphi_y,\varphi_z} = \frac{M_{x,y,z}}{\varphi_{x,y,z}} \quad \text{(N.mm/rad)}, \tag{2}$$

where k_{φ_x}, k_{φ_y}, k_{φ_z} are equivalent rotational stiffness about x-, y-, z- axis, respectively,
φ_x, φ_y, φ_z are maximum angular deflection about x-, y-, z- axis, respectively,
M_x, M_y, M_z are loading moments acting about x-, y-, z- axis, respectively.

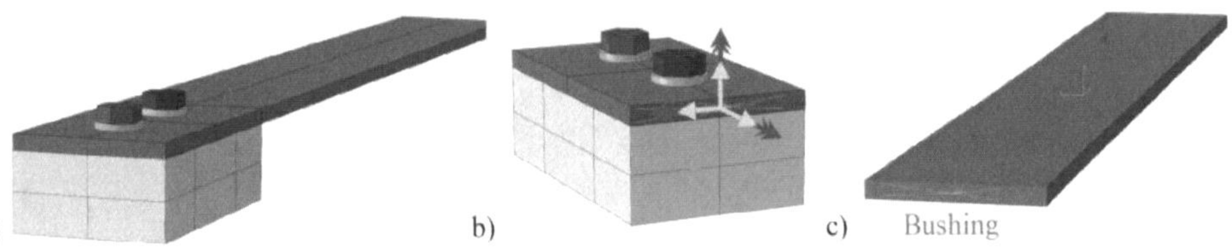

Fig. 1: a) The complete initial FEM model, b) Part of the model used to obtain stiffness parameters, c) The equivalent model with bushing connector.

Table 1: Stiffness parameters of the bushing.

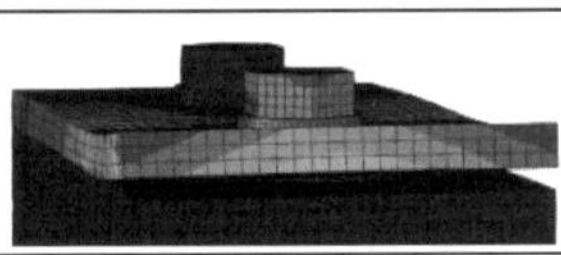	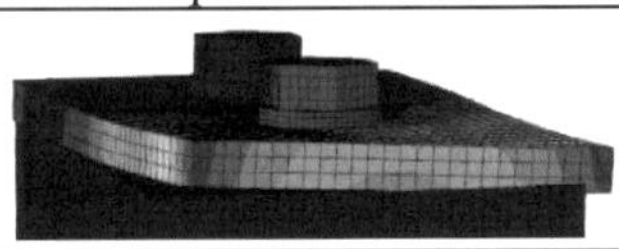	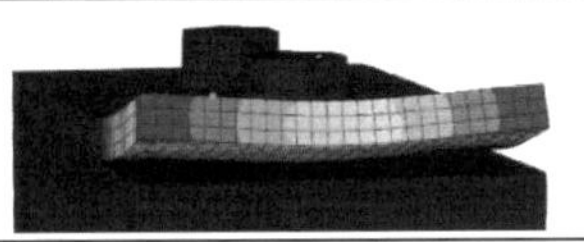
k_x = 397e3 N/mm	k_y = 101e3 N/mm	k_z = 9.8e3 N/mm
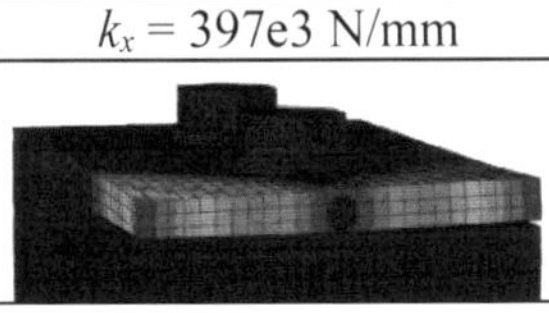	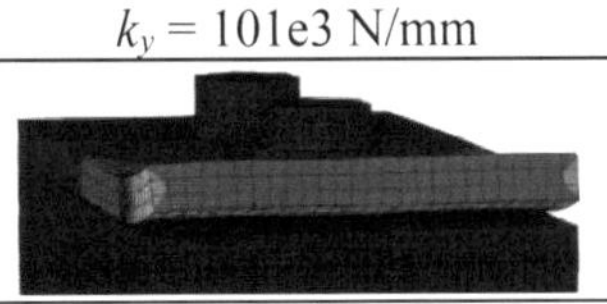	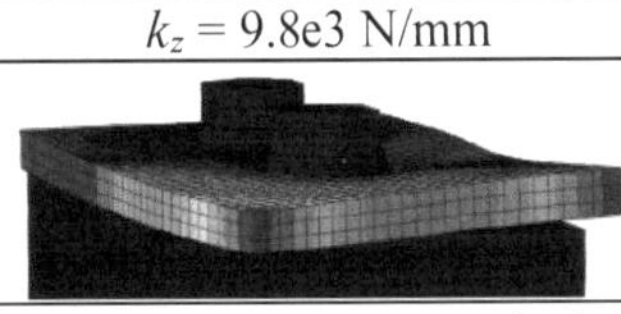
$k_{\varphi x}$ = 1.547e6 N.mm/rad	$k_{\varphi y}$ = 0.972e6 N.mm/rad	$k_{\varphi z}$ = 45.455e6 N.mm/rad

Both FE models, the complete initial model (Fig. 1a) and the equivalent model (Fig. 1c), were subjected to modal analysis to compare them. Their natural frequencies are listed in Table 2. Two selected mode shapes can be seen in Fig. 2

Table 2: Natural frequencies of compared FE models in Hz.

Initial model	Equivalent model	Initial model	Equivalent model
37.6	37.7	1975.4	2058.6
235.5	239.7	2129.2	2194.7
332.6	339.4	2624.3	2610.5
511.2	505.4	3173.7	3256.5
658.7	676.6	3765.5	3736.1
1289.7	1329.7	4111.3	4107.4
1547.5	1533.3	4424.3	4494.1

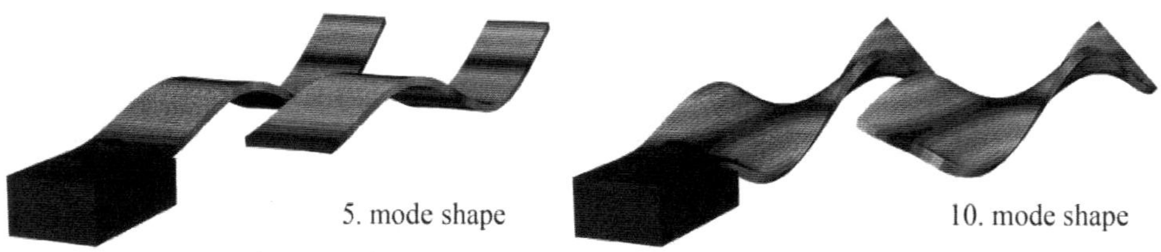

Fig. 2: Selected mode shapes of compared models.

The results obtained show very good agreement, which means that the equivalent model response corresponds to that of the complete initial model. In addition to maintaining dynamic behavior, the model has been significantly simplified. The number of elements decreased from 59747 to 15148.

The second equivalence method

This method uses the values of natural frequencies determined by experimental modal analysis (EMA) to calculate the stiffness parameters. The parameters are determined in the optimization process, which is based on frequency FEM analysis, where the objective function is the known natural frequencies. By changing the stiffness of the bushing, the model is retuned, i.e. to change its modal parameters. The shape, dimensions and material properties of the structure must be known.

Fig. 3: Experimental modal analysis of the beam.

The method is presented on a similar beam model as used in the previous case. The rectangular steel beam was fixed to the rigid base on one side by means of two bolt connections (Fig. 3). Its natural frequencies and mode shapes were determined by a modal test, in which the beam was excited with a Bruel & Kjaer 8206 impact hammer and its responses were measured with a Polytec PDV100 laser vibrometer. The measurement and the evaluation was performed using a Bruel & Kjaer Pulse system. In the frequency range up to 3.2 kHz, 7 modes were identified, of which 5 were bending and 2 torsion mode shapes of vibration (Fig. 4). Their natural damped frequencies and damping ratios are listed in Table 3.

Table 3: Modal parameters of the beam.

Mode	Damped frequency (Hz)	Damping ratio (%)	Mode shape
1	46.95	2.02	1. bending
2	289.44	0.33	2. bending
3	660.99	0.21	1. torsion
4	810.21	0.17	3. bending
5	1567.15	0.08	4. bending
6	1998.37	0.08	2. torsion
7	2562.31	0.27	5. bending

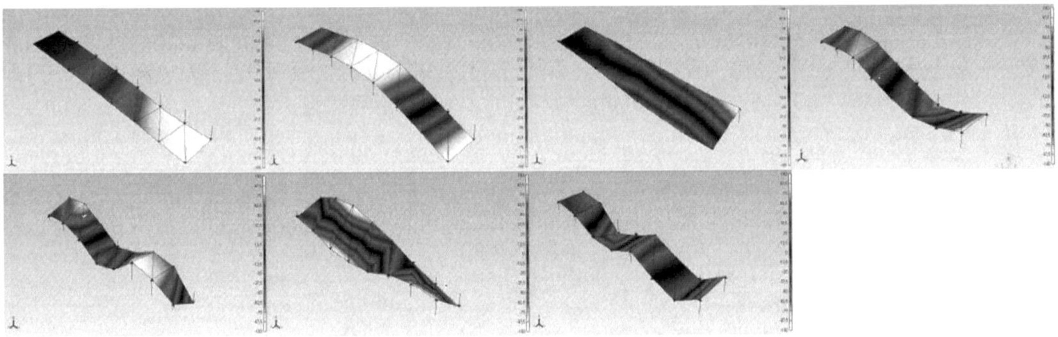

Fig. 4: Mode shapes of the beam.

The parametric optimization was performed in NX Nastran software that uses gradient-based numerical optimization algorithm [12]. The FE model (Fig. 5) consists only of the free part of the beam. The removed part representing the beam attachment was replaced by bushing elements connecting the section plane of the beam to the ground.

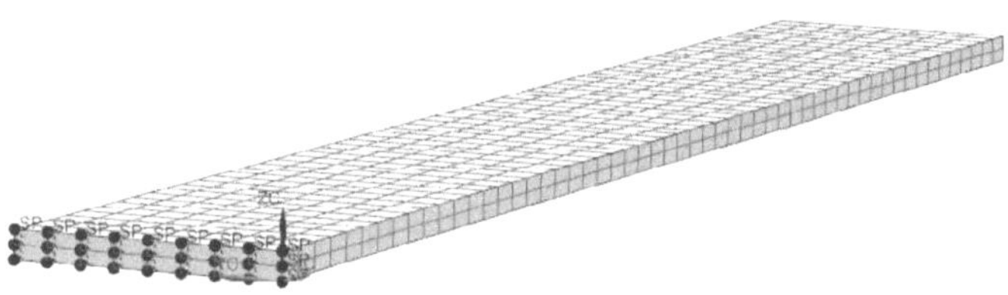

Fig. 5: FE model of the beam.

Since NX allows only one objective function to be defined, this function was the frequency of the first mode. Other frequencies were defined as optimization constraints with a tolerance of +/- 5%. It should be noted that the lateral mode shapes of vibration were not identified by the modal test. However, they occur in FEM analysis (mode 3 and mode 8). For this reason, these modes have not been included among the optimization constraints. The parameters k_x, k_y, k_z representing the equivalent translational stiffness of the bushing in x-, y-, z- direction, respectively, were consider as the design variables.

The optimization parameters were as follows:

```
Design Objective:
Target Model Frequency Mode 1, Target value = 47.000 Hz

Design Constraints:
Model Frequency, Mode 1, Upper limit = 49.000 Hz
Model Frequency, Mode 1, Lower limit = 45.000 Hz
Model Frequency, Mode 2, Upper limit = 304.00 Hz
Model Frequency, Mode 2, Lower limit = 275.00 Hz
Model Frequency, Mode 4, Upper limit = 694.00 Hz
Model Frequency, Mode 4, Lower limit = 628.00 Hz
Model Frequency, Mode 5, Upper limit = 851.00 Hz
Model Frequency, Mode 5, Lower limit = 769.00 Hz
Model Frequency, Mode 6, Upper limit = 1645.0 Hz
Model Frequency, Mode 6, Lower limit = 1488.0 Hz
Model Frequency, Mode 7, Upper limit = 2098.0 Hz
Model Frequency, Mode 7, Lower limit = 1898.0 Hz
Model Frequency, Mode 9, Upper limit = 2690.0 Hz
Model Frequency, Mode 9, Lower limit = 2434.0 Hz

Design Variables:
kx (N/mm), Initial value = 6e6, Lower limit = 500, Upper limit = 10e6
ky (N/mm), Initial value = 6e6, Lower limit = 500, Upper limit = 10e6
kz (N/mm), Initial value = 6e6, Lower limit = 500, Upper limit = 10e6
```

Fig. 6 shows how the value of the objective function has changed in the individual optimization cycles. Fig. 7 shows the changes of the design variables.

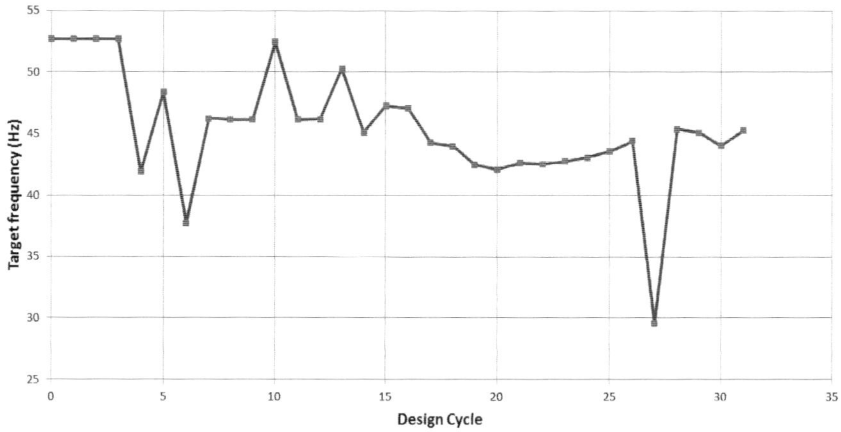

Fig. 6: Objective function plot.

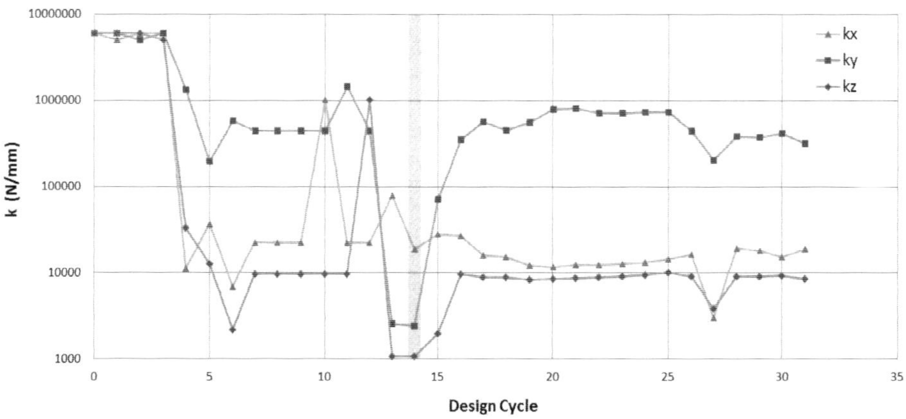

Fig. 7: Design variables plot.

Since not all optimization constraints were met in any cycle, it was necessary to choose a solution where the estimate was closest to meeting them. In the 14. design cycle, five of the seven conditions were met, so this cycle is therefore considered to be the result of optimization. Stiffness parameters in this cycle are k_x = 18654.71 N/mm, k_y = 2428.06 N/mm, k_z = 1068.86 N/mm. For comparison, Table 4 lists the natural frequencies of the beam model with bushing with these parameters, the frequencies of the clamped-free beam, and the frequencies obtained by measurement. The experimentally determined frequencies are considered as reference. Percentage deviations are calculated with respect to them.

Table 4: Natural frequencies of the beam obtained by EMA and FEM.

Mode	EMA	FEM Beam with bushing	FEM Clamped-free beam
1	46.95	45.09 (–3.97%)	52.743 (12,34%)
2	289.44	291.07 (0.56%)	330.11 (14,05%)
3	-	411.26	516.60
4	660.99	682.50 (3.25%)	724.09 (9,55%)
5	810.21	829.92 (2.43%)	924.37 (14,09%)
6	1567.15	1682.42 (7.35%)	1812.71 (15,67%)
7	1998.37	2064.50 (3.31%)	2192.14 (9,69%)
8	-	2463.03	2992.52
9	2562.31	2741.52 (6.99%)	2998.71 (17,03%)

It is evident that the percentage differences are significantly smaller in the case of the model with bushing and increase the accuracy of the FE model compared to the clamped beam with a perfectly rigid constraint. Therefore, it can be stated that the estimation of the stiffness parameters is correct. A more accurate estimate could be obtained by multicriteria optimization.

Conclusions

In the paper, the methods to increase the accuracy of the boundary conditions of the FE model are described. Their basis is the replacement of the boundary condition, bond or part of the model by bushing or elastic connectors with adequate stiffness. The paper presents two approaches to determine these parameters. The first is based on a purely numerical calculation and is applicable when a real physical model does not exist. This approach is very accurate and effective. The second approach is based on experimental modal analysis and parametric optimization using FEM. Its accuracy depends on several factors, such as natural frequencies, material properties, dimensions and geometry of the structure, on the basis of which the FE model is created. From this point of view, it is an estimation method. However, the test results indicate that the estimate is relatively accurate and the procedure is practically applicable.

Acknowledges

This research was supported by project VEGA 1/0355/18.

References

[1] V. Ferreira, L. P. Santos, M. Franzen, et al., Improving FEM crash simulation accuracy through local thickness estimation based on CAD data, Advances in Engineering Software, Volume 71, 2014, Pages 52-62, ISSN 0965-9978.

[2] S. Soni, S. K. Pradhan, Improving crash worthiness and dynamic performance of frontal plastic automotive body components, Materials Today: Proceedings, 2019, ISSN 2214-7853.

[3] A. Kaľavský, R. Huňady, Fatigue analysis of a notched beam in frequency domain, AIP Conference Proceedings 2198, 020007 (2019).

[4] A. Kaľavský, R. Huňady, M. Kicko, M. Kučinský, Calculation of the Fatigue Life of a Dynamically Loaded Component, American Journal of Mechanical Engineering, 2017, Vol. 5, No. 6, 252-256.

[5] M. Goletti, V. Mussi, A. Rossi, M. Monno, Procedures for Damping Properties Determination in Metal Foams to Improve FEM Modeling, Procedia Materials Science, Volume 4, 2014, Pages 233-238, ISSN 2211-8128.

[6] F. S. Iglesias, A. F. López, Rayleigh damping parameters estimation using hammer impact tests, Mechanical Systems and Signal Processing, Volume 135, 2020, ISSN 0888-3270.

[7] D. Siano, M. Viscardi, R. Aiello, Sensitivity Analysis and Correlation Experimental/Numerical FEM-BEM for Noise Reduction Assessment of an Engine Beauty Cover, Energy Procedia, Volume 81, 2015, Pages 742-754, ISSN 1876-6102.

[8] M. Pastor, M. Binda, T. Harčarik, Modal Assurance Criterion, Procedia Engineering, Volume 48, 2012, Pages 543-548, ISSN 1877-7058.

[9] M. Brehm, V. Zabel, Ch. Bucher, An automatic mode pairing strategy using an enhanced modal assurance criterion based on modal strain energies, Journal of Sound and Vibration, Volume 329, Issue 25, 2010, Pages 5375-5392, ISSN 0022-460X.

[10] R. Huňady, P. Pavelka, P. Lengvarský, Homogenization of Material Properties of a Combined rubber/steel Sheet using Modal Analysis, In: EAN 2019 - Conference Proceedings, 2019, p. 146-150.

[11] A. Kaľavský, R. Huňady, P. Pavelka, Determining the young's modulus of a specimen using experimental modal analysis. In: EAN 2018 - Conference Proceedings, 2018, p. 176-83.

[12] Design Sensitivity and Optimization User's Guide (NX Nastran help), Siemens, URL: docs.plm.automation.siemens.com/data_services/resources/nxnastran/10/help/en_US/tdocExt/pdf/optimization.pdf

EAN 2020
58th conference on experimental stress analysis

Leak Localization by Using Time Delay Estimation of Negative Pressure Wave Signal

IŽOLD R.[1,a], JANČO R.[1,b]

[1]Institute of Applied Mechanics and Mechatronics, Faculty of Mechanical Engineering, Slovak University of Technology in Bratislava, Námestie slobody 17, 812 31 Bratislava 1, Slovak Republic

[a]roman.izold@stuba.sk, [b]roland.janco@stuba.sk

Keywords: Delay estimation, Generalized cross-correlation, Leak localization, Negative pressure wave

Abstract. The aim of this paper is to investigate usability and accuracy of different time delay estimation techniques for purposes of pipeline leak localization. The generalized cross correlation with various weighting functions and time domain estimation techniques are compared. Limited measurements of pressure waves initiated by artificial leak are utilized for brief accuracy analysis of mentioned estimation techniques.

Introduction

Basic though of acoustic leak detection techniques is assumption that a leak is an acoustic source. Occurrence of a pipeline leak leads to local rarefaction and initiate a negative pressure wave propagated toward both ends of a pipeline. Using of a pair of sensors on both sides of the leak allow to sense a negative pressure wave. If a speed of sound wave propagated in the pipeline is known, time delay estimation calculated between two sensor signals can be used for leak localization.

This paper compares time delay estimation (*TDE*) techniques used for localization of the artificial leak on the experimental water-filled steel pipe. The negative pressure wave (*NPW*) initiated by the artificial leak is sensed by a pair of hydrophones placed on both sides of the leak. Subsequently, leak position is calculated by using of *TDE* techniques in time domain and the generalized cross-correlation in frequency domain. The basic cross-correlation (*BCC*), the average square difference function (*ASDF*), and the average magnitude difference function (*AMDF*) described by Jacovitti [1] and the generalized cross-correlation presented by Knapp [4] and reinterpreted by Gao [2] are investigated for the purpose of leak detection. In the first chapter are presented basic assumptions of sampled mutually delayed signals which represent inputs to the cross-correlation and TDE techniques. Equations of time domain techniques are introduced and various weighting functions obtained from listed literature for the GCC purposes are presented. The second chapter investigates performance of individual time domain techniques and GCC functions through variance of delay estimation. The third chapter shows use of most effective methods for delay estimation of negative pressure waves sensed through the experimental pipeline setup.

1 Theoretical development

Based on the assumption of two sensors placed on both sides of the pipeline leak, discretized signals x_1 and x_2 can be expressed as

$$x_1(k) = s_1(\mathrm{k}) + n_1(k)$$
$$x_2(k) = \alpha s_1(\mathrm{k} + \mathrm{D}) + n_2(k) \tag{1}$$

where k is number of samples, D is sample delay, s_1 is focused leak signal in the form of negative pressure wave and n_1, n_2 present additive noise.

Time domain techniques. Jacovitti [1] presents the direct cross-correlation method in time domain, also known as sliding dot product for sampled signal in the form

$$R_{x1x2}{}^{D}(\tau) = \frac{1}{N}\sum_{k=1}^{N} x_1(k)x_2(k+\tau) \tag{2}$$

where T is sampling interval and N is number of samples. Said author also specified the average magnitude difference function (*AMDF*) and the average square difference function (ASDF) in the following form.

$$R_{x1x2}{}^{M}(\tau) = \frac{1}{N}\sum_{k=1}^{N} |x_1(k) - x_2(k+\tau)| \tag{3}$$

$$R_{x1x2}{}^{S}(\tau) = \frac{1}{N}\sum_{k=1}^{N} [x_1(k) - x_2(k+\tau)]^2 \tag{4}$$

Delay estimation $\widehat{D}$ of true delay D is presented by Jacovitti [1] as argument of maxima or minimum for the specific function.

$$\widehat{D}^{D} = \mathrm{argmax}\left[R_{x1x2}{}^{D}(\tau)\right]$$
$$\widehat{D}^{M} = \mathrm{argmin}\left[R_{x1x2}{}^{M}(\tau)\right] \tag{5}$$
$$\widehat{D}^{S} = \mathrm{argmin}\left[R_{x1x2}{}^{S}(\tau)\right]$$

In respect to the computational time efficiency, the AMDF and the ASDF are more convenient than the direct correlation due to products of original and lagged signal are not required. The AMDF and the ASDF involves only subtracts, respectively their squares [1].

Frequency domain techniques. Gao [2] investigated using of the generalized cross-correlation (*GCC*) function for the purpose of leak detection on buried plastic pipes. He defines the GCC function as the inverse Fourier transform of the cross-spectral density in the following form.

$$R^{g}{}_{x1x2}(\tau) = F^{-1}\{\varphi^{g}(f)S_{x1x2}(f)\} \tag{6}$$

The function $\varphi^{g}(f)$is defined as $\varphi^{g}(f) = H_1(f)H_2(f)$. The weighting function $\varphi^{g}(f)$ introduces filters $H_1(f)$and $H_2(f)$ into calculation in order to facilitate estimation of time delay [4]. Gao [2] reinterpreted study by Knapp [4] for purpose of leak detection on plastic pipes. His study is focused on estimation of time delay by using the GCC with leak noise as input signal. Gao [2] compares five different forms of weighting functions (*PHAT, WIENER,*

SCOT, ML, ROTH). Weighting functions stated by Knapp [4] and used in this study for purpose of negative pressure wave delay estimation are presented in the Table 1.

Table 1: Weighting functions

BCC	Roth	SCOT	PHAT	Eckart	HT
1	$\frac{1}{S_{x1x1}(f)}$	$\frac{1}{\sqrt{S_{x1x1}(f)S_{x2x2}(f)}}$	$\frac{1}{\lvert S_{x1x2}(f)\rvert}$	$\frac{S_{s1s1}(f)}{S_{n1n1}(f)S_{n2n2}(f)}$	$\frac{\frac{S_{s1s1}(f)}{S_{n1n1}(f)S_{n2n2}(f)}}{\frac{S_{s1s1}(f)}{S_{n1n1}(f)}+\frac{S_{s1s1}(f)}{S_{n2n2}(f)}+1}$

2 Variance of delay estimation

In accordance with Knapp [4] and Jacovitti [1] variance of delay estimation is introduced as representative value of the mentioned GCC and time domain techniques. In order to investigate usability of GCC functions for leak detection by delay estimation of negative pressure wave, artificially generated synthetic data are used. Generation of synthetic negative pressure waves is based on real measured negative pressure waves presented in previous work of the author [3]. Synthetic data presented on the Fig.1 are in accordance with the Eq.1. The signal $s_1$is represented by damped sine wave. Additive noises n_1, n_2 are introduced as white Gaussian noise with the specified S/N ratio. Signals x_1, x_2 with various S/N ratios in range (-10 dB, 50 dB) are used as inputs for each cross-correlator. The variance is calculated from 500 sample delay estimations for each S/N ratio step.

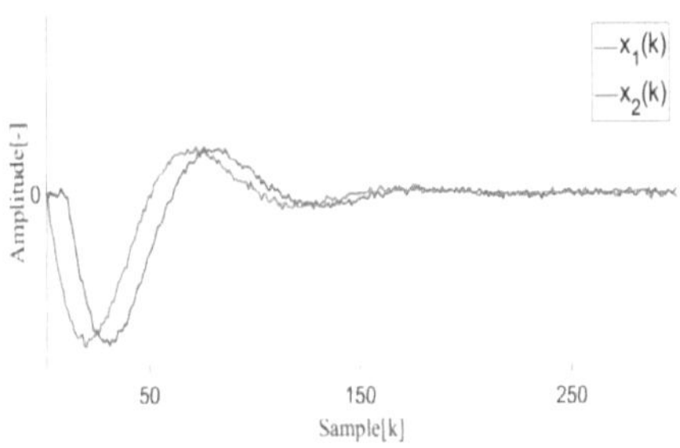

Fig. 1: Synthetic negative pressure wave SNR=40 dB

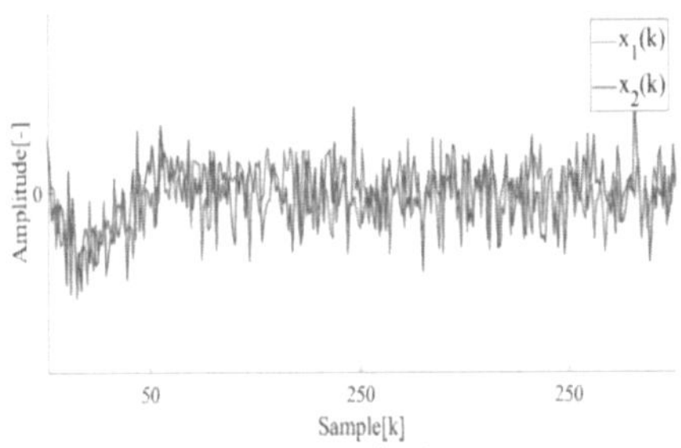

Fig. 2: Synthetic negative pressure wave SNR=10 dB

Delay variance of time domain techniques. The Fig. 3 represents variance as a function of SNR for S/N ratio step 0.5 dB. In each step is calculated 500 time delay estimations using the BCC method, the average magnitude difference function and the average square difference function listed in the previous chapter. Despite of possibility of the calculation BCC in both frequency and time domain, all of figures show the basic cross-correlation method based on the Fourier transform which is computationally more efficient than the dot sliding product mentioned in previous chapter. Due to computational efficiency and simplicity the BCC serves as reference method for comparison with other time domain methods and GCC functions. Detailed variance of delay estimation at the interesting SNR range between 0 dB and 10 dB is presented on the Fig. 5. Although the Fig. 5 shows less variance of delay estimation for the AMDF and the ASDF for low S/N ratio (0 dB-3 dB), the BCC method provides with increasing S/N ratio faster 'convergence' to the theoretical zero variance. With increasing true delay between signals x_1, x_2 vanishes 'convergence' advantage of the BCC method in comparison with the AMDF and the ASDF (Fig. 7), (Fig. 9).

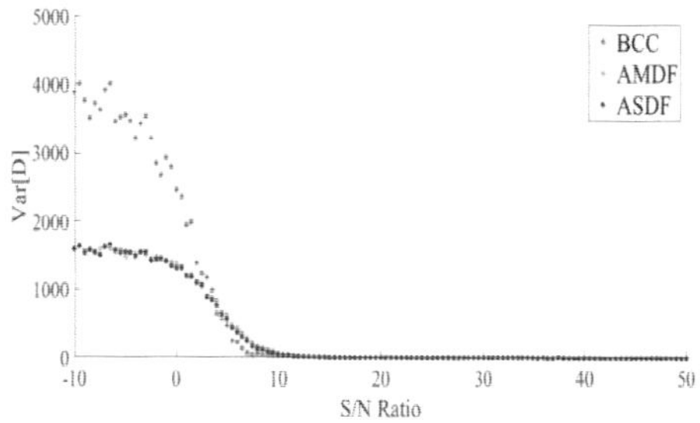

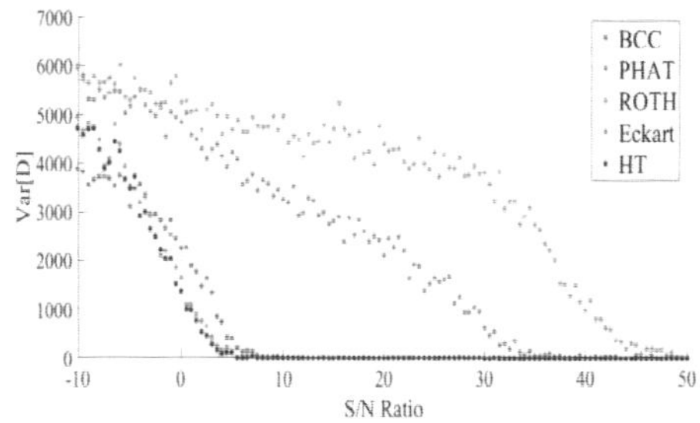

Fig. 3: Variance of delay estimation for time domain techniques (True Delay=0)

Fig. 4: Variance of delay estimation for GCC functions (True Delay=0)

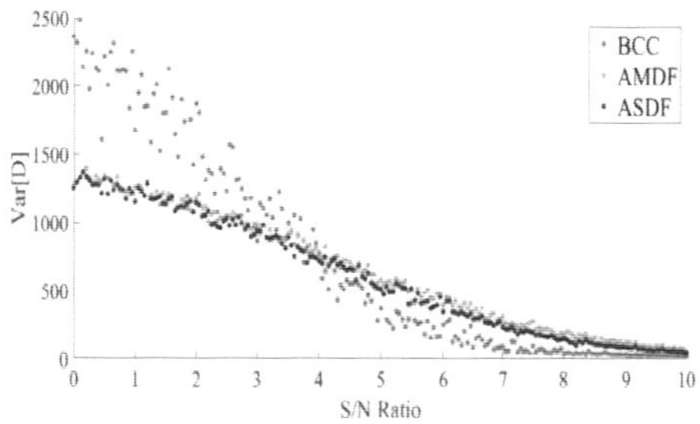

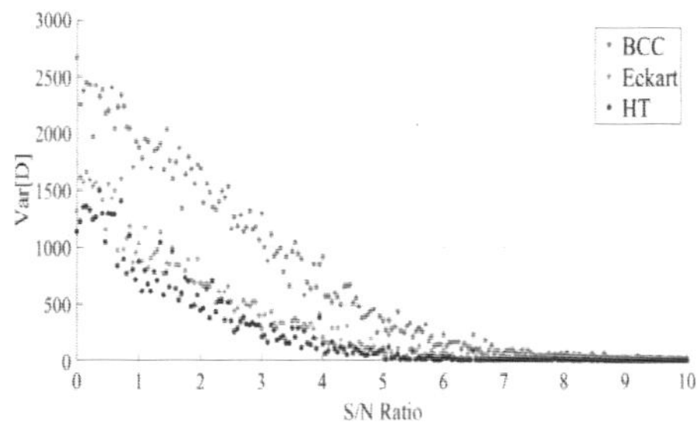

Fig. 5: Detailed variance (SNR= [0 dB,10 dB]) of delay estimation for time domain techniques (True Delay=0)

Fig. 6: Detailed variance of delay estimation for GCC functions (True Delay=0)

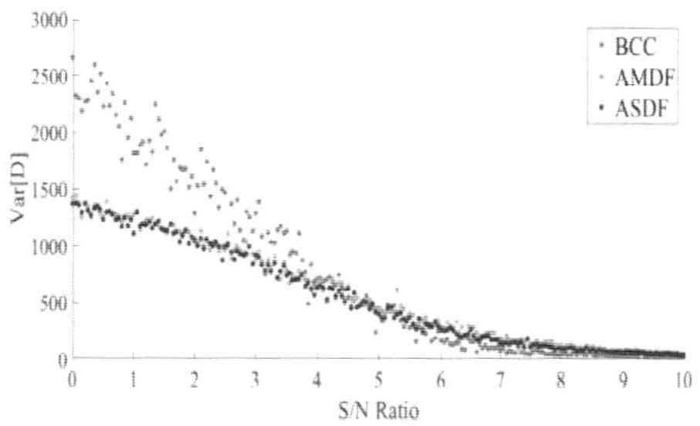

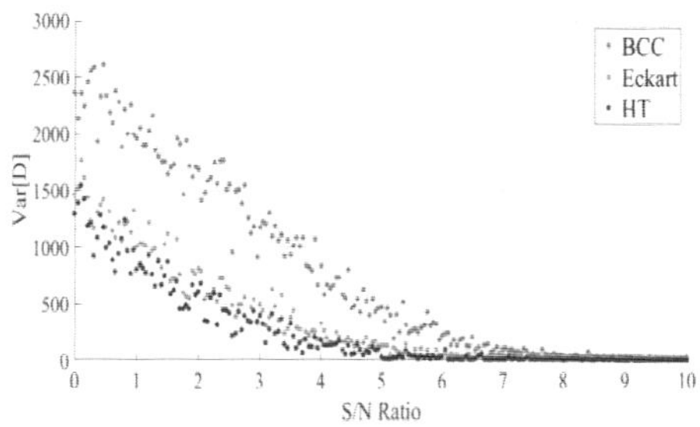

Fig. 7: Detailed variance of delay estimation for time domain techniques (True Delay=10)

Fig. 8: Detailed variance of delay estimation for GCC functions (True Delay=10)

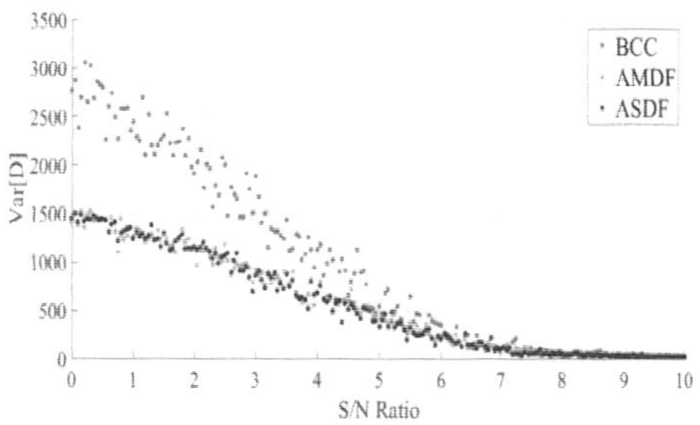

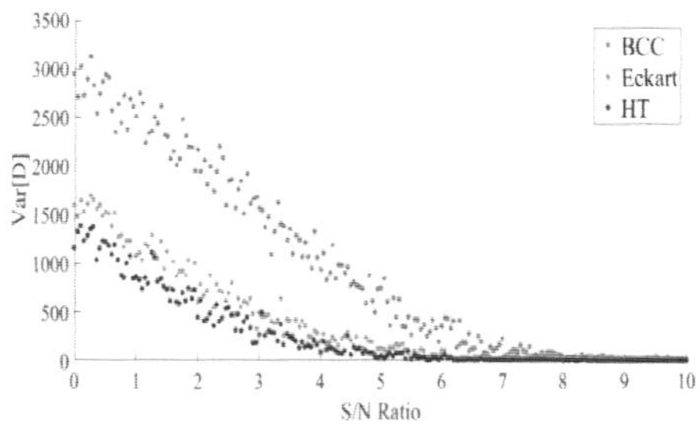

Fig. 9: Detailed variance of delay estimation for time domain techniques (True Delay=30)

Fig. 10: Detailed variance of delay estimation for GCC functions (True Delay=30)

Delay variance of GCC methods. Fig. 4 shows variance of delay estimation for five GCC weighting functions mentioned in previous chapter, also presented by Knapp [4] and reinterpreted for purpose of leak detection by Gao [2]. If we assume variance od delay estimation as representative value of GCC methods performance, the BCC, the Eckart function and the HT (or ML) clearly outperform remaining GCC methods (Fig. 4). For this reason, only the Eckart and the HT are compared in the following analysis. The Eckart and the HT weighting functions provide better delay estimation performance than the BCC method (Fig. 6). It is caused by involving spectral densities of additive white noise into the weighting function (Tab.1). These methods suppress frequency bands of high noise [4]. With increasing value of true delay, the Eckart and the HT hold approximately equal 'convergence' to the zero variance, while the BCC worsen its delay estimation performance (Fig. 10). Of course, the PSD of noise signal can be easily obtained from artificially generated synthetic data. However, in real application, signal s_1 and noise n_1, n_2 are usually unknown and need to be estimated.

3 Experimental results and Discussion

The laboratory setup (Fig. 11) constructed for experimental investigation of leak detection and localization methods is presented in detail by Izold [3]. The setup consists by the water-filled steel pipe with welded branches where hydrophones are placed. The artificial leak (ball valve) is assumed as a source of the acoustic signal. The ball valve opening initiate a local rarefaction which is followed by a formation of the negative pressure wave propagated toward both hydrophones (Fig. 12).

Fig. 11: Experimental setup for sensing of negative pressure wave

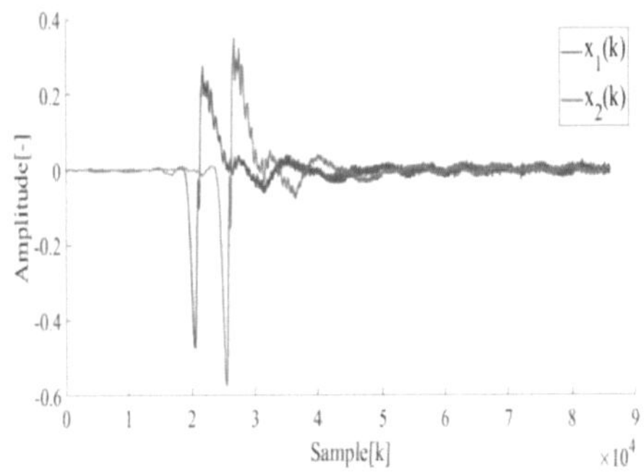

Fig. 12: Negative pressure wave sensed by pair of hydrophones

The Fig. 12 shows sensed negative pressure wave initiated by opening of the ball valve. Due to short length of the experimental setup and high speed of sound in water, real measured signals are artificially delayed for better illustration.

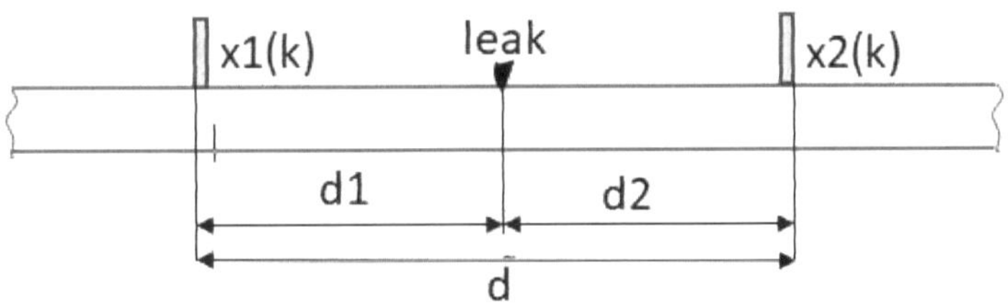

Fig. 13: Experimental setup scheme

Leak position estimation. If a pressure distribution is assumed as uniform across the cross-section of the pipe and a speed of sound wave is approximately constant across the pipe

length, estimation of the leak position depends only on time (sample) delay estimation of pressure signals sensed by pair of hydrophones [5].
Leak position estimation in the form of distance from the Sensor No.1 is then defined by the following formula.

$$d_1 = \frac{d - c\widehat{D}}{2} \tag{7}$$

Where c[m/s] is speed of sound in water affected by pipeline calculated in accordance with Liu [6]. Based on the analysis stated in the previous chapter, five different techniques are used for leak position estimation. Negative pressure wave is actuated by valve opening for 10 times. Sensed pressure signals are used as inputs to the five estimation methods showed in Fig. 14. Aim of the calculation is to investigate delay estimation of two negative pressure waves actuated by valve open not delay estimation of leak noise. Therefore, for the calculation only segment of distinct negative pressure wave is selected. Noise spectra required for the Eckart and the HT weighting functions are estimated from segments of input signals where noise dominates.

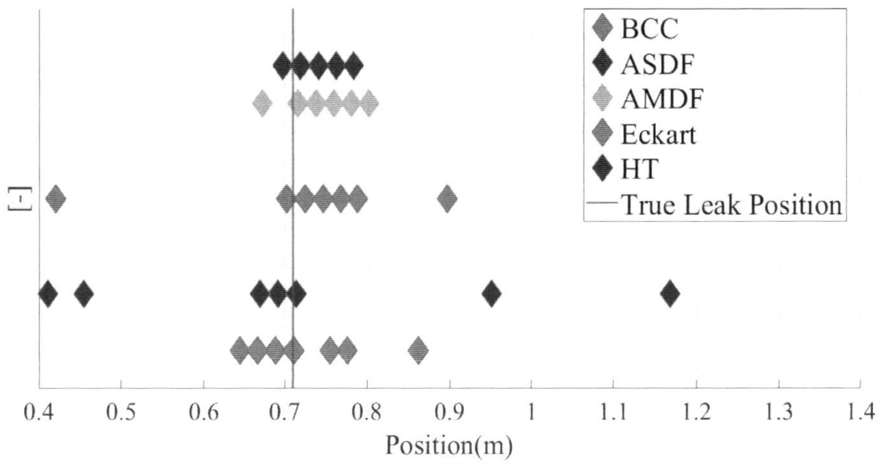

Fig. 14: Leak position estimation

The Fig. 14 surprisingly shows comparatively better estimation performance of time domain AMDF and ASDF techniques. Both provides better leak position estimate than the BCC and both GCC methods. The worsening of the Eckart and the HT weighting functions is probably caused by inaccurate estimation of the noise spectra. In addition, use of the GCC with the HT weighting function for experimental data provides in contrast with previous chapter less accuracy of leak position estimation than other compared methods. Mentioned results present only briefly suggestion of usability of listed techniques for experimental leak localization. For more relevant result, larger quantity of experimental data is required.

Conclusions

Investigation of significantly larger experimentally obtained data of negative pressure waves is necessary for relevant analysis of usability and accuracy of mentioned time estimation techniques for purpose of leak localization. Analysis with artificially generated inputs suggests more accurate delay estimation of the GCC with the HT and the Eckart functions only under the assumption of accurate noise spectra estimation. Analysis of experimentally obtained data approves using of mentioned TDE techniques.

References

[1] G. Jacovitti, G. Scarano, Discrete Time Techniques for Time Delay Estimation, IEEE Transactions on Signal Processing, 41 (1993), pp. 525–533.

[2] Y. Gao, M.J. Brennan, P.F. Joseph, A comparison of time delay estimators for the detection of leak noise signals in plastic water distribution pipes, Journal of Sound and Vibration, 292 (2006), pp. 552-570.

[3] R. Izold, R. Janco, Propagation of Negative Pressure Wave in Water-Filled Steel Pipe: Proceedings of the APLIMAT 2020, Bratislava, Slovakia, 2020, pp. 629–635.

[4] Ch. H. Knapp, C. G Carter, The Generalized Correlation Method for Estimation of Time Delay, IEEE Transactions on Acoustics, Speech, and Signal Processing, Vol. ASSP-24, No. 4, 1976.

[5] Y. Gao, M.J. Brennan, P.F. Joseph, J.M. Muggleton, O. Hunaidi, A model of the correlation function of leak noise in buried plastic pipes, Journal of Sound and Vibration, 277 (2004), pp. 133-148.

[6] Liu, H., Pipeline Engineering. CRC Press LLC, 2003, ISBN 0-203-50668-5.

EAN 2020
58th conference on experimental stress analysis

An Experimental Investigation of Dynamic Response of Bladed Disk

JAMRÓZ T.[1,a], MALÍNEK P.[1,b]

[1] Aviation Division - Czech Aerospace Research Centre;
Beranových 130, 19905 Praha, Czech Republic

[a]jamroz@vzlu.cz, [b]malinek@vzlu.cz

Keywords: simulation of the operational conditions, blade vibration, impeller

Abstract

This work focuses on controlling of the dynamic behavior of turbomachines which requires an experimental validation phase to ensure their safety. The application of an innovative experimental techniques to measure the modal properties of a rotating bladed integral disk (blisk) is described in this paper. The aim of this analysis is to simulate a correct harmonic component of engine order (EO) that excites the blisk in an engine. The blisk is tested without rotation as static part and an excitation of the impeller is performed by rotating of a jetting air.

This work focuses on the study of the dynamic behavior of a centrifugal compressor impeller in view to preparing of an operational test in an engine requiring precise characterization of modal frequencies and nodal diameters.

Introduction

Turbomachinery designers have a constant preoccupation: improve performance while reducing masses. One main consequence, often underestimated, is the effect on the blade stress design and dynamic behaviour. Static loads due to the rotation speed, the temperature and the steady aerodynamic forces are now so important that the ability of the material to accept extra dynamic stresses is dramatically reduced. Moreover, the intensity of the sources of excitation in turbomachines has increased: smaller axial gaps between stators and rotors, smaller tip gaps inducing rotor/casing contact risks, etc... as was recalled in [1].

Nowadays, the analysis of failure problems under operating condition show that the majority of problems encountered is connected to High Cycle Fatigue (HCF). These problems are often very sudden because a huge number of stress cycles can be performed in a very short time. The margins for dynamic loads being smaller and smaller, the tolerance to vibration have become a key point.

That is the reason why the measuring of the modal parameters of the bladed disk is very useful for predicting their response to aerodynamic forces, for mistuning identification and FE model updating.

Although the measurement of rotating vibrating structures has been addressed by numerous authors but in spite of a new approach was found. Standard procedure is a challenging task especially with hight rotation speed, the surrounding environment, gyroscopic effect, boundary conditions and the possible presence of sensors where the dynamic response of a rotating impeller is measured inside an evacuated chamber [2].

This work focuses on the study of the dynamic behavior of a centrifugal compressor impeller in view to preparing of an operational test in an engine requiring precise characterization of modal frequencies and nodal diameters.

Experimental set up

General description of a test rig:

A general diagram of the test rig is shown in Fig. 1. The system consists of two main parts. The first part is an excitation frame (position 05) and the second is a frame (position 07) for the test piece. These frames are dynamically separated so that the measurement is not affected. The centrifugal compressor impeller is fixed to the frame by a flange. Excitation of the centrifugal compressor impeller is performed by rotating the jetting air (position 02). Specific harmonic excitation is defined by the number of nozzles on the exciter rotor. The excitation rotor is driven by electric motor (position 08). A rotor speed range is from 0 to 11000 rpm.

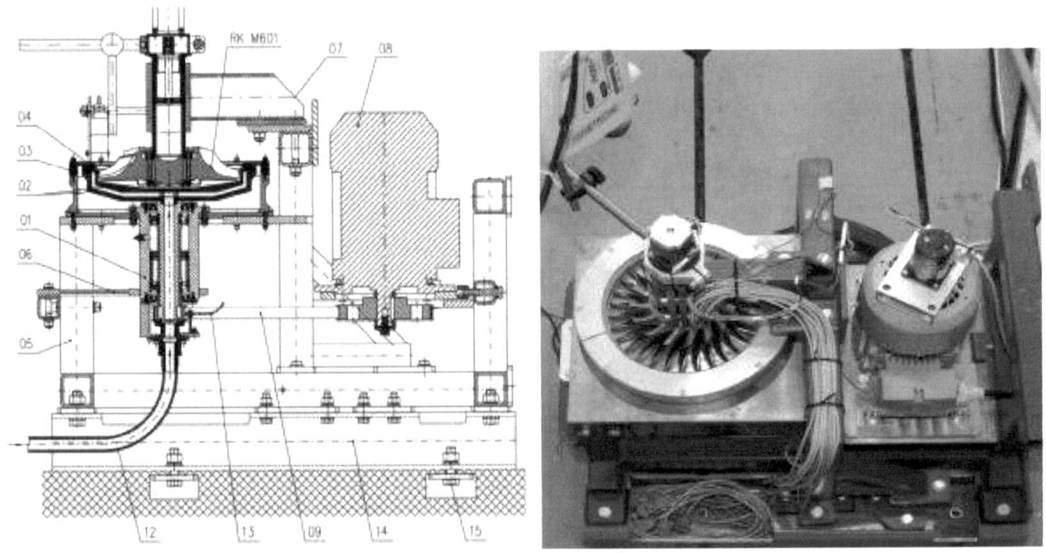

Fig. 1: Test-rig in general view and with labels

Measurements were performed by three means: the first one by accelerometers that were located on disk and on the root of main blades, the second one by strain gauges using an inductive telemetry system; the third by a Laser Doppler Vibrometer (LDV) Ometron VH300+.

The test conditions were monitored using an accelerometer, a thermocouple and a pressure gauge. The tests were performed at ambient temperature.

Definition of harmonic excitation

The sources of excitation in the engine are numerous and most of them are harmonic. Effectively, blades excitations are mainly due to non-uniformity of the upstream pressure field. When rotating, the blades are loaded by a fluctuating pressure field at a frequency connected to the rotation speed. The test rig simulates this aerodynamic dynamic load caused by interaction between rotor blades and diffuser vanes. The sources of excitation can be by the upstream/downstream stator (with Es blades) generating excitation frequencies equal to the rotation frequency (ω) times the number of stator blades plus higher harmonic ($Es.\omega$, $2Es.\omega$, $3Es.\omega$, etc…) with decreasing intensity. Turbines can be excited also by the number of injectors (Ei) in the combustion chamber at a frequency equal to $Ei.\ \omega$ and upper harmonics. Moreover, low engine orders are present in the engine, due to static parts non perfect symmetries, such as not perfectly circular casings, not perfectly constant stator vanes or injector spacing, etc…

The rotor of the centrifugal compressor impeller has maximal speed 40 000 rpm. The excitation of the rotor in the test rig has a limited speed, as mentioned above. Therefore, it is important to find another harmonic excitation that excites the same operational shapes in the same frequencies as is the case of the engine.

Bladed disk with *Er* blades that is excited by *Es-th* harmonic, can oscillate by modal shape of bladed disk with x nodal diameter (ND). The number of ND is defined by following equation

$$x = |h \cdot Es - k \cdot Er| \quad (1)$$

Where h is harmonic index $\langle 1,2,3,\ldots \rangle$, k is integer $\langle 0,1,2,\ldots \rangle$, *Es* is the number of stationary elements (it is also EO) which define harmonic of speed $h \cdot Es$, *Er* is the number of rotating blades.

In this specific case, the number of ND is determined the number of rotation blades $Er=30$ and the number of stationary elements (diffusor vanes) $Es=25$. For given number of stationary elements *Es* is excited five ND ($x_{min}=5$ for harmonic index $h=1$). The maximum emphasis is put on minimal values of x (harmonic index $h=1$). All x values for a particular harmonic index h can be used as harmonic excitations of impeller and all these harmonic excitations will generate operation shape with the same number of ND, which is given by the value x_{min}.

Table 1: Number of ND which is excited by diffusor vanes

k	*h*	
	1	2
0	25	50
1	5	20
2	35	10

Numerical Analysis of the centrifugal compressor impeller

In order to get an idea about basic vibration characteristics a FE model segment is derived from the ideal design of centrifugal compressor with main blade and splitter blade as is shown in Fig. 2a. A sector of impeller was used to launch FE calculations under cyclic symmetry boundary conditions as is shown in Fig. 2a. The sector of impeller includes three structures (disk, main blade and splitter blade) and represents one fifteenth of the whole.

Additionally, the modes can be distinguished using the blade percentage of the total blisk strain energy as a reliable indicator Table 2. Hence, several blade-modes mode families can be easily identified. These mode families occur as nearly horizontal lines. In contrast to this, disk-modes and coupled modes show more rising frequencies with an increasing number of nodal diameters caused by increasing disk stiffness, as mentioned above can be found in [3,4].

Table 2: Distinguish of blisk modes

	Disk mode	Coupled mode	Blade mode
Blade percentage of total blisk strain energy	<50%	50%÷75%	>75%

Based on this model an eigenvalue analysis with the effect of rotational speed included is carried out, firstly to derive the nodal diameter plot Fig. 2b and with that to get an overview about relevant blade mode families and disk dominated modes as well. A first overview of the possible resonances in operating range and a suitable point for measurement of the response by strain gauge or LDV were obtained based on this analysis. The strain gauges are the most common technique used to measure the dynamic behaviour. The effectiveness of these techniques relies on the pertinent placement of the strain gauges or the laser target point.

The numerical analysis shows a lot of structural vibration modes in engine operating range. Some of them are close to operating modes of the engine. The operating modes of the engine are 60% speed (Idle), 94% speed (Continuous mode) and 100% speed (Starting mode). The engine operating range for the 25 EO is represented by a red line in Fig. 2b. The critical point is between the continuous mode and the starting mode of the engine because there is structural vibration mode which is excited by 25 EO.

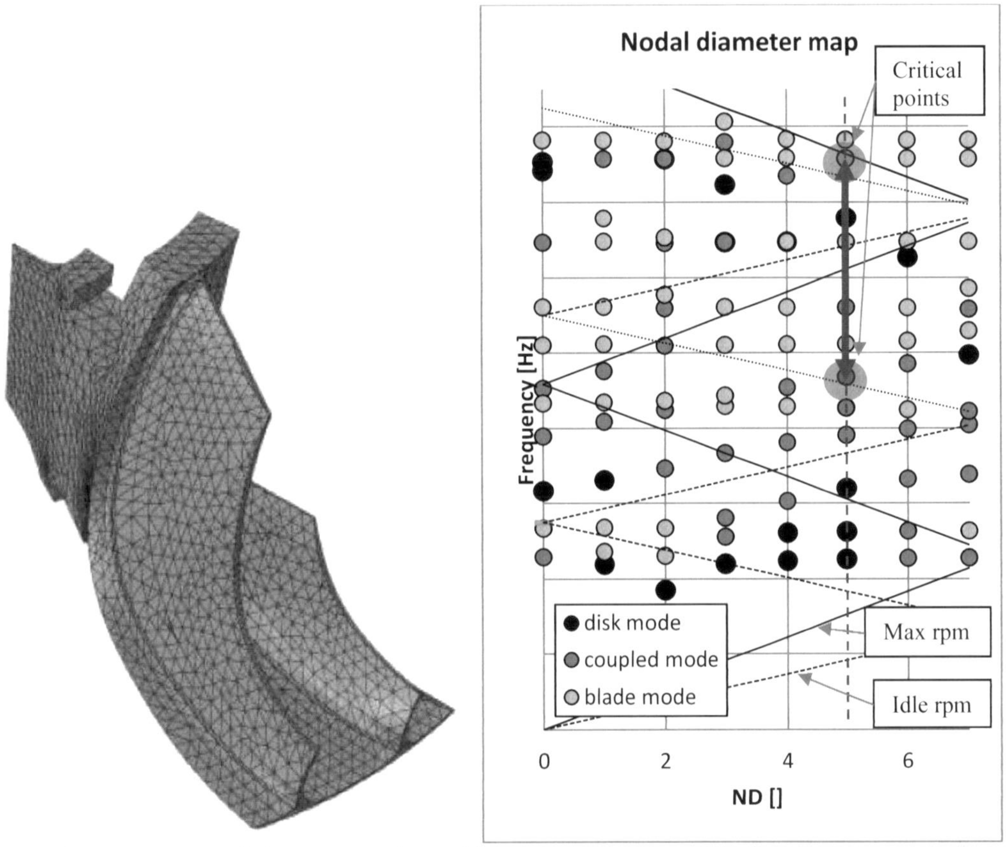

Fig. 2: a) FE model segment of centrifugal compressor and b) SAFE diagram

Experimental method

The impeller was fixed to the frame, as already mentioned. The impeller did not rotate and therefore it was not loaded by static force such as centrifugal force. The measurement was performed using two approaches.

- The first approach determine operational modal shapes. The speed of the excitation of rotor was set up outside any resonance. The impeller was essentially excited by wide-band noise via air turbulence. Responses levels of chosen points were measured by LDV OMETRON VH300+. The chosen points for identification of operational modes are shown in Fig. 3. The Auto and Cross-correlation functions were processed to prepare an input to the modal extraction algorithm in order to analyse natural frequency, damping ratio, and modes shapes.
- The second approach defined the resonance regions in the operating range. The excitation rotor was continually accelerated up to 8600 rpm, which is approximately 108% of maximum operating speed of the compressor rotor. Response levels of chosen points on blades were measured by strain gauges and accelerometers. Harmonic (125th

EO) extraction was performed in off-line mode by digital tracking filter with a bandwidth 0.25h and time step 0.05s.

Excitation of the impeller was performed by rotating of a jetting air. Specific harmonic excitation is defined by the number of nozzles on the exciter rotor. A definition of the number of nozzles was described in the previous chapter. The impeller was excited by 125 nozzles. The mass flow of the jetting air was 172 kg/h.

The responses were measured by LMS SCADAS III frontend with 32 input channels and 2 tacho inputs. The dynamic signal measurement resolution is 16 or 24 bits. Frequency bandwidth was set to 25600 Hz, frequency resolution 2.5 Hz.

Experimental results

Operation modal analysis

Modal identification was performed to verify the excitation method. The identification of modes by operational modal analysis was performed outside any region of resonance (0.81 norm frequency). Fig. 3 shows the selected points on disk, main blade and splitter for the measurement of the mode shapes. The measurement did not include all main blades, splitters and all point on disk. The cause is a frame which holds the impeller. On-line data reduction has been used for modal data measurement. Cross and Auto Spectra were measured for each measured point on the structure, linear averaging of 10 spectra was used. A miniature accelerometer (mass 0.2 g) placed on the disc served as a reference signal measuring in axis of rotation. Operational disk modal shape has expected the number ND, which was excited, as is shown in Fig. 4.

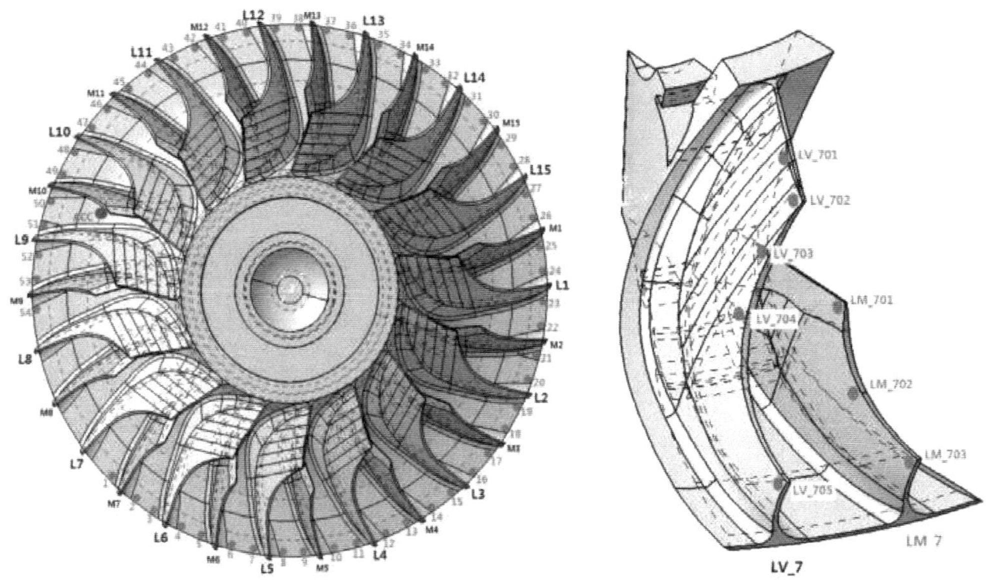

Fig. 3: Points for identification of operational modal shapes

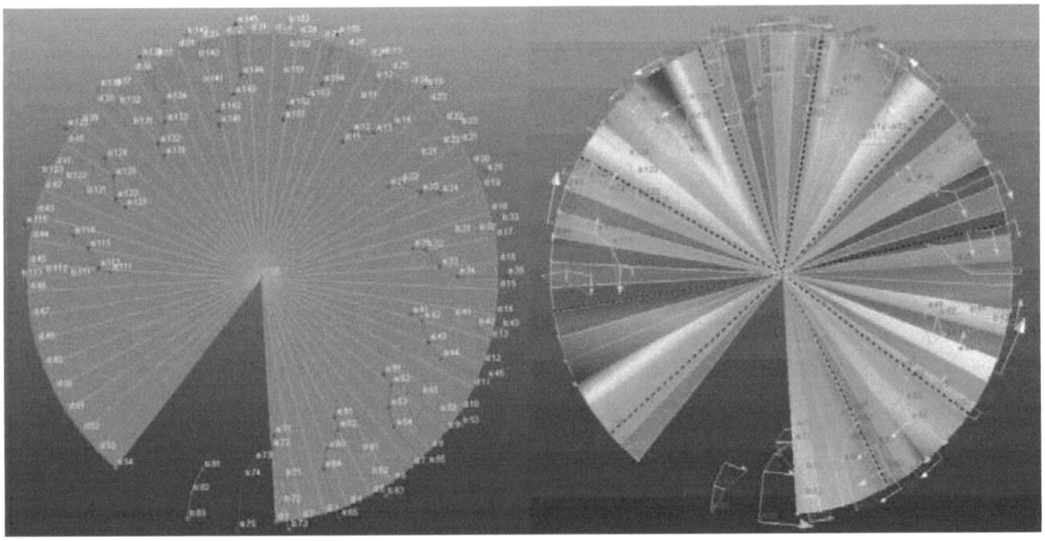

Fig. 4: Model for visualization of the shapes

The first two operational modal shapes have expected the number ND, which was excited, as is shown in Fig. 4. On the other hand the higher operational modal shapes have twice the number of ND. The cause of this phenomenon is the opposite phase of an oscillation of the main blades and the splitters. The mistuning of the blades of the impeller worsens the identification of the mode shapes. Some modes cannot be identified due to the localization of the mode, so-called "blade modes".

Resonance regions

An example of frequency response measured using strain gauges is shown in Fig. 6 for a particular gauge on all blades at theirs trailing edges as is shown in Fig. 5. This was obtained using a slow acceleration during a run-up test in order to improve the quality of the low signal of strain gauges.

Fig. 5: Placement of the strain gauges on the trailing edge of the impeller blades and accelerometers on root of leading edges

As all the blades were measured simultaneously by their specific strain gauges, the operation deflection shape (in term of strain) with their nodal diameter numbers could be determined automatically as described in the previous chapter. Then, a nodal diameter number (ND) could be attributed to each of the peaks on the response spectrum Fig. 6.

Fig. 7 shows the magnitude of response spectrums from measurement by a mini accelerometers PCB 352A73. The excitation of the centrifugal compressor impeller is the same as was with the strain gauges. The high sensitivity of the accelerometers provides better signal quality. This method was limited by the number of accelerometers. The measurement had to be repeated twice. The accelerometers were placed at the root of the leading edge of the blades as is shown in Fig. 5. A small frequency shift is noticeable. This phenomenon is due to the added mass of accelerometers.

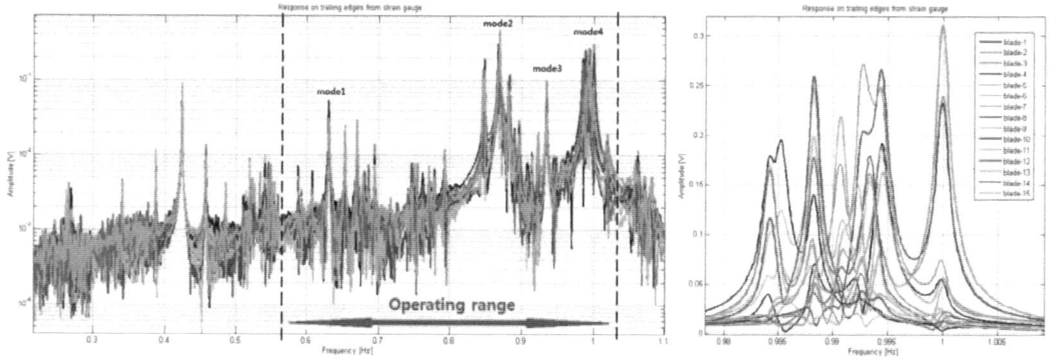

Fig. 6: Spectra measured with strain gauges at trailing edges

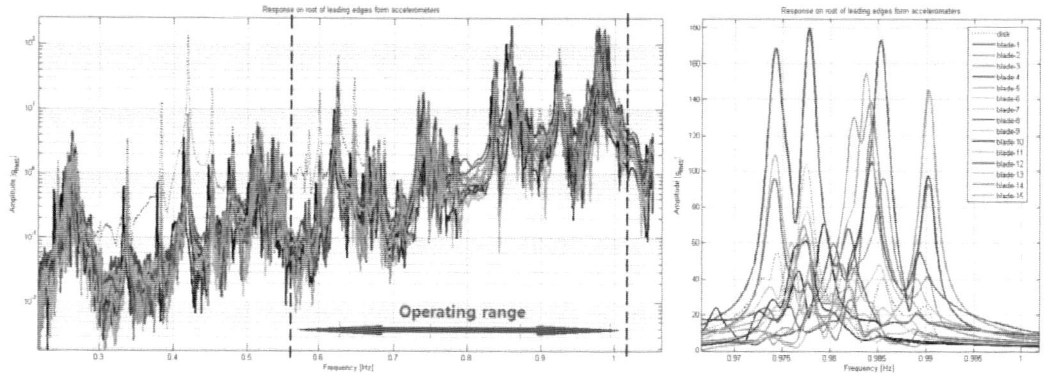

Fig. 7: Spectra measured with accelerometers at the root of leading edges

The mistuning of the system is well obvious in the resonance spectra. Generally, the mistuning of a bladed disk assembly results in several fundamental effects such as splitting of double mode and mode localization [4]. A splitting of double modes is more pronounced in the bladed modes. Disk modes and coupled modes are not susceptible to the mistuning effect as is the case with blade modes [5].

The mistuning is caused by the inaccuracy in the manufacturing. The vibration modes of blade integrated disks are very sensitive to individual blade mistuning. The structure has very low structural damping. The mistuning effect leads to the formation of local zoning of the vibration modes (mode localization) [6]. These local zones have a significantly higher dynamic response and reduce of reserve for the high cyclic fatigue (HCF) of the system.

The instrumentation of strain gauge is an exacting task and in terms of time (tens of hours). It is necessary to perform FEM analysis at first to identify suitable points. On the other hand it is not difficult to install accelerometers. Acceleration sensors have good sensitivity and resolution. The accelerometer is suitable for initial measurement of dynamic behaviours, but the accelerometers cause shifts of natural frequencies though their weight is small (0.2g) and they are placed in the roots of the leading edges of the blades.

Conclusions

This article shows a simple way to experimental validation of the dynamic properties of the bladed integral disk. This process is shown on the centrifugal compressor impeller. The centrifugal compressor impeller was analysed using strain gauges, the accelerometers and the LDV system in order to predict its dynamic behaviours and the influence of certain parameters. A simple test rig was introduced that simulates a specific dynamic load of the impeller (interaction between rotor blades and diffuser vanes) and the definition of the excitation vector. A methodology dealt with searching the correct placement of the strain gauges and advantages and drawbacks of each system were mentioned. The advantages and drawbacks of strain gauges and accelerometers were evaluated, based on the experiment on this specific application.

The measurement of the operational modal shapes was performed by LDV and LMS SCADAS III. The measurements have shown that it is possible to visualize the modal shape oscillation of the particular excited resonance. There was demonstrated that the mistuning of the blades has a useful property. Modal shapes are fixed to the geometry of the impeller and the shapes of oscillation do not change their position. Blades response for a particular resonance is always the same.

The test rig was designed primarily to search out the resonances of the blisks (bladed integral disks). Another function should be a reliable identification of the mode localization, the corresponding blades, and the maximum amplitude magnification. These are not only the prerequisite for dimensioning of aeroengine compressors but also for the optimum selection of blades to be instrumented with strain gauge, in particular, in case of lowdamped blisk rotors. Determination of suitable blades is important because the number of measurement channels is limited. A bad choice of the blades gives a misleading information, as is evident from the amplitude-frequency responses in Fig. 6 and Fig. 7. In general blade vibration modes govern the maximum operational lifetime of compressor rotors caused by the high strain levels of the blades.

Acknowledgement

This result was achieved with the help of institutional funding of the Czech Ministry of Industry and Trade in the programme dedicated to the development of research organizations.

References

[1] Seinturier, E., Forced Responce Computation for Bladed Disks Industrial Practices and Advanced Methods, 12th IFToMM World Congress, Besançon, France, June18-21, 2007.

[2] P. Almeida, C. Gibert, X. Leblanc, J.-P. Ousty, F. Thouverez, Experimental and numerical investigations on a rotating centrifugal compressor, In: ASME Turbo Expo Paper No. GT2012-69760, pp. 1133-1142 (2012)

[3] Strehlau, U. & Kühhorn, A., Experimental and Numerical Investigations of HPC Blisks With a Focus on Travelling Waves. ASME Turbo Expo2010: Power for Land Sea and Air, Glasgow, UK, June 14–18, ASME Paper No. GT2010-22463 (2010).

[4] A. V. Srinivasan, Flutter and Resonant Vibration Characteristics of Engine Blades", R Journal of Engineering for Gas Turbines and Power, 119, pp. 741–774, 1997.

[5] Jamróz, T., Study of Localized Vibrations of Mistuned Blades of Centrifugal Compressor, Engineering Mechanics 2016 - 22nd International Conference, pp. 241-244. (ISBN: 978-80-87012-59-8), Academy of Sciences of the Czech Republic.

[6] T. Klauke, A. Kü hhorn , B. Beirow & M. Golze: "Numerical Investigations of Localized Vibrations of Mistuned Blade Integrated Disks (Blisks)", Journal of Turbomachinery 131, 031002, 2007.

EAN 2020
58th conference on experimental stress analysis

Development of 3D Printed Insoles

JANDOVÁ S.[1,a], MENDŘICKÝ R.[2,b], JAŠUREK M.[2,c]

[1]Deparment of Applied Mechanics, Faculty of Mechanical Engineering, Technical University of Liberec, Studentska 2, 461 17 Liberec, Czech Republic

[2]Deparment of Manufacturing Systems and Automation, Faculty of Mechanical Engineering, Technical University of Liberec, Studentska 2, 461 17 Liberec, Czech Republic

[a]sona.jandova@tul.cz, [b]radomir.mendricky@tul.cz, [c]milan.jasurek@tul.cz

Keywords: Plantar pressure, Scanning, Modelling

Abstract

The main objective of the study was to print a pair of customized insoles based on foot scanning. At first plantar pressure distribution while walking in an average group of 51 young adults (21 males and 30 females, age 22 ± 2 years, height 1.72 ± 0.1 m, weight: 70 ± 16 kg) was measured. One of these participants who had an European foot type was selected for future feet scanning and modeling. For plantar scanning a scanner type ATOS III TripleScan was used and then a 3D model of the feet was made. These models were used for the subsequent 3D printing of the insoles. Developed insoles were printed with an OBJET Connex 500 printer (technology PolyJet Matrix) and the blend of Agilus30 and VeroBlack material was used. Initial plantar pressure measurement results with the insole prototype indicated a positive effect to the gait cycle when a subject´s foot contact area was changed. Our experience and knowledge may be used in the future to influence gait cycle through adjustment of the customized insoles in a patients with a different foot deformities.

Introduction

In designing of footwear and the components of the shoe, it is very useful to perform a comprehensive study of the foot. As human feet vary widely from person to person, it is beneficial to understand and recognize trends which will then lead to the creation of shoe and sole designs that will be optimized for the average foot in terms of comfort and style. As such, the advancement of 3D scanning to analyze and measure the human foot accelerates the process, becoming a very useful tool for research on this design.

As one of the primary considerations of selecting the proper footwear, comfort is a complex parameter in the design of footwear. The proper fit may be dictated by several factors such as shape, size, weight, materials among others and it is also necessary to take into consideration measurements such as heel width, heel to ball width, heel to ball length, toe box space among others. And as such, the three-dimensional shape of the foot is critical in understanding and designing for the proper fit of the shoe [1].

Foot scanning methodology

Prior to the more robust and accurate methods of measuring the foot using 3D scanning, many techniques have also been used for analyzing the structure of the foot, including manual measurements of the lengths using calipers or Brannock devices [2,3,8,9], as well as other conventional methods such as determination from digital footprints or the Harris mat also known as the force foot imprinter kit [8]. However, 3D scanning equipment prove to provide accurate and precise results as compared to these other methods in collecting anthropometric data, where consistency in the measurement is important while measuring across different foot types [8]. Usually, subjects are asked to stand up and loaded foot is scanned. The output is a 3-dimensional cloud of data points which represents the surface of the foot.

Model of the foot

As the foot is a flexible structure, there are several measurements taken of the foot as the parameters for differentiating foot types and determining the structure of the foot. Measurement of foot length, foot width, horizontal foot breadth, instep length, instep height, ball angle, medial ball length, lateral ball length, anatomical heel width, hallux and toe height are used as parameters to classify the foot shape or for customization of the shoe as needed. According to these measurements three major clusters of European foot types are identified [4,5]. The most common is foot characterized by low instep length, average anatomical ball width, low heel width, long medial ball length and large ball angle. The aim of our study was to develop a 3D printed customized insoles for this type of foot.

Methods

The first objective was to measure plantar pressure distribution while walking in an average group of 51 young adults (21 males and 30 females, age 22 ± 2 years, height 1.72 ± 0.1 m, weight: 70 ± 16 kg). Dynamographic records of all participants′ gait at a normal speed (v = 5 $km.h^{-1}$) using force plates (Emed®-c50, Novel, Munich, Germany) were performed. During their walk, every third step in the sequence was measured. In each participant, a total of 10 steps (5 left, 5 right) were analysed, from which the average values of the foot length, forefoot width, heel width, plantar angles and Arch index were analysed (Fig. 1). Based on these results, participant with the typical European foot type without any deformities like flatfoot or highfoot was selected for future scanning and foot model developing.

All experiments were performed with the approval granted by the institutional Review Board. The experimental work conforms to the highest standards of safety and ethics, with respect to the Declaration of Helsinki and to the national laws. The informed consent of all people participating in the research has been obtained.

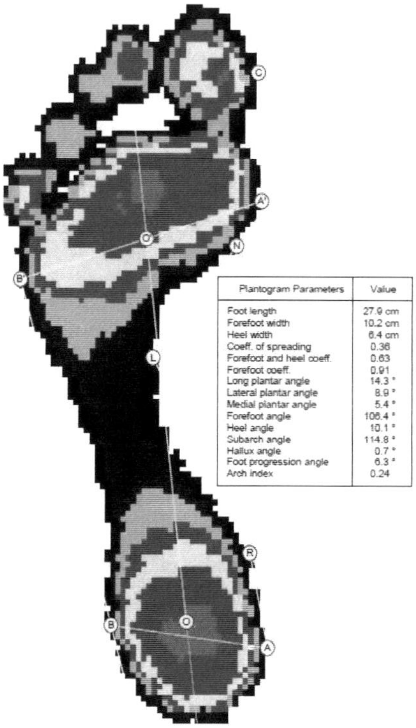

Plantogram Parameters	Value
Foot length	27.9 cm
Forefoot width	10.2 cm
Heel width	6.4 cm
Coeff. of spreading	0.36
Forefoot and heel coeff.	0.63
Forefoot coeff.	0.91
Long plantar angle	14.3 °
Lateral plantar angle	8.9 °
Medial plantar angle	5.4 °
Forefoot angle	106.4 °
Heel angle	10.1 °
Subarch angle	114.8 °
Hallux angle	0.7 °
Foot progression angle	6.3 °
Arch index	0.24

Fig. 1: Plantar pressure distribution in a subject with European foot type

The next step was to scan the foot with the scanner type *ATOS III TripleScan* (Fig. 2).

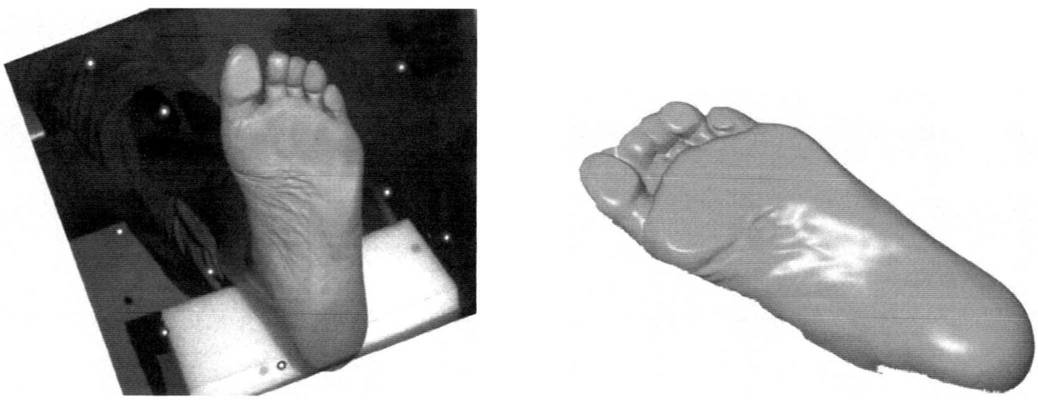

Fig. 2: Scanner view and model of the foot

Based on the plantar scanning (scanner type *ATOS III TripleScan*) a 3D model of the feet was made. These models were used for the subsequent 3D printing of the insoles. Developed insoles were printed with a *OBJET Connex 500* printer (technology PolyJet Matrix) and the blend of *Agilus30 and VeroBlack* material was used (Fig. 3).

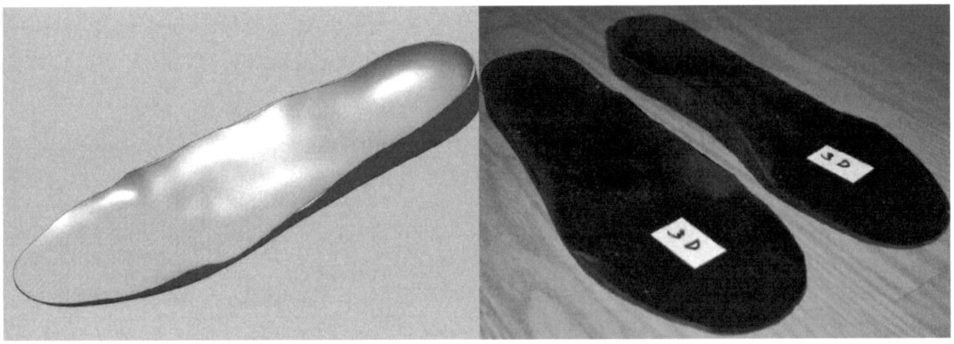

Fig. 3: Development of printed insoles

Resulting printed insoles were tested by a subject while walking. Plantar pressure analysis carried out while walking barefoot and using these insoles showed a plantar pressure (PP) differences that can cause a positive effect of the use of insoles in terms of balance and posture stabilization during stance phase. Relative peak pressure ($PP_{rel.}$) normalized to 1 kg of body weight (PP_{max}/kg, where *kg* is the body weight) decreased with the use of customized printed insoles from 4.46 (without customized insoles) to 3.94. In absolute values it was 45 ± 8 kPa.

Conclusions

The objective of the study was to print a pair of customized insoles for one of the three main foot types based on foot scanning. As a suitable material for 3D printed insoles was chosen the blend of *Agilus30* and *VeroBlack* material. Initial plantar pressure measurement results with the insole prototype indicated a positive effect to the gait cycle when a subject´s foot contact area was changed. Our experience and knowledge may be used in the future to influence gait cycle through adjustment of the customized insoles in a patients with a different foot deformities.

Acknowledgement

This publication was written at the Technical University of Liberec as part of the project "Research, development and application of advanced methods and technologies in technical mechanics, biomechanics and strength and elasticity" with the support of the Specific University Research Grant, as provided by the Ministry of Education, Youth and Sports of the Czech Republic in the year 2020.

References

[1] R.S. Goonetilleke, A. Luximon, Designing for Comfort: A Footwear Application, in Proceedings of Computer-Aided Ergonomics and Safety Conference, 2001.

[2] A. Luximon, R.S. Goonetilleke, Foot shape modelling, Human Factors 46(2) 2004 pp. 304-315.

[3] E.S. Song, W. Yoon, H. Son, S. Yu (2018) Foot Measurement Using 3D Scanning Model, International Journal of Fuzzy Logic and Intelligent Systems 18(3) 2018 pp. 167-174.

[4] I. Krauss, S. Grau, M. Mauch, C. Maiwald, T. Horstmann, Sex-related differences in foot shape, Ergonomics 51(11) 2008 pp. 1693-1709.

[5] I. Krauss, C. Langbein, T. Horstmann, S. Grau, Sex-related differences in foot shape of adult Caucasians – a follow-up study focusing on long and short feet. Ergonomics 54(3) 2011, pp. 294-300.

[8] Y. Lee, G. Lin, M.J. Wang, Comparing 3D foot scanning with conventional measurement methods. Journal of Foot and Ankle Research 7(44) 2014.

[9] S. Telfer, J. Woodburn, The use of 3D surface scanning for the measurement and assessment of the human foot. Journal of Foot and Ankle Research 3(19) 2010.

EAN 2020
58th conference on experimental stress analysis

Examination of Dynamical Parameters of the Production Line Manipulator

JÍROVÁ R.[1,a], PEŠÍK L.[1,b]

[1]Deparment of Machine Parts and Mechanism, Faculty of Mechanical Engineering, Technical University of Liberec, Studentská 2, 461 17 Liberec, Czech Republic

[a]radka.jirova@tul.cz, [b]lubomir.pesik@tul.cz

Keywords: Linear rolling guides, Dynamical parameters

Abstract. In the automotive industry, manipulators are essential for production lines and production processes. Manipulators ensure transport of materials, parts and tools, and are frequently based on linear rolling guides. However, the failure of any linear guide may cause stopping the production and may lead to significant production losses. Therefore, firms demand their sufficient reliability and prediction of possible failures. For ensuring these demands, at first, we need to identify and verify the dynamical parameters of production manipulators that are necessary for computing the operational loads of linear guides. The paper describes a method for identifying dynamical parameters of production manipulators on a specific manipulator. Dynamical parameters were determined by experimental and numerical methods, comparing the measured and simulated accelerations at two points on the manipulator. The measured data were cleared by finite impulse response filtering and explored by fast Fourier transformation for getting the natural oscillation frequency. This oscillation frequency could be compared with frequencies of a dynamical model at different mass and stiffness parameters. Results showed a sufficient match between measured and simulated acceleration values.

Introduction

Nowadays, the reliability of production machines is highly required. To prevent possible losses, producers demand prediction of failures. Production Line Manipulators often use linear rolling guides for linear motion. As well as rolling bearing in rotational machines, linear guides are a bottleneck of a manipulator design.

For ensuring reliability, the failure prediction of linear rolling guides is needed. The failure prediction is related to a dynamical load of linear guides [1], which is connected with dynamical parameters, the effect of mass inertia and operating conditions of production manipulators [2]. Therefore, the objective of this paper is to identify and verify the dynamical parameters of a manipulator.

Clamping Frames Manipulator

Dynamical parameters are identified and verified on a specific manipulator for clamping frames. The manipulator horizontally transports clamping frames of car bodies during the welding process. Clamping frames are connected to the cart through telescopic parts. For transporting, the manipulator uses linear rolling guides and an electric motor. The power from the electric motor is transmitted and transformed through a rack gear. For finding dynamical parameters, the acceleration was measured at two points of the manipulator and in two directions – x and y [3]. Fig. 1 shows a 3D model of the manipulator with marked positions of acceleration sensors.

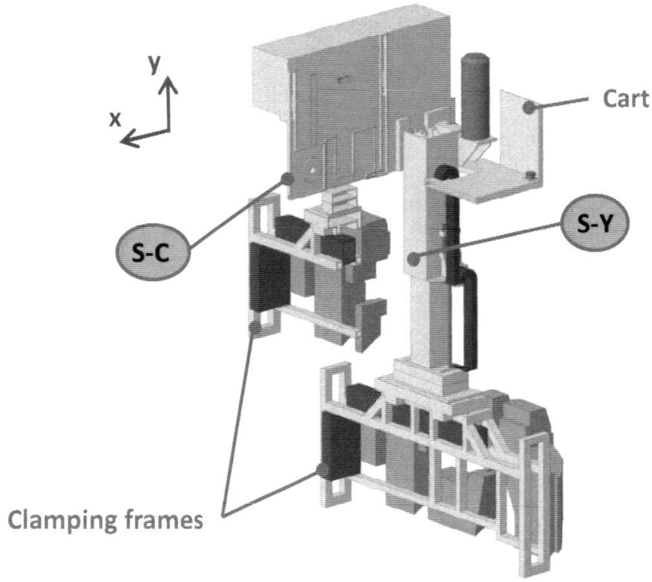

Fig. 1: Clamping frames manipulator, positions of sensors

The first acceleration sensor "S-C" is placed nearby the electric motor; the second one "S-Y" is placed on the telescopic part.

FIR Filtering

For getting a pure acceleration of motion at two chosen points, measured data were filtered by the FIR filter. Measured and filtered data in x and y directions are shown in Fig. 2 - Fig. 5.

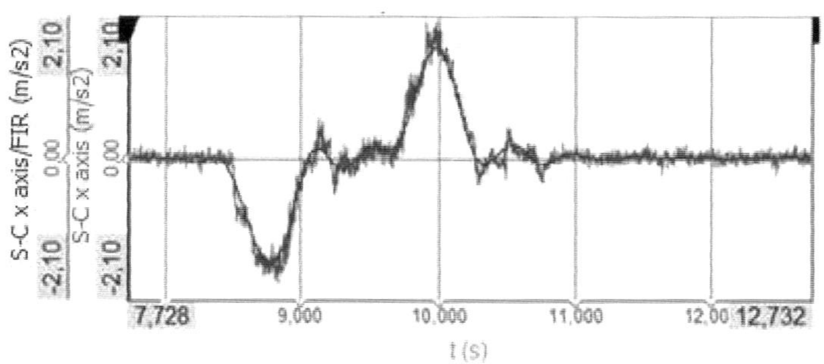

Fig. 2: Acceleration in x axis direction at the first point

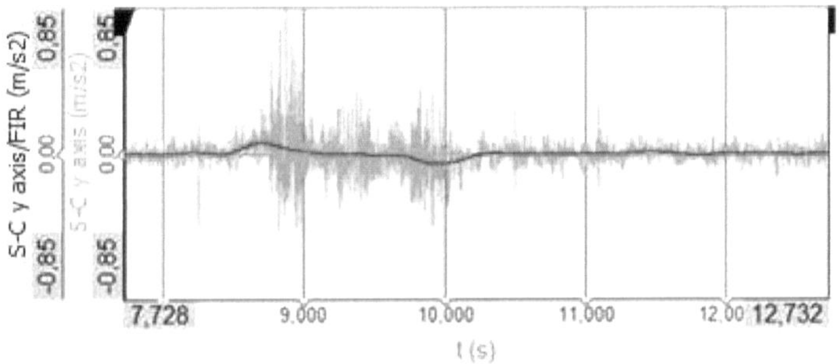

Fig. 3: Acceleration in y axis direction at the first point

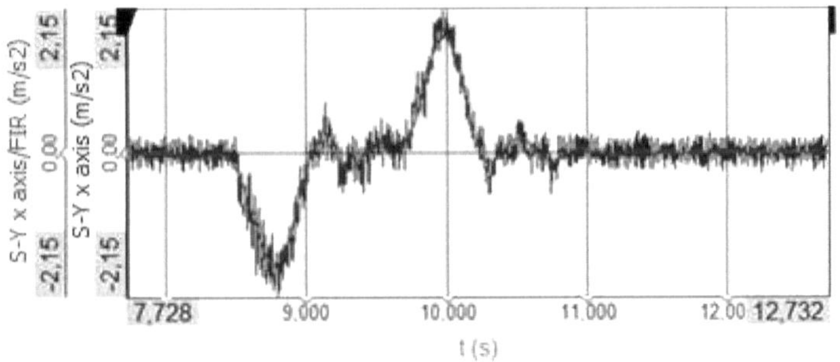

Fig. 4: Acceleration in x axis direction at the second point

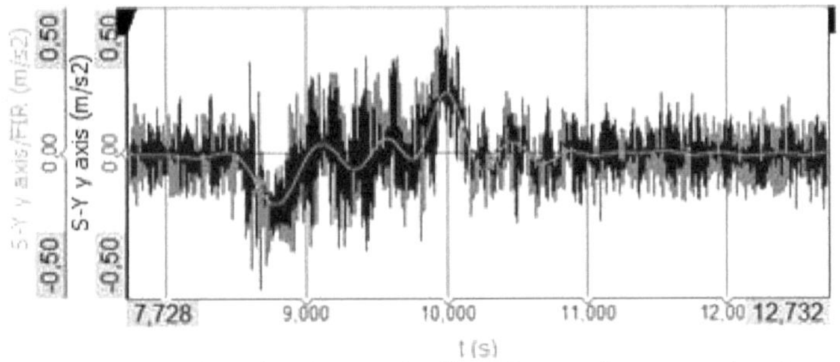

Fig. 5: Acceleration in y axis direction at the second point

The filtered data shows the oscillation of the manipulator during the linear motion. This oscillation is the natural oscillation of the manipulator related to the inertia effect of its mass.

FFT Analysis

The natural oscillation frequency is evaluated by Fast Fourier Transformation (FFT). The results of the FFT Analysis are summarised in Fig. 6.

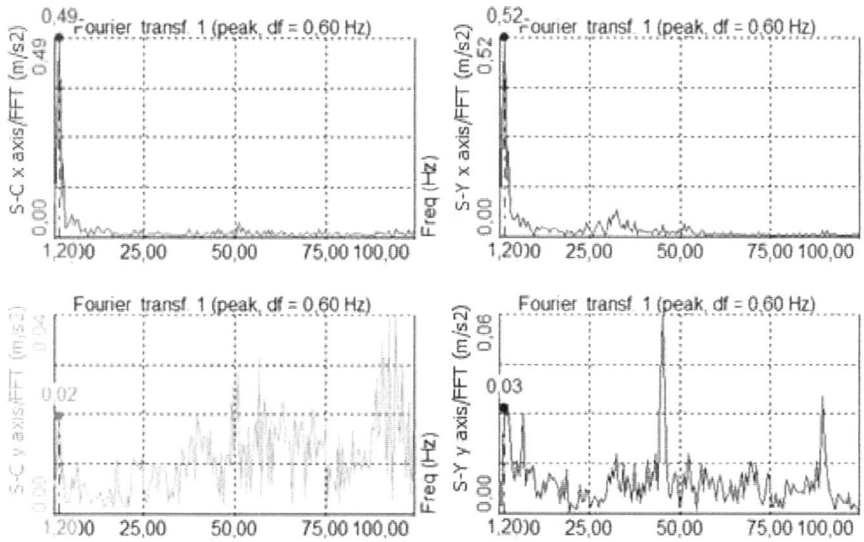

Fig. 6: Frequency spectrum

The highest peak at the lowest frequency in each spectrum corresponds to the natural oscillation frequency 1,2Hz that is marked in each graph.

MBS Analysis

The linear motion of the manipulator was simulated using a multibody system analysis. Where, linear rolling guides were substituted by elastic and damping connections (Fig. 7). The contact point of the rack gear was substituted by a pin and slider connection with assigned kinematic excitation at L point [4,5].

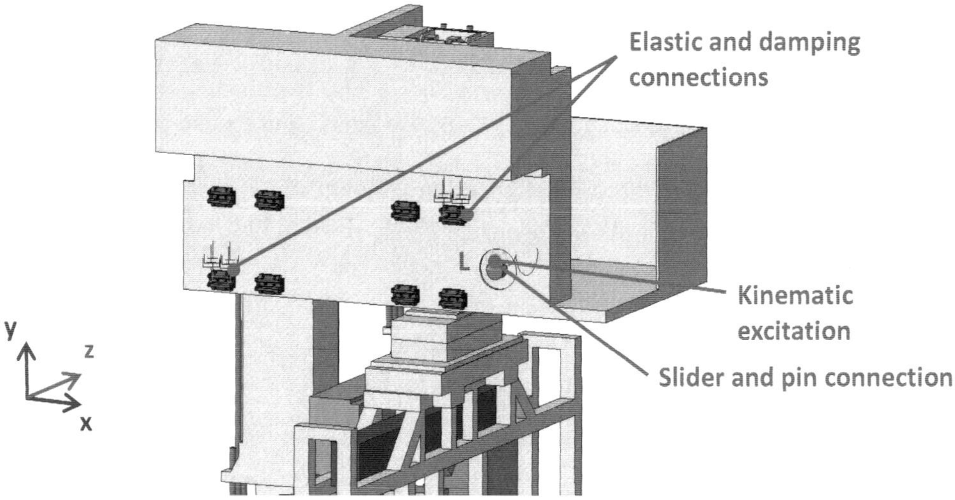

Fig. 7: Dynamical model

The velocity data used for kinematic excitation were taken from the data of an electric motor frequency converter. Stiffness and damping coefficient may be calculated using the evaluated natural oscillation frequency. The stiffness is

$$k_y = \frac{J_z(2\pi f_z)^2}{\sum_{j=1}^{n} l_j^2} \tag{1}$$

where J_z is the moment of inertia related to the z axis, f_z is the natural frequency related to the z axis, l_j a perpendicular distance between L point and damping and elastic connection and n a number of damping and elastic connections.

The damping coefficient equals

$$b_y = \frac{2\sqrt{J_z k_y \sum_{j=1}^{n} l_j^2}}{\sum_{j=1}^{n} l_j^2} b_{rel} \tag{2}$$

where b_{rel} is the relative damping coefficient.

Results

The measured and simulated values of acceleration are compared in Fig. 8 – Fig. 11.

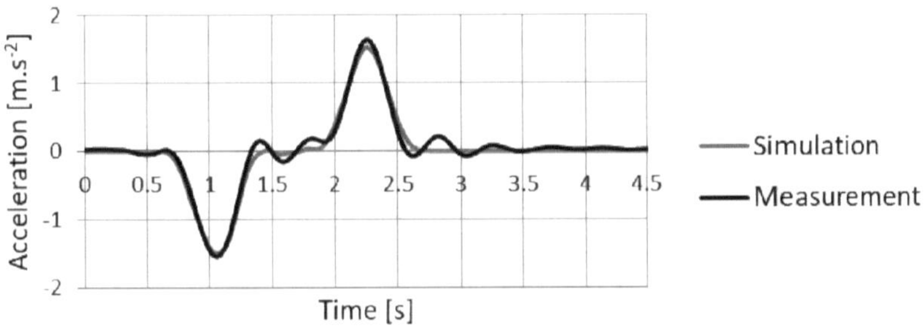

Fig. 8: Acceleration in x axis direction at the first point

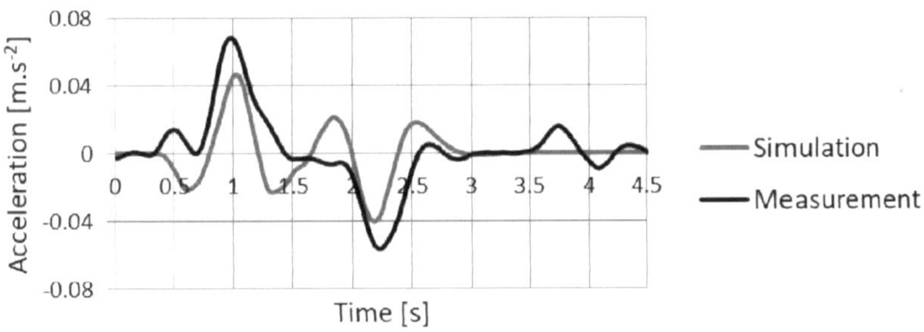

Fig. 9: Acceleration in y axis direction at the first point

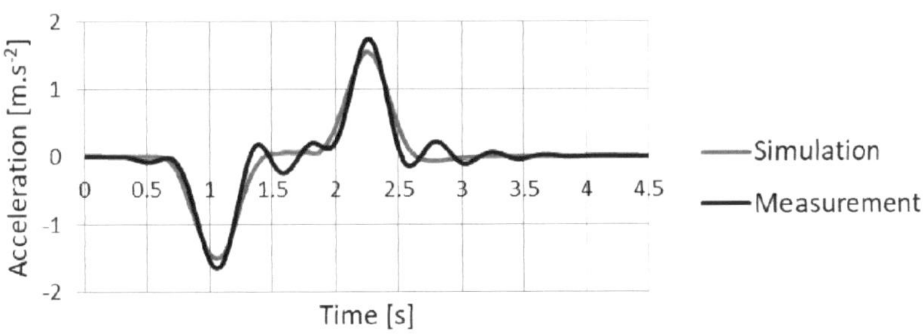

Fig. 10: Acceleration in x axis direction at the second point

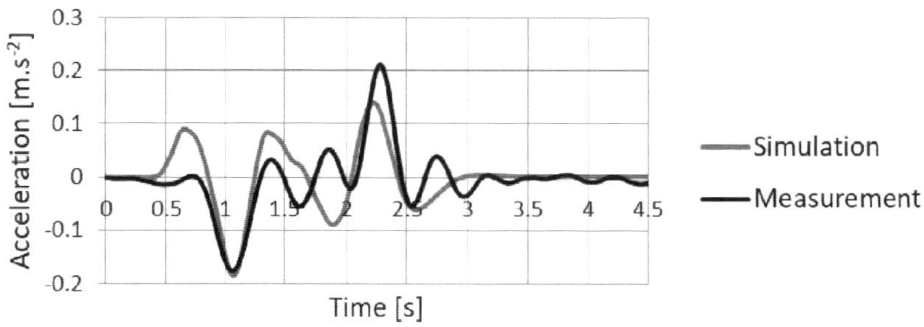

Fig. 11: Acceleration in y axis direction at the second point

Results show a sufficient match between measured and simulated values of acceleration. The identified dynamical parameters, especially stiffness and damping coefficient, may be further used for calculation of the dynamical load of manipulator linear rolling guides.

Conclusions

The paper describes the identification of dynamical parameters of manipulators. The proposed method uses a comparison between measured values of acceleration at two defined points and acceleration values calculated by numerical simulation. The results, demonstrated on the specific manipulator of clamping frames, show a relatively good match. Identified dynamical parameters might be used for calculation of the dynamical load of linear rolling guides. In granted patent [1], the authors pointed out the negative effect of linear rolling guide load on identifying their wear. For evaluating this negative effect to the failure detection in practice, dynamical parameters of the operated machine need to be sufficiently identified at first.

Acknowledgements

This work was supported by the Student Grant Competition of the Technical University of Liberec under the project No. SGS-2019-5036.

References

[1] Škoda Auto a.s. Lineární valivé vedení s integrovaným diagnostickým zařízením. Patentový spis CZ 308 232 B6, 11.3.2020

[2] E. Tomeh, Technická diagnostika. Vibrační diagnostika strojů a zařízení. Technická univerzita v Liberci. Liberec, 2015. ISBN 978-80-7494-174-0.

[3] M. Navrátil, Měření mechanického kmitání. Úvod do teorie snímačů. SNTL. Praha, 1981.

[4] V. Zeman, Z. Hlaváč Z, Kmitání mechanických soustav. Západočeská univerzita v Plzni. Plzeň, 2010. ISBN: 978-80-7043-337-9.

[5] H. Dresig, F. Holyweißig, Maschinendynamik, 8. auflage, Springer. Berlin, 2008.

EAN 2020
58th conference on experimental stress analysis

Spectral Fatigue and Probability Density Function

KAĽAVSKÝ A.[1,a], HUŇADY R.[1,b], PAVELKA P.[1,c]

[1]Deparment of Applied Mechanics and Mechanical Engineering, Faculty of Mechanical Engineering, TUKE – Technical University of Košice, Letná 9, 042 00 Košice, Slovak Republic

[a]adam.kalavsky@tuke.sk, [b]robert.hunady@tuke.sk, [c]peter.pavelka@tuke.sk

Keywords: spectral fatigue, Dirlik, Rayleigh, Zhao-Baker, Lalanne-Rice, Tovo-Benasciutti

Abstract. The article deals with the influence of the choice of the probabilistic model in the numerical resp. analytical calculation of fatigue life in the frequency domain. In the introduction, the authors present modern probabilistic models that are suitable for application in the mentioned problem. The article also includes an experiment where an experimental modal analysis and a lifetime test are performed on 10 test-specimens. The results of the experiment are compared at the end of the paper with the numerical solution.

Introduction

Structural and mechanical systems are often exposed to irregular loads. If these loads are known in advance (for example, they are obtained experimentally), a time domain analysis based on the values of the rainflow matrix and consequently the accumulation of linear damage is generally used to determine fatigue damage and system life [1].

In fact, structures such as a car drives on rough road or a wind turbine are exposed to random loads (e.g. road surface, wind speed). Such a random load can be seen as performing a random Gaussian process, which can be described in the frequency domain using power spectral density (PSD) representing the propagation of the mean quadratic amplitude in the frequency range [2]. Working with power spectral density has proven to be particularly beneficial when working with complicated finite element models, where the frequency response calculation is much faster than transient time domain dynamic analysis [3].

One approach is the development of frequency domain fatigue assessment methods that offer a direct link between spectrum power density and damage intensity or load cycle distribution, respectively. Therefore, most authors consider the rainflow method to be the most accurate, the frequency domain methods seek to obtain the distribution of cycles by rainflow counting over a time domain [4].

At present, the analysis of frequency fatigue is dealt with by many authors who describe relatively complex load models as multiaxial loads [5-7] resp. random non-Gaussian signals[7-9].

Theory

The paper describes methods related to the stationary Gaussian process. These methods are divided into narrowband and broadband processes, of which narrowband allows direct derivation of cycle distribution. For the broadband process, the relationship of peak distribution and cycle amplitude is much more complex. Several empirical solutions (e.g. Dirlik [4] and Zhao-Baker [10]) have been proposed, but very few completely theoretical solutions (Bishop [11]). The theoretical solution presented by Bishop is based on the Markov process theory and is computationally demanding. However, there is a slight improvement in accuracy over the Dirlik method, which is usually the preferred method.

In 2004, Benasciutti and Tovo [3] compared a group of methods in the frequency domain (Wirsching-Light [12], Zhao-Baker [10], Dirlik [4], empirical α 0.75 and Tovo-Benasciutti [3]) and found that the Tovo-Benasciutti method matches the accuracy of the Dirlik method in terms of numerically simulated power spectral densities. In Table 1 some methods with corresponding formulas are mentioned.

Dirlik model. Dirlik created an empirical expression for calculating the probability density function, which was obtained using extensive computer simulations using the Monte Carlo technique. Although, this model is more complicated than other methods, it is still only a function of four spectral moments. However, this solution has a wide range of applicability, thus surpassing other available methods. The formula for calculating the assumed cumulative damage by the Dirlik method is

$$p_D(S) = \frac{\left(\frac{D_1}{Q} \cdot e^{\frac{-S_i}{Q}} + \frac{D_2 \cdot S_i}{R^2} \cdot e^{\frac{-S_i^2}{2R^2}} + D_3 \cdot S_i \cdot e^{\frac{-S_i^2}{2}} \right)}{\sqrt{m_0}} \tag{1}$$

where D_1, D_2, D_3, Q, R are constants calculated from spectral moments m_0, m_1, m_2, m_4 and skewness γ . S_i in (1) represents normalized stress amplitude.

Table 1: Overview of others formulas for solving the probability density function.

Model	Formula
Rayleigh	$p_R(S) = \frac{S_i}{m_0} \cdot e^{\frac{-S_i^2}{2m_0}}$
Zhao-Baker	$p_{ZB}(S) = \frac{w \cdot \alpha \cdot \beta \cdot \hat{S}_i^{(\beta-1)} \cdot e^{\left(-\alpha \cdot \hat{S}_i^{\beta}\right)} + (1-\alpha) \cdot \hat{S}_i \cdot e^{\left(-\frac{\hat{S}_i^2}{2}\right)}}{\sqrt{m_0}}$
Lalanne-Rice	$p_L(S) = \sqrt{\frac{1-\gamma^2}{2 \cdot \pi \cdot m_0}} \cdot e^{\left(\frac{-S_i^2}{2 \cdot m_0 \cdot (1-\gamma^2)}\right)} + \frac{\gamma \cdot S_i}{2 \cdot m_0} \cdot e^{\left(\frac{-S_i^2}{2 \cdot m_0}\right)} \cdot \left[1 + \operatorname{erf}\left(\frac{\gamma \cdot S}{\sqrt{2 \cdot m_0 \cdot (1-\gamma^2)}} \right) \right]$
Tovo-Benasciutti	$p_B(S) = B \cdot \rho + (1-B) \cdot K$

Cumulative damage is calculated based on the formula

$$E[D] = \frac{E[P].T}{k} \cdot \int_0^{\infty} S^b . p(S).dS \tag{2}$$

where $E[P]$ represents the peak rate, T is a time duration, k, b are material constants.

Practical part

This part consists of an experiment and a numerical simulation combined with an analytical calculation using the Matlab program.

Experiment. It consisted of an experimental modal analysis, in which the eigenmodes, eigenfrequencies and damping of the measured specimens were determined, which were used to calculate and refine the numerical simulation. The second part of the experiment was a fatigue testing at stochastic excitation. The valid measurement was performed on 10 specimens made of aluminium alloy AL5052-H32 sheet of 4 mm thickness (Fig 1.).

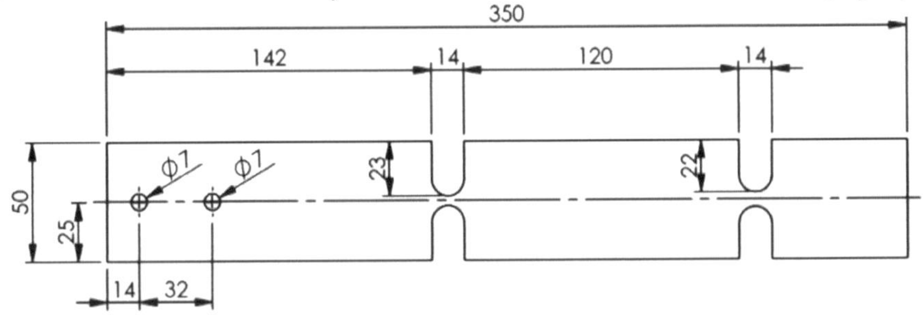

Fig. 1: Testing specimen with dimensions.

Two accelerometers were placed on the specimen. The sensor placed on the flange was acquiring the real excitation signal generated by the vibro-exciter, and the second sensor was used to measure the response of the examined specimen near the notch. A measuring configuration is shown in the Fig 2.

Experimental modal analysis (EMA) was performed by Bruel&Kjaer Pulse system. A modal hammer Type 8206 with an aluminum tip without an additional weight was used to excite the specimen. Responses were measured using a Polytec PDV100 laser vibrometer. The results of EMA are shown in Table 3 and Table 4.

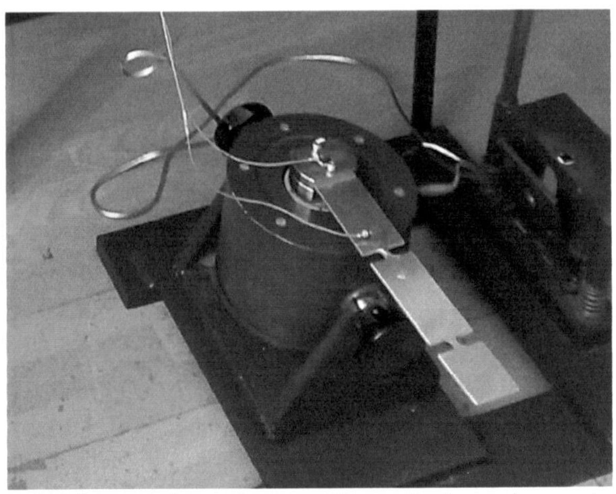

Fig. 2: Measured specimen with applied accelerometers.

Broadband white noise (1 - 1200 Hz) was used for analysis. The measurement was performed using Pulse LAN XI 3050-B-060 module and processed in LabShop software. Acceleration time histories were recorded and used in numerical analysis, the results of which were later correlated with the experiment.

Table 2: Failure times.

Specimen	Time to failure
Specimen No. 3	1040 s (17 min 20 s)
Specimen No. 5	1445 s (24 min 05 s)
Specimen No. 6	1000 s (16 min 40 s)
Specimen No. 7	815 s (13 min 35 s)
Specimen No. 8	1022 s (17 min 02s)
Specimen No. 9	820 s (13 min 40 s)
Specimen No. 10	1132 s (18 min 52 s)
Specimen No. 11	1150 s (19 min 10 s)
Specimen No. 12	1470 s (24 min 30 s)
Specimen No. 13	1058 s (17 min 38 s)

Table 2 shows the results of 10 valid fatigue tests. The time to failure ranged from 815-1470 seconds. The higher scatter of values was caused by earlier failure of two specimens (spec. No. 7 and 9), whose surface around the notch was roughened due to the application of an inappropriate technological procedure in the notch production. Therefore, these specimens were not included in the statistical processing. The remaining 8 specimens with a smooth surface were broken in the time from 1000 to 1470 seconds. Specimen after the fatigue test is shown in the Fig. 3.

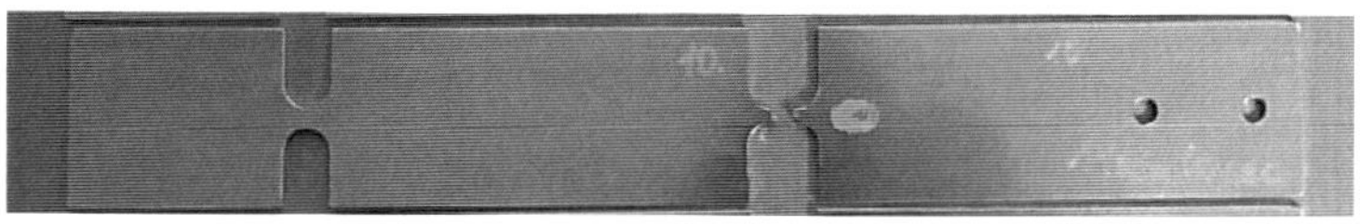

Fig. 3: Specimen after failure.

Numerical simulation. The pre-simulation was carried out in SolidWorks program, where the primary calculation containing the PSD response was performed. The PSD response was obtained by the linear dynamic random vibration module.

Since the specimen was firmly attached in the flange, the model had taken all degrees of freedom at the point of contact of the specimen with the flange. The PSD function, which can be seen in the Fig. 5, was used as an excitation. The PSD was estimated using a periodogram in the Matlab program from the actually measured history of acceleration as mentioned above. The excitation was performed in a direction perpendicular to the largest specimen area.

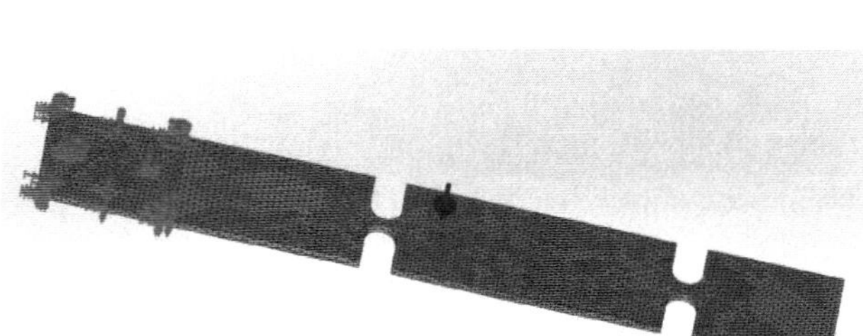

Fig. 4: Numerical model prepared to simulation

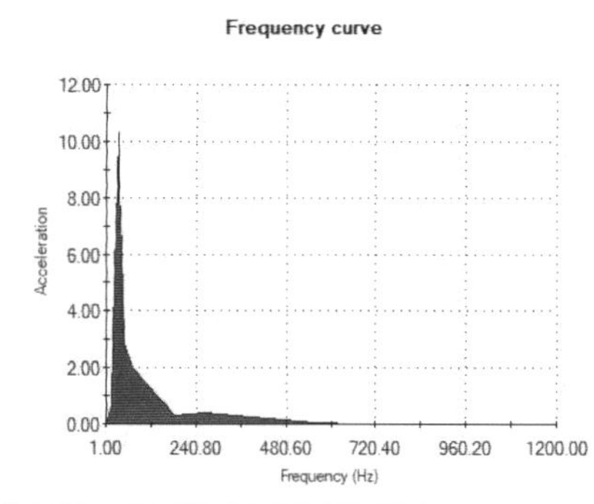

Fig. 5: PSD response

Results comparison of the numerical modal analysis with EMA can be seen in Table 3 and Table 4. The results of the numerical analysis are slightly higher. It can be caused due to the idealization of boundary conditions (perfect rigid couplings that do not exist in reality). Nevertheless, it is possible to state on the basis of the figures from Table 3 that the eigenmodes (also the numerical model of the beam) correspond to reality.

Table 3: Comparison of bending eigenmodes.

Mode	Experimental modal analysis	Simulation
1.	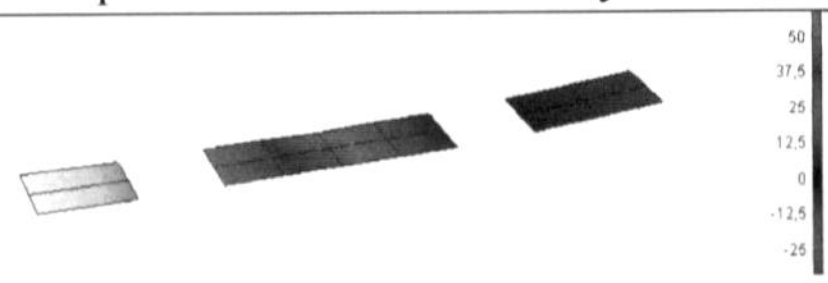	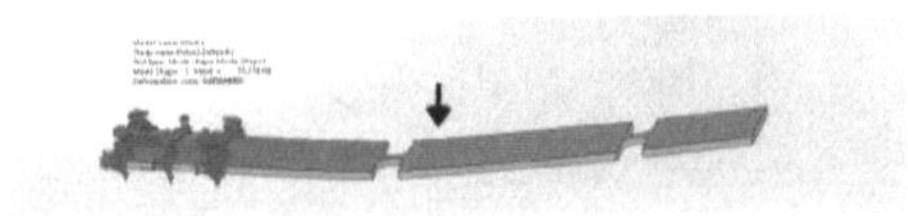
3.	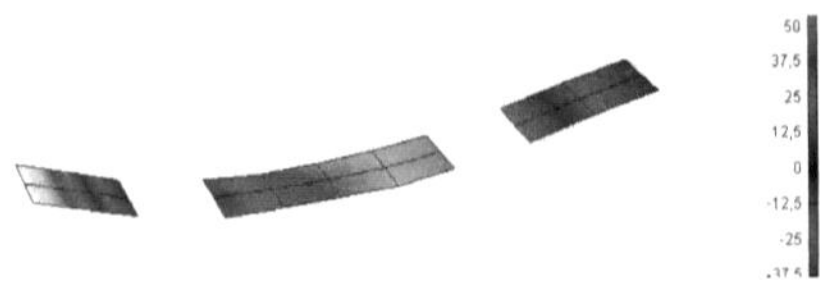	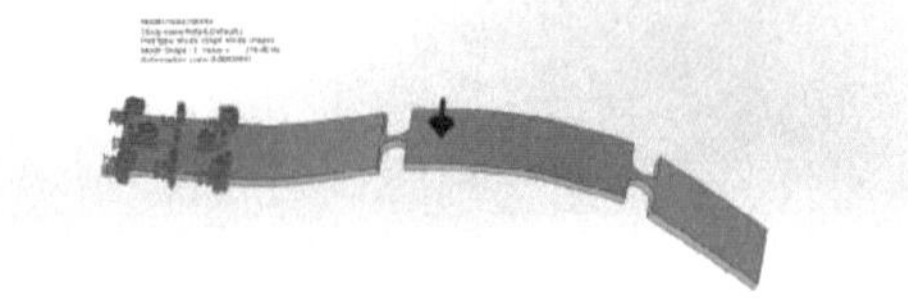
5.		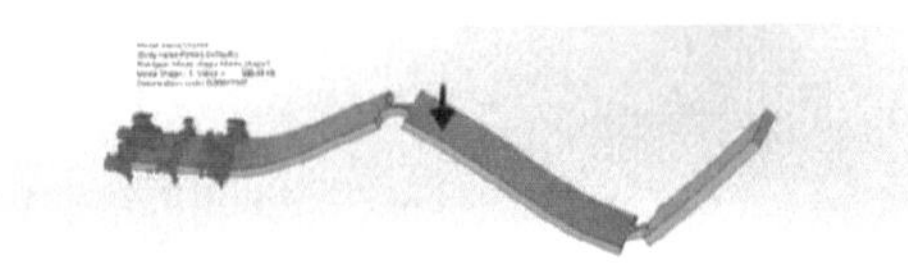
8.	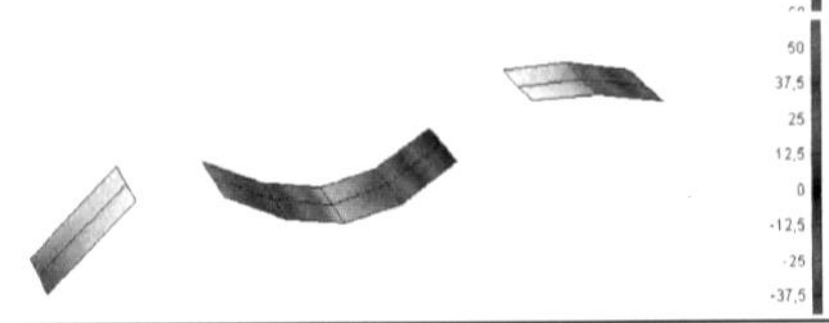	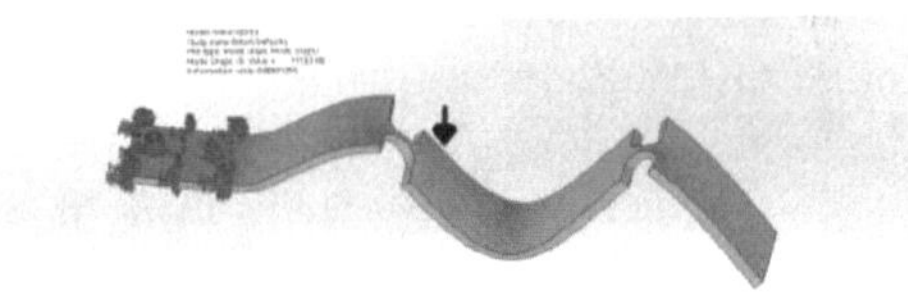

Table 4: Modal parameters of the in numerical and experimental analysis.

Mode	Numerical eigenfrequency [Hz]	Experimental eigenfrequency [Hz]	Damping [%]	Mode complexity
1.	31,92	31,80	1,22684	0,01680
2.	62,21	59,32	0,51345	0,02140
3.	218,48	191,68	0,45291	0,00278
4.	283,22	279,79	0,03545	0,04141
5.	500,64	472,87	0,22234	0,00074
6.	735,78	728,28	0,31693	0,07025
7.	856,51	853,03	0,05126	0,05126
8.	1119,12	1081,9	0,09551	0,00378

Damping, based on Rayleigh damping coefficients, was used to refine the computational PSD response model.

RAYLEIGH DAMPING COEFFICIENTS:

$$\beta = \frac{2\cdot\zeta_1\cdot\omega_1 - 2\cdot\zeta_m\cdot\omega_m}{\omega_1^2 - \omega_m^2} = \frac{2\cdot 0,0122684\cdot 31,8 - 2\cdot 0,0009551\cdot 1081,9}{31,8^2 - 1081,9^2} = 1,0999394\cdot 10^{-6}$$

$$\alpha = 2\cdot\zeta_1\cdot\omega_1 - \beta\omega_1^2 = 2\cdot 0,0122684\cdot 31,8 - 1,0999394\cdot 10^{-6}\cdot 31,8^2 = 0,7791579$$

The result of the simulation is a PSD stress field (von Misses), which is variable depending on the frequency (Fig. 6). The most endangered point (node) was selected from the critical area and this node was further subjected to a spectral fatigue analysis in the Matlab program. A PSD response of this node has been exported from SolidWorks software.

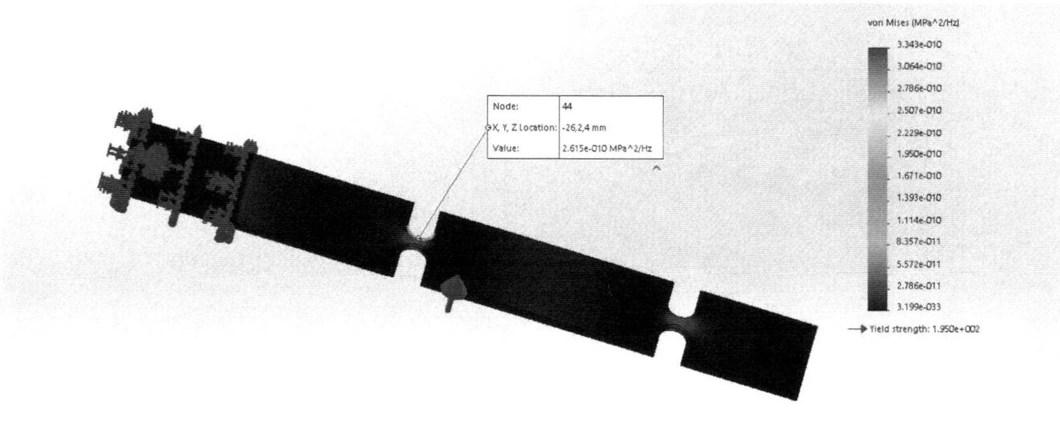

Fig. 6: Field of (von Misses) PSD stresses

Analytical calculation. A Matlab script has been programmed to determine the frequency domain lifespan using the Dirlik, Rayleigh, Zhao-Baker, Lalanne-Rice and Tovo-Benasciutti methods.

The input data for the script were the PSD response curve and the S-N curve. The program uses a common interpretation of the S-N curve based on the Basquin equation. The points of the S-N curve corresponded to real tests performed on notched standardized specimens of the AL 5052-H32 material, which was also used to produce specimens subjected to the fatigue test.

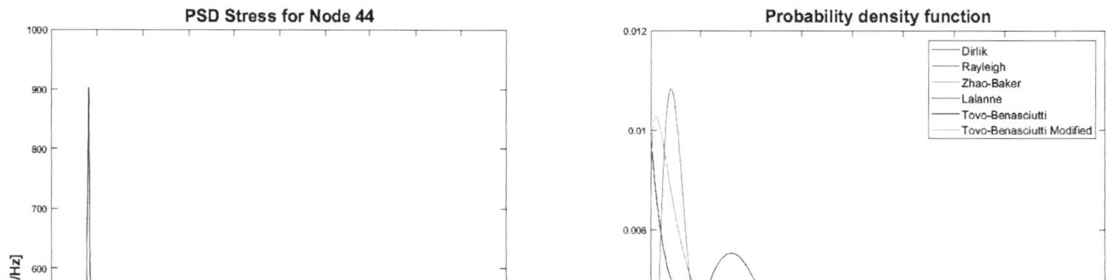

Fig. 7: PSD response for most critical node (left plot), probability density functions of all probability models (right plot)

Results are summarized in Fig. 7 and Table 5, where it is possible to see individual differences in histories of probability density functions and fatigue life estimate.

Table 5: Results of spectral fatigue analysis.

Model	Damage	Time to failure
Dirlik	0,00089622	1116 s (18 min 36 s)
Rayleigh	0,0014895	671 s (11 min 11 s)
Zhao-Baker	0,00073885	1353 s (22 min 33 s)
Lalanne	0,00092158	1085 s (18 min 05 s)
Tovo-Benasciutti	0,00090137	1109 s (18 min 29 s)
T-B Modified	0,00090163	1109 s (18 min 29 s)

Discussion

Experiment. As mentioned above, the failure of 8 smooth surface specimens occurred in the time from 1000 to 1470 s. The standard deviation in this interval was 175.9 s. In the case of specimens 5 and 12, the time to failure was much longer than in the case of other specimens. If these specimens were also excluded from the measurement, the time variance would be significantly smaller (1000 - 1150 s). When the standard deviation at this interval decreased to 55.4 s, the mean time to failure was determined based on the times of only 6 specimens (3, 6, 8, 10, 11 and 13). Thus, it can be stated that the average time to fatigue failure of the tested specimens at vibration is 1067 s (17 min 47 s).

Numerical + analytical solution. If the average failure time of the specimen is experimentally set at 1067 s, it can be stated that according to the results in Table 5, the fatigue life was the most accurately estimated by the Lalanne model. As the Dirlik and Tovo-Benasciutti models provide estimates with an error of up to 5%, this can also be taken as a positive result. From the PDF histories it can also be stated that the most damaging amplitudes were located somewhere in the range of 100-250MPa, where it can be see how the Rayleigh model is clearly higher and the Zhao-Baker model is just below the other models.

Conclusions

The paper compares models of the probability density function (Dirlik, Rayleigh, Zhao-Baker, Lalanne-Rice and Tovo-Benasciutti). The experimentally obtained average time to damage under random variable loading was 1067 s. Fatigue life was most accurately predicted by the Lalanne (error 1,68%), Tovo-Benasciutti (error 3,94%) and Dirlik (error 4,59%) models. The worst prediction of fatigue life was estimated by the Rayleigh model (error 37,11%). The differences in the results can be attributed to the ideal rigid of the specimens in the FEM program, the approximation of the PSD signal, the damping of the individual eigenmodes, and the accuracy of the S-N curves.

Acknowledges

This research was supported by project VEGA 1/0355/18.

References

[1] F. Trebuňa, Odolnosť prvkov mechanických sústav. 1. vyd. Košice: Emilena, 2004. 980 s. ISBN 978-80-8073-148-9.

[2] P. Stoica, R.L. Moses, Spectral Analysis of Signals. [s.l.]: Pearson Prentice Hall, 2005. 490 p. ISBN 978-0-13-113956-5.

[3] D. Benasciutti, R. Tovo, Spectral methods for lifetime prediction under wide-band stationary random processes. In Int. J. Fatigue. 2005. Vol. 27, no. 8, pp. 867–877.

[4] T. Dirlik, Application of computers in fatigue analysis. Coventry, England: University of Warwick, 1985.

[5] D. Benasciutti, R. Tovo, Comparison of spectral methods for fatigue analysis of broad-band Gaussian random processes. Probabilistic Engineering Mechanics. 2006. Vol.21, no.4, pp.287-299. doi: https://doi.org/10.1016/j.probengmech.2005.10.003

[6] A. Nieslony et al., Fatigue life prediction for broad-band multiaxial loading with various PSD curve shapes. In Int. J. Fatigue. Vol.44. 2012. pp. 74–88. doi: https://doi.org/10.1016/j.ijfatigue.2012.05.014

[7] A. Nieslony, E. Macha, Spectral Method in Multiaxial Random Fatigue. Opole: Springer, 2007. pp. 152.

[8] M. Palmieri et al., Non-Gaussianity and non-stationarity in vibration fatigue. In Int. J. Fatigue. 2017. Vol.97, pp. 9-19. Doi: https://doi.org/10.1016/j.ijfatigue.2016.12.017

[9] F. Cianetti et al., The effort of the dynamic simulation on the fatigue damage evaluation of flexible mechanical systems loaded by non-Gaussian and non stationary loads. In Int. J. Fatigue. 2017. Vol. 103, pp. 60–72. doi: https://doi.org/10.1016/j.ijfatigue.2017.05.020

[10] W. Zhao, M.J. Baker, On the probability density function of rainflow stress range for statonary Gaussian processes. In Int. J. Fatigue. 1992. Vol. 14, no. 2, pp. 121–135.

[11] N.W.M. Bishop, The use of frequency domain parameters to predict structural fatigue. University of Warwick, 1988

[12] P.H. Wirsching, A.M. Shehata, Fatigue under wide band random stresses using the rain-flow method. In Published in: The Journal of Engineering Materials and Technology, July 1977, pp. 205-211.

EAN 2020
58th conference on experimental stress analysis

Experimental Assessment of Active Wheelset Steering System Using Scaled Roller Rig

KALIVODA J.[1,a], BAUER P.[1,b]

[1] Czech Technical University in Prague, Faculty of Mechanical Engineering, Department of Automotive, Combustion Engine and Railway Engineering, Technická 4, 166 07, Prague 6, Czech Republic

[a]Jan.Kalivoda@fs.cvut.cz, [b]Petr.Bauer@fs.cvut.cz

Keywords: Active wheelset steering, Roller rig, Wheel-rail contact forces

Abstract. The paper is created within the project which aim is to design a system of active wheelset steering for an electric four-axle locomotive. The wheelset steering system enables reduction of forces acting in the wheel-rail contacts in a curved track and consequently a reduction of wear and maintenance costs of both vehicles and rails is achieved. The project consists of three main parts: computer simulations, scaled roller rig experiments and field tests. The paper is focused on the second part on the project. The scaled roller rig has been innovated in order to simulate bogie run in a curved track with uncompensated value of lateral acceleration and instrumented with a system of measurement of lateral wheel-rail forces. The experimental bogie has been equipped with systems of active wheelset steering and measurement of axle-box forces. The experiment setup, newly developed and applied systems of forces measurement and wireless signal transmission, and results of the first experiments are described in detail.

Introduction

Force interaction in between rails and railway wheels is one of the most important issues in the development of the new rolling stock. Todays' effort to build an economic and environment friendly railroad brings a general and sustained demand to reduce wheel-rail contact forces below legislative limits as much as possible. Particular attention is paid to the lateral component of wheel-rail contact forces during passing a curved track, also called guiding forces. Conventional methods of reduction of guiding forces are based on the optimization of suspension characteristics [1], or on mechanic or hydraulic linkages between the various components of the running gear. Because the possibilities of conventional methods are increasingly encountered at their limits, ideas of the utilization of active controlled elements in the wheelset guidance and railway vehicle suspension occur [2, 3]. One of the first practical utilization of such systems are active yaw dampers developed by Liebher and offered as an option for Siemens Vectron locomotives [4] (Fig. 1).

In order to compare the effectiveness of individual methods for reducing guiding forces, a simplified multibody simulation (MBS) model of an electric locomotive has been created (Fig. 2). The model consists of 7 rigid bodies (car body, 2 bogie frames, 4 wheelsets) that are connected by linear force elements. Wheel-rail contact respects non-linear characteristics of S1002 wheel and UIC 60 rail profiles, forces acting in the wheel-rail contacts are calculated using FASTSIM method [5].

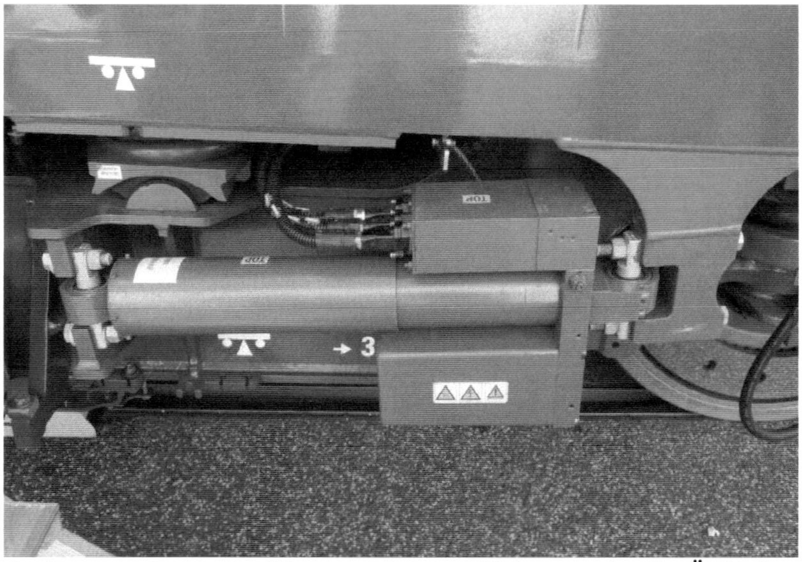

Fig. 1: Active yaw damper on Siemens Vectron MS locomotive for ÖBB, Innotrans fair 2018 (photo author)

Fig. 2: Graphical representation of the simulation model

Using this model, a simulations of vehicle run in a low speed in the curved track of radius $R = 150$ m considering the friction coefficient in the wheel rail contact $f = 0.4$. The quasistatic value of guiding forces in a constant curvature track without irregularities Y_{qst} was evaluated on all wheels. A vehicle setup and parameters has been sought in which the value of Y_{qst} is minimized. The maximum value of Y_{qst} is typically reached on the outer wheel of the first wheelset. Simulations were performed for 6 different vehicle setups:

- STD – Standard, the suspension parameters corresponds to the standard 4-axle electric locomotive with flexi-coil type secondary suspension.
- YFS – Yaw Flexible Suspension, the characteristics of the primary suspension and wheelset guidance are modified in order to soften the yaw stiffness of the connection between wheelsets and bogie frame.
- MBC1 – Mechanical Bogies Connection Type 1, one of the classical methods of reducing guiding forces based on the direct mechanical connection of bogie frames [6].
- MBC2 – Mechanical Bogies Connection Type 2, mechanical connection of bogie frames by a mechanism. This method woks on the similar principle like MBC1, but has

less space demands. Thus MBC2 can be utilized also on asynchronous locomotives, which have usually a large transformer located between the bogies.

- AYD – Active Yaw Dampers. Method based on active controlled yaw torque acting between car body and bogies. The torque is generated by a couple of linear actuators acting in between each bogie frame and car body [7].
- AWS – Active Wheelset Steering, yaw angle of wheelsets towards a bogie frame is actively controlled.

The parameters of mechanical bogie connections MBC1 and MBC2 (ie. stiffness and preload od coupling elements) were optimized in order to achieve the best performance in a 150 m radius as well as the force produced by AYD actuators and the wheelsets yaw angle for AWS. The Fig. 3 and Table 1 thus expresses the maximum possible effect of reducing the guiding forces, which can be achieved by individual methods. It is important to note that:

- Contribution of mechanical bogie connections MBC1 and MBC2 to the guiding force reduction will be lower then calculated values 23 respectively 10.5 pro cent. The parameters of mechanical bogie connections should be compromised in the wide range of curve radiuses.
- For AYD the impact of forces in the actuators on the secondary suspension deflections was not taken into an account. To avoid undesired large deflections of secondary suspension in the lateral, direction and transmitting forces via lateral bump-stops the power of the actuators would probably have to be lower than considered in the simulation. Consequently, a reduction in guiding forces will be lower then calculated 25.7%.
- The highest reduction of guiding forces (75%) shows YFS. However, such reduction is achieved for zero value of the yaw stiffness of the wheelset guidance which drastically affect the stability and lower the maximum speed of the vehicle.

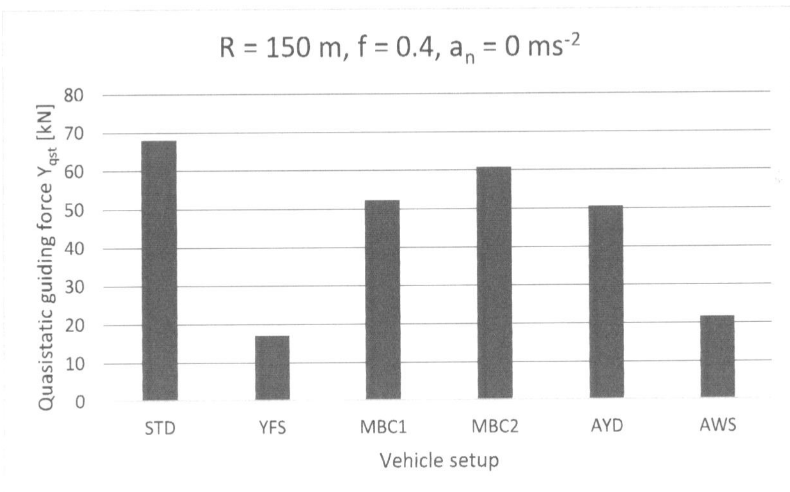

Fig. 3: Quasistatic guiding force Y_{qst} acting on outer wheel of the 1st wheelset in curve with radius R = 150 m for various vehicle configurations

The AWS appears to be the most effective and practically implementable method for the reduction guiding forces, because exhibits significant reduction of guiding forces (68.1%) and do not deteriorate other important properties of a vehicle such as forces in the secondary suspension, maximum speed, stability, ride comfort, etc. Moreover, in the performed simulation all four wheelsets were steered to the same yaw angle towards the bogie frame. The performance of AWS could be improved by individual wheelset steering and thus even distribution of guiding forces on all wheels and further reduction of guiding forces could be achieved.

Table 1: Quasistatic guiding force Y_{qst} for different vehicle setups

Vehicle setup	STD	YFS	MBC1	MBC2	AYD	AWS
Y_{qst} [kN]	67.9	17.0	52.3	60.8	50.4	21.6
Y_{qst} reduction [%]	0.0%	75.0%	23.0%	10.5%	25.7%	68.1%

In view of the above, a project with the aim of design a system of active wheelset steering for an electric four-axle locomotive was launched. The project is divided into the three main stages:

I. computer simulations,
II. scaled roller rig experiments (Fig. 4),
III. on track tests.

The goal of Stage I is composing and verification of the detailed simulation model including wheelset steering actuators and control loop and optimize the wheelset steering control algorithm considering various vehicle speeds and track conditions.

The Stage II is focused on the verification of computer simulations and demonstration of the benefits of AWS using a scaled laboratory test device.

Fig. 4: Scaled roller rig of the Czech Technical University

The Stage III includes implementation of AWS system on an existing locomotive and performing the track tests. Prior to the first field tests of active wheelset steering system it needs to be thoroughly tested to meet the requirements of safety. Roller rigs can be advantageously used for these tests, because they allow testing of vehicle running behaviour in laboratory conditions, where it is possible to simulate extreme situations without a risk of railway accident [8, 9]. The principle of roller rig is in the replacement of a track by rotating rollers with a rail profile on their circumference. The tested vehicle is longitudinally fixed. Nevertheless, the creep conditions and forces in the wheel-roller contact points are analogical to the conditions in wheel-rail contacts of a vehicle running in a real track. In order to fulfil the goals of the Stage II and perform laboratory tests for the assessment of the impact of AWS on guiding forces the CTU test rig and experimental bogie had to be considerably modified. The paper focuses on the test stand design changes and the first tests that were performed to verify its function.

Roller rig setup

The CTU scaled roller and experimental bogie have been considerably improved in order to simulate bogie negotiation of arbitrary shaped track and to measure force interaction between vehicle and track. The rig capability of a curved track simulation has been extended by simulation of the uncompensated lateral acceleration. It is based on tilting of the entire rig [10]. The completely new main frame of the rig, that is supported on four rollers, has been designed, manufactured and assembled. (Fig. 5). Thus the CTU roller rig is capable to simulate vehicle run in straight, transition and constant curvature track up to radius of 15 m and rail cant deficiency up to 200 mm.

Special attention was paid to measurements of forces in the wheel-roller contacts and between wheelsets and bogie-frame. Lateral components of the forces acting in the wheel-roller contacts are measured by strain gauge measurement of roller disc deformation [12]. The own telemetry system of wireless signal transmission from the rotating roller has been developed (Fig. 4, left) [10].

Fig. 5: Principle and design solution of the roller rig tilting

Fig. 6: Strain gauge instrumented roller with wireless transmitter module

The instrumented rollers have been calibrated by loading by directly measured lateral force. The loading was gradually performed at 16 points on the roller. The maximum error is 2.5% in the range of 0.1 – 1 kN. Although the roller rotates in the speed up to 1000 RPM, the calibration constants are not influenced by rollers speed. The measured signal is wirelessly transmitted in 1 kHz sampling rate to the standard PC and further processed in LabView software. The signal transmission is realized independently for each roller, time synchronisation of the signals is done by global positioning system (GPS) clock signal. The wireless signal transmission has been broadly tested, the number of lost samples is less than 0.05%. The measured signal is in parallel saved without lost samples to the SD card which is attached to the transmitter on the roller.

Experimental bogie setup

The experimental bogie is 1:3.5 scaled, but it does not correspond to any specific bogie of a real vehicle [13]. Its design is based on the goals of experimental research. In order to achieve high geometrical accuracy, the most of the main structural parts are made of aluminium by CNC machining. The connections of the mutually movable components are provided by roller and linear roller bearings, dry friction joints are avoided.

Wheelset steering mechanism.

The wheelsets are steered by an actively controlled steering mechanism (Fig. 7). The actuator is a permanent magnet synchronous servomotor (item 1) with rated torque 2.5 Nm. The actuator torque is transmitted via toothed belt (item 2) to the steering rod (item 3) and then to the wheelset by pair of linkages (item 5). Each wheelset is controlled independently to a desired value of yaw angle between wheelset and bogie frame by analog voltage signal connected to the servomotor controller.

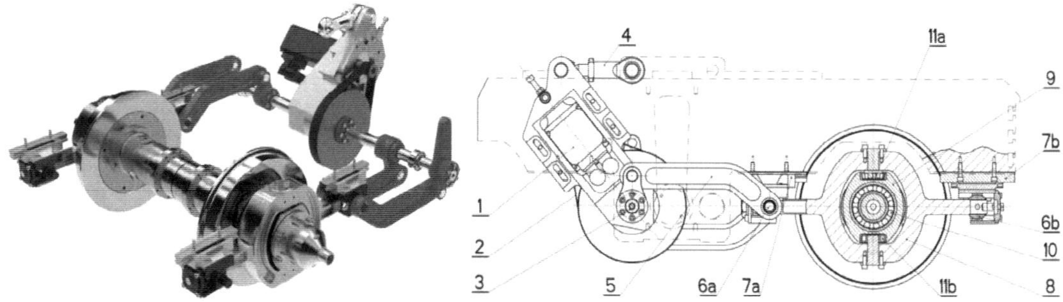

Fig. 7: Wheelset steering mechanism

Axle-box forces measurement.

The experimental bogie is equipped with a system of measurement axle-box forces, i.e. forces transmitted between axle-boxes and bogie frame.

Forces between each axle-box and bogie frame are transmitted via a stirrup (item 8 on Fig. 7), that was optimized for the strain gauge placement and serves as 3-axis load cell. Each stirrup is instrumented by 36 strain gauges connected in 3 Wheatstone bridges (Fig. 8). Thus independent measurement of longitudinal, lateral and vertical component of axle-box force is achieved [11].

The systems of wheel-roller contact force measurement and axle-box forces measurement has been successfully implemented, calibrated and tested on the rig.

Setup of experiment

The main goal of the first experiments was to verify the function of the simulation of the curved track, the test of active wheelsets steering and position control, while verifying the systems of measurement of axle-box forces and guiding forces. Two types of experiments were performed:

1. Vehicle run in superelevated straight track. During this test was the wheelset steering switched off, the bogie behaved as a standard passive suspension bogie. Straight track was simulated, whilst the rail superelavation p was continuously changed from zero to $p = 28.6$ mm ($p = 100$ mm. in a full scale).
2. Test of active wheelset steering. Test of active wheelset steering test was conducted in a curved track of a radius R = 60 m that corresponds to 210 m in a full scale. The yaw angle towards the bogie frame of both wheelsets was continuously changed from zero 0.22 deg. The yaw angle of both wheelsets was the same but in the opposite orientation.

Fig. 8: Instrumented stirrup

In the experiments the yaw angle of both wheelsets, track radius and superelevation of rails were directly measured by potentiometric transducers. At the same time the guidance and axle-box forces were measured. The experiments were done in a low speed of rollers N = 50 RPM.

Results of the first experiments

Vehicle run in canted straight track. The cant angle of the track β could be expressed by:

$$\beta \cong \sin\beta = \frac{p}{2s} \tag{1}$$

where $2s$ is the distance of the rails in the lateral direction and p is superelevation of rails. Then the lateral a_n acceleration is given by the component of the gravitational acceleration as:

$$a_n = g\sin\beta = g\frac{p}{2s} \tag{2}$$

Lateral force F_y acting on mass m is proportional to the rail superelevation by formula:

$$F_y = ma_n = mg\frac{p}{2s} \tag{3}$$

The results plotted on Fig. 9 are fully consistent with above assumption. The red line shows the sum of the lateral component of axle-box forces across all four wheels, whilst the black line shows the sum of wheel-rail guiding forces. Both are proportional to the superelevation of rails. The difference between them corresponds to the mass of wheelsets.

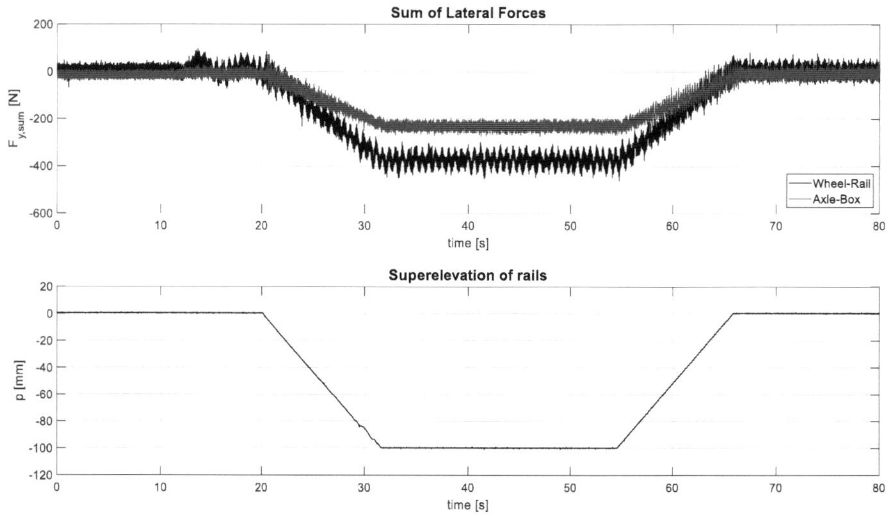

Fig. 9: Measured data – vehicle run in a canted straight track

Test of active wheelset steering. The test of active wheelset steering showed the functionality of wheelset steering mechanism and capability to steer both wheelsets to the desired yaw angle towards the bogie frame. Simultaneously the strong dependence of guiding forces on the yaw angle of wheelsets and possibility to considerably reduce guiding forces by active wheelset steering were confirmed. The plots on Fig. 10 show the time development of guiding force on the outer wheel of the 1st wheelset. This wheel exhibits the highest value of guide forces of all wheels of a bogie. With the increasing value of the wheelset yaw angle fell guiding force there below 30% of its original value.

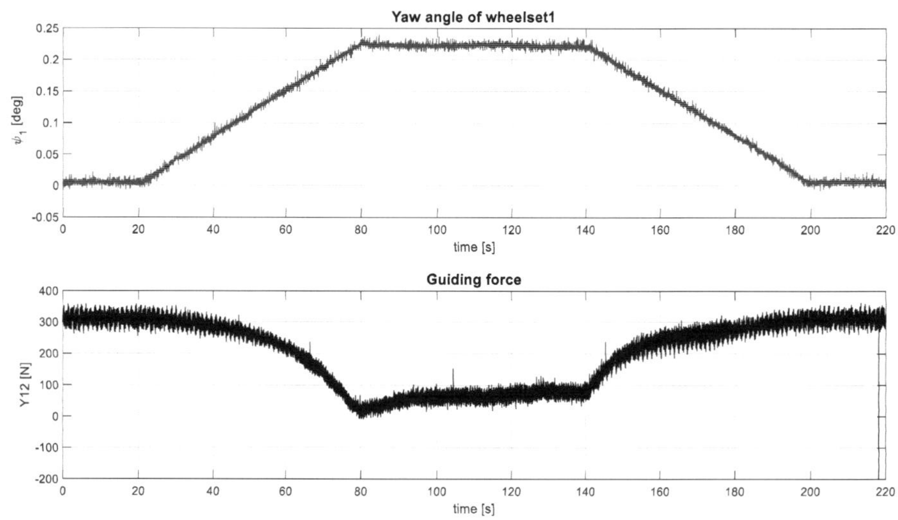

Fig. 10: Measured data – active wheelset steering test

Conclusions

The guiding forces acting in the wheel-rail contacts while a vehicle passes a curved track play very important role in the development of the new types of rolling stock. Reduction of guiding forces by conventional methods based on tuning suspension parameters and mechanical connections between bogies is not very effective. According simulations results the active wheelset steering is very effective and practically applicable method for reduction of guiding forces. Czech technical university together with industrial partner develops a system of wheelset steering for an electric locomotive. The paper focuses on the 2nd stage of the project – measurement of the performance of the system using experimental scaled roller rig. The experimental bogie has been equipped with the actuated mechanism that tis capable to steer wheelsets positions in the yaw direction and measure forces transmitted between axle-boxes and bogie frame. The roller rig was modified to fully simulate vehicle run in a curved track including effects of centrifugal acceleration and equipped with measurement of guiding forces. The performance of the experimental device was successfully tested during initial tests. The device is ready for conducting the experiments with varying speed, track radius, uncompensated lateral acceleration and control algorithms of the wheelset steering. Such experiments are scheduled as the next step in the near future.

Acknowledgement

This research has been realized using the support of Technological Agency, Czech Republic, programme National Competence Centres, project # TN01000026 Josef Bozek National Center of Competence for Surface Vehicles. This support is gratefully acknowledged.

References

[1] I. Klimenko, J. Kalivoda, L. Neduzha, Parameter optimization of the locomotive running gear, in: Transport Means - Proceedings of the International Conference, 2018, pp. 1095-1098.

[2] B. Fu, R. L. Giossi, R. Persson, S. Stichel, S. Bruni, R. Goodall, Active suspension in railway vehicles: a literature survey, Railway Engineering Science 28 (2020) 3–35. DOI: 10.1007/s40534-020-00207-w.

[3] N. Farhat, C. P. Ward, R. M Goodall, R. Dixon, The benefits of mechatronically-guided railway vehicles: A multi-body physics simulation study, Mechatronics 51 (2018) 115-126. DOI: 10.1016/j.mechatronics.2018.03.008.

[4] Liebherr to supply active yaw damper systems for Vectron locomotives, [online], 2017, [cited 2020-04-02], available at: https://www.liebherr.com/en/usa/latest-news/news-press-releases/detail/liebherr-to-supply-active-yaw-damper-systems-for-vectron-locomotives-news.html#lightbox

[5] J. J. Kalker, A fast algorithm for the simplified theory of rolling contact, Vehicle System Dynamics 11 (1982) 1-13.

[6] J. Bek et al., Elektrické lokomotivy, NADAS, 1976, in Czech.

[7] T. Michálek, J. Zelenka, Reduction of lateral forces between the railway vehicle and the track in small radius curves by means of active elements, Applied and Computational Mechanics 5 (2011) 187–196.

[8] A. Jaschinski, H. Chollet, S. Iwnicki, A. Wickens, J. Von Würzen, The application of roller rigs to railway vehicle dynamics, Vehicle System Dynamics 31 (1999) 345-392. DOI: 10.1076/vesd.31.5.345.8360.

[9] S. Myamlin, J. Kalivoda, L. Neduzha, Testing of Railway Vehicles Using Roller Rigs, Procedia Engineering 187 (2017) 688-695, DOI: 10.1016/j.proeng.2017.04.439.

[10] J. Kalivoda, P. Bauer, Scaled Roller Rig to Assess the Influence of Active Wheelset Steering on Wheel-Rail Contact Forces, Advances in Dynamics of Vehicles on Roads and Tracks, pp. 82-89, Springer Int. Publishing, 2020. DOI: 10.1007/978-3-030-38077-9_10.

[11] P. Bauer, J. Kalivoda, System of axle-box force measurement for experimental railway bogie, in: Experimental Stress Analysis - 56th International Scientific Conference EAN 2018 - Conference Proceedings, Harrachov, Czech Republic 2018, pp. 9-16.

[12] J. Kalivoda, P. Bauer, System of axle-box force measurement for experimental railway bogie, in: Experimental Stress Analysis - 57th International Scientific Conference, EAN 2019 - Conference Proceedings, Luhačovice, 2019, pp. 194-201.

[13] J. Kalivoda and P. Bauer, Mechatronic Bogie for Roller Rig Tests, in: The Dynamics of Vehicles on Roads and Tracks: Proceedings of the 24th Symposium of the International Association for Vehicle System Dynamics (IAVSD 2015), Graz, Austria, 2015.

EAN 2020
58th conference on experimental stress analysis

Investigation of Mechanical Behaviour of Commercially Available Polymeric Materials and their Suitability as Impact Absorber

KAŇÁKOVÁ S.[1,a], KOTTNER R.[2,b], KRYSTEK J.[2,c], BOŃKOWSKI T.[3,d], HECZKO J.[2,e]

[1]Department of Mechanics, Faculty of Applied Sciences, University of West Bohemia, Univerzitní 8, 301 00 Pilsen, Czech Republic

[2]NTIS - New Technologies for the Information Society, Faculty of Applied Sciences, University of West Bohemia, Technická 8, 301 00 Pilsen, Czech Republic

[3]Department of Biomechanical Human Body Models, New Technologies – Research Centre, University of West Bohemia, Univerzitní 8, 301 00 Pilsen, Czech Republic

[a]kanaksan@kme.zcu.cz, [b]kottner@kme.zcu.cz, [c]krystek@kme.zcu.cz,

[d]tomasz@ntc.zcu.cz, [e]jheczko@ntis.zcu.cz

Keywords: Drop test, Elastomer, Foam, Impact, Polymer

Abstract. A drop test was carried out to investigate mechanical behaviour of commercially available polymeric materials, namely expanded foams, polyurethane elastomers and urethane plastic. The Shore A hardness of presented material was verified. Moreover, the influence of the impact from higher position was investigated. Selected materials were further exposed to the higher temperatures 40 and 50 °C and subsequently tested. The obtained data were evaluated, promising damping material for use in personal protective equipment was selected.

Introduction

The variety of materials used in the industry nowadays is immense. Metals are gradually substituted by other types of plastic-like materials e.g. polyurethanes, polypropylenes, and silicones [1]. Unfortunately, the mechanical behaviour of many of these plastics is not sufficiently described. Therefore, their potential is not fully utilized.

One particularly problematic aspect of such materials is the influence of the temperature. A change of few degrees Celsius can significantly change mechanical behavior. This creates essential problem for personal protective equipment as shown in [2]. In the cited paper, the presented issue is overcomed by an additional composite layer.

This works deals with the comparison of different types of polymeric materials as a material suitable for impact-damping components. Obtained data will serve for the identification of selected material model parameters. Furthermore, the selected material will be used for the improvement of personal protective equipment as shown in [3].

Experiment

A drop test was carried out to evaluate the ability of the material to dampen the force during an impact. Three different types of materials were used, expanding foams (FF-it 8, FF-it 10, FF-it 14, and FF-it 25), polyurethane elastomers (CF 50, VF 60, PMC 770) and solid urethane plastic (T 13a14). Samples were approximately 10 mm high and had 55 mm in diameter. All presented materials were commercially available in form of two-component suspensions and the creation process of the final samples followed the distributor instruction.

According to the distributor, the last two digits in the label of the foams and elastomers refer to their hardness on the Shore A scale. This fact was verified using Wolpert Wilson Durometer (see Fig. 1a). Obtained data are shown in table 1. The Shore A value for the plastic T 13a14 was also measured. Values were obtained from five points on the sample, one centre point and four points around the diameter. The scheme of these points is displayed in Fig. 1b. Measured values correspond to the material labels. The biggest difference was in the case of PMC 770, all measured points gave lesser value than 70. The average Shore A values equal or slightly exceeded material label values. The difference could be caused by the material production process, when the presented materials are developed by a chemical reaction between two different types of components. The samples were produced by the distributor.

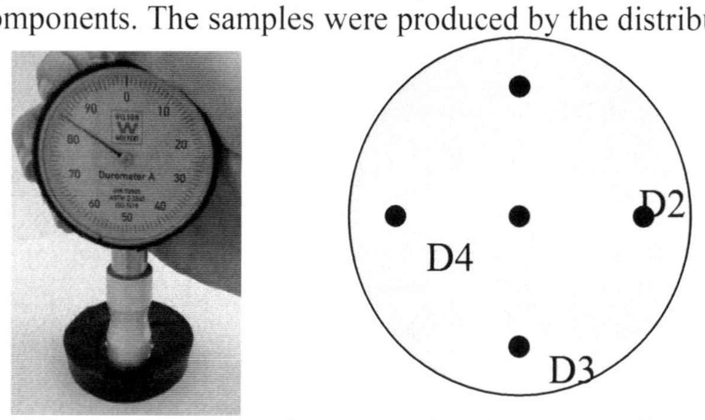

Fig. 1: a) Verification of Shore A value, b) measured positions

Table 1: Shore A values

	Shore A values					
Sample	C	D1	D2	D3	D4	Average
FF-it 8	8	9	11	9	9	9
FF-it 10	12	13	14	13	12	13
FF-it 14	15	14	15	14	14	14
FF-it 25	24	26	27	26	26	26
CF 50	51	52	52	52	53	52
VF 60	60	61	62	61	62	61
PMC 770	67	67	66	66	66	66
T 13a14	85	84	85	85	85	85

The experiment was divided into two parts. In the first part, cylindrical samples were placed on an anvil with a spherical head. The samples were approximately 1 cm high and their diameter was 5.5 cm. A flat head impactor was released from 25 cm height above the anvil. The impactor weight was 5 kg. The temperature during the experiment was 23 ± 2 °C. The transmitted force was measured using the KISTLER 9351B load cell and the impactor displacement was measured by the two Micro-Epsilon optoNCDT 2300-50 lasers on each side of the impactor crossbeam railing. The aim of this experimental part was to determine the damping ability of the material.

The second part of the experiment focused only on the outstanding materials from the first part. These materials were further investigated, using the described drop test with the higher 50 cm release position of the impactor.

The influence of the higher temperature was also investigated in the second part of the experiment. The best samples were conditioned in the temperatures 40 ± 2 °C and 50 ± 2 °C for one hour prior the drop test. The impactor release height was 25 cm.

Results

Fig. 2 shows maximal transmitted forces *F* in the first part of the experiment, the drop test from 25 cm and in temperature of 23 ± 2 °C. FF-it 25 transmitted the lowest force from not only the investigated expanding foams but also elastomers and solid urethane. The best polyurethane elastomer was CF 50, transmitting lower force than the two other elastomers with higher hardness. The maximal transmitted forces increase their values with the increasing hardness of the material. The only solid urethane T 13a14 has the worst ability to dampen the force during the impact.

The foam FF-it 25 and polyurethane elastomer CF 50 displayed the best force damping ability, therefore, were selected for further investigation in the second part of the experiment.

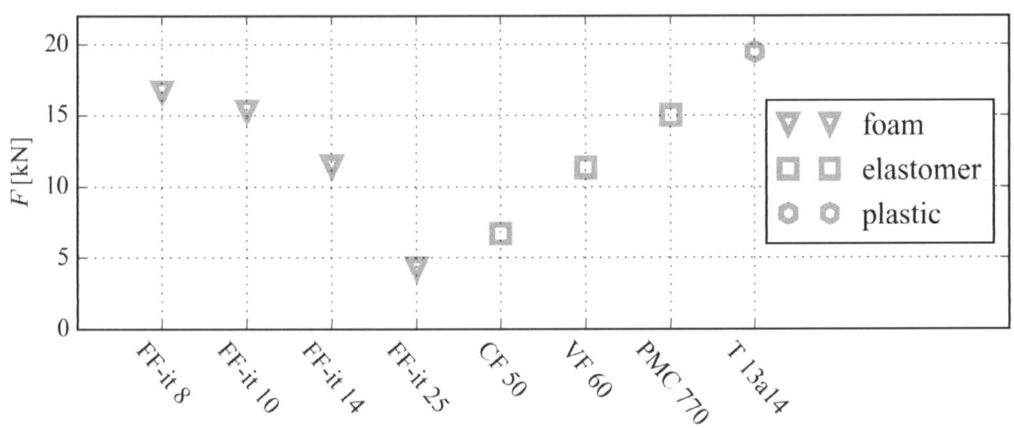

Fig. 2: Maximal transmitted forces

Fig. 3a shows the minimal height *u* of the sample during the drop test (maximal compression occurred). The 10 mm value equals uncompressed sample, 0 mm equals collision of the impactor head and the anvil. Fig. 3b shows the maximal transmitted forces *F*. The change of the minimal height and the maximal transmitted force in dependency on the impactor releasing height are similar for both FF-it 25 and CF 50. The maximal force transmitted by FF-it 25 is lower than the maximal transmitted force by CF 50. The increase of the height may cause ultimate compression of the sample and direct contact between anvil and impactor.

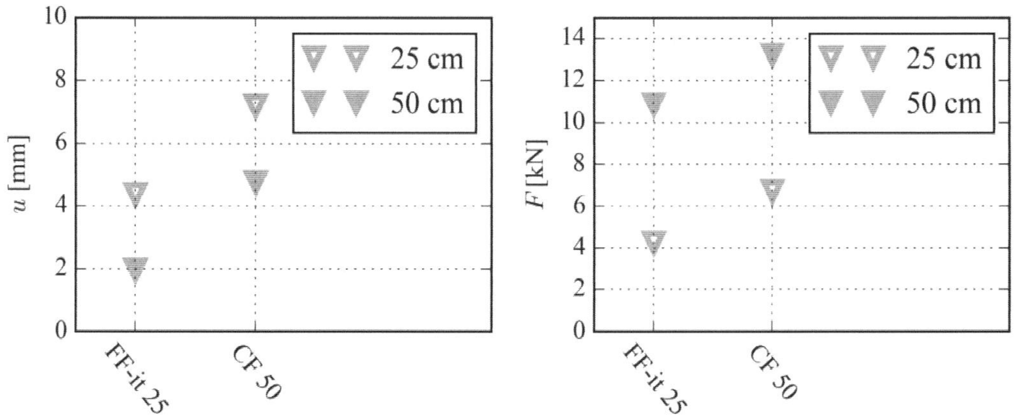

Fig. 3: a) Minimal height of the sample during the drop test, b) maximal transmitted force

Fig. 4a shows the influence of the temperature on the minimal sample height u and Fig. 4b shows the influence of the temperature on the maximal transmitted force F. Fig. 5 and 6 show the comparison of the impactor displacement and the transmitted force in time for all tested temperatures.

The sample fell of the spherical anvil after the first landing in case of the FF-it 25 - 40 °C and the CF 50 - 50 °C drop test, therefore, the impactor collided directly with the anvil, thus obtained data after the first anvil landing were excluded from further work.

The increase of the temperature had negative influence on the mechanical behaviour of the investigated materials. The samples exposed to higher temperature were more compressed during the drop test and transmitted higher forces. This trend is more evident in the case of expanding foam FF-it 25. The impact damping ability of both the expanding foam FF-it 25 and the polyurethane elastomer CF 50 was reduced in dependency on the increasing temperature. Therefore, the impactor jumps reached higher positions.

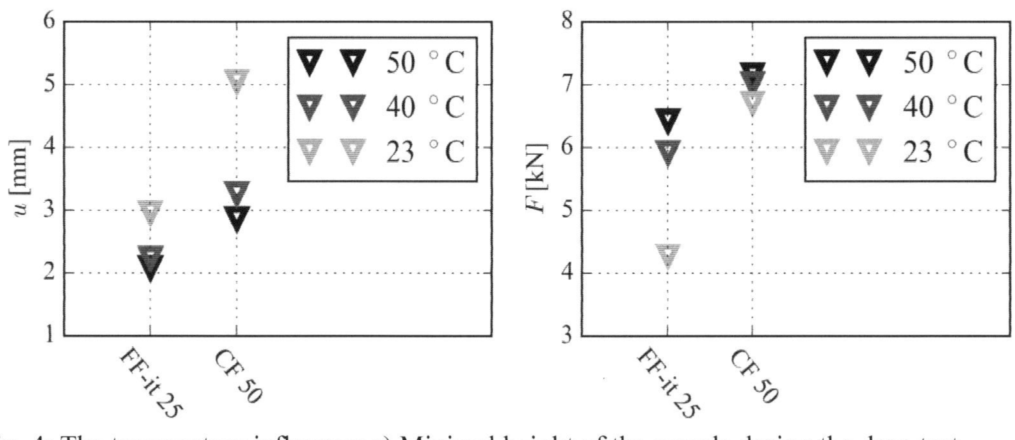

Fig. 4: The temperature influence: a) Minimal height of the sample during the drop test, b) maximal transmitted force

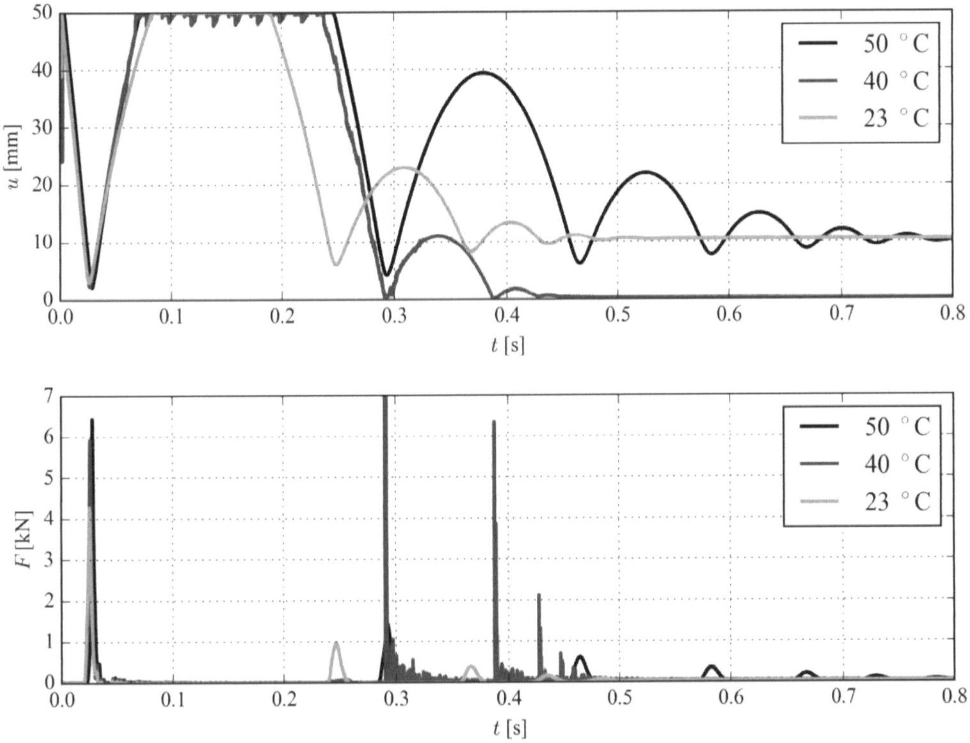

Fig. 5: Temperature influence on FF-it 25

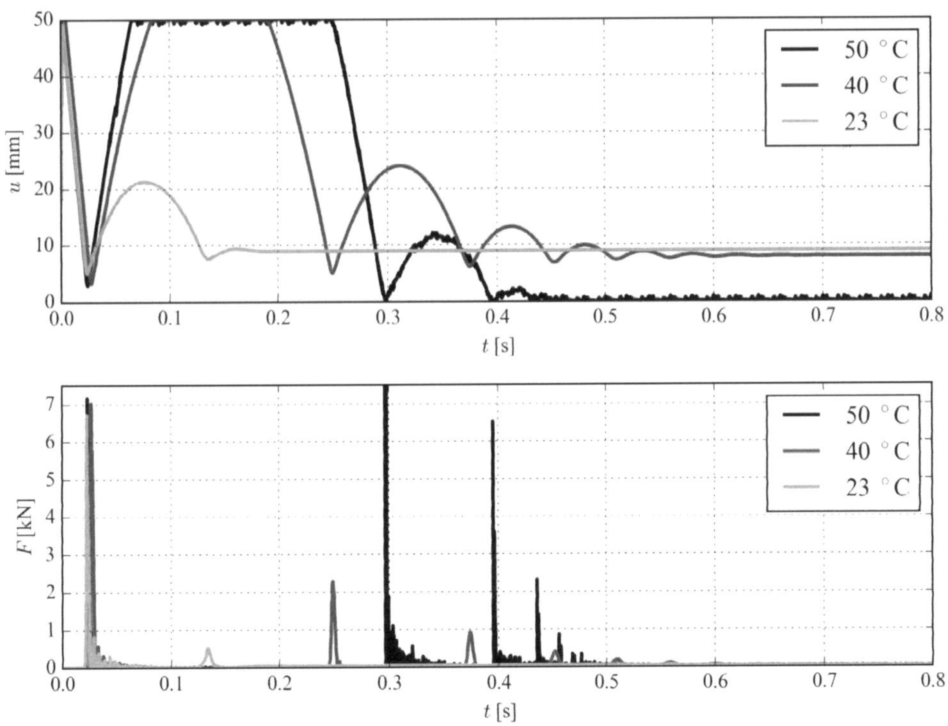

Fig. 6: Temperature influence on CF 50

Conclusions

The hardness specified by the distributor (the label of the materials) corresponded to the Shore A measurements.

The best impact damping ability for 23 °C had the expanded foam FF-it 25. This foam transmitted lower forces for drop test from 25 cm and 50 cm. The CF 50 polyurethane elastomer was better damper than other two elastomers with higher hardness from the Shore A scale. The worst damper was the only solid urethane T 13a14.

The influence of the elevated temperature was negative for FF-it 25. The differences between individual temperatures in case of CF 50 were not significant. Therefore, this material is suitable for an impact damping components, such as motorcycle protectors, where are higher temperatures common matter.

Acknowledgements

This work has been supported by the internal grant project SGS-2019-002 "Biomechanical models of human body, biological tissues and biomechanical processes with application in industry and medicine" and by the project LO1506 of the Czech Ministry of Education, Youth and Sports under the program NPU I.

References

[1] B. Kratochvíl, V. Švorčík, D. Vojtěch. Introduction to study of materials. 1st ed. ISBN 80-7080-568-4. 2005.

[2] R. Kottner, S. Kaňáková, T. Bońkowski, R. Yeung, A. Pukaro Improvement of impact protection by KORDCARBON-CPREG-200-T-3K-EP1- 4-A composite. Materials Today: Proceedings. 10.1016/j.matpr.2020.02.084.

[3] B. J. Ramirez, V. Gupta. Evaluation of novel temperature-stable viscoelastic polyurea foams as helmet liner materials. Materials & Design, Vol. 137, 5 January 2018, pp. 298-304.

Strength Calculation of Prestressed Bolted Connections Analysis using Available Computational Software and FEM method

KANAVAL J.[1,a], CÉZOVÁ E.[1,b]

[1]Department of Designing and Machine Elements; Faculty of Mechanical Engineering; Czech Technical University in Prague; Technická 4, 166 07 Prague 6 - Dejvice, Czech Republic

[a]jan.kanaval@fs.cvut.cz, [b]eliska.cezova@fs.cvut.cz

Keywords: Bolted Connection, Strength Calculation, KissSoft, MITCalc, FEM Analysis

Abstract. This paper deals with analytical calculations and FEM strength analysis of prestressed bolted connections as well as their experimental verification. Design engineers use in the strength calculation technical recommendation of fasteners sellers and fasteners manufacturers when designing of bolted joints. In industrial practice, it is now very often required to calculate the prestressed bolted connections according to the standard *VDI 2230*. [1] The calculation is usually performed using *KissSoft* or *MITCalc* programs on a personal computer.

Introduction

Actual tensile bolts prestress after *controlled tightening* ensures correct functioning of bolted joints during operation. There is defined the concept of controlled tightening of the presstresed bolted joints in the technical support of the company *Bossard*. [2] This tightening method of the presstresed bolted joints respects the variable value of the friction coefficient in the thread and the *stress relaxation* of the prestressed bolted connections after assembly.

Strength Calculation of Prestressed Bolted Joints Using MITCalc Software

There are presented parts of the strength calculation of the prestressed bolted connection for bolt of size *M20*, strength class *12.9*, loaded by static axial force of *55 kN*, in Fig. 1, Fig. 2, Fig. 3 and Fig. 4. [3] The strength calculation was performed using the program *MITCalc*.

?	Input section			
1.0	☑ **Loading of the connection, basic parameters of the calculation.**			
1.6	**Loading of the connection**			
1.7	Maximum axial force	F_{amax}	55000,00	[N]
1.8	Minimum axial force	F_{amin}	55000,00	[N]
1.9	Maximum radial force	F_r	0,00	[N]
2.0	☑ **Operational and mounting parameters of the connection.**			
2.1	Desired coefficient of tightness (prestressing) of the connection	q_a	1,500	
2.2	Desired safety against side shift	q_r	1,000	
2.3	Required residual prestressing of clamped parts of the connectio	F_{2min}	82500,00	[N]
2.4	Desired safety of the bolt at the yield point	n_S	1,200	
2.5	Friction coefficient in threads	μ_t	0,130	

Fig. 1: Part of the strength calculation of the prestressed bolted connection for bolt of size *M20*, strength class *12.9*, loaded by static axial force of *55 kN* (*MITCalc* program) [3]

2.6	Friction coefficient in seating face of the head (nut) of the bolt	μ_c	0,130	
2.7	Friction coefficient between he connected surfaces	μ_a	0,200	
2.8	Consider additonal bending stresses		No	
2.9	Deviation of perpendicularity of the bolt head seating surface	δ	0,100	[°]
2.10	Assembly temperature	T_0	20,0	[° C]
2.11	Consider effects of operational temperature to connection prestressing		No	
2.12	Operational temperature of the bolt	T_b	148,9	[° C]
2.13	Operational temperature of the clamped parts	T_m	148,9	[° C]
2.14	☐ The connection is loaded only at the operational temperature			
2.15	Consider reduction of mounting prestressing using deformation of the joint		Yes	
2.16	Plastic pernament deformation (settlement) of the connection	ΔL	0,01400	[mm]
2.17	**Factor of implementation of operational force**			
2.18	○ Coefficient of implementation of the operational force	n	0,800	
2.19	◉ Distance of the point of action of the force from bolt head	L_{F1}	8,800	[mm]
2.20	Distance of the point of action of the force from the nut	L_{F2}	8,800	[mm]
3.0	☑ **Design, dimensions and material of connected parts.**			
3.1	Design of connected parts		A ... Plate	
3.2	Number of clamped parts	i	2	
3.5	Total height of the clamped parts	L	88,000	[mm]

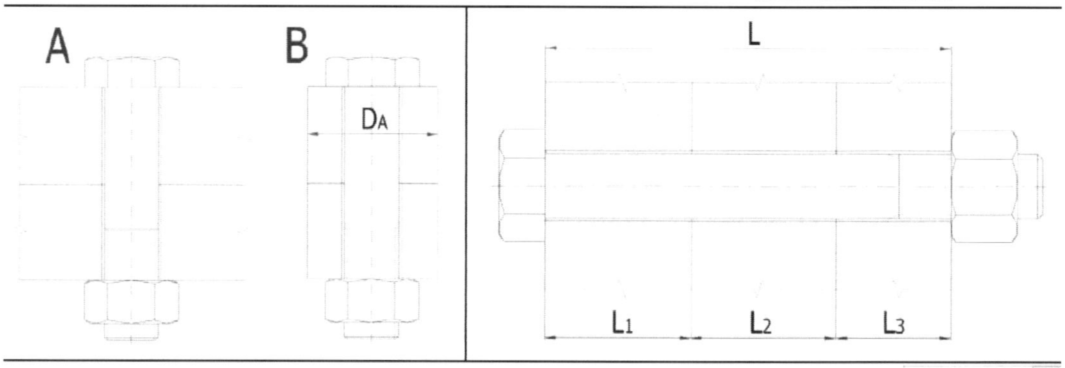

3.6	L_i	E	α	p_A	Material	EN
Part 1	55,000	169000	10,5	800	Ductile cast iron EN-GJS-500-7 [5.3200]	
Part 2	33,000	207000	11,1	490	Structual steel S235JR [1.0038]	

4.3	**Material of the bolt**			
4.4	Strength class (Material) of the bolt		Class 12.9	
4.5	Modulus of elasticity in tension	E	211000	[MPa]
4.6	Ultimate tensile strength	S_u	1220	[MPa]
4.7	Yield strength	S_y	1100	[MPa]
4.8	Heat expansion coefficient	α	11,5	[10^{-6}/°C]
4.9	Density	ρ	7850	[kg/m^3]
4.11	**Thread parameters**			
4.12	Thread type		Metric thread - Coarse series	
4.13	Automatic bolt design			
4.14	Thread size		M20	
4.15	Basic major diameter	d	20,0000	[mm]
4.16	Thread pitch	p	2,5000	[mm]
4.17	Minor diameter	d_r	16,9330	[mm]
4.18	Pitch diameter	d_m	18,3760	[mm]

Fig. 2: Part of the strength calculation of the prestressed bolted connection for bolt of size *M20*, strength class *12.9*, loaded by static axial force of *55 kN* (*MITCalc* program) [3]

Results section

5.0	☑ **Prestressing, force conditions and operational diagram of the connection.**			
5.1	**Stiffness constants of the connection**			
5.2	Stiffness of the connecting bolt	c_b	516875,36	[N/mm]
5.3	Stiffness of the clamped parts	c_m	3905885,41	[N/mm]
5.4	Resulting stiffness od the group of surcharged parts of the joint	c_1	503868,37	[N/mm]
5.5	Resulting stiffness of the group of relieved parts of the joint	c_2	4852461,57	[N/mm]
5.6	**Assembly preload of the bolt connection**			
5.7	Maximum axis component of the operational force	F_a	55000	[N]
5.8	Maximum radial component of the operational force	F_r	0	[N]
5.9	Minimum needed clamping force for transfer of the radial force	F_{cmin}	0	[N]
5.10	Part of the operational force additionally loading the bolt	ΔF_1	5173,83	[N]
5.11	Part of the operational force relieving the clamped parts	ΔF_2	49826,17	[N]
5.12	Minimum needed assembly preload of the joint	F_{0min}	138717	[N]
5.13	Assembly preload of the joint	F_0	138717,0	[N]
5.14	Tightening torque	M	452,06	[Nm]
5.15	**Force conditions of the loaded connection**			
5.16	Change of prestresssing due to heating of the connection	ΔF_{0T}	0,00	[N]
5.17	Loss of prestressing due to deformation of the connection	ΔF_{0L}	-6390,57	[N]
5.18	Operating prestressing of the joint	F_0'	132326,43	[N]
5.19	Residual prestressing of clamped parts of the connection	F_2	82500,26	[N]
5.20	Resulting internal axis force in the bolt	F_1	137500,26	[N]
5.21	Coefficient of tightness (prestressing) of the connection	q_a	1,500	
5.22	Safety against side shift	q_r	0,000	

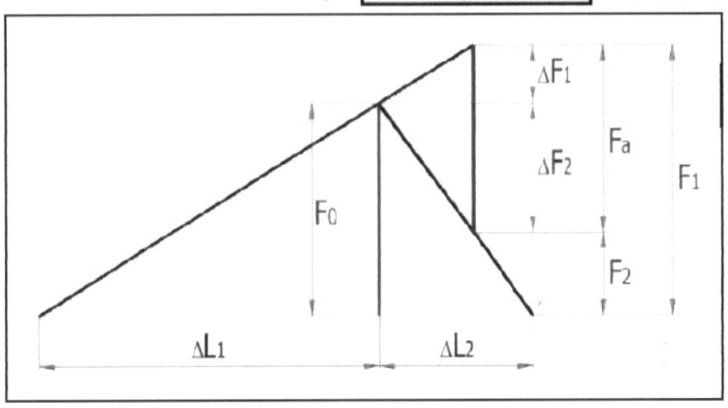

6.0	☑ **Strength checks of statically loaded bolt connections.**			
6.1	**Strength check of connections in the working state**			
6.2	Internal axis force in the bolt	F_1	137500,26	[N]
6.3	Tensile stress in bolt core from the axis force	σ	561,70	[MPa]
6.4	Torsional stress in bolt core from tightening moment	τ	205,60	[MPa]
6.5	Additional bending stress	σ_b	0,00	[MPa]
6.6	Resulting reduced stress in the bolt core	σ_{red}	589,24	[MPa]
6.7	Yield point of the bolt material	S_y	1100	[MPa]
6.8	Safety at yield point	n	1,87	
6.9	**Strength check of connections in the assembly state**			
6.10	Assembly preload of the joint	F_0	138717,00	[N]
6.11	Tensile stress in bolt core from the assembly preload	σ	566,67	[MPa]
6.12	Resulting reduced stress in the bolt core	σ_{red}	669,27	[MPa]
6.13	Allowable stress (90% Sy)	σ_A	990	[MPa]

Fig. 3: Part of the strength calculation of the prestressed bolted connection for bolt of size *M20*, strength class *12.9*, loaded by static axial force of *55 kN* (*MITCalc* program) [3]

No.	Parameter	Symbol	Value	Unit
6.14	**Check of pressure in seating face of the bolt head**			
6.15	Pressure in the bolt head (nut) seating face	p	381,42	[MPa]
6.16	Permitted pressure in the marginal clamped part	p_A	490	[MPa]
6.17	**Strength check of connections for maximum prestressing**			
6.18	Maximum operating prestressing of the joint	$F_0'_{max}$	138717,00	[N]
6.19	Maximum internal axis force in the bolt	F_{1max}	143890,83	[N]
6.20	Tensile stress in bolt core from the maximum axis force	σ_{max}	587,80	[MPa]
6.21	Resulting reduced stress in the bolt core	σ_{red}	614,18	[MPa]
6.22	Maximum pressure in the bolt head (nut) seating face	p_{max}	399,14	[MPa]
8.0	**Assembly parameters of the connection.**			
8.1	**Assembly preload of the joint**			
8.2	Minimum assembly preload	F_{0min}	138717,0	[N]
8.3	Tightening factor	α_A	1,70	
8.4	Maximum assembly preload	F_{0max}	235818,9	[N]
8.5	**Tightening torque**			
8.6	Friction coefficient in threads (min/max)		0,100 / 0,160	
8.7	Friction coefficient in seating face of the head (nut) of the bolt (min/max)		0,100 / 0,160	
8.8	Minimum possible tightening torque	M_{min}	360,31	[Nm]
8.9	Maximum possible tightening torque	M_{max}	924,65	[Nm]
8.10	**Strength check of connections in the assembly state**			
8.11	Tensile stress in bolt core from the assembly preload	σ_{max}	963,34	[MPa]
8.12	Torsional stress in bolt core from tightening moment	τ_{max}	410,56	[MPa]
8.13	Resulting reduced stress in the bolt core	σ_{red}	**1197,37**	[MPa]
8.14	Allowable stress (90% Sy)	σ_A	990	[MPa]
8.15	**Strength check of connections in the working state**			
8.16	Operating prestressing of the joint	$F_0'_{max}$	229428,3	[N]
8.17	Internal axis force in the bolt	F_{1max}	234602,2	[N]
8.18	Tensile stress in bolt core from the axis force	σ_{max}	958,37	[MPa]
8.19	Resulting reduced stress in the bolt core	σ_{red}	1022,19	[MPa]
8.20	Yield point of the bolt material	S_y	1100	[MPa]
8.21	Safety at yield point	n	**1,08**	
8.22	**Check of pressure in seating face of the bolt head**			
8.23	Pressure in the bolt head (nut) seating face	p_{max}	**650,77**	[MPa]
8.24	Permitted pressure in the marginal clamped part	p_A	490	[MPa]
8.25	**Fatigue check of the connection**			
8.26	Medium cycle stress in the thread core	σ_m	958,37	[MPa]
8.27	Cycle stress amplitude in the thread core	σ_a	0,00	[MPa]
8.28	Max. fatigue strength of the bolt for the given course of loading	σ_A	17,38	[MPa]
8.29	Dynamic safety in tension	n_σ	---	

Fig. 4: Part of the strength calculation of the prestressed bolted connection for bolt of size *M20*, strength class *12.9*, loaded by static axial force of *55 kN* (*MITCalc* program) [3]

The following user settings of *MITCalc* program have been set: calculation of plate stiffness by the method of pressure cones according to *VDI 2230*; the torsional stress reduction coefficient was chosen to be 0,5; the calculated thread cross-section was chosen as the critical cross-section of the bolt (thread). The static external loading force *55 kN* for this strength calculation using *MITCalc* program was chosen so that the resulting internal force and tension in the bolt correspond to the results of the FEM analysis for the bolt *M20*, strength class *12.9*, see reference [4]. Subsequently, the results of the strength calculation performed in the *MITCalc* program were compared with the results of the stress analysis performed by the

FEM method. Creating FEM model of a bolted joint, the choice of main characteristics of the finite element model and more details about the performed FEM analysis are also described in reference [4].

FEM Analysis of the Prestressed Bolted Connections

It is possible to verify of the prestressed bolted connection design in two ways:
1) with the using of FEM analysis of the prestressed bolted connection (see Fig. 5),
2) by experimental evaluation or experimental testing - e.g. using a special sensors.

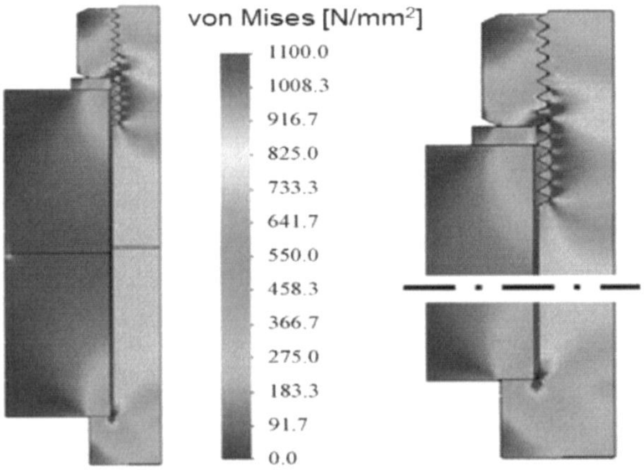

Fig. 5: Examples of FEM analysis of the prestressed bolted connection for bolt of size *M20*, strength class *12.9*, loaded by external static axial force of *55 kN* [4]

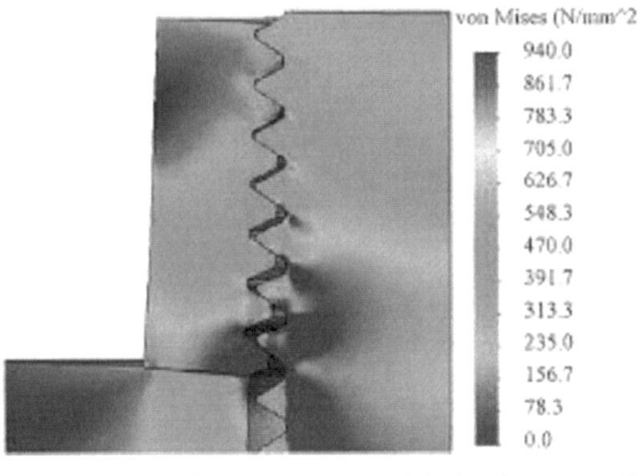

Fig. 6: Examples of FEM analysis of the prestressed bolted connection for bolt of size *M24*, strength class *10.9*, loaded only by assembly static preload of *312 kN* [4]

There is presented equivalent stresses in the bolted connection for bolt of size *M20*, strength class *12.9*, loaded by external static axial force of *55 kN*, in Fig. 5. The stress scale maximum was set to 1100 N/mm^2. And there is shown the radial deformation of the loaded nut in the bolted connection for bolt of size *M24*, strength class *10.9*, loaded only by assembly static preload of *312 kN*, in Fig. 6.

FEM analysis was performed at our Department of designing and machine elements for other cases of prestressed bolted connections recently (see Fig. 7).

The FEM full-model was created in SW *ANSA (Beta CEA systems)* and evaluation of the calculated data was created in SW *META (Beta CAE systems)*. FEM calculations were performed using school license SW *ABAQUS*. *ANSA* is an advanced multidisciplinary *CAE pre-processing tool* that provides all the necessary functionality for full-model build up, from CAD data to ready-to-run solver input file, in a single integrated environment. *META* is a powerful, intuitive *CAE post-processing tool* that works seamlessly with the leading finite element analysis solvers.

The CAD data of the bolt and nut did not contain a detailed modeled thread part. For the purposes of the FEM analysis, it was necessary to model the threaded part manually. The bolt and nut were created using *TETRA SOLID* elements of size 2 mm. In the area of the threads and under the screw head, the size of the elements is reduced to 1 mm. The connected boards are formed using *HEXA* and *PENTA SOLID* elements of size 5 mm. In the area of the screw hole, the size is locally softened to a size of 2 mm. Boundary conditions were realized by removing all degrees of freedom using the *BOUNDARY FIXED* function on the edges of the joined boards. During assembly and under loading, the screw connection elements interact with each other. In the FEM model, these interactions are modeled using the *CONTACT_PAIR* function with the corresponding values of the friction coefficients in the threads (μ_Z), under the screw head (μ_H) and under the nut (μ_M). The preload was introduced into the bolt using the *PRE-TENSION_SECTION* function. In the first step of the calculation, the required assembly preload of the prestressed bolted connection was entered. The area of external operating force was 6 mm below the surface of the joined boards (see Fig. 7). The external operating force was thus introduced symmetrically in these places under the screw head and under the nut. The magnitude of the force was budgeted to the appropriate number of nodes. For the case of static external force, the value of F_A was considered. The calculation consists of two separate steps. In the first step, the specified mounting preload was introduced into the bolt. In the second step, the preload was locked in the bolt and an external operating force was introduced into the bolt.

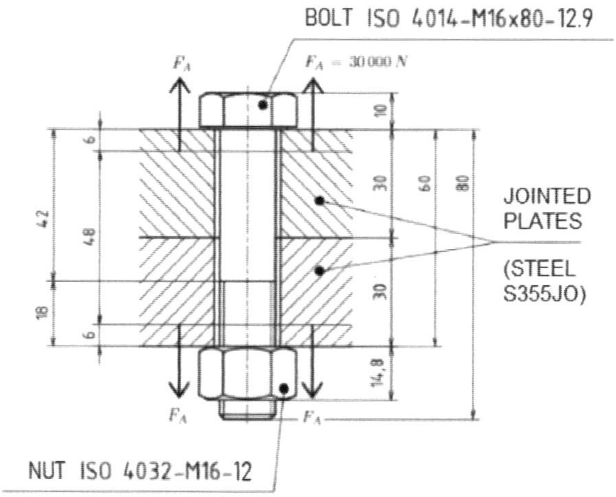

Fig. 7: Design of the prestressed bolted connection with bolt of size *M16*, strength class *12.9*, loaded by external static axial force of *40 kN* as a model for FEM stress analysis [Author]

There is presented equivalent stresses in the bolted connection for bolt of size *M16*, strength class *12.9*, loaded by external static axial force of *40 kN*, in Fig. 8 and Fig. 9. The stress scale maximum was set to 600 N/mm^2, (respectively 460 N/mm^2).

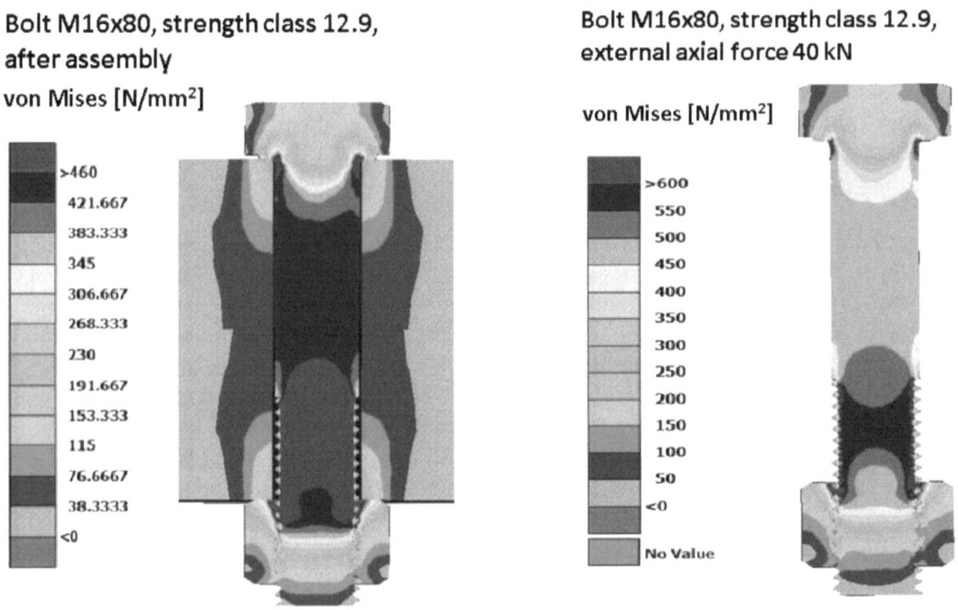

Fig. 8: Examples of FEM analysis of the prestressed bolted connection for bolt of size *M16*, strength class *12.9*, loaded by external static axial force of *40 kN*; equivalent stresses in the bolted connection for bolt of size *M16*, strength class *12.9*, loaded by external static axial force of *40 kN* [Author]

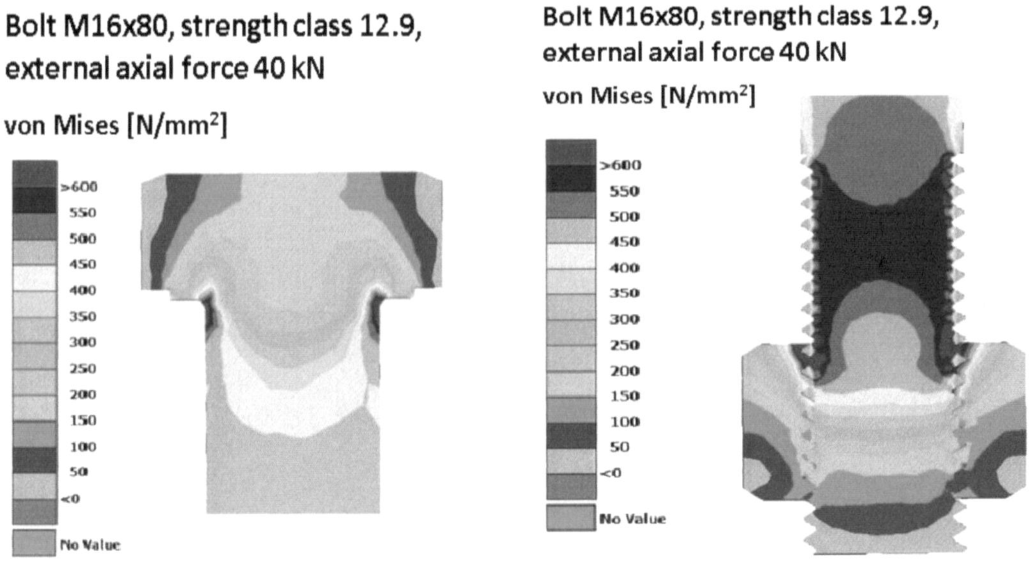

Fig. 9: Examples of FEM analysis of the prestressed bolted connection for bolt of size *M16*, strength class *12.9*, loaded by external static axial force of *40 kN* (detailed view); equivalent stresses in the bolt of size *M16*, strength class *12.9*, loaded by external static axial force of *40 kN* [Author]

Testing Equipment for Experimental Analysis of the prestressed bolted connection

Comparing the results of strength calculations of prestressed bolted joints using available software and stress analysis performed by the FEM method and experimental verification of the results is necessary in terms of refinement of design methods and strength control of prestressed bolted joints and in terms of development of modern fasteners.

In the development of threaded fastener materials, surface finishes, plating, coatings, and thread locking adhesives, it is necessary to measure and control the friction coefficients on both the threads and underhead regions of the fasteners. [1, 5]

Also, to verify the assembly process of bolted connections and to verify strength of the bolted connections is necessary to perform experiments on special test equipment. There is presented typical laboratory fastener test machine in Fig. 10. [5]

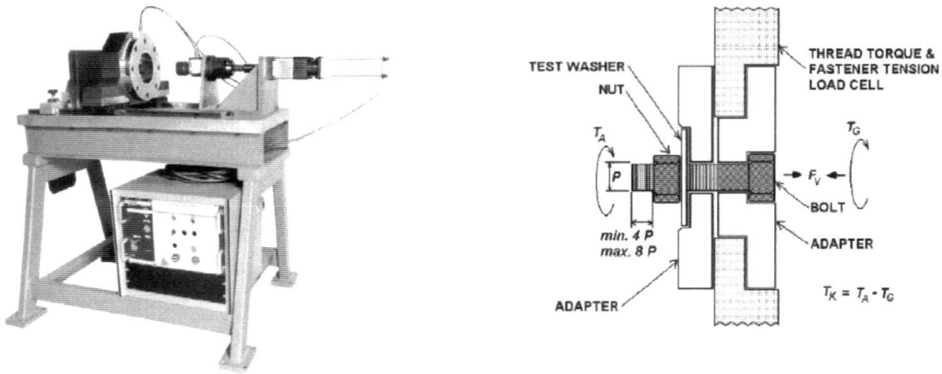

Fig. 10: Laboratory fastener test machine, typical, (on the left side). Torque-tension research head, (on the right side) [5]

Testing equipment for complex analysis of screw fasteners was designed, manufactured and is tested currently in the laboratory of our Department of designing and machine elements to verify the correctness of strength calculations and FEM analysis (see Fig. 11).

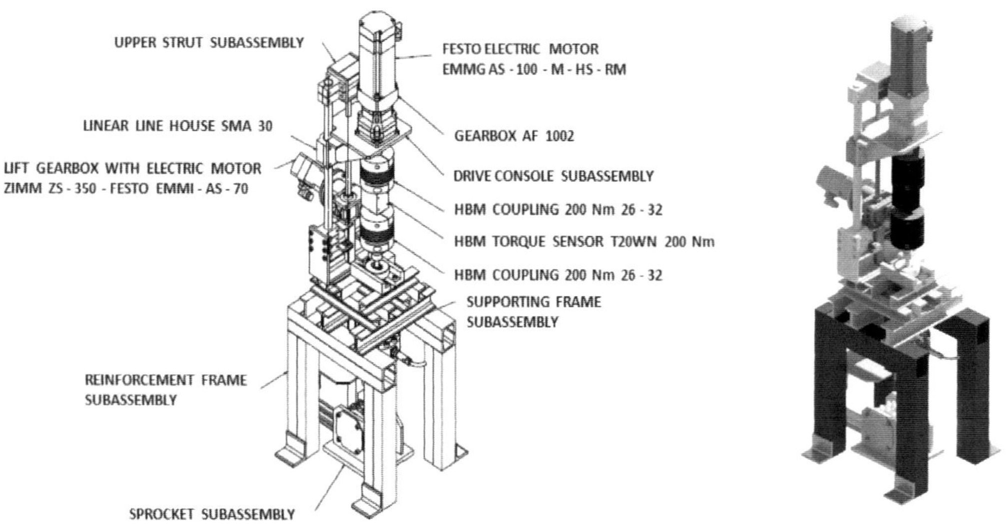

Fig. 11: Testing equipment for complex analysis of screw fasteners [Author]

On this testing equipment can be detected:

- value of the friction coefficient in the thread,
- value of the friction coefficient under bolt head (under nut),
- tension axial prestressed reduction (stress relaxation) after assembly in time,
- to evaluate the characteristics of individual fastener tightening processes, (Torque-angle curves),
- to verify the strength of the bolt connection.

Technical parameters of this testing equipment:

- maximum testing torque moment	185 Nm,
- maximum testing axial force	100 kN,
- maximum testing rotation speed	97 min^{-1},
- maximum testing stroke	300 mm.

Conclusions

Strength calculations of selected prestressed bolted connections were realized using available calculation software (*MITCalc* program) and FEM analysis of selected prestressed bolted connections were realized too. Results of strength calculations of prestressed bolted connections using *MITCalc* program and results of stress FEM analysis for prestressed bolted connection for bolt of size *M20*, strength class *12.9*, loaded by external static axial force of *55 kN* were compared.

The following results were obtained by calculating the prestressed bolted connection in the assembly state using the *MITCalc* program: tensile stress in bolt core from the assembly preload $\sigma_{max} = 963{,}34\ N/mm^2$, torsional stress in bolt core from tightening moment $\tau_{max} = 410{,}56\ N/mm^2$, reduced stress in the bolt core $\sigma_{red} = 1197{,}37\ N/mm^2$. And the following results were obtained by calculating the prestressed bolted connection in the working state using the *MITCalc* program: tensile stress in bolt core from the axis force $\sigma_{max} = 958{,}37\ N/mm^2$, torsional stress in bolt core from torque $0{,}5 \cdot \tau_{max} = 205{,}28\ N/mm^2$, reduced stress in the bolt core $\sigma_{red} = 1022{,}19\ N/mm^2$ (see Fig. 4). If we compare the calculated reduced stresses with the results of the stress analysis performed by the FEM method, then the values of the reduced stress at the most exposed place of the bolt (place of first supporting thread) do not differ significantly. The difference between the reduced stresses determined by the FEM stress analysis and the reduced stresses determined using the *MITCalc* program was estimated at a maximum of 5 percent (see Fig. 4 and Fig. 5). Experimental verification of strength calculations and FEM analysis results of prestressed bolted connections will be performed in the laboratory of our Department using the developed testing equipment for complex analysis of screw fasteners in the near future (see Fig. 11).

Acknowledgements

Acknowledgments belong to the student of the combined form of bachelor's study, Mrs. Mgr. Monica Blaškovič for performing FEM analysis of prestressed bolted connections.

References

[1] Guideline VDI 2230, Systematic calculation of high duty bolted joints, Part 1, Joints with one cylindrical bolt, 2003.

[2] Bossard CZ s.r.o. Technical support. Brno. [2020-03-08]. Available from https://www.bossard.com/cs/assembly-technology-expert/technical-information-and-tools/technical-resources/.

[3] MITCalc - Mechanical, Industrial and Technical Calculations. [2020-03-08]. Available from https://www.mitcalc.com.

[4] Molnar, L. et al. Stress Analysis of Bolted Joints Part I. Numerical Dimensioning Method. Department of Machine and Product Design, Budapest University of Technology and Economics, Budapest, Hungary, 2014.

[5] R. Shoberg, Engineering fundamentals of threaded fastener design and analysis, MI 48335, USA: RS Technologies, a Division of PCB Load & Torque, Inc., 2010.

EAN 2020
58th conference on experimental stress analysis

The Modified Elastomagnetic Sensor Intended for a Quick Application on an Existing Prestressed Concrete Structures

KLIER T.[1,a], MÍČKA T.[1,b], POLÁK M.[2,c], PLACHÝ T.[2,d], HEDBÁVNÝ M.[3,e], KREJČÍKOVÁ L.[3,f]

[1]Pontex, spol. s r. o., Bezová 1658, 147 14, Prague 4, Czech Republic

[2]Faculty of Civil Engineering, Czech Technical University in Prague, Thákurova 7, 166 29, Prague 6, Czech Republic

[3]Freyssinet CZ a. s., Zápy 267, 250 01, Brandýs nad Labem, Czech Republic

[a]klier@pontex.cz, [b]micka@pontex.cz, [c]polak@fsv.cvut.cz, [d]plachy@fsv.cvut.cz, [e]hedbavny@freyssinet.cz, [e]krejcikova@freyssinet.cz

Keywords: Magnetoelastic method, Elastomagnetic sensor, Tensile force, Prestressed strand, Prestressed cable

Abstract. The fully equipped elastomagnetic sensor is assembled from six basic components which are namely the sensor body, the primary coil, the secondary coil No. 1, the system of Hall probes, the secondary coil No. 2 and the steel shield of the sensor. Its additional installation on a tension steel element of an existing structure (as is typically a prestressed cable of a concrete bridge) involves the in situ process of the winding of its three coils that is relatively laborious and time-consuming. There is usually a large number of the tension elements on the investigated existing structure which should be investigated during the detailed construction inspection. The development of the modified elastomagnetic sensor that is described in this paper was based on an idea to reduce the laboriousness of the in situ process and also to obtain an effective tool for the quick detection and selection of which ones of the tension elements are most important for further detailed analysis and for installation of the fully equipped elastomagnetic sensors. In order to fulfil this idea, the modified elastomagnetic sensor was designed so that its each part can be easily mounted and then removed from the investigated tension element.

Introduction

In civil engineering practice, nowadays there are more often designed and built special engineering and building structures in which various structural elements are used that are loaded by large tensile forces as are, for example, internal or external cables of prestressed concrete bridges, stays of cable stayed bridges, hangers of arch bridges, suspension cables of suspended roofs, tensioning cables of tensile fabric membrane structures, ground and rock anchors. The forces in these elements, in many cases, significantly influence the reliability of the whole structure and they usually fluctuate considerably in time. Correct information about the actual force value is very important for accurate assessment of the structural reliability at any of the life cycle stages of the entire structure both during its construction or operational state.

There are also many similar existing structures with the same problem. The prestressed concrete bridges with internal cables or strands that are more than twenty years old form a special group. The knowledge of forces in internal prestressed reinforcements of these bridges, that is accurate enough, is important for thorough preparation of their reconstruction.

In civil engineering practice, five experimental techniques are generally used for determination of axial tensile forces in the important structural elements. The two of them (the direct measurement of the force by a pre-installed load cell and the approach based on a strain measurement with strain gauges) can be used only for evaluation of the total value of the tensile force in the observed structural elements during an experiment by which the applied sensors were installed before the elements were activated.

However, the next three of them (the vibration frequency method [1,2], the force determination in a flexible structural element based on the relation between the transverse force and the caused transverse displacement [3,4], and the magnetoelastic method [5–11]) are usable for new experiments on existing structures that are for some time already in service. It means that the observed structural elements remain activated during the experiment preparation, realization and also after the experiment completion.

The another advantages and disadvantages of all five mentioned techniques are discussed in more detail in the reference [8].

Motivation and objectives of the study

The fully equipped elastomagnetic sensor was introduced in detail in references [5,6]. It is assembled from six basic components that are the sensor body, the primary coil, the secondary coil 1, the system of Hall probes, the secondary coil 2 and the steel shield of the sensor.

The parts of the sensor body are printed on a 3D printer. The sensor proportions are always adapted to particular dimensions of a studied structural element. If the elastomagnetic sensor is to be installed on an already activated structural element on an existing structure, its body is consequently assembled, in a relatively easy way, on the element during preparation phase of the experiment.

However, the subsequent steps of the elastomagnetic sensor production procedure involves the in situ process of the winding of the three coils that is relatively laborious and time-consuming which last for about three hours in relatively comfortable conditions.

Moreover, there is usually a large number of the structural elements on the investigated existing structure (prestressed cables or strands on a prestressed concrete bridge, for example) which should be potentially investigated during the detailed construction inspection.

The fundamental motivation for the development of the modified elastomagnetic sensor that is described in this paper was based on an idea to reduce the laboriousness of the in situ production process.

The next reason for the research was to obtain also an effective tool for the quick detection and selection of which ones of the studied structural elements are most important for further detailed analysis and for installation of the fully equipped elastomagnetic sensors. For example, it means to select the elements with the smallest or largest axial tensile force.

In order to fulfil these ideas, the modified elastomagnetic sensor was designed so that its each part can be easily mounted and then removed from the investigated structural element and that no one coil has to be winded in situ.

The brief description of the developed elastomagnetic sensor

Two variants of the removable elastomagnetic sensor (see Fig. 1) were developed and studied to fulfil the above mentioned motivation and objectives of the study.

The first variant of the removable sensor (see Fig. 1) applies two permanent neodymium magnets as the source of variable magnetic field. The parameters of the magnetic field are then measured using four Hall probes. The sensor consists of two basic parts. The first one is intended for attachment to the investigated structural element (see Fig. 2) by four disposable cable ties (see Fig. 1) and the second part is the moving one (see Fig. 2).

The four Hall probes are fixed in the first stationary part of the sensor. Two Hall probes are situated in its centre (see Fig. 2) and they measure the properties of magnetic field in a direction parallel with the element axis. The next two Hall probes are located on axes of the magnets (see Fig. 2) and they monitor the magnetic field properties in a direction perpendicular to the element axis.

Fig. 1: Two studied variants of the removable modified elastomagnetic sensor with two permanent neodymium magnets (on left) and with the portable primary coil (on right) which are installed on the prestressed strand

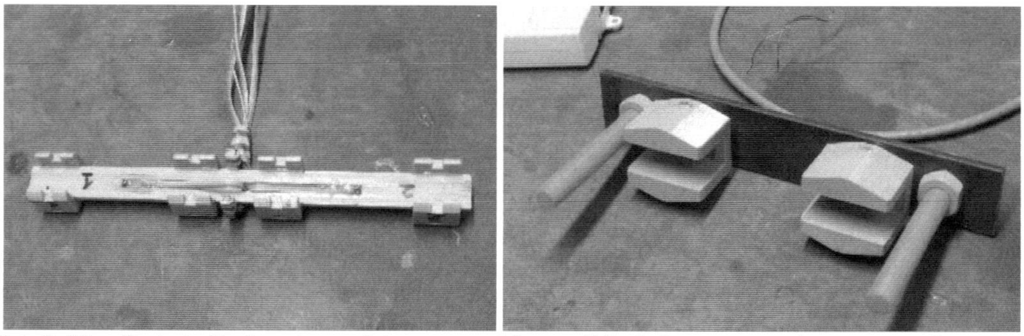

Fig. 2: The first studied variant of the removable modified elastomagnetic sensor – the sensor part intended for attachment to the element (on left) and the moving part with two permanent neodymium magnets (on right)

The second moving part of the sensor includes two permanent neodymium magnets and it can move in the direction perpendicular to the investigated element and in this way can change the magnetic field. The movement is controlled by two orange screws (see Fig. 1 and Fig. 2) in the case of the applied primary prototype of the sensor.

The second variant of the removable sensor (see Fig. 1) applies a portable primary coil wrapped around a steel horseshoe-shaped core (see Fig. 3) for developing sufficiently intense variable electromagnetic field. The parameters of the magnetic field are measured applying two pole secondary coils and two Hall probes. The pole secondary coils are winded in orange cases that are put on the both ends of the steel core (see Fig. 1 and Fig. 3) and they measure the magnetic field properties in a direction perpendicular to the element axis.

The two Hall probes (see Fig. 3) are fixed in a black holder that is mounted on the studied structural element so that the Hall probes are situated in the centre plane of the sensor perpendicular to the element (see Fig. 2 and Fig. 3) and they measure the properties of magnetic field in a direction parallel with the element axis.

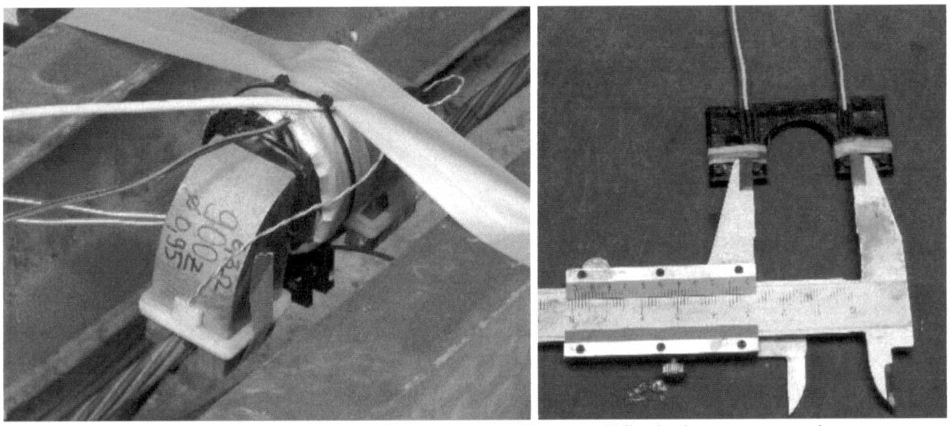

Fig. 3: The second studied variant of the removable modified elastomagnetic sensor – the steel horseshoe-shaped core with the wrapped primary coil and the pole secondary coil put on it (on left) and two Hall probes fixed in the part of their holder (on right)

The description of the experiment arrangement

The both variants of the removable sensor were verified by practically identical experiments (see Fig. 4). A prestressed strand was fastened in a steel test frame and next the particular variant of the removable sensor was attached to the strand. Then a calibrated hollow hydraulic jack was used to apply the tension force into the strand in several force steps and simultaneously the influence of the mechanical stress to the magnetoelastic characteristics of the prestressed strands were observed and evaluated.

Fig. 4: View of arrangement of the experiments with the hollow hydraulic jack, the steel test frame, the fastened prestressed strand and the first variant of the removable elastomagnetic sensors attached to the strand (on left) and the second one (on right)

The obtained results

During the results evaluation, the experiments based on the modified magnetoelastic method, which is described in more details in references [5–8], apply usually measured hysteresis loops, that include also their sections with the technical saturation points. This procedure allows to realize experiments independently on application of particular elastomagnetic sensors or other devices.

For a deeper understanding of relationships, the both developed variants of the removable sensor were also analysed theoretically. Two created theoretical models were built in the software Ansys Maxwell 3D (see Fig. 6, Fig. 9 and Fig. 10) and they both correspond to the real arrangement and conditions of the both experiments.

The default boundary conditions that are available in Ansys Maxwell 3D software were used on boundaries of the resolved regions for both theoretical models. And the flux normal conditions were, of course, applied on the plane of symmetry. The resolved regions were about three times larger than investigated sensors in each dimension. Likewise, the default parameters of the material applied in the models were adopted from the software material library. Namely, the ferromagnetic „Steel_1010“ was used as a strand material. Suitability of this material and its properties for the purpose of the realized theoretical analysis was verified in the course of some former authors' works [5,6].

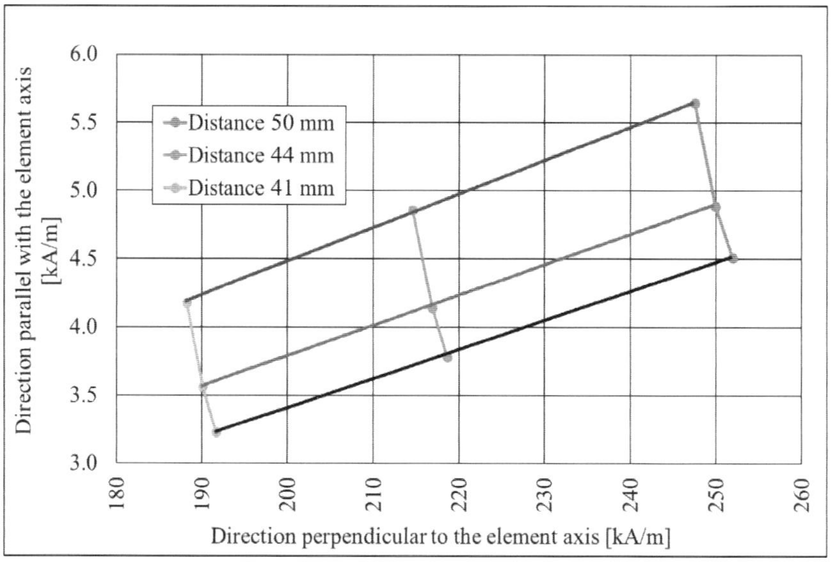

Fig. 5: The results obtained by application of the first studied variant of the removable modified elastomagnetic sensor – the dependence between measured values of the magnetic field intensity "H" for three tension force steps and three magnet positions

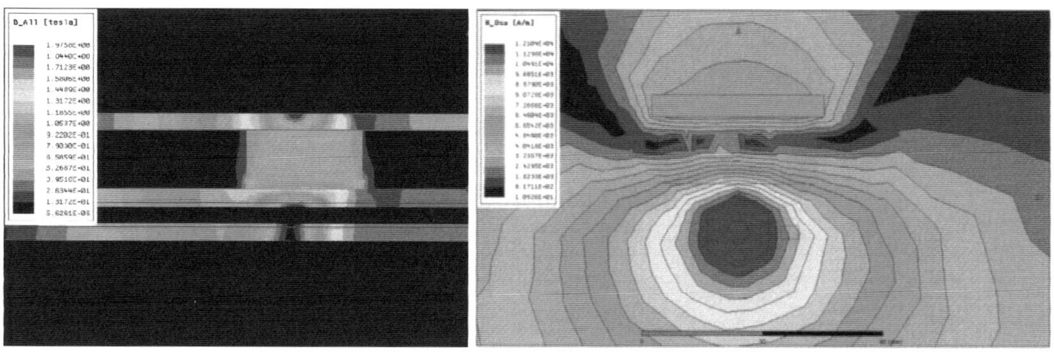

Fig. 6: The magnetic induction field "B" (on left) and the magnetic field intensity "H" (on right) determined on the theoretical model of the first variant of the sensor (the variant with two permanent neodymium magnets) installed on the studied prestressed strand

The experimental results obtained by application of the first variant of the removable elastomagnetic sensor (the variant with two permanent neodymium magnets) show that there is an applicable dependence between measured parameters of the magnetic field and the value of tension force in the strand as is demonstrated in Fig. 5.

The dependence between values of the magnetic field intensity "H" measured by the Hall probes for three tension force steps and three magnet positions relative to the strand is depicted in Fig. 5 where the "H" values, which were measured in the direction perpendicular to the strand axis in the points located on axes of the magnets, are plotted on the horizontal axis and the "H" values, which were measured in the direction parallel with the strand axis in the sensor centre, are plotted on the vertical axis.

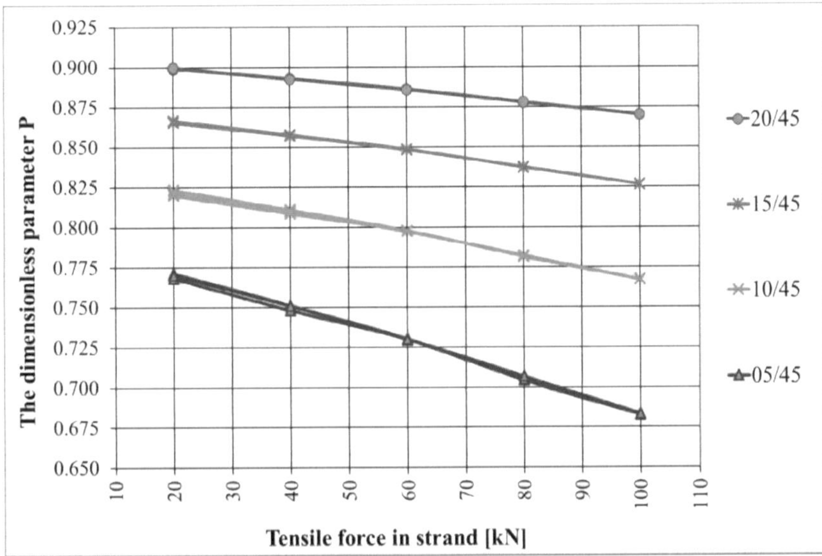

Fig. 7: The results obtained by application of the second studied variant of the removable modified elastomagnetic sensor - comparison of the different resultant dimensionless parameter P

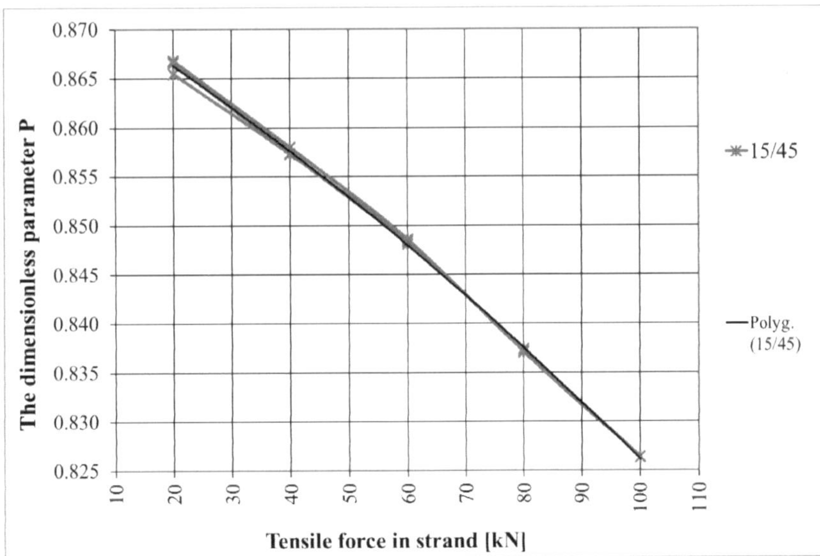

Fig. 8: The results obtained by application of the second studied variant of the removable modified elastomagnetic sensor - the relation between the tensile force in the observed strand and the chosen resultant dimensionless parameter P 15/45

The three lines in Fig. 5 (the red, green and black one) correspond to three specific set values of tensile force in the strand. The red line is related to the force 100 kN, the green line to 60 kN and the black one to 20 kN.

However, the required state of the studied strand, when its technical saturation would be reached in its central section, was not achieved. Moreover, the theoretical analysis revealed that this state could not be also obtained by the application of stronger neodymium magnets in the course of the experiment.

In the course of the application of the second variant of the removable elastomagnetic sensor (the variant with the portable primary coil) the required state of the studied strand with its technical saturation was achieved. The experimental results revealed that the second sensor variant can be used for the intended purpose to select quickly which tension elements are most important for further detailed analysis. However, the precision of the gained results is lower than for the fully equipped elastomagnetic sensors.

The influence of the defined stress in the investigated strand on the parameters of the magnetic field was studied for five steps of the tension force that were 20 kN, 40 kN, 60 kN, 80 kN and 100 kN. For each used force step, the hysteresis loop was measured and determined, results presented in Fig. 7 were evaluated based on the experimental data about magnetic field intensity "H" measured by two applied Hall probes located in the centre plane of the sensor and about magnetic induction “B” monitored by the two pole secondary coils.

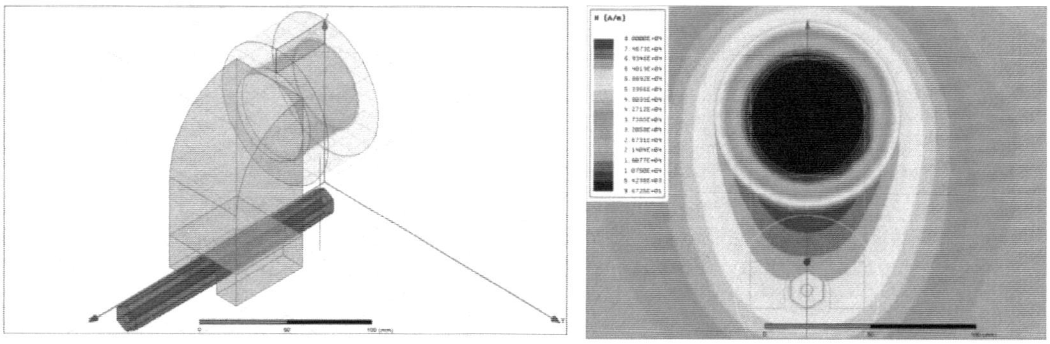

Fig. 9: The view of the theoretical model of the second sensor variant installed on the studied strand (on left) and the magnetic field intensity “H” calculated on this model (on right)

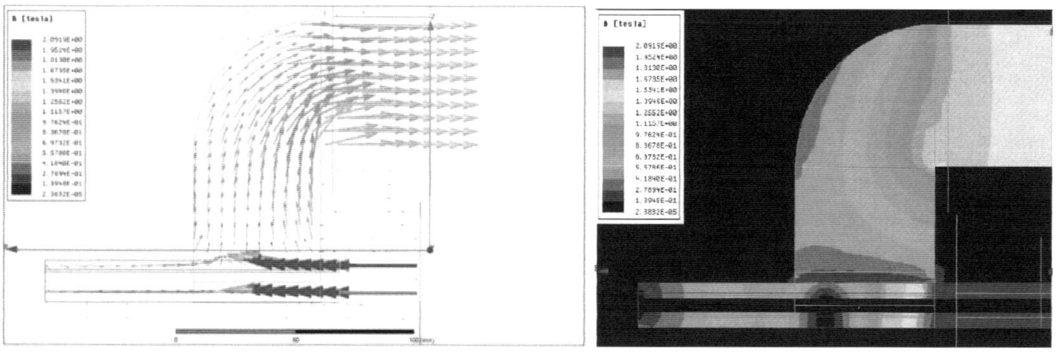

Fig. 10: The magnetic induction field "B" determined on the theoretical model of the second variant of the sensor – the vectors in the longitudinal plane of symmetry (on left) and intensity (on right)

The dimensionless parameter P, that is shown for the realized experiment in Fig. 7, is standardly evaluated for the purpose of a practical application of the modified magnetoelastic method [5,7,8]. It is used to convert a complex measured shape of the hysteresis loop, that depends on the actual force magnitude, to one simple numeric value. The fractions in Fig. 7 describe points of the hysteresis loop taken into account as the key node points for the conversion of the loop shape to the particular parameter P. The numerator of the fractions is the most important value for evaluation of the parameter P indicating the level of the magnetic field intensity "H" in the main node point. The lower values of this indicator indicate the preference of the portion of the hysteresis loop close to the remanence (the intersection with the vertical axis in the B–H curve). On the contrary, its higher values prefer the loop portion near to the saturation. The more exact definition of the parameter P is an industrial secret.

During the realized verification experiment using the second sensor variant, four various dimensionless parameters P were evaluated that represented in a simple way the character and shape of measured hysteresis loops. The sensitivity of the evaluated parameters P to the tension force was studied (see Fig. 7). For advance analysis, the parameter P "15/45" was chosen as the most suitable one because of its sufficient sensitivity and satisfactory stability of obtained results.

For the chosen parameter P "15/45" the detailed analysis of its sensitivity to the tensile force in the studied strand was realized, the results of this analysis are shown in Fig. 8. The resultant regression fitting curve (the black curve in Fig. 8) was determined using methods of mathematical analysis and statistics.

Conclusions

The fundamental results of two verification experiments that were focused on application of two variants of the removable elastomagnetic sensor are described in this paper.

The published results show that the both variants of the removable sensor can be used for a relatively quick experiment intended for the evaluation of the total value of the tensile force in the structural elements activated before the start of the experiment.

However, the second variant of the modified elastomagnetic sensor (the variant with the portable primary coil) is more suitable for this purpose than the first one. Most particularly because, the required state of the studied strand with its technical saturation, that is important for experiments realized using the modified magnetoelastic method, can be achieved applying the second variant of the sensor only.

As it was expected at the beginning of the investigation, the precision of the results obtained by application of the second sensor variant is distinctly lower than for the fully equipped elastomagnetic sensors.

On the other hand, the experimental results revealed that the second sensor variant can be used for the intended purpose to select quickly with sufficient accuracy which tension elements from their group are most important for further detailed analysis and so for installation of the fully equipped elastomagnetic sensor.

The utilization of the both removable and fully equipped modified elastomagnetic sensor is suitable especially for applications on structural elements with short free length, as are, for example, the prestressed cables permanently embedded in the monolithic concrete construction, that are exposed for the purpose of the experiment to the minimum necessary length. For structural elements with a long free length, the other methods mentioned above in this paper can be applied more effectively.

The results presented in this paper are outputs of the research project FV 30457 "Utilization of a Magnetoelastic Method for Increasing the Reliability and Durability of Existing and Newly Built Prestressed Concrete Structures" supported by Ministry of Industry and Trade of the Czech Republic.

References

[1] M. Polák, T. Plachý, Determination of Forces in Roof Cables at Administrative Center Amazon Court, Procedia Engineer. 48, (2012) 578-582.

[2] M. Polák, T. Plachý, Experimental Evaluation of Tensile Forces in Short Steel Rods, Appl. Mech. Mater. 732, (2015) 333-336.

[3] P. Fajman, M. Polák, Measurement of structural cable of membranes, in: Proceedings of the 50th Annual Conference on Experimental Stress Analysis, EAN 2012, Tábor, Czech Republic, 2012, pp. 61-64.

[4] P. Fajman, M. Polák, J. Máca, T. Plachý, The Experimental Observation of the Prestress Forces in the Structural Elements of a Tension Fabric Structure, Appl. Mech. Mater. 486 (2014) 189-194.

[5] T. Klier, T. Míčka, M. Polák, T. Plachý, M. Hedbávný, L. Krejčíková, New Information about Practical Application of the Modified Magnetoelastic Method, MATEC Web Conf. 310 (2020) 00026.

[6] T. Klier, T. Míčka, M. Polák, T. Plachý, M. Hedbávný, R. Jelínek, F. Bláha, Application of the Modified Magnetoelastic Method and an Analysis of the Magnetic Field, Acta Polytechnica CTU Proceedings 15 (2018) 46-50.

[7] T. Klier, T. Míčka, T. Plachý, M. Polák, T. Smeták, M. Šimler, The verification of a new approach to the experimental estimation of tensile forces in prestressed structural elements by method based on the magnetoelastic principle, MATEC Web Conf. 107 (2017) 00015.

[8] T. Klier, T. Míčka, M. Polák, T. Plachý, M. Šimler, T. Smeták, The in Situ Application of a New Approach to Experimental Estimation of Tensile Forces in Prestressed Structural Elements by Method Based on the Magnetoelastic Principle, in: Proceedings of the 55th Conference on Experimental Stress Analysis 2017, Košice, Slovakia, 2017, pp. 122-132.

[9] M. Chandoga, P. Fabo, A. Jaroševič, Measurement of Forces in the Cable Stays of the Apollo Bridge, in: Proceedings of the 2nd fib Congress, Naples, Italy, 2006, pp. 674-675.

[10] A. M. Sarmento, A. Lage, E. Caetano, J. Figueiras, Stress measurement and material defect detection in steel strands by magneto elastic effect. Comparison with other non-destructive measurement techniques, in: Proceedings of the 6th International Conference on Bridge Maintenance, Safety and Management IABMAS 2012, Stresa, Lake Maggiore, Italy, 2012, pp. 914-921.

[11] H. J. Wichmann, A. Holst, H. Budelmann, Magnetoelastic stress measurement and material defect detection in prestressed tendons using coil sensors, in: Proceedings of 7th International Symposium on Non-Destructive Testing in Civil Engineering NDTCE'09, Nantes, France, 2009.

EAN 2020
58th conference on experimental stress analysis

Monotonic Testing of 3D Printed SS316L under Multiaxial Loading

KOŘÍNEK M.[1,a], HALAMA R.[1,b], FOJTÍK F.[1,c], PAGÁČ M.[2,d], KRČEK J.[3,e], ŠEBEK F.[4,f], KRZIKALLA D.[1,a]

[1]Deparment of Applied Mechanics, Faculty of Mechanical Engineering, VŠB – Technical University of Ostrava, 17. listopadu 2172/15, 708 00 Ostrava, Czech Republic

[2]Department of Machining, Assembly and Engineering Metrology, Faculty of Mechanical Engineering, VŠB – Technical University of Ostrava, 17. listopadu 2172/15, 708 00 Ostrava, Czech Republic

[3]Department of Mathematics and Descriptive Geometry, Faculty of Mechanical Engineering, VŠB – Technical University of Ostrava, 17. listopadu 2172/15, 708 00 Ostrava, Czech Republic

[4]Institute of Solid Mechanics, Mechatronics and Biomechanics, Faculty of Mechanical Engineering, Brno University of Technology, Technická 2896/2, 616 69 Brno, Czech Republic

[a]michal.korinek@vsb.cz, [b]radim.halama@vsb.cz, [c]frantisek.fojtik@vsb.cz, [d]marek.pagac@vsb.cz, [e]jiri.krcek@vsb.cz, [f]sebek@fme.vutbr.cz, david.krzikalla@vsb.cz

Keywords: Stainless steel S316L, 3D printing of metal, Multiaxial monotonic loading, Digital Image Correlation Method, Hill yield criterion, isotropic hardening, FEM simulation

Abstract. The aim of this work was to monitor the behaviour of 316L stainless steel produced by 3D printing. The material was tested in the state "as printed". Digital Image Correlation measurements were used for 4 types of notches. The behaviour of these notches under monotonic loading was investigated in two loading paths – tension and torsion. Based on experimental data, two material models were used in finite element analyses. Subsequently, the load-deformation responses of simulations and experiments were compared. Ductility of 3D printed specimens in "as printed" state is also compared with 3D printed machined samples and samples produced by conventional methods.

Introduction

The austenitic stainless steel AISI 316L is one of the most utilized constructional material for various parts in the power industry and beyond. It is usually used in the conventional wrought state [1] while it has been loaded in tension, torsion, and even combination of both. Nevertheless, it is increasingly utilized in the additively manufactured form [2], as it opens new possibilities. It may be optimized and applied in an organic shape or it can even serve as a custom made part or machine element utilized in the repair or reconstruction of structure, where the commercial products are not available or hardly producible by a conventional manufacturing like the machining. Various process parameters used during the additive manufacturing of 316L have been examined [3, 4]. One of the important outputs are the mechanical properties [5, 6] or porosity [7]. The building direction also plays a vital role [8] and the final surface roughness is of a particular interest [9].

This paper presents new results of the deformation response obtained during monotonic multiaxial loading of the samples made from Stainless Steel 316L (SS316L) produced by

Selective Laser Melting (SLM) technology in "as printed" state. Due to the character of the specimens that contain notches, the Digital Image Correlation Method (DICM) was used. The DICM is a progressive optical-numerical method suitable for 3D analysis of structural components, under uniaxial and multiaxial loading in the full-field [10, 11]. Averaged characteristics gained in this experimental study with DIC measurements were used for validation of a numerical model based on Finite Element Method (FEM).

Experiment description

For the test were used the samples created with SOLIDWORKS 2019 (Dassult Systemes SoliDWorks, France) and made using Powder Bed Fusion 3D printing technology– Selective Laser Melting. The machine used was 3D printer Renishaw AM400 (Renishaw, Great Britain, 2016) and material was atomized SS316L powder. This is additive manufacturing technology, where the laser scans and selectively melts the atomized metal powder particles, bonding them together and building a model layer-by-layer [12, 13]. In the beginning of the process, the build chamber is filled with inert gas argon, to minimize the oxidation of metal powder. The layer thickness was set to 50 µm and Chessboard strategy was used. Strategy translates by 5 mm in horizontal direction X and Y and rotates for optimum homogeneous distribution of stress [14, 15]. Other 3D Printing parameters are shown in Table 1 (QuantAM, SW made by company Renishaw) [16]. Building time was 76 hours. The part orientation and the position in chamber, the 3D printing preview and chessboard strategy preview in cross-section are in the Figure 1.

Table 1: 3D Printing parameters [15, 17]

3D Printer:	Renishaw AM400
Powder description:	SS Powder AISI 316L (DIN 1.4404)
Powder Particle Size:	15–45 µm
Layer Thickness:	50 µm
Focus Size:	70 µm
Print Strategies:	Chessboard
Border Power:	110 W
Border Exposure Time:	100 µs
Border Point Distance:	20 µm
Hatches Power:	200 W
Hatches Exposure Time:	80 µs
Hatches Point Distance:	60 µm
Jump speed:	5000 mm·s-1
Dosing time:	7 s
Melting range:	1371 °C to 1399 °C
Concentration of Oxygen:	< 0,1 % O_2
Inert Gas:	Argon
Purity:	5.0 (99.998 %)

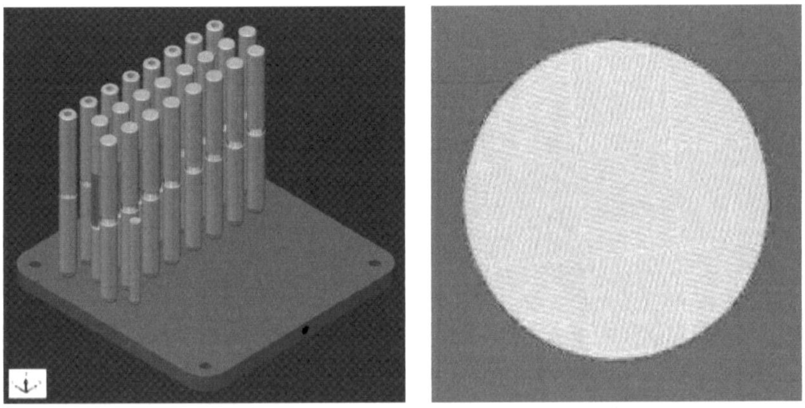

(a) (b)

Figure 1: The part orientation and the position in chamber: (a) the 3D printing preview, (b) chessboard strategy preview in cross-section

The specimens were not further machined (outer surface), or heat treated ("as printed") therefore had naturally a high surface roughness. The geometry of notches considered in this study is shown in Fig. 2. The only one was tubular, other were solid bodies. Tubular (specimens A) had to be drilled on required internal diameter. Each specimen was 160 mm long and outer diameter was 15 mm. In addition, the standard tensile test has been performed. The solid specimens were loaded only in tension. The tubes were subjected to two different loading modes: tension and torsion. Each measurement was repeated four times. Testing machine LabControl 100kN/1000Nm was used. Multiaxial tests were done under deformation control under 2 millimetres per minute for tension and under 0.157 radians per minute for torsion. The results of tests were evaluated in the form of force (torque) vs elongation (twist) diagrams.

DIC measurements were used to monitor the deformation. This method is characterized by the creation of a light area with dark points, also known as pattern. Represented pattern is shown in Fig 3. Two optical sensors were used for this measurement. Sensors are high resolution cameras. The principle of DIC measurement is, that two images of the specimen are compared at different loading state by using a pixel. Advantages of this method are ability to monitor deformation of very complex shaped areas and determination of Young's modulus and Poisson's ratio [18]. MERCURY RT system provided by Sobriety company was used for all DIC measurements. This software was also used for configuration and calibration cameras. The optical probe must be set up in this software before starting the measurements. This probe must be aligned for both cameras. The optical probe provides initial length and must cover the whole area examined.

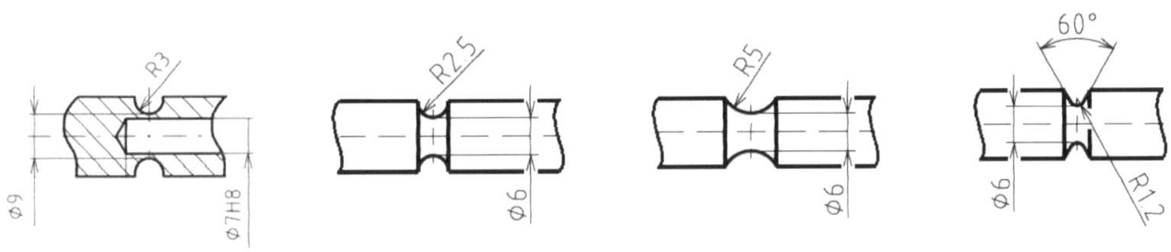

A B C D

Fig. 2: Notched specimens: The tubular specimen with notch for multiaxial loading (A), and solid specimens with notch for axial loading (B), (C) and (D).

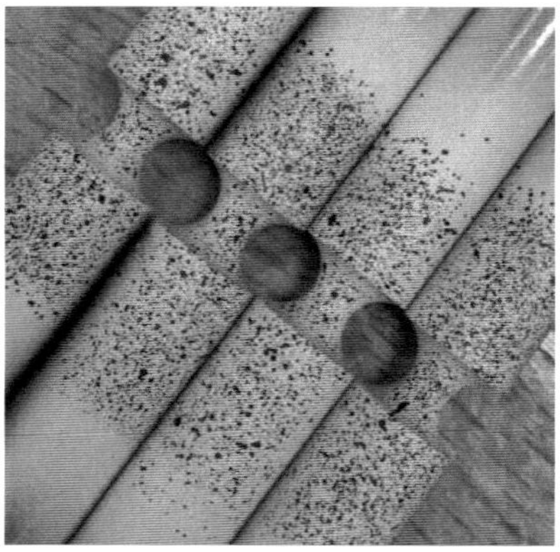

Fig. 3: Representative pattern for DIC measurement

Numerical Modelling

Finite element modelling approach, material models used, and presentation of results are discussed next.

Finite Element Modelling Approach. Since additive manufacturing technology is getting increasingly popular, it is important to examine if FEM approach can predict sufficiently accurate results with respect to experimental response of material. Number of analyses were thus performed in this study to validate finite element model response under several loading conditions. Validation of FEM was performed in terms of comparison of results from FEM with results from experiments.

For purposes of FEM, specimens were modelled as cut-outs of length equal to initial length measured by a probe during experiments. Various levels of symmetry were utilized to reduce computation time. For tensile analyses 1/8 symmetry was used and for torsion analyses half symmetry was used. Models were meshed using linear hexagonal elements (SOLID185). Usage of mapped mesh and sizing settings ensured regular and sufficiently sized mesh to capture stress and strain gradients accurately (example of mesh is in Fig. 4. All analyses were prescribed in form of macros in Ansys Parametric Design Language for easy and fast running and automatic post-processing of desired results.

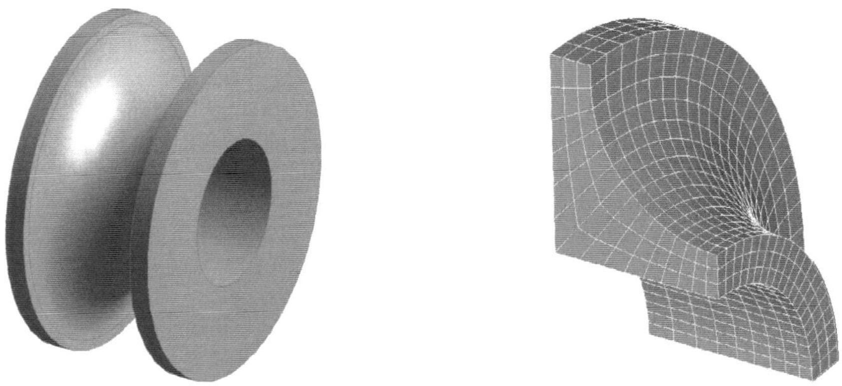

(a) *(b)*

Fig 4.: (a) Model of full specimen A cut-out, and (b) meshed specimen utilizing 1/8 symmetry

Boundary conditions were set in accordance with experimental loading conditions and symmetry assumptions. For simulations of tensile test, nodal displacement and symmetry plane boundary conditions were used. Those were applied upon nodes within appropriate faces. See

Fig. 5a for an example of boundary conditions for tensile test simulations as it is similar for all specimens.

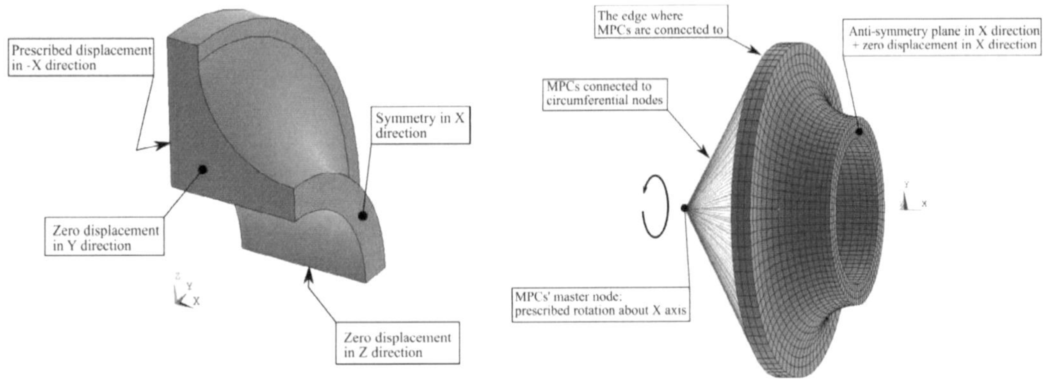

(a) *(b)*

Fig 5: Description of boundary conditions for tensile test simulation (a), and for torsion test simulation (b)

For simulation of torsion test of specimen A, structural multipoint constraints (MPC184) were utilized to load the specimen in torsion. Then nodal displacement in "x" direction and anti-symmetry plane in "x" direction were used (see Fig. 5b). Fix of displacement of nodes in "x" direction was sufficient since Ansys's anti-symmetry plane formulation constraints - in this case - all other degrees of freedom. Observing boundary conditions, one can see that all simulations were displacement controlled. Correctness of applied boundary conditions for both tensile and torsion test simulations were confirmed by simulations of full models.

Material Model. To simulate the material response of specimens under conditions of tension and torsion, a suitable material model was needed. The chosen material model captures nonlinear isotropic hardening together with either Hill yield criterion or von Mises yield criterion. The linear elastic part of the model obeys Hooke's law for three-dimensional problem denoted in Table 2 and requires two input parameters of Young's modulus E and Poisson ration μ. Values of the parameters are denoted in Table 3.

Table 2: Hooke's law for three-dimensional problem

$$\varepsilon_x = \frac{1}{E}\left[\sigma_x - \mu\left(\sigma_y + \sigma_z\right)\right],\ \varepsilon_y = \frac{1}{E}\left[\sigma_y - \mu\left(\sigma_x + \sigma_z\right)\right],\ \varepsilon_z = \frac{1}{E}\left[\sigma_z - \mu\left(\sigma_y + \sigma_x\right)\right],$$
$$\gamma_{xy} = \frac{\tau_{xy}}{G},\ \gamma_{yz} = \frac{\tau_{yz}}{G},\gamma_{xz} = \frac{\tau_{xz}}{G},$$

where σ_i - stress components, ε_i- normal strain components, γ_{ij}- shear strain components, τ_{ij} – shear stress components and G- shear modulus.

Table 3: Elastic parameters

Parameter	Value	Unit
E	183	[GPa]
μ	0.3	[-]

Isotropic hardening is suitable for this study since the loading was monotonic. Isotropic hardening during plastic deformation causes a uniform increase of the yield surface. This results in increased yield stress. Thus, the yield criterion takes the form:

$$f(\sigma) - Y = 0 \quad (1)$$

where $f(\sigma)$ is function of stress and Y is yield stress (or size of yield surface). For description of isotropic hardening, the Voce law was used. Voce law is, however, combination of linear and nonlinear isotropic variables and has the form:

$$Y = \sigma_Y + R \quad (2)$$

where σ_Y is initial yield stress and R is a new internal variable. The evolution of R is done by superposition of two parts:

$$dR = dR_1 + dR_2, \qquad dR_1 = R_0\, dp\,, \qquad dR_2 = b(R_\infty - R_2)dp \quad (3)$$

By integration of Eq. (3) with zero initial values of p, R_1 and R_2 respectively and use of Eq. (2), a constitutive equation is obtained:

$$Y = \sigma_Y + R_0\, p + R_\infty(1 - e^{-bp}) \quad (4)$$

where R_0 is slope of the saturation stress, p is accumulated equivalent plastic strain, R_∞ is the difference between the saturation stress and the initial yield stress and b is hardening parameter that governs the rate of saturation of the exponential term. Values of the parameters are denoted in Table 4.

Table 4: Voce law parameters

Parameter	Value	Unit
σ_Y	575	[MPa]
R_0	950	[MPa]
R_∞	60	[MPa]
b	125	[-]

Hill yield criterion and von Mises criterion were used in this material model to compare their accuracy with respect to real additive manufactured specimens' responses. Hill criterion is anisotropic, independent of hydrostatic pressure and depends on the orientation of the stress relative to the axis of anisotropy, thus suitable for materials in which microstructure influences the macroscopic behaviour of the material, what is the case for additive manufactured steels [19, 20]. Hill yield criterion was, in this study, utilized for modelling of yield strength anisotropy based on build direction [21 - 24]. Hill yield criterion's stress function has the form:

$$f(\boldsymbol{\sigma}) \equiv F(\sigma_{22} - \sigma_{33})^2 + G(\sigma_{33} - \sigma_{11})^2 + H(\sigma_{11} - \sigma_{22})^2 + 2L\sigma_{23}^2 + 2M\sigma_{31}^2 + 2N\sigma_{12}^2 = \sigma_y^2 \quad (5)$$

where F, G, H, L, M, N are coefficients which are functions of the ratio of the scalar yield stress parameter and yield stress in each of the six stress components. The coefficients and ratios are denoted in Table 5 [25 - 27].

Table 5: Hill yield criterion coefficients and ratios

$F = \frac{1}{2}\left(\frac{1}{R_{22}^2} + \frac{1}{R_{33}^2} - \frac{1}{R_{11}^2}\right), G = \frac{1}{2}\left(\frac{1}{R_{33}^2} + \frac{1}{R_{11}^2} - \frac{1}{R_{22}^2}\right), H = \frac{1}{2}\left(\frac{1}{R_{11}^2} + \frac{1}{R_{22}^2} - \frac{1}{R_{33}^2}\right),$
$L = \frac{3}{2}\left(\frac{1}{R_{23}^2}\right), M = \frac{3}{2}\left(\frac{1}{R_{13}^2}\right), N = \frac{3}{2}\left(\frac{1}{R_{12}^2}\right),$
$R_{11} = \frac{\sigma_{11}^y}{\sigma_y}, R_{22} = \frac{\sigma_{22}^y}{\sigma_y}, R_{33} = \frac{\sigma_{33}^y}{\sigma_y},$
$R_{12} = \sqrt{3}\frac{\sigma_{12}^y}{\sigma_y}, R_{23} = \sqrt{3}\frac{\sigma_{23}^y}{\sigma_y}, R_{13} = \sqrt{3}\frac{\sigma_{13}^y}{\sigma_y},$

where the directional yield stress ratios R_{ii} and R_{ij} are related to the isotropic yield stress parameter, and σ_i^y is the yield stress in the direction given by the value of subscript *i*. Almost all directional yield stress ratios for uniaxial and torsional loading are equal to 1. The only different one is the directional ratio R_{33} which is equal to 0.87. The ratio R_{33} corresponds to axial direction of all specimens, thus introduces the effect of build direction into the material model.

Von Mises yield criterion is isotropic, independent of hydrostatic pressure and commonly used for metals, polymers, etc. Independence of hydrostatic pressure can limit its applicability to micro structured materials. In this study, accuracy of the von Mises yield criterion is examined by comparing to the real material response. Von Mises stress function takes the form:

$$f(\boldsymbol{\sigma}) = \sqrt{\frac{(\sigma_{11} - \sigma_{22})^2 + (\sigma_{22} - \sigma_{33})^2 + (\sigma_{11} - \sigma_{33})^2}{2}} \quad (6)$$

where σ_{ii} are principal stresses.

Results

All specimens were subjected to loading as described above. Each test was deformation controlled. Values of applied force or torque were recorded by testing machine and values of deformation were recorded by DICM system. Because experimentally measured data embodies natural oscillations, the presented force vs. elongation diagrams were smoothed using functions of the Curve Fitting Toolbox in Matlab. The combination of moving average and smoothing splines seemed to lead to sufficient results. Optimal smoothing parameters were chosen with respect to the size and character of data sets.

Presentation of simulation results and their comparison with respect to experimental response of specimens follow next. The comparison is performed in form of plots comprised of experimental smoothed responses in dashed lines and simulation results in full lines (see Fig. 7 - 11).

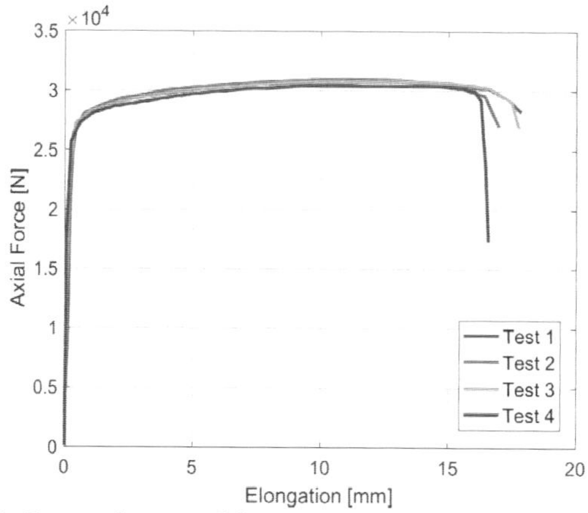

Fig. 6: Dependency of force on elongation for tensile test

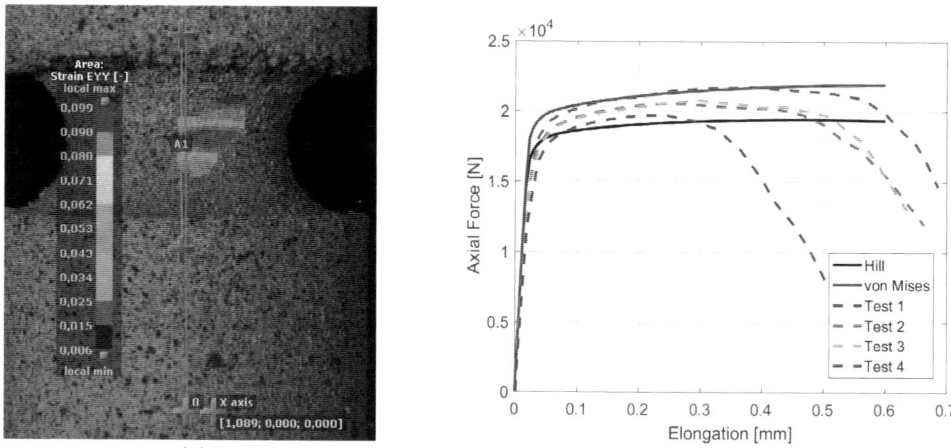

(a) *(b)*

Fig. 7: Longitudinal strain distribution (a), and dependency of force on elongation for specimen type A compared to experimental and simulation results (b)

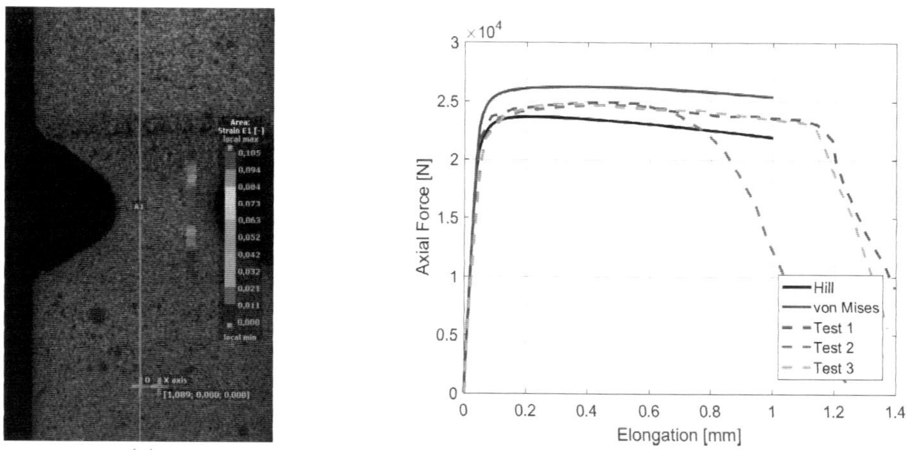

(a) *(b)*

Fig. 8: Maximal principal strain distribution (a), and dependency of force on elongation for specimen type B compared to experimental and simulation results (b)

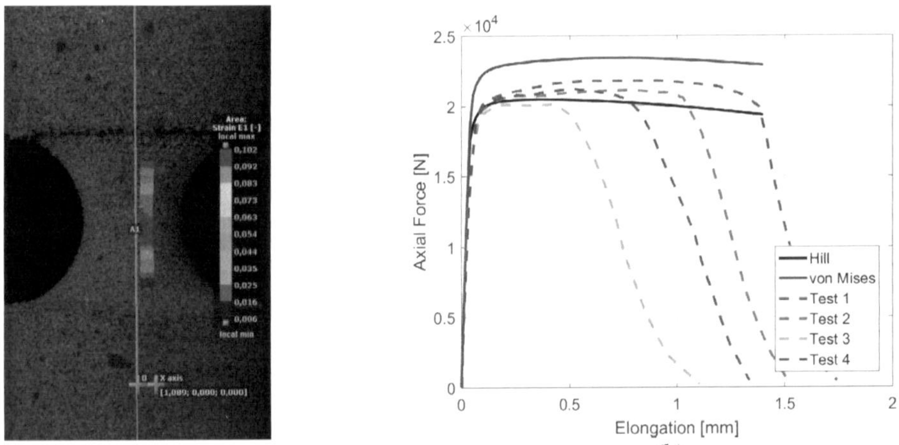

(a) *(b)*

Fig. 9: Maximal principal strain distribution (a), and dependency of force on elongation for specimen type C compared to experimental and simulation results (b)

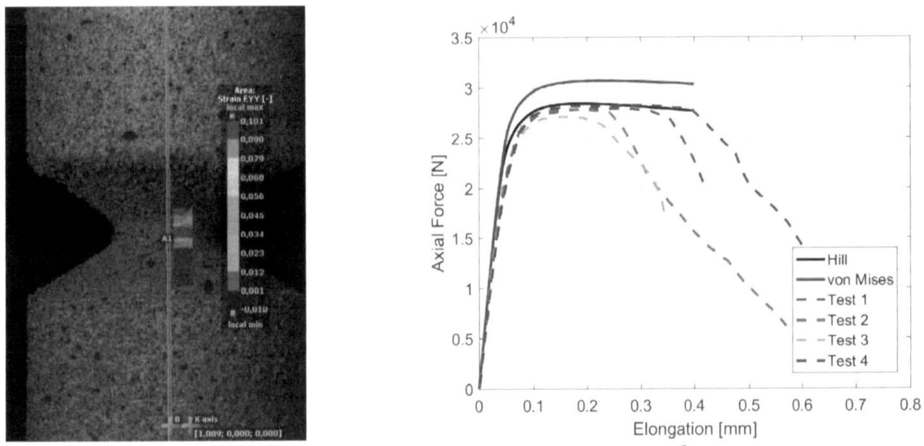

(a) *(b)*

Fig. 10: Longitudinal strain distribution (a), and dependency of force on elongation for specimen type D compared to experimental and simulation results (b)

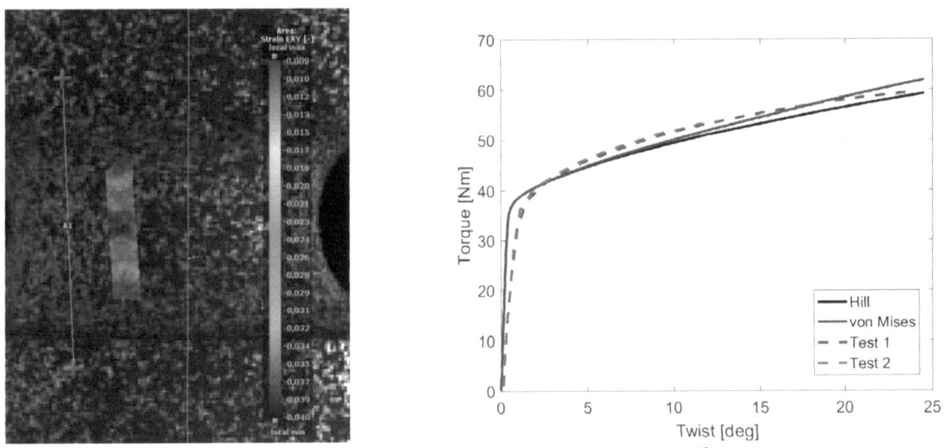

(a) *(b)*

Fig. 11: Shear strain distribution for torsional test (a), and dependency of torque on twist for specimen type A compared to experimental and simulation results (b)

Comparison of experimental results with results from FEM analysis revealed that simulation based on Hill yield criterion gives more accurate results than simulation based on isotropic yield criterion. Usage of isotropic yield criterion leads to stiffer response of simulation model for each specimen type. Thus, usage of Hill yield criterion could be crucial for obtaining reasonably accurate results from FEM analysis once simulating additive manufactured steels such as SS316L. Depending on directional yield stress ratios (which were, in this study, correlated from experimental material response), the Hill yield criterion can give more compliant response than the real material response is. Such a phenomenon was observed for majority of responses in this study (see Fig. 6 - 11). In this case, one should be aware that once having a displacement-controlled simulation the stresses are likely to be underestimated.

Ductility was calculated from tensile test results for specimen without notch (Fig. 6).

$$A = \frac{L_u - L_0}{L_0} \cdot 100\ [\%], \qquad (7)$$

where L_u is the final length of the specimen after the test and L_0 is initial length of the specimen. Specimens without machined outer surface had ductility from 42 % to 45 %. "As printed" specimens show about 2/3 higher ductility than machined specimens. The machined samples had ductility 13 % to 15 %. The printed SS316L reveals surprisingly good ductility even printed in vertical direction (43 % in comparison with 60 % of the conventional SS316L), but just for "as printed" variant.

Conclusions

Objective of this paper was also to study the effect of machining. Four types of specimen were investigated. The geometry of the specimens is show in Fig 2. In addition, the standard tensile test has been performed. The specimens were subjected to different loading modes, as stated above. All specimens were made from Stainless Steel 316L produced by Selective Laser Melting Technology in "as printed" state. DICM was used to monitor deformation. DIC method is a good alternative for complicated shape areas such as notches. Averaged characteristics gained in this experimental study were used for validation of a numerical model based on FEM.

Acknowledgement:

This work was supported by The Ministry of Education, Youth and Sports from the Specific Research Project (SP2020/23), by the Czech Science Foundation (GACR), grant No. 19-03282S and has been done in connection with the DMS project reg. no. CZ.02.1.01/0.0/17_049/0008407 financed by Structural Funds of Europe Union.

References

[1] F. Šebek, P. Kubík, J. Hůlka, J. Petruška, Strain hardening exponent role in phenomenological ductile fracture criteria, Eur. J. Mech. Solids A 57 (2016) 149–164.

[2] M. Marya, V. Singh, S. Marya, J.Y. Hascoet, Microstructural development and technical challenges in laser additive manufacturing: Case study with a 316L industrial part, Metall. Mater. Trans. B 46 (4) (2015) 1654–1665.

[3] J. Metelkova, Y. Kinds, K. Kempen, Ch. de Formanoir, A. Witvrouw, B.V. Hooreweder, On the influence of laser defocusing in selective laser melting of 316L, Addit. Manuf. 23 (2018) 161–169.

[4] G. Tapia, S. Khairallah, M. Matthews, W.E. King, A. Elwany, Gaussian process-based surrogate modeling framework for process planning in laser powder-bed fusion additive

manufacturing of 316L stainless steel, Int. J. Adv. Manuf. Technol. 94 (2018) 3591–3603.
[5] M.S. Pham, B. Dovgyy, P.A. Hooper, Twinning induced plasticity in austenitic stainless steel 316L made by additive manufacturing, Mater. Sci. Eng. A 704 (2017) 102–111.
[6] X. Chen, J. Li, X. Cheng, B. He, H. Wang, Z. Huang, Microstructure and mechanical properties of the austenitic stainless steel 316L fabricated by gas metal arc additive manufacturing, Mater. Sci. Eng. A 703 (2017) 567–577.
[7] Z.'E.E. Tana, J.H.L. Pang, J. Kaminski, H. Pepin, Characterisation of porosity, density, and microstructure of directed energy deposited stainless steel AISI 316L, Addit. Manuf. 25 (2019) 286–296.
[8] A. Leicht, U. Klement, E. Hryha, Effect of build geometry on the microstructural development of 316L parts produced by additive manufacturing, Mater. Charact. 143 (2018) 137–143.
[9] Y. Kaynak, O. Kitay, The effect of post-processing operations on surface characteristics of 316L stainless steel produced by selective laser melting, Addit. Manuf. 26 (2019) 84–93.
[10] F. Trebuňa, R. Huňady, M. Hagara. Experimentálne metódy mechaniky - Digitálna obrazová korelácia. Košice: Technická univerzita v Košiciach, 2016. ISBN 978-80-5532-346-6.
[11] Z. Paška, R. Halama, F. Fojtík, P. Gajdoš, Strain response determination in notched specimens under multiaxial cyclic loading by DICM. In: Proceedings of EAN 2017 - 55th Conference on Experimental Stress Analysis 2017, Novy Smokovec, Slovakia, May 30 - 1 June, 2017, p. 143-149.
[12] I. Gibson, D. Rosen, B. S. Tucker, Additive manufacturing technologies: rapid prototyping to direct digital manufacturing. New York: Springer, c2010, xxii, 459 p. ISBN 1441911200.
[13] 3D Hubs. 2020. Introduction To Metal 3D Printing | 3D Hubs. [online] Available at: <https://www.3dhubs.com/knowledge-base/introduction-metal-3d-printing/> [Accessed 16 June 2020].
[14] J. Hajnyš, M. Pagáč, J. Mesicek, J. Petrů, M. Krol, Influence of scanning strategies parameters on residual stress in SLM process according to bridge curvature method for stainless steel AISI 316L. Materials. 2020, 13, 1527. doi: 10.3390/ma13071659.
[15] J. Hajnyš, M. Pagáč, O. Kotera, J. Petrů, S. Scholz, Influence of Basic Process Parameters on Mechanical and Internal Properties of 316L Steel in SLM Process for Renishaw AM400. MM Science Journal, 2019. Pgs: 2790-2794. DOI: 10.17973/MMSJ.2019_03_2018127
[16] RENISHAW, Magics training: Material profile editing. RENISHAW: apply innovation. United Kingdom: Renishaw, 2016. Available from: https://www.renishaw.com. [Accessed 16 June 2020].
[17] Data sheets - Additive manufacturing. Object moved [online]. Copyright © [cit. 16.06.2020]. Available at: https://www.renishaw.com/en/data-sheets-additive-manufacturing--17862. [Accessed 16 June 2020]
[18] D. Lecompte et al., „Quality assessment of speckle patterns for digital image correlation", Opt. Lasers Eng., roč. 44, č. 11, s. 1132–1145, 2006, DOI: 10.1016/j.optlaseng.2005.10.004.
[19] L. Hitzler, J. Hirsch, B. Heine, M. Merkel, W. Hall, A. Öchsner, 2017. On the anisotropic mechanical properties of selective laser-melted stainless steel. Materials, 10(10), p.1136. DOI: 10.3390/ma10101136
[20] P. Bian, J. Shi, Y. Liu, Y. Xie, Influence of laser power and scanning strategy on residual stress distribution in additively manufactured 316L steel, Optics & Laser Technology, Volume 132, 2020,106477, ISSN 0030-3992, DOI: 10.1016/j.optlastec.2020.106477.

[21] N. Zhang, H. Li, X. Zhang, D. Hardacre, 2016. Review of the fatigue performance of stainless steel 316L parts manufactured by selective laser melting. Available from: https://hdl.handle.net/10356/84380

[22] S. Afkhami, M. Dabiri, S. H. Alavi, T. Björk, A. Salminen, 2019. Fatigue characteristics of steels manufactured by selective laser melting. International Journal of Fatigue, 122, pp.72-83. DOI: 10.1016/j.ijfatigue.2018.12.029

[23] X. Song, S. Feih, W. Zhai, C. N. Sun, F. Li, R. Maiti, J. Wei, Y. Yang, V. Oancea, L. R. Brandt, A. M. Korsunsky, 2020. Advances in additive manufacturing process simulation: Residual stresses and distortion predictions in complex metallic components. Materials & Design, p.108779. DOI: 10.1016/j.matdes.2020.108779

[24] P. Ferro, F. Berto, L. Romanin, 2020. Understanding powder bed fusion additive manufacturing phenomena via numerical simulation. Frattura ed Integrità Strutturale, 14(53), pp.252-284. DOI: 10.3221/IGF-ESIS.53.21

[25] R. B. Colby, 2013. Equivalent plastic strain for the Hill's yield criterion under general three-dimensional loading (Doctoral dissertation, Massachusetts Institute of Technology).

[26] T. Altan, A. E. Tekkaya, eds., 2012. Sheet metal forming: fundamentals. Asm International.

[27] M. Bertin, F. Hild, S. Roux, 2017. On the identifiability of Hill-1948 plasticity model with a single biaxial test on very thin sheet. Strain, 53(5), p.e12233. DOI: 10.1111/str.12233

EAN 2020
58th conference on experimental stress analysis

Non-Destructive Evaluation of Small Aircraft Wing Adhesive Joints using Pulse Infrared Thermography Method

KOSTROUN T.[1,a], DVOŘÁK M.[2,b]

[1]Department of Aerospace Engineering, Faculty of Mechanical Engineering, Czech Technical University in Prague, Karlovo nám. 13, 121 35 Prague 2, Czech Republic

[2]Department of Mechanics, Biomechanics and Mechatronics, Faculty of Mechanical Engineering, Czech Technical University in Prague, Technická 4, 160 00 Prague 6, Czech Republic

[a]tomas.kostroun@fs.cvut.cz, [b]milan.dvorak@fs.cvut.cz

Keywords: NDE, Infrared Thermography, IRNDT, composite, GFRP

Abstract. In this article we examine possibility of using active infrared thermography as a non-traditional non-destructive evaluation method for an adhesive joints testing. The attention was focused on the load bearing wing structure and related structural joints, specifically the adhesive joints of wing spar caps and skins on the wing demonstrator of a small sport aircraft made mainly of carbon composite. The pulse thermography method using flash lights optical excitation was tested. A modified differential absolute contrast method was used to process the measured data to reduce the effect of heat source inhomogeneity and surface emissivity. The tested method proved a very good ability to detect defects in the adhesive joints. The achieved results are easy to interpret and use for qualitative and quantitative evaluation of the adhesive joints of thin composite parts.

Introduction

The aim of an experimental work was to evaluate bonding quality of the small composite aircraft wing adhesive joints. The all-composite wing is assembled from the pre-cured carbon/epoxy composite (CFRP) parts using the Hysol 9394 two component epoxy adhesive system. The work was focused on the joints of wing skin to spars and ribs.

One of the less common Non-Destructive Evaluation (NDE) methods was used, namely Pulse infrared Thermography (PT). This method allows inspection of specific areas of the tested object with direct graphical output. It is suitable for the finding of flaws and voids located close to the surface. In the case of the performed tests, defects of the adhesive joints were expected at a depth of 0.5-2 mm below the tested surface. They would be difficult to detect using other standard NDT methods.

Experimental specimen

The test sample was the right half of the wing of the small composite aircraft demonstrator. The wing was used for a development static structural strength testing. Its structure consists of the spars and ribs made of CFRP. The wing skins are made of a sandwich structure consisting of a carbon sheet and a foam core. The outer skins are additionally covered with a thin layer of surfacer made of glass/epoxy composite (GFRP). Wing skins are glued to the load bearing structure using the Hysol 9394 epoxy adhesive. The scheme of the bonded joints configuration in the tested areas is pictured in Fig. 1.

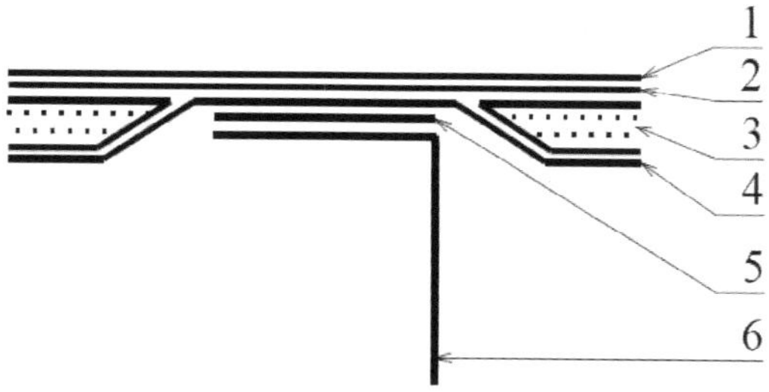

no.	description	material
1	surfacer	AeroGlide (CFRP)
2	outer skin	TeXtrem (GFRP)
3	foam	Rohacel
4	inner skin	TeXtrem (GFRP)
5	adhesive	Hysol 9394
6	spar (rib)	GFRP

Fig. 1: Bonded joint configuration

Experimental method description

PT method is based on the principle of heating the sample from one side by a short thermal pulse (for example halogen light flash) and subsequent monitoring of the cooling curve at each point of the surface using an infrared (IR) thermal camera. By sending a pulse, the heat wave begins to propagate through the material. The surface cools due to heat wave propagation (conduction) into the material in the depth direction, but also due to convection and radiation losses. If there is a defect underneath the surface with different thermal effusivity than the base material (delamination, cavity or void in an adhesive joint), the heat wave will be reflected back to the surface and the cooling process will change at this point. This behaviour of surface cooling curves is demonstrated in Fig. 2. Defects that occur at a greater depth will appear on the thermogram with time delay [1, 2]. Fig. 3 shows an example of the time evolution of thermograms for different moments after the excitation pulse. The figure shows a time-sequential drawing of deeper layers of the adhesive joint. At time $t = 5\ s$, the poor quality of the joint can be seen due to inappropriate technology (the adhesive bead wasn't compressed and spread sufficiently). At the same time, it can be seen that due to lateral diffusion, thermograms lose their sharpness with increasing time.

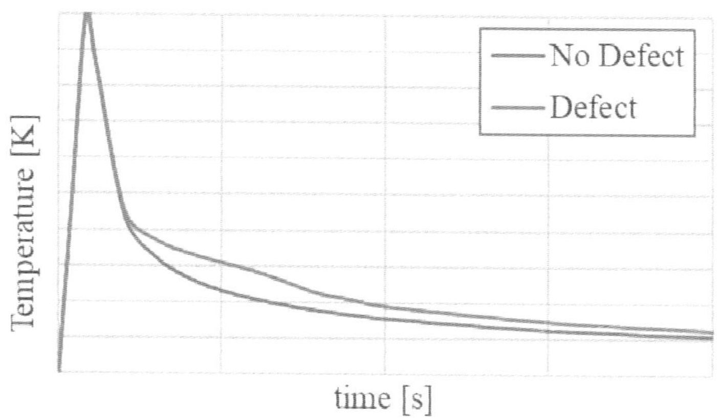

Fig. 2: Surface cooling curves

The disadvantage of this method is the sensitivity to the unevenness of the heat source and the distribution of emissivity on the surface, which can be partially eliminated by subsequent postprocessing.

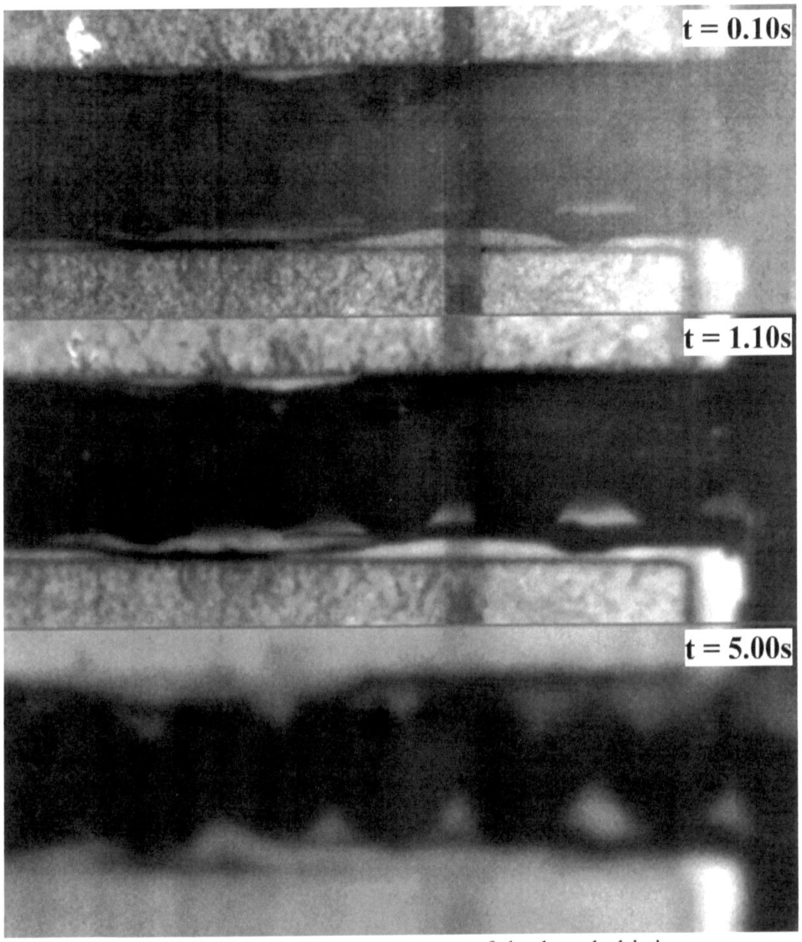

Fig. 3: Sequence of thermograms of the bonded joint

Experimental system description

A modular test system was designed and used for the PT non-destructive testing (NDT) method. The basic hardware elements of this system consist of a FLIR A325SC bolometric uncooled IR camera (resolution 320x240 pixels, NETD <50 mK, maximum scanning frequency 60 Hz), instrument unit equipped with PC for test control and data recording and two flash lights (2x 1200 Ws).

The instrument unit works as the communication interface between the PC, IR camera and excitation lights. It consists mainly of the cDAQ measuring and control system from National Instruments, equipped with analog output cards (for excitation light control) and digital input/output, as well as other necessary auxiliary electronics.

A program created for this purpose in the LabView environment from the National Instruments was used to control the tests process and record the measured data. Processing and evaluation of the obtained data was performed in the MATLAB environment from the MathWorks company.

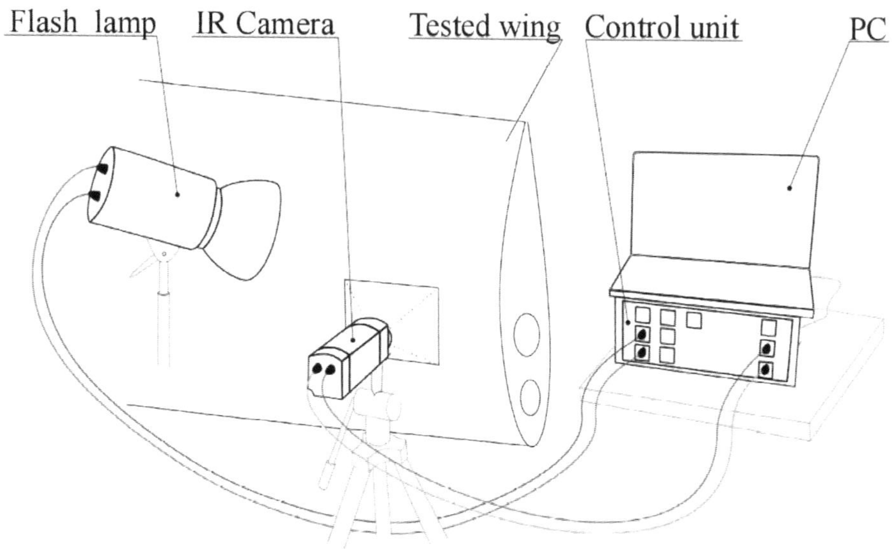

Fig. 4: Experimental configuration

Experimental data evaluation

A modified Differential Absolute Contrast (DAC) method was used to process the measured data. It allows partial elimination of heat source unevenness and of emissivity distribution on the surface. The DAC method compares the temperature of the tested place with the defect with the theoretical value of the temperature in case the defect does not occur in the tested place. This theoretical temperature is calculated on the basis of the 1D form of the Fourier equation of heat conduction in a semi-infinite medium from the measured temperature at a time when the temperature defect does not manifest itself [3]. The standard DAC method works with the temperature in time just before the manifestation of the defect, which, however, usually requires manual intervention of the test operator. The modified DAC method, on the other hand, uses the temperature taken from the end of the measurement. The thermal contrast is calculated according to (1), which describes relation of temperature $T_{(t)}$ at the observed time t and the temperature $T_{(t')}$ at steady state at the end of the measurement at time t' for each individual pixel of the record.

$$\Delta T_{DAC}(t) = T(t) - \sqrt[b]{\frac{t'}{t}}T(t')\ [K] \tag{1}$$

Parameter b represents the steepness of the cooling curve shape correction factor, which is given by the properties of the examined material, especially the rate of surface cooling after excitation (standard DAC method uses $b = 2$). This parameter must be determined experimentally in order to gain the greatest possible contrast. [1]

Results and Discussion

The following figures show examples of the test results. Each individual image covers the tested area with a size of approx. 320x240 mm, which at a given resolution of the IR camera represents a resolution of 1x1 mm for each pixel of the image. The final images were adjusted so that the grayscale range covers the entire range of evaluated data (from white to black). Fig. 5 and 6 show a representative selection of images on which the test results are demonstrated with an explanation of the individual indications.

The area of the adhesive joint of wing tip rib (right) and the front wing spar cap on the lower wing skin side is pictured in Fig. 5. The red lines mark the area of the front wing spar for the PT NDT evaluation process. The blue lines mark the area of the sandwich foam core reinforcements of the wing skin. From the point of view of the adhesive joint evaluation, the critical places are represented by the lighter shade of corresponding colour (A), which can be interpreted as the voids in the adhesive joint. The indication marked with the D in the figure represents an overflow of excessive adhesive outside the joint area. Indication B is caused by overlapping of the top layers of the surfacer and thus its local doubling. Indication C represents the region of resin accumulation in the place where foam core ends and sandwich skins are joined together.

Fig. 6 represents the area of the front wing spar at the fuel tank location on the upper wing skin side. Again, the places with the voids in adhesive layer are clearly visible (A). The use of two types of different adhesives is visible in the bonded area (B). This is because the need for the increased resistance to the influence of fuel in the fuel tank area (use of C-resin type). Indication C represents the local reinforcement in the area of fuel tank lid by adding one layer of fabric to the outer skin lay-up.

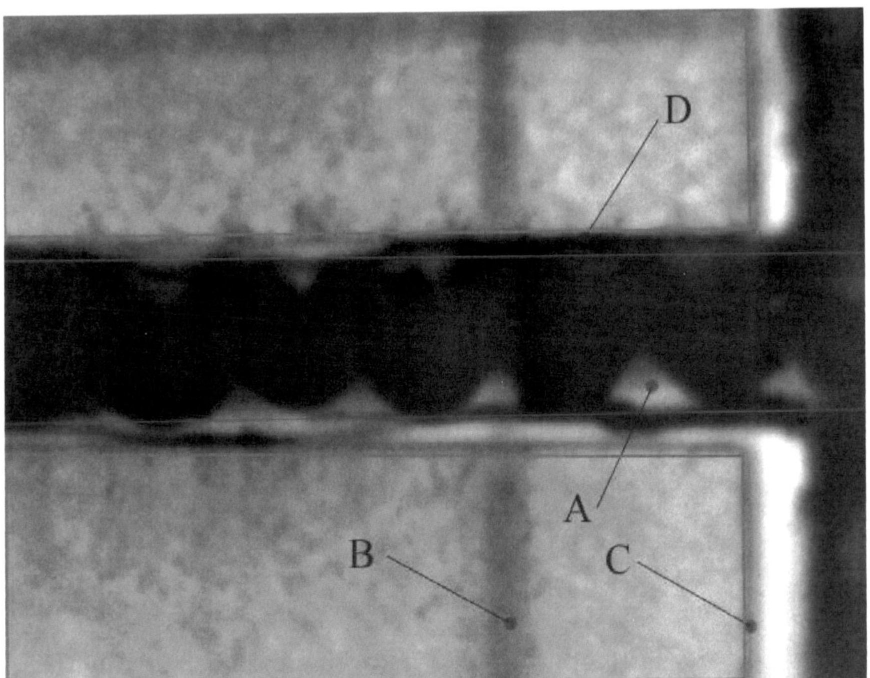

Fig. 5: Results of PT NDE method – adhesive joint of the front wing spar and the wing tip rib on the lower wing skin

Due to the fact that no etalons with artificial defects were available for testing, an additional comparison of the NDT findings with the actual condition of the adhesive joint at the failure area was performed after the static strength test of the wing demonstrator. In the area of the main wing spar, the CFRP wing skin was removed to the depth of the adhesive joint. Fig. 7 shows a comparison of the NDT findings with the actual condition of the adhesive joint. The individual color circles mark the corresponding defects of the joint. The performed NDT measurement shows a good agreement with the actual condition of the joint. The smallest detected defect (marked in red) is about 3 mm in diameter. Due to the resolution of the IR camera, this dimension can be considered as the smallest detectable defect in the adhesive joint for a given test configuration. Except for defects in the adhesive layer, the use of two adhesives (Hysol and C-resin) is clearly visible in the picture. In the lower half of the picture, two vertical cracks caused by a failure during the strength test of the wing demonstrator can be also seen.

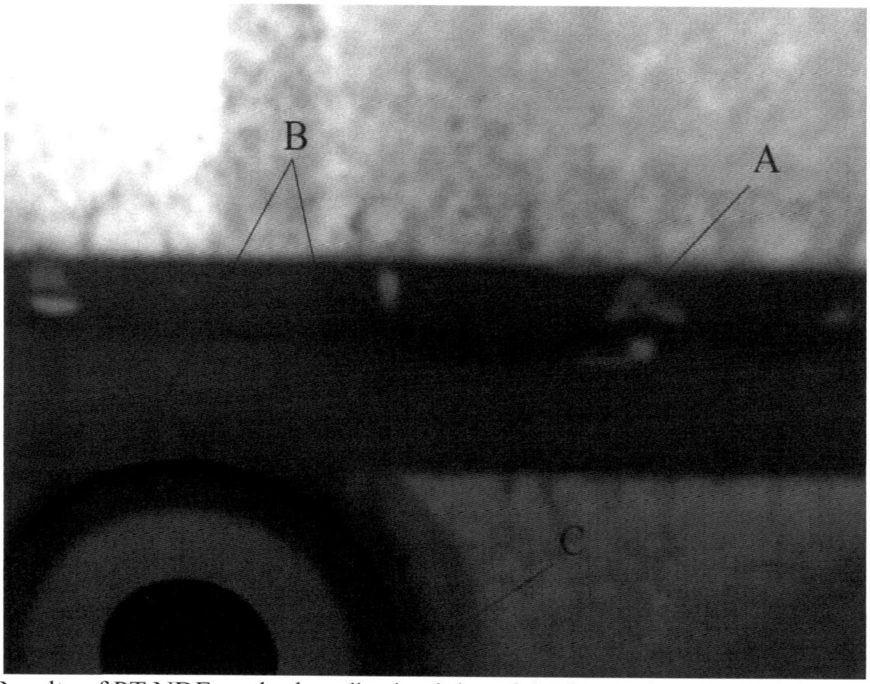

Fig. 6: Results of PT NDE method – adhesive joint of the front wing spar and the upper wing skin in the fuel tank area

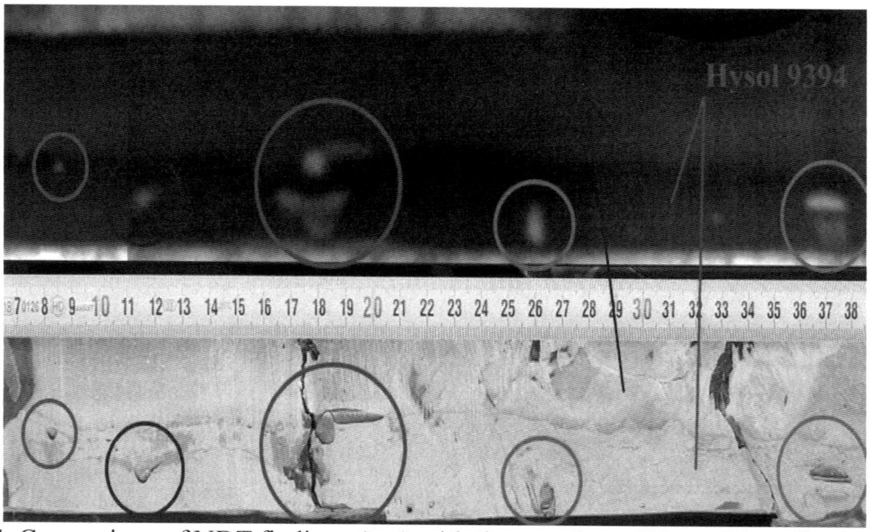

Fig. 7: Comparison of NDT findings (top) with the actual condition of the adhesive joint (bottom)

Fig. 8 shows an overview image (top and bottom view of the right wing) from the test of the entire wing, composed of individual images. The tested area is limited to adhesive joints only. The dark vertical areas represent the adhesive joints of the spar caps to the wing skin. The horizontal areas represent the adhesive joints of the ribs to the wing skin. Areas with insufficient adhesive coverage (brighter areas in the adhesive joints) can be seen practically along the entire length of the adhesive joints of the spar caps with the wing skin. Furthermore, the areas of overlapping of the outer layers of the wing skin or reinforcements of the structural openings in the skin are clearly visible.

Fig. 8: Results of PT NDE method – overview image of the tested wing area (lower wing side on the left, upper wing side on the right)

Conclusions

Described NDE testing method proved very good usability for detection of the flows in the adhesive joints of the wing skins, ribs and the load-bearing beam structure made of thin GFRP. The achieved results are clearly interpretable and usable for qualitative and quantitative evaluation of adhesive joints. Besides defects in an adhesive layer, manufacturing technology defects and inaccuracies such as adhesive overflow, foam inserts misalignments or composite layers overlaps are detectable.

This work has been supported by project No. SGS20/162/OHK2/3T/12 of the Grant Agency of the Czech Technical University in Prague.

References

[1] J. Pilař, T. Kostroun, T. Čenský, Design of active infrared thermography NDT system with optical excitation, Technical report TZP-ULT-88-2016, ČVUT v Praze. 2016

[2] T. Kostroun, J. Pilař, M. Dvořák, Design of active infrared thermography NDT system with optical excitation, in: Proceeding of 49th International conference DEFEKTOSKOPIE 2019 / NDE for Safety, VUT v Brně, 2019, p. 75-85

[3] H.D. Benítez, C. Ibarra-Castando, A. Bendada, X. Maldague, H. Loaiza, E. Caicedo, Definition of a new thermal contrast and pulse correction for defect quantification in pulsed thermography, in: Infrared Physics & Technology, Volume 51, Issue 3, Elsevier 2008, p. 75-85, ISSN: 1350-4495

EAN 2020
58th conference on experimental stress analysis

Investigation of Mechanical Behaviour of LLDPE Foil

KOTTNER R.[1,a], KUČEROVÁ K.[2,b], KOCHOVÁ P.[1,c], BOŃKOWSKI T.[3,d]

[1]NTIS - New Technologies for the Information Society, Faculty of Applied Sciences, University of West Bohemia, Technická 8, 301 00 Pilsen, Czech Republic

[2]Department of Mechanics, Faculty of Applied Sciences, University of West Bohemia, Technická 8, 301 00 Pilsen, Czech Republic

[3]Department of Biomechanical Human Body Models, New Technologies – Research Centre, University of West Bohemia, Univerzitní 8, 301 00 Pilsen, Czech Republic

[a]kottner@ntis.zcu.cz, [b]sykoka@students.zcu.cz, [c]kochovap@ntc.zcu.cz, [d]tomasz@ntc.zcu.cz

Keywords: LLDPE foil, Tensile test, Impact test, Drop tower, Nanobag

Abstract. LLDPE foil is commonly used for packing. Since this material allows large plastic deformations, it was suggested to use it as the main part of a new concept of safety systems for the automotive industry. A tensile test and an impact test with a spherical impactor were performed to investigate the mechanical behaviour of the foil. The force-displacement curves of the foil, and deceleration of the impactor and displacement of the foil with the impactor at the bottom dead centre were evaluated to obtain data for the material model.

Introduction

Low-density polyethylene (LDPE) is a frequently used material for packing. This material can be also incorporated with biological materials, such as different starch sources, to obtain partially biodegradable product [1]. The LDPE itself can be recycled and thus reduce the plastic waste [2]. This plastic has been found to be promising e.g. in asphalt mixtures and plastic bonded sand blocks [2, 3].

When comparing low-density polyethylene (LDPE) and linear low-density polyethylene (LLDPE), the strength and the elongation (energy to break a unit volume of material) is higher in the case of LLDPE [4]. Therefore, LLDPE was suggested to use it as the main part of Nanobag system [5] - a new concept of safety systems for the automotive industry. The suitability of the LLDPE foil for the Nanobag safety system should be assessed using numerical simulations of car crashes such as [6].

Both LDPE and LLDPE foils are anisotropic. The highest strength is in the longitudinal (machine) direction [4]. The crystallographic deformations were studied in [4] using the scanning electron microscopy and differential scanning calorimetry techniques for lamellar structure and orientation. The LLDPE structure shows more randomly orientated and thicker lamellae than that of the LDPE foil. The result of this structure is a more balanced toughness.

The foil produced by TICHELMANN was used for the study. The thickness was 12 μm, the density was 0.091-0.092 g/cm^3, as provided by the material data-sheet from the producer.

This work aimed to investigate the mechanical behaviour of the LLDPE foil and subsequently to obtain material data for the numerical simulations. The mechanical behaviour was investigated using a tensile test and an impact test.

Tensile test

The 574LE2 TestResources machine was used for the uniaxial tensile test of rectangular samples having the initial length $l_0 = 5$ mm and the width $w = 10$ mm. Tensile samples were cut from the foil in two perpendicular directions (see Fig. 1a), commonly referred to as machine (M) direction and transverse (T) direction, related to the foil manufacturing process [7]. Displacements of mechanical grips Δl (see Fig. 1b) were prescribed and force F was measured. Three loading velocities were applied, namely 0.2 mm/s; 20 mm/s; 100 mm/s. Six samples were tested per each velocity group. The engineering tensile strength and the ultimate elongation were determined from the force-displacement curves. The Mann-Whitney U-test was used to compare the resultant data between velocity groups.

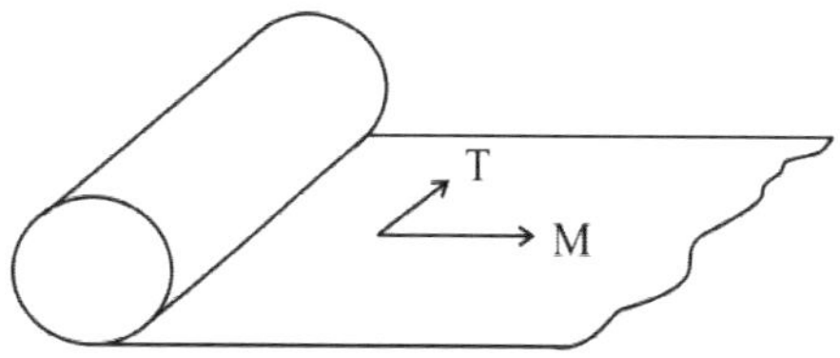

(a)

(b)

Fig. 1: a) Material directions, b) a tensile sample in the mechanical grips

Impact test

A drop tower designed by the authors was used to test the foil. The measured foil was n-times ($n = \{8, 9, 10\}$) wrapped in M direction around a specially designed frame imitating the clamping of the foil in the Nanobag safety system (see Fig. 2). A spherical impactor having the diameter of 149 mm and the weight of 10.72 kg dropped from the heights $h_1 = 1$ m and $h_2 = 1.5$ m. The deceleration of the impactor a was measured by KISTLER 8742A5 accelerometer. The displacement of the impactor at the bottom dead centre w_{max} was measured using the Micro-Epsilon optoNCDT 2300-50 laser. The maximum measurable value of the displacement was set to $w_{limit} = 275$ mm, as displacements exceeding this value are not suitable for the Nanobag safety system. The sampling frequency was 26 kHz.

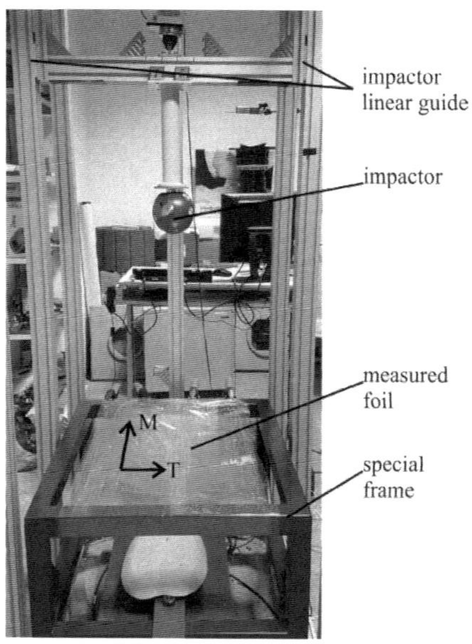

Fig. 2 The impact test

Results

The force-displacement curves of the tensile test are shown in Fig. 3. Evaluated mean values of the engineering tensile strength and ultimate elongation are listed in Table 1 and Table 2. The tensile strength of the foil is approximately two times higher in the M direction than in the T direction. Conversely, the elongation is approximately two times lower in the M direction when compared with the T direction. The influence of the loading velocity on the force-displacement curves was not significant except for the strength in the M direction when the strength was reduced in the case of the fastest velocity (54 MPa at 100 mm/s).

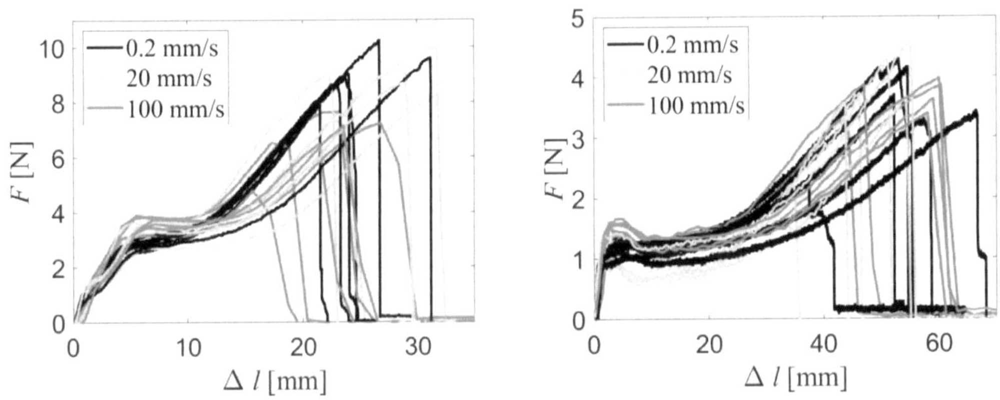

(a) (b)

Fig. 3: The force-displacement curves: a) the M direction, b) the T direction

Table 1: The mean values of the engineering ultimate strength

Loading velocity [mm/s]	Strength in M direction [MPa]	Strength in T direction [MPa]
0.2	74.5	28.3
20	72.3	30.0
100	54.0	29.9

Table 2: The mean values of the ultimate elongation

Loading velocity [mm/s]	Elongation in M direction [%]	Elongation in T direction [%]
0.2	499	1065
20	555	950
100	427	1091

The impactor deceleration a and the impactor displacement w in relation to the number of the foil coatings n are shown in Fig. 4 and Fig. 5, respectively. It can be seen that a higher number of the foil coatings n resulted in lower value of the impactor displacement at the bottom dead centre w_{max} (see Table 3). In the case of the drop height h_2: when $n < 10$, the foil did not stop the impactor before the displacement w reached the limit $w_{limit} = 275$ mm and when $n = 8$, the impactor even hit the drop tower frame. Therefore, the maximum deceleration value was 837 ms^{-2}. It follows from the above that the necessary number of the foil coatings was $n = 10$ to comply with the w_{limit} limit for both drop heights. When $n = 10$, maximum deceleration values were 154 ms^{-2} and 342 ms^{-2} for the drop heights $h_1 = 1$ m and $h_2 = 1.5$ m, respectively.

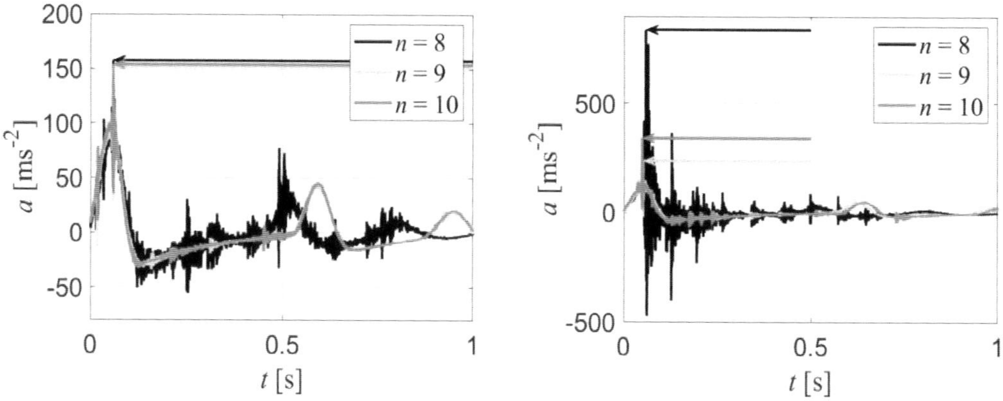

(a) (b)

Fig. 4: The deceleration in time: a) h_1 = 1 m, b) h_2 = 1.5 m

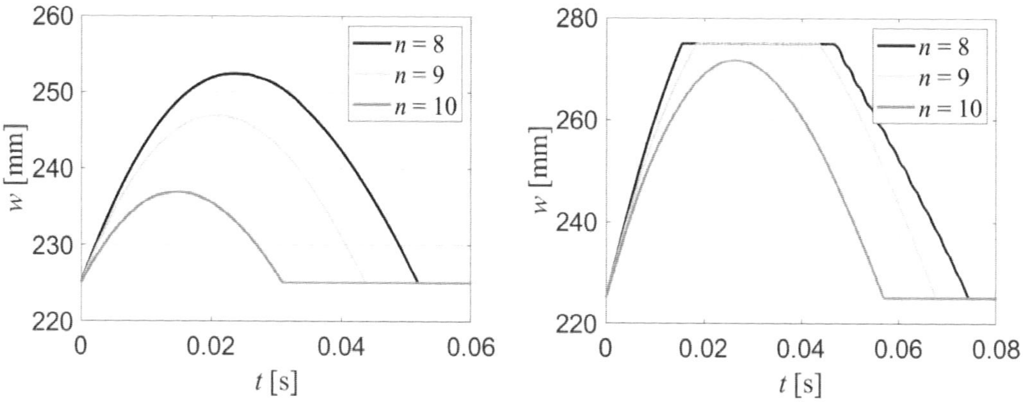

(a) (b)

Fig. 5: The displacement of the impactor in time: a) h_1 = 1 m, b) h_2 = 1.5 m

Table 3: The displacement of the impactor at the bottom dead centre

Number of foil coatings N	w_{max} [mm] with drop height h_1 = 1 m	w_{max} [mm] with drop height h_1 = 1.5 m
8	252.4	over 275
9	247.0	over 275
10	236.9	271.6

Conclusions

The material data suitable for the crash numerical simulations involving LLDPE foil were obtained. The tensile curves in both material directions did not significantly depend on the used loading velocity, except for the strength in the machine direction which was reduced at the fastest velocity. Both tensile and impact tests have shown the promising energy-absorbing properties of the foil. This support the usage of the foil not only for packing but also for further application as the part of Nanobag system, where the energy-absorbing properties are crucial. The ten-times wrapped foil complied with the impactor displacement limit for both tested drop heights.

Acknowledgement

This publication was supported by the project LO1506 of the Czech Ministry of Education, Youth and Sports under the program NPU I, from European Regional Development Fund-Project *Application of Modern Technologies in Medicine and Industry* (No. CZ.02.1.01/0.0/0.0/17 048/0007280), and by the internal grant project SGS-2019-002 *Biomechanical models of human body, biological tissues and biomechanical processes with application in industry and medicine*.

References

[1] S. Kormin, F. Kormin, M.D.H. Beg, M.B.M. Piah, Physical and mechanical properties of LDPE incorporated with different starch sources, IOP Conf. Ser.: Mater. Sci. Eng. 226 (2017) 012157, 1–9.

[2] P.O. Awoyera, A. Adesina, Plastic wastes to construction products: Status, limitations and future perspective, Case Studies in Construction Materials 12 (2020) 1-11.

[3] A. Kumi-Larbi, D. Yunana, P. Kamsouloum, M. Webster, D.C. Wilson, Chr. Cheeseman, Recycling waste plastics in developing countries: Use of low-density polyethylene water sachets to form plastic bonded sand blocks, Waste Management 80 (2018) 112-118.

[4] K. Mezghani, S. Furquan, Analysis of dart impact resistance of low-density polyethylene and linear low-density polyethylene blown films via an improved instrumented impact test method, Journal of Plastic Film and Sheeting (2012) 1-17.

[5] A. Hanuliak, Safety Restraint System for Motor Vehicles, PV 2017-307 (CZ), B60R 21/08 (2006.01), WO2018/219371, valid from 19 November 2018, signed 06 December 2018.

[6] J. Špička, A. Hanuliak, L. Hynčík, L. Kovář, Virtual assessment of new safety systems for future mobility modes, in: Proceedings of the 2nd International Symposium on Future Mobility Safety Science and Technology, Plzeň, Czech Rep., 2019, pp. 25.

[7] G. H. Michler, F. J. Baltá-Calleja, Nano- and micromechanics of polymers, Hanser, München, 2012.

Mechanical Properties of Coated Fabric by Comparing Computational Model to Experiment

KOVÁČIK M.[1,a], FUSEK M.[1,b]

[1]Deparment of Applied Mechanics, Faculty of Mechanical Engineering, VŠB – Technical University of Ostrava, 17. listopadu 2172/15, 708 00 Ostrava, Czech Republic

[a]michal.kovacik@vsb.cz, [b]martin.fusek@vsb.cz

Keywords: Fabric, Tensile testing, Crimp interchange, Finite element method, Curve fitting

Abstract. Coated fabrics are used in many applications such as fuel bladders for example. When subjected for testing, accurate constitutive data have to be found. However testing for strength as well as the resistance to stretching of such coated fabrics come with specific difficulties. For once, the thickness as well as the volume ratios of the fibre and coating cannot be stated very accurately. Secondly, the resilience, e.g. the stiffness of the stretching is generally low compared to metallic materials, which causes difficulties in simplifications like when using engineering stresses and strains. Both of these issues can be overcome using correlation to a finite element analysis of the actual experiment, which is uniaxial tensile testing in this case. At last, coated fabric materials are generally not isotropic, which requires tensile testing in different directions. However, an accurate finite element representation can be used to find more accurate values of constitutive parameters such as elastic moduli for each direction, and even the Poisson ratio. The strength estimates are however more difficult to represent accurately using a single material as they are still composite materials with more layers which should be represented accurately using for example the Hill criterion.

Introduction

Coated fabrics are used in many flexible applications such as tensile structures in architecture, belts for transport or flexible containers in this particular case. The fabric in this study is used in bladders for water storage. A woven textile coated in polyvinylchloride is suitable for such purpose since it is water-tight and non-toxic. However, since two different materials are used to form one inseparable material, it is basically a composite.

It is quite easy to find a constitutive model for a composite by linear mixing rules by volume fractions [1]. For example, Young's modulus for a composite E can be found by a linear combination of the Young's moduli of both materials E_T, E_C (T stands for textile, C for coating) multiplied by their volume fractionsV_T, V_C:

$$E = E_T V_T + E_C V_C \tag{1}$$

However, fabric can be looked upon as a special case. The textile in a fabric has a pattern caused by the weave of the textile, which causes non-uniform thickness of the whole, which causes problems in evaluating the fractions.

There is a possibility to disassembly the fabric into yarns and measure the properties of the yarns for purposes like in work [2] and verify the data by measuring the fabric properties. In such case, the data are not usually abstracted to variables as elastic modulus, since the geometry of the yarns is still uncertain.

These data can be theoretically used for coated fabrics with linear superposition of layer properties; however such procedure is ineffective for general properties such as the elastic modulus. The coating layer is usually small and can be considered isotropic in many cases such as in work [3]. It can be observed from figure 2 that the coated fabric is clearly made using the plain weave consisting of warp and weft yarns, which allows to consider these two axis of material symmetry. For an orthotropic description, one additional direction needs to be measured for the shear modulus. In fabrics, this is known as the bias direction which is tilted 45° to the warp and weft direction.

There is also the crimp interchange phenomenon (a term used in work [3]) present in constitutive relationships in the study. This effect is associated to the specific properties of fabrics caused by the weave which is originally fabricated with the warp yarns being tensioned and weft yarns being woven around them causing the so-called crimp. However, when weft yarns are being tensioned, they straighten first and bend the warp yarn around which is the so called crimp interchange. Since the flexural stiffness of the yarns is very low, this phase imposes virtually no additional forces and influences the deflection only. This can be seen in the evaluation part of this paper.

There is an extra difficulty with the scale of the stiffness of the fabric. The changes in geometry due to extension in length and contraction in width affects the difference of true and engineering stresses and strains significantly. This difficulty along with the uncertainty in the thickness is addressed here by making a computational model of the specimen with average width and thickness subjected to same forces as the actual experiment. By varying the Young's modulus and the Poisson ratio of the computational model, one can find an accurate displacement corresponding to the experiment.

The strength of a composite is usually can be usually expressed in terms of layers such as maximum stress criterion or in fractions such as the Hill criterion [5]. However, since the strength of each layer is not known here, the strength is evaluated only for the material as a whole.

Test setup

The experiment was carried out on a tensile testing device. The dimensions should be as large as possible to minimize the effect of the pattern caused by the weave; however the dimensions of the specimen are limited, because they need to fit in the clamps used for the test. Although this is not a metallic material, the geometry can be set the same as for metallic specimens for reference, therefore, the specimen geometry was acc. to ISO 6892-1 [6] for all three directions. The size is limited since there is a limited supply of the material. For each direction, three specimens are made. The total number of specimens is then 9.

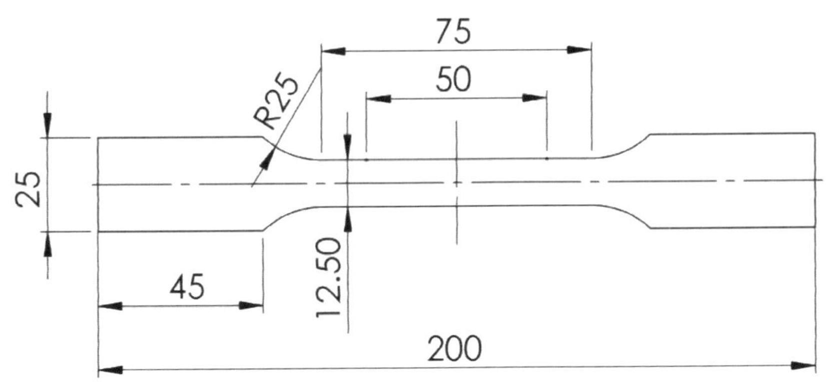

Fig. 1: Specimen geometry

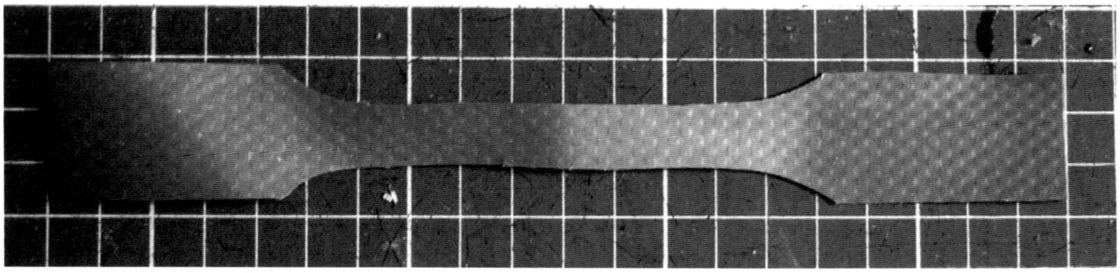

Fig. 2: Specimen example cut on the warp direction

Computation

The computation was done using the finite element method. The geometry was set according to the specimen, with the thickness set as an average of the measured thickness in different spots of the specimen. The actual average of the thickness of the specimen is around 1mm so unit thickness was used for the calculation.

The calculation was carried out using linear 2D membrane elements in MSC.Marc with updated large deformations and strains. The clamp is simulated using multi-point constraints with prescribed velocity 20mm/min according to the experiment. Only the quarter of the model is modelled with accounted symmetry. The forces can then be read as force reactions. A linear elastic constitutive model was used, where the Young's modulus and Poisson ratio constants were varied until behaviour of the model similar to the actual specimen was obtained.

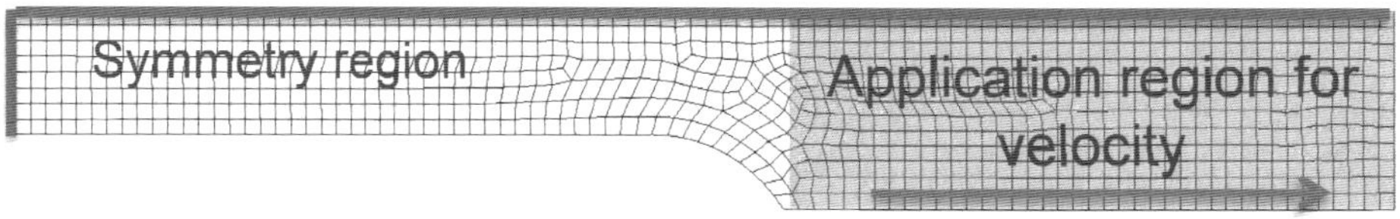

Fig. 3: Computation model with boundary conditions

Results

The test results are summarized by force vs. deflection relationships. The same data were computed by the simulation and both relationships were compared below for the warp direction, where values of 340 MPa for the elastic modulus and 0.3 for Poisson ratio used for the calculation.

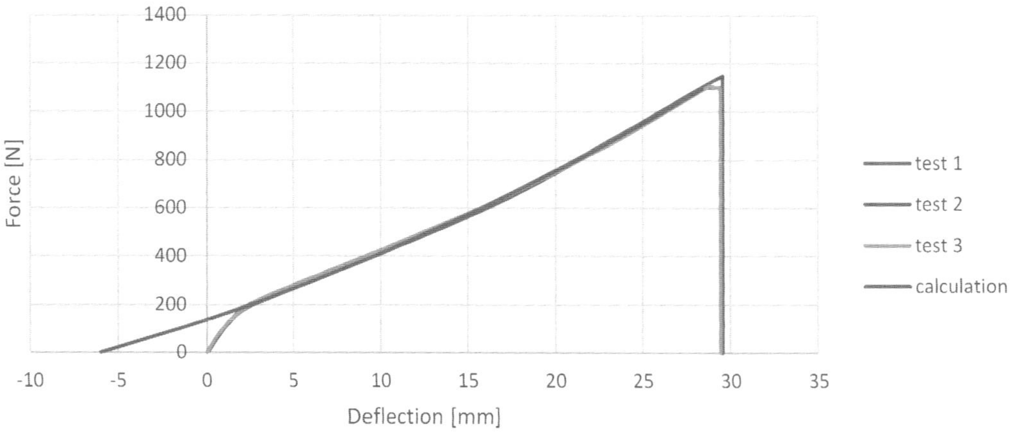

Fig. 4: Comparison of the experimental and computed values for the warp direction

For the weft direction, the same values of the elastic modulus and the Poisson ratio were used to fit the experimental data properly.

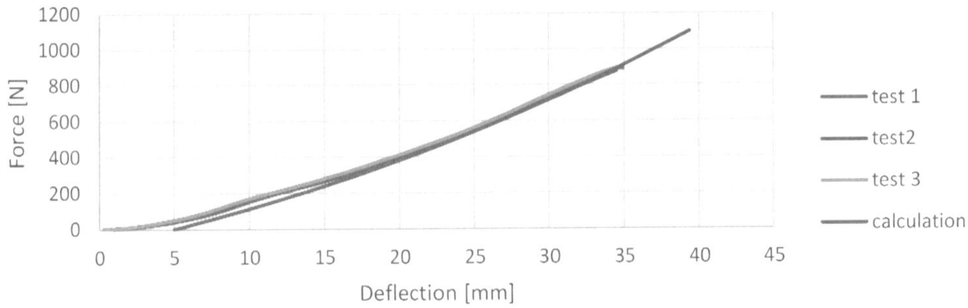

Fig. 5: Comparison of the experimental and computed values for the weft direction

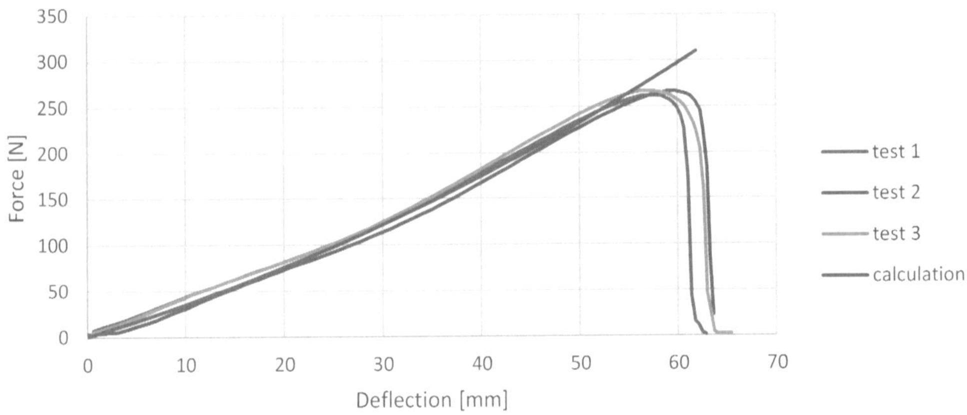

Fig. 6: Comparison of the experimental and computed values for the bias direction

The bias direction used a value of the elastic modulus 50 MPa. The results of the computation show stress levels after curve fitting on figure 4 corresponding to the maximal stress at the failure for the warp direction

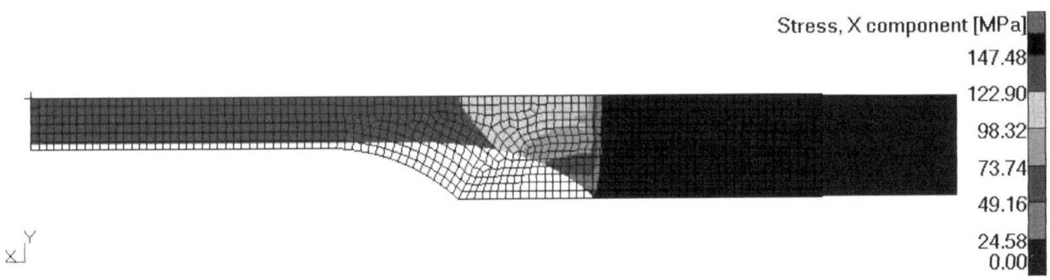

Fig. 7: Stress levels for computation in the warp direction

The failure occurred in all cases in the narrower part of the specimen at the beginning of the radius which corresponds to the maximal stress observed in the calculation. An example of a failure in the warp direction specimen can be observed below.

Fig. 8: Failure example in the warp direction

Conclusions

Based on the figure 7, the actual uniaxial strength of the material for the material as a whole in the warp direction can be associated to value somewhere around 147 MPa and 120 MPa for the weft direction. The strength here however does not account failure of single layers of the composite. Such case requires further study and experiments. Bias direction specimens have no continuous yarns from one and to the other and are held only together by the weave and the coating. Therefore, their stiffness and strength which is around 36 MPa is quite low compared to the warp and weft directions relationships.

The standard procedure in determining the Young's modulus uses a conversion of the force to engineering stresses using the initial cross-sectional area and a conversion of the deflection to engineering strains using the initial length, where regression between these two measures can be made to find the value. Such procedure would be highly inaccurate in this case due to excessive elongations and contractions which can be observed in figure 7. The excessive elongations and contractions cause the force-deflection relationships to curve concavely even though a linear stress-strain constitutive model was used, since the difference in true and engineering values differ greatly in such scale of stiffness. Since the curvature of the relationship is dependent not only on the elongation, but contraction as well, it is partly influenced by the value of the Poisson ratio, where value of 0.3 showed accurate enough for proper fitting of the curvature in the experimental data.

Unlike the bias direction relationship (fig. 6) the warp and weft directions (figure 4 and 5) show different initial slopes for the curve. These parts of the curve were omitted in the fitting process since they can be associated to the so-called crimp interchange phenomenon. The pattern of the initial slopes for both warp and weft direction seem to correspond to the shapes described in works [3,4]. The bias direction is not affected by it because the phenomenon is present only in the warp and weft directions.

References

[1] D. Hull, T. W Clyne. An introduction to composite materials. Cambridge Press, 1996.

[2] Y. Duan, M. Keefe, T.A. Bogetti, B.A. Cheeseman, B. Powers. A numerical investigation of the influence of friction on energy absorption by a high-strength fabric subjected to ballistic impact. International Journal of Impact Engineering, 2006

[3] K. Carbonez, Finite element simulation of tensioned membrane structures in deployable systems (Master dissertation), Universiteit Gent, 2013.

[4] K. Vandenboer, Numerical assessment of material models for coated fabrics in foldable tent structures (Master dissertation), Universiteit Gent, 2011.

[5] MSC Software, Marc® 2011 Volume A: Theory and User information, 2011

[6] ISO 6892-1:2019 Metallic materials – Tensile testing – Part 1: Method of test at room temperature

EAN 2020
58th conference on experimental stress analysis

Experimental Testing of the Effect of PEEK Surface Treatments on the Shear Strength of Glued Joints

KRÁL M.[1,a], BOHÁČIK P.[2,b], KRUML J.[1,c], UHER O.[1,d]

[1]Czech Aerospace Research Centre, Design and Development Department, Space Division, Beranových 130,199 05 Praha - Letňany, Czech Republic

[2]Czech Aerospace Research Centre, Chemical Technologies Department, Space Division, Beranových 130,199 05 Praha - Letňany, Czech Republic

[a]kral@vzlu.cz, [b]bohacik@vzlu.cz, [c]kruml@vzlu.cz, [d]uher@vzlu.cz

Keywords: Polyetheretherketon, Apparent shear strength, Glued Joints, Single lap-shear test

Abstract. Adhesive bonding is a way of joining two components together. Bonded joints have many advantages in comparison with other types of joints (bolts, rivets, welds, etc.), such as low weight, more uniform stress distributions, easy manufacture and the possibility to join dissimilar materials, possibility to join very thin adherents and very large surfaces. On the other hand, bonded joints also have a number of disadvantages. They are not suitable for structures subjected to fatigue, require surface preparation and have limited stability to elevated temperatures. With the rapid development of the properties of adhesives (higher peel and shear strength and also allowable ductility up to failure), their use in all areas of industry has expanded.

Introduction and motivation

One of the areas of research activities in the Design and Development Department of Space Division in the Czech Aerospace Research Centre is 3D printing focused on Polyetheretherketon (PEEK). PEEK is a semi-crystalline organic thermoplastic with glass transition temperature of approximately 143 °C and melting temperature of around 343 °C. Due to its properties, PEEK is used in aerospace, chemical and pharmaceutical industry and is expected to be used in more important areas. PEEK is also used for ultra-high vacuum applications thanks to very low outgassing. Most plastics contain volatile organic compounds which, when vaporized, are poisonous chemicals. Direct contact with these compounds can cause serious health problems. High temperatures usually accelerate outgassing. Since many industrial applications involve high temperatures, plastic outgassing becomes a critical factor in choosing plastic materials. PEEK is also used for good mechanical properties. The yield strength of PEEK is approximately 110 MPa, Young modulus is 4400 MPa and specific mass is 1.32 g/m^3 [1, 2].

PEEK is also suitable for 3D printing. However, performance of most of the 3D printers is not sufficient to melt PEEK. In comparison to other materials widely used in 3D printing, PEEK has a higher melting point (343 °C). To solve this problem, it is necessary to apply a nozzle heated to about 400 °C [2]. In the Design and Development Department of Space Division in the Czech Aerospace Research Centre, 3D printing using PEEK based on the FFF technology is used for the development of spacecraft elements. A typical example is presented in Fig. 1. To connect this honeycomb with skins based on thermoset matrix composite, it is necessary to use some adhesive. Using of adhesives is necessary in more applications where

PEEK parts are bonded together or to other materials. This is the main motivation for the development of surface treatments in order to improve the properties of bonded joints of PEEK 3D printed parts. Also PEEK parts made by injection are used in the design of aerospace structures, but this technology does not offer so big freedom in the choice of the shape.

Fig. 1: 3D printed honeycomb for sandwich panel made of PEEK with integrated places for plastic inserts

As already mentioned structural bonding is frequently used today for connecting two components together. This connection must ensure a satisfactory structural strength and good durability. A good design of bonded connection is a necessary condition for reliable function. Design of bonded connections with many useful information are described in [3].

It is generally known that bonding of plastics can cause problems especially in the case of plastics with low surface energy [4]. Today adhesives for these types of plastics are available. But there are not many information in terms of bonding of PEEK with PEEK or with other plastics. Mechanical properties of these connections of PEEK are not described on a satisfactory level.

Description of test specimens, used adhesive and methods

Test specimens were not designed in the line with recommendations of commonly used standards. This approach is possible thanks to the fact, that only comparison of individual surface treatments was expected. Tests were not performed in order to obtain design allowables, where the conformity with standards is necessary.

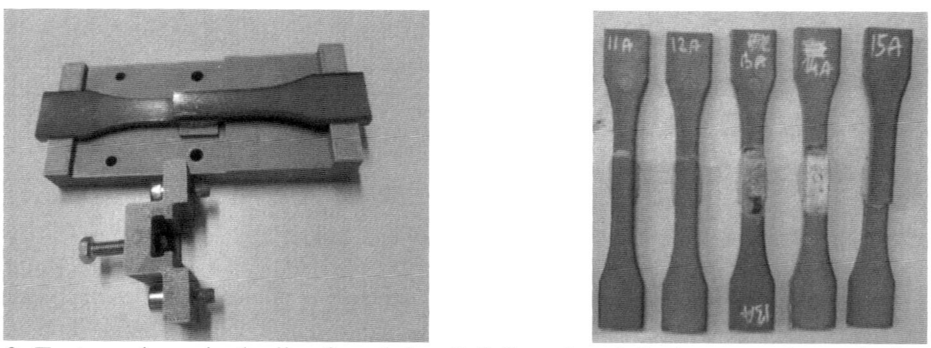

Fig. 2: Test specimen in the jig after gluing (left figure) and final test specimens before testing (right figure)

Test specimens have a typical "dog bone" shape. To ensure a good repeatability of test specimens during the process of gluing a special jig was designed and manufactured using 3D printing, see Fig. 2. Trial tests started with specimens made by injection. The motivation for this was an effort to compare results with other results available in the literature. Bonding of PEEK 3D printed parts is discussed in the literary sources only minimally.

During trial tests, three surface treatments in combination with one type of adhesive were tested. 3M™ Scotch-Weld™ Structural Plastic Adhesive DP8005 was used for the preparation of test specimens. It is a two-part acrylic adhesive specially formulated to bond low surface energy plastics without priming. Mechanical properties can be found in [5]. Based on manufacturer information, see [5], typical strength reached in single lap shear test is 16.3 MPa for joint of two FRP adherends, where cohesion failure was reached. Other example is a joint of two ABS adherends where the strength is 6.7 MPa and adhesion failure was reached.

In terms of surface treatments, three types including acetone degreasing (immersion in a bath for 5 minutes, acetone, puriss. p.a.), sandblasting with acetone degreasing (blasting agent Alu. oxide 120 mesh, pressure 2 bar, distance 0.3 m, blasting angle 45 ° and then immersion in a bath for 5 minutes, acetone, puriss. p.a.) and nickel plating (activating bath 20% H_2SO_4 was used for 5 minutes prior to chemical Ni plating – 20 minutes in 10% $NiSO_4.6H_2O$ with 5% NaH_2PO_2 water solution at 50 °C) were tested till now.

Test specimens were subjected to a single lap-shear test according to the standard ASTM D3163-01(2014) [6]. Test set-up is shown in Fig. 3. Test speed was 1,3 mm/s. Since only baseline screening tests were performed, three test specimens of each type of surface treatment were tested with some exceptions.

Fig. 3: Single lap-shear test - the test set-up

Discussion of the results

The results are presented in Table 1. The evaluation was performed on the basis of ASTM D3163-01. The original idea was to test three test specimens for each surface treatment type during trial tests. This plan was followed only in the case of sandblasting, see Tab. 1. The results for nickel plating are determined only on the basis of two specimens and therefore have only informative character. More tests will be performed to obtain a satisfactory statistical sample. It can be seen that the coefficient of variation is very low. This is probably caused by the good repeatability of test specimens thanks to the jig for their preparation and also good repeatability of the surface treatments quality. The highest apparent shear strength was reached in the case of sandblasting surface treatment. This finding is in the line with generally known fact, that sandblasting has a positive effect on the shear strength of glued joints.

Good results were also obtained in the case of nickel plating surface treatment although the apparent shear strength is significantly lower. Acetone degreasing without any other treatment shows the lowest strength and the highest scatter in results, which was expected.

Test specimens after testing are presented in Fig. 4. Failure mode of the specimens can be defined as cohesive failure of the adhesive. In some cases the combination of cohesive failure of adhesive and the adhesion failure of adhesive was determined. Cohesive failure of adhesive is a valid failure mode. More information about valid failure modes during testing of shear strength of glued joints can be fined e.g. in [7].

Tab. 1: Results of apparent shear strength testing with statistical evaluation

	Sandblasting and Acetone degreasing [MPa]	Acetone degreasing [MPa]	Nickel plating [MPa]
1	6.37	3.89	4.92
2	6.86	3.70	4.80
3	6.93	4.03	-
4	-	3.31	-
Mean [MPa]	6.72	3.73	4.86
Standard dev. [MPa]	0.305	0.313	0.085
Coef. of variation [%]	4.54	8.37	1.75

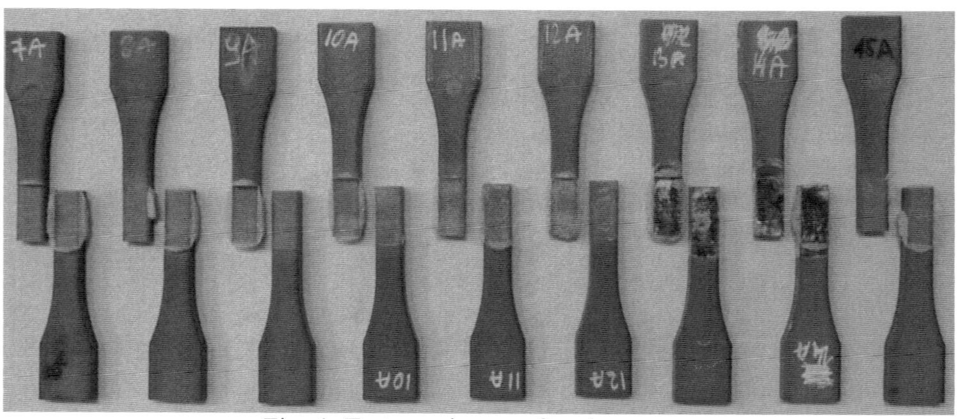

Fig. 4: Test specimens after the test

Conclusions

Presented results show that it is possible to realize the glued joint between PEEK adherends using common adhesive. Appropriate surface treatment is a necessary step before gluing. Although the statistical sample was low in these baseline experiments, the results show, that it is possible to affect the properties of PEEK surface before gluing and get a satisfactory results.

The results show a very low scatter in apparent strength values. In terms of experimental determination of design allowables it is a positive finding. For the calculation of design allowables, both mean value and standard deviation are necessary. With increase in standard deviation, decrease in design allowables is obtained. Generally, low scatter during measuring of mechanical properties is a good expectation for the design of reliable structure. Single lap shear test is used for the comparison of the effect of surface treatments. In order to get more precise strength values, double lap shear will be performed.

In other step of the research, another surface treatment using plasma will be tested. It is generally known fact that plasma surface treatment has a positive effect on the surface properties in terms of glued joints.

In terms of tested adhesives it is expected to test some foil adhesive. Foil adhesives are used for the manufacturing of sandwich panels. The use of these panels in the design of aircraft and spacecraft structures is very frequent.

References

[1] S.A. Mushin, P. V. Hatton, A. Johnson, N. Sereno, D. J. Wood, Determination of Polyetheretherketone (PEEK) mechanical properties as a denture material, The Saudi Dental Journal (2019) 382-391, 31(3), ISSN 10139052.

[2] Ch. Yang, X. Tian, D. Li, Y. Cao, F. Zhao, Ch. Shi, Influence of thermal processing conditions in 3D printing on the crystallinity and mechanical properties of PEEK material, Journal of Materials Processing Technology (2017) 1-7, ISSN 09240136.

[3] L.J Hart-Smith, Design Methodology for Bonded-Bolted Composite Joints, Volume I. Analysis Derivations and Illustrative Solutions, 1982

[4] N.G.C. Barbosa, R.D.S.G. Campilho, F.J.G. Silva et al, Comparison of different adhesively-bonded joint types for mechanical structures, Applied Adhesives Science (2018)

[5] 3M Scotch Weld Structural Plastic Adhesive DP 8005, manufacturer datasheet available from https://multimedia.3m.com/mws/media/79749O/3m-scotch-weld-structural-plastic-adhesive-dp-8005.pdf

[6] ASTM D3163-01(2014), Standard Test Method for Determining Strength of Adhesively Bonded Rigid Plastic Lap-Shear Joints in Shear by Tension Loading, ASTM International, West Conshohocken, PA, 2014.

[7] F.C. Campbell, Adhesive Bonding and Integrally Cocured Structure: A Way to Reduce Assembly Costs through Parts Integration, Manufacturing Processes for Advanced Composites (2004) 241-301, Elsevier.

Inflation-Extension Behaviour of the Human Vena Saphena Magna

KRATOCHVÍL O.[1,a], CHLUP H.[1,b], HORNÝ L.[1,c] ŠPAČEK M.[2,d], MĚŘIČKA P.[3,e],

[1]Faculty of Mechanical Engineering, Czech Technical University in Prague, Technická 4, 160 00 Prague, Czech Republic

[2]2nd Department of Surgery – Department of Cardiovascular Surgery, 1st Faculty of Medicine, Charles University in Prague and General University Hospital in Prague, U nemocnice 2, 120 00 Prague, Czech Republic

[3]Tissue Bank, University Hospital Hradec Králové, Sokolská 581, 500 05 Hradec Králové, Czech Republic

[a]ondrej.kratochvil@fs.cvut.cz, [b]chynek@seznam.cz, [c]lukas.horny@fs.cvut.cz, [d]miroslav.spacek@vfn.cz, [e]pavel.mericka@fnhk.cz

Keywords: Axial prestretch, inversion point, Pressurization, Saphenous vein

Abstract. Saphenous vein grafts are still frequently used in both coronary and peripheral bypass surgery. Although much work is now being done to deepen our knowledge of the mechanobiology of graft remodelling, these processes are still not completely understood. To contribute to an effort aimed at filling this gap, a pressurization experiment with a tubular sample of the human great saphenous vein was carried out. Our study particularly focused on the so-called inversion point, known from arterial mechanics. Inversion point is the value of the axial prestretch at which a pressurized tube does not move axially. The experiment confirmed that axial inversion takes place in the mechanical response of the prestretched saphenous vein. This means that there is a value of the axial prestretch at which pressure-induced elongation changes to pressure-induced shortening of the inflated tube.

Introduction

Saphenous vein grafts are still frequently used in both coronary and peripheral bypass surgery especially when alternatives like the internal thoracic artery or the radial artery are not available [1,2]. In contrast to their arterial counterparts, properties of vein grafts, however, are optimized for a mechanical environment very different from arterial conditions. An obvious consequence of this fact is that immediately after surgery, remodelling processes are triggered and the vein adapts to the elevated blood pressure, flow rate, and oscillatory wall shear stress [3]. Although much work is now being done to deepen our knowledge of the mechanobiology of graft remodelling, these processes are still not completely understood [3–6].

In particular, a fully nonlinear constitutive description, validated experimentally at multiaxial stress states, is rare in the scientific literature [4–6]. To contribute to an effort aimed at filling this gap, a pressurization experiment with a tubular sample of the human great saphenous vein was carried out.

It is well known that arteries sustain significant axial prestretch in their in situ positions [7–9]. Since venous grafts are placed into an arterial position during bypass surgery, we propose to study how their mechanical response is changed by axial prestretching. To the best of our knowledge, our study is the first that focuses on the effect of axial prestretching on the biomechanical response of the human great saphenous vein.

Methods

Instead a simple inflation experiment, an inflation-extension test was carried out. In this experimental technique, a tubular sample is mounted into the pressurization setup vertically. The upper end of the tube is fixed, whereas the lower end can freely move in an axial direction. The axial prestretch was induced by hanging a weight of constant mass at lower end of the tube. The whole set-up is illustrated in Fig. 1. The loading protocol comprised of 8 quasi-static pressurization cycles from 0 to 20 kPa at a specific axial weight.

The mechanical response of the sample was recorded onto a PC. It consisted in digitization of the pressure signal measured with pressure transducer Cressto (0 – 30 kPa, 500 Hz) and in the recording of digital photographs of the deformed sample (two 5MPx cameras at 20 Hz). In the data post-processing, the deformation of the sample was determined with the help of an in house developed image analysis procedure, based on edge detection implemented in Matlab. In a final step, both pressure signal and measured deformation were synchronized.

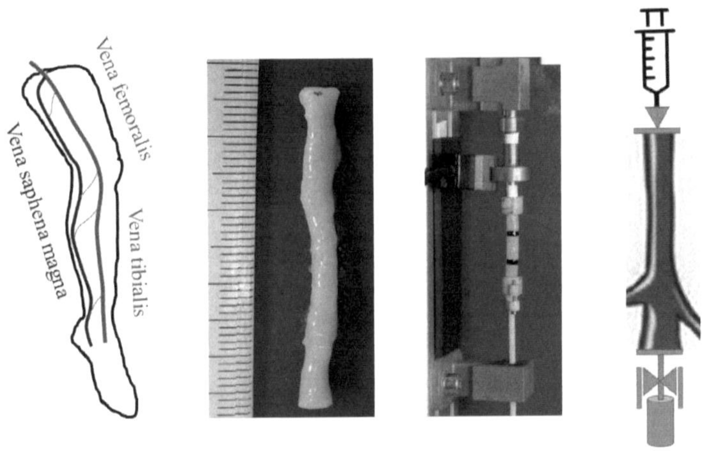

Fig. 1: Left panel shows a sketch of the lower extremity with a position of the great saphenous vein. The two middle panels display the sample of the great saphenous vein before and after mounting to the experimental setup. Finally, the right panel depicts a scheme of the inflation-extension experiment (pressurization and hanging mass).

Results

Figure 2 shows the results of the experiment. The first panel (from left) shows the pressure–circumferential stretch (the ratio of current to reference radius of the inflated tube) dependences obtained under five different values of axial preload (0, 10, 50, 200, and 500 g). The second panel depicts pressure–axial stretch (the ratio of the current to reference length of the pressurized tube) curves obtained in the same experiment. All depicted curves correspond to the loading part of the eighth cycle, when the sample exhibited a preconditioned response.

Two mechanical properties are clearly visible on the graphs: First, in axial response, it is a presence of the inversion phenomenon, at which pressure-induced elongation changes to pressure-induced shortening (cf. red curve, 0 g, and yellow curve, 50 g, in the second panel). It is worth noting that at the inversion point, which can be roughly identified with orange curve (10 g), no work is done on axial displacements, because the tube does not axially move. It follows that all the mechanical work of the loading pressure is carried out by radial distension of the tube at such a point.

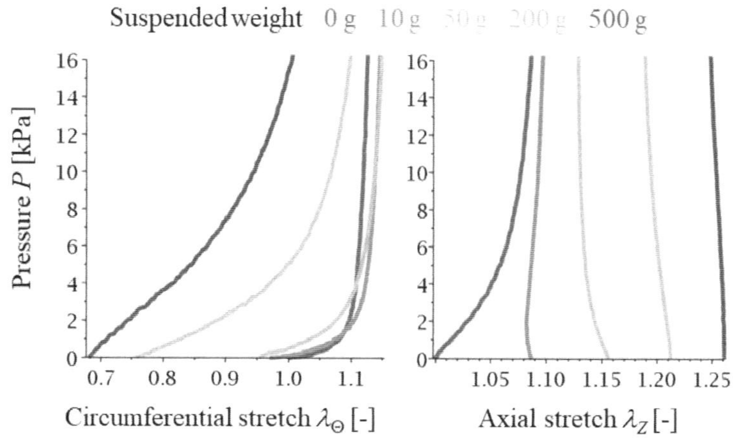

Fig. 2: Left panel shows pressure–circumferential stretch curves and the right panel displays pressure–axial stretch relationships obtained with different axial load. A colour indicates weight used as described above the image.

Table 1: Circumferential stiffness at 10 kPa

Suspended weight [g]	Circumferential stiffness [kPa]
0	740.9
10	695.8
50	416.3
200	132.8
500	70.57

The second interesting phenomenon is related to a stiffness of the pressurized tube. When curves on the left panel are compared with each other, notices that their tangents made at a given pressure have lower slopes when created on prestretched responses. In Figure 2, compare for instance green curve (500 g) and red curve (0 g) at 10 kPa. For more convenience, specific numerical values are listed in Table 1. It is clear that there is a value of the pressure, for which the axially prestretched tube exhibits more compliant response than the non-prestretched one. Such behaviour is known from arterial mechanics [8], but has not been discussed in the context of venous physiology so far.

Conclusions

Our results suggest that axially prestretched veins show similar phenomena to prestretched arteries. In particular, the prestretch reduces their stiffness during pressurization, and pressure-induced elongation may change to pressure-induced shortening. It follows that axial prestretch could offer a mechanism for how the response of venous grafts could be pre-modulated.

Acknowledgement

Czech Science Foundation has supported this study via project GA18-26041S entitled “Effect of axial prestretch on mechanical response of nonlinearly elastic and viscoelastic tubes”.

References

[1] M. Gaudino, D. Taggart, H. Suma, J.D. Puskas, F. Crea, M. Massetti, The choice of conduits in coronary artery bypass surgery, J. Am. Coll. Cardiol., 66 (2015) 1729–37.

[2] F.-J. Neuman, M. Sousa-Uva, A. Ahlsson, F. Alfonso, A.P. Banning, U. Benedetto, et al., 2018 ESC/EACTS guidelines on myocardial revascularization, Eur. Heart J., 40 (2019) 87–165.

[3] X. Berard, S. Déglise, F. Alonso, F. Saucy, P. Meda, L. Bordenave, et al., Role of hemodynamic forces in the ex vivo arterialization of human saphenous veins, J. Vasc. Surg., 57 (2013) 1371–82.

[4] J. Veselý, L. Horný, H. Chlup, T. Adámek, M. Krajíček, R. Žitný, Constitutive modeling of human saphenous veins at overloading pressures, J. Mechan. Behav. Biomed. Mater. 45 (2015) 101–108.

[5] W. Li, Biomechanical property and modelling of venous wall, Prog. Biophys. Mol. Biol. 133 (2018) 56–75.

[6] W. Li, Constitutive laws with damage effect for the human great saphenous vein, J. Mechan. Behav. Biomed. Mater. 81 (2018) 202–213.

[7] L. Horný, T. Adámek, M. Kulvajtová, A comparison of age-related changes in axial prestretch in human carotid arteries and in human abdominal aorta, Biomech. Model. Mechanobiol. 16 (2017) 375–383.

[8] L. Horný, M. Netušil, T. Voňavková, Axial prestretch and circumferential distensibility in biomechanics of abdominal aorta, Biomech. Model. Mechanobiol. 13 (2014) 783–99.

[9] L. Horny, T. Adamek, H. Chlup, R. Zitny, Age estimation based on a combined arteriosclerotic index, Int. J. Leg. Med., 126 (2012) 321–6.

EAN 2020
58th conference on experimental stress analysis

Identification of MTB Enduro Bicycle Frame Behaviour

KROPÍK B.[1,a], MALÁ A.[1,b], SCHMIDOVÁ N.[1,c], DOUBRAVA K.[1,d] , PONÍŽIL T.[2,e] , DVOŘÁK M.[1,f], MAREŠ T.[1,g]

[1] Department of Mechanics, Biomechanics and Mechatronics, Faculty of Mechanical Engineering, Czech Technical University in Prague. Technická 4, Praha 6, Czech Republic

[2]Compo Tech PLUS, spol. s r.o. Nová 1316, Sušice, Czech Republic

[a] Bohumil.Kropik@fs.cvut.cz, [b] Anna.Mala@fs.cvut.cz, [c] Nikola.Schmidova@fs.cvut.cz, [d] Karel.Doubrava@fs.cvut.cz, [e] tomas.ponizil@compotech.com, [f] Milan.Dvorak@fs.cvut.cz, [g] Tomas.Mares@fs.cvut.cz

Keywords: Bicycle frame, CFRP, Filament winding, MTB enduro, Integrated joint

Abstract. This paper deals with the design of a bicycle frame. The experimental part is focused on identification and evaluation of the front triangle down tube behaviour with a significant effect of the shock absorber. Evaluation of extreme values and assessing the frequency of their occurrence with focus on selecting important measuring places for future tests was done. Part dedicated to the finite element method is focused on new version of bicycle frame front triangle with focus on finding of equivalent horizontal force acting into front fork related to measured strain and detection of critical places of bicycle frame front triangle. Critical places together with equivalent force were identified. Also bending and tension/compression component of strain distribution along the front triangle down tube was evaluated. Important measured channels were selected, and all these data are valuable for future testing of new version of bicycle frame front triangle. Laboratory test and comparison with finite element (FE) analysis was done.

Introduction

Joining of composite tubes is big a design challenge. Current joining techniques are mostly based on bonding in of additional parts or in bicycle industry commonly used hand lamination over the joined tubes. Objective of project is to develop cost-efficient high-performance joint for frame structures demonstrated on mountain bicycle frame. This contribution deals with an experimental identification of behaviour of MTB enduro bicycle frame made fully from Carbon Fibre Reinforced Polymer (CFRP) filament wound tubes with Integrated Loop Technology (ILT) joints. Motivation is thus to identify applied loads to the frame tubes and joints during testing which is crucial and provide crucial information for design of the strong, stiff and durable joint.

Bicycle design. The main triangle of the bicycle frame is made of CFRP filament wound tubes with ILT joints as shown in Fig. 1. The ILT is described in detail in [1]. The single pivot full suspension mountain bike frame consists of two parts: the front triangle and the swing arm. These two parts are connected by the main pivot and a shock absorber. The front triangle tubes made including the integrated joints are bonded together to form the base of bicycle frame. The scheme of the bicycle frame is shown in Fig. 2.

Fig. 1: CFRP filament wound tubes with ILT joint

Measurement. Most complex load state on the bicycle frame is on front triangle down tube due to shock absorber connection. Critical in terms of safety and performance of the bicycle is the front triangle Down Tube (DT) and Head Tube (HT) joint. Therefore, 1- axis linear strain gauges (SGs) were installed on top and bottom sides of DT (see Fig. 2. place a – d) and shear SGs were installed on both sides (place e). Combination of quarter, half and full bridge connection was used for identification of axial, bending and torque load. The set of HBM QuantumX measurement units carried in a backpack was used for measurement. Some basic load cases were recorded at first, followed by measuring of routine driving on tracks in terrain as shown in Fig. 3.

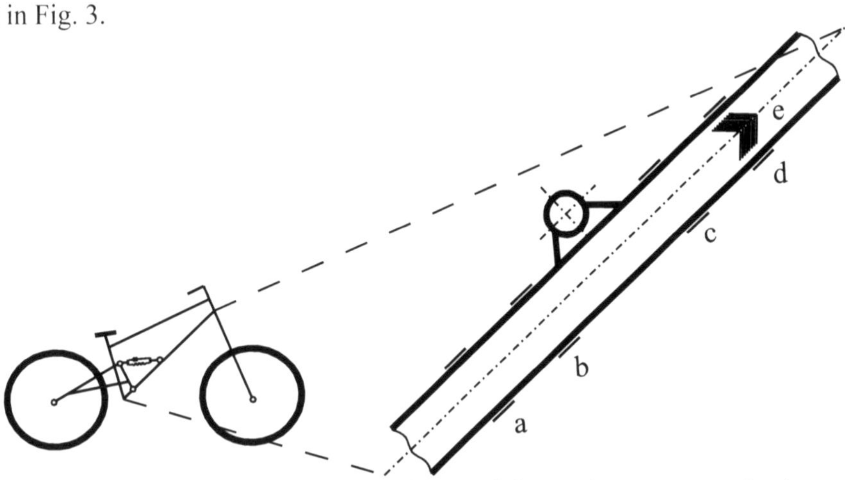

Fig. 2: Scheme of the frame and the positions of the strain gauges on the down tube

Fig. 3: One of the load cases

Evaluation. With distances of positions of SGs (see Fig. 4 a – e) and shock absorber (place f) related to the maximal length of down tube, bending and tension/compression component of strain from SGs connected as quarter bridges (places a and d) was evaluated. Together with bending strain from half bridges (places b and c), they were used for extrapolation to the position of shock absorber. From this, a distribution of bending and tension/compression component of strain behaviour was obtained in time during load cases. All strain values in this paper are related to the maximal measured strain value during experimental measurement. Example of distribution measured during simple load state – single jump – is shown in Fig. 5 for static state with sitting rider, in Fig. 6 during take-off, in Fig. 7 for flying part and in Fig. 8 during landing. Another example of distribution shown in Fig. 9 was observed during routine driving on tracks. This evaluation gives us an idea of load effects on the bicycle frame front triangle down tube and influence of shock absorber. Torsion SGs in position e connected as full bridges gives us an idea of torsion behaviour near DT and HT connection.

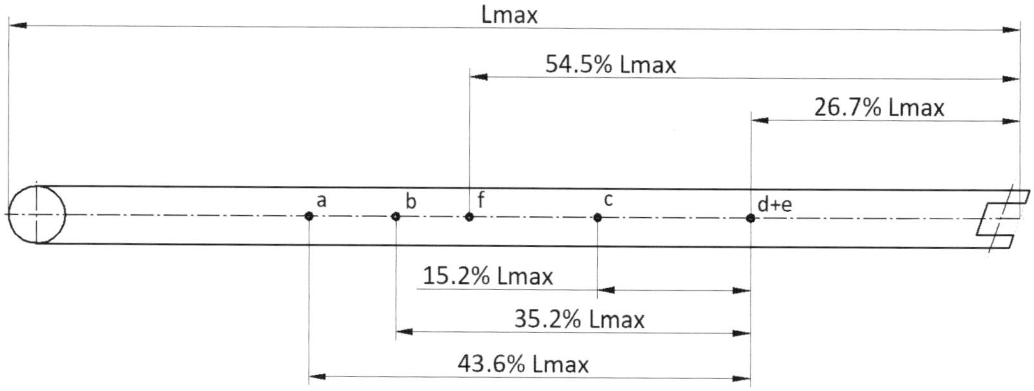

Fig. 4: Distances of positions of SGs and shock absorber on down tube

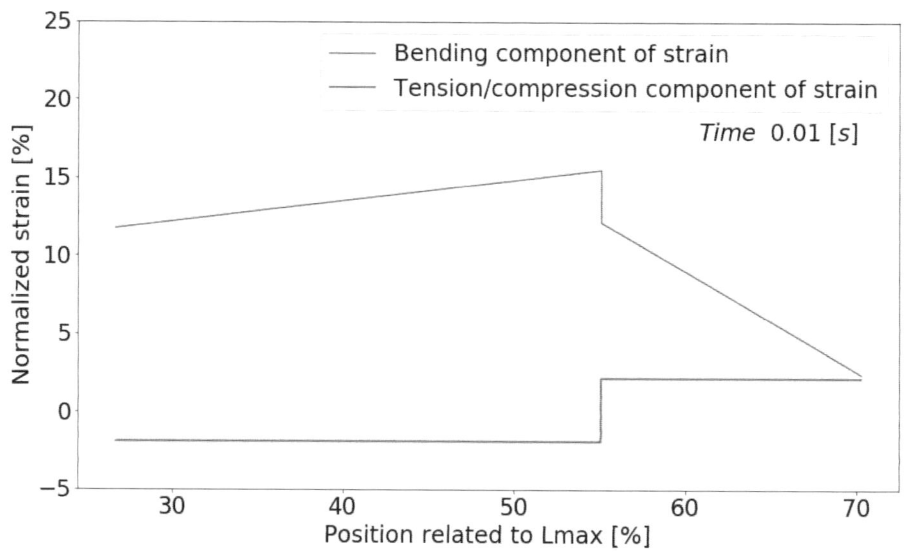

Fig. 5: Distribution of bending and tension/compression component of strain along DT in static state mode with sitting rider

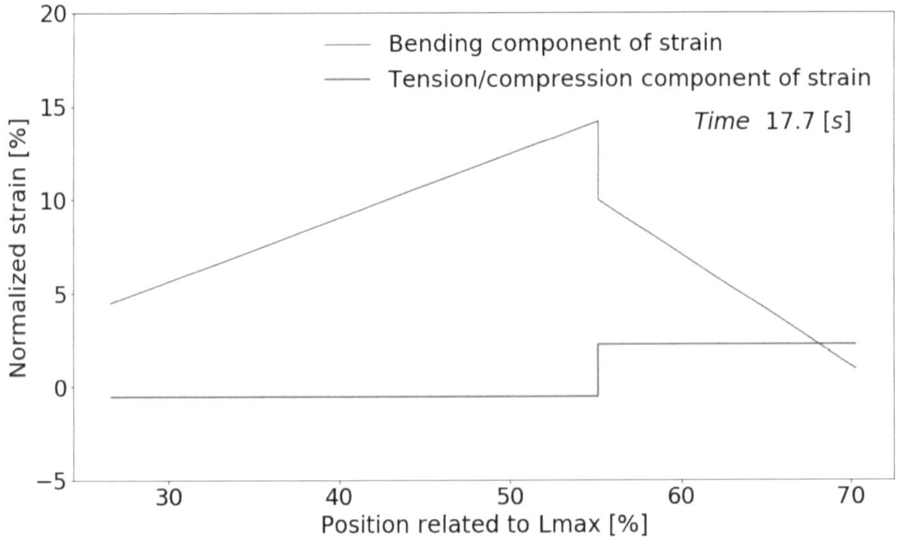

Fig. 6: Distribution of bending and tension/compression component of strain along DT during take-off

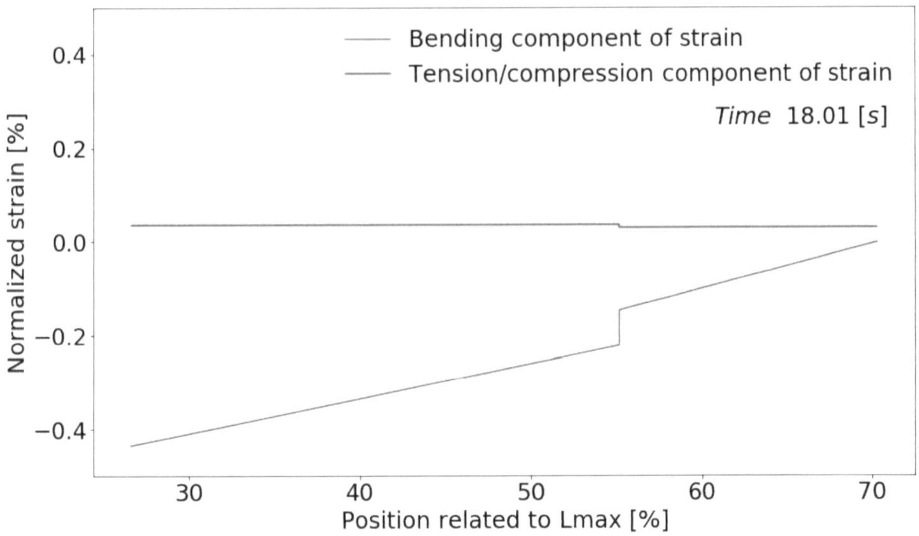

Fig. 7: Distribution of bending and tension/compression component of strain along DT during fly

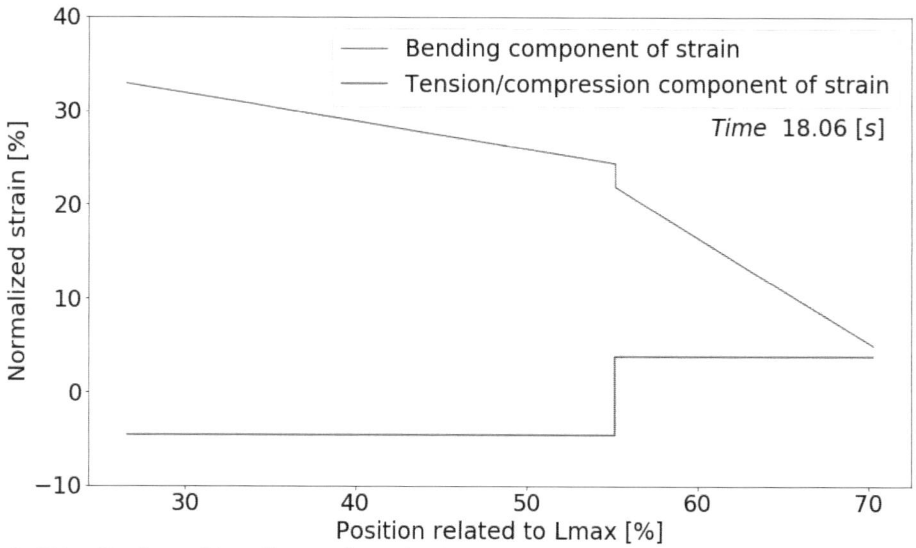

Fig. 8: Distribution of bending and tension/compression component of strain along DT during landing part

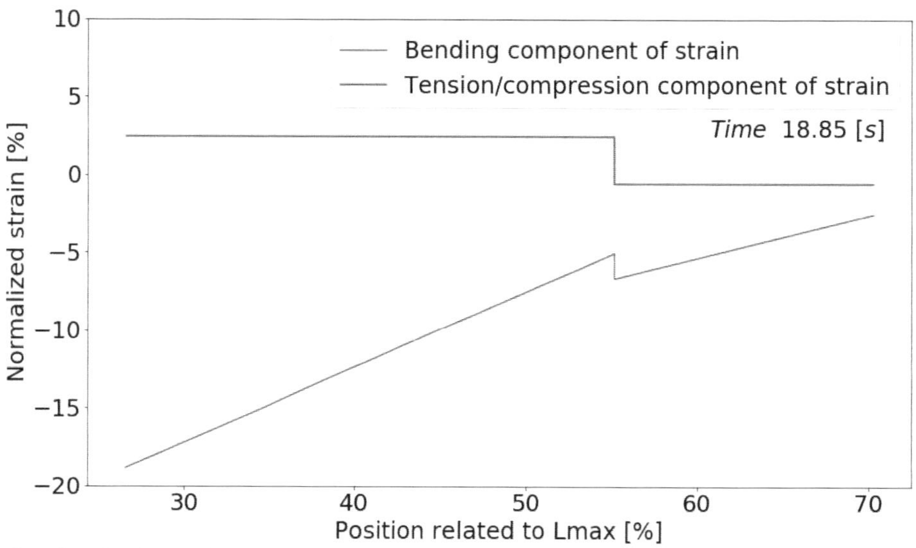

Fig. 9: Distribution of bending and tension/compression component of strain along DT during routine driving

Minimal and maximal values of measured normalized strain were evaluated as shown in Tab. 1. These values were used for the finite element (FE) model to find equivalent force which is described below in FE analysis chapter. Sorting of measured strain values into predefined levels was also done in order to find how often these and close values are achieved. This evaluation shown that most of these values are corresponding to extreme cases, such as bad landing. Extreme values from SGs, torsion behaviour and distribution of bending and tension/compression component of strain along DT were also used for selection of important positions for future tests. Linear SGs in position c and d in a quarter bridge configuration were chosen for future testing. Based on the evaluation can be said that the measured data were valid and may be used for further analysis as inputs for cases of fatigue load or identification of specific behaviour of the frame according to individual load cases such as jump, etc.

Tab. 1: Evaluated strain extremes in % of maximal strain

Position	Minimum [%]	Maximum [%]
a_top	-15.3	10.7
a_bottom	-5.6	25.7
b	-8.4	44.9
c	-19.8	82.8
d_top	-100	39.8
d_bottom	-30.7	75.9
e	-14.9	14.4

Laboratory test stand based on ČSN EN ISO 4210-6 [2] was built in order to compare different versions of bicycle frame and with regard to future fatigue tests necessary for approval of the bicycle frame. Test stand is shown in Fig. 10 and corresponds to the FE analysis simulation. Test arrangement with corresponding FE model was designed for the first comparison of bicycle frames in quasi-static load cases and to validate FEM simulations within the beginning of the project. Both should give us basic idea of bicycle frame durability during the future fatigue test. First quasi-static test with tension force acting on front fork of newer version of bicycle frame was done and comparison of measured values and FE analysis is given in the end of FE section.

Fig. 10: Laboratory test stand for quasi-static and fatigue load cases

FE model. As mentioned before, the FE model was used to find equivalent force in tension and compression in quasi-static load case and for comparison between quasi-static laboratory test and simulation. Development of the ILT [1] and the bicycle frame is parallel. Improved design of the ILT to the one used for the frame analysed in the field was developed based on the analyses in [1]. It was decided to focus on gradual development of more complex FE model of the frame made with the last version of the ILT due to good correspondence of FE model results and test coupons results in [1]. First simple FE model was made within the beginning of the project with focus on basic design and comparison and it is described here. FE model is being continuously modified to simulate load cases more realistically and to simulate more complex load cases closer to real riding behaviour. Test configuration with rotary binding was used for bottom bracket and sliding binding for front fork which was also horizontally pushed or pulled (see red arrow) is shown in Fig. 11. Model was equipped with selected SGs sensors modelled and bonded to appropriate positions.

Fig. 11: FE model of bicycle frame according to ČSN EN ISO 4210-6 [2] with force oriented in tension

This analysis gave us tension and compression forces acting on the front fork in the horizontal plane corresponding to the measured strain extremes. The list of values is shown in Tab. 2. A discussion of used model is essential for interpretation of these values. Bicycle frame was made from shell elements with oriented layers of CFRP according to new version of bicycle frame. Analysis was performed as strictly linear, which means that damaged places still retains the same mechanical properties, model is geometrically linear and do not consider plasticity in the first step. Since the progressive damage model was not used, it is expected that these values are bigger than in a real quasi-static test. Even so, this model serves well to detect the first points of failure and to detect weak places of the front triangle design. It should be mentioned that since the model does not fully correspond to the measured bicycle frame these values are only informative.

Tab. 2: FE analysis data

Position	Force in tension [kN]	Force in compression [kN]
c	6.9	-1.6
d	5.4	-2.2
d_top	6.4	-2.5
d_bottom	4.5	-1.8

As a representation of damage failure in this FE model a failure index Inverse Reverse Factor (IRF) is used. Value of IRF equal to 1 or over means that the element is damaged. IRF with tension force of 2 kN is shown in Fig. 12. Maximal IRF value is in this case lower than 1 so there is no visible damage yet. In Fig. 13 is shown the same with 4 kN force and similarly in Fig. 14 with 6 kN tension force. In these figures IRF is equal or over 1 so and damage on areas near head tube was observed.

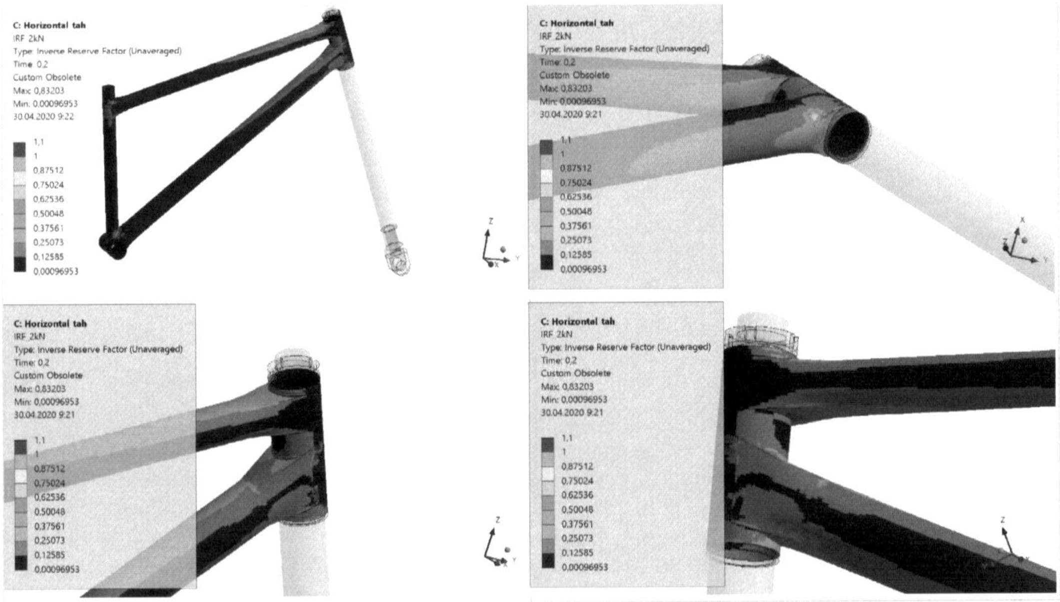

Fig. 12: IRF factor from Ansys analysis for 2 kN tension force

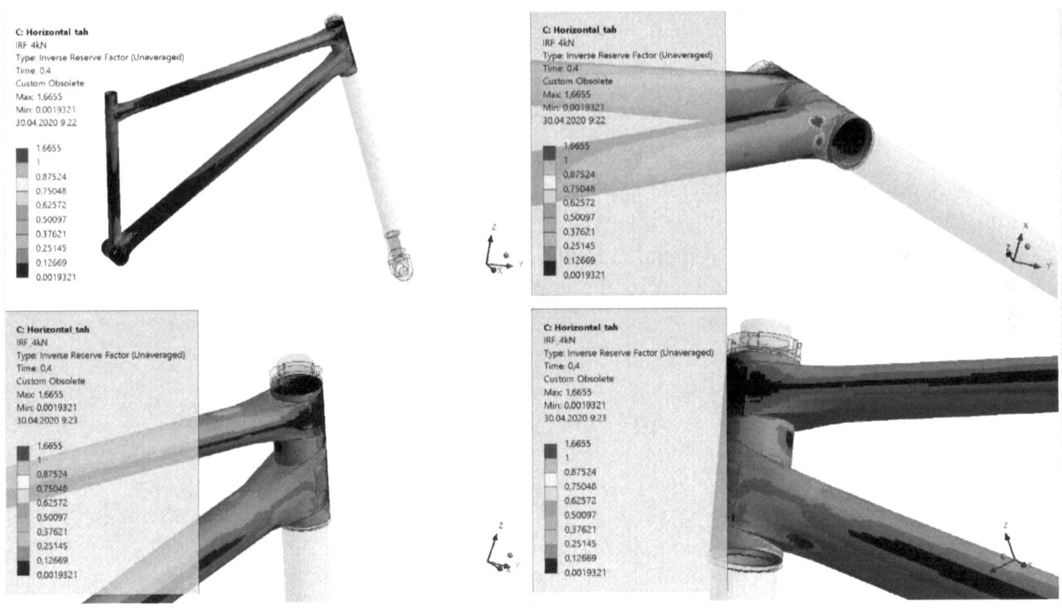

Fig. 13: IRF factor from Ansys analysis for 4 kN tension force

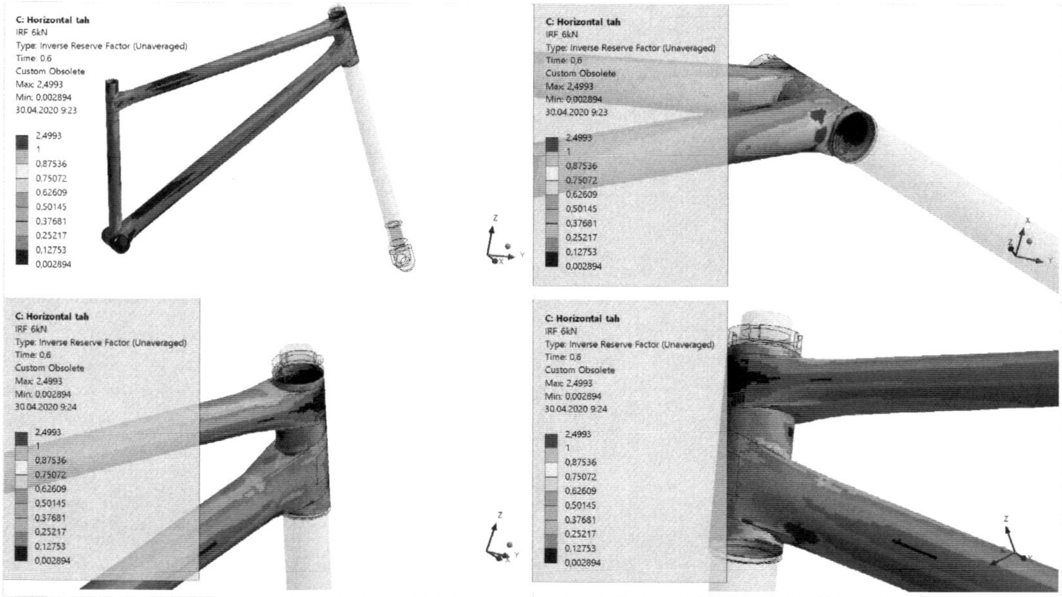

Fig. 14: IRF factor from Ansys analysis for 6 kN tension force

FE analysis also shown damage close to the rest of integrated joints of bicycle frame front triangle with rising load force, but these areas were not that large as joint of head tube to down and top tube. This analysis gives us horizontally oriented equivalent force related to measured strain and information about potentially weaker places of bicycle frame which is useful for future improvement and prediction of maximal load during laboratory test. Progressive damage model will be used in future for improvement of force prediction.

As mentioned before, FE model was also used for comparison between analysis and experimentally measured data in laboratory test stand of newer version of bicycle frame. All forces in next figures are related to maximal measured value. Since the measured frame and analysis had SGs in the same positions there is comparison between force and displacement (taken from hydraulic actuator) dependence in Fig. 15 and comparison between force and strain dependence of d_top SG. Due to the visible damage of bicycle frame joints from previous rides it is expected that FE simulation is more rigid and comparison look sufficiently for simple analysis.

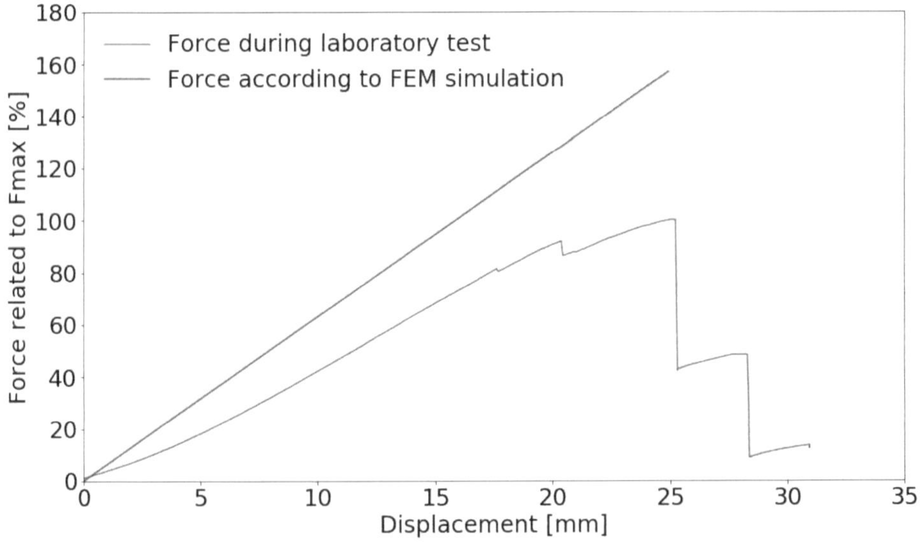

Fig. 15: Comparison of FEM analysis and experimentally measured dependence between force and displacement

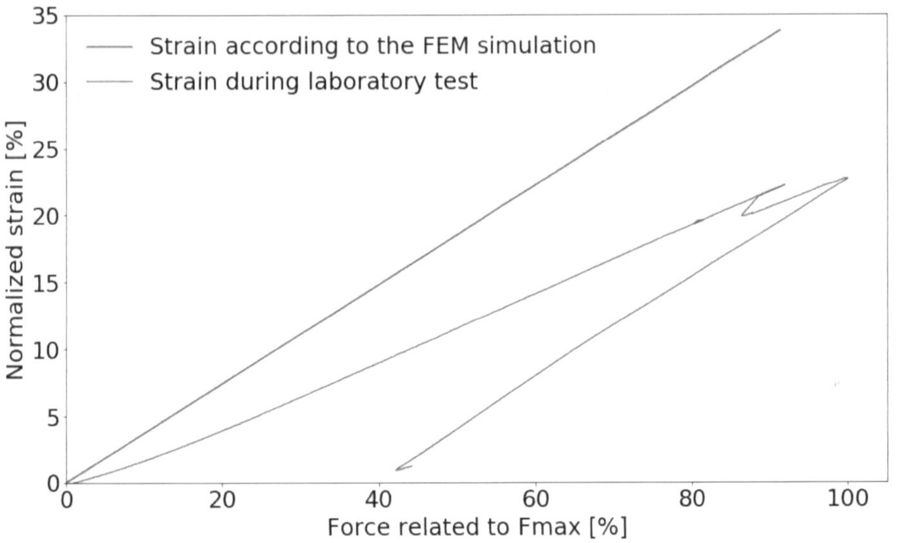

Fig. 16: Comparison of FEM analysis and experimentally measured dependence between force and strain

Conclusions

Complex strain behaviour of the down tube was identified. Significant impact of the shock absorber on its behaviour was confirmed. The measured data were used in FE analysis for finding the equivalent force and identification of potentially critical places. Also, reduction of measured channels to 4 quarter bridges in positions of c and d was done, which allowed usage of miniature measurement unit for new version of bicycle frame front triangle which will not restrict the rider so the full load range during a regular enduro race could be measured. Also, quasi-static laboratory test was done with comparison with FEM simulation. First generation of FE model provides expected results and will be improved according to laboratory and field measurements.

Acknowledgments

This work has been supported by project No. SGS18/175/OHK2/3T/12 of the Grant Agency of the Czech Technical University in Prague and by project No. TJ02000252 of the Technological agency of the Czech Republic.

References

[1] A. Malá, V. Kulíšek, N. Schmidová, T. Zámečníková, T. Ponížil, T. Mareš. Finite element analysis of composite tubes with integrated loop connections. In: PROCEEDINGS OF COMPUTATIONAL MECHANICS 2019. Plzeň: University of West Bohemia, 2019. ISBN 978-80-261-0889-4.

[2] ČSN EN ISO 4210-6. Jízdní kola - Bezpečnostní požadavky na jízdní kola - Část 6: Zkušební metody pro rám a vidlici. 2. vydání. 2016.

[3] Ansys® Academic Research Mechanical, Release 19.2, Help System, ANSYS, Inc.

EAN 2020
58th conference on experimental stress analysis

Influence of Position in 3D Printing Process on Mechanical Properties of Metal Prints

KRYSTEK J.[1,a], KAŇÁKOVÁ S.[1,b], KROUPA T. [1,c], ZEMČÍK R. [1,d], ZIKMUND, P. [2,e], ROUS, P.[3,f], STEINER, F.[3,g], VAŇKOVÁ T. [1,h]

[1]NTIS – New Technologies for the Information Society, Faculty of Applied Sciences, University of West Bohemia, Technická 8, 301 00, Plzeň, Czech Republic

[2]METAL 3D, s.r.o., Dobříš 1989, 263 01 Dobříš, Czech Republic

[3]University of West Bohemia, Faculty of Electrical Engineering, Regional Innovation Centre for Electrical Engineering (RICE), Univerzitní 8, 301 00 Plzeň, Czech Republic

[a]krystek@kme.zcu.cz, [b]kanaksan@kme.zcu.cz, [c]kroupa@kme.zcu.cz, [d]zemcik@kme.zcu.cz, [e]petr.zikmund@metal3d.cz, [f]rousp@ket.zcu.cz, [g]steiner@ket.zcu.cz, [h]vankovat@kme.zcu.cz

Keywords: Experiment, Metal 3D print, Tensile test, computed tomography

Abstract. This paper researches the influence of the placement during the 3D printing process on the quality of the final product. The tensile tests were carried out on standard dog-bone specimens, which were printed both horizontally or vertically with three orientation angles 0°, 45°, 90° to the main direction of the plate. The main goal of the measurements was to determine the influence of the printing orientation and placement of the specimens on the mechanical properties. Obtained data were evaluated and used for the determination of the best printing orientation.

Introduction

The growing popularity of 3D printing and its availability significantly influenced the manufacturing process. Nowadays, an application of 3D printing technology is immense, from products with simple decorative purposes to medical implants as shown in [1], or even whole buildings [2]. However, this technology also presents many aspects that can potentially affect the mechanical properties of the final product, e.g. a printing layer thickness, the orientation of the product against the printing equipment or a curing process.

The *Renishaw AM250* machine with the 200 W fiber laser and argon atmosphere were used. Printing parameters were developed by Metal 3D s.r.o. to get the best possible homogeneity and minimal deformation after the process of cutting the specimens from the base plate [4, 5]. Pure stainless steel powder with spherical particles was used. It was produced with gas atomization to get the desired shape of particles.

At the beginning of the process, the build chamber is closed, vacuumed and filled with argon 5.0 to get less than 1000 ppm of oxygen. The build file consists of several hundreds of layers. The whole build is made layer by layer. The base plate is stepping down in 50 μm steps, while a special mirror device is taking care of precise laser beam guidance, during the printing process. Every layer has its own parameters and strategy which the beam follows. As soon as the chamber is ready and the build file is loaded, the building starts. The base plate is covered with the powder, which is then melted with the laser. Next, the coating process follows as soon as the base plate moves down for another 50 μm. This is how the process continues until the maximum part height is reached.

X-ray tomography, also known as computed tomography (CT), allows capturing spatially large objects with their internal structure in the required accuracy. The use of X-ray technology in the form of film shots of two-dimensional images has been used in medicine for a long time. Instruments for medical 3D-tomography have been offered since the 1970s. In the early 1990s, X-ray tomography began to be used also for the inspection of technical objects, i.e. non-destructive diagnostics.

The use of CT allows checking whether there are any cracks, voids, and any other inhomogeneities, anomalies or disturbances in specimens. The second area of application is metrology. By means of CT it is possible to determine the dimensions of individual parts of analyzed objects. These dimensions (3D model) can be further compared with other samples (e.g. "gold" sample). Software tools for computed tomography provide the possibility of comparison with models created in CAD systems.

This work investigates the influence of a specimen's orientation and placement during the printing process. Thin bone-shaped specimens were placed throughout the whole base plate and printed in two ways (plane perpendicular to the thickness direction is parallel or perpendicular to the base plate, further these types of printing will be designated also as horizontal or vertical) with three different orientations made of *316L stainless steel*. The capabilities of using computed tomography scan to verify print quality have been tested.

The aim of this work was to analyze Young's modulus dependency on the specimens' position, the printing orientation, and specimens' placement on the base plate. A similar problem was presented in [3] for Polylactic Acid thermoplastic material. In the cited article, the printing orientation notably influenced the tensile strength, its value increased with increasing of the printing angle, measured from the transverse axis of the specimen.

The obtained results will be used during the future printing processes, determining the best printing position of designed products.

Tensile test

The experimental specimens were made from *316L stainless steel* (15 - 45 microns particle distribution) with a layer thickness of 50 μm, using a chessboard strategy. The 3D printing of the specimens was performed in cooperation with company Metal 3D s.r.o. The experimental specimens (Fig. 1) were printed horizontally (the specimens LA, LB and LC) and vertically (SA, SB, SC) with three orientation angles 0°, 45°, 90° throughout the whole area of the base plate. All specimens were made at one build plate with production time exceeding 30 hours. Additional 14 hours were used for a wire EDM cutting to remove the specimens from the base plate. A traditional cheaper and faster band saw could be used, but the aim was for a precise cut. At the last step the remaining supporting structures were removed.

Geometric parameters (width W and thickness H) and weight M of the specimens are presented in Table 1. Total length of specimens was $L = 100$ mm. Specimens were prepared according to ASTM E8 / E8M Standard Test Methods for Tension Testing of Metallic Materials. The specimens were exposed to a tensile quasi-static loading in the longitudinal direction. The loading velocity of the crosshead was $v = 1$ mm/min and temperature was $T = 21$ °C. The universal testing machine *Zwick/Roell Z050* was used. A uniaxial extensometer was used for measuring the elongation (the gage length was $L_0 = 10$ mm). The force–displacement dependencies were obtained from the tensile test. The stress–strain dependencies (Fig. 2) were calculated using

$$\sigma = \frac{F}{W \cdot H}, \quad and \quad \varepsilon = \frac{\Delta L}{L_0}. \tag{1}$$

The elastic modulus was identified on the interval of stress $\sigma \in \langle 250{,}300 \rangle\, MPa$. Elastic modulus, yield strength and ultimate stress (strength of material) are presented in Table 2.

The influence on the resulting mechanical properties of the way the specimens were 3D printed was found. Horizontally printed specimens have higher values of elastic modulus, yield strength and maximum stress than vertically printed specimens (Table 2).

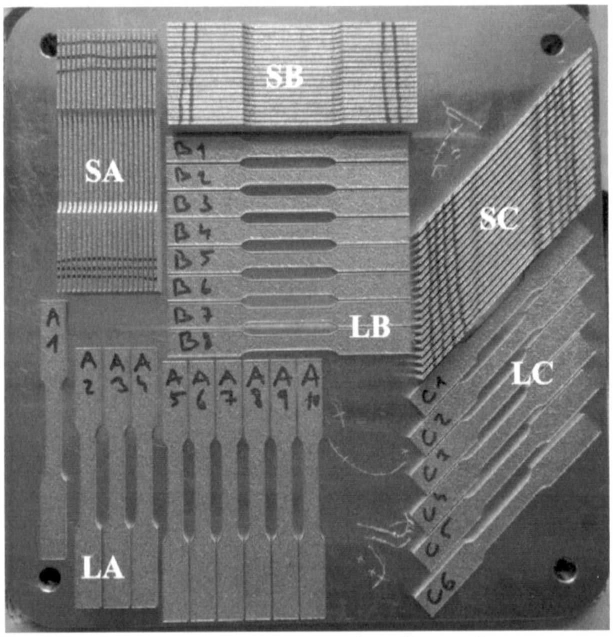

Fig. 1: The placement of each specimen

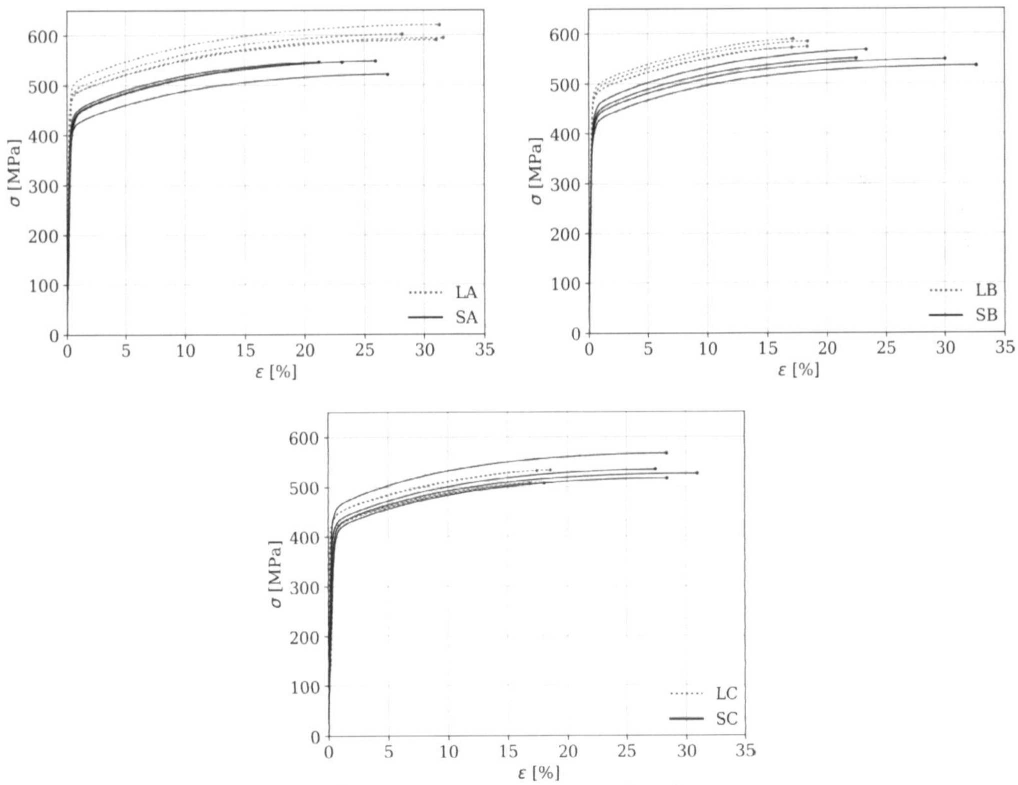

Fig. 2: Stress–strain dependencies

Table 1: Geometric parameters and weight of specimens

Specimen name	W [mm]	H [mm]	M [g]	Specimen name	W [mm]	H [mm]	M [g]
LA_1	6.14	1.17	7.158	SA_1	6.40	1.24	7.207
LA_2	6.20	1.16	6.889	SA_2	6.48	1.22	7.182
LA_3	6.17	1.16	6.935	SA_3	6.57	1.28	7.220
LA_4	6.19	1.25	7.635	SA_4	6.55	1.22	7.252
LB_1	6.16	1.29	7.519	SB_1	6.42	1.21	7.252
LB_2	6.14	1.29	7.550	SB_2	6.41	1.28	7.281
LB_3	6.13	1.26	7.576	SB_3	6.53	1.23	7.321
LB_4	6.16	1.28	7.624	SB_4	6.61	1.23	7.286
LC_1	6.24	0.97	5.579	SC_1	6.42	1.29	6.994
LC_2	6.25	0.99	5.718	SC_2	6.30	1.28	7.135
LC_3	6.19	0.96	5.627	SC_3	6.37	1.30	7.094
LC_4	6.18	0.98	5.701	SC_4	6.36	1.31	7.115

Table 2: Elastic modulus (E), yield strength ($Rp_{0.2}$), ultimate strength (σ_{max})

Specimen name	E [GPa]	$Rp_{0.2}$ [MPa]	σ_{max} [MPa]	Specimen name	E [GPa]	$Rp_{0.2}$ [MPa]	σ_{max} [MPa]
LA_1	140	495	621	SA_1	114	424	546
LA_2	121	487	591	SA_2	94	422	549
LA_3	178	478	595	SA_3	116	407	522
LA_4	155	483	602	SA_4	110	432	547
mean value	148.5	485.8	602.3	mean value	108.5	421.3	541.0
coefficient of variation [%]	14.1	1.3	2.0	coefficient of variation [%]	8.0	2.2	2.1
LB_1	152	472	574	SB_1	121	438	570
LB_2	134	474	573	SB_2	112	404	536
LB_3	163	485	590	SB_3	130	425	551
LB_4	112	480	585	SB_4	121	421	549
mean value	140.3	477.8	580.5	mean value	121.0	422.0	551.5
coefficient of variation [%]	13.8	1.1	1.3	coefficient of variation [%]	5.3	2.9	2.2
LC_1	105	414	509	SC_1	134	441	567
LC_2	123	413	508	SC_2	114	413	535
LC_3	165	433	533	SC_3	99	411	527
LC_4	132	435	535	SC_4	75	401	518
mean value	131.3	423.8	521.3	mean value	105.5	416.5	536.8
coefficient of variation [%]	16.6	2.5	2.5	coefficient of variation [%]	20.5	3.6	3.5

Computed tomography

A Phoenix v|tome|x s 240 inspection equipment was used for analysis. The device is equipped with a 240 kV 320 W micro-focal open X-ray tube and 180 kV high-performance nano-focal X-ray tube. The computed tomography system with a conical X-ray source enables inspection of samples up to 260 mm in diameter, 420 mm in height and 10 kg in weight with a high resolution of up to 1 μm.

Fig. 3: Phoenix v|tome|x s 240 inspection equipment - 240 kV 320 W micro-focal open X-ray tube and 180kV high-performance nano-focal X-ray tube

Fig. 4: Defect distribution for specimen LA_1 after tensile test

After scanning the 3D model, it is also necessary to analyze the acquired data. *VGSTUDIO MAX* (high-end software for analysis and visualization of CT data) was used. In order to obtain the necessary information, the extension module of the system (Porosity / Inclusion Analysis Module) was used.

Specimens LA_1 (after tensile test) and LA_10 (non-tested in tensile test) were analyzed by means CT. A defect value ratio of specimen LA_1 ($v_d = 0.44$ %) was 4 times greater than specimen LA_10 ($v_d = 0.11$ %).

Conclusions

The influence on the resulting mechanical properties of the way the specimens were 3D printed was found. Horizontally printed specimens have higher values of elastic modulus, yield strength and maximum stress than vertically printed specimens. It is worth to note, that specimens, whose longitudinal axis is not parallel to any of the edges of the base plate also show the lowest values of mechanical properties. Furthermore, the capabilities of the computed tomography were shown.

Future work will be aimed at finding a correlation between mechanical properties and volume fraction, spatial distribution and/or orientation of the voids in the 3D printed specimens. This will need a new set of specimens with more suitable dimensions for CT scanning.

Acknowledgement

This publication was supported by the project LO1506 of the Czech Ministry of Education, Youth and Sports and project of the Technology Agency of the Czech Republic no. TN1000007 (subproject no. 1, segment no. 1, working package no. 5).

References

[1] A. Jíra, P. Hájková, L. Řehounek, Trabecular structures as efficient surface of dental implants, in: Proceedings of the 57th International Scientific Conference on Experimental Stress Analysis. Luhačovice, Czech Republic, 2019, pp. 169–174.

[2] I. Hager, A. Golonka, R. Putanowicz, 3D printing of buildings and building components as the future of sustainable construction?, International Conference on Ecology and new Building materials and products, ICEBMP 2016, pp. 292–299.

[3] Y. Zhao,Y. Chen, Y. Zhou, Novel mechanical models of tensile strength and elastic property of FDM AM PLA materials: Experimental and theoretical analyses, Materials and Design, Vol. 181, 5 November 2019. ISSN 0264-1275.

[4] V. Novotny, M. Vitvarova, M. Kolovratnik, B. B. Stunova, V. Vodicka, J. Spale, P. Zikmund, M. Drasnar, and E. Schastlivtseva. Design and Manufacturing of a Metal 3D Printed kW Scale Axial Turboexpander, in: Proceedings of the ASME Turbo Expo 2019: Turbomachinery Technical Conference and Exposition. Volume 8: Microturbines, Turbochargers, and Small Turbomachines; Steam Turbines. Phoenix, Arizona, USA, 2019. https://doi.org/10.1115/GT2019-91822

[5] V. Novotny, J. Spale, B. B. Stunova, M. Kolovratnik, M. Vitvarova and P. Zikmund, 3D Printing in Turbomachinery: Overview of Technologies, Applications and Possibilities for Industry 4.0, in: Proceedings of the ASME Turbo Expo 2019: Turbomachinery Technical Conference and Exposition. Volume 6: Ceramics; Controls, Diagnostics, and Instrumentation; Education; Manufacturing Materials and Metallurgy. Phoenix, Arizona, USA, 2019. https://doi.org/10.1115/GT2019-91849

EAN 2020
58th conference on experimental stress analysis

A Wear Resistance Analysis of Orthopedic Instruments

KUBÁŠOVÁ K.[1,a], SEDLÁČEK R.[1,b], HORNÝ L.[1,c]

[1] Czech Technical University in Prague, Faculty of Mechanical Engineering, Department of Mechanics, Biomechanics and Mechatronics, Technická 4, 160 00 Praha 6, Czech Republic

[a]Kristyna.Kubasova@fs.cvut.cz, [b]Radek.Sedlacek@fs.cvut.cz, [c]Lukas.Horny@fs.cvut.cz

Keywords: Milling cutters, Drilling, Wear, Experiment,

Abstract. This paper reports on an investigation into the wear problem of 5 types of orthopedic instruments. We conducted a series of experiments based on a proven methodology with modified parameters for the design and development of new instruments. A total of 3 wear cycles for 40 drill holes (respectively for 8 drill holes) per instrument were performed and supplemented by 4 measured cycles. The main monitored parameters were the resistive force and the torque. These parameters can evaluate the penetration of the instrument into the material. We then used statistical methods to evaluate the significance of the detected differences. In most cases, it was statistically proven that the wear did not change the monitored parameters. This implies that drilling 120 holes (respectively 40 holes) does not compromise the sharpness of the instrument.

Introduction

In surgical procedures, especially in the repair of complex fractures (with implanted screws) there is a need to use instruments that can penetrate through hard cortical bone. These instruments are reused (after sterilization). Repeated use of these instruments leads to a blunt cutting edge. An important factor influencing the healing of the surrounding tissue is the heat generated by the instrument in the drilling area – a blunt instrument generates more heat than a sharp instrument. This work employs a proven methodology to evaluate the wear rate of 5 sets of orthopedic instruments [1].

Methodology

The methodology was based on controlled drilling by instruments into beech blocks. The original methodology was adapted to the requirements of MEDIN, the company that produces the instruments [2, 3].

First, one instrument from each group was subjected to microscopy. Before using tools from group D, respectively group E, a hole with a diameter of 2.5 mm, respectively 3.5 mm was pre-drilled. During the experiment, instruments were inserted into the pre-drilled holes. Drilling was carried out at a constant speed of feed [$mm \cdot min^{-1}$], at a constant rpm [min^{-1}] to a depth of 5 to 12 mm. Three wear cycles were performed with 40 holes, respectively 8 holes in each cycle. After each wear cycle, we performed a control measurement cycle using MTS 858 Mini Bionix. The measured value was the resistive force [F = N] required to penetrate the instrument into the material. Due to the specially designed fixture, it was also possible to measure the torque [M = N·m]. The measured value is the torque required to compensate the rotation of the workpiece in the cutter/drill direction [3, 4].

The control measurement cycle was carried out with new tools and with worn tools. The instrument wear was carried out under the same cutting conditions (as the measurement cycle) on the vertical cantilever milling machine. The control measurement drilling was performed using a BOSCH hand drill attached to the frame of the MTS 858.02 Mini Bionix system (MTS, Minnesota, USA) with a force sensor (MTS, 440 N) and a torque sensor (MTS, 5.7 Nm). The figure below shows one cycle of our experiment (see **Chyba! Nenalezen zdroj odkazů.**).

The table (see Table 1) contains information on the speed of the feed, the drilling depths, the rpm and the number of wear cycles in one period. Three wear cycles were performed with all tools after the initial force and torque measurements. Each period was followed by a measurement cycle of the force and the torque.

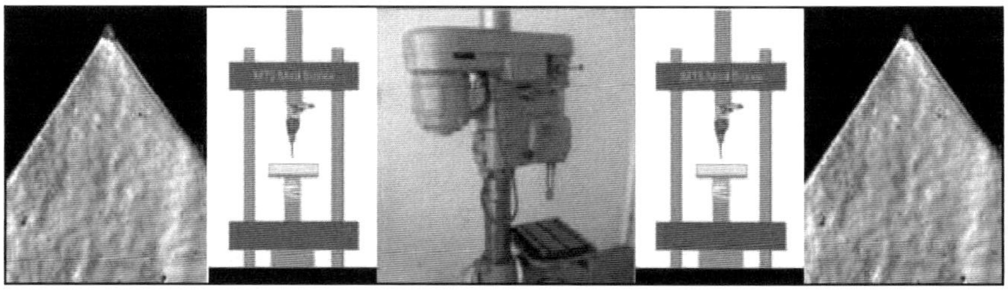

Fig. 1: One cycle of the experiment: 1. Olympus LEXT OSL 3000, confocal microscope, photo and profile; 2. MTS 858.02 Mini Bionix, measured *F, M*; 3. Defined drilling wear, 40 cycles or 8 cycles; 4. MTS 858.02 Mini Bionix, measured *F, M*; 5. Olympus LEXT OSL 3000 confocal microscope, photo and profile

Table 1: Drilling and wear parameters

	RPM n [min^{-1}]	Speed of feed u [mm·min^{-1}]	Hole depth u [mm]	Number of cycles per period	Number of wear periods	Diameter of the hole [mm]
A	360	63	5	40	3	-
B	360	63	10	8	3	-
C	360	63	5	40	3	-
D	360	63	12	40	3	2.5
E	360	63	10	40	3	3.5

The resistive force during drilling and the temperature of the cutting instrument were both measured. The temperature was measured in each control measurement cycle on the MTS Mini Bionix, i.e. in the initial measurement of the intact instrument and then after each wear period. The temperature was also measured during the first wear cycle. The temperature measurement was always performed before drilling and then immediately after the instrument was removed from the workpiece. The monitored value was the temperature difference, i. e. the increase in the temperature at the tip of the instrument due to drilling. The measurement was carried out using two devices. The FLUKE 574 three-point infrared thermometer was used during the control measurement cycles, and the FLIR E60 thermal camera was used in the first wear cycle [4].

Due to insufficient temperature measurement accuracy, the measurement was terminated after evaluating the first wear cycle. We continued to measure temperature only during the control measurement cycles.

The cutting surfaces of the instruments were subjected to microscopic examination, during which micrographs and height profiles of the material of the instrument were taken. Microscopy was performed using an Olympus LEXT OSL 3000 laser confocal microscope.

Evaluation and Results

Time, displacement, resistive force and torque were recorded during loading. The data were plotted – the dependence of the resistive force (see Fig. 2) and the torque (see Fig. 3) on the feed (for example, the graphs of sample B3). The data were fitted by a polynomial of the 6th degree (purple curve).

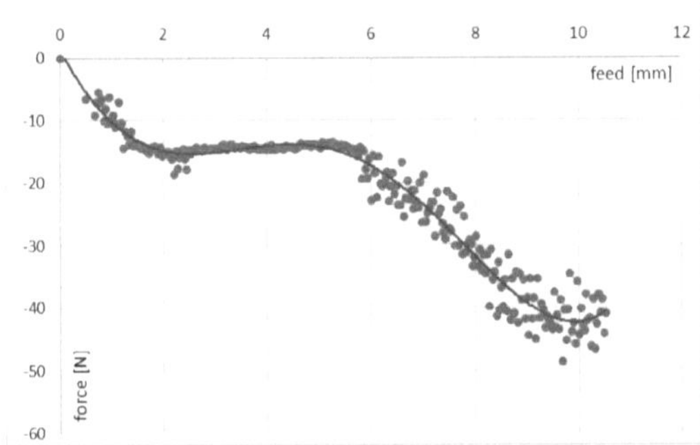

Fig. 2: The dependence of the resistive force on the feed – 1st wear cycle – sample B3

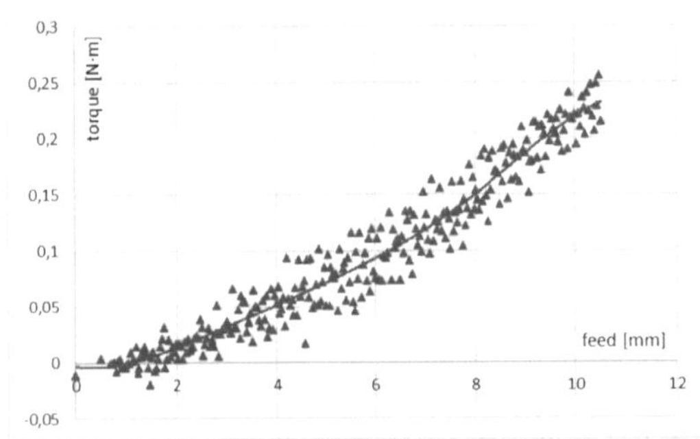

Fig. 3: The dependence of the torque on the feed – 1st wear cycle – sample B3

Histograms and boxplots (see Fig. 4) were also compiled for a better visual comparison of the differences between the measured data. The data were subjected to a statistical evaluation. For each instrument, the hypothesis that the wear affected the measured resistive forces and torques was verified using the Kruskal-Wallis (KW) test. We used the Dunn (D) test to determine the differences between the groups [5].

We presumed that a worn instrument increases the absolute value of the resistive force, and that a blunt instrument transmits more torque to the workpiece.

Thus, if the instrument had been worn out, the resistive force and the torque would have increased gradually over the wear cycles. However, if the mean values of the forces (torques) fluctuate, the random effects prevail over the effects of wear. The results were divided into 3 classification categories:

level 2: The increase is monotonic – damage is accumulated – the force/torque measurements show a statistically significant increase after all three wear periods relative to the values for new instruments (each value is greater than the previous value)

level 1: There is an increase without the accumulation of damage – the force/torque measurements show a statistically significant increase in wear against the values for new instruments (each value is greater than the initial value, not greater than the previous value)

level 0: There is no increase – there are fluctuating results – the force/torque measurements show fluctuating results, there was no significant effect of wear damage on the cutting conditions

Fig. 4 shows the boxplots of the resistance force and the torque for the selected instrument of Group E3.

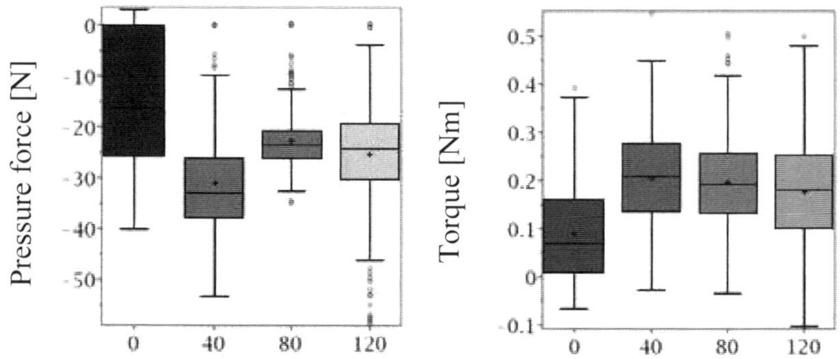

Fig. 4: Boxplots of instrument group E3 across all wear cycles

Table 2: Results for each group of instruments by levels

Group	Measured parameter	Sample number					Total
		1	2	3	4	5	
A	*F*	0	0	0	0	0	**0**
	M	0	0	0	0	**1**	**1**
B	*F*	0	**1**	**1**	0	0	**2**
	M	0	0	0	0	0	**0**
C	*F*	0	0	0	0	0	**0**
	M	0	0	0	0	0	**0**
D	*F*	0	0	0	0	0	**0**
	M	0	0	0	**1**	0	**1**
E	*F*	0	**1**	**1**	0	0	**2**
	M	0	0	**1**	0	0	**1**

Note: 0 – no increase; 1 – an increase without accumulation of damage; 2 – the increase is monotonic

On seven occasions (14 %), wear was demonstrated by the measured parameters (see Table 2) – on four occasions (16 %) in the resistive force measurement, and on three occasions (12 %) in the torque measurement. Level 2, i.e. monotonically increasing force or torque relative to wear did not appear anywhere. In several cases, an increase in force or in torque relative to the new instrument was observed. However, there were no differences among the results for the worn-out instruments.

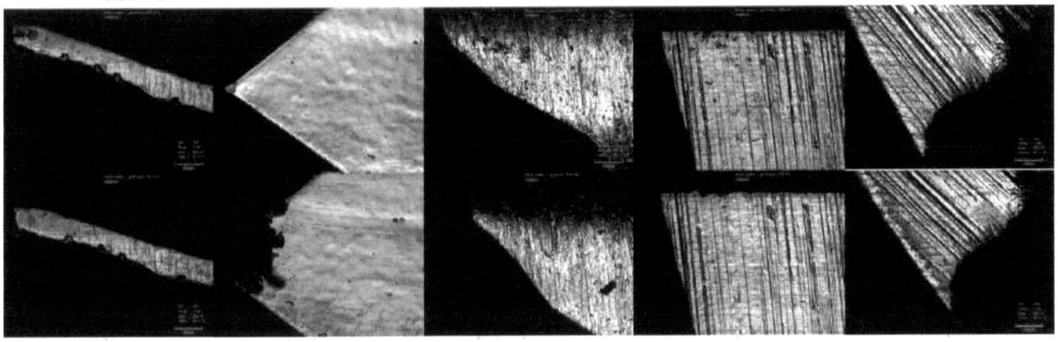

Fig. 5: Confocal microscope images – above, intact instruments, below, the instruments after the second wear cycle

Photographs of the blades taken by an Olympus LEXT OSL 3000 confocal microscope are shown above (Fig. 5). Intact instruments are shown above, and the instruments after the second wear cycles are shown below. In the case of instrument A, and especially in the case of instrument B1, we can see some signs of wear. In the case of instrument B1 there was major tip breaking. For instruments C1, D1 and E1, a visual inspection was inconclusive. The scale for A1 and B1 is 80 μm, and for C1, D1 and E1 the scale is 160 μm.

Conclusions

In total, we performed 4 control measurement cycles and 3 wear cycles for 5 types of instruments (5 items per group). We measured the resistive force and the torque for each instrument. The statistical tests showed that a monotonic increase was never achieved. There was no systematic permanent increase in resistive force and in torque during the 3 wear cycles (120 holes, respectively 24 holes). We can state that there was no significant change in the measured parameters representing resistive effects. This conclusion is consistent with the microscope observations. It was therefore possible to use the instruments up to 120 times (respectively 24 times) without signs of wear. Although the temperature measurements were not accurate, and temperatures were not measured during the wear cycles, the temperature measurement during the control measurement cycle never exceeded the limit of 50° C, a value which doctors have found to be the maximum acceptable level.

Acknowledgements

The work has been supported by grant-funded project FV30348 of the Ministry of Industry and Trade of the Czech Republic and by the Grant Agency of the Czech Technical University in Prague, grant number SGS19/155/OHK2/3T/12.

References

[1] K. Kubášová, R. Sedláček, L. Horný, Evaluation of Wear of Milling Cutters for Drilling of Nail Holes Used for Osteosynthesis of Long Bones of Lower Limbs, in: Strojnícky časopis – Journal of Mechanical Engineering, SjF STU Bratislava, 2019, 69(3), 69–74.

[2] T. Staroveski, D. Brezak, T. Udiljak, Drill wear monitoring in cortical bone drilling, 2015, 37(6), 560-566

[3] M. Litschmannová, Martina Litschmannová: Katedra aplikované matematiky, Ostrava: WordPress, Graphene Theme., 2019

[4] E. Jantunen, A summary of methods applied to tool condition monitoring in drilling, International Journal of Machine Tools and Manufacture, 2002, 42(9), 997-1010

[5] P. Nestrojil, Léčení zlomenin dlouhých kostí, SANQUIS: Časopis o umění, medicíně a zdravém životním stylu, Brno: Audabiac, o. s., 2002, 2002(22), 52

EAN 2020
58th conference on experimental stress analysis

Al/Cu Laminated Wire Conductors; Effect of Stacking Sequence on Residual Stress

KUNČICKÁ, L.[1,2a], KOCICH, R.[2,b]

[1]Institute of Physics of Materials, Academy of Science of Czech Republic, Brno, 61662, Czech Republic

[2]Faculty of Materials Science and Technology, VŠB-Technical University of Ostrava, 70833 Ostrava 8, Czech Republic

[a]kuncicka@ipm.cz, [b]radim.kocich@vsb.cz

Keywords: rotary swaging, clad composite, residual stress, microstructure

Abstract. The presented study deals with preparation of Al-Cu clad composite wires with two different stacking sequences via the intensive plastic deformation technology of rotary swaging (RS). The aim of the work was to provide detailed characterization of the effects of RS on the development of structure and residual stress within the composites and their components. The results showed that fine more of less equiaxed grains with no prevailing preferential orientations were present within all the Al and Cu composites' components of the final 5 mm thick wires, which points to the occurrence of dynamic recovery/recrystallization during processing. The analyses of grains misorientations revealed that residual stress was locally present primarily in the Cu components; the Al components did not exhibit substantial presence of residual stress, which is in accordance with the observed structure relaxation. The tensile tests of the swaged composites revealed both the stacking sequences to exhibit comparable ultimate tensile strength, however, the plasticity differed notably.

Introduction

Research and development in virtually all the industrial fields has led to the emergence of numerous modern components, as well as to the introduction of a variety of innovative materials, such as plasma and laser melted alloys [1], (pseudo)alloys manufactured via powder metallurgy [2], high entropy alloys (HEAs) or multi-principal element (MPE) alloys [3,4], hierarchical and hybrid materials [5,6], functionally graded materials [7], and composites [8]. The latter can be of various types; the most characteristic are the Metal Matrix Composites (MMCs) [9] offering the advantages of the individual component metals and, at the same time, benefiting from the additions of dispersed particles/fibres. Nevertheless, laminated and clad composites consisting of several layers of various metals, each one of which introduces characteristic properties to create their unique combination within the final product, are also popular, since composites combining several metals together feature enhanced combinations of properties when compared to the individual components [10–13].

Among the widely researched composite systems consisting of bulk metallic components, i.e. layers, is the Al/Cu one [14]. Copper is favoured for its excellent electric and heat conductivity, while aluminium is preferred for its high conductivities and favourable combination of low density and reasonable strength [15]. Beside these (and other) advantages, both the metals have a wide applicability, e.g. in the automotive and electrotechnics. However, they both feature the tendency to form intermetallic compounds when in contact, i.e. to form intermetallics at mutual interfaces, especially at high (elevated) temperatures.

Clad composites consisting of multiple layers bond together at mutual interfaces are typically produced by welding, however, welding can introduce local structure modifications and formation of brittle intermetallics [16]. Nevertheless, clad composites can also be fabricated at room temperature via methods of (intensive) plastic deformation, which provides an advantageous solution for production of clad composites at room temperatures, or under cold conditions. Considering the layered Cu/Al composites, even post-process heat treatment at temperatures above 300°C has been documented to introduce the development of intermetallics, which makes post-processing of these composites unfavourable despite the fact that it also supports structure restoration and grain refinement within both the metals [17]. Methods of severe plastic deformation (SPD) are especially advantageous for preparation of clad composites since they can typically be performed at low/room temperatures. Among the widely researched and/or used SPD methods are, e.g., equal channel angular pressing (ECAP) [18,19] and its modifications (ECAP-PBP [20], TCAP [21], TCMAP [22], etc.), high pressure torsion (HPT) [23], accumulative roll bonding (ARB) [24], and rotary swaging (RS) [25–27], which was used to manufacture the herein studied composites.

Rotary swaging (RS) is a versatile industrially applicable intensive plastic deformation technology enabling to manufacture various axially symmetrical products via a repeated action of a set of rotating dies, which incrementally affects the surface of the work-piece with compressive radial forces and thus introduces a favourable combination of shear and compressive strains [28]. The nature of the process enables to avoid large stress gradients and supports progressive grain refinement, which introduces improvement of mechanical and utility properties [29].

The presented study deals with preparation of Al-Cu clad composite wires with two different stacking sequences via room temperature rotary swaging. The aim of the work was to provide a detailed characterization of the effects of RS on the development of structure and residual stress within the 5 mm swaged composite wires.

Experimental

The originally used metals were commercially pure (CP) copper (with 0.002 wt.% O, 0.015 wt.% P, and 0.002 wt.% Zn), and CP electro-conductive aluminium (with 0.20 wt.% Si, 0.25 wt.% Fe, and 0.05 wt.% Cu). The clad composites fabricated from the Cu and Al were gradually swaged from the original 30 mm diameter to the final wires having 5 mm in diameter. Both the stacking sequences originally consisted of a sheath with 30 mm in diameter, and 19 reinforcing wires having 3 mm in diameter each. The *stacking sequence I* consisted of Al sheath and Cu wires, whereas the *stacking sequence II* consisted of Cu sheath and Al wires. To acquire the idea of the composite design, Fig. 1 depicts a cross-sectional cut through the *stacking sequence I* swaged composite with 10 mm in diameter.

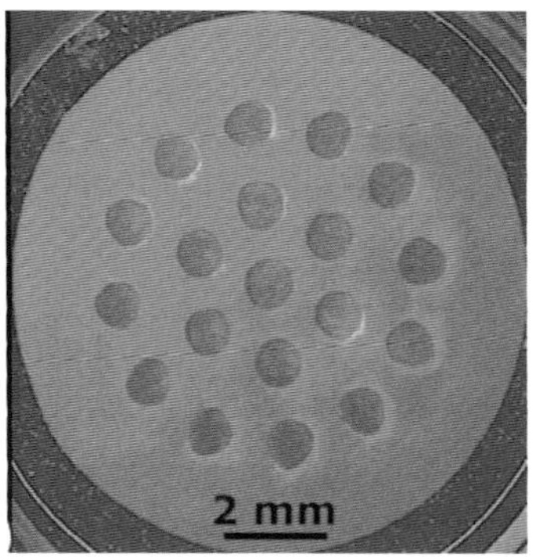

Fig. 1: Example of cross-sectional cut through investigated composites (sheath and 19 wires)

Detailed structure analyses of the swaged wires were performed by scanning electron microscopy (SEM); Tescan Lyra 3 FIB/SEM microscope equipped with a NordlysNano detector for electron back-scattered diffraction (EBSD) analyses was used. The scanned images were evaluated with the help of OXFORD Instruments [30] and LEM3 Laboratories [31] software. Microhardness measurements were performed using Zwick/Roell equipment; the average HV values for the Al/Cu sheath were calculated from a set of ten individual values measured across the particular composite cross-sectional cut (excluding the wires), while the average HV values for the wires were calculated from a set of ten individual values measured throughout all the wires. The samples for all the analyses were prepared via manual grinding on SiC papers and final diamond suspension polishing. For EBSD analyses, the samples were eventually polished electrolyticaly.

Results and Discussion

Mechanical properties. The original values of microhardness for the Cu and Al composite components were 84.5 HV, and 25.7 HV, respectively. The results of HV measurements of the 5 mm room-temperature swaged composites of *stacking sequence I* showed that the average HV value of the Al sheath was 55.1 HV, while for the Cu wires it was 106.9 HV. The HV measurements of the 5 mm room-temperature swaged composites of *stacking sequence II* then resulted in the average HV value of the Cu sheath to be 108.3 HV, while for the Al wires it was 34.4 HV.

Whereas the HV values of the Cu components were comparable for both the stacking sequences, the average HV values of the Al component differed significantly. This phenomenon can be explained by the mutual effect of the following two factors: 1) the RS process affects primarily the peripheral regions of the swaged piece, by the effect of which the imposed strain was primarily "consumed" by the prevailing volume of the Al sheath for *stacking sequence I* [32]; and 2) the maximum work hardening ability of the Al sheath was achieved during swaging (given by the intrinsic properties of Al, especially the stacking fault energy) [15].

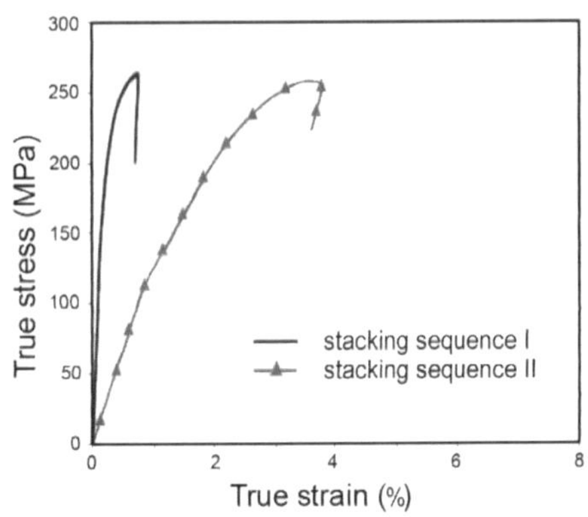

Fig. 2: Tensile stress-strain curves for both 5 mm swaged composite wires

The stress-strain curves resulting from the tensile tests of both the 5 mm swaged composite wires are depicted in Fig. 2. The tensile tests of the swaged composites revealed both the stacking sequences to exhibit comparable ultimate tensile strengths, however, the plasticity differed notably. Whereas *stacking sequence II* composite exhibited the elongation to failure of~ 3.5%, composite with *stacking sequence I* exhibited the plasticity of less than 1%. The stress-strain curve for *stacking sequence I* featured a steep increase at quite small deformations, i.e. *stacking sequence I* exhibited rapid work hardening at the expense of plasticity. The 5 mm *stacking sequence II* also exhibited rapid work hardening. Nevertheless, its intensity was not as high as for *stacking sequence I* (the slope of the stress-strain curve was less steep).

The relatively high strength of the swaged composites can primarily be attributed to the high volume fraction of Cu for *stacking sequence II*, and to the maximum achieved work hardening of the Al sheath for *stacking sequence I* (this supposition corresponds to the increased microhardness, as mentioned above). On the other hand, the imposed shear strain supports increase in plasticity for the Cu [33]. This factor was most probably the reason for the increased plasticity of *stacking sequence II*, whereas the plasticity of *stacking sequence I* was exhausted given by the major portion of work-hardened Al.

Structure analyses. The structure observations revealed the presence of fine more of less equiaxed grains within all the Al and Cu composites' components; the examples document Figs. 3a and 3b depicting the orientation image maps (OIMs) of structures of the Al sheath and Cu wire (both taken from the peripheral wires' layer radius) of *stacking sequence I*, respectively. The OIMs also depict the low and high angle grain boundaries (LAGBs and HAGBs), featuring the misorientations lower/higher than 15°, as red and black lines, respectively.

The observed grain refinement points to the occurrence of recovery/recrystallization imparted by the severe imposed strain. The grains within the Al components were evidently smaller than within the Cu components, which is primarily given by the fact that the activation energy needed to promote relaxation processes within Al is smaller than the activation energy of Cu [15]. The supposition of occurring recrystallization was confirmed also by the HAGBs analyses (Figs. 3a and 3b show their prevailing portion for both the composite components). The volume fractions of HAGBs, i.e. the boundaries the misorientation angles of which are larger than 15°, were 54% for Cu and 82% for Al within *stacking sequence I*, and 62% for Cu and 79% for Al within *stacking sequence II*. Also, the orientation image mapping showed that the fine grains within all the investigated composite components exhibited no prevailing preferential orientations (i.e. no colour depicting single preferential grains' orientation prevails in the Figures), which confirms the presence of relaxation processes. Fig. 3b also shows that the Cu wires of *stacking sequence I* exhibited the presence of bimodal structure. This phenomenon can be attributed to the effects of the following phenomena: as confirmed above by the observed work hardening and exhaustion of plasticity, the Al sheath of *stacking sequence I* tended to consume the majority of the imposed energy (given by the nature of the RS process and intrinsic properties of Al [15]). Also, the energy needed for grain growth (i.e. secondary recrystallization) is generally up to ten times lower than the energy needed for primary recrystallization [33]. The mutual effect of these phenomena imparted substructure development and grain growth at the expense of further recrystallization during the last swaging pass, which resulted in the formation of bimodal structure within the Cu wires.

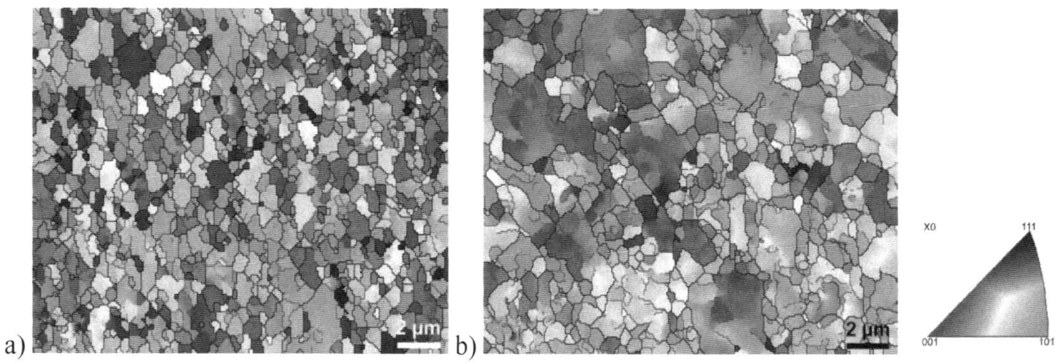

Fig. 3. OIMs with LAGBs depicted in red and HAGBs depicted in black for: Al sheath of *stacking sequence I* (a); Cu wire of *stacking sequence I* (b).

Residual stress. The internal grains misorientations depicting the presence of residual stress within the structures of the Al sheath and Cu wire (both taken from the peripheral wires' layer radius) of *stacking sequence I* are shown in Figs. 4a and 4b, respectively, whereas the internal grains misorientations within the structures of the Cu sheath and Al wire (both taken from the peripheral wires' layer radius) of *stacking sequence II* are shown in Figs. 4c and 4d, respectively.

The results of residual stress analyses showed the grains misorientations, i.e. residual stress, to be locally present in the Cu components of both the stacking sequences. Considering the results of structure analyses, i.e. the presence of bimodal structure within Cu, this phenomenon was primarily caused by the imparted substructure development as the misorientations were mostly observed within the growing grains. The Al components within both the stacking sequences did not exhibit substantial presence of residual stress, which also confirms the suppositions of structure relaxation and occurring recrystallization.

All the structure phenomena are inevitably affected by the two following processing factors. The deformation force affecting the processed material during rotary swaging - consisting of two individual components, radial and axial [2]. The designed composite sequence and the used composite materials – featuring their individual intrinsic properties, such as stacking fault energy, work hardening ability, activation energy for recovery, recrystallization, precipitation, etc. [15]. The plastic flow imparted by the swaging force can non-negligibly affect not only the mechanical properties, but also substructure development and the presence of residual stress within the swaged composites. As documented by the results, the imposed shear strain was sufficient to introduce grain refinement and structure relaxation within the Al components of both the stacking sequences. On the other hand, the Cu components, the activation energy necessary for relaxation processes for which is higher than for Al, exhibited the presence of residual stress. In other words, the occurring recovery/recrystallization was not complete after the last swaging pass for the Cu component, regardless the stacking sequence. Nevertheless, both the final 5 mm composite wires exhibited sufficient bonding of both the components, the present residual stress originating primarily from incomplete substructure development, did not negatively affect any of the composite wires as a whole product.

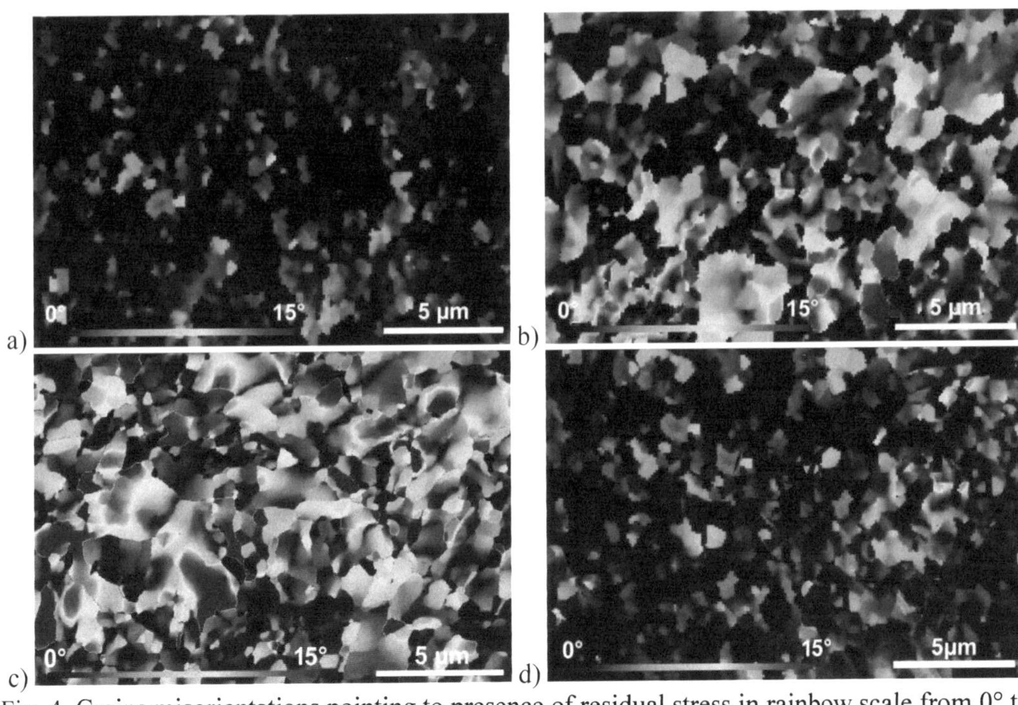

Fig. 4. Grains misorientations pointing to presence of residual stress in rainbow scale from 0° to 15° for: Al sheath of *stacking sequence I* (a); Cu wire of *stacking sequence I* (b); Cu sheath of *stacking sequence II* (a); Al wire of *stacking sequence II* (b).

Conclusions

The presented study proved that the technology of rotary swaging is applicable for production of long electro-conductive Al-Cu clad composite wires; both the swaged composite wires exhibited sufficient bonding of the component metals. The results revealed that, for sufficient swaging ratios, the stacking sequence of the individual metallic components had negligible effects on the substructure development and residual stress within the composite components; both the investigated stacking sequences featured Al components having very fine equiaxed grains with minor presence of residual stress, whereas the Cu components featured bimodal structures with a local presence of residual stress, which was primarily caused by substructure development, i.e. incomplete relaxation during the last swaging pass. Nevertheless, differences between the stacking sequences were observed during mechanical testing. Due to significant work hardening (microhardness over 55 HV) and exhaustion of plasticity of the Al sheath, the composite consisting of Al sheath and Cu wires exhibited very low plasticity (<1%), however, the maximum strength (276 MPa) was comparable to the composite consisting of Cu sheath and Al wires.

References

[1] S.L. Sing, Characterisation of Titanium-Tantalum Lattice Structures Fabricated by Selective Laser Melting, in: Sel. Laser Melting Nov. Titanium-Tantalum Alloy as Orthop. Biomater., 2019: pp. 87–95. doi:10.1007/978-981-13-2724-7_6.

[2] R. Kocich, L. Kunčická, D. Dohnalík, A. Macháčková, M. Šofer, Cold rotary swaging of a tungsten heavy alloy: Numerical and experimental investigations, Int. J. Refract. Met. Hard Mater. 61 (2016) 264–272. doi:10.1016/j.ijrmhm.2016.10.005.

[3] T. Hori, T. Nagase, M. Todai, A. Matsugaki, T. Nakano, Development of non-equiatomic Ti-Nb-Ta-Zr-Mo high-entropy alloys for metallic biomaterials, Scr. Mater. (2019). doi:10.1016/j.scriptamat.2019.07.011.

[4] X. Huang, J. Miao, A.A. Luo, Lightweight AlCrTiV high-entropy alloys with dual-phase microstructure via microalloying, J. Mater. Sci. 54 (2019) 2271–2277. doi:10.1007/s10853-018-2970-4.

[5] Y. Yan, S. Priya, Multiferroic Magnetoelectric Composites/Hybrids, in: C.-S. Kim, C. Randow, T. Sano (Eds.), Hybrid Hierarchical Compos. Mater., Springer International Publishing, Cham, 2015: pp. 95–160. doi:10.1007/978-3-319-12868-9_4.

[6] Y. Zhao, L. Peng, G. Yu, Electrochemical Hierarchical Composites, in: C.-S. Kim, C. Randow, T. Sano (Eds.), Hybrid Hierarchical Compos. Mater., Springer International Publishing, Cham, 2015: pp. 239–286. doi:10.1007/978-3-319-12868-9_7.

[7] F. Watari, A. Yokoyama, M. Omori, T. Hirai, H. Kondo, M. Uo, T. Kawasaki, Biocompatibility of materials and development to functionally graded implant for bio-medical application, Compos. Sci. Technol. 64 (2004) 893–908. doi:10.1016/j.compscitech.2003.09.005.

[8] S.. Tjong, Z.. Ma, Microstructural and mechanical characteristics of in situ metal matrix composites, Mater. Sci. Eng. R Reports. 29 (2000) 49–113. doi:10.1016/S0927-796X(00)00024-3.

[9] T.W. Clyne, P.J. Withers, An introduction to metal matrix composites, New York, NY, USA, 1993.

[10] L. Kunčická, R. Kocich, K. Dvořák, A. Macháčková, Rotary swaged laminated Cu-Al composites: Effect of structure on residual stress and mechanical and electric properties, Mater. Sci. Eng. A. 742 (2019) 743–750. doi:10.1016/j.msea.2018.11.026.

[11] M. Vaezi, H. Seitz, S. Yang, A review on 3D micro-additive manufacturing technologies, Int. J. Adv. Manuf. Technol. 67 (2012) 1721–1754. doi:10.1007/s00170-012-4605-2.

[12] J. Luo, S. Zhao, C. Zhang, Microstructure of aluminum/copper clad composite fabricated by casting-cold extrusion forming, J. Cent. South Univ. Technol. 18 (2011) 1013–1017. doi:10.1007/s11771-011-0796-1.

[13] R. Kocich, L. Kunčická, C.F. Davis, T.C. Lowe, I. Szurman, A. Macháčková, Deformation behavior of multilayered Al-Cu clad composite during cold-swaging, Mater. Des. 90 (2016) 379–388. doi:10.1016/j.matdes.2015.10.145.

[14] L. Kunčická, R. Kocich, K. Dvořák, A. Macháčková, Rotary swaged laminated Cu-Al composites, Effect of structure on residualstress and mechanical and electric properties, Mater. Sci. Eng. A. 742 (2019) 743–750.

[15] A. Russell, K.L. Lee, Structure-Property Relations in Nonferrous Metals, 1st ed., John Wiley & Sons, Inc., New Jersey, 2005.

[16] A.N. Cherepanov, V.I. Mali, I.N. Maliutina, A.M. Orishich, A.G. Malikov, V.O. Drozdov, Laser welding of stainless steel to titanium using explosively welded composite inserts, Int. J. Adv. Manuf. Technol. 90 (2017) 3037–3043. doi:10.1007/s00170-016-9657-2.

[17] L. Kunčická, R. Kocich, Deformation behaviour of Cu-Al clad composites produced by rotary swaging, IOP Conf. Ser. Mater. Sci. Eng. 369 (2018) 012029. doi:10.1088/1757-899X/369/1/012029.

[18] R. Kocich, M. Kursa, I. Szurman, A. Dlouhý, The influence of imposed strain on the development of microstructure and transformation characteristics of Ni–Ti shape memory alloys, J. Alloys Compd. 509 (2011) 2716–2722. doi:10.1016/j.jallcom.2010.12.003.

[19] M. Suresh, A. Sharma, A.M. More, R. Kalsar, A. Bisht, N. Nayan, S. Suwas, Effect of equal channel angular pressing (ECAP) on the evolution of texture, microstructure and mechanical properties in the Al-Cu-Li alloy AA2195, J. Alloys Compd. 785 (2019) 972–983. doi:10.1016/J.JALLCOM.2019.01.161.

[20] L. Kunčická, R. Kocich, J. Drápala, V.A. Andreyachshenko, FEM simulations and comparison of the ecap and ECAP-PBP influence on Ti6Al4V alloy's deformation behaviour, Met. 2013 22nd Int. Met. Mater. Conf. (2013) 391–396.

[21] R. Kocich, J. Fiala, I. Szurman, A. Macháčková, M. Mihola, Twist-channel angular pressing: effect of the strain path on grain refinement and mechanical properties of copper, J. Mater. Sci. 46 (2011) 7865–7876. doi:10.1007/s10853-011-5768-1.

[22] R. Kocich, L. Kunčická, A. Macháčková, Twist Channel Multi-Angular Pressing (TCMAP) as a method for increasing the efficiency of SPD, IOP Conf. Ser. Mater. Sci. Eng. 63 (2014) 012006. doi:10.1088/1757-899X/63/1/012006.

[23] S.V. Dobatkin, J. Gubicza, D.V. Shangina, N.R. Bochvar, N.Y. Tabachkova, High strength and good electrical conductivity in Cu–Cr alloys processed by severe plastic deformation, Mater. Lett. 153 (2015) 5–9. doi:10.1016/J.MATLET.2015.03.144.

[24] R. Kocich, A. Macháčková, F. Fojtík, Comparison of strain and stress conditions in conventional and ARB rolling processes, Int. J. Mech. Sci. 64 (2012) 54–61. doi:10.1016/j.ijmecsci.2012.08.003.

[25] Y. Yang, J. Nie, Q. Mao, Y. Zhao, Improving the combination of electrical conductivity and tensile strength of Al 1070 by rotary swaging deformation, Results Phys. 13 (2019) 102236. doi:10.1016/J.RINP.2019.102236.

[26] W.M. Gan, Y.D. Huang, R. Wang, G.F. Wang, A. Srinivasan, H.G. Brokmeier, N. Schell, K.U. Kainer, N. Hort, Microstructures and mechanical properties of pure Mg processed by rotary swaging, Mater. Des. 63 (2014) 83–88. doi:10.1016/j.matdes.2014.05.057.

[27] R. Kocich, L. Kunčická, P. Král, P. Strunz, Characterization of innovative rotary swaged Cu-Al clad composite wire conductors, Mater. Des. 160 (2018) 828–835. doi:10.1016/j.matdes.2018.10.027.

[28] L. Kunčická, A. Macháčková, N.P. Lavery, R. Kocich, J.C.T. Cullen, L.M. Hlaváč, Effect of thermomechanical processing via rotary swaging on properties and residual stress within tungsten heavy alloy, Int. J. Refract. Met. Hard Mater. 87 (2020) 105120. doi:10.1016/j.ijrmhm.2019.105120.
[29] L. Kunčická, R. Kocich, C. Hervoches, A. Macháčková, Study of structure and residual stresses in cold rotary swaged tungsten heavy alloy, Mater. Sci. Eng. A. 704 (2017) 25–31. doi:10.1016/j.msea.2017.07.096.
[30] O. Instruments, Providing leading-edge tools for SEM, TEM &FIB - Nanoanalysis -, (2018). https://nano.oxinst.com/ (accessed December 24, 2018).
[31] B. Beausir, J.J. Fundenberger, Analysis Tools for Electron and X-ray diffraction, ATEX - software, www.atex-software.eu, (2017).
[32] R. Kocich, L. Kunčická, A. Macháčková, M. Šofer, Improvement of mechanical and electrical properties of rotary swaged Al-Cu clad composites, Mater. Des. 123 (2017) 137–146. doi:10.1016/j.matdes.2017.03.048.
[33] F.J. Humphreys, M. Hetherly, Recrystallization and Related Annealing Phenomena, 2nd ed., Elsevier Ltd, Oxford, 2004.

Measurement of Gearbox Efficiency by Strain Gauges: Direct Application

LOPOT F.[1,a], DUB M.[1,b], HAVLÍČEK M.[1,c], ŠTOČEK O.[1,d], VESELÝ T.[2,e], KUČERA L.[2,f]

[1]Department designing and machine elements, Faculty of mechanical engineering, Czech Technical University in Prague, Technická 4, 160 00 Prague 6, Czech Republic

[2]Department of Information and Communication Technologies in Medicine, Faculty of Biomedical Engineering, Czech Technical University in Prague, nám. Sítná 3105, 272 01 Kladno 2, Czech Republic

[a]frantisek.lopot@fs.cvut.cz, [b]martin.dub@fs.cvut.cz, [c]martin.havlicek@fs.cvut.cz, [d]ondrej.stocek@fs.cvut.cz, [e]tomas.vesely@fbmi.cvut.cz, [f]lukas.kucera@fbmi.cvut.cz

Keywords: strain-gauge, efficiency, notch, FEM, calibration

Abstract.
Experimental determination of gearboxes efficiency provides important information. The use of specialized torque-sensors is, however, often complicated. Thus, we engage in experimental determination of the efficiency by direct strain gauges application. To achieve meaningful results, the calibration quality and strain gauges location are crucial. There is also an effort to increase the overall apparatus sensitivity by placing strain gauges in notches vicinity. To make the apparatus as light as possible, the amplifiers were replaced by 24 bits AD-converter. Places for the strain gauges installation are selected using FEM analysis. This analysis is used as a "virtual" calibration of the apparatus at the same time. Currently tested gearbox for city rail-vehicle was mounted in the measuring stand in full compliance with its real installation. Strain gauges were installed directly at the input shaft and at the output both, on the shaft and on Hirt teething on coupling hub. The presented method demonstrates reliable interpretation of results from measurements with conventional foil wire strain gauges far beyond their commonly declared sensitivity at the level of about 2 MPa. Furthermore, it brings a proposal how to define realistic uncertainities of the measured data considering calculated directly from raw data set without any filtering or other processing to eliminate the noise or in contrary highlight the points of interest.

Introduction

Experimental determination of gearboxes efficiency provides important information. Use of special torque sensors is, however, often complicated mainly due to spatial reasons. Moreover, their installation will affect the araangement of the whole set. Thus, we engage in experimental determination of the efficiency by direct application of strain gauges on the shafts of the measured gearboxes or on the hubs of the shaft couplings used. Also, the measuring apparatus is usually placed directly on measured rotating shafts. To achieve meaningful results, the quality of calibration and location of strain gauges are crucial [2, 3, 4, 6, 7, 9, 10]. There is also an effort to increase the overall apparatus sensitivity by convenient placing of strain gauges in notches vicinity. The use of notches for strain gauges is quite common, especially in the field of sensors [4, 5, 6].

A fundamental prerequisite for the use of a notch is clear linear relationship between mechanical stress and deformation. Because the stress is concentrated in the notches, the proportionality limit (yield point) can be exceeded locally. This condition is limiting for their use for strain gauge measurements, although it cannot be an obstacle for the operation of the component. Therefore, it is very important to subject the notch area to a thorough analysis.
The second fundamental output of this analysis is the determination of the area in which the searched local maximum of mechanical stress is located. Its position depends not only on the shape of the notch, but also on its size, geometry of the surroundings, surface roughness and many other factors. The influence of these factors varies depending on parameters of the material, production technologies, etc. [5, 11]. For this reason, it is appropriate to accompany the computational analysis with an experiment. The experience of our team so far indicates that the geometry of the notch and its surroundings has an overwhelming influence on steel shafts produced by turning and grinding.

Methods

Our measurement of efficiency by direct application of strain gauges is a complex process of series of successive steps. These steps can vary in specific design according to the situation. In any case, however, these steps are perfomred according to the following specifications:

- analysis of drawing documentation
- FEM analysis
- Preparation of installation foils
- Installation of strain gauges and apparatus
- Measurement
- Data processing using virtual calibration through processed FEM analysis
- Interpretation of results

The analysis of the drawing documentation represents an input step. In frame of it, places for installation of strain gauges and apparatus are selected. Secondly, the drawing documentation is used for preparation of installation foils. Usually, and in this case also, we use to work with standard foil strain gauges, LY series from HBM.

FEM analysis is a key step. First, it is used to select locations for installation of the strain gauges. A criterion for the final selection is the highest achieved stress value at a location, which must be available and large enough to install the strain gauge. As a rule, subjects of an evaluation are stress values in the network nodes. The smallest possible strain gauges are used for the measurement and it is thus not necessary to determine the average stress value over their length usually. In other words, the error we introduce into the result in this way is well below the limit of measurability. However, depending on the software used, it is possible to use various beam-type elements for strain gauge simulation. The above mentioned measurement error can thus be nearly completely eliminated. In order to be sure of the correctness of analysis results, the FEM calculation is repeated several times with various degrees of local refinement of the network and various settings of other marginal conditions. As the subject of monitoring, there are values in selected places of the calculated geometry. These values are compared with each other for various versions of the calculation. The result of the FEM analysis is considered correct when an invariable result is achieved. It is practically impossible to achieve a completely unchanging result. Obviously, results obtained by the FEM calculation after stabilization fluctuate around actual value (Fig. 1). It is therefore necessary to work with differences that are understood to be negligible.

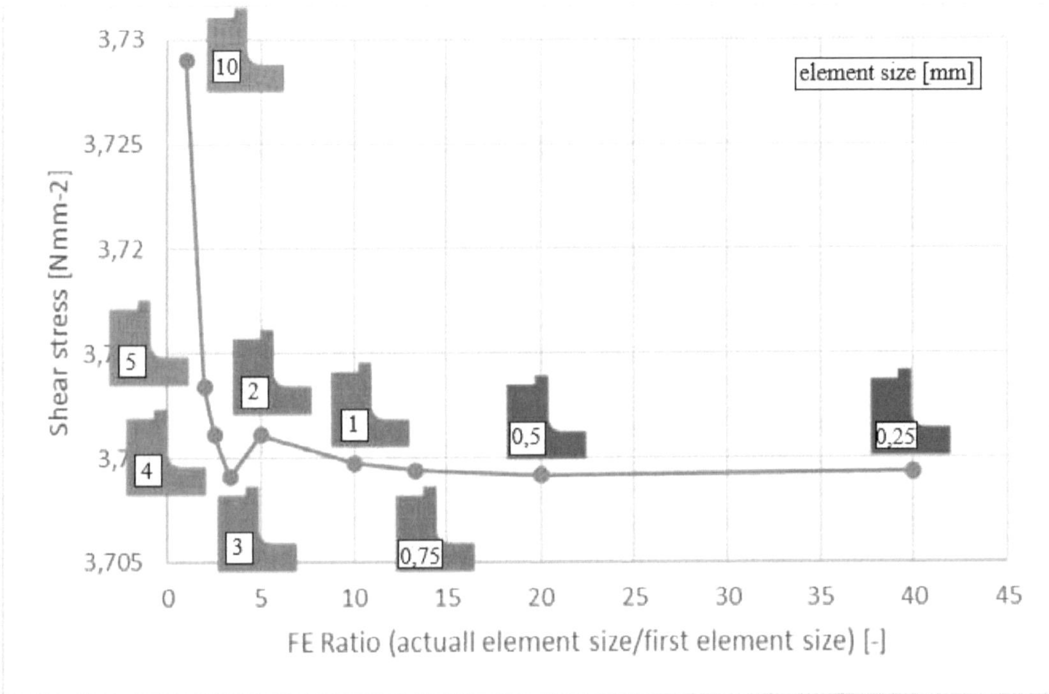

Fig. 1: Result of FEM analysis in strain gauge site related to element size

To calculate the stress value which will be comparable with acquired signal from strain gauge, we calculated presented values as average from results included in site of the strain gauge bordered by its active length [11]. This is the reason why in presented graph at FE ratio 5 (2 mm element size) is discontinuity which is caused by the partition of the area under the strain gauge active length. The number of elements at the edge of the model representing the strain gauge is 2 till the element size of 3 mm. Then, there is the jump in the number of elements from 2 to 4. The following curve then continues in the expected trend. With regards to the accuracy of the detection of actual values by practical measurements, we set this limit at a difference of 1% in our study. Once it is clear where the strain gauges are to be glued, we return to the drawing documentation to determine the specific geometry in specified places as a starting point for the next phase of work. Second, FEM analysis is also used for "virtual" calibration of the apparatus [4] during the processing of the measured data.

Preparation of the installation foils and jigs for the apparatus assembly is a routine step based on projection of measured geometry into the design of assembly jigs for the apparatus and on the use of developed shapes of appropriate geometry for printing of self-adhesive foils. On these foils, the strain gauges are placed before installation. In this way, we are able to position the strain gauges very accurately. The error caused by this way is then immeasurable for us.

The installation of strain gauges and apparatus already takes place at the installation site of the measured transmission set. During this phase, we follow common recommendations for the installation of strain gauges - thorough surface preparation for gluing, strain gauge gluing and wiring processes, etc. With regards to the need of the determination of the current torque on specified shafts using measurement, we work almost exclusively with strain gauges intended for torsion sensing [9]. These are installed into a full bridge due to temperature compensation. Usually, the apparatus is supplemented by other strain gauges for sensing of bending in two perpendicular planes. Alternating bending is usually an undesirable phenomenon in transmission sets. The function of these strain gauges is thus usually rather checking one. Nevertheless, the information is available. For its application in the calculation of required efficiency, it is not possible to describe a simple universal solution. It is always strongly dependent on each specific task. In general, it can only be stated that in cases where bending is present, there is an outflow of transmitted energy. This outflow is reflected in a decrease of the efficiency of the unit. At the end of this phase of the work, the installed measuring system is tested and the software bridge balancing, or tara zeroing only, is performed [8]. To make the apparatus as light as possible, the amplifiers are not used. Instead, a system developed by our team only works with a 24 bits AD converter. Using a battery with 1500 mAh capacity, we can measure continuously for up to 18 hours. Practically, this phase is performed twice - once for the input shaft and once for the output shaft. In both places, we worked with standard foil strain gauges from, the LY type.

Both subsystems communicate with each other within the overall measuring system via a selected protocol (x-bee, wifi, etc.). They also communicate with the control computer using the same protocol. Data can be read online while being stored on the storage media of both subsystems. We do not normally use online data reading due to high energy demands of this operation. Instead, we send data to the control computer with a significantly lower frequency (around 10 Hz). We display these data online on the monitor to check the operation of both subsystems and the whole complete system. Very often, subsystems are wrapped in a protective foil in the final phase of their installation (Fig. 2).

Fig. 2: Installed subsystem wrapped in a protective foil
(see also the shaft coupling with rubber blocks)

Data processing using virtual calibration through processed FEM analysis is performed after measurement finishing. Obtained data are not filtered or otherwise reduced. For the purpose of determining the efficiency of measured set, data processing is focused on reliable determination of mean values of measured waveforms. The question then is how to express the uncertainty of assessed value. We used standard deviation obtained from data set for mentioned average curve. The searched for efficiency is then determined according to the common definition. The gear ratio is considered as constant, because its fluctuations due to the unevenness of the teeth engagement are negligible for involute gearing. Thus, it has practically no effect on the result. An important aspect of data processing is correct signal zeroing. This can be complicated, because the gluing of the strain gauges takes place on the installed equipment during the agreed downtime. Usually, it takes place without any practical possibility of eliminating any residual stresses that remain in the shafts after the machine has been shut down. Noteworthy, the presence of these residual stress is specifically reflected in the measured signal by a step change of recorded values after starting the unit and its approach to prescribed test mode.

Interpretation of results is the final phase of the work. In frame of it, obtained results are arranged into comprehensible outputs, most often into so-called efficiency maps.

Efficiency maps clearly show a connection between efficiency and important parameters that directly or indirectly define it. In case of transmissions, it is primarily the speed and the transmitted torque that define the degree of load on the transmission under investigation. These are therefore 3D graphs, where the parameter is most often the operating temperature in the case of transmissions. This is given mostly by temperature of the filling (oil). The influence of this parameter on the form of the map varies depending on construction of the device and oil used, from very low to completely decisive.

For needs of our research in the field of industrial gearboxes for relatively large powers (from hundreds of kW to MW units), the most commonly used type is a projection of the efficiency map determined for the operating temperature of given design into a speed and torque plane. The speed and torque plane is a plane from which the connection of the investigated gearbox with the degree of its load can be seen clearlierst.

Furthermore, other important information can very easily be obtained from the efficiency maps, such as the development of power dissipation depending on a load of examined transmission, etc.

All the above-mentioned partial steps have been a subject of long-term development. As a part of this development, the methodology of tensometric measurement of the efficiency of transmission sets was certified by TÜV NORD Czech, s.r.o. [1].

Experiment

The measurement was performed using the above-mentioned methodology in a cooperation with Škoda Electric a.s. and Wikov MGI et al. The tested gearbox for city rail vehicle was installed in the measuring stand in full compliance with its real installation (Fig. 3).

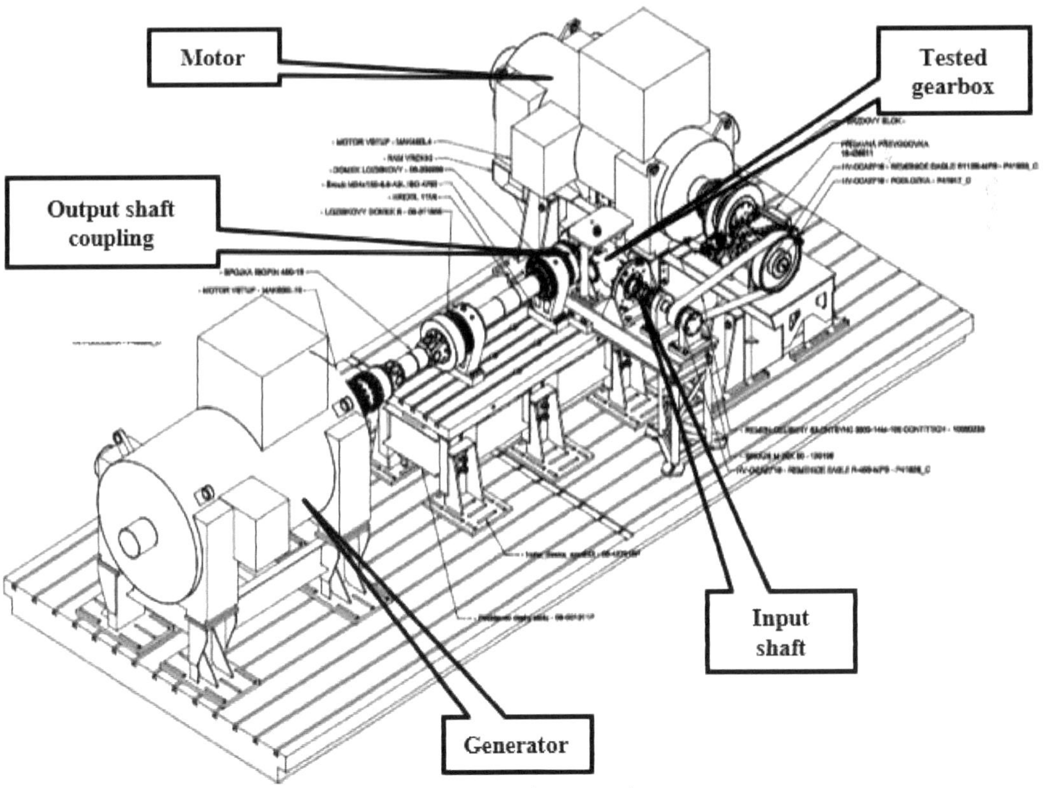

Fig. 3: Testing stand

The measurement was performed according to the following specification (Tab. 1):

Tab. 1: Assignment

Mode	Input speed [1/min]	Mk on the input [Nm]
1	10	10
2	10	330
3	10	660
4	10	1080
5	1840	10
6	1840	330
7	1840	660
8	1840	1080
9	2670	10
10	2670	358
11	2670	710
12	2670	1080
13	4200	10
14	4200	250
15	4200	500
16	4200	775

Based on the analysis of the drawing documentation, places for position of strain gauges and apparatus were selected. The outputs of this phase are clear from Fig. 4.

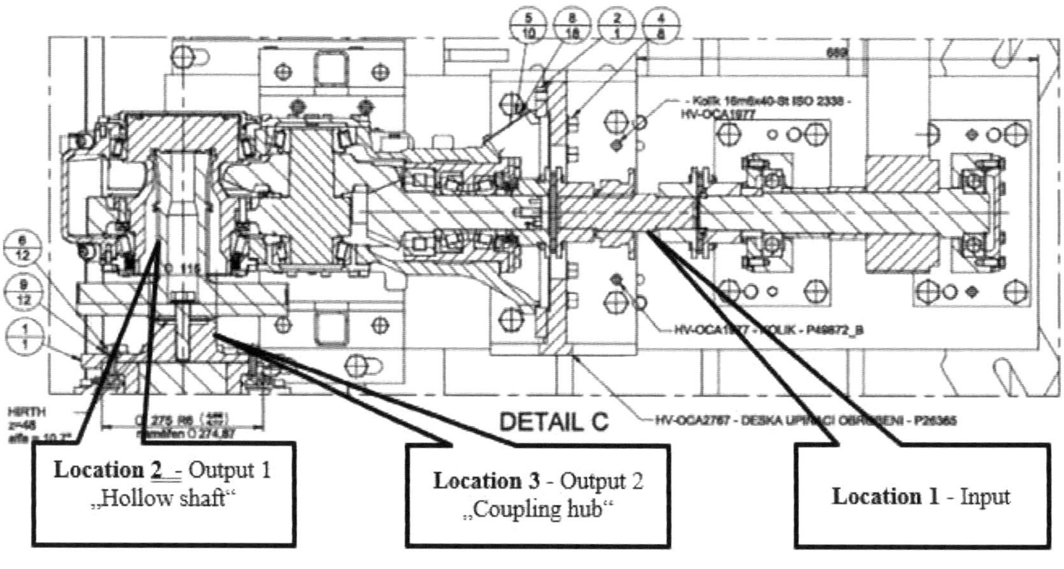

Fig. 4: Drawing with determination of places for application of strain gauges

The second step was the calculation of expected stress for individual load levels and FEM analysis of complicated shapes with notches to ensure virtual calibration even in places for which the analytical calculation does not provide reliable data. Because all the selected locations except for the first third (Fig. 4) are located on smooth circular cross-sections, FEM analysis was applied only to the hub of the used Hirt toothed coupling. We worked with Abaqus software. The FE model is designed as axisymmetric. It brings many benefits in case of meshing and computation time. There are used axisymmetric stress elements CGAX8R with reduced integration and quadratic shape functions. Quadratic elements are chosen to obtain the most precise results.

The FE size sensitivity analysis was carried out to ensure the correctness of the result shear stress at measured place. The obtained values are averaged across the strain gauge active length. The dependency of the result shear stress on the FE size ratio can be seen in the graph. It is evident that the results are very stable and converges quickly to the „exact" solution. The final FE size is set to 0,5 mm on the surface, the global seed size is 4 mm.

The calibration constant for the Hirt joint measured point is calculated as a ratio of the shear stress of the smooth shaft and the shear stress on the modelled part with realistic shape. The result calibration constant is equal to K=0.902 (Fig. 5).

There is also the inner thread M24 placed in the centre line of the shaft. The influence of this thread to the surface shear stress is negligible.

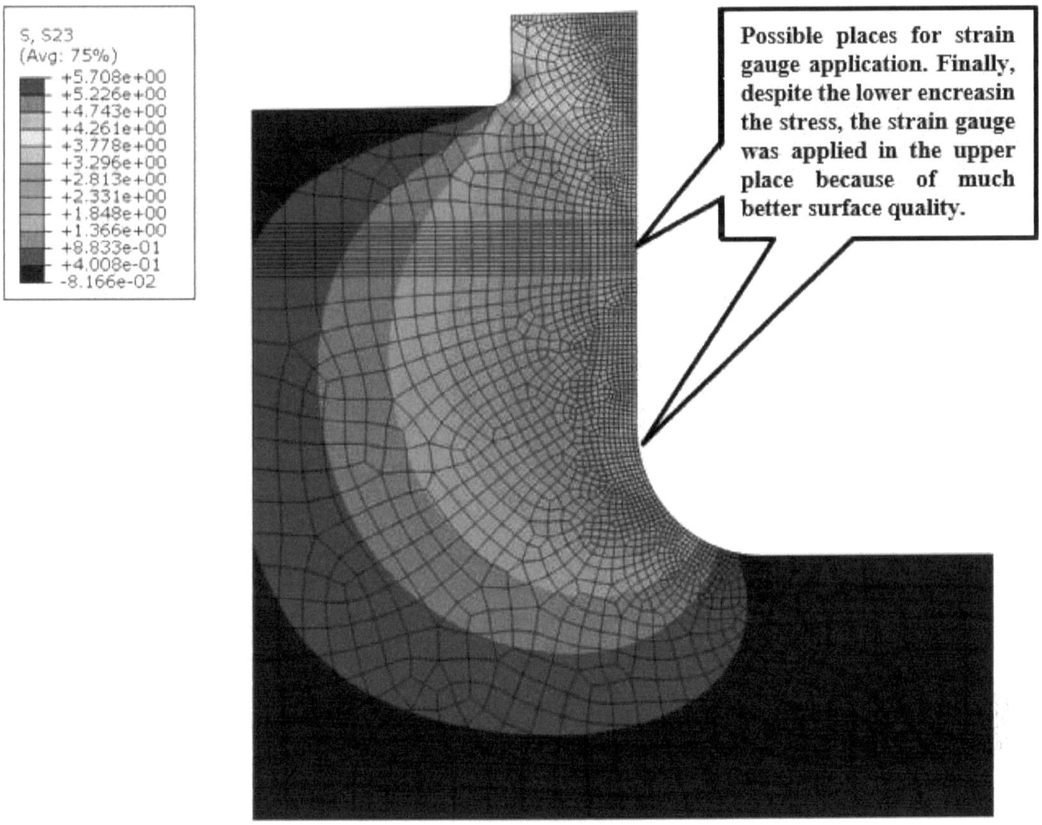

Fig. 5: FEM analysis use

The outputs of this calculation were subsequently used also for the above mentioned final calibration during data processing.

After confirming that selected places are well usable for the installation of strain gauges, the phase of preparation of installation foils took place on the basis of the processing of local geometry. This is followed by the installation of strain gauges on site. In this case, the output shaft was varnish coated and it was necessary to remove the varnish at gluing place for the strain gauge. The outputs of these phases can be seen in Figs. 6 and 7.

Fig. 5: Applied application foil with strain gauges on the input shaft (Location 1 in Fig. 4)

Fig. 6: Applied measuring strain gauge after removing the foil and installing the soldering fields on the hollow output shaft (Location 2 in Fig. 4)

After wiring, the parts were handed over to the mechanics and laboratory technicians of Wikov MGI company. They installed all the measuring parts prepared into the tested stand. The installation of data loggers and batteries followed. In order to ensure permanently unchanged position of data loggers and accumulators on the surface of rotating measured shafts, a strong adhesive tape with textile fibers in the adhesive line was applied (Figs. 7 and 2).

Fig. 7: Completed installation of the apparatus on the input shaft

The measurement was performed without physical presence of workers in the area of the gearbox measured. Progress of the measurement was monitored by means of video camera on computer monitor in a control room of the testing laboratory. The setting of required operating modes was performed by an experienced employee of Wikov MGI company. Tab. 2. provides an overview of the measured modes. These have been implemented within the possibilities offered by the laboratory equipment, respecting of operating conditions of the gearbox operation (especially the oil temperature).

Tab. 2: Overview of implemented measured modes

Mode	Input speed [ot/min]	M_k on the engine [Nm] (indicatively)	No. of measurement	Measurement start	Measurement time [s]	Oil temperature [°C]
1	12,2	10	9	16:56	35	48
2	12,2	330	10	16:57	35	46
3	12,2	660	11	16:59	35	45
4	12,2	1080	12	17:00	20	45
4	12,2	1080	12a	17:54	20	42
5	1840	10	1	16:04	40	44
6	1840	330	2	16:09	40	45
7	1840	660	3	16:11	40	48
8	1840	1080	4	16:14	40	49
9	2670	10	5	16:17	40	50
10	2670	358	6	16:19	40	53
11	2670	710	7	16:22	40	55
12	2670	1080	8	16:24	40	56
13	4200	10	13	17:58	30	42
14	4200	250	14	17:59	30	44
15	4200	500	15	18:00	30	45
16	4200	775	16	18:01	30	46

Mode no. 4 was repeated because control of operation of the unit reached limit of laboratory equipment. Moreover, online monitored control data did not show stable character. The delay between modes 12 and 13 was caused by oil cooling to operating temperature. Data on input torque in Tab. 2 is for guidance only - this data come from motor control converter. RPM reading is accurate. It was obtained from an external speed sensor located on the transmission input shaft.

Measurement time in Tab. 2 is for guidance only. It is a time for which the unit worked exclusively in the operating mode set.

Data recording was provided by data loggers with 24-bit AD converters. These convertors stored data on an SD card with sampling frequency of 1000 Hz per channel. At the same time, these data were sent via wireless interface X-Bee with a frequency of 10 Hz to a notebook. These data were there displayed online to check the operation of the apparatus.

The purpose of data processing before final evaluation is to identify and eliminate artifacts and errors which have been introduced during the measurement. Furthermore, readings from strain gauges have to be converted to Nm.

The actual data processing was preceded by detailed determination of dimensions of measured components at the strain gauge application site.

Data obtained after removal of artifacts and conversion to Nm using the data of gear ratios are presented graphically for individual modes according to tab. 2 (an example in Fig. 8).

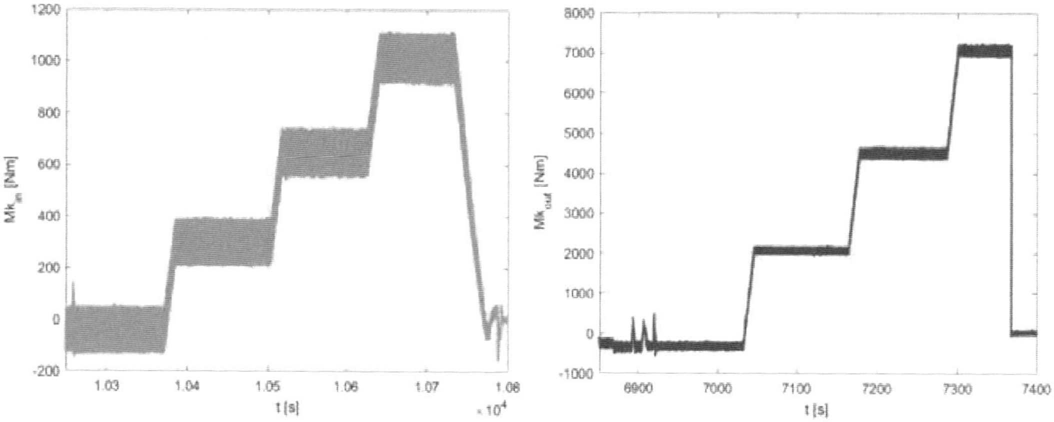

Fig. 8: Exapmple of torque recording for modes 9, 10, 11 and 12
(input - green, outputs - red, blue)

From the presented total course, the sections of the constant load in individual working modes are excluded for the determination of the mean values of the acting moments (example in Fig.9).

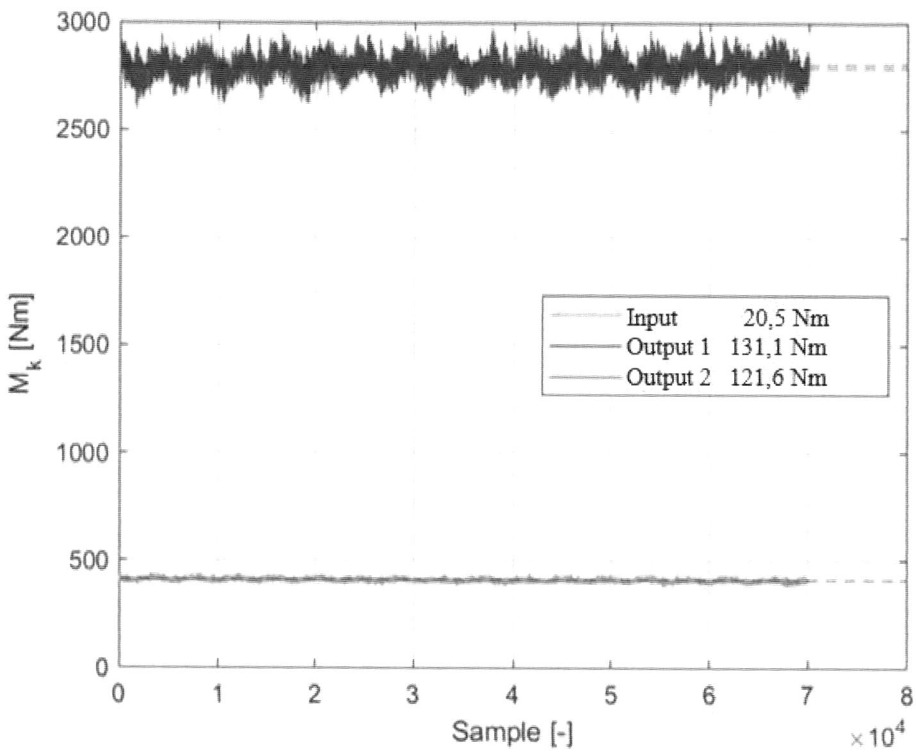

Fig. 9: Example of evaluated torque record for one mode
(input - green, outputs - red, blue)

The data are interesting in terms of frequency also but from current point of view, we use them primarily to calculate the searched efficiency. The interpretation of reached results is clear from Tab. 3 and 4. They give an overview of torque values for the input and output hollow shaft and actuall transmission efficiency in studied modes. The relative deviation of efficiency evaluation of the tested transmission is assessed using a theory of error folding as the square root of squares of relative errors of the input and output torque in each mode. Relative error means the ratio of the standard deviation and the average value of the monitored quantity.

Tab. 3: Evaluated data for hollow output shaft

mode	Input			Hollow shaft			
	Speed	Average	Std. dev.	Average	Std. dev.	Efficiency	Std. dev.
	1/min	Nm	Nm	Nm	Nm	%	%
1	12.2	20.46	0.620	131.07	3.609	**89.89**	17,898
2	12.2	409.95	0.750	2808.87	4.258	**96.14**	1,055
3	12.2	756.13	0.701	5169.80	3.698	**95.93**	0,498
4	12.2	1106.13	0.758	7607.35	4.233	**96.50**	0,389
5	1840	-39.94	2.972	-347.03	3.623	**82.02**	-11,733
6	1840	271.82	2.940	1853.23	4.221	**95.66**	1,892
7	1840	595.86	3.034	4128.51	6.444	**97.22**	1,195
8	1840	1007.57	3.019	7018.25	5.688	**97.73**	0,639
9	2670	-35.93	4.049	-322.31	2.876	**79.45**	-13,823
10	2670	304.77	4.076	2071.48	2.959	**95.37**	1,653
11	2670	651.44	4.101	4509.53	5.155	**97.13**	1,011
12	2670	1016.67	4.188	7080.37	5.719	**97.72**	0,697
13	4200	-54.89	1.327	-507.55	2.478	**77.08**	-5,121
14	4200	181.76	1.452	1147.52	2.599	**88.58**	1,638
15	4200	428.56	1.452	2888.15	2.903	**94.56**	0,757
16	4200	700.97	1.466	4803.84	3.011	**96.16**	0,478

In modes 5, 9 and 13, the power flow reversed in the direction from the generator to the motor. This was reflected in a change in the polarity of the signal from the measured components. To calculate the efficiency in these cases, the output of the gearbox was therefore considered as an input and, conversely, the input was considered as an output.

Tab. 4 presents similar data for the input shaft and the hub of the output shaft coupling.

Tab. 4: Evaluated data for coupling hub in the output shaft

mode	Input			Clutch hub – Hirt			
	Speed	Average	Std. dev.	Average	Std. dev.	Effieciency	Std. dev..
	1/min	Nm	Nm	Nm	Nm	%	%
1	12.2	20.46	0.620	121.57	4.082	**83.37**	20,180
2	12.2	409.95	0.750	2789.57	4.752	**95.48**	1,174
3	12.2	756.13	0.701	5144.14	4.203	**95.46**	0,564
4	12.2	1106.13	0.758	7570.35	4.731	**96.03**	0,433
5	1840	-39.94	2.972	-386.87	4.463	**73.57**	-13,425
6	1840	271.82	2.940	1813.00	4.859	**93.58**	2,089
7	1840	595.86	3.034	4083.77	6.839	**96.16**	1,256
8	1840	1007.57	3.019	6963.65	6.206	**96.97**	0,685
9	2670	-35.93	4.049	-345.10	3.672	**74.20**	-15,213
10	2670	304.77	4.076	2043.42	3.319	**94.07**	1,725
11	2670	651.44	4.101	4478.72	5.388	**96.47**	1,039
12	2670	1016.67	4.188	7038.89	6.058	**97.14**	0,724
13	4200	-54.89	1.327	-520.35	4.499	**75.18**	-8,545
14	4200	181.76	1.452	1125.83	4.368	**86.91**	2,532
15	4200	428.56	1.452	2859.89	4.544	**93.63**	1,113
16	4200	700.97	1.466	4772.37	4.673	**95.53**	0,699

A graphical representation of the development of efficiency of the tested gearbox in all operational modes is presented as a fundamental output of the measurement (Fig. 10 - without presented uncertainties/errors).

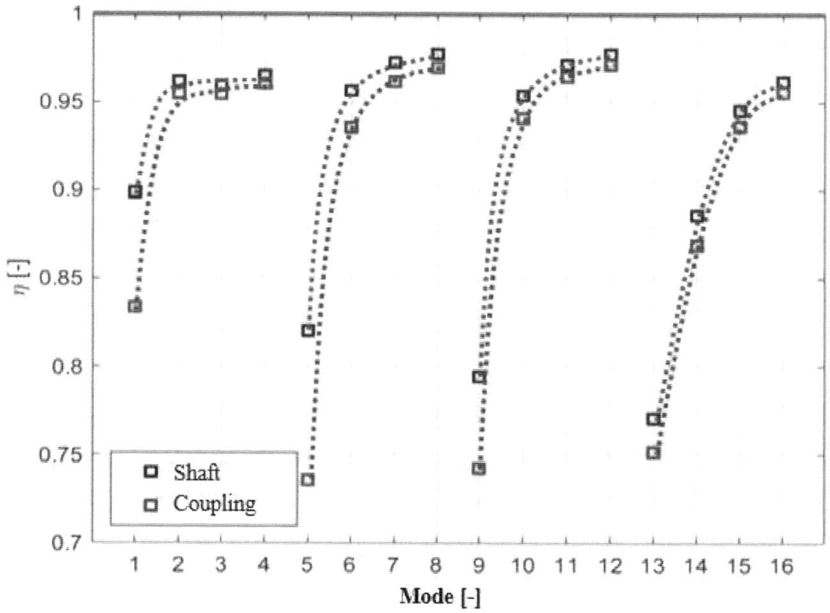

Fig. 10: Gearbox efficienty – trend curves for constant RPM

Smooth trend lines of marked points were added to the graph in Fig. 10 to show the development of efficiency of tested gearbox for individual measured modes in connection with the growth of the gearbox load at constant RPM.

If trend lines of marked points corresponding to each other across the modes are added to the same graph, it is possible to monitor the development of the efficiency of the tested gearbox in connection with the increase of RPM for a constant load level (Fig. 11).

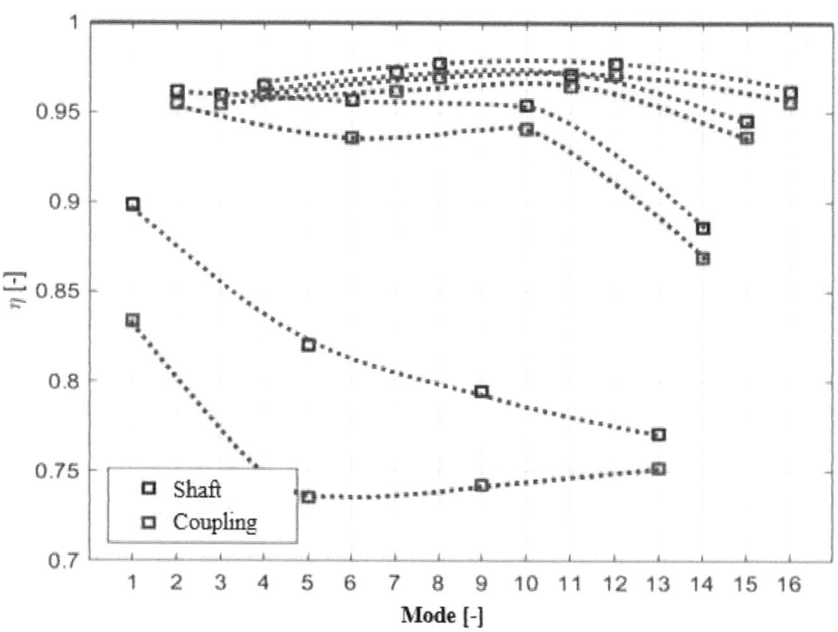

Fig. 11: Gearbox efficienty – trend curves for constant load level

With reference to the results presented above, it can be stated in general that the presence of flexible coupling on the output shaft (Figs. 2 and 7) which reduces the overall efficiency of the tested gearbox compared to the efficiency measured directly on the hollow shaft inside the gearbox.

Conclusions

The purpose of this study was to present a functional approach to measuring the efficiency of transmissions by direct application of conventional foil strain gauges. We aimed to draw attention to pitfalls that accompany it. To clearly document described phases of the whole process, the procedure was presented on an example of concrete measurement. In connection with this, it can be stated that the efficiency map above corresponds well with expectations of the gearbox manufacturer and, above all, it matches well with theoretical assumptions.
An important outcome of our work is the confirmation that with a suitable constellation of apparatus and the use of notches for the installation of strain gauges where possible, it can be obtained reliable interpretation of results from measurements with conventional foil wire strain gauges far beyond their commonly reported sensitivity at the level of about 2 MPa. Furthermore, it brings a proposal how to define realistic uncertainities of the measured data considering calculated directly from raw data set without any filtering or other processing to eliminate the noise or in contrary highlight the points of interest.

Acknowledgements

The paper was made with the support of Wikov MGI company which provided its laboratory for the placement of the measuring stand.

References

[1] O. Berka, M. Dub, V. Dynybyl, M. Janda, F. Lopot, Měření účinnosti převodovek s ozubenými koly [Applied Certified Methodology] 2016.

[2] R. L. Hannah, S. E. Reed, Strain Gage Users‘ Handbook. Springer Science & Business Media, 1992

[3] K. Hoffmann, An Introduction to Stress Analysis using Strain Gauges. HBM Germany, Dresden. 2017.

[4] J. Koch, J. Strain gauges: theory and application. Philips technical library. University of California. 1952

[5] F. Lopot, M. Dub, P. Kubový, Improvement in accuracy of strain gauge measurements for using in high precision measurements of gearboxes efficiency. EAN 2019 Proceedings. Česká společnost pro mechaniku, 2019.

[6] H. K. Neubert, P. Strain gauges: kinds and uses. Macmillan, University of Michigen. 1968

[7] J. Němec, Odporové tenzometry v praxi. SNTL – Nakladatelství technické literatury. Praha. 1967

[8] M. Riedel, Calibration with Strain gauge and bridge measurements. imc Meßsysteme GmbH, 2015 (available on www.imc-berlin.com)

[9] R. Schicker, G. Wegener, Measuring Torque Correctly. HBM Germany, Dresden. 2020.

[10] L. A. Window, Strain Gauge Technology. Springer Netherlands, 1992

[11] N. T. Younis, B. Kang, Averaging effects of a strain gage. Journal of Mechanical Science and Technology 25/1, Springer. 2011.

EAN 2020
58th conference on experimental stress analysis

Development of Smart Insoles for Gait Monitoring of Patients after Low Extremity Orthopedic Treatment

LUFINKA A.[1,a], JANDOVÁ S.[2,b], PETRIKOVÁ I.[2,c]

[1]Deparment of the Design of Machine Elements and Mechanism, Faculty of Mechanical Engineering, Technical University of Liberec, Studentska 2, 461 17 Liberec, Czech Republic

[2]Deparment of Applied Mechanics, Faculty of Mechanical Engineering, Technical University of Liberec, Studentska 2, 461 17 Liberec, Czech Republic

[a]ales.lufinka@tul.cz, [b]sona.jandova@tul.cz, [c]iva.petrikova@tul.cz

Keywords: Gait Monitoring, Foot loading, Measuring insole, Measuring system

Abstract. The aim of our study was to develop a new measuring system that enables to measure foot loading and temporal variables during stance phase while walking. This device is not supposed to compete with expensive professional measuring systems, but it should be a system that will be available to patients in rehabilitation after articular lower limb operations.

Introduction

Walking is the most common human locomotion and it is essential for self-service. Walking stereotype is changing during life and in the elderly, walking can be affected by a long-term overload of the musculoskeletal system. Very often, knee and hip joints are overloaded and joint replacement is indicated. According to the Institute of Health Information and Statistics of the Czech Republic (ÚZIS), about 11 145 (43.78%) knee replacements and 14 312 (56.22%) hip replacements per year are performed. The results of treatment are assessed in several ways; clinical examination results may not always be consistent with X-ray images and computed tomography results [1]. Nevertheless, it is quite common that despite the satisfactory results of clinical examination, abnormalities in the gait pattern after surgery remain noticeable. Therefore, some authors attempt to use other instrumentation methods as for example pedobarography to evaluate treatment outcomes [2]. This method enables to measure foot loading at various time interval after the joint replacement. Research institutes and university researchers very often have the opportunity to use extremely expensive commercially available measuring systems with high resolutions but these systems are not accessible to be used by common patients after the joint replacement. This is the reason why one of this study objectives is to develop a pressure-sensing insole with high resolution while maintaining a low cost. This could be achieved through the research and understanding of flexible pressure sensors regarding their electro-mechanical response.

The measuring insole principle

The insole will be intended for patients after surgery of the lower limb joints. Being part of the shoe, it is clear that it will not be reusable for multiple patients. This is the biggest difference compared to, for example, cardiac monitoring systems, which are only lent to patients during the monitoring period and can be reused for other patients after simple disinfection. Therefore, the most important requirement for the design of the insole measurement system was the low cost of the resulting solution to make the insoles available for single use. It follows from this requirement that the number of measuring points in the insole must be optimized so that the number of sensors is minimized while maintaining the possibility to measure the pressure distribution over the entire foot area with sufficient accuracy.

The design of sensors distribution in the insole is based on the pressure distribution measured by Emed forceplate (Novel, Munich, De) during one step (see Fig. 1).

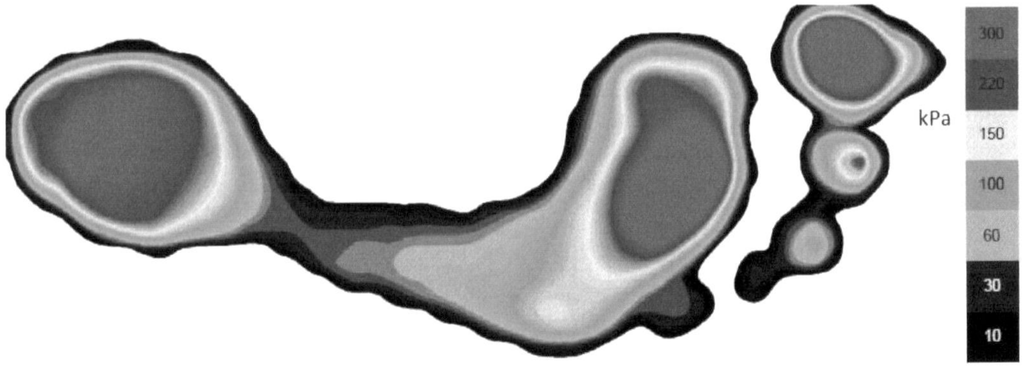

Fig. 1: Foot pressure distribution example

The foot pressure distribution shows three significant load areas [3]. If sensors are placed in these areas the entire measurement area will be covered. As a good compromise between the number of sensors and the resulting price, it was proposed to place eight pressure sensors in the insole area. The design of their location is shown in Fig. 2.

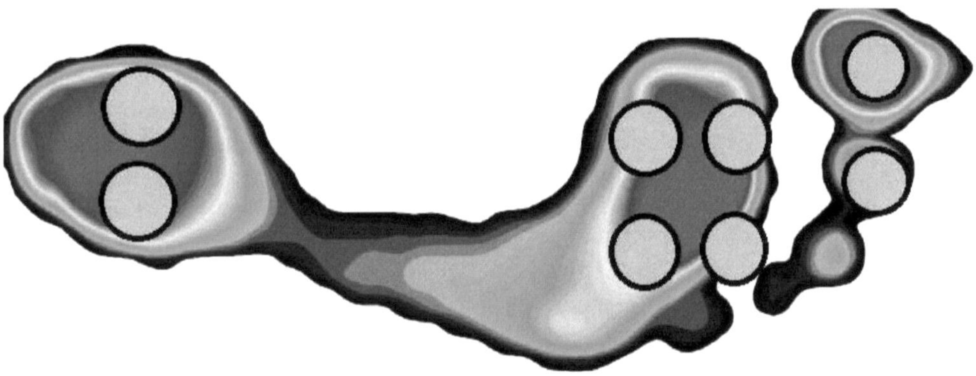

Fig. 2: The design of sensors location

Signals from the sensors will be led outside the insole into a microcomputer, which can be placed on top of the shoe including the power supply. The final solution assumes a separate microcomputer on each shoe, which is wirelessly connected, for example, to a mobile phone. The user application will evaluate data from both insoles and compare the load of lower limbs.

The pressure sensor and measuring insole design

The sensor is based on an elastic layer with electrodes on both sides. Electrodes and the elastic layer form a capacitor with the capacity according to Equation 1.

$$C = \varepsilon_0 * \varepsilon_r * \frac{S}{d} \quad (1)$$

where S is the area and d the electrodes distance, ε_0 and ε_r the vacuum and layer permittivity.

The capacitor is connected to an oscillator whose frequency depends on its capacitance. The electrodes distance and thus the capacitance and oscillator frequency are changed when the elastic layer is compressed by force. The sensor principle is shown in Fig.3.

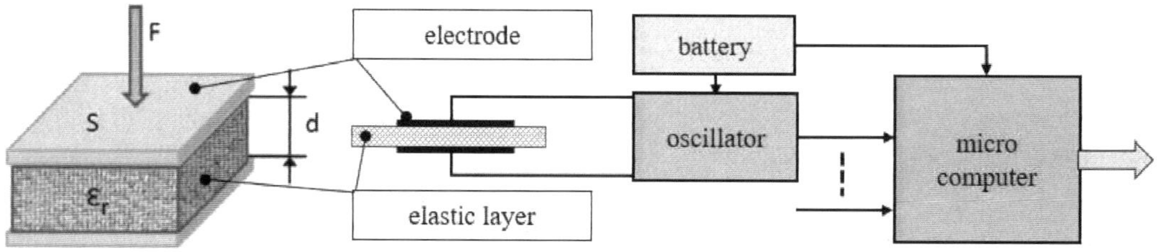

Fig. 3: The design of sensors location

Obviously, the capacitor's capacity with respect to the sensor dimensions considered is very small, in the order of picofarads. The theoretical capacitance and its corresponding oscillator frequency for electrode area 150 mm^2 and varying electrodes distance are shown in Table 1.

Table 1: Capacity and frequency dependence on electrodes distance

d[mm]	1.8	1.6	1.4	1.2	1	0.8	0.6	0.4	0.2
C [pF]	0.74	0.83	0.95	1.11	1.33	1.66	2.21	3.32	6.64
f [kHz]	174	172	168	164	158	150	139	120	86

It is evident from Equation 1 that the dependence of capacitance on electrode spacing is non-linear. The area where the sensitivity of the frequency to the distance change is good is indicated in red in the table. For the insole construction, it follows that the optimum initial thickness of the elastic layer is approximately 1.2 mm and must be compressed to 0.4 mm or less at maximum pressure. Then the frequency change will be sufficient to evaluate the pressure with the required accuracy. The nonlinearity of the dependence does not matter, because the computing power of the microcomputer enables realization of linearization in real time.

Due to the small capacitance of the active capacitor, it is necessary to minimize additional parasitic capacitances at the oscillator input. These parasitic capacitances (connected in parallel to the active capacitor) affect the frequency of the oscillator and reduce the sensitivity of the frequency to the capacitance change. Therefore, the length of the connecting wires between the capacitor and the oscillator must be minimized, the oscillator must be located as close as possible to the active capacitor. The oscillator integrated circuit will therefore be located directly in the insole. The insole must, of course, be flexible so that it can be used in the shoe during walking. The capacitor electrodes will therefore be formed on a flexible printed circuit board, on which the integrated circuits of the oscillators will also be placed. They will be in SMD design to keep their size as small as possible. The possible arrangement of the insole is shown in Fig 4.

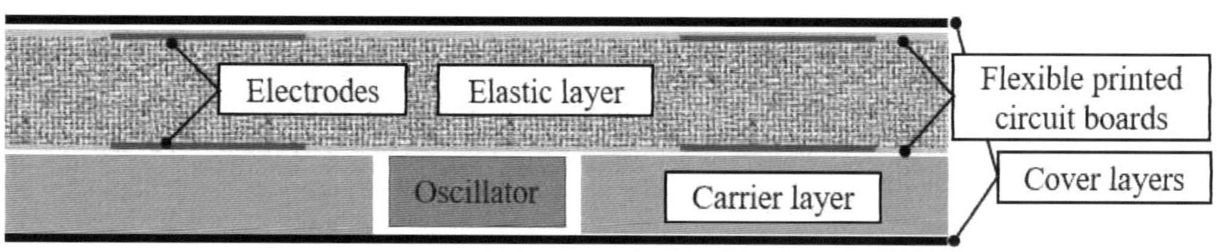

Fig. 4: The possible design of the measuring insole

The carrier layer forms the basis of the measuring insole. In the carrier layer there are openings for integrated circuits which are on the underside of the flexible printed circuit, on which there are also the lower electrodes. The printed circuit is laid on the carrier layer. This is followed by an elastic layer and a second flexible printed joint with the upper electrodes. The cover layers are then on both sides of the insole. If the elastic layer has a thickness of 1.2 mm, the overall thickness of the insole is less than 4 mm, which is an acceptable result for using the insole in footwear.

The pressure sensor prototype

The properties of the elastic layer will have a major influence on the functionality of the entire measuring insole. The sensor prototype was therefore designed to measure the properties of various materials of the elastic layer. For this prototype a classic hard printed circuit instead of flexible was used, because basic verification of material properties will not be done directly in the shoe and therefore flexibility is not necessary. Electrode dimensions, oscillator circuit type, and its location on the underside of the printed circuit board have been retained. The plastic rigid pad also replaced the insole support layer so that the sensor with the elastic layer tested can be inserted to the testing device. The rigid pad ensures that only the elastic layer is compressed during loading. The sensor prototype on the plastic pad and the sensor with elastic layer are shown in Fig. 5.

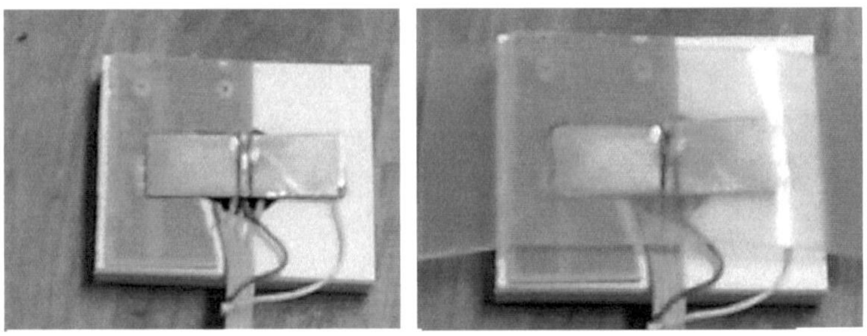

Fig. 5: The sensor prototype

Measurement of elastic layers properties

A simple device for measuring the properties of the elastic layer has been built (see Fig. 6). The elastic layer is compressed by the tightening nut. The sensor oscillator frequency, the achieved force and the elastic layer compression value are measured.

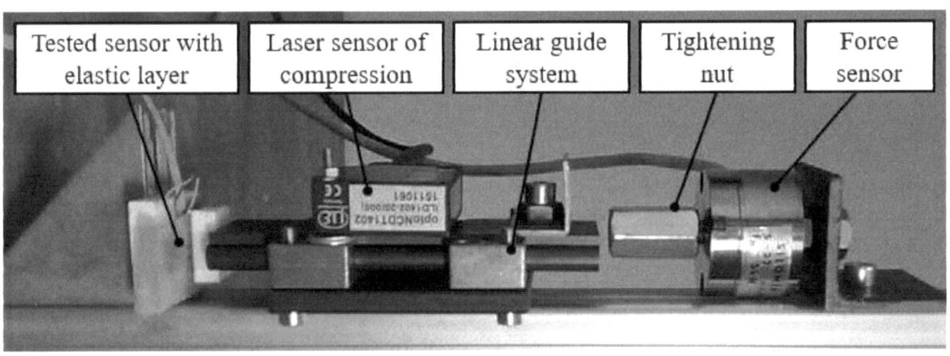

Fig. 6: The elastic layer testing device

The elastic layer is placed on the sensor and the top electrode is placed on the elastic layer. The whole is then inserted in the testing device. The oscillator is connected to the power supply and all signals are connected to the Dewe 5000 measurement unit. The elastic layer is compressed manually by a tightening nut. The progress of all signals is recorded in the measuring unit. An example of a load cycle for one type of elastic layer is shown graphically in Fig. 7.

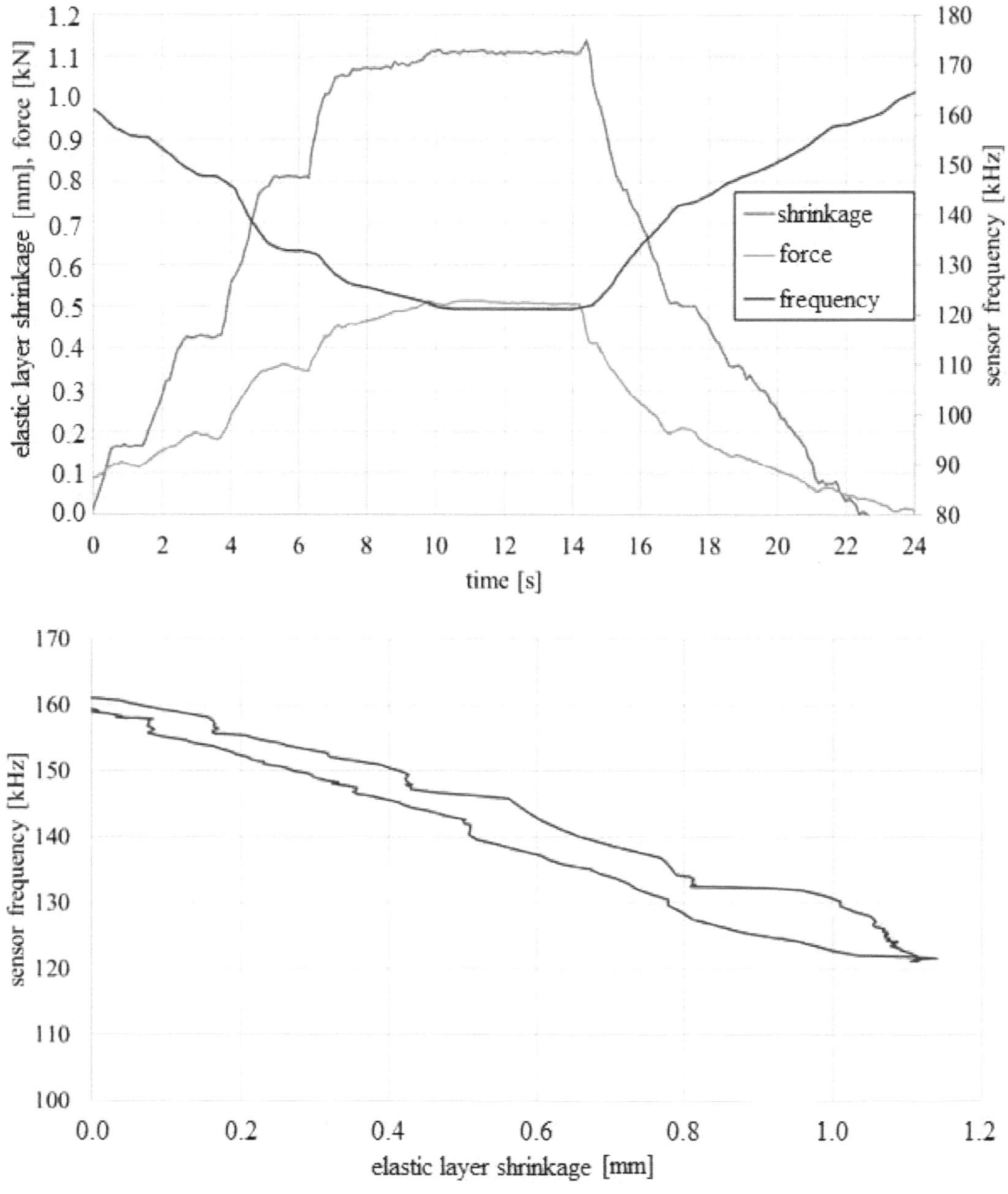

Fig. 7: Example of elastic layer load cycle (time record at the top, frequency dependence on compression at the bottom)

The cycle shown shows that the load with the hand-tightened nut is not smooth and even. However, this system is sufficient to verify the properties of the various materials of the elastic layer. The lower graph shows the hysteresis of the elastic layer measured.

The measurements made so far show that the properties of the elastic layer are absolutely crucial for the reliable operation of the entire measuring insole. The material of the elastic layer must have minimal hysteresis and its stiffness must be such that the compression under walking load is within the required range.

Conclusions

The goal of device development is not to compete with expensive professional measuring systems, but to develop a system that will be affordable to patients in rehabilitation after lower limb joint operations. Due to the intended use, the initial measurement results with the prototype insole indicate satisfactory accuracy and repeatability of the measurement. Of course, the device does not reach the parameters (especially in the number of measured points) of expensive professional systems, but for a fraction of the price of these systems, it provides results corresponding to the intended use. This will allow its single-use by the patient during rehabilitation.

Acknowledgment

This publication was written at the Technical University of Liberec as part of the project "Research, development and application of advanced methods and technologies in technical mechanics, biomechanics and strength and elasticity" with the support of the Specific University Research Grant, as provided by the Ministry of Education, Youth and Sports of the Czech Republic in the year 2020. The role of students was to put together overview about previous studies of the topic.

References

[1] A. Hirschmüller, L. Konstantinidis, H. Baur, et al. Do changes in dynamic plantar pressure distribution, strength capacity and postural control after intra-articular calcaneal fracture correlate with clinical and radiological outcome? Injury. 2011; 42(10):1135-1143.

[2] S. Jandová, J. Pazour, M. Janura, Comparison of Foot Load in walking after two different surgical treatment of calcaneal fracture. The Journal of Foot & Ankle Surgery. 2019; 58: 260-265.

[3] S. Jandová, Foot strike pattern in sport shoes with different construction. Journal of Mechanical Engineering, 2018; 15(2): 29-39.

EAN 2020
58th conference on experimental stress analysis

Testing of Glued Joints on 3D Printed Flexible Materials Made by FFF Technology

MACHALLA V.[1,a], SUDER J.[2,b], FOJTÍK F.[1,c], ZEMAN Z.[2,d]

[1]Deparment of Applied Mechanics, Faculty of Mechanical Engineering, VŠB – Technical University of Ostrava, 17. listopadu 2172/15, 708 00 Ostrava, Czech Republic

[2]Deparment of Robotics, Faculty of Mechanical Engineering, VSB – Technical University of Ostrava; 17. listopadu 2172/15, 708 00 Ostrava, Czech Republic

[a]vojtech.machalla@vsb.cz, [b]jiri.suder@vsb.cz, [c]frantisek.fojtik@vsb.cz, [d]zdenek.zeman@vsb.cz

Keywords: 3D print, Tensile test, Glued joints, Flexible plastic, FFF technology,

Abstract. The article deals with testing of glued joints of plastic parts manufactured by 3D printing. In the first section of this article is introduction of elastic plastic materials used for 3D-printing, followed by an overview of selected glues used to glue plastic samples. The glued plastic samples were tested on tensile testing machine accordingly to standard DIN EN 1465 (standard for testing of the glued samples). For each type of glued joint, the maximal and minimal breaking force, the dependence of loading force on displacement and the maximal and minimal stress are evaluated. The aim of this article is to evaluate and compare tensile tests of glued joints for different types of glues used on different types of plastic materials.

Introduction

3D printing is a process of making three dimensional solid objects from a digital file. This technology is based on adding thin layers of material on top of each other in order to create desired solid object. [1,2]

There are many advantages to this method of creating 3D objects (speed, price, shape diversity, weight, etc), but also some disadvantages as well. One of them being a limited print area of 3D printers to make objects in. Therefore, bigger objects must be divided into several smaller parts that fits into that print area. The glue method is offered to connect individual parts.

This paper deals with testing of the strength of glued joints made by using various glues, where the glued parts are made from flexible plastic TPE (Thermoplastic elastomer). Research done in this area shows, that the glued joints between flexible plastics are not stronger than the plastic material itself. While testing, all glued samples made from TPE material broke in the glued joint (the glued surfaces separated from each other). This can be caused either by bad adhesion of the glue to the surface, or by high elasticity and elongation of the TPE material compared to the hardened adhesive layer that has lower elasticity, thus when the material is elongated the adhesive layer tears itself off the surface. [2-6].

Description of samples – Each glued sample is marked by two letters. First letter represents material, from which the sample is made. Second letter is the type of adhesive glue used to glue parts of the sample together.

First letter (type of material):

A) Fiberflex 30D.

B) TPE 88

Second letter (type of adhesive glue):
G) PECKALEP GUMMI
L) LOCTITE 4850
B) BLACK BOLT
Z) ZAP-RT CA

Samples are made of two identical printed parts of dimensions 70 x 25 x 1 mm (length x width x thickness) glued together.
Parts were connected by the glue across all their width in length of 12.5 mm see at Fig. 1 (blue area marks the glue), which corresponds to DIN EN 1465 (the standard for testing glued samples by tensile test) [3].

Surface preparation and gluing – The top surface of the test samples was treated before gluing. The surface was first mechanically sanded with a medium-thick sandpaper and then subsequently cleaned with acetone and degreaser. After that, the samples were manipulated only with gloves, to prevent smearing of surface. A thin layer of glue was applied on the samples and then the two parts were pressed together as shown at Fig. 1. [7, 8]

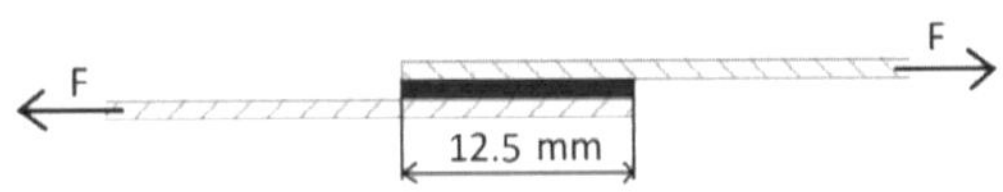

Fig. 1: Example of bonding samples

A fixture for precise sample placement during gluing was created to make sure the parts of the sample were always glued in the same position.

Tensile test
All samples were subjected to a tensile test on the Testometric M500/50CT machine (Fig. 2) in the laboratory. The load force F was gradually increased until the glued joint failed and the samples separated. The load force F (N) and elongation Δ_L (mm) were recorded. The pic. 2 shows ripped sample after tensile test. The sample is clamped in testing device.

Fig. 2: Sample clamped in to testing device after tensile test,

While testing, sample is loaded to tension, whereas the glued joint is loaded to shear. Due to a little misalignment of glued parts, some load to bending occurs. This phenomenon causes removal of adhesive glue, but because in this case misalignment is minimal, this influence can be neglected.

Results

Material Fiberflex 30D (A thermoplastic elastomer with a hardness of 30D Shore) Dependence of the loading force on sample deflection for material Fiberflex 30D is shown in Fig. 3, 4. Five samples tested for each group (AG, AL, AB, AZ).

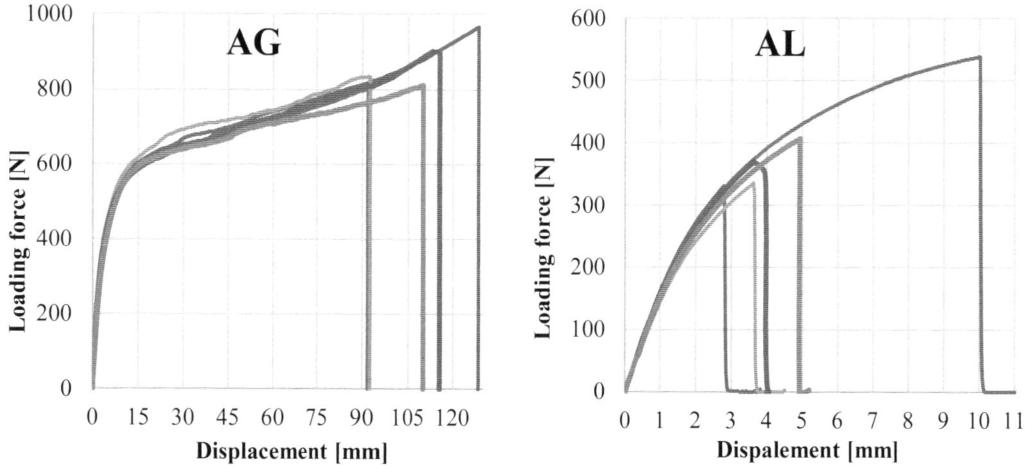

Fig. 3: Measurement – Dependence of loading force on displacement:
(a) Samples marked AG, (b) Samples marked AL

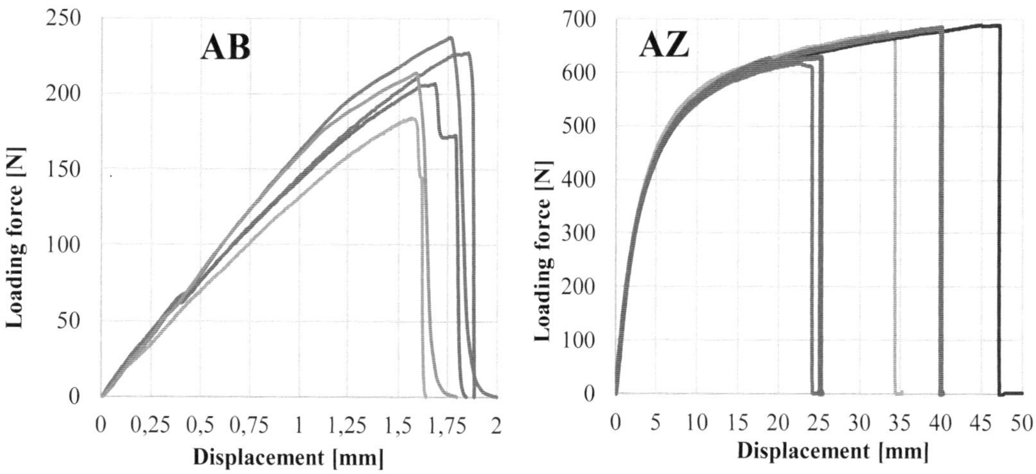

Fig. 4: Measurement – Dependence of loading force on displacement:
(a) Samples marked AB, (b) Samples marked AZ

Material TPE 88 (A flexible material like a rubber, the designation TPE 88 indicates the degree of elasticity). Dependence of the loading force on sample deflection for material TPE 88 is shown in Fig. 5, 6. Five samples tested for each group (DG, DL, DB, DZ).

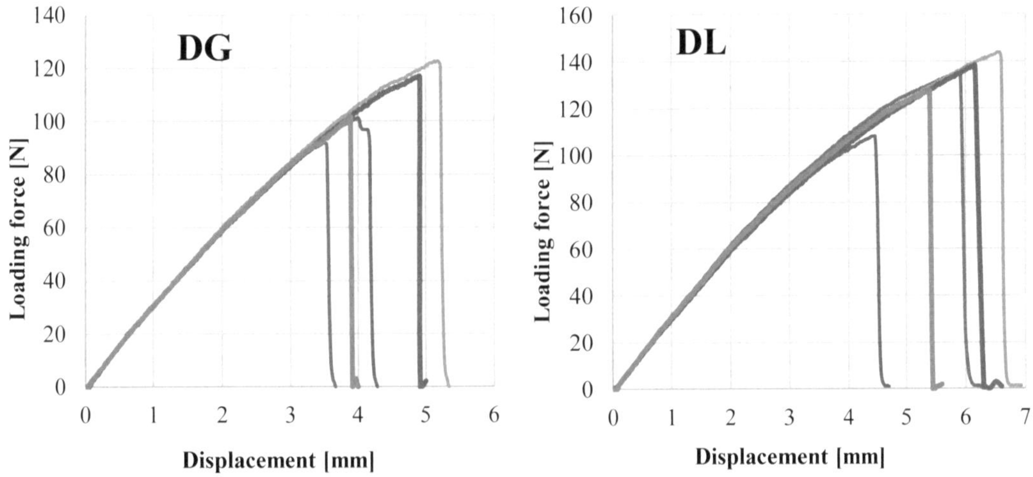

Fig. 5: Measurement – Dependence of loading force on displacement:
(a) Samples marked DG, (b) Samples marked DL

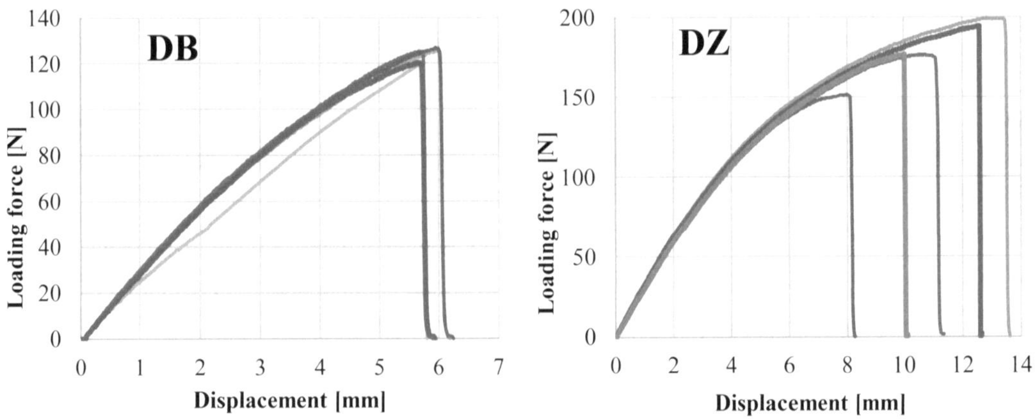

Fig. 6: Measurement – Dependence of loading force on displacement:
(a) Samples marked DB, (b) Samples marked DZ

The maximal shear stress calculated as

$$\tau_{S_{MAX}} = \frac{F_{MAX}}{S} = \frac{F_{MAX}}{a \cdot b} \tag{1}$$

the minimal shear stress calculated as

$$\tau_{S_{MIN}} = \frac{F_{MIN}}{S} = \frac{F_{MIN}}{a \cdot b} \tag{2}$$

where a is with of sample (25 mm), b is length of glued joint (12,5 mm) and F_{MAX}, F_{MIN} is maximal and minimal loading force.

Table 1: Comparison of test results

Mark of sample [-]	Maximal force F_{MAX} [N]	Minimal force F_{MIN} [N]	Stochastic definitions force [N]	Maximal Shear Stress $\tau_{S_{MAX}}$ [MPa]	Minimal Shear Stress $\tau_{S_{MIN}}$ [MPa]	Mean value of Shear Stress [MPa]
AG	**966**	**802**	$\mathbf{863^{+103}_{-61}}$	**3.09**	**2.57**	**2.76**
AL	538	330	396^{+142}_{-66}	1.72	1.06	1.27
AB	237	184	213^{+24}_{-29}	0.76	0.59	0.68
AZ	689	616	657^{+32}_{-41}	2.2	1.97	2.1
DG	122	92	107 ± 15	0.39	0.29	0.34
DL	144	108	131^{+13}_{-23}	0.46	0.35	0.42
DB	127	120	124^{+7}_{-4}	0.41	0.38	0.4
DZ	**199**	**151**	$\mathbf{179^{+20}_{-28}}$	**0.64**	**0.48**	**0.57**

The next Fig. 7 (a), (b), shows box plots of maximal and minimal shear stress. The individual boxes show the material and type of glue used on the sample, compared to maximal a minimal shear stress needed to rip the sample.

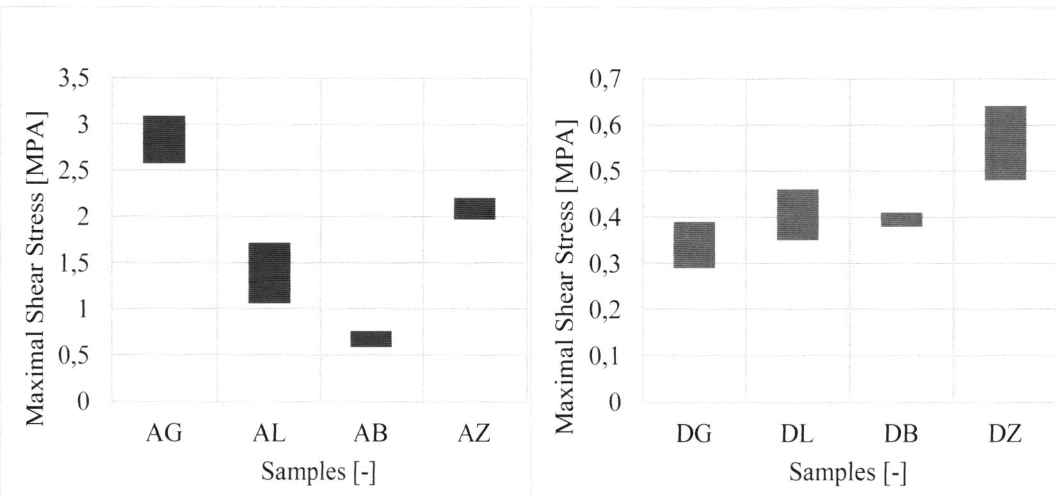

Fig. 7: Boxplots of maximal shear stress: (a) Material Fiberflex 30D, (b) Material TPE80E

Conclusions

This paper deals with gluing flexible plastic TPE and testing the glued joint. The results of tensile tests show the strength of glued joints of individual glue types used. The choice of glue type, depending on the material, leads to a bigger strength of glued joint.

Peackalep Gummi (marked G) is the most suitable for gluing parts made of material Fiberflex 30D (marked A). Five samples were created for each type of glue. In the worst case the glued joint broke at load of 802 N (minimum force to break sample) see at Fig. 3(a) and

Tab. 1. In the best case the glued joint broke at load of 966 N was needed, see at Fig. 3(a) and Tab. 1. According to equations (1) and (2), the maximal and minimal shear stress was determined, see. Tab. 1. The maximal and minimal shear stress for each type of glued joint are clearly displayed in the boxplots, see. Fig at 7(a). The worst type of glue for this material is Black bolt (marked B), glued joint made with this type of glue broke under a load of 184-237 N, see at Fig. 4(a). The results of other glued joints are shown in Tab. 1.

Zap-RT CA (marked Z) is the most suitable for gluing parts made of material TPE 88 (marked D). Five samples were created for each type of glue. In the worst case the glued joint broke at load of 151 N (minimum force to break sample) see at Fig. 6(b) and Tab. 1. In the best case the glued joint broke at load of 199 N, see at Fig. 6(b) and Tab. 1. According to equations (1) and (2), it was determined maximal and minimal shear stress, see. Tab. 1. The maximal and minimal shear stress for each type of glued joint are clearly display in the boxplots, see at Fig. 7(b). The worst type of glue for this material is Peckalep Gummi (marked G), glued joint made with this type of glue broke under a load of 92-122 N, see at Fig. 5(a). The results of other glued joints are shown in Tab. 1.

Acknowledgement

The authors gratefully acknowledge the funding from the Czech projects SP2020/23 and CZ.02.1.01/0.0/0.0/17_049/0008441.

References

[1] J. Lipina et al., "Shaped Glued Connection of Two Parts Made by Rapid Prototyping Technology", Applied Mechanics and Materials, Vol. 555, pp. 541-548, 2014.

[2] E. M. Petrie, Handbook of Adhesives and Sealans, McGraw-Hill, New York USA, 2000, ISBN 0-07-049888-1.

[3] J. Drkoš. Bonding of composite materials. Brno, 2018. Bachelor thesis. Brno university of technology.

[4] 3D Printer Original Prusa i3 MK3S. Prusa Research [online]. Praha: Prusa Research, 2019 [cit. 2019-03-17]. Available at: https://shop.prusa3d.com/cs/3d-tiskarny/181-3d-tiskarna-original-prusa-i3-mk3s.html#1

[5] J. Folta, Evaluation of the strength of glued joints in the construction of buses. Pardubice, 2018. Bachelor thesis. University of Pardubice.

[6] J. Ruth, R. Kleer and F. Piller. Predicting the future of additive manufacturing: A Delphi study on economic and societal implications of 3D printing for 2030. Technological Forecasting and Social Change. 2017, 117, 84-97. DOI: 10.1016/j.techfore.2017.01.006. ISSN 00401625. Available at: https://linkinghub.elsevier.com/retrieve/pii/S0040162517300276.

[7] B. M. MALYSHEV, R.L. SALGANIK. The strength of adhesive joints using the theory of cracks. International Journal of Fracture Mechanics. 1965, 1-1(2). DOI: 10.1007/BF00186749. ISSN 0020-7268. Available at: http://link.springer.com/10.1007/BF00186749.

[8] Loctite-406. 2010. Online: http://polymerteknik.com/doc/Loctite-406.pdf.

EAN 2020
58th conference on experimental stress analysis

Verification of the Hyperplastic Material Model with Damage for Woven Composites by Experimental Measurements Using Digital Image Correlation

MANDYS T.[1,a], LAŠ V.[1,b], KROUPA T. [1,c], LOBOVSKÝ L.[1,d]

[1]University of West Bohemia in Pilsen, NTIS – New Technologies for Information Society, Technická 8, 301 00 Pilsen, Czech Republic

[a]tmandys@ntis.zcu.cz, [b]las@ntis.zcu.cz, [c]kroupa@ntis.zcu.cz, [d]lobo@ntis.zcu.cz

Keywords: Woven composite, Digital image correlation, Experiment, Damage, Elasto-plastic

Abstract. The paper is aimed at testing the newly developed elasto-plastic material model with damage for the description of non-linear mechanical behaviour of the woven composites. Capabilities of the material model are verified by the usage of the experimental testing of the specimens with the geometry providing the complex combination of the plastic deformation and material damage during the tensile loading. The results of the numerical simulations are compared to the experimental data analysed by means of digital image correlation.

Introduction

Composite materials have found widespread applications in the entire sector of transport industry in the current time. It is due to their advantages such as high strength to weight ratio, high stiffness to weight ratio, corrosion resistance and flexibility in the final product design. The main disadvantage of these materials is their susceptibility to the occurrence of damage that causes the irreversible reduction of their stiffness during the loading and affects negatively the lifetime of the composite product. Based on the experimental analysis of the fiberglass woven composites, their non-linear behaviour is most evident during the in-plane shear testing but it can be also found in warp and weft direction during the tensile loading too [1]. A suitability of the usage of composite materials for new industrial applications is usually verified by numerical simulations. Material models used for the modelling of the behaviour of the composite materials are most often based on the linear elastic models of the orthotropic material in commercial finite-element software. Extending the knowledge and using more advanced material models, that are more consistent with the real behaviour of the composite materials, will lead to other expansion of their application.

The hyperplastic material model with damage based on the thermodynamic principles was developed by the authors in order to describe the non-linear behaviour of the woven composites. The mathematical background of the used material model with the relations and the equations including the computational principles and principles of material parameters determination are published in [2]. This material model considers the additive decomposition of the total logarithmic strain tensor to the elastic and the plastic strain tensor in the finite strain theory in accordance with the work [3]. The degradation functions used in the material model and dependent on the elastic strains were proposed for description of the damage states of the modelled material. The applied damage evolution equations in the material model are in accordance with the principles of continuum damage mechanics [4]. Authors proposed material model was then implemented into the finite-element software Abaqus 6.14 using material subroutines UMAT and VUMAT written in Fortran code too.

Woven composite

The examined woven composite of the total thickness 0.9 mm was made of three layers of plain weave fiberglass fabric with the material density 816 g/m^2 and epoxy resin with designation Epicote MGS 385LR. Experimental cyclic static tensile and compressive tests were performed on the material specimens in accordance with ASTM standards. Material parameters of the woven composite material were identified using mathematical optimization, the principles and the real execution of that are described in [2]. The material parameters are summarized in Tables 1 and 2 [2].

Table 1: Elastic and strain-hardening curve parameters

E_{11} [GPa]	E_{22} [GPa]	E_{33} [GPa]	G_{12} [GPa]	G_{13} [GPa]	G_{23} [GPa]	ν_{12} [-]	ν_{23} [-]	ν_{31} [-]	R_0 [MPa]	K_R [MPa]	n_R [-]
25.30	25.79	8.00	4.50	4.50	2.75	0.337	0.337	0.280	35	265	0.4

Table 2: Material damage parameters, units [-]

$\varepsilon^0_{11_T}$	$\varepsilon^0_{11_C}$	$\varepsilon^0_{22_T}$	$\varepsilon^0_{22_C}$	$\varepsilon^0_{33_T}$	$\varepsilon^0_{33_C}$	γ^0_{12}	γ^0_{23}	γ^0_{13}
0.0030	0.0007	0.0040	0.0076	0.0	0.0076	0.0048	0.0	0.0
$\varepsilon^R_{11_T}$	$\varepsilon^R_{11_C}$	$\varepsilon^R_{22_T}$	$\varepsilon^R_{22_C}$	$\varepsilon^R_{33_T}$	$\varepsilon^R_{33_C}$	γ^R_{12}	γ^R_{23}	γ^R_{13}
0.0084	0.0500	0.0060	0.0500	0.0001	0.0500	0.0102	0.0001	0.0001
$\varepsilon^F_{11_T}$	$\varepsilon^F_{11_C}$	$\varepsilon^F_{22_T}$	$\varepsilon^F_{22_C}$	$\varepsilon^F_{33_T}$	$\varepsilon^F_{33_C}$	γ^F_{12}	γ^F_{23}	γ^F_{13}
0.0125	0.2000	0.0137	0.2000	0.0233	0.2000	0.0391	0.0350	0.0350
a_{11_T}	a_{11_C}	a_{22_T}	a_{22_C}	a_{33_T}	a_{33_C}	a_{44}	a_{55}	a_{66}
0.0125	0.2000	0.0137	0.2000	0.0233	0.2000	0.50	0.50	0.50
b_{11_T}	b_{11_C}	b_{22_T}	b_{22_C}	b_{33_T}	b_{33_C}	b_{44}	b_{55}	b_{66}
0.0	10.0	0.0	10.0	100.0	10.0	100.0	100.0	0.8550
$K^0_{11_T}$	$K^0_{11_C}$	$K^0_{22_T}$	$K^0_{22_C}$	$K^0_{33_T}$	$K^0_{33_C}$	K_{44}	K_{55}	K_{66}
1.0	1.0	1.0	1.0	0.0	1.0	0.0	0.0	0.76
$K^0_{11_T}$	$K^0_{11_C}$	$K^0_{22_T}$	$K^0_{22_C}$	$K^0_{33_T}$	$K^0_{33_C}$	K_{44}	K_{55}	K_{66}
1.0	1.0	1.0	1.0	0.0	1.0	0.0	0.0	0.76
$D^U_{11_T}$	$D^U_{11_C}$	$D^U_{22_T}$	$D^U_{22_C}$	$D^U_{33_T}$	$D^U_{33_C}$	D^U_{44}	D^U_{55}	D^U_{66}
0.1290	0.7500	0.1210	0.75	1.0	0.75	1.0	1.0	1.0
D^U_{12}	D^U_{13}	D^U_{23}	$D^{max}_{ij_T}$	$D^{max}_{ij_C}$				
1.0	1.0	1.0	1.0	0.75				

Testing of the capabilities of the material model

The proposed material model was verified using tests on complex geometries. The planar specimens from woven composite with the total dimension of 160 x 50 mm and five circular perforations were prepared and subjected to tensile loading. All dimensions of the specimen geometry and setup from experimental testing are displayed in Fig. 1. The perforation of the specimens causes a complex combination of the damage states and the plastic deformation of the material during the loading. The investigated area of the specimens is bounded by the extensometer arms. Four types of specimens differing by the angle φ representing the angle between the loading direction and the material direction 1 (warp fibers direction) were tested. The denoting of the specimens was realized according to the value of $\varphi = 0°$, 15°, 30° and 45° as TH-0, TH-15, TH-30 and TH-45. The testing machine Zwick Roell Z050 was used to exert the tensile loading with the velocity $v = 1.0$ mm/min. The loading state was performed until the visual rupture of the specimens and decrease of the loading force to 20% of its maximal value.

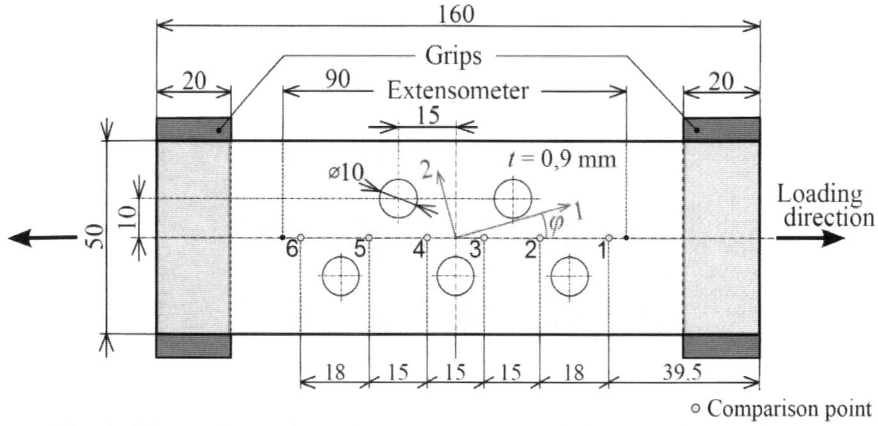

Fig. 1: The perforated specimen geometry and the experimental setup

Distribution of the principal true strain along the perforated specimen's surface during the tensile loading was analyzed by the non-contact and non-destructive technique of digital image correlation (DIC). The DIC optical system consist of the four digital high resolution cameras (4 MPix) was used for the strain distribution analysis based on image captured with frequency of 0.2 Hz. Alternatively, a strong source of light was applied in order to backlight illuminate the fiberglass specimens which are translucent and the permanent structural changes and damage occurred in the material during the tensile loading was captured by the cameras. Three of each type of the perforated specimens were tested and analyzed. The obtained results were compared with results of the corresponding numerical simulations. The element deletion and the finite strain theory were assumed in the case of numerical simulation.

The force-displacement curves obtained from experimental measurement and numerical simulations are compared for the perforated specimens TH-0 and TH-15 in Fig. 2. The same is shown for the specimens TH-30 and TH-45 in Fig. 3. The experimentally recorded force-displacement dependencies showing the relation between the loading force and displacement from the extensometer (u) were not post-processed. The black curves correspond to the experimental runs when DIC analysis was applied. The grey curves display the experimental results when the backlight illumination of the specimen was employed.

The strain distribution during the loading was analyzed in the form of the true principal strains (logarithmic strains) along the entire surface of the specimen. In addition, strain-displacement relations were post-processed at six selected points on the surface of the perforated specimen. The location of these comparison points is depicted in Fig. 1. The strain-displacement curves are shown for the selected point no. 2 (see Fig. 1) for all tested specimens TH0 - TH45 in Fig. 4 and Fig. 5. Results for both the maximum (ε_1) and minimum (ε_2) true principal strain are presented. The value of displacement reflects the total displacement of the grips of the testing machine (u_t) (which is different from displacement u measured by the extensometer and used in the force-displacement plots. The numerical results were computed at the selected point using the interpolation from the neighbourhood nodes of the computational mesh. The Fig. 6 – Fig. 13 show a visual comparison of the strain distribution along the specimen surface as a result of the tests with backlight illumination, the tests with the DIC analysis and numerical simulation for all specimen types TH0 - TH45. The comparison is provided always for two cases of loading states of each specimen type. Namely the state when the specimen is loaded by the maximal loading force and the state when the rupture of the specimens occurs. The first true principal strain ε_1 (maximum true principal strain, logarithmic strain 1) is shown in all cases of visual comparisons.

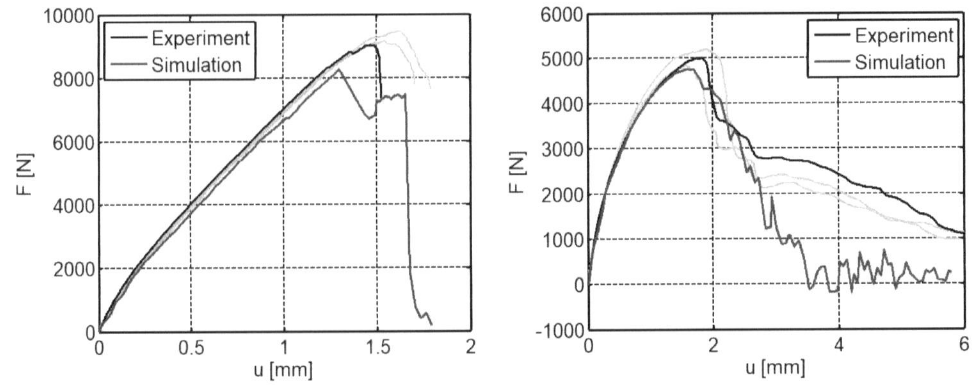

Fig. 2: Force-displacement diagram comparison between experimental and numerical results for specimens TH-0 (left) and TH-15 (right)

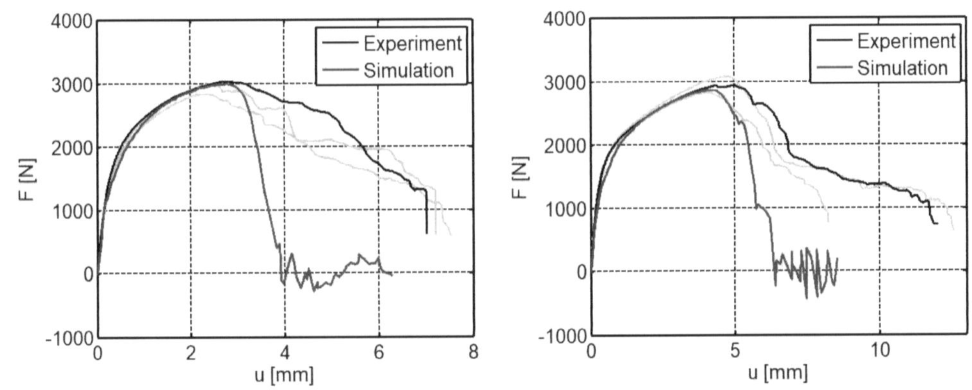

Fig. 3: Force-displacement diagram comparison between experimental and numerical results for specimens TH-30 (left) and TH-45 (right)

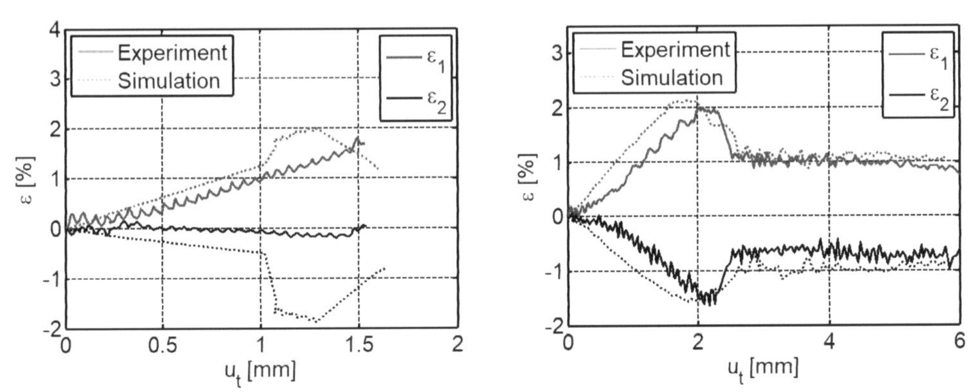

Fig. 4: Relation between the true principal strains (maximum ε_1 and minimum ε_2 true principal strain) and the total displacement for specimens TH-0 (left) and TH-15 (right), comparison between experimental and computational results at point no. 2

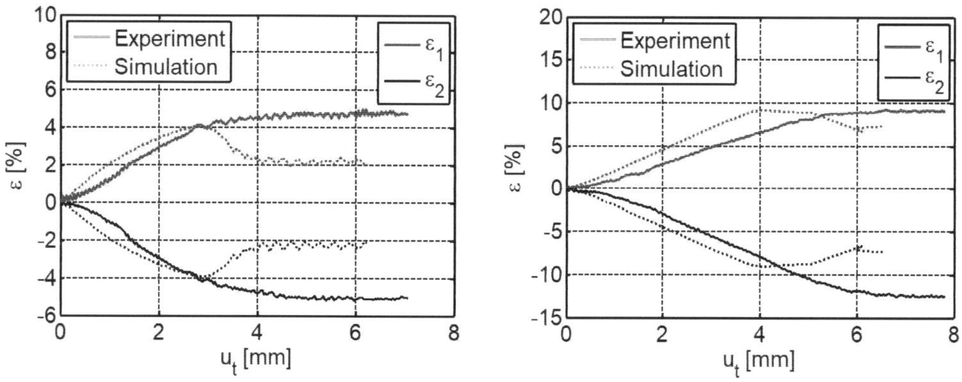

Fig. 5: Relation between the true principal strains (maximum ε_1 and minimum ε_2 true principal strain) and the total displacement for specimens TH-30 (left) and TH-45 (right), comparison between experimental and computational results at point no. 2

Conclusion

The proposed hyperplastic material model with damage for woven composites was verified based on the experimental measurements of the specimens with complex geometry. The material model is implemented into commercial FEM software Abaqus and provides results which correspond well to the recorded experimental data. The force-displacement curves, the strain-displacement relations at selected points, the contour plots of the strain distribution along the specimens surface and the material damage were evaluated and the computational results reflects well experimental data. The force-displacement curves show minimal difference between the experiments and numerical simulations for loading up to the maximal force. The total ruptures of the specimens occur in the case of the numerical simulations but not in the case of experiment, where the damaged specimens show the fiber bridging between individual damaged parts of the specimens. This fact is also evident on the plots of the strain-displacement relation, where the trend diversities occur after the maximal force is reached. The visual comparison of the contour plots of the true strain and of the locations with the material damage provide a qualitative as well as quantitative agreement between experimental and numerical data for all types of the tested specimens.

Acknowledgment

This publication was supported by the project LO1506 of the Czech Ministry of Education, Youth and Sports and by the project SGS–2019–009.

References

[1] T. Kroupa, K. Kunc, R. Zemčík, T. Mandys, The non-linear finite element simulations of tensile tests of textile composites, Materiali in Technologie 49 (2015), 509-514.

[2] T. Mandys, T. Kroupa, V. Laš, L. Lobovský, Hyperplastic material model with damage for woven composites, Composite Science and Technology 183 (2019), online.

[3] M.Á. Caminero, F.J. Montáns, K.J. Bathe, Modeling large strain anisotropic elasto-plasticity with logarithmic strain and stress measure, Computers & Structures 89 (2011), 826-843.

[4] S. Murakami, Continuum Damage Mechanics – A continuum Mechanics Approach to the Analysis of Damage and Fracture, Springer Science + Business Media press, 2012.

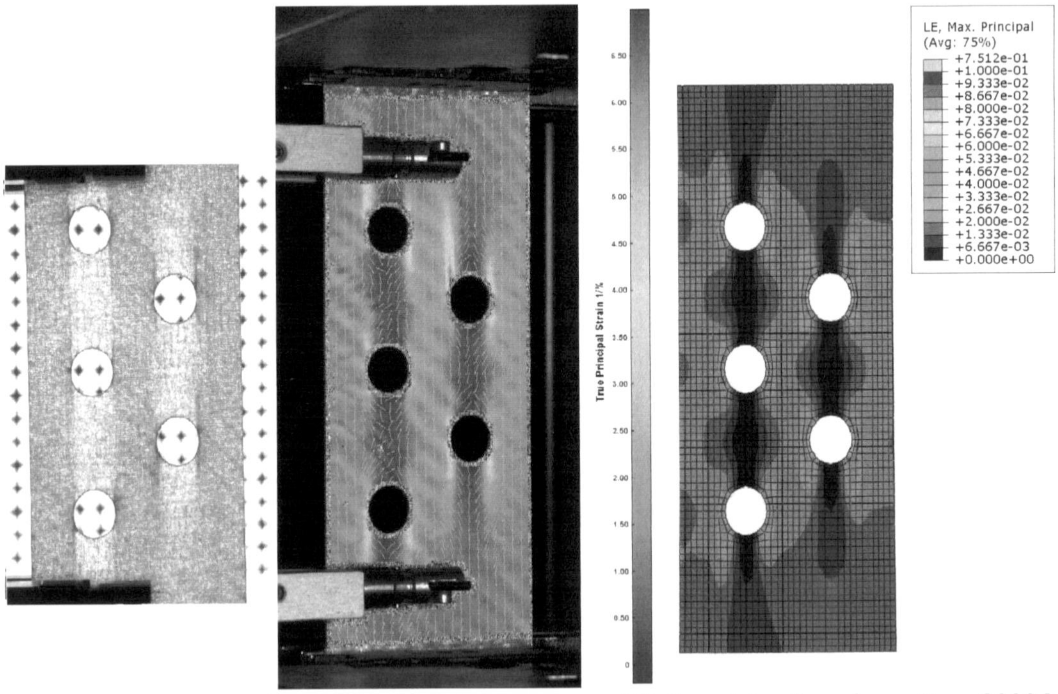

Fig. 6: Visual comparison of results for specimen TH-0 and loading force F = 8000 N (u = 1.2 mm): the backlight illuminated specimen (left), the maximum true principal strain as resolved by the DIC system (center) and as computed by the numerical simulation (right)

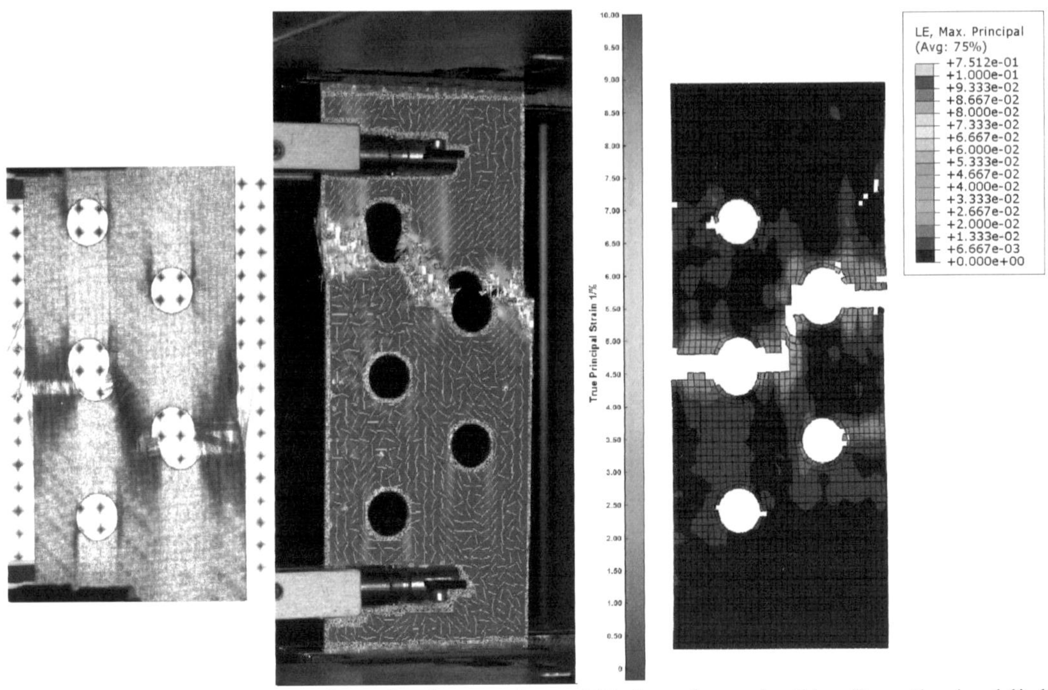

Fig. 7: Visual comparison of results for specimen TH-0 at the end of loading: the backlight illuminated specimen (left), the maximum true principal strain as resolved by the DIC system (center) and as computed by the numerical simulation (right)

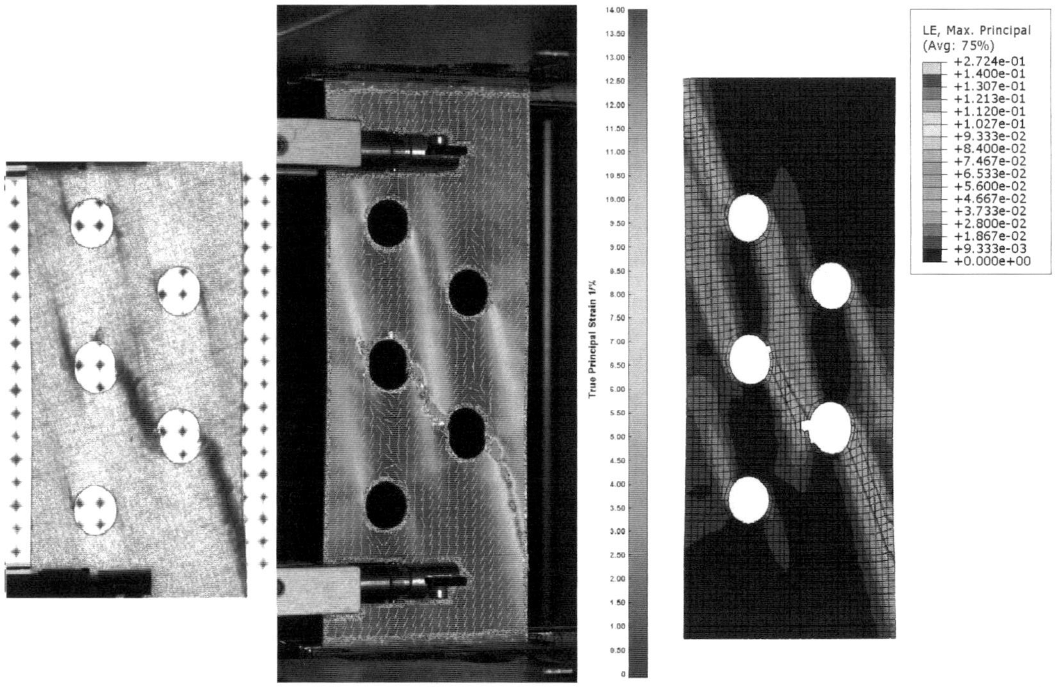

Fig. 8: Visual comparison of results for specimen TH-15 and displacement $u = 2.0$ mm: the backlight illuminated specimen (left), the maximum true principal strain as resolved by the DIC system (center) and as computed by the numerical simulation (right)

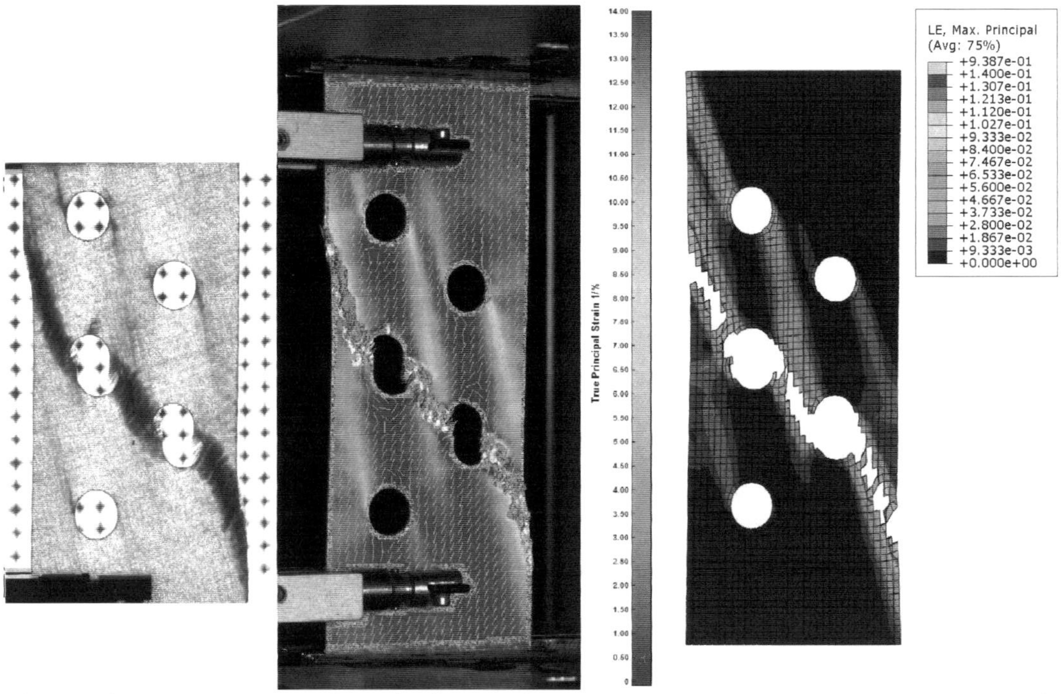

Fig. 9: Visual comparison of results for specimen TH-15 and displacement $u = 4.0$ mm: the backlight illuminated specimen (left), the maximum true principal strain as resolved by the DIC system (center) and as computed by the numerical simulation (right)

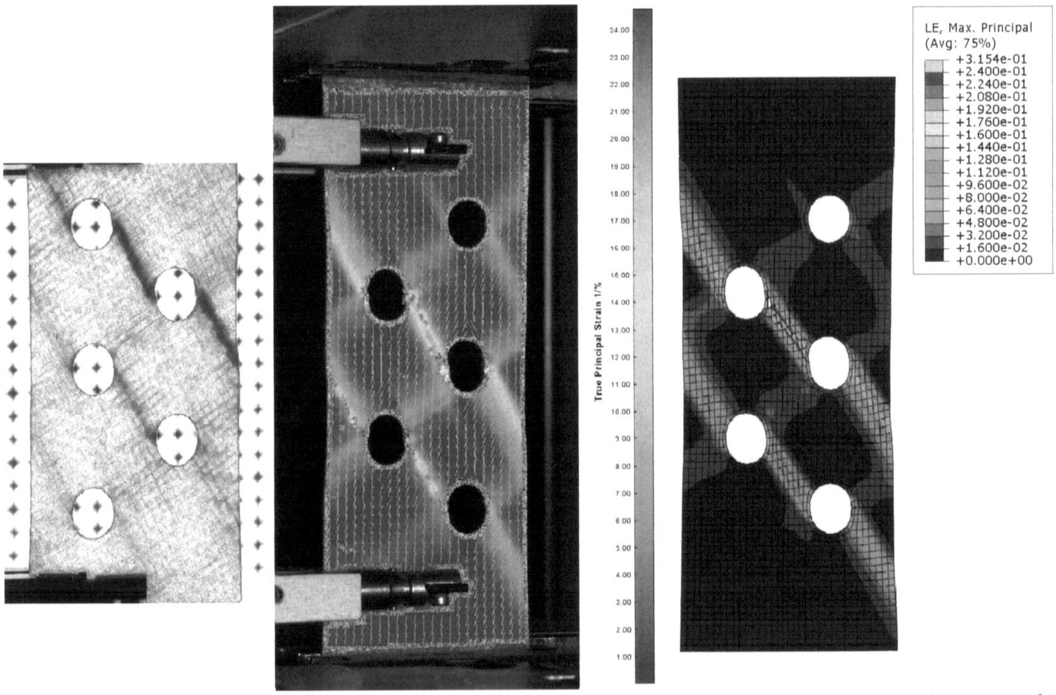

Fig. 10: Visual comparison of results for specimen TH-30 and displacement $u = 3.0$ mm: the backlight illuminated specimen (left), the maximum true principal strain as resolved by the DIC system (center) and as computed by the numerical simulation (right)

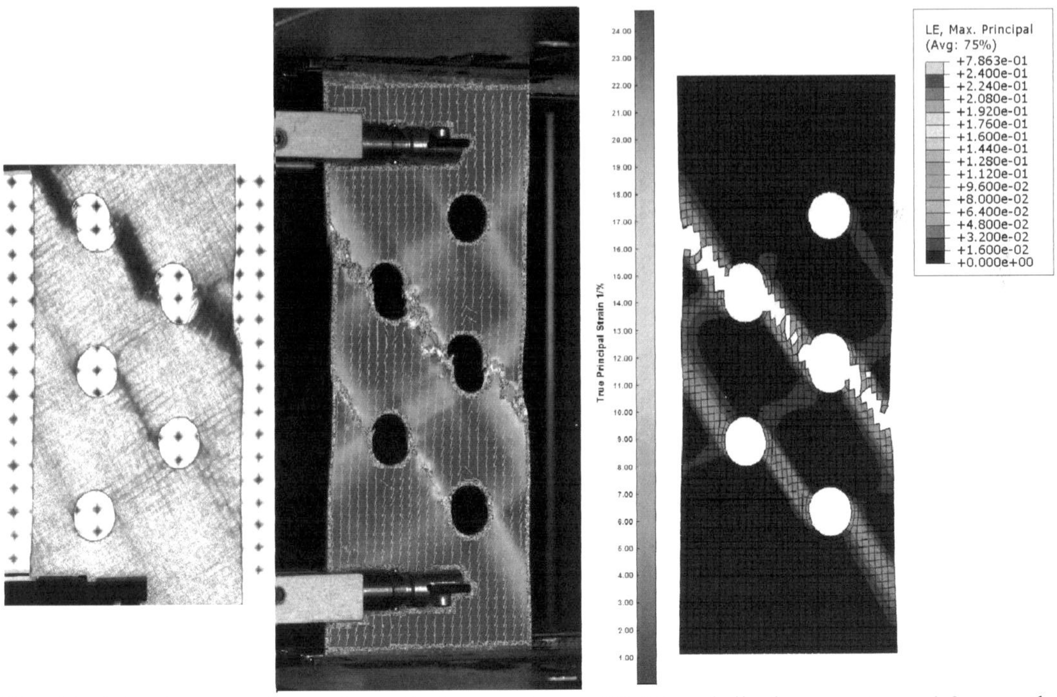

Fig. 11: Visual comparison of results for specimen TH-30 and displacement $u = 4.0$ mm: the backlight illuminated specimen (left), the maximum true principal strain as resolved by the DIC system (center) and as computed by the numerical simulation (right)

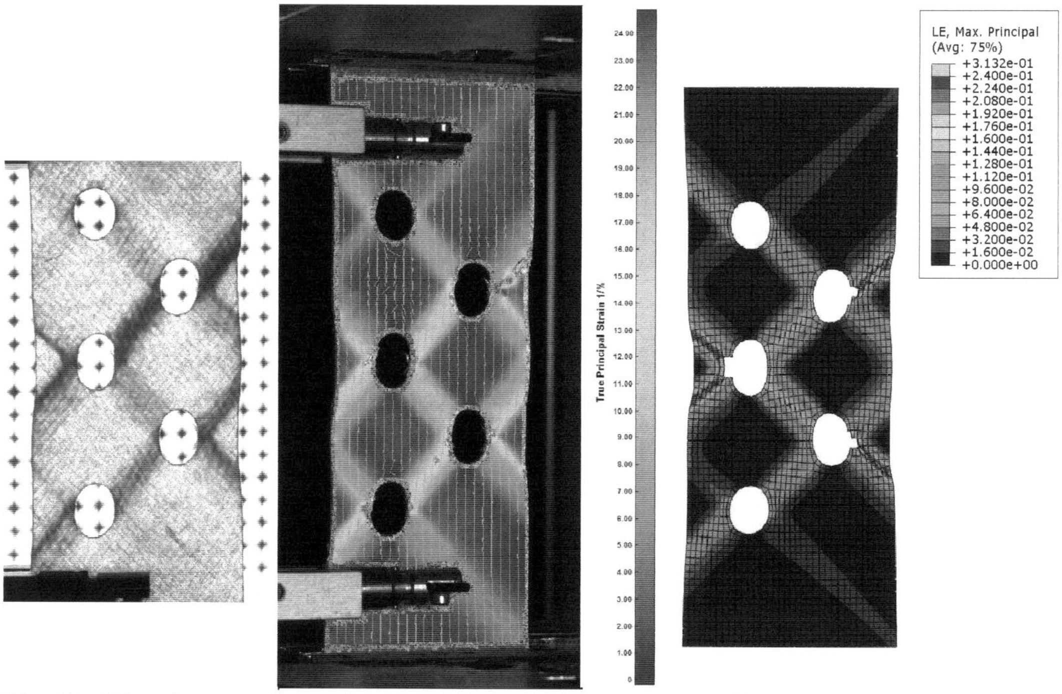

Fig. 12: Visual comparison of results for specimen TH-45 and displacement $u = 5.0$ mm: the backlight illuminated specimen (left), the maximum true principal strain as resolved by the DIC system (center) and as computed by the numerical simulation (right)

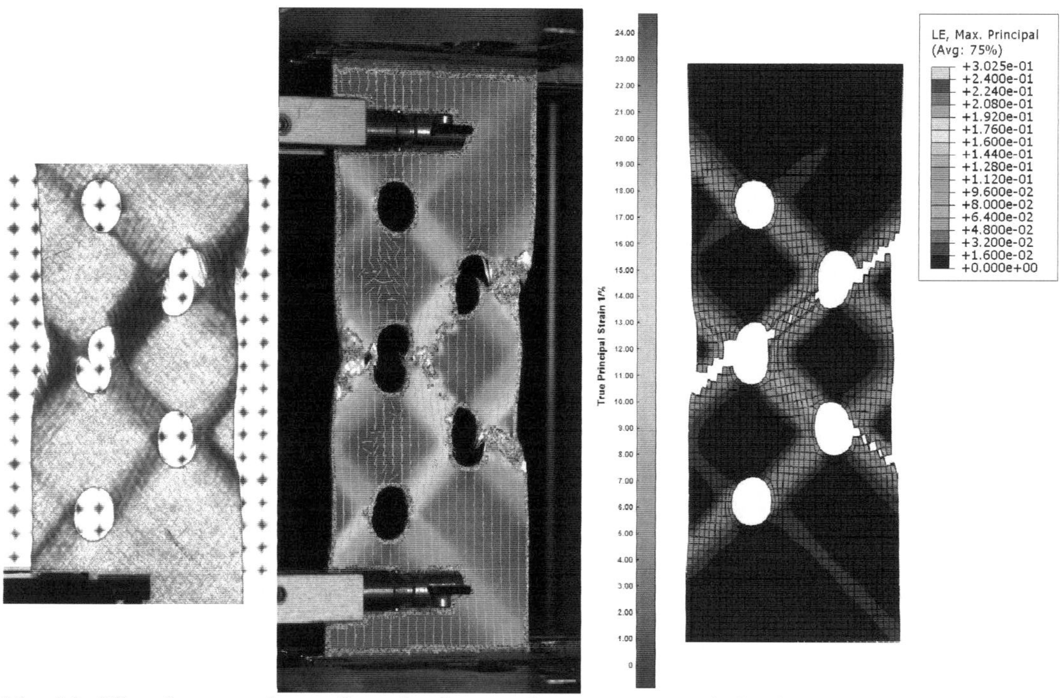

Fig. 13: Visual comparison of results for specimen TH-45 and displacement $u = 7.0$ mm: the backlight illuminated specimen (left), the maximum true principal strain as resolved by the DIC system (center) and as computed by the numerical simulation (right)

EAN 2020
58th conference on experimental stress analysis

Wireless Signal Transmission at Testing Vehicle Response to Kinematic Excitation

MELCER J.[1,a]

[1]Deparment of Structural Mechanics and Applied Mathematics, Faculty of Civil Engineering, University of Žilina; Univerzitná 8215/1, 010 26 Žilina, Slovak Republic

[a]jozef.melcer@uniza.sk

Keywords: Wireless transmission, Moving load, Roadway, Vehicle response, Experiment

Abstract. The article is devoted to monitoring the real effects of the moving load on the structure of the roadway. It solves the problem of wireless transmitting the measured signal from a moving vehicle to a stationary registration device. It describes the way of realization of experiment and specifies the parameters of individual devices. It presents the results of experimental measurements of vehicle response and did conclusions for real application in practice.

Introduction

The moving load is the predominant component of the load in the case of transport structures. It is important to know the real load and the change of the load over time for the design of the road structure and the issues related to its service life and reliability [1]. In order to know the real dynamic load, the experimental way is the best way to obtain the necessary information. Since moving loads are represented by transport means, it is the best to use absolute sensors during experiment, preferably acceleration sensors. Offer of developed acceleration sensors that could be used for this purpose is wide and the choice is really rich. So it is not a technical question but a financial one. The technical problem to be solved is the transmission of the signal from the sensor to the registration device. One way is to realize wireless signal transmission and take advantage of the Wi-Fi system. The transmitted signal can then be archived in digital form on a computer and then processed using suitable software. The results thus obtained can be used to verify numerical simulations.

Experimental technique and measuring line

The measurement was performed on a T815 lorry. The aim of the measurement was to detect vertical accelerations of the vehicle axles and consequently dynamic road loads. The measuring line consists of the following components: Sensor B&K 4507 or 4508, A/D interface NI 9234, Wi-Fi system NI cDAQ-9191, receiver Wi-Fi MOXA AWK-3121, operative computer PC-hp with software National Instruments (LabVIEW, Measurement Studio, Signal Express, DIAdem, MATRIXx, Developer Suite, VI Logger) [2].

For measurement of acceleration, it is advantageous to use sensors from Brüel & Kjær type BK 4507 or 4508 [3]. Miniature CCLD Accelerometers Types 4507 and 4508 are specifically designed to withstand the rough environment of the automotive industry. A combination of high sensitivity, low mass and small physical dimensions make them ideal for modal measurements, such as automotive body and power-train measurements, as well as for modal analysis on aircraft, trains and satellites. These are sensors of small dimensions of about 1 cm with a weight of 4.8 g, which operate in the frequency range 0.8 - 8000 Hz at temperatures –54 - +100 ° C. They have a high sensitivity of 100 mV/ms^{-2}. They are attached to a plastic base which is glued to the structure. The sensor itself and its installation on the front and rear axle of the vehicle is shown in Fig. 1.

Fig. 1: Sensor BK 4508 and its installation on the front and rear axle of the vehicle

The voltage signal from accelerometer is routed via coaxial cable to the National Instruments NI 9234 module which also works as amplifier and an A/D interface. The NI 9234 is a 4-channel, ±5 V, 24 Bit, C series sound and vibration input module for making high-accuracy measurements from integrated electronic piezoelectric (IEPE) and non-IEPE sensors such as accelerometers, tachometers, and proximity probes. The NI-9234 is also compatible with smart TEDS (Transducer Electronic Data Sheets) sensors. The NI 9234 delivers 102 dB of dynamic range and incorporates Integrated Electronics Piezoelectric (IEPE) signal conditioning at 2 mA constant current for accelerometers and microphones. The four input channels simultaneously acquire at rates up to 51.2 kS/s. It operates at temperatures –54 - +70 °C. In addition, the module includes built-in anti-aliasing filters that automatically adjust to your sampling rate. Compatible with a single-module USB carrier and NI CompactDAQ and CompactRIO hardware, the NI 9234 is ideal for a wide variety of mobile or portable applications such as industrial machine condition monitoring and in-vehicle noise, vibration, and harshness testing.

The wireless transmission of the signal is provided by the National Instruments NI cDAQ-9191 Wi-Fi system. The cDAQ-9191 is a Wi-Fi chassis designed for small, remote, or distributed sensor measurement systems. The chassis controls the timing, synchronization, and data transfer between C Series I/O modules and an external host. You can use this chassis with a combination of modules to create a mix of analog I/O, digital I/O, and counter/timer measurements. You can send data from the cDAQ-9191 to a host PC over Ethernet or IEEE 802.11 Wi-Fi. The chassis also has four 32-bit general-purpose counters/timers built in. You can access these counters through an installed, hardware-timed digital C Series module for applications that involve quadrature encoders, PWM, event counting, pulse train generation, and period or frequency measurement. Antenna (packaged with the cDAQ-9191 chassis). Antenna Connector Bulkhead RP-SMA connector Electrical performance VSWR Maximum 2.0 (2.4 GHz to 2.5 GHz) Impedance 50 Ω nominal Directivity Omni Maximum gain 2.0 dBi (2.4 GHz to 2.5 GHz).

The MOXA AWK-3121 can be used to receive wireless signals. The AWK-3121 Access-Point/Bridge and AP Client is ideal for applications that are hard to wire, too expensive to wire, or use mobile equipment that connects to a TCP/IP network. The AWK-3121 can operate at temperatures ranging from 0 to 60°C for standard models and -40 to 75°C for extended temperature models, and is rugged enough for any harsh industrial environment. Installation is easy, with either DIN-Rail mounting or wall mounting in distribution boxes. The DIN-Rail/wall mounting capability, wide operating temperature range, and IP30 housing with LED indicators make the AWK-3121 a convenient yet reliable solution for any industrial wireless application.

Swivel-type antennas (2dBi, RP-SMA, 2.4&5GHz). The modules described above [4-6] are shown in Fig. 2.

Fig. 2: A/D interface NI 9234, Wi-Fi NI cDAQ9191, receiver MAXA AWK 3121

The rod antenna of the NI cDAQ-9191 module can be replaced by a panel antenna of the JPA-9 type, for example, of 123 x 123 x 13 mm, Fig. 3, [7]. The antenna operates in the 2.4 - 2.485 GHz frequency band, horizontal or vertical polarization, radiation angle of 65 ° in both planes. Both the NI 9215 and the cDAQ-9191 require a power supply. It is possible to use 14.4 V LiTon batteries, 83.5 kWh, max current 3A.

Fig. 3: Panel antenna JPA.9

Results of experimental testing

Measurements according to the above methodology were performed on the Tatra T815 lorry. The road section on which the vehicle response was registered was 120 m long. The passage time of the vehicle along this section was measured. The average speed of the vehicle was calculated from the measured time and the known length. The measurements were carried out in the campus of the University of Žilina on the newly built roadway, which can be classified in category A according to ISO 8608 [8]. The record of acceleration on the vehicle front axle at the speed of 18 km/h is shown in Fig. 4 and the record of the acceleration on the rear axle is shown in Fig. 5.

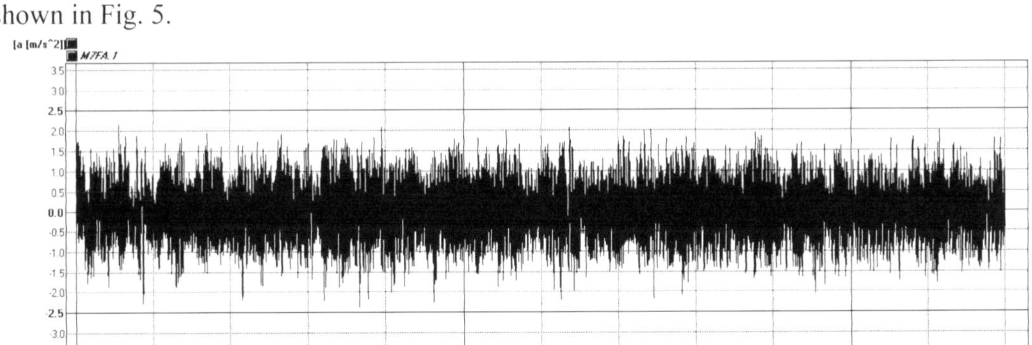

Fig. 4: Acceleration record on the vehicle front axle

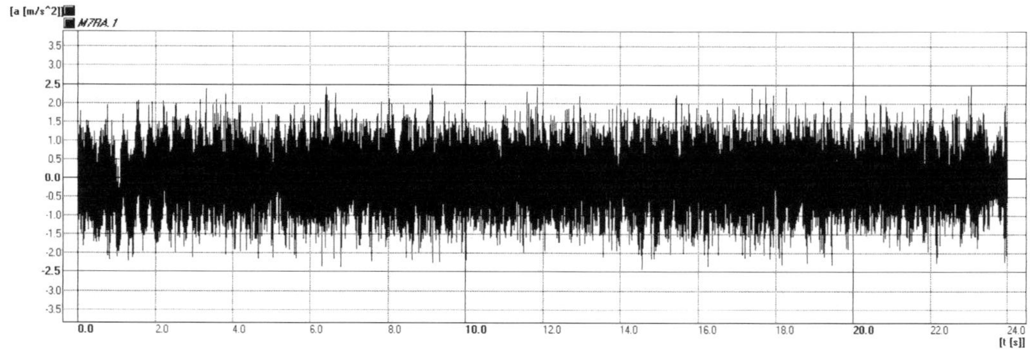

Fig. 5: Acceleration record on the vehicle rear axle

Conclusions

The wireless transmission of the signal from the measured point to the registration device is advantageous, especially when measured on vehicles. For signal transmission it is possible to use WiFi system and e.g. devices of the firm National Instruments. As long as there is no barrier between the transmitter and receiver antennas ("the antennas see each other") the methodology described above gives good results and is reliable. Problems may arise if the antennas "do not see each other". The methodology can also be used to measure the response of bridges. It spares pulling a large number of cables.

Acknowledgements

This work was supported by the Grant National Agency VEGA of the Slovak Republic, grant No. 1/0006/20.

References

[1] J. Melcer et al., Dynamics of Transport Structures (in Slovak), first ed., EDIS, Žilinská univerzita v Žiline, 2016.

[2] https://www.ni.com/cs-cz/support/downloads/software-products.html.

[3] https://www.bksv.com/en/products/transducers/vibration/Vibration-transducers/accelerometers/4508-B-002.

[4] http://www.ni.com/pdf/manuals/373779a_02.pdf.

[5] http://www.ni.com/pdf/manuals/374048c.pdf.

[6] http://www.ni.com/pdf/manuals/AWK-3121_UM.pdf.

[7] http://cz.jirous.com/anteny-2,4ghz/jpa-9.

[8] ISO 8608, Mechanical vibration – road surface profiles – reporting of measured data. International standard, 1995.

EAN 2020
58th conference on
experimental stress analysis

On a Possible High-Resolution Residual Strain/Stress Measurements by Three Axis Neutron Diffractometer

MIKULA P.[1,a], ŠAROUN J.[1,b],RYUKHTIN V. [1,c], ROGANTE M.[2,d]

[1]Deparment of Neutron Physics, Nuclear Physics Institute ASCR, v.v.i., 250 68 Řež, Czech Republic

[2]Rogante Engineering Office, Contrada San Michele, n. 61, 62012 Civitanova Marche, Italy

[a]mikula@ujf.cas.cz, [b]saroun@ujf.cas.cz, [c]ryukhtin@ujf.cas.cz, [d]main@roganteengineering.it

Keywords: Neutron diffraction, Three axis setting, High resolution, Bent crystal monochromator, Bent crystal analyzer

Abstract. New unconventional high-resolution neutron diffraction three axis set-up for strain/stress measurements of rather large bulk polycrystalline samples is presented. Contrary to the conventional two axis set-ups in this case the strain measurement on the sample situated on the second axis is carried out by rocking the bent perfect crystal (BPC) analyzer situated on the third axis of the diffractometer. The neutron signal is registered by a detector. This new set up provides a considerably higher resolution (~5x) which, however, requires much longer measurement time. Therefore, it can be effectively used, namely, for bulk gauge volumes and/or plastic deformation studies on the basis of analysis of diffraction line profiles.

Introduction

Residual stresses are typical phenomena associated e.g. with welding of any kind of structural material. Stresses are generally responsible for the deformation of welded structures during production and subsequently influence the behavior of these structures during service [1-3]. However, they also occur when external load or some shape forming is applied on the sample.

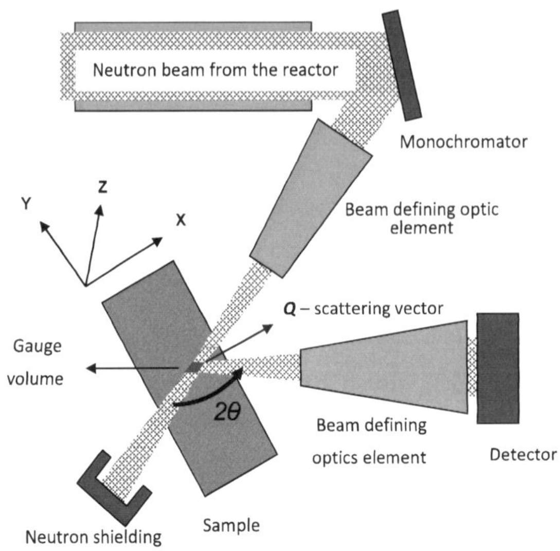

Fig. 1: Schematic drawing of the conventional strain scanner.

In our contribution, a new unconventional high-resolution neutron diffraction three axis set-up for strain/stress measurements of rather large bulk polycrystalline samples is presented. Contrary to the conventional two axis strain/stress scanner (see Fig. 1) three axis set-ups carried out by rocking the bent perfect crystal (BPC) analyzer situated on the third axis of the diffractometer and the neutron signal registered by a point detector (see Fig. 2) [4,5]. By a proper adjustment of the curvature of the analyzer the three axis set-up exploits focusing in momentum space. The high-resolution determination of the lattice changes is achieved even on large irradiated gauge volumes (see Fig. 2b.

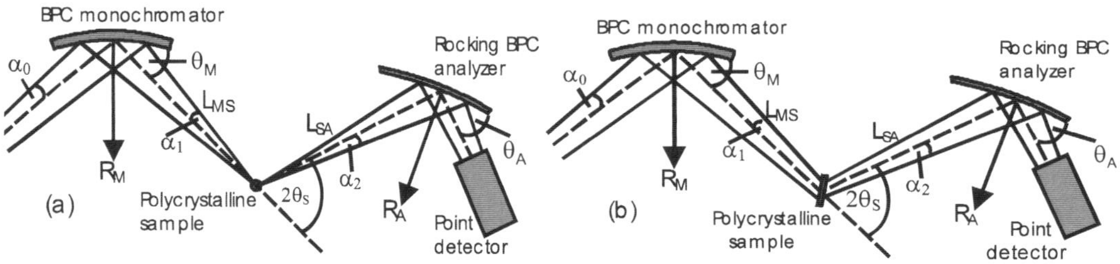

Fig. 2: Scheme of the three axis diffractometer performances employing BPC monochromator and analyzer operating with bar samples in vertical – (a) and/or in horizontal position – (b) with a point detector (R_M, R_A - radii of curvature, θ_M, θ_A - Bragg angles).

Experimental resolution

The experiment was carried out on the three axis neutron optics diffractometer installed at the Řež research reactor LVR-15. Si(111)-monochromator and Si(311)-analyzer single crystals had the dimensions of 200x40x4 mm^3 and 20x40x1.3 mm^3 (length x width x thickness), respectively. After the adjustment of the diffractometer setting for an optimum resolution as the first step, the resolution of the experimental setting was studied for different thicknesses of the standard samples (see Fig. 3). The width of the slit situated just before the sample was 10

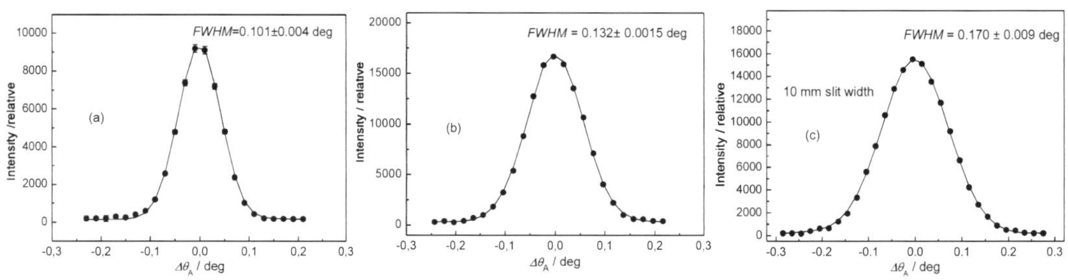

Fig. 3: Analyzer rocking curves for the virgin α-Fe(110) samples of (a) - ϕ=5.1 mm and (b) - ϕ=8 mm situated in the vertical position and the sample of (c) - ϕ=8 mm in the horizontal position (see Fig. 2b).

mm and in the case of Fig 3c represents the irradiated width of the horizontally installed sample. This trick can be used for the sake of comparison, because the vertical dimension of the sample does not play a principal role with respect to the resolution of the experimental setting. It can be seen from Fig. 3 that the width of the sample (represented by the diameter or by the slit width) plays an important role, however, in all cases it is sufficiently high for observation not only peak shifts resulting from elastic strains, but also for microstructural (dislocation density, mean grain size) studies of plastically deformed samples on the basis of diffraction profile analysis. By using the PSD detector instead of the point detector, it is possible to reconstruct the 2D profile of the neutron beam after the analysis by the analyzer crystal (see Fig. 4). In this case both the $\Delta\theta_A$ and Channel number scales are relative.

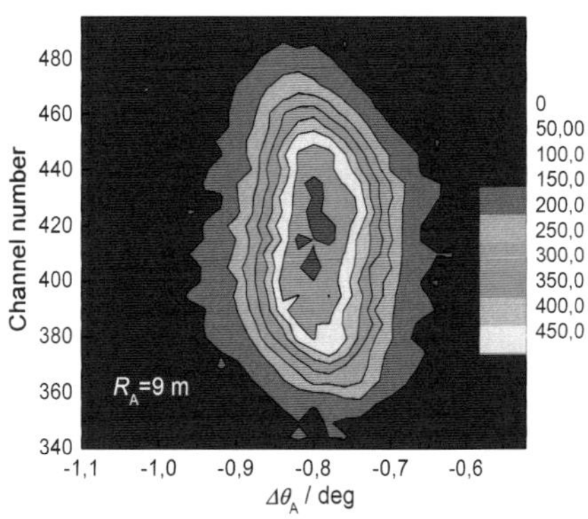

Fig. 4. 2D imaging of the analyzed beam for the virgin sample of ϕ=8 mm used in vertical position.

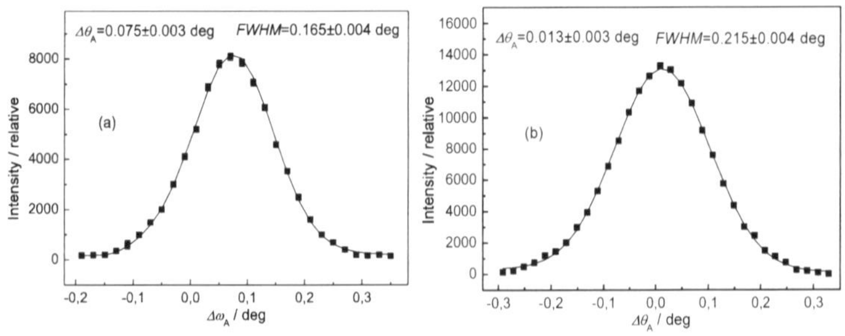

Fig. 5. Analyzer rocking curves for the deformed α-Fe(110) steel wire installed in the vertical – (a) and horizontal position – (b).

In the next step a plastically deformed α-Fe(110) steel wire (shear deformation 23 %, drawing deformation 23.2 %, ϕ=4.28 mm) was used. The corres-ponding experimental rocking curves are shown in Fig. 5. If we compare them with the curves shown in Fig. 3, it can be seen that the effect of plastic de-formation on the *FWHM* of the rocking curve and the dif-fraction profile itself is clearly visible and sufficient for a possible application of the diffraction profile analysis. If the samples are situated at the diffractometer with a high accuracy, the relative change of the lattice constant $\Delta d_S/d_0$ (strain) can be derived from the shift $\Delta\theta_A$ of the rocking curve (see Fig. 5). With the help of the Bragg condition in our case the values $\Delta\theta_A$=0.075±0.003 deg and $\Delta\theta_A$=0.013±0.003 deg correspond to $\Delta d_S/d_0$=(-1.50±0.03)x10^{-3} (radial component) and $\Delta d_S/d_0$=(-0.26±0.03)x10^{-3} (longitudinal component), respectively (d_0 is the lattice spacing of the virgin sample).

Conclusion

New alternative high-resolution neutron diffraction method which can be successfully exploited namely, in the strain/stress measurements on bulk samples exposed to an external load (e.g. in tension/compression rig, in aging machine etc.) is presented. The bulk samples e.g. of the diameter of several millimetres can be investigated in the vertical as well as horizontal position. In comparison with the conventional two axis neutron diffraction strain/stress scanners [6-8] the three axis alternative provides a considerably higher $\Delta d/d$ resolution permitting also microstructure studies (root-mean-square microstrains as well as the effective grain size) as a function of macroscopic strain from applying shape analysis of the neutron diffraction peak profiles [9,10]. On the other hand the presented alternative is much more time consuming. However, contrary to conventional two axis neutron diffraction bulk gauge volumes of the samples e.g. of the width of 5-10 mm can be studied (when keeping a requested high resolution) and can be successfully used, namely, in measurements on samples under the thermomechanical load (including the oscillating loads). It can be stated that the presented three axis neutron diffraction setting can offer an additional support to complement the information achieved by using the other conventional characterization methodologies.

Acknowledgements

Measurements were carried out in the CANAM NPI CAS Řež infrastructure (MŠMT project no. LM2015056). The presented results were also supported in the frame of LM2015074 infrastructure MŠMT project "Experimental nuclear reactors LVR-15 and LR-0". Furthermore, J. Šaroun and V. Ryukhtin acknowledge support from the Czech Academy of Sciences in the frame of the program "Strategie AV21, No.23" and P. Mikula acknowledges support from ESS participation of the Czech Republic – OP (CZ.02.1.01/0.0/0.0/16_013/0001794). We thank Ms. B. Michalcová for significant assistance in measurements and basic data processing.

References

[1] I.C. Noyan, J.B. Cohen, Residual stress: Measurement by diffraction and interpretation, 1st edn; 1987, New York, Springer-Verlag.

[2] M.T. Hutchings, A. D. Krawitz (Eds.), Measurement of residual and applied stress using neutron diffraction, NATO ASI Series, Applied Sciences 26, Kluwer Academic Publisher, 1992.

[3] V. Stelmukh, L. Edwards, J.R. Santisteban, S. Ganguly, M.E. Fitzpatrick, Weld stress mapping using neutron and synchrotron x-ray diffraction, Materials Sc. Forum, 404-407 (2002) 599-604.

[4] M. Vrána, P. Lukáš, P. Mikula J. Kulda, Bragg diffraction optics in high resolution strain measurements, Nucl. Instrum. Methods in Phys. Res. A 338 (1994) 125-131.

[5] P. Mikula, J. Šaroun, J. Stammers, V. Em, High-resolution analysis with bent perfect crystal (BPC) in powder diffraction - I, Journal of Surface Investigation: X-ray, Synchrotron and Neutron Techniques, 2020, (in press)

[6] P. Mikula, J. Šaroun, J. Stammers, V. Em, High-resolution analysis with bent perfect crystal (BPC) in powder diffraction - II, Journal of Surface Investigation: X-ray, Synchrotron and Neutron Techniques, 2020, (in press)

[7] P. Mikula, M. Vrána, P. Lukáš, J. Šaroun, P. Strunz, H.J. Ullrich, V. Wagner, Neutron diffractometer exploiting Bragg diffraction optics - A high resolution strain scanner. In Proc. of ICRS-5, June 16-18, 1997, Linköping, T. Ericsson, M. Odén and A. Andersson (Eds.), 2 (1997) 721-725.

[8] B.S. Seong, V. Em, P. Mikula, J. Šaroun, M.H. Kang, Unconventional performance of a highly luminous strain/stress scanner for high resolution studies, Materials Sc. Forum, 681 (2011) 426-430.

[9] R. Delhez, T.H. de Keijser, E.J. Mittemeijer, Determination of crystallite size and lattice distortions through X-ray diffraction line profile analysis, Fresenius' Zeitschrift für analytische Chemie, 312(1) (1982) 1-16.

[10] V. Davydov, P. Lukáš, P. Strunz, R. Kužel, Single-line diffraction profile analysis method used for evaluation of microstructural parameters in the plain ferritic steel upon tensile straining, Materials Sc. Forum, 571-572 (2008) 181-188.

The Effect of the Heat Treatment at 450 °C on Distribution of Residual Stresses of Modified Cr-Mo Steel Welds

MRÁZ Ľ.[1a], HERVOCHES CH.[2,b], MIKULA P.[2c], KOTORA J.[3d]

[1]Welding Research Institute – Industrial Institute SR, Bratislava, Slovak Republic

[2]Dept. of Neutron Physics, Nuclear Physics Institute ASCR, 250 68 Řež, Czech Republic

[3]SES Tlmače, Slovak Republic

[a]lmrazl@ @mail.t-com.sk, [b]hervoches@ujf.cas.cz, [c]mikula@ujf.cas.cz, [d]jan_kotora@ses.sk

Keywords: Steels, Welding, Stress measurements, Neutron diffraction, Neutron optics

Abstract. The effect of low-temperature long-term heat treatment on distribution of residual stresses on the modified chromium-molybdenum steel of the type 7 CrMoVTiB10-10 and known as T24 steel which was studied by using neutron diffraction method, is presented. The investigation of residual stress measurement by neutron diffraction was concentrated on the identification of the effect of low temperature heat treatment on stress distribution close to the circumferential weld of the steel tube. It has been recognized that the applied heat treatment did not remarkable change local properties (hardness) of the weld. Despite of this the drop of residual stresses after heat treatment compared to as welded condition was remarkable. According to the obtained results it is assumed that low temperature heat treatment can influence final behaviour of circumferential welds in service.

Introduction

The modified chromium-molybdenum steel type 7 CrMoVTiB10-10 steel is a material that has attracted manufacturers′ attention for a long time. This steel belongs to the group of modified Cr-Mo low-alloyed creep resistant steels. The reasons for development of some properties of these steels is to increase the efficiency of thermal power equipment using higher media parameters in membrane walls of the super heater up to the creep zone. Post Weld Heat Treatment (PWHT) is usually applied when the conventional Cr-Mo steels are used for membrane walls. The intention is to develop and produce new steels, especially for super heaters, which themselves also in an as-welded condition would achieve better properties than the parent 10CrMo9-10 steel and be comparable to 9 Cr steel.

The leak of welds in pressure boiler systems for supercritical parameters represents an unexpected phenomenon which does not occur in conventional boilers during the development and testing. On the other hand, in certain cases, the defects, cracks, in welded joints were observed in huge components of membrane walls. The same can be stated also in case of circumferential welded joints in tubes of different dimensions. The leakage of welds in Cr-Mo steel did not occur everywhere, as e.g. in the region where the material of membrane walls is subject to maximum load. Even workshop or on-site welds has not to avoid the leakage. The cause of the leakage can be briefly summarised as: The quality and the type of the material, hydrogen diffusion effects and extremely high mechanical stresses. Moreover, the consequences of potential secondary hardening of welded joints in the early stages of their operation at working temperatures have to be considered. We assume that dealing with the above-mentioned large-sized leakage of the welds in the Cr-Mo steel could provide more information about this material and some experience with its welding as well as the manufacture of pressure parts of this steel.

The limit for maximum hardness in the vicinity of welds (HAZ) is associated with the potential risk of cold crack formation in the HAZ. The 2.25Cr1Mo steels with the maximum thickness of 15 mm are generally recommended for welding at the preheat temperature of 100 °C when diffusible hydrogen content is below 5 ml/100 g and of 150 °C when hydrogen content is from 5 to 10 ml/100 g. Due to the presence of vanadium, titanium and especially boron, these preheat temperatures can be considered as minimum ones and in some cases seem to be even insufficient.

Cold cracks, especially in the weld metal of long submerged arc welds which are oriented in transverse direction to the longitudinal weld axis, were also observed [1]. Hydrogen seems to be the most important factor for the cold cracking behaviour.

Remarkable secondary hardening for Cr-Mo steel after 100 hour exposure at 500 °C was identified. This phenomenon is typical for vanadium hardened steels during tempering [2]. According to these results, it is supposed that PWHT should be carried out in order to avoid a low ductility and low toughness properties mainly during operation.

The objective of the paper is to present the effect of low-temperature and long-term heat treatment on distribution of residual stresses using neutron diffraction method. The distribution of residual stresses in the vicinity of circumferential welded joints was measured in as welded conditions and subsequently, after low-temperature heat treatments at 450 °C for 48 hours.

Experimental part

Circumferential weld of two tubes made of 7 CrMoVTi10-10 creep resistant steel diameter of ∅ 38 mm and wall thickness of 6.3 mm has been prepared for this investigation. The TIG process, medium alloyed filler material WZ, CrMo2VTi/Nb (Union I P24) (typical content Cr = 2.2 %, W = 1.7 %) and argon as shielding gas have been used for welding. The heat input was in the range of 8.0 to 13.0 kJ cm, preheat temperature from 150 °C to 200 °C and interpass temperature of max. 250 °C have been applied.

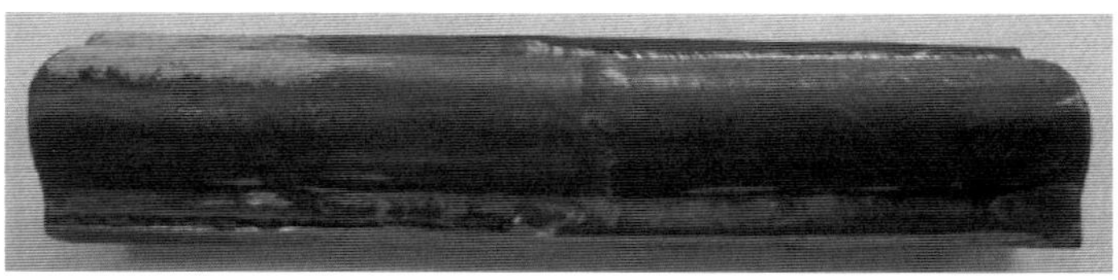

Fig. 1: Circumferential test weld of 7 CrMoVTiB10-10 creep resistant steel

Hardness distribution across butt weld has been measured by Vickers method at the load of 98.1 N (HV10). Neutron diffraction has been performed in order to identify residual stress distribution in the vicinity of the circumferential weld. Hoop, radial and axial stresses have been estimated for as/welded conditions and after long term heat treatment at the temperature 450 °C for 48 hours.

Principles of the neutron diffraction method

The principle of the neutron diffraction method is quite simple. It consists in the precise determination of the d_{hkl}-spacing of particularly oriented crystal planes [3,4]. In neutron and X-ray diffraction the angular positions of the diffraction maxima are directly related to the values of the lattice spacing through the Bragg equation $2d_{hkl} \cdot \sin \theta_{hkl} = \lambda$ (d_{hkl} -lattice spacing, θ_{hkl}- Bragg angle, λ - the neutron wavelength) and thus offer a unique non-destructive technique for investigation of stress fields. When a specimen is strained elastically, the lattice

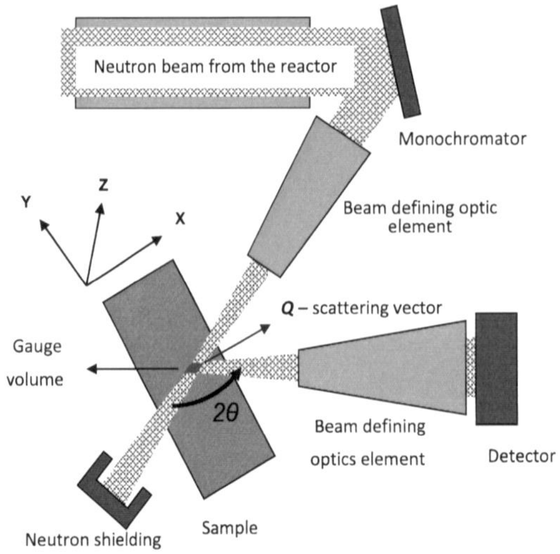

Fig. 2: Schematic illustration of a reactor source based diffractometer for strain measurement in parallel diffraction geometry.

spacing changes. Then, when defining the strain ε as $\varepsilon = \Delta d/d_{0,hkl}$ ($d_{0,hkl}$ is the lattice spacing of the strain-free material) it is related to a change in the lattice spacing, i.e. to a component parallel to the scattering vector $\boldsymbol{Q}$ perpendicular to the reflecting set of planes. Therefore, the knowledge of the $d_{0,hkl}$ value is a crucial task [3]. Then by differentiation of the Bragg condition we arrive at $\varepsilon = -\mathrm{cotg}\ \theta_{hkl} \cdot \Delta\theta_{hkl}$. The relation for the strain ε indicates that it gives rise to a change in the scattering angle $2\theta_{hkl}$ resulting in an angular shift $\Delta(2\theta_{hkl})$ of the peak position for a particular reflecting plane illuminated by a fixed wavelength. In such a way, the shift in the Bragg angle (relative to that of the stress-free material) permits the determination of the average lattice macrostrain over the irradiated gauge volume (see Fig. 2). The conversion of strains to stresses is carried out by means of the relation

$$\sigma_x = \frac{E_{hkl}}{(1-2\nu_{hkl})(1+\nu_{hkl})}\left[(1-\nu_{hkl})\varepsilon_x^{hkl} + \nu_{hkl}(\varepsilon_y^{hkl} + \varepsilon_z^{hkl})\right] \qquad (1)$$

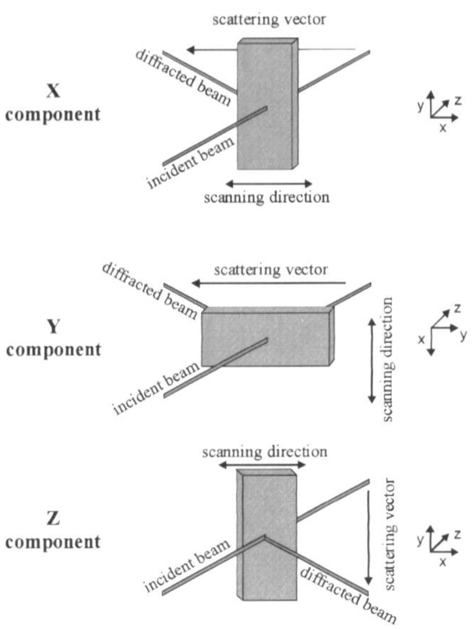

Fig. 3. Sketch of the sample setting for determination of three strain/stress components.

where $\varepsilon_{x,y,z}^{hkl}$ is the x,y,z-component of the lattice strain measured at the hkl crystal lattice planes, E_{hkl} and ν_{hkl} are the diffraction elastic Young modulus and diffraction Poisson ratio, respectively. For the determination of the stress tensor in this case of steel samples, three strain components should be determined as schematically shown in Fig. 3. The residual strain/stress experiment was carried out on the dedicated strain/stress scanning diffractometer (http://neutron.ujf.cas.cz/en/hk4) installed at the 10 MW medium power research reactor LVR-15 and operating at the neutron wavelength of 0.235 nm. Diffractometer uses optimally bent perfect crystal monochromator which provides necessary resolution and luminosity required for the experimental measurements [5,6].

Experimental results

Hardness distribution. Hardness distribution accross the circumferential test weld prior and after low temperature heat treatment is summarized in Fig. 4. No remarkable change has been noticed due to applied heat treatment.

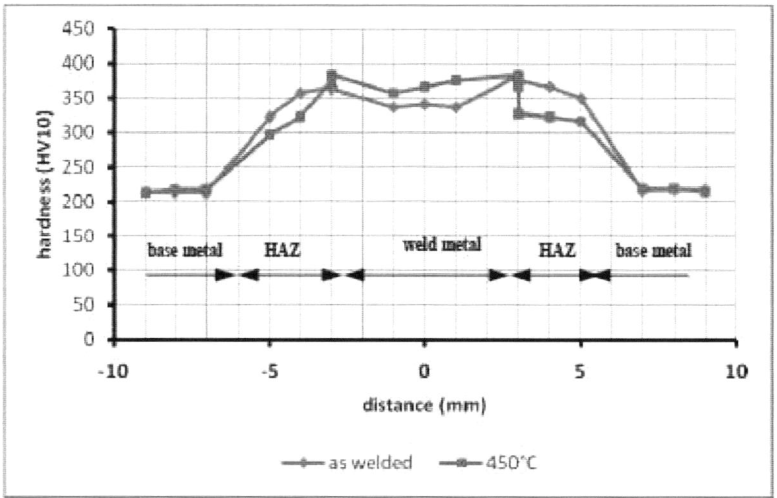

Fig. 4: Hardness distribution accross test weld prior and after heat treatment

Neutron diffraction measurements

Axial, radial and hoop stresses have been estimated in the vicinity of the test weld (see Figs. 5 to 7). The results show that all stress components of weld residual stresses clear dropped after low temperature long term heat treatment. The irradiated gauge volume had the dimensions of about 3x3x3 mm^3. In this case the measurements were carried out on the α-Fe(110) lattice planes.

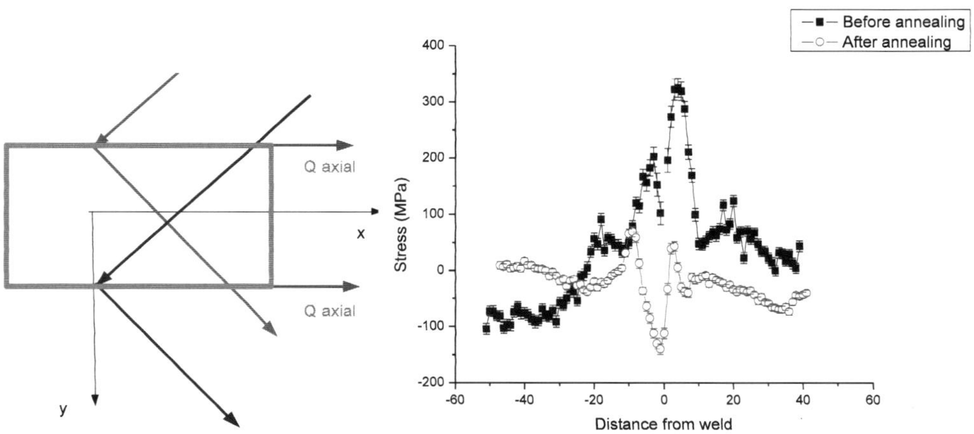

Fig. 5: Axial stresses (parallel to the z-axis) in the vicinity of the test weld; scheme of the measurement and the experimental results.

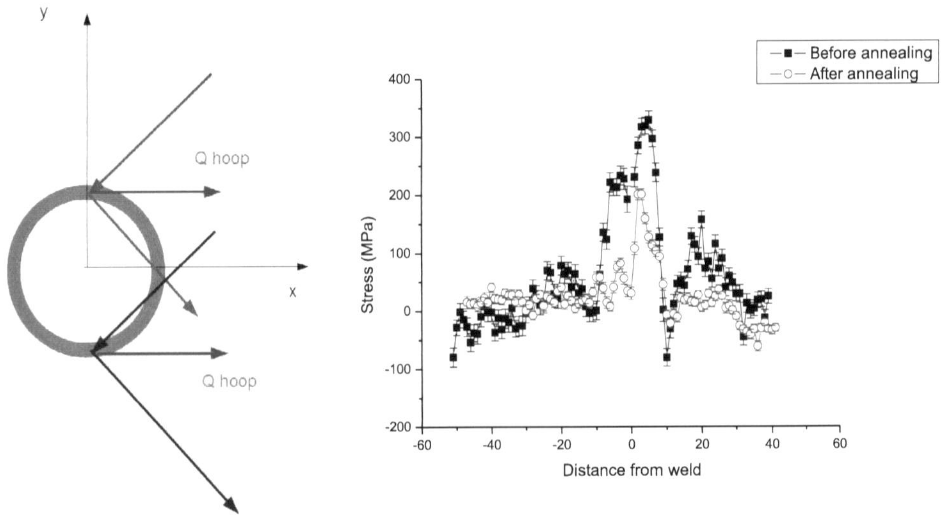

Fig. 6: Hoop (tangential) stresses in the vicinity of the test weld; scheme of the measurement and the experimental results.

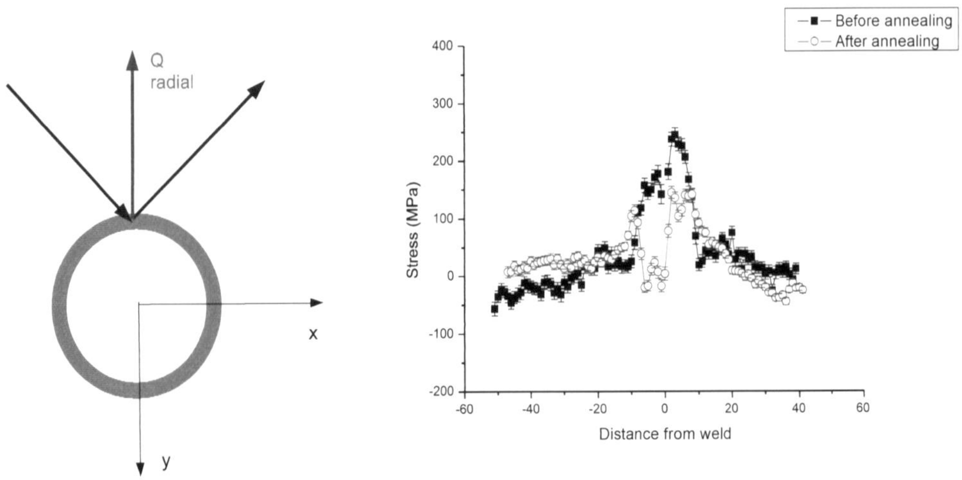

Fig. 7: Radial stresses in the vicinity of the test weld; scheme of the measurement and the experimental results.

Discussion

The investigation of residual stress measurement by neutron diffraction has been concentrated on the identification of the effect of low temperature heat treatment on stress distribution close to the circumferential weld of T24 steel tube. It has been recognized that the applied heat treatment did not remarkable change local properties (hardness) of the weld. Despite of this the drop of residual stresses after heat treatment compared to as welded condition is clear and remarkable. According to these results we assume that low temperature heat treatment can influence final behaviour of circumferential welds in service. We expect that the results will contribute to the extension of knowledge about the behaviour of welded joints of the creep resistant T24 steel during the start-up period of power stations.

Conclusion

The contribution summarizes the results of neutron diffraction measurements for the estimation of residual stresses distribution in the circumferential weld of creep resistant Cr-Mo steel. The results show that the even low-temperature and long-term heat treatment can have an influence on the level of residual stresses. It can be supposed that this effect can play an important role in the behaviour of the weld during the initial period of a power station start up.

Acknowledgements

Measurements were carried out in the CANAM NPI CAS Řež infrastructure (MŠMT project no. LM2015056). The presented results were also supported in the frame of LM2015074 infrastructure MŠMT project "Experimental nuclear reactors LVR-15 and LR-0". Furthermore, P. Mikula acknowledges support from ESS participation of the Czech Republic – OP (CZ.02.1.01/0.0/0.0/16_013/0001794). The authors thank Ms. B. Michalcová for a significant help in the measurements and basic elaboration of the data.

References

[1] Ľ. Mráz, F. Király, M. Paľo, Analysis of T24 steel welded joints – etapa IIb, October 2011, Technical report 222/2000, ME095-IIb.

[2] P. Mohyla, V. Foldyna, Improvement of reliability and creep resistance in advanced low-alloy steel, Material Sci. Eng. A 510-511 (2009) 234-237.

[3] M.T. Hutchings, A. D. Krawitz (Eds.), Measurement of residual and applied stress using neutron diffraction, NATO ASI Series, Applied Sciences 26, Kluwer Academic Publisher, 1992.

[4] V. Stelmukh, L. Edwards, J.R. Santisteban, S. Ganguly, M.E. Fitzpatrick, Weld stress mapping using neutron and synchrotron x-ray diffraction, Materials Science Forum 404-407 (2002) 599-604.

[5] P. Mikula, M. Vrána, P. Lukáš, J. Šaroun and V. Wagner, High-resolution neutron powder diffractometry on samples of small dimensions, Materials Science Forum, 228–231 (1996) 269–274.

[6] P. Mikula, M. Vrána, P. Lukáš, J. Šaroun, P. Strunz, H.J. Ullrich, V. Wagner, Neutron diffractometer exploiting Bragg diffraction optics - A high resolution strain scanner, in: Proc. of ICRS-5, Linköping, T. Ericsson, M. Odén and A. Andersson (Eds.), 2 (1997) 721–725.

EAN 2020
58th conference on experimental stress analysis

Mechanical Properties of Rammed Earth with High Water-Clay Ratio in Tensile Bending Test

MUŽÍKOVÁ B.[1,a], PLAČEK OTCOVSKÁ T.[1,b], PADEVĚT P.[1,c]

[1]Department of Applied Mechanics, Faculty of Mechanical Engineering, CTU – Czech Technical University in Prague, Thákurova 2077/7, 160 00 Praha, Czech Republic

[a]barbora.muzikova@fsv.cvut.cz, [b]tereza.otcovska@fsv.cvut.cz, [c]pavel.padevet@fsv.cvut.cz

Keywords: Rammed earth, Water-clay ratio, Tensile strength in bending, Modulus of elasticity

Abstract. The paper is focused on mechanical properties of rammed earth with illitic clay and high water-clay ration in tensile bending test. The modulus of elasticity and tensile strength in bending were measured and evaluated. Three different mixture of rammed earth were designed and tested. The amount of water and binder is one of the key properties of the rammed earth, the amount of the water is expressed by the water-clay ratio. Mechanical properties of the earth material highly depend on the composition of sand, clay and water. The prescription AGL III with 80 % of sand, 20 % of clay and 0.4 water-clay ratio seem to be optimal.

Introduction

The rammed earth is nowadays considered as a sustainable way for modern building. The article is focused on illitic rammed earth with high water-clay ratio. The mechanical properties that were searched are modulus of elasticity and tensile strength in bending.

Fig. 1: Producing the mixture and final specimens

Producing the mixture. The material is a mixture of sand, clay and water. The amount of water and binder is one of the key properties of the rammed earth, it is similar to concrete, so the amount of the water is expressed by the water-clay ratio. The type of the clay is another important factor. In this case all prescriptions were made with illitic clay, so that the volume of water-clay ration can be evaluated just to one clay. Three different prescriptions were designed and tested for determining mechanical properties of rammed earth with high water-clay ratio. The AGL sets III (80 % of sand, 20 % of clay and water-clay ratio 0.400), V (85 % of sand, 15 % of clay and water-clay ratio 0.370 – set with minimum amount of clay), and IX (80 % of sand, 20 % of clay and water-clay ratio 0.450 – set with the highest water-clay ratio), are all with illitic clay.

Table 1: Tested specimens and their prescription and results

Prescription	Sand/clay – water-clay ratio	Modulus of Elasticity [MPa]	Tensile Strength in Bending [MPa]
AGL III	80/20 – 0.400	68.742 ± 28.224	0.059 ± 0.002
AGL V	85/15 – 0.370	49.056 ± 15.171	0.028 ± 0.003
AGL IX	80/20 – 0.450	27.038 ± 7.597	0.051 ± 0.001

Building of rammed earth. The process of building rammed earth wall is as follows: first of all, the framework is built, then the first layer of rammed earth is filled in. Secondly the layer of moist earth is compressed by a tamper. Than next layers of moist earth are added and compressed up to the top of framework. Finally, the framework is removed. [1, 2] In the same way laboratory specimens were made. This process can be seen in Fig. 1.

Fig. 2: A rammed earth specimen during a tensile bending test

Evaluation of Measured Data

The evaluated data are shown in the table 1. In the columns there are stated the name of the tested set, a composition of the mixture (ratio of sand and clay – in summary it gives 100 % and water-clay ratio), the mean value of modulus of elasticity, a standard deviation of the data and the mean value of tensile strenght in bending and standard deviation of the data.

The maximal force in the test and then tensile stenght in bending was measured. The modulus of elasticity was evalueted from the force displacement curve. The curve was croped at the beging and the it was possible to evalueted the modulus. [5, 6] The maximal and minimal values are shown in the table 1., they represent the mean values and their standart deviation, in the text the standart deviation are missed out. The maximum of tensile strength in bending is 0.059 MPa (AGL III) and the maximum modulus of elasticity is 68.742 MPa (AGL III). The set AGL III have the maximal values of both observed mechanical properties. The minimum of tensile strength in bending is 0.028 MPa (AGL V) and the minimum modulus of elasticity is 27.038 MPa (AGL IX). In contrast with the same set AGL III with both maximal values, the minimal values differs for modulus of elasticity and tensile strenght in bending. From this point of view the prescription AGL III with 80 % of sand, 20 % of clay and 0.400 water-clay ratio seem to be the best composition.

In comparison to the other tests that were already done [3,4], the values of tensile strength in bending is lower at the prescription with high water-clay ratio.

Discussion

As a problem that seem to be characteristic for this material is a quite big range of measured data, big value of the standard deviation – especially in the case of modulus of elasticity up to 41 %. The manufacturing of the specimen is very exacting and several types of tests are made, three pieces for the bending test are used. One reason can be the small number of tested specimens and secondly, the other reason can be that the rammed earth is a quite brittle material. On the other hand, the standard deviation of the tensile strenght is only up to 9 % and very simillar to all prescription

Conclusion

The mechanical properties of the earth material highly depend on the composition of sand, clay and water. The amount of water and binder is one of the key properties of the rammed earth, it is similar to concrete, so the amount of the water is expressed by the water-clay ratio. The water-clay ratio 0.400 seems to be optimal with the composition of 80 % sand and 20 % clay.

Acknowledgements

This research was financial supported by the Czech Science Foundation (GAČR 18-10884S) and the Faculty of Civil Engineering at CTU in Prague (SGS project No. 16/201OHK1/3T/11).

References

[1] G. Minke, Building with Earth - Design and Technology of Sustainable Architecture. Berlin, 2006 pp. 11-18, 158-160.

[2] I. Žabičková, Earth Constructions. Brno, 2002. pp. 5-14.

[3] T. Otcovska, P. Padevet. Dependence of tensile bending strength of rammed earth on used clay composition and amount of mixture water. Modern Methods of Experimental and Computational Investigations in Area of Construction II pp. 48–53, 2017.

[4] T. Otcovska, P. Padevet. Dependence of compressive strength of rammed earth on used clay composition. Experimental Stress Analysis 2016.

[5] S. Shah Surendra P., S. Stuart E., O. Chenhsheng, Fracture Mechanics of Concrete, New York, John Wiley and Sons, 1995.

[6] M. Jirasek, J. Zeman. Deformation and failure of materials: Creep, plasticity, fracture and damage, CTU in Prague, 2006.

EAN 2020
58th conference on experimental stress analysis

Rammed Earth and its Properties for Design of Construction

PADEVĚT P.[1,a], MUŽÍKOVÁ B.[1,b], OTCOVSKÁ T.[1,c]

[1]Deparment of Mechanics, Faculty of Civil Engineering, CTU – Czech Technical University in Prague, Thákurova 7, 166 29 Prague, Czech Republic

[a]pavel.padevet@fsv.cvut.cz, [b]barbora.muzikova@fsv.cvut.cz, [c]tereza.otcovska@fsv.cvut.cz

Keywords: Rammed earth, Clay, Compression test, Compression strength, Tensile strength in bending, Volumetric weight

Abstract. Two types of clays are compared in the article. The first type of clay is illitic, the second is ilitic - kaolinitic clay. Material properties such as compressive strength and flexural tensile strength are compared for different types of manufactured material mixtures. The test specimens were made of rammed earth, and that of defined mixtures. The dependence of material properties on the amount of clay, the type of clay and the amount of mixing water was monitored. The results of the comparison of properties were summarized in the conclusion.

Introduction

The design of building structures depends on the required shape of the structure, the type of use and the selected material design. Some constructions can be built only from suitable building materials, others can be built from any materials. If the above assumptions permit, rammed unfired earth can be used for building structures [1].

Material selection

Unfired clay is suitable for structures that are not exposed to direct water. In this case, the use of concrete is the most suitable solution. The rammed construction of clay allows to create a relatively free shape, which is limited in the case of using masonry elements. Like other building materials, rammed earth has its specific properties which make it suitable for use in selected building constructions. The paper presents the properties of unburnt earth, and it will be seen that the use of this material will be particularly suitable for vertical structures [2].

Rammed earth is a material consisting of three basic components, water, sand and clay, which has the function of a binder. The results of two types of clay are presented here. The first is Ilitic and the second is an Illitic-Kaolinitic clay. The mixtures labeled "KR" are prepared from Illitic- Kaolinitic clay [3]. On the other hand, "AGL" mixtures were prepared from Illitic clay. These two types of clays belong to the most frequently offered types of clays that are available in the Czech Republic. At the same time, these are mineralogically clearly defined types of clays.. The sand to clay weight ratio was the same for all mixtures. The difference was in the amount of water used in the prepared mixture. The amount of water is based on the weight of clay for all of the present compositions. This ratio is referred to as the water factor. This factor is called the water / clay ratio and is captured in the last column of Table1.

Preparation of specimens for testing

The specimens were manufactured by placing and ramming into moulds of 40 × 40 × 160 mm. Approximately 3 layers of material were compacted to produce of one specimen. The load direction of all specimens was perpendicular to the compaction direction. The ratio of base width to specimen height was 0.25.

Acquisition of the strength of the rammed earth is carried out by evaporating the mixing water from the body and drying it out. The drying process must be natural, as no cracks have formed since drying. Too fast drying of the material does not occur in the case of laboratory conditions. The clay itself is able to bind water, and gradually release it in the drying process. The specimens could dry out and gain strength for more than a month. Rammed earth is able to saturate the air humidity very well. All specimens were tempered to 20° C at 50% relative humidity before of testing. This was done to ensure that the saturation of the material was uniform [4]. The same method of compaction and the same procedure in the preparation of the bodies for testing gave very comparable bulk densities of the prepared mixtures, see Fig.6.

Composition of the all tested mixtures is described in Table 1. The water / clay ratio was between 0.25 and 0.45. A total of 12 specimens were prepared for each mixture. Of these, 3 were tested in a compression test and 3 bodies were intended for the bending tensile test. The remaining specimens were used for other experiments.

Table 1: Tested rammed earth in compression.

Description	Clay	Sand	Water / clay
AGL 3	20	80	0.4
KR 2	20	80	0.37
AGL 5	20	80	0,37
AGL 8	20	80	0.29
AGL 9	20	80	0,45
KR 7	20	80	0,25

Material tests

Compressive strength tests were performed by recording force and axial deformation using two extensometers and axial deformation of whole height of prism, see Fig.1. A typical working diagram in compression is presented in Fig. 2. The displayed deformation is recorded in the deformation of the entire height of the specimen.

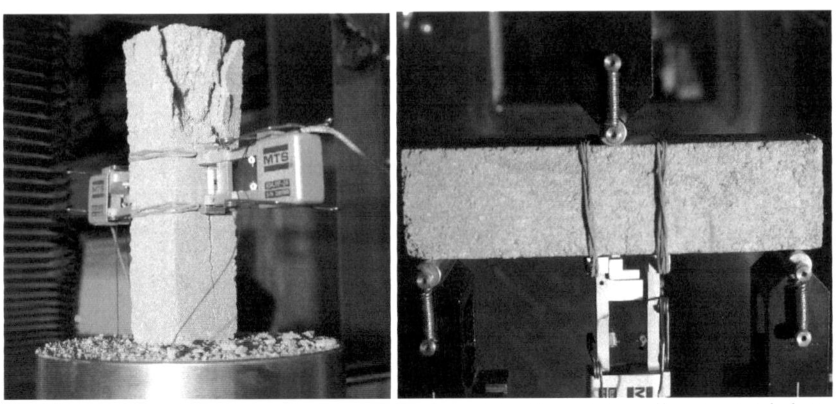

Fig. 1: Instrumentation of compression test (left) and bending test (right).

As can be seen from the working diagram, the ascending branch contains an elastic and later also a plastic area. The descending branch of the diagram is not steeply decreasing, as we would expect for more fragile materials.

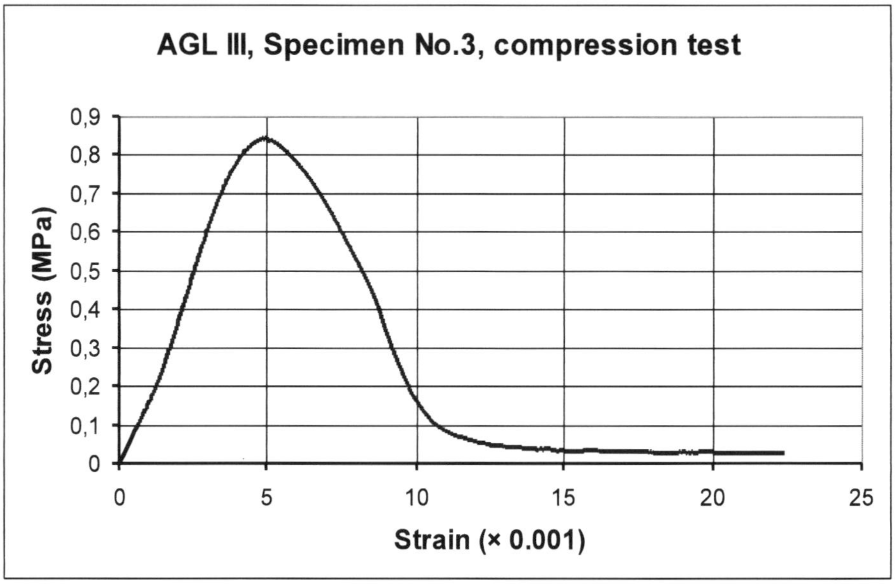

Fig. 2: Selected graph of compression test of mixture prepared from illitic clay.

The bending tensile test was instrumented by one extensometer on the lower surface of the bent specimen. It measured the opening of the tensile crack. Furthermore, the force and deflection of the prism in the middle of the span were recorded. The distance of the supports was 140 mm for all tested specimens. The test was performed in a three-point bend, as shown in Fig.1.

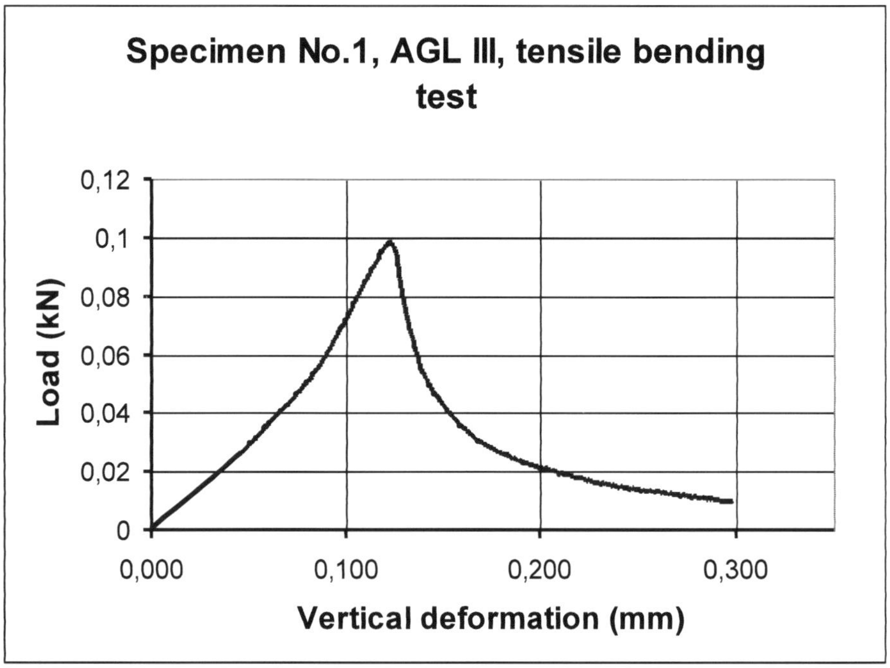

Fig. 3: Selected graph of bending test of mixture prepared from illitic clay.

The bending tensile behaviour of the material is interesting in that the descending branch of the working diagram is rapidly descending. The rapid decrease in force subsequently slows down and the deflection of the body increases considerably during the development of the main crack. The decrease in force slows down as the crack develops, as can be seem from Fig.3.

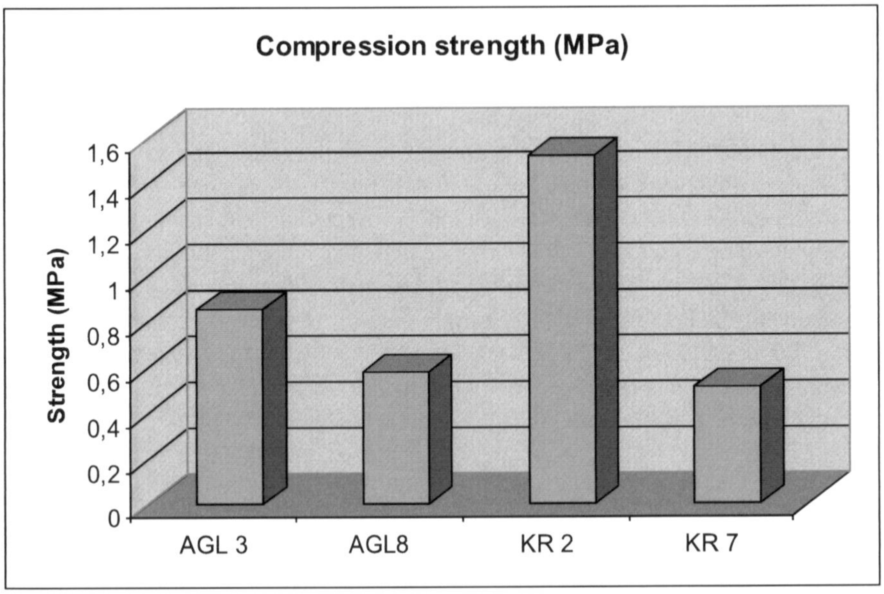

Fig. 4: Compression strength of specimens prepared from Illitic and Illitic-Kaolinitic clay.

Fig.4. provides an overview of the achieved compressive strengths. The achieved strengths of the material made of Illitic clay were 0.58 and 0.86 MPa. The specimens prepared from Illitic-Kaolinitic clay reached compressive strengths of 0.51 and 1.52 MPa. It should be noted that the KR7 mixture contained the least mixing water. This was clearly reflected in the lowest compressive strength achieved.

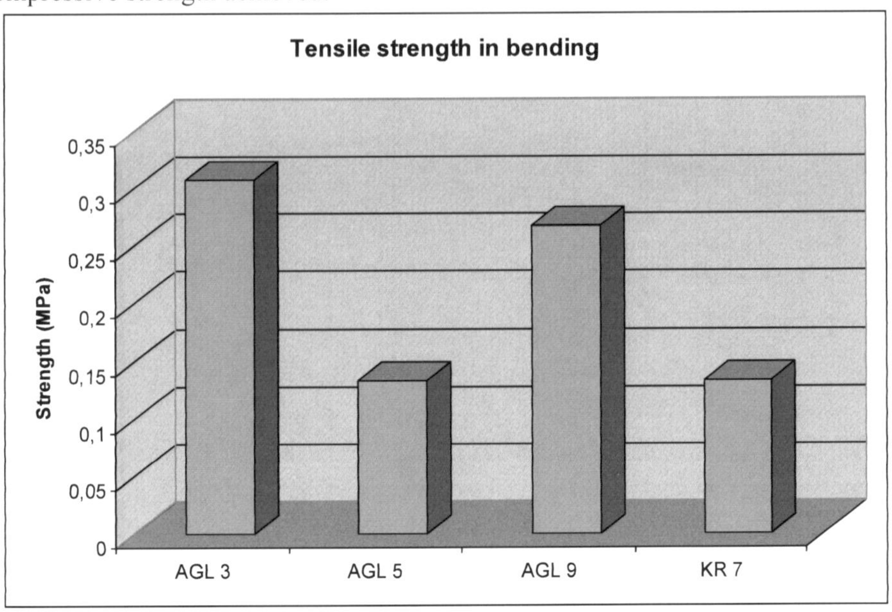

Fig. 5: Tensile strength in bending of specimens prepared from Illitic and Illitic-Kaolinitic clay.

On the contrary, the KR2 mixture did not contain the most mixing water, but its amount was given by a factor of 0.37. AGL 8 contained more mixing water than KR7. Its strength was higher. AGL3 also contained more water than KR2, but its strength was comparatively lower. It follows that for the compressive strength, the limit values of the water coefficient were 0.25 and 0.4. The most suitable water coefficient in terms of the highest compressive strength was a ratio of 0.37.

The lowest value of compressive strength is evident in connection with the lowest bulk density, as is possible see on Fig. 6. The density of the rammed clays was close to 2100 kg / m³. Only the KR7 set showed a lower strength value of 2020 kg / m³. For both types of used clays it is evident that with decreasing amount of mixing water the compressive strength decreases. The compressive strengths obtained, see Fig. 3, show that by means of an Ilitic-Kaolinitic clay a significantly higher strength can be achieved, while maintaining comparable proportions of water and clay than with an Illitic clay.

The graph in Fig. 5 shows a comparison of three materials prepared from Illitic clay with one material from Illitic-Kaolinitic clay. The comparison is made in flexural tensile strength. The range of water coefficient of Illitic clay materials was from 0.37 to 0.45. In this case, the highest flexural strength was achieved by a mixture with a water to clay ratio of 0.4. The AGL9 mixture, which was prepared with a water coefficient of 0.45, reached a flexural strength of 0.27 MPa. However, a single KR7 blend reached almost the same bending strength value of 0.13 MPa as the AGL 5 blend. The AGL5 blend had a water content of 0.37 and the KR7 blend was prepared with a water content of 0.25. The results show that the higher content of mixing water was favorable for achieving better values of flexural strength.

The volumetric density of all mixtures ranged between 2006 and 2098 kg / m³, see Fig.6. Two mixtures of AGL5 and KR7 had the lowest values. The two blends also had comparable flexural strength results. Other materials prepared from different blends had very similar bulk densities. These ranged from 2086 to 2098 kg / m³, although the amount of mixing water was between 0.29 and 0.45.

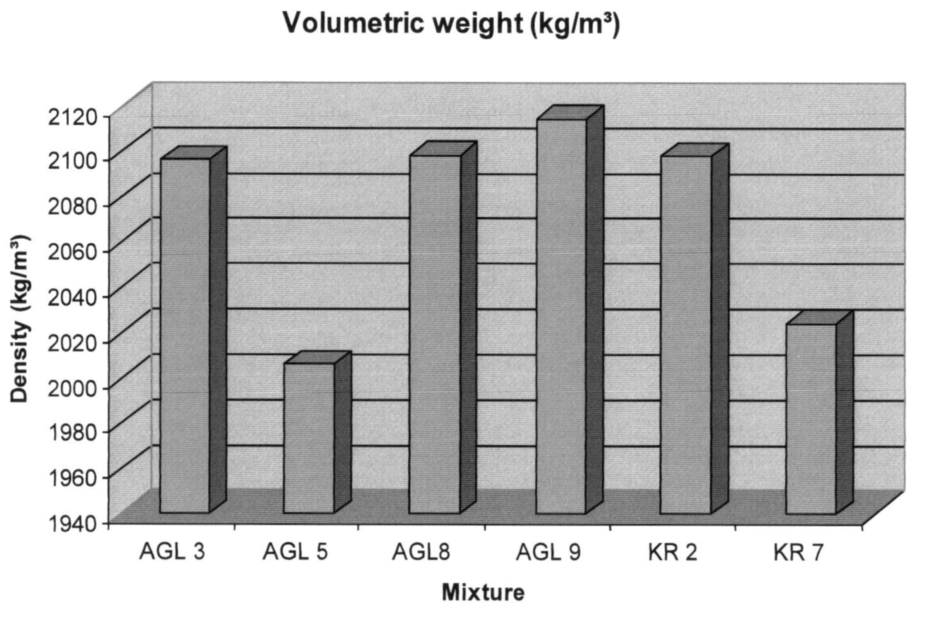

Fig. 6: Volumetric weight of all tested mixtures prepared from Illitic and Illitic-Kaolinitic clay.

Conclusions

Tests performed show that the compressive strength is affected by the amount of mixing water. In conclusion, it should be noted that the test specimens were tested after stabilization of their weight in a temperature-humidity chamber.. The optimal amount of mixing water was 0.37. A material prepared from Illitic kaolinitic clay (KR) seems to be more suitable. Since only one mixture of Illitic-Kaolinitic clay was tested in bending tensile strength, it cannot be confirmed with certainty that materials made of this clay are also more advantageous from the point of view of flexural strength.

As follows from the presented properties, unfired earth is suitable for structures loaded with pressure. Suitable applications can be found in vertical load-bearing and non-load-bearing structures. The results of compressive strength and their dependence on selected parameters differ from rheological properties [5]. Finding optimal properties will depend on comparing the results of multiple material properties.

Acknowledgement

This work was supported by Grant Agency of Czech Republic GACR No. 18 – 10884S "Unfired and rammed clay for construction".

References

[1] G. Minke, Building With Earth [online]. 2006, [published August 18, 2015]. ISBN 978-3-0346-0822-0. Available from: http://archive.org/details/Gernot Minke-Building With Earth.

[2] I. Žabičková, Clay buildings [online]. Brno, Era 21, 2002 [vid. August 18, 2015], ISBN 80-86517-21-7.

[3] P. Padevět, T. Otcovská, B. Mužíková, Material Properties, of Rammed Earth Prepared from Illitic - Kaolinitic Clay, WSEAS Transaction on Applied And Theoretical Mechanics, Vol. 13, 2018, pp. 181 – 186.

[4] T. Plaček Otcovská, B. Mužíková, P. Padevět, Mechanical Properties of Rammed Earth with Respect to Clay Mixture Composition, Acta Polytechnica, Volume 59, Issue 4, 2019, pp.372-383.

[5] P. Padevět, P. Bittnar, T. Otcovská, Shrinkage of Rammed Earth Based on Illitic Clay, 56th International Scientific Conference on Experimental Stress Analysis EAN 2018, 2018, Harrachov, Czech Republic, pp. 321 – 325.

Numerical and Experimental Analysis of C/PPS Y-Shaped Profile

PADOVEC Z.[1,a], ZÁMEČNÍKOVÁ T.[1,b], SEDLÁČEK R.[1,c]

[1]Deparment of Mechanics, Biomechanics and Mechatronics, Faculty of Mechanical Engineering; Czech Technical University in Prague; Technická 4, 160 00 Praha, Czech Republic

[a]zdenek.padovec@fs.cvut.cz, [b]tereza.zamecnikova@fs.cvut.cz, [c]radek.sedlacek@fs.cvut.cz

Keywords: Composite structure, FEM, Experiment

Abstract. Presented paper deals with testing and its FE simulation of Y shaped composite profile designed for joining of aircraft construction. This type of profile was derived and designed according to previous design, testing and simulation of T profile manufactured from same material and with the same lay-up. Results from both experiments were compared with each other with conclusion that new Y shape design can withstand higher loading than T profile.

Introduction

Main goal of this work was testing, and simulation of Y-shaped profile manufactured from C/PPS (carbon fiber, polyphenylene sulphide matrix, fabric with 5H satin weave 0.33 mm thick) composite which will be used for connection of aircraft keel beam with the ribs. Profile was developed in two construction versions – T (which was tested and analysed in [1]) and Y. Results from both tested profiles were evaluated and compared with each other. Both profiles were manufactured by thermoforming process from three plates with $[[(0/90)/(\pm45)]_2]_s$ lay-up.

Experiment. Experiment was done on TIRA 2300 universal testing machine with loading speed of 2 mm/min (respectively 0.5 mm/min). Tensile load was realized through the screws in the web of the element jointed with the jaws of the machine. Relationship between loading force and displacement for three tested Y and five tested T specimens can be seen in Fig. 1.

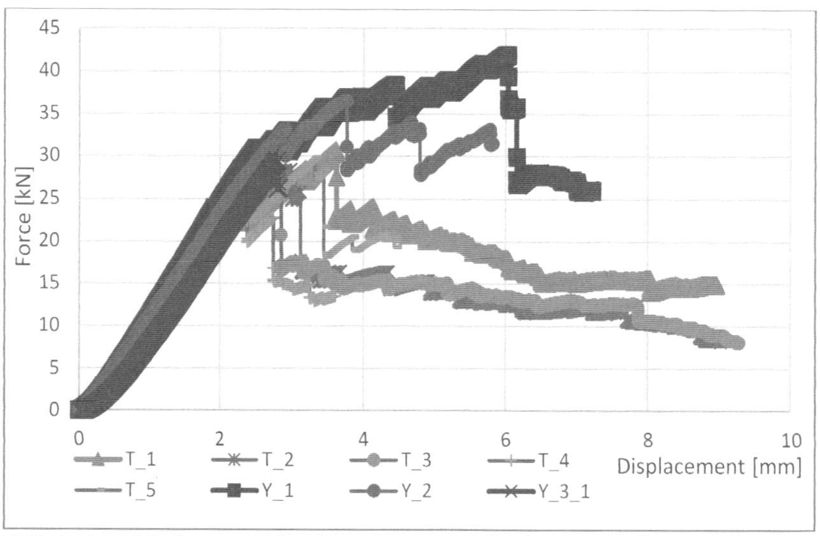

Fig. 1 Relationship between loading force and displacement for tested specimens

First two Y specimens were tested until total failure, third one was tested until first failure occurred. Then the specimen was removed and put into CT scan to find out the area of the failure (see Fig. 2). This failure occurs under the loading force 29.5 kN.

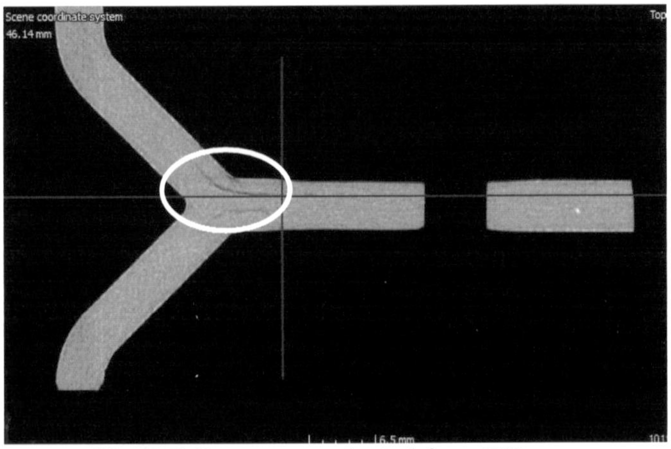

Fig. 2 Failure area evaluated from CT scan

FEM. Simulation of the experiment was done in similar way as simulation of T profile with the use of Abaqus software [1]. The main goal of the simulation was determination of through thickness stress direction in the failed area taken from CT scan. That´s why layered solid elements were used instead of shell or continuum shell elements which can´t be used because of plane stress assumption. Loading was distributed through reference points in the web holes coupled with their area. Material parameters (taken from technical sheet for certified material used in aircraft industry [3]) of each layer were entered as engineering constants (see Table 1), area of pure matrix was modeled as isotropic material (see Table 2). Material orientation can be seen in Fig. 3.

Table 1: Material parameters for composite layer [3]

$E_1 = E_2$ [MPa]	E_3 [MPa]	G_{12} [MPa]	$G_{23} = G_{13}$ [MPa]	ν_{12} [-]	$\nu_{23} = \nu_{13}$ [-]
58 000	10 000	4 100	3 500	0.046	0.33

Table 2: Material parameters for pure matrix

E [MPa]	ν [-]
3 800	0.33

FE model of the experiment, stress field on whole profile and stress field in detail in the failed area can be seen in Figures 4 – 6.

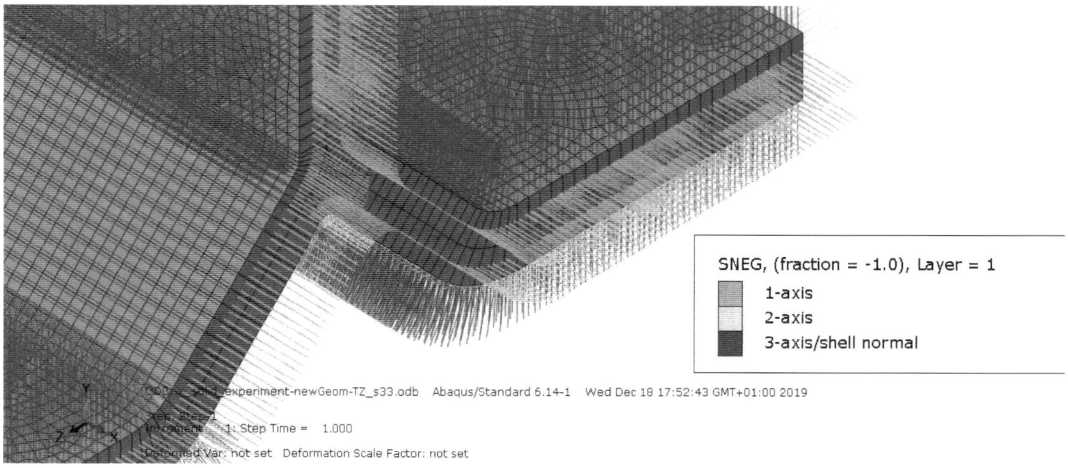

Fig. 3 Material orientation in the beam

Fig. 4 FE model of the experiment

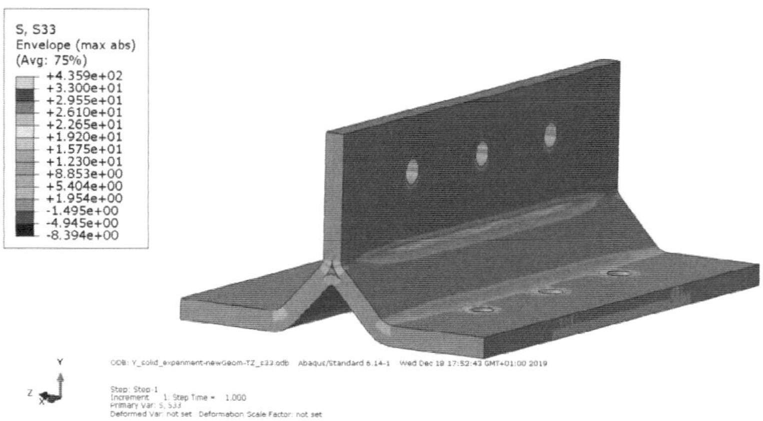

Fig. 5 S33 stress field in whole profile

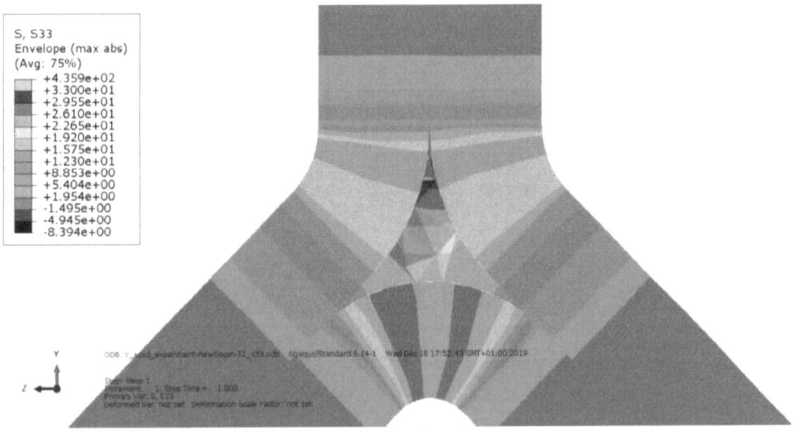

Fig. 6 S33 stress field in the failed area

Conclusions

If we compare maximal value of stress from FE simulation which is 33 MPa and presumed allowed value from stress taken from material sheet [2] which is 39 MPa we see that the agreement between simulation and experiment is very good. Failure theories for composites implemented in Abaqus (maximum stress, maximum strain, Tsai-Hill, Tsai-Wu and Azzi-Tsai-Hill) can't be used because the shell/continuum shell elements which assume plane stress condition (S33 = 0) were not used in this analysis. Also, the area of the failure from CT scan compared with the maximal stress area from FE simulation shows good agreement.

If we compare results from the experiment of T profile and Y profile (Figure 1) we can see that the first decrease of the loading force (which is crucial for fatigue) in tested T profiles occurs between 14 and 16 kN. In tested Y profiles this decrease occurs between 24 and 34 kN which means circa 71 – 112 % improvement compared to T profile results.

Acknowledgements

This work has been supported by project FV30033 of the Ministry of Industry and Trade of the Czech Republic and by the Grant Agency of the Czech Technical University in Prague, under grant No. SGS18/175/OHK2/3T/12.

References

[1] Z. Padovec, R. Sedláček, T. Zámečníková, Testing of C/PPS Element for Aircraft, in: EAN 2019 57th conference on experimental stress analysis CONFERENCE PROCEEDINGS, Luhačovice, Czech Republic, 2019, pp. 361-364.

[2] TenCate Cetex® TC1100 PPS Resin System, Product Data.

[3] AIMS05-09-002, Airbus Material Specification Carbon Fabric, 285 g/m^2 fiber area mass with 43% PPS Resin or Equivalent resin Material Specification.

EAN 2020
58th conference on experimental stress analysis

Prediction of Residual Stresses using FEM before 3D printing by Selective Laser Melting of Stainless Steel AISI 316L

PAGÁČ M.[1,a], HALAMA R.[1,b], ALDABASH, T.[1,c], HAJNYŠ J.[1,d], MĚSÍČEK, J.[1,e], ŠTEFEK, P. [1,g], JANČAR, L. [1,g], JANSA, J. [1,h]

[1]Faculty of Mechanical Engineering, VŠB – Technical University of Ostrava, 17. listopadu 2172/15, 708 00 Ostrava, Czech Republic

[a]marek.pagac@vsb.cz, [b]radim.halama@vsb.cz, [c]tariq.aldabash.st@vsb.cz, [d]jiri.hajnys@vsb.cz, [e]jakub.mesicek@vsb.cz, [f]petr.stefek@vsb.cz, [g]lukas.jancar@vsb.cz, [h]jan.jansa1@vsb.cz

Keywords: Residual Stresses, SLM, 316L, Additive Manufacturing, FEM

Abstract. This paper deals with experimental analysis of stress prediction and simulation before 3D printing by Selective Laser Melting method (SLM) and subsequent separation of printed sample from substrate (building board) in Simulation Additive (MSC Software) and Additive Manufacturing Simulation (ANSYS). Practical verification of the simulation was performed on a 3D printed topologically optimized part made of 316L steel (DIN 1.4404). The paper summarizes new knowledge in the field of stress analysis and simulation of 3D printing metallic alloys. The paper also summarizes current trends in the area of simulation software for additive production and reflects their weaknesses and strengths not only with regard to their use in science and research, but also in practice.

Introduction

3D printing (additive manufacturing - AM) of precise metal models includes three technological phases: pre-process (simulation, prediction and data preparation), process (3D printing) and post-process (heat treatment, support removal, machining, etc.). These processes largely affect the reduction or even elimination of stresses and deformations, reducing scrap and manufacturing costs. Selective Laser Melting (SLM) belongs to the Powder Bed Fusion (PBF) additive manufacturing technology, whose principle consists in melting or sintering atomized metal powder by laser. The SLM method is characterized by a thermal process taking place in the so-called melt pool, where the metal powder melts [1]. Due to the local heat input, the printed part does not cool homogeneously, and thus thermal gradients are created, which generally lead to thermal stresses around the melt pool [2]. Heat transfer in the SLM process has a great influence on the final mechanical properties of the print. To obtain a fully dense component without pores, it is necessary to completely melt the metallic particles of the powder. Therefore, it is advisable to use a high laser power, but this brings negative thermal effects, such as a balling effect and internal stress [3] causing distortion of the part or crack.

Several researchers have already researched distortion prediction [4, 5], most of them based on simulation of a model for welding using the FE method. This simulation method consists in the application of a mathematically modeled heat source to the thermomechanical FEA model [4]. This method has been mainly applied in a process where the input material is powder or additional wire. However, these simulations are very limited when the geometry of the part is complex. Chiumenti et al. [5] performed a fully-coupled thermo-mechanical numerical simulation including phase-change phenomena defined in terms of both latent heat release and shrinkage effects. Thus, they formulated a new activation methodology for simulating layer deposition. This method was followed by Lundbäck et al. [6], when it was successfully tested to simulate the deposition of individual layers. In addition, they have improved it by calibrating the inflow and heat input, to predict the correct temperature and distortion. They compared the results experimentally. Hussein et al. [7] developed a 3D FE model for the prediction of melt pool size and temperature gradients that are affected by the scan speed parameter during single layer deposition, this model is valid only for AISI 316L material. He concluded that higher speeds cause a smaller melt pool size in the width and depth parameters, but the pool length parameter is higher. Another conclusion of the publication was that the highest temperature gradients were reached with the first layer and then rapidly decreased with each subsequent layer (at all scan speeds tested). Zhang et al. [8] used a 3D FE model to investigate the dependence of temperature gradients on laser power and scanning speeds. It has been shown that high laser power and low scanning speed lead to significant heat input and high maximum temperature in the SLM process. However, these simulation approaches require high PC performance and a lot of computational time. To speed up simulations, some researchers have begun to use the inherent strain approach, which is commonly used in the field of welding for large components. Keller et al. [9] used the inherent strain approach on a macro scale and successfully managed to significantly reduce the computation times of the simulations, however, the question arises how accurate these simulations are because they have not been experimentally verified. A similar approach based on FE and inherent strain was developed by the team around Afazov et al. [10], when they succeeded in simulating an industrial model of an impeller blade on a macro scale. The data obtained from the simulation were successfully experimentally verified with a high degree of accuracy. Song et al. [11] pointed out in their publication an important fact that needs to be taken into account when designing the simulation, namely that for accurate simulation it is necessary to take into account the surrounding powder around the printed part, when heat is transferred to the surrounding powder.

This study focuses on the prediction and evaluation of deformations using two solutions by ANSYS Additive Suite and MSC Simufact. In both cases, a 3D non-linear finite element model based on thermo-mechanical fields was used. Furthermore, simulations and predictions of temperature distribution and thermal stress were performed. The aim and motivation of this article is to realistically verify and compare the results of advanced commercial software, even experimentally.

Methods

Both ANSYS Additive Suite (AAS) and MSC Simufact work on the principle of Finite Element (FE) simulation and are directly designed for SLM technology. The primary task of these software is to avoid unnecessary cost and time and improve product quality. These tools allow, thanks to the prediction of stresses and deformations in the AM process, to choose the optimal design and orientation of the part in the build chamber. Another function is compensation and distortion prediction, which allows the part to be pre-deformed. Both simulation programs use a layered discretization with voxel elements (cube with a defined edge length). AAS even offers layered tetrahedral elements for better adaptation to the shape of the part. In the AAS program, it is recommended to set 10-20 times the actual height of the print layer for layer simulation [12]. For MSC Simufact, the manual recommends setting the voxel size to 1.8 mm, according to the literature [13] it is clear that the smaller the voxel size is set, the longer the simulation time takes. In general, the simulation procedure can be described by a flowchart, see Fig. 1.

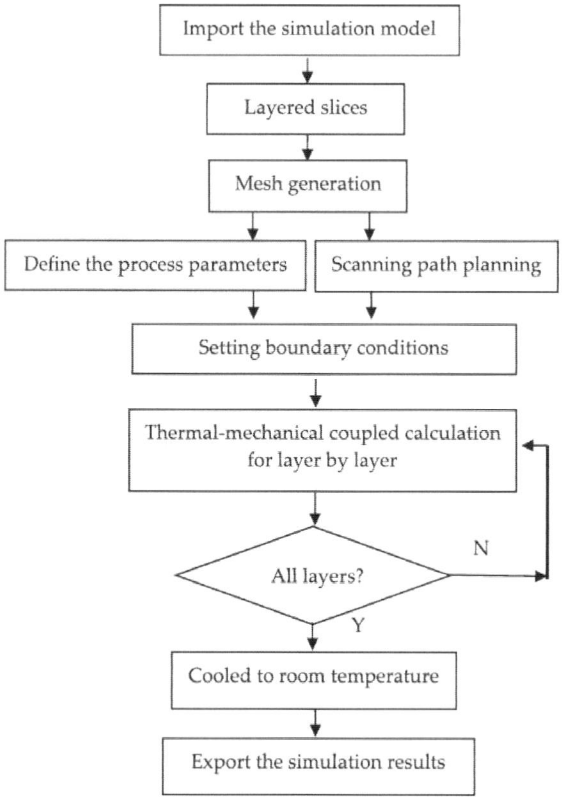

Fig. 1: Flowchart of simulation model [14]

Simulation approach

ANSYS Additive Suite. AAS is fully adapted for work in the ANSYS Workbench environment. The component is first completely meshed to individual layers of either voxel or tetrahedral elements. Each FE layer represents a number of real metal powder layers, assuming thermal continuity of the next layer. The program does not take into account a moving heat source, for example, when the thermal gradient in the build direction dominates over the thermal gradient in the plane due to its effect on residual deformations. Unlike MSC Simufact, AAS requires more input parameters. To simulate the process, it is necessary to specify the basic properties of the material, which are temperature dependent, or inherent strain calibration is also possible. However, the first approach was considered in this study. These parameters include elastic modulus, orthotropic thermal conductivity, Poisson's ratio, density, elastic bilinear curves of plastic stress and strain, coefficient of thermal expansion, and specific heat capacity up to the melting point of the material. In this study, the material properties for AISI 316L (AM), which is predefined in ASS, were set for the simulation. Other process parameters necessary for print simulation are given in Table 1. Meshed was also build plate with hexahedral element.

For this study was used ANSYS Inc. ANSYS Mechanical Workbench, version 2020R1.

Table 1: Process parameter for input data AAS

Parameter	Value
Element size	0.75 mm
Method	Cartesian
Hatch Spacing	0.13 mm

Scan Speed	650 mm/s
Preheat Temperature	22 °C
Layer thickness	50 μm
Scan strategy	Meander

MSC Simufact. MSC Simufact is a program developed specifically for Power Bed Fusion Additive Manufacturing simulations. Selective Laser Melting (SLM), Electron Beam Melting (EBM), Laser Beam Melting (LBM) and Direct Metal Laser Sintering (DMLS) methods can be analyzed. One of the advantages of the software is the coverage of the simulation of the whole AM process, such as printing simulation, support cutting, heat treatment, Hot Isostatic Pressing (HIP) and of course the evaluation of residual stresses and distortions. Simufact offers only a hexahedral element for voxel mesh. After creating the voxel mesh, a volumetric mesh structure is created, which provides a 3D network, then a surface mesh is generated. Voxel and surface mesh complement each other and thus guarantees coverage of the entire part. Since the program offers work with inherent strain, it is necessary to calibrate using cantilever, when a set of cantilever was first printed on the printer in various positions, these were then cut lengthwise, thus activating residual stresses and entering the measured values into software. Material curves and constants have already been predefined for the AISI 316L material as well as for AAS by the manufacturer. The only necessary input from the user is the input of the machine type (for setting the working space), layer thickness, the direction of cutting the part from the printing pad and the voxel size.

For this study was used MSC Software Inc. Simufact version 2019. Main considered parameters in the simulation are stated in Tab.2.

Table 2: Process parameter for input data MSC Simufact

Parameter	Value
Element size	1 mm
Method	Inherent Strains
Hatch Spacing	0.11 mm
Scan Speed	650 mm/s
Preheat Temperature	Ambient
Layer thickness	50 μm
Scan strategy	Meander

Test geometry. For evaluation and validation purposes, was chose a practical component called the Shifting thumb, which is used in the gearbox of trucks. The original shape was topologically optimized for maximum use of the 3D printing potential, see Fig. 2.

Topology optimization has become a powerful area since its algorithms can be applied in many design problems in different physical disciplines such as solid mechanics, fluid dynamics, as well as thermal dynamics. Using topology optimization methods and related algorithms, a designer can generate innovative design ideas especially in engineering fields. Instead of using trial and error methods in designing engineering products, designers use different topology optimization methods to generate conceptual designs.

In topology optimization, it is usually done by creating a geometry that only considers the constraints of the part ranging from supports to the vacant area that the body can move around within the assembly. However, since that part has almost no space within the assembly, the original geometry has been considered in this study with a slight modification for simplicity of calculations and some fitment issues, where the radius at the tip of the thumb was not perfectly matching when attaching it into the main assembly. ANSYS Discovery Live software was used for topological optimization. Final reduction of volume was 63%. The verification of design has been performed in ANSYS Workbench leading to 214.2MPa of Von Mises stress (232.3MPa in ANSYS Discovery Live) what is comparable with 221.5MPa obtained for the original design. Young's modulus E=195GPa and Poisson's ratio ν=0.3 were used in the simulations.

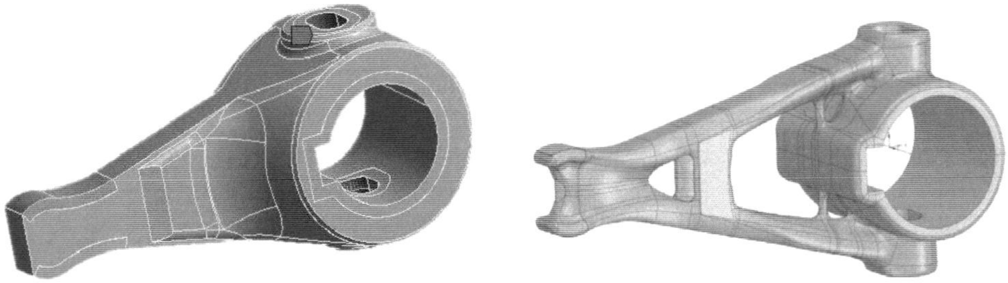

(a) (b)

Fig. 2 Original geometry of main shifting thumb (a); Topologicaly smoothed shifting thumb (b)

SLM printer RenAM400. A Renishaw AM400 3D printer was used to manufacture and test the part. It is a device for 3D printing of metal parts from the English industrial company Renishaw. This is an improved version of the AM250 printer. The AM400 features an improved optics control system, redesigned inert gas flow, a window protection system and a 400 W optical system with a fiber laser with a wavelength of 1070 nm, which provides a beam with a diameter of 70 μm at the melting point. Argon, which has a higher atomic number than nitrogen, was used to make the experimental sample, displacing more oxygen in the printing chamber and keeping the oxygen level below 500 ppm.

Digital Image Correlation. Testing was performed using Digital Image Correlation (DIC) which is an optical method. Mercury®RT system provided by Sobriety company was used to capture displacement field. The optical contrast coating was created on the printed part (generally known as pattern). The main goal was to investigate the deformation of the part after support removal and subsequent comparison with the deformation predicted by FEM.

The software used in DIC measurement is Mercury RT which is capable of three-dimensional measurements. Two high accuracy cameras (2x2.3Mpx@40Hz) where fixed in front of the investigated part during the cutting process. The cameras record the deformation of the part once the cutting process has begun. With the capabilities of the software deformation contours were gained for comparison with the numerical prediction.

Results and discussion

Total displacement simulation in AAS. The software offers an orientation map feature where it simulates all possible angles and calculate build time, distortion tendency, and supports. After the selection of the appropriate orientation the model then undergoes an additive manufacturing thermal analysis to simulate the printing process and predict structural defects such as deformation due to shrinkage during cooling time. Using ANSYS Additive Manufacturing Wizard, the following project setup would automatically generate. AM thermal analysis meshing is displayed in Fig. 3a, also AM thermal analysis results are shown in Fig. 3b which confirms that the effective thermal stress proportionally related to the increase of layer height from bed surface.

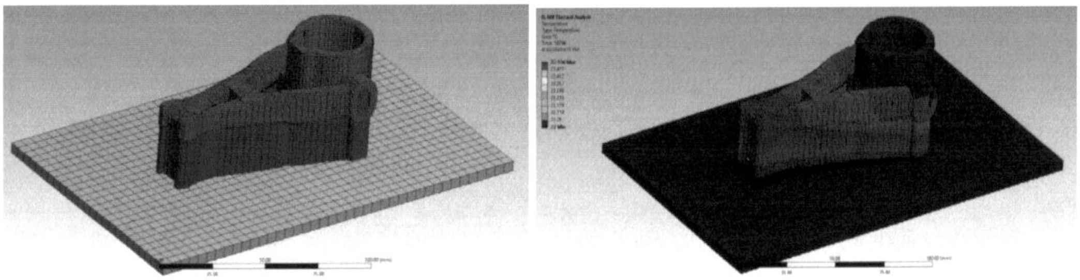

(a) **(b)**

Fig. 3: AM meshing for thermal analysis (a); thermal analysis results (b)

After thermal analysis, structural AM analysis was performed. From the nature of using Cartesian mesh, we can get an unacceptable result. Therefore, a subsequent finite element analysis (submodeling) is required to obtain finer and more accurate results by selecting finer mesh. Such process was performed with suppressed base plate and ignored nonlinear effect as well as initial strain. After obtaining the evaluation of the simulated printing process in Fig. 4, theoretical results are then verified by performing the printing process and perform a deformation test during cutting the part from its base plate. Maximum displacement is about 0.83 mm. The largest deviation from the original shape is shown on the "nose" of the part. Here the greatest deformation occurs due to the fact that the coarser part does not deform and the rest bends due to thermal stress and also the corners of the part act as stress concentrators.

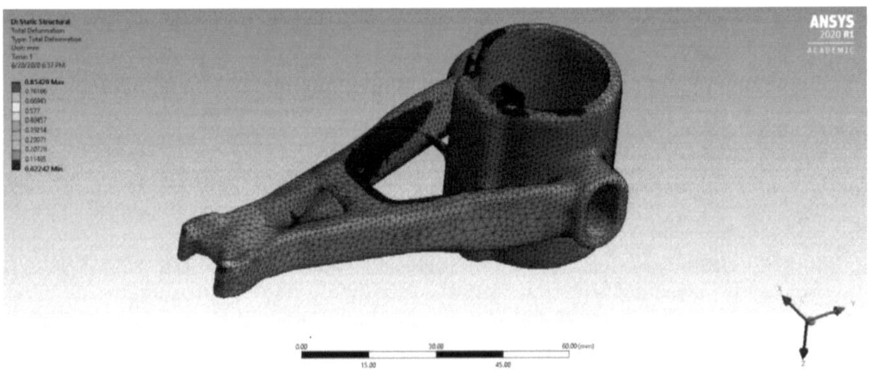

Fig. 4: Standalone structural analysis by ASS

Total displacement simulation in MSC Simufact. Performing a simulation in the Simufact program consists of eight steps. After importing the geometry, support elements are generated or imported. In the third step, the manufacturing parameters are defined, see Table 2. By selecting the AM printer, the dimensions of the working environment are automatically set, then the distance of the Z-axis direction is manually set. In the fourth step, a surface and voxel mesh are formed, see Fig. 5a. The surface mesh is set to 2 mm by default for each analysis. Surface mesh is the type of mesh that is transferred to the shifting thumb part after the voxel mesh. The voxel element size was set to 1 mm. Voxel mesh then directly analyzes the deformations, see Fig. 5b. The main reason for creating the surface and voxel mesh is to bring the part as close as possible to the actual shape. The fifth step analyzes the results and shows a preview, in the sixth step the results are graphically displayed, which are used to calculate the pre-deformed model and can be further analysed. The penultimate step consists in removing the supporting elements and in the last step the cutting of the part from the build plate is realized.

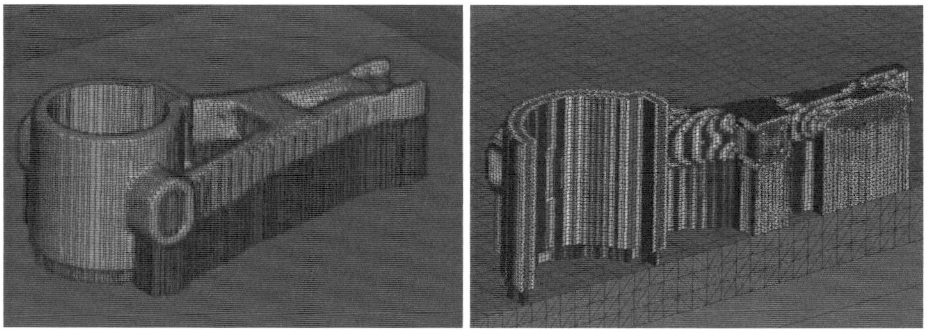

(a) (b)

Fig. 5: Voxel and Surface mesh (a); Voxel mesh analysis (b)

In this study, the layer height was set to 0.05 mm and the inherent strains ε_{xx} = -0.0036, ε_{yy} = -0.0036, ε_{zz} = -0.03 at distribution uniform. These values were determined by calibration using a reference structure for the equipment used. From the results of the simulation with the printing plate, a total displacement in the range of 0.6-0.7 mm was obtained. The surface deformation deviation was 0.58 mm and, as with the ASS software, the largest deformation on the "nose" of the component was evaluated, see Fig. 6.

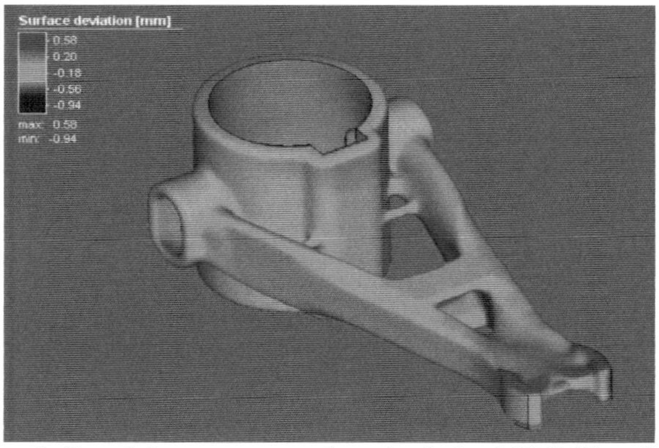

Fig. 6: Standalone structural analysis by MSC Simufact

Validation of the results. The part was printed on RENISHAW AM400 machine using Stainless Steel AISI 316L as a print material. After the completion of printing process, excess powder material was then safely removed for recycling into future jobs. Then, the printed part goes through cutting from base plate process which was investigated to observe if any deformation occurs in the printed part.

Testing was performed using Digital Image Correlation (DIC). It is observed in Fig. 7, that the resulting deformation is a slightly higher than the simulated deformation in Fig. 4 and Fig. 6. There are many factors that might have a role in such difference. For instance, computational tools have different approaches and input parameters and that might result in different values. Also, the factor that can be easily pointed out is the material properties that were implemented into the simulation process where ANSYS and MSC material library was the source of the material properties assigned. Material properties of the printed part differ from the material properties available in the library. Even more, meshing of the bodies might have a significant impact since the calculations where constrained to limited number of elements to run the calculations Moreover, heating due to slicing could be a factor for additional deformation of the part.

Cutting from the base plate was realized on saw blade and the maximum displacement was measured to 1.137 mm.

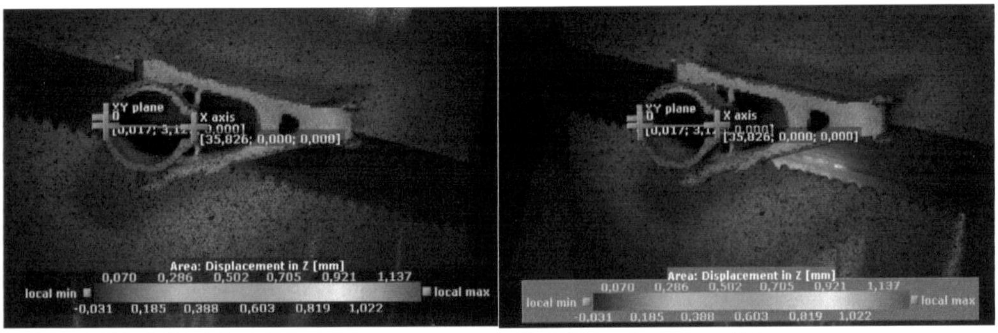

(a) **(b)**

Fig. 7: Maximum displacement on the left camera (a) and on the right camera (b)

Conclusions

Simulations play an important role in the additive manufacturing process as well as in welding. Thanks to them, we are able to virtually design and verify the entire process. This saves material and costs. We are able to eliminate the trial-and-error method, and, in addition, we can benefit from benefits such as pre-defining the shape, which straightens into the desired shape due to residual stresses. Simulations unlock the potential of AM technology.

This study summarized new knowledge in the field of stress analysis and simulation of 3D printing metallic alloys. The paper also summarizes current trends in the area of two commercial simulation software ANSYS AAS and MSC Simufact. A real component was selected, which underwent topological optimization and subsequently simulations were performed in both programs. For both SW, the default setting of material properties supplied by the manufacturer of the SW was used (it was not the same for both simulations). The part was also experimentally printed using the SLM method to verify the development of deformations during removal from the set plate. Thanks to the performed simulations and real verifications, several conclusions were drawn:

- All performed simulations provide a great basis for the prediction of stresses and strains. Both programs showed very similar deviation values.
- When comparing the simulation with real measurements, a deviation of ± 0.5 mm is achieved. However, the DIC measurement could be affected by the heating of the material at the cutting point, which probably caused the part to deflect towards the cameras, which corresponds to the fact that the measured maximum displacement is larger than in the simulations. The actual shape of the part will be verified by scanning for a possible magazine article.
- Generated support elements in both software can be successfully used instead of generated supports in slicer software.

Further research could compare other software as well as focus on multiple scanning strategies and the use of multiple different print settings.

Funding

This paper has been completed in association with project Innovative and additive manufacturing technology – new technological solutions for 3D printing of metals and composite materials, reg. no. CZ.02.1.01/0.0/0.0/17_049/0008407 financed by Structural Funds of the European Union and project.

References

[1] Q. Chen, X. Liang, D. Hayduke, J.K. Liu, L. Cheng, J. Oskin, R. Whitmore, A.C. To, An inherent strain based multiscale modeling framework for simulating part-scale residual deformation for direct metal laser sintering, Additive Manufacturing, 2019, Vol. 28. pp. 406-418.

[2] M.J. Ansari, D.-S. Nguyen, H.S. Park, Investigation of SLM process in terms of temperature distribution and melting pool size: modeling and experimental approaches, Materials 12, 2019

[3] J. Hajnys, M. Pagáč, J. Měsíček, J. Petru, M. Król, Influence of scanning strategy parameters on residual stress in the SLM process according to the bridge curvature method for AISI 316L stainless steel, (2020) Materials, 13 (7), art. no. 1659

[4] L. Lindgren, Computational Welding Mechanics, Woodhead, Cambridge, 2007.

[5] M. Chiumenti, M. Cervera, A. Salmi, C. Agelet de Saracibar, N. Dialami, K.Matsui, Finite element modeling of multi-pass welding and shaped metal deposition processes, Comput. Methods Appl. Mech. Eng. 199 (2010). pp. 2343–2359

[6] A. Lundbäck, L. Lindgren, Modelling of metal deposition, Finite Elem. Anal. Des. 47 (2011). pp. 1169–1177.

[7] A. Hussein, L. Hao, C. Yan, R. Everson, Finite element simulation of the temperature and stress fields in single layers built without-support in selective laser melting, Mater. Des. 52 (2013). pp. 638–647.

[8] D. Zhang, Q. Cai, J. Liu, L. Zhang, R. Li, Select laser melting of W-Ni–Fe powders: simulation and experimental study, Int. J. Adv. Manuf. Technol. 51 (2010). pp. 649–658.

[9] N. Keller, V. Ploshikhim, New method for fast predictions of residual stress and distortion of AM parts, in: Annual International Solid Freeform Fabrication Symposium –An Additive Manufacturing Conference, The University of Texas in Austin, USA (2016). pp. 1229–1237

[10] S. Afazov, A.D.D. Willem, L.T. Borja, A. Holloway, A. Yaghi. Distortion prediction and compensation in selective laser melting. Additive Manufacturing, 2017, 17, pp. 15-22

[11] X. Song, S. Feih, W. Zhai, et al. Advances in additive manufacturing process simulation: Residual stresses and distortion predictions in complex metallic components. Materials & Design, (2020), 193

[12] T. Mayer, G. Brändle, A. Schönenberger, R. Eberlein. Simulation and validation of residual deformations in additive manufacturing of metal parts. Heliyon. 2020, 6(5), pp. 1-13

[13] A. Çelebi, E. Z. Appavuravther. Analyzing the Effect of Voxel Mesh and Surface Mesh Application on Residual Stress by Simufact Additive Software. Düzce Üniversitesi Bilim ve Teknoloji Dergisi 2018, 6(4). pp. 930-940

[14] P. Bian, J. Shi, Y. Liu, Y. Xie. Influence of laser power and scanning strategy on residual stress distribution in additively manufactured 316L steel. Optics & Laser Technology (2020), 132

EAN 2020
58th conference on experimental stress analysis

Damage Parameters During Thermomechanical Fatigue

PAŘEZ J.

Výzkumný a zkušební letecký ústav, a.s., Czech Aerospace Research Centre; Beranových 130, 199 05, Prague - Letnany, Czech Republic

parez@vzlu.cz

Keywords: Thermo-mechanical fatigue, Damage parameter, TMF prediction

Abstract. Presented paper deals with damage parameters in prediction of thermomechanical fatigue. The objective of the paper is to compare the input parameters and evaluate the results. Showed thermomechanical fatigue analysis is based by the Nagode calculation with the damage parameter according to the SWT methodology based on the fatigue coefficients and the power relation. Further according to Zamrik's and Ostergren's methodology. Showed methodologies were applied to experimental data, where regression coefficients of respective models were obtained to obtain uniform dependence on the number of half cycles and plots of the damage parameter were plotted.

Introduction

Most methods of predicting fatigue life when considering a combination of thermal and mechanical stress are focused on iso-thermal fatigue, which is characterized by variable mechanical stress of the component in a steady-state temperature field. On the other hand, when the temperature field changes, the stress gradients change due to temperature gradients and during thermal cycling the component is subjected to "thermal fatigue".

A more complicated type is when a component is subjected to a variable temperature field and variable mechanical stress at the same time. One such component is a turbine blade that is under different thermomechanical cycles depending on the blade position. i.e. at the root, leading edge or trailing edge. Thus, there is an interaction of these effects collectively referred to as the "thermo-mechanical fatigue". In general, the time courses of mechanical and thermal stresses can be independent of one another. This type of load typically occurs under different transition modes of thermally stressed machine parts. With regard to turbine engines, these are especially the engine start and stop modes, respectively transition state when operating mode changes, there are two model types of cycles, in-phase (IP) and out-phase (OP).

Variable thermal and mechanical loads may have a much greater damaging effect than mere isothermal stress if improperly combined. Damage, type of damage, eventually its individual modes, which are formed during loading and depend mainly on the type of material and its properties, amplitude and strain rate, temperature and, last but not least, the mutual phase between the course of thermal and mechanical loading. A characteristic feature of TMF is often significant amplitude of plastic deformation, which evokes the problem of low cycle fatigue with life prediction in several thousand cycles.

The following chapters describe the damage parameters and their effect on the prediction of fatigue from mechanical stresses. The methodology of the calculation of thermo-mechanical fatigue was developed within the project to increase the utility properties of the GE H75 turboprop engine with modifications necessary for use in a training aircraft capable of advanced aerobatics. The object of the project is the development of methodology and software for the prediction of thermo-mechanical fatigue and calculation of damage growth on critical parts. By determining the life of individual critical parts, the knowledge of the whole engine is deepened, individual parts can be structurally optimized and maintenance needs can be appropriately determined. Validation of this method will be used to evaluate the possibility of transitioning from a scheduled maintenance system to an actual condition maintenance system.

Damage parameter

The damage parameter is a parameter quantifying the key fatigue damage control variables. The original damage parameter proposed in the publications, sees for example [3], [5] or [6], is the classic SWT parameter. In addition to this, other parameters are also used for the target area of complex thermo-mechanical stress. The parameters differ in the consideration of other variables and their different interdependencies. Thus, in addition to the SWT parameter, the Ostergren and Zamrik damage parameters are newly considered in the prediction.

Parameter SWT

The SWT parameter is a classical damage parameter used mainly in the low-cycle isothermal fatigue. The parameter is constructed as a combination of stress amplitude, total strain and elastic modulus according to (1).

$$P_{SWT} = \sqrt{\sigma_a \cdot \varepsilon_a \cdot E} \tag{1}$$

Regression analysis over the experimental data is able to be performed in various ways. A common way is to gain the P_{SWT} parameter by substituting regression relationships for each parameter element. Substituting the equations σ_a and ε_a showed in (2, 3) into (1) gives (4).

$$\sigma_a = \sigma_f'(2N)^b \tag{2}$$

$$\varepsilon_a = \frac{\sigma_a}{E} + \varepsilon_f'(2N)^c \tag{3}$$

$$P_{SWT} = \sqrt{\sigma_a \cdot \varepsilon_a \cdot E} = \sqrt{\sigma_f'(2N)^b \cdot E \cdot \left(\frac{\sigma_f'(2N)^b}{E} + \varepsilon_f'(2N)^c\right)} \tag{4}$$

Another method of using regression fitting of experimental data may be polynomial fitting according to formula (5). It is also used in the Ostergen and Zamrik damage models as shown below.

$$P_{SWT} = A(N)^B \tag{5}$$

Damage parameter based on Ostergren model

Another model of damage factor is model proposed by Ostergren in [7]. The model was used in [8] and [9]. The main parameters of the model are the tensile part of inelastic hysteresis energy and cycling frequency. According to [8], the model was used by combination of plastic deformation amplitude and the maximum stress as shown in Eq. (6).

$$P_{OST} = \Delta\varepsilon^{pl} \cdot \sigma_{max} \tag{6}$$

Regression analysis over the experimental data is usually performed to obtain regression coefficients as shown in Eq. (7).

$$A = N_f^B \cdot P_{OST} = N_f^B \cdot \Delta\varepsilon^{pl} \cdot \sigma_{max} \tag{7}$$

Damage parameter based on Zamrik model

Zamrik proposed in his work [9] a modification of the Ostergren model by using the maximum tensile energy amplitude instead of the non-elastic energy amplitude, ie. by including the highest tensile stress and temperature, and by including or eliminating the effect of pressure strokes. The model can be expressed according to [9] in the form (8).

$$N_f = A(\Delta W)^B \cdot \left(h(t)\right)^C \cdot \left(r(T)\right)^D \tag{8}$$

In Eq. (8) ΔW represents an energy function, $h(t)$ represents a creep or environmental effect due to hold-time and $r(T)$ represents an elevated temperature effect.
Parameter of deformation energy amplitude ΔW can be expressed as the strain energy ratio during the fatigue test W_f and the deformation energy from the static test W_s. Parameter of amplitude of strain energy is then able to simplified to form (9) where the parameter is expressed as the dependence maximum stress, strain amplitude of tensile strain of the hysteresis loop in the middle of life, the strength limits and the extension of the static tensile test. The function of time-dependent phenomena is designed to take into account the effect of hold-time, as shown in Eq. (10).

$$\Delta W = \frac{W_f}{W_s} \approx \frac{\sigma_{max} \cdot \varepsilon_{ten}}{R_m \cdot A_f} \tag{9}$$

$$h(t) = 1 + \frac{t_h}{t_c} \tag{10}$$

Temperature-dependent parameters are included through activation energy to change the failure mechanisms. Temperature-dependent phenomena $r(T)$ are expressed by the Arhenius equation in the following Eq. (11)

$$r(T) = exp\left(\frac{-Q}{R(T_{max} - T_0)}\right) \tag{11}$$

The regression analysis over the experimental data is designed according to [9] to obtain the regression coefficients A and B in the form (12).

$$N = A(P_{ZAM})^B = A\left(\frac{\sigma_{max} \cdot \varepsilon_{ten}}{R_m \cdot A_f}\right)^B \tag{12}$$

Partial damage

Total damage consists of contributions from partial damage of various damage sources (mechanisms). In addition to the most significant type of mechanical fatigue damage caused by the cyclic change in plastic deformation, in particular in the case of thermally stressed parts, the further significant damage is the time-dependent effect of the temperature field. The stress under the influence of temperature induces in the components another damaging mechanism, namely creep. However, the process assumes that mechanical fatigue is the main quantitative damaging effect.

The Prandtl-Ishlinskii hysteresis model is predominantly used to determine damage from mechanical fatigue. Using the hysteresis model in case of damage, the impact from the previous load is introduced into the damage calculation at the current point of solution. This procedure was presented in [2].

Implementation of the prediction scheme into the TMF_PRED program

The following chapter describes the prediction software for the prediction of thermomechanical fatigue based on available knowledge of the issue. After running the program, the program runs sequentially along the axis illustrated in the following Fig. 1. The calculation can be divided into the preparation phase and the damage calculation itself.

In the first step, fatigue curves and cyclic deformation curves for individual temperatures are discretized. The following is the identification of the closed hysteresis loop to the solved point and then the determination of the increase in damage size of the solved point.

For the subsequent calculation part, the input is a mechanical temperature history of loading. Input stress is a set of incremental time values of stress and temperature distribution. The input can be either results from the FEM program, or input directly using MATLAB functions.

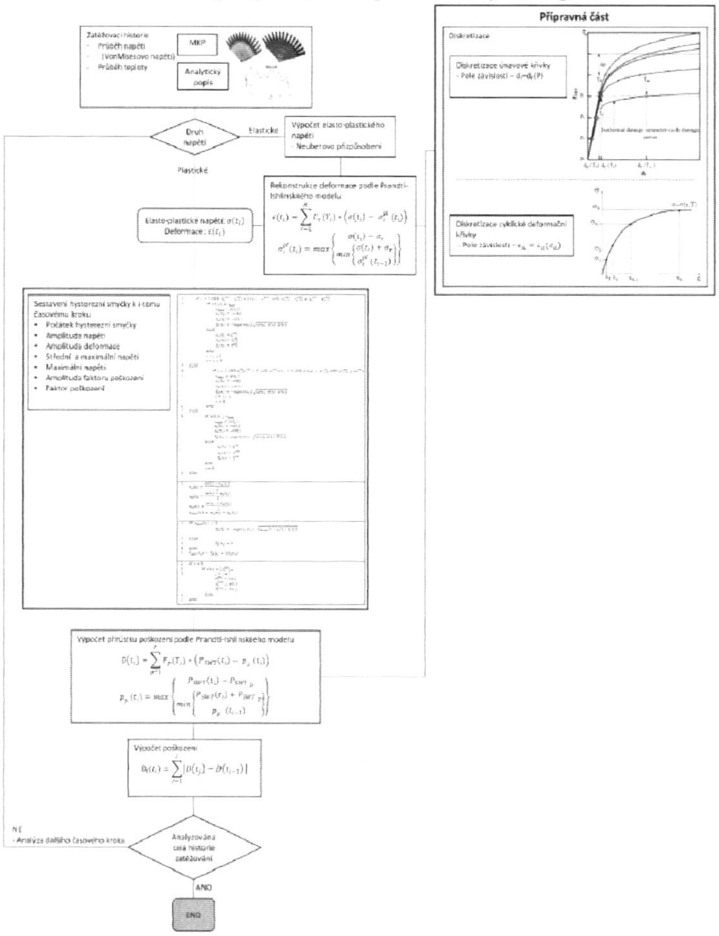

Fig. 1: TMF_PRED software scheme

Validation of a method of prediction thermo-mechanical fatigue and increase of damage

In the article [13], the results of damage calculation for the proposed basic stress spectra are published. It is a solution of combination of thermomechanical loading with variable temperature and stress. In the paper, it is possible to clearly distinguish values of individual peaks of stress and temperature of the load history. The 4 load histories listed in Table 5 are selected. The fatigue data of this material are taken from the article in graphical form of dependence $\sigma_a = \sigma_a(N)$. Values are used to correct the equivalent stress amplitude $\sigma_c = 360[MPa]$ for R=-1 and $\sigma_c = 305[MPa]$ for R=0.

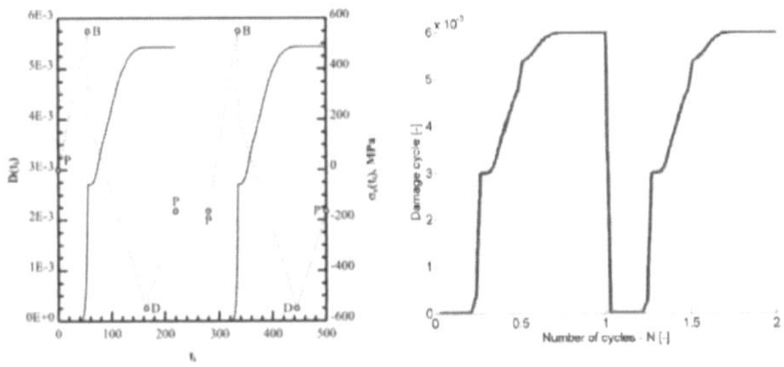

Fig. 2: Comparison of published results with the total damage for in-phase load (IP)

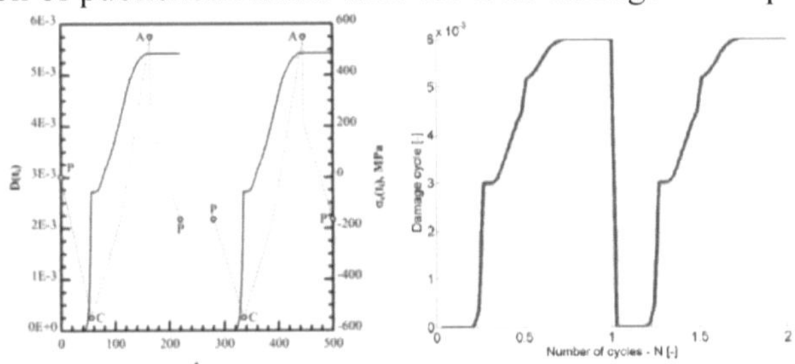

Fig. 3: Comparison of published results with total damage for out-phase load (OP)

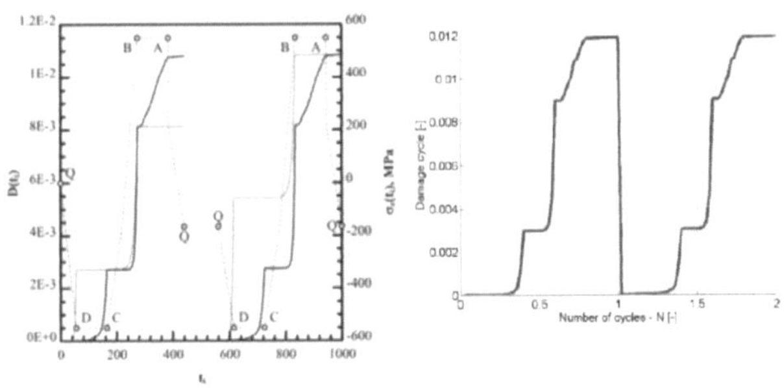

Fig. 4: Comparison of published results with total damage for thermo-mechanical cycling, ver. 1 ("clockwise")

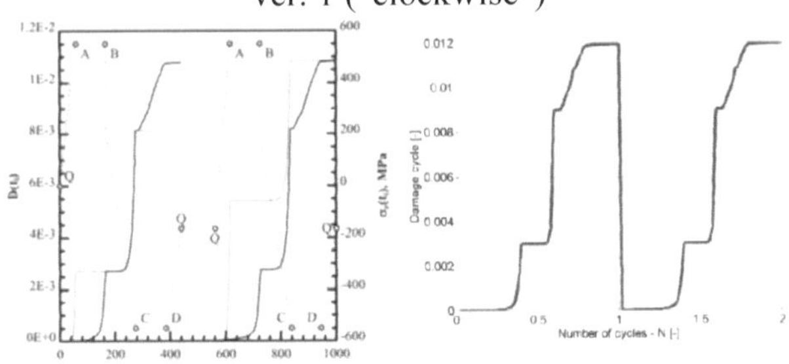

Fig. 5: P Comparison of published results with total damage for thermo-mechanical cycling, ver. 2 ("anti-clockwise")

Table 1 compares the published results ("PUB") and the results of the actual implementation ("PRED").

Table 1: Comparison of published results

Load history	Published result		Calculated result		Deviation $1-\frac{PUB}{PRED}$	
	1. cycle	2. cycle	1. cycle	2. cycle	1. cycle	2. cycle
Load in-phase	5,427 E-03	5,431 E-03	5,994 E-03	5,998 E-03	9,5%	9,4%
Load out-phase	5,427 E-03	5,431 E-03	5,994 E-03	5,998 E-03	9,5%	9,4%
Cycling ver. 1	1,078 E-02	1,085 E-02	1,192 E-02	1,198 E-02	9,5%	9,5%
Cycling ver. 2	1,081 E-02	1,085 E-02	1,194 E-02	1,198 E-02	9,4%	9,5%

Aplication of experimental data

Measured data were compiled parameters of damage, according to the Eq. in Tab. 2. The SWT damage parameter was used based on the Eq. 4, the Ostergren damage parameter based on the Eq. 6, and the Zamrik model based on the Eq. 12.

In the case of Zamrik's model of damage parameter, the damage parameter in a simplified form Eq. 12 is used, since the loading did not contain hold-time according to experimental reports and the failure mechanism was not changed based on the results. The resulting summary of damage factors for each model is given in Table 4 below.

Table 2: Damage parameter summary

Name	**Model**	
SWT	$P_{SWT} = \sqrt{\sigma_a \cdot \varepsilon_a \cdot E}$	(4)
Ostergren	$P_{OST} = \Delta\varepsilon^{pl} \cdot \sigma_{max}$	(6)
Zamrik	$P_{ZAM} = \frac{\sigma_{max} \cdot \varepsilon_{ten}}{R_m \cdot A_f}$	(12)

The regression coefficients of the respective models were obtained by regression. Pair of regression coefficients for respective dependence were gradually determined by linear regression. Individual relationships and corresponding regression coefficients are summarized in Tab. 3. All fatigue models listed in Tab. 5 are expressed uniformly depending on the number of half cycles. The main reason is a uniform implementation of the damage calculation and a uniform approach to discretization of fatigue curves.

When processing regression dependencies in which plastic deformation occurs, the measurements usually do not take into account those measurements in which the magnitude of plastic deformation was less than a certain minimum value. This practice is related to the low accuracy of small plastic deformation determination. Usually the minimum value is used $\varepsilon_{0\,min} = 10^{-4}$.

Table 3: Fatigue models and regression coefficients

Model	Regression coefficients		
$\sigma_a = K'\left(\varepsilon_a^{pl}\right)^{n'}$	K'	n'	
$\varepsilon_a^{pl} = \varepsilon_f'\left(2N_f\right)^{c}$	ε_f'	c	(3)
$\sigma_a = \sigma_f'\left(2N_f\right)^{b}$	σ_f'	b	(2)
$P_{SWT\ IF\ const} = \sqrt{\sigma_f'(2N)^b \cdot E \cdot \left(\frac{\sigma_f'(2N)^b}{E} + \varepsilon_f'(2N)^c\right)}$	$\sigma_f', b, \varepsilon_f', c$		(4)
$P_{SWT\ POWER} = A_{SWT}\left(2N_f\right)^{B_{SWT}}$	A_{SWT}	B_{SWT}	(5)
$P_{OST} = A_{OST}\left(2N_f\right)^{B_{OST}}$	A_{OST}	B_{OST}	(6)
$P_{ZAM} = A_{ZAM}\left(2N_f\right)^{B_{ZAM}}$	A_{ZAM}	B_{ZAM}	(12)

Interleaving of material models according to Tab. 3 experimental data is shown in the following figures. Fig. 6 to Fig. 12 shows an example of model fitting with experimental data for T = 500 °C. The figures show the fit of the cyclic strain curve, Eq. 19, Manson-Coffin stress curve, Eq. 3, and the strain fatigue curve of the Basquin curve, see Eq. 2.

The course of the SWT damage factor, which is composed of the regression coefficients obtained from the deformation and fatigue curve according to Eq. 4, is shown in Fig. 9. The SWT parameter with power description according to Eq. 5 is shown in Fig. 10. The Ostergren model of the damage factor according to Eq. 6 is shown in Fig. 11, respectively. The Zamrik model of Eq. 12 is shown in Fig. 12.

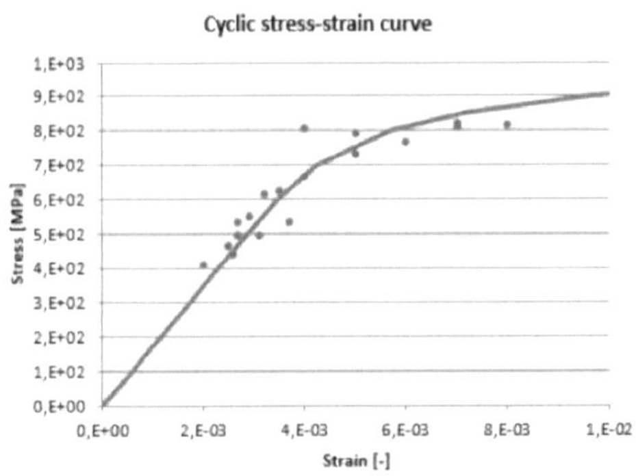

Fig. 6: Stress-strain curve, T=500°C

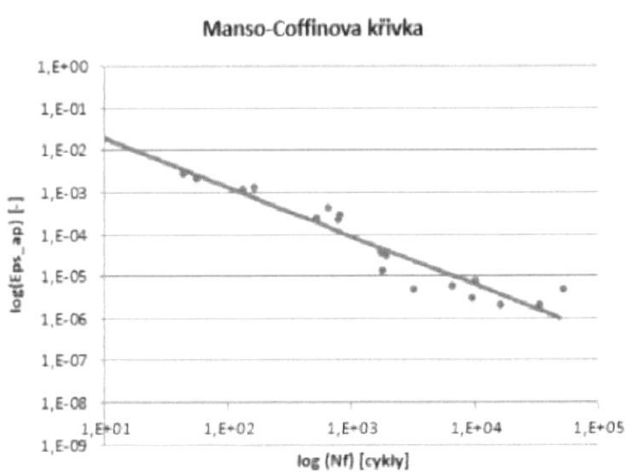

Fig. 7: Deformation fatigue curve, T=500°C

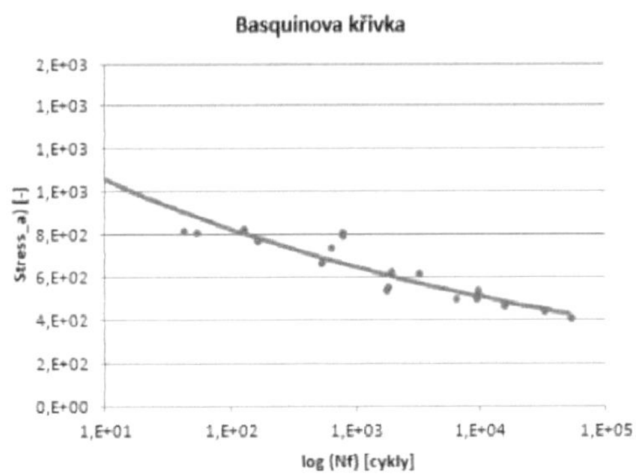

Fig. 8: Stress fatigue curve, T=500°C

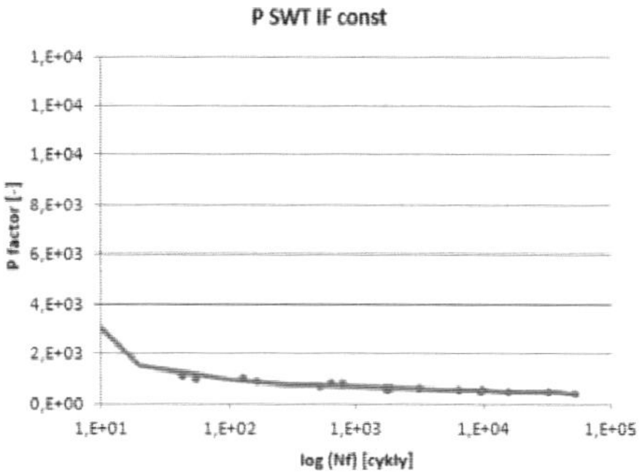

Fig. 9: SWT damage factor dependence on the number of cycles to failure, P factor composed of fatigue coefficients, T=500°C

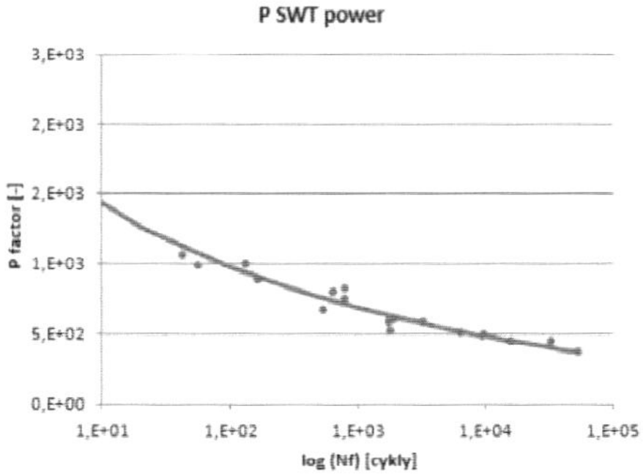

Fig. 10: SWT damage factor dependence on the number of cycles to failure, P factor composed of power model, T=500°C

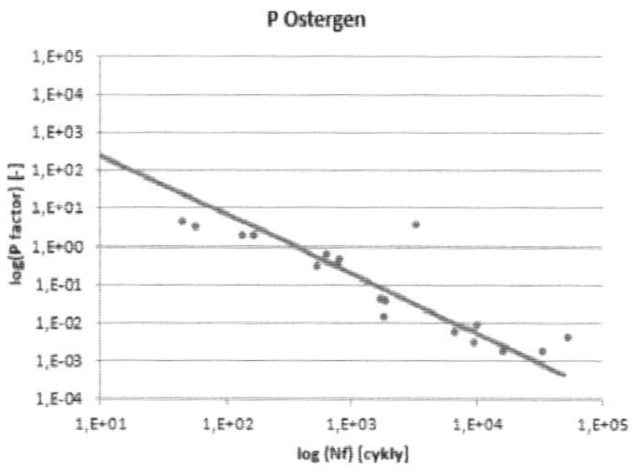

Fig.11: Ostergren damage factor dependence on the number of cycles to failure, T=500°C

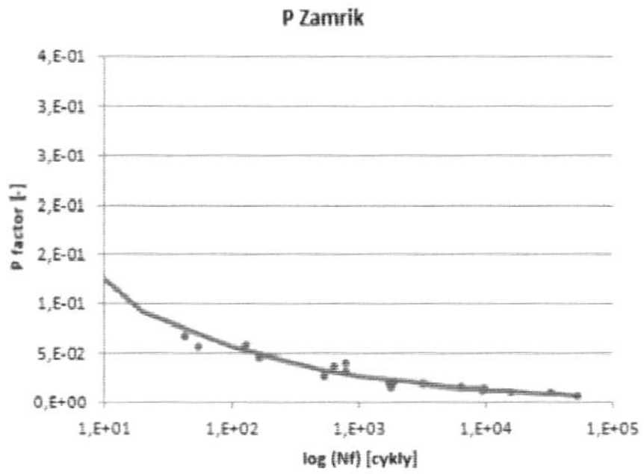

Fig.12: Zamrik damage factor dependence on the number of cycles to failure, T=500°C

In the figures, Fig. 13 to Fig. 19 is a comparison of cyclic strain curve, stress and strain fatigue curve, SWT, Ostergren and Zamrik damage factors for each processed temperature.

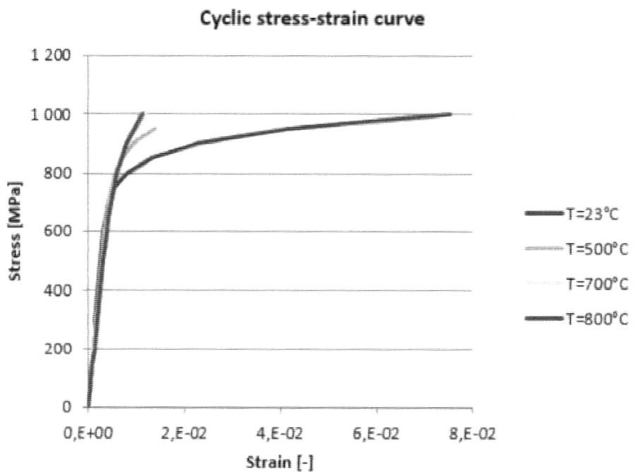

Fig. 13: Cyclic deformation curves for individual temperatures

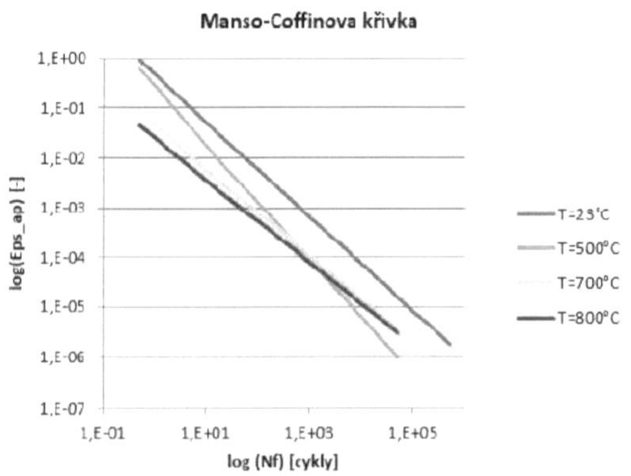

Fig.14: Deformation fatigue curves temperatures

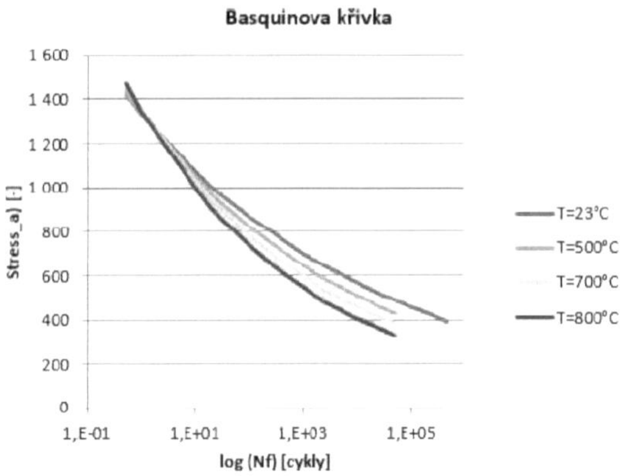

Fig.15: Stress fatigue curves according to according to temperatures

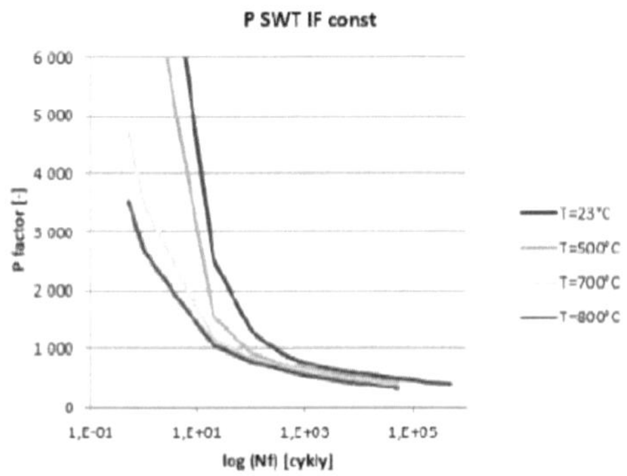

Fig.16: SWT damage factor depending on the number of failure cycles for individual temperatures, P factor based on fatigue coefficients

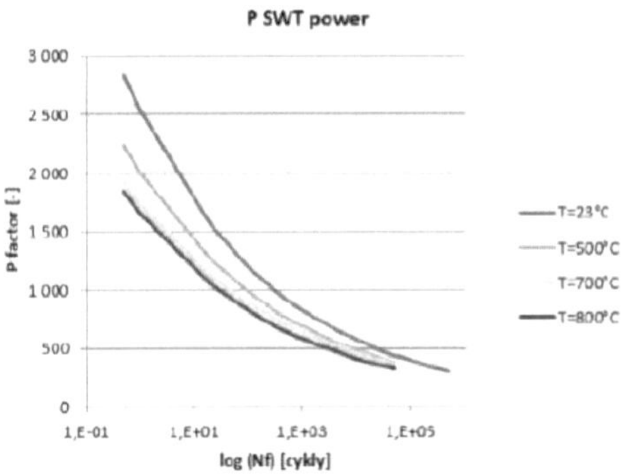

Fig.17: SWT damage factor depending on the number of failure cycles for individual temperatures, power model of P factor

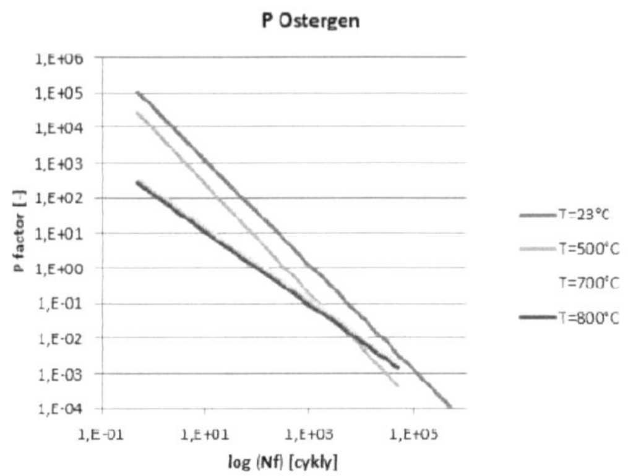

Fig.18: Ostergren damage factor depending on on the number of failure cycles for individual temperatures

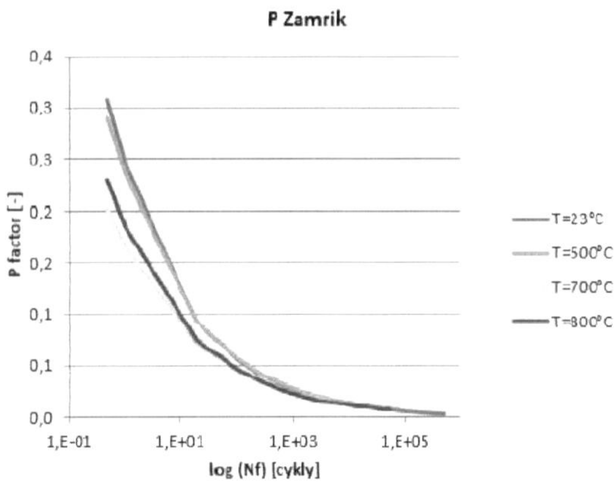

Fig.19: Zamrik damage factor depending the number of failure cycles for individual temperatures

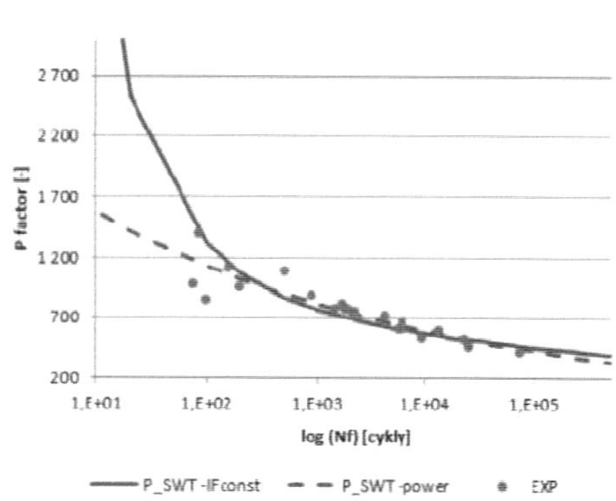

Fig.20: Comparison of SWT damage factor determined by fatigue coefficient and power model, temperature T = 23°C

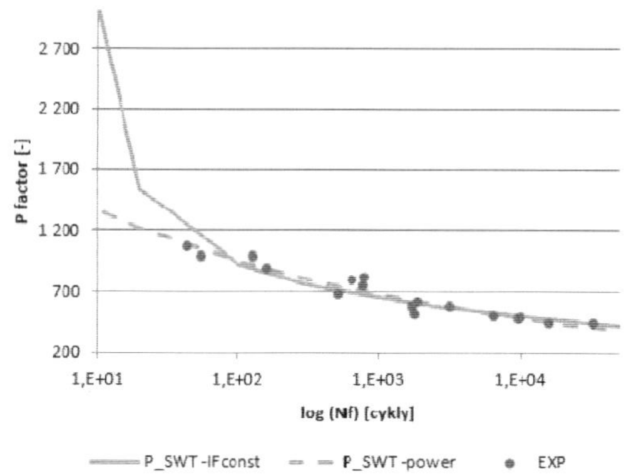

Fig.21: Comparison of SWT damage factor determined by fatigue coefficient and power model, temperature T = 500°C

A similar comparison for all temperatures is shown in Fig. 22. In this case, the biggest difference between the two ways of describing the SWT parameter is in the case of lower temperatures, namely T=23°C and 500°C. In the case of higher temperatures T=700°C and 800°C the difference is not so significant.

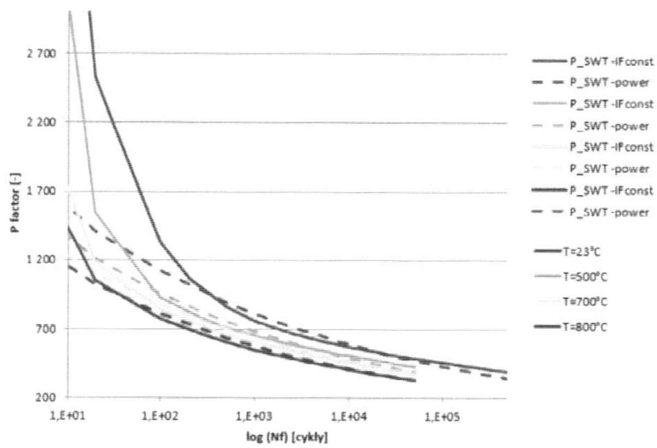

Fig. 22: Comparison of SWT damage factor determined by fatigue coefficients and power relations for individual temperatures

Conclusions

Damage parameters used for prediction of thermomechanical fatigue were presented in this paper. These methodologies are based on published sources and the object of the work was to compare input parameters and evaluate different results. TMF_PRED software for thermomechanical fatigue prediction was used to evaluate the output values. The whole prediction metodology is validated on the basis of available experimental data and based on published reports by prof. Nagodeho.

The damage parameter was compared in the SWT methodology in both the fatigue coefficient model (SWT IF const.) and the power relation (SWT power). Furthermore, the Ostergren methodology was compared and the last methodology was the Zamrik methodology. These methodologies were applied to experimental data where regression coefficients of the respective models were obtained to obtain uniform dependence on the number of half cycles.

Comparative CDK curves, stress and strain fatigue curves for individual temperatures were plotted for each methods. It can be seen from the graphs that the SWT method shows a significant difference in the damage factor between models with fatigue coefficients and power eq. form. For the model with fatigue coefficients, the damage factor increases significantly with decreasing number of half cycles at low temperatures, at higher temperatures the growth is not so significant and the difference between the parameter of fatigue coefficient damage and power relation decreases with temperature. Furthermore, for all methods it can be stated that above 1e+3 half cycles there is no significant difference in the damage parameter with temperature, this fact corresponds to the fact that below this area, which is not the area of low-cycle fatigue and fatigue prediction is inappropriate.

It is also apparent from the comparison that the damage predicted by the SWT power factor described by the power eq. form (SWT power) is very close to the results of the Zamrik damage factor. In the case of the Ostergren factor, significantly less damage is predicted than other damage factors.

Based on the data presented in [6], where the fatigue life determined by the OP TMF is slightly longer than the IP TMF test, the difference between the prediction model and the test results is greater in higher cycles. This result shows that the Ostergren model is not suitable for predicting TMF life using isothermal fatigue test data. Zamrik modified the Ostergren model and developed a new model of life prediction for OP TMF.

Life predictions according to the Ostergren model are not effective in high cycle fatigue areas, life predictions according to the Zamrik model are more effective in these areas than according to Ostergren. It can also be seen that the Zamrik model is not suitable for predicting the life of an OP.

The main reason for the difference between the two prediction models is to incorporate the amplitude of the elastic stress as part of the damage parameter in addition to the inelastic strain deformation. Unlike Zamrik's model, the Ostergren model does not include the elastic stress amplitude as part of the damage parameter. However, in the case of materials with high strength and limited ductility, such as nickel-based superalloys, the elastic stress amplitude is significant, and therefore the elastic stress amplitude should be considered for TMF life predictions. In addition, the amplitudes of the inelastic deformations of LCF and TMF are small, and therefore small measurement errors and process utilization can cause large errors in results.

Discussion

The paper demonstrated different course and dependence of damage factor results as a quantifier of key fatigue damage control variables in thermomechanical fatigue. The above mentioned relations show the difference of input parameters for individual methodologies and under variable conditions. This raises the question of the suitability of specific methodologies for specific conditions affecting fatigue behaviour and lifetime prediction. The influence of temperature, LCF and HCF assumption, stress loading and material characteristics should be considered as control parameters of these specific conditions.

For example, the graph shows the suitability of the Zamrik methodology under varying temperature conditions or with a higher number of cycles. It is also appropriate to consider the quality of conservative and non-conservative estimates. Based on the previously studied VZLU data, these comparisons suggest that the closest prediction of LCF was achieved using the SWT parameter constructed using the power relationship in the experimental isothermal campaigns tested. Using the Ostergren parameter, a non-conservative estimate was predicted on average. Using other parameters a conservative prediction is achieved. Moreover, in the case of the Ostergren parameter of damage, the cycle numbers are predicted with the greatest variance. The variations for the other damage parameters were comparable.

Acknowledgement

This work was supported by Výzkumný a zkušební letecký ústav, a.s. in the project dealing with TMF prediction, project IMotor.

References

[1] J. Had, Implementation of interacting creep and fatigue under thermo-mechanical loading. *Engineering mechanics 2017*. Svratka, Czech Republic, 23rd International Conference.

[2] J. Had, Modelling of thermo-mechanical fatigue on engine generator turbine. *Engineering mechanics 2016*. Svratka, Czech Republic, 22nd International Conference.

[3] M. Nagode, M. Fajdig, Coupled elastoplasticity and viscoplasticity under thermo-mechanical loading. Fatigue & Fracture of Engineering Materials & Structures. (2006) 510-519.

[4] M. Nagode, D. Šeruga, M. Hack, E. Hansenne, Damage Operator-Based Lifetime Calculation Under Thermomechanical Fatigue and Creep for Application on Uginox F12T EN 1.4512 Exhaust Downpipes Strainl. (2012) 198-207.

[5] M. Nagode, Continuous damage parameter calculation under thermo-mechanical random loading. MethodsX. (2014) 149-160.

[6] M. Nagode, M. Hack, M. Fajdiga, Low cycle thermo-mechanical fatigue: damage operator approach. Fatigue & Fracture of Engineering, Materials & Structures. (2010) 149-160.

[7] W. J. Ostergren, A Damage Function and Associated Failure Equations for Predicting Hold Time and Frequency Effects in Elevated Temperature, Low Cycle Fatigue. Journal of Testing and Evaluation, (1976) 327–339.

[8] D. Lee, J. Lee, Y. Kim, J. M. Koo, C. S. Seok, Y. J. Kim, Thermo-mechanical fatigue life prediction of Ni-based superalloy IN738LC. International Journal of Precision Engineering and Manufacturing, 18(4), (2017) 561-566.

[9] S. Y. Zamrik, M. L. Renauld, Thermo-Mechanical Out-of-Phase Fatigue Life of Overlay Coated IN-738LC Gas Turbine Material. ASTM Special Technical Publication, 1371, (2000) 119–137.

[10] D. Holländer, D. Kulawinski, M. Thiele, C. Damm, S. Henkel, H. Biermann, U. Gampe, Investigation of isothermal and thermo-mechanical fatigue behaviour of the nickel-base superalloy IN738LC using standardized and advanced test methods. Materials Science and Engineering: A, 670, (2016) 314-324.

[11] D. Lee, I. Shin, Y. Kim, J. M. Koo, C. S. Seok, A study on thermo mechanical fatigue life prediction of Ni-base superalloy. International Journal of Fatigue, (2014) 62, 62-66.

[12] E. L. Robinson, Effect of temperature variation on the long-time rupture strength of steels. Trans. Am. Soc. Metals, (1952) 74, 777-781.

[13] M. Nagode, M. Hack, M. Fajdiga, High cycle thermo-mechanical fatigue: Damage operator approach. Fatigue and Fracture of Engineering Materials and Structure. 32.6 (2009) 505-514

EAN 2020
58th conference on experimental stress analysis

Experimental Verification of Stress State in Selected Steel Load-Bearing Elements of Silo

PÁSTOR M.[1,a], TENGLER, M.[2,b], FRANKOVSKÝ P.[1,c], HAGARA M.[1,d], LENGVARSKÝ P.[1,e]

[1]Deparment of Applied Mechanics and Mechanical Engineering, Faculty of Mechanical Engineering, Technical University of Košice, 042 00 Košice, Slovakia

[2]Ing. Miloslav Tengler, Aut. Ing. - diagnostika, 811 03 Bratislava, Slovakia

[a]miroslav.pastor@tuke.sk, [b]tengler@tengler.sk, [c]peter.frankovsky@tuke.sk, [d]martin.hagara@tuke.sk, [e]pavol.lengvarsky@tuke.sk

Keywords: Silo, Measurement, Tensometry, Safe Operation

Abstract. When evaluating the service life of already operated silos, information about the history of the current load is needed. Although we have sufficient information about the history of the operating load, with unforeseen stresses, its impact on future operational safety needs to be evaluated. Such a case is the assessment of silo S1 in the immediate vicinity of which the adjacent silo S2 collapsed. Although the structure of silo S1 did not show any signs of damage at first sight, the operator decided to assess its future operational safety and reliability through the analysis of experimental measurements. This paper describes strain gauge measurements in selected areas. The aim was to compare measured values with results of numerical modelling and to draw conclusions about future operational safety of the silo based on such comparison.

Introduction

Among other vessels, steel tanks or silos are used to store liquids, gases or bulk materials. Their shells are formed by thin sheets reinforced with vertical steel reinforcing elements [1,2]. The total breakdown of steel structures is preceded by the occurrence of local plastic deformation, which causes the loss of stability of the entire object. Published analyses of possible causes of breakdowns document that such deformation often occurs not only due to incorrect operation but also due to incorrect construction [3,4,5]. In rare cases, local plastic deformation allows further operation without the need for major structural repair [6]. However, there are many cases where, due to overloading, the occurring local deformation subsequently caused a global loss of stability of the entire structure. An example hereof is the collapse of the supporting roof structure of a hot water storage tank due to underpressure, Fig. 1 [7]. In this case, a structural modification was necessary to ensure its future operational safety. The last and most critical situation occurs when a load-bearing structure breaks down completely. Such situation often occurs with silos, Fig. 2. Silos are used in agriculture to store grain or fermented feed known as silage. The load is caused by the weight of the stored material. When designing silos, several design situations are considered [8,9]. In addition to loads in the state of complete filling, the following limit states are also evaluated:

- maximum normal stress on the vertical wall of the silo,
- maximum friction thrust (traction) on the vertical wall,
- maximum vertical pressure on the bottom of the silo,
- maximum load of the discharge hopper.

Fig. 1: a) View of storage tanks, b) collapsed roof of the storage tank

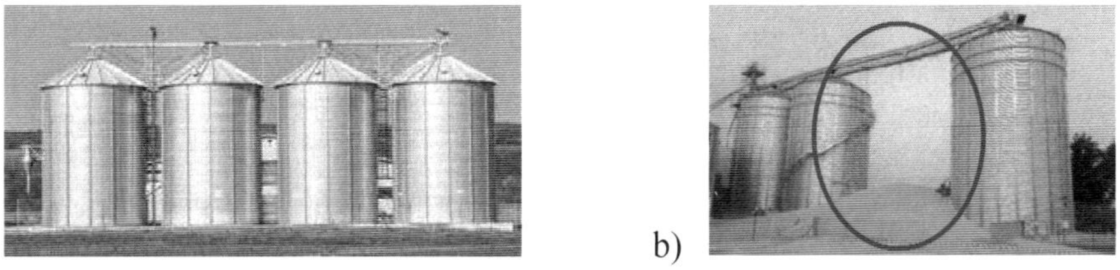

Fig. 2: Steel silos a) without failure, b) collapse of a silo for the storage of approximately 120 t of wheat

The paper presents the procedure of experimental measurement on steel supporting structure of the silo S1 (Fig. 3a) located near the collapsed silo S2 (Fig. 3b). As presented in Fig. 3b, the collapse of silo S2 also caused the breakdown of the adjacent silo S3. Although the structure of silo S1 did not show any signs of damage at first sight, the operator decided to assess safety and reliability of its future operation through the outputs of the analysis of experimental measurements. The circumferential walls of the silo are made of steel profiled sheets, the corrugated profiles of which run in the horizontal direction. The structure is reinforced around the perimeter with vertical stiffening profiles. The stiffeners are evenly distributed around the circumference of the silo along the entire height of the cylindrical part.

Basic dimensions:

- diameter: 32.08 m,
- total high: 32.00 m (cylindrical part 22.88 m, conical roof 9.12 m),
- volume 20 924 m^3.

The measurement was done under operational load in order to validate the results of numerical modelling performed by the facility operator. The aim of the first stage was to evaluate the impact of temperature changes on the steel structure of the empty silo during one day. In the second stage, the stress inside the steel shell of the silo was monitored as the silo was gradually filled until its subsequent complete emptying.

Fig. 3: View of the analysed silo S1, a) before breakdown, b) after breakdown

Strain measurement and determination of the size of stress components

The resistance strain gauge method was chosen for the experimental stress analysis, while in application areas of strain gauges' solo devices 1-LY11-6/120, k = 2.05 were used, also regarding technical possibilities of their application. The strain gauges were applied in places with assumed maximum exploitation of the cross-section near the lower part in the "second field" at a height of approx. 1.7 m using a concrete support. The "first field" above the concrete support was excluded because of possible local influences due to placement, i.e. due to uneven expansion causing possible uneven distribution of forces and measurement distortion or uneven stress distribution in monitored elements. The areas of strain gauge application were located on the south and the north side of the steel structure (opposite each other). Fig. 4a shows a location diagram of strain gauges. For experimental determination of strain components by means of the strain gauge method in locations of strain gauges, the gauges were connected to half-bridges, while the active branch was formed by a strain gauge applied to the silo structure. The compensation gauge connected in the second branch of the bridge was applied to a steel plate fixed at the measuring point. Fig. 4b shows the orientation of active grids of strain gauges.

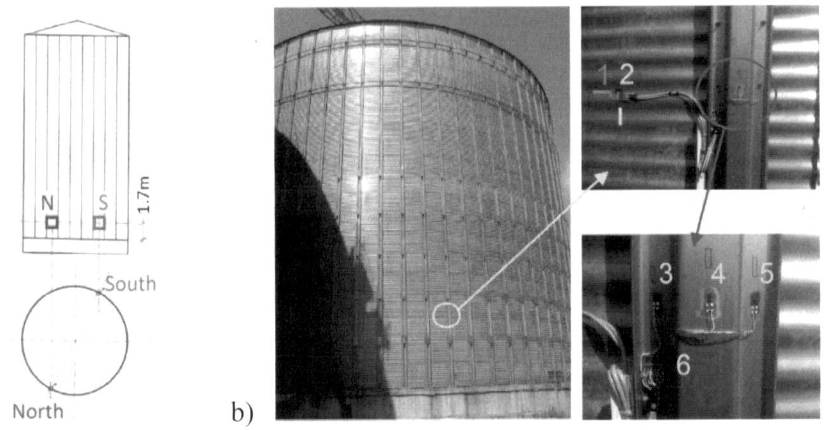

Fig. 4: a) Location diagram of strain gauges, b) orientation of grids

Fig. 4b shows a detailed view of the applied strain gauges on a corrugated sheet (areas 1 and 2) and on a stand (areas 3 to 5). In addition to the mentioned strain gauges, a temperature sensor type FNA611L0100 was used in the vicinity of strain gauges.

The application of strain gauges to an already existing empty structure allows us to measure strain increments under operating load after connecting the strain gauges to the measuring apparatus. Calibration of the used measuring chain was performed before and after the strain gauge measurement. Strain gauge measurements provided us with strain values at locations of applied strain gauges during filling or discharging of the silo according to operator's instructions. The proposed methodology assumed that the regulation of operating parameters is sufficiently accurate, and data provided by the operator are relevant to actual values. From strain values measured by strain gauges, stress increments since the change of load caused by internal pressure as well as stress increments due to temperature changes were determined using constitutive equations.

As already mentioned, the measurement was divided into two stages. The focus of the first stage was to determine the effect of temperature changes on the steel structure of the empty silo within one summer day with intense sunlight. The measurement started at 3:25 o'clock. when all strain gages were balanced. The recording of values during the day was at about 3 hours' intervals.

An overview of the values of recalculated stresses is given in Graph 1 and 2. The given values imply that during that day both tensile and compressive stresses were recorded. The maximum values were measured on the south side at noon, as expected.

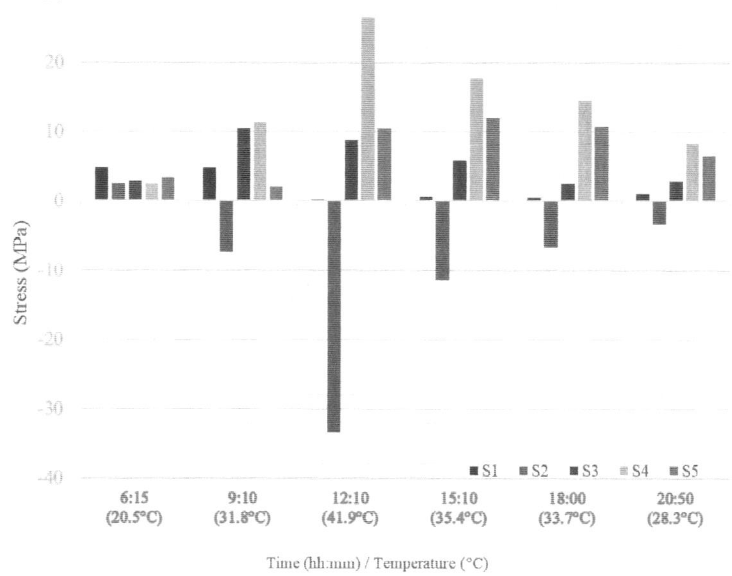

Graph 1: Stress values depending on the influence of ambient temperature within one day – south side

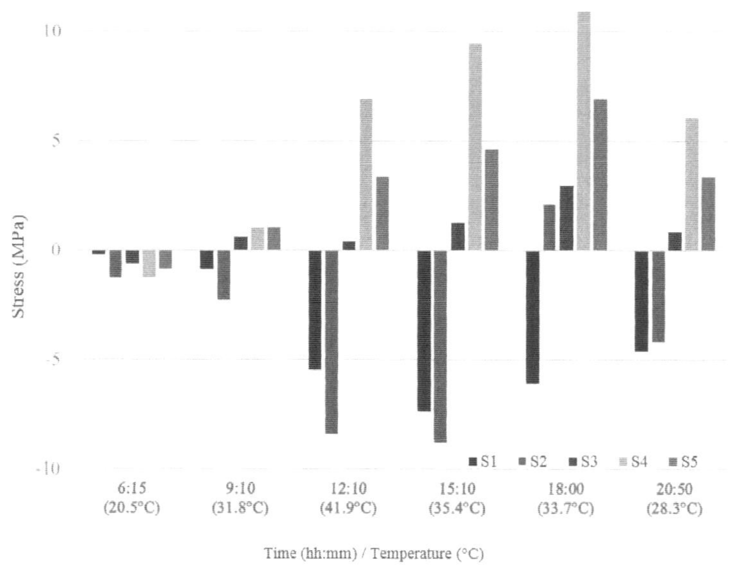

Graph 2: Stress values depending on the influence of ambient temperature during one day – north side

The aim of the second stage was to monitor and then assess stress inside the steel shell of the silo as the silo was gradually filled until its subsequent complete discharge during normal operation. It was a long-term measurement. The first measurement was performed on November 12 and the last on August 31 of the following year. The values of strains were not registered permanently but only in 7 characteristic stages. An overview of the values of recalculated stresses is given in Tables 3 and Table 4.

Table 3: Stress values during filling and discharging of the silo – south side

Date / time	Mass [t]	S-I [MPa]	S-II [MPa]	S-III [MPa]	S-IV [MPa]	S-V [MPa]	S-VI [°C]
		Measurement area					
12.11. / 10:35	7099	52.08	-58.17	7.98	31.50	-0.42	30.18
14.11. / 14:15	10800	73.08	-47.04	6.51	38.01	3.36	16.32
27.11. / 14:05	12164	57.33	-54.81	-6.51	26.04	-15.54	6.84
29.12. / 12:35	15574	119.91	-78.75	-25.41	19.74	-38.01	2.77
21.07. / 15:35	12050	73.92	-73.92	-9.87	45.15	-16.17	36.40
05.08. / 11:55	9055	67.83	-45.99	6.93	49.77	-13.44	38.40
31.08. / 08:15	0	18.7	-2.1	-8.8	-30.2	-29.4	23.84

Table 4: Stress values during filling and discharging of the silo – north side

Date / time	Mass [t]	N-I [MPa]	N-II [MPa]	N-III [MPa]	N-IV [MPa]	N-V [MPa]	N-VI [°C]
		Measurement area					
12.11. / 10:35	7099	58.17	-57.75	2.10	10.71	-10.08	10.80
14.11. / 14:15	10800	74.97	-72.87	-0.63	16.80	-12.81	11.30
27.11. / 14:05	12164	99.96	-98.49	-6.72	17.43	-24.99	4.62
29.12. / 12:35	15574	118.65	-129.99	-18.69	18.69	-39.69	1.72
21.07. / 15:35	12050	125.72	-132.93	-36.33	-30.24	-39.06	31.67
05.08. / 11:55	9055	105.48	-101.22	-18.48	-31.08	-21.21	28.66
31.08. / 08:15	0	27.94	-30.3	-9.7	-17.7	-15.1	14.10

Strain gauges applied in the meridional direction from both sides of the structure enabled the assessment of a possible change in shape (geometry) due to possible eccentricity of the stored material. The correctness of the proposed methodology of experimental measurements was confirmed by different values of stresses corresponding to excessive deformations of the whole structure identified even by the naked eye. The values of compressive normal stresses in vertical direction did not exceed 40 MPa (Table 4). In addition to compressive stresses, tensile stresses due to bending of the structure, which was visible to the naked eye at the time of measurement, were also recorded in the same areas. The maximum values of these tensile stresses in vertical direction did not exceed 50 MPa (Table 3). After complete discharge of the silos, non-zero stress values were recorded in the measurement areas, which is documented by the fact that permanent structural deformations occurred during operating load. The task of the authors was not to draw conclusions on future operational safety of the silo, but to provide the operator with information about stress inside the structure for further analysis performed by the operator. It should be noted that the measuring points as well as stress directions were determined according to operator's requirements.

Conclusions

When assessing service life of steel tanks (silos), information, among other things, about the history of the current load is necessary. Based on verified analyses, it can be stated that the greatest influence on damage accumulation have stress amplitudes which exceed the fatigue limit of the supporting element in the examined area. In this case, in order to evaluate operational safety and reliability of silo S1, an experimental stress investigation was performed in selected areas on both the southern and the northern side of the silo during its filling or discharging. It was a long-term measurement lasting about 1 year in total. Based on the measured strain, stress values were recalculated. Based on the measured data, it can be stated that the limit safety values of the evaluated structure were not exceeded in any of the analysed areas. It should be noted, however, that these were local values which were forwarded to a further analysis of the supporting structure carried out by an operator who subsequently examined the silo and its safety.

Acknowledgment

This paper was supported by project VEGA 1/0290/18 and ITMS 26220220182.

References

[1] STN EN 1991-4 Eurocode 1. Actions on structures. Part 4: Silos and tanks.

[2] V. Křupka, P. Schneider, Konstrukce aparátů. PC DIR, Brno, 1998.

[3] A. Dogangun, Z. Karaca, A. Durmus, H. Sezen, Cause of Damage and Failures in Silo Structures Cause of Damage and Failures in Silo Structures, Journal of Performance of Constructed Facilities 23 (2) 2009.

[4] A. M. Ghasemi, F. Nourai, A framework for minimizing domino effect through optimum spacing of storage tanks to serve in land use planning risk assessments. Journal of Safety Science, 97 (2017) 20-26.

[5] M. Abo-Elkhier, K. Muhammad, Failure analysis of an exploded large-capacity liquid storage tank using finite element analysis. Engineering Failure Analysis, 110 (2020) 104401.

[6] F. Trebuňa, et al., Possible causes of initiation of plastic deformation in the containers used in fo od industry, Applied Mechanics and Materials Vol. 816 (2015) 195-203.

[7] F. Trebuňa, et al., Determination of failure reasons and proposal for restricted operation of storage tank, Acta Mechanica Slovaca. Vol. 10, No. 3 (2006) 15-28.

[8] STN EN 1993-1-10 Eurocode 3. Design of steel structures. Part 1-10: Material toughness and through-thickness properties.

[9] C. K. Shedd, W. A. Foster, Silo construction, Bulletin: Vol. 16, No. 189, (2017) Article 1.

EAN 2020
58th conference on experimental stress analysis

Residual Stresses Measurement Around Welds By Optical Methods

PÁSTOR M.[1,a], HAGARA, M.[1,b], ČARÁK P.[1,c]

[1]Deparment of Applied Mechanics and Mechanical Engineering, Faculty of Mechanical Engineering, Technical University of Košice; 04200 Košice, Slovakia
[a]miroslav.pastor@tuke.sk, [b]martin.hagara @tuke.sk, [c]peter.carak@tuke.sk

Keywords: Residual Stresses, Welding Joints, Photostress, DIC

Abstract. The aim of this paper is the analysis levels of residual stress created by the welding process. The paper focuses on the evaluation of residual stresses. Residual stresses in welded structures affect the quality of the structure and therefore their quantification plays an important role. We can improve the mechanical properties of structures with the proper method of evaluation and with the help of suitable technology of removing residual stresses in welded joints, which can lead to increased durability and reliability of the structure. The residual stresses were evaluated using the hole drilling method with strain gauge rosettes, as well as the optical Photostress method using photosensitive coatings and digital image correlation (DIC).

Introduction

Among the most commonly used method for the permanent joining of materials is welding. It is widely used and is currently one of the preferred methods. Welded joints in structures present a certain risk because they can act as stress concentrators. The residual stresses introduced into the materials have a negative effect on the quality of the whole construction. At present, there is a great effort to evaluate residual stresses. The most common source of residual stresses is the machining of materials. Almost all machining technologies cause residual stresses in the workpiece. Another source of residual stresses may be the modification of the mechanical elements of the mechanism during its operation. In other cases, they may arise during installation of the element, assembly of the structure, occasional loading. According to its origin, the ways of residual stress formation can be divided into three categories, i.e. mechanical, temperature and metallurgical ones [1,2].

The cause may be one of these factors, or their combination, which results in residual stress. For example, during grinding, all three factors are combined (Fig. 1). The residual stresses in the structures can reach relatively high values. Especially if they are close to cracks, welds, cavities, etc. In these cases, residual stresses are difficult to observe and can cause damage at a lower load than the maximum permissible load. [3].

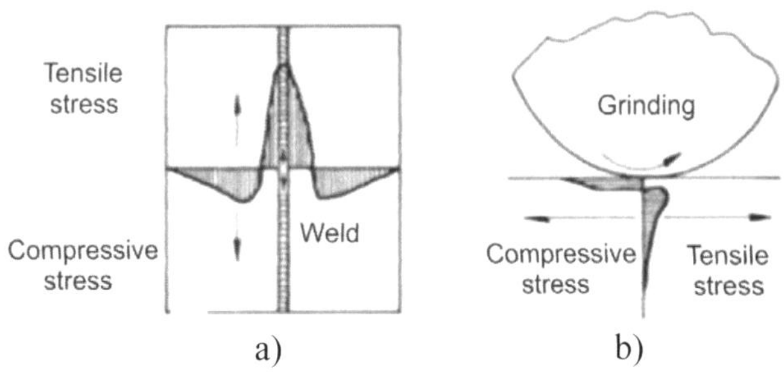

Fig. 1: Residual stress created by a) welding and b) grinding [3]

Fastening technology is a very important part of the industry. Welding of metals is the best known and most commonly used connection technology [4]. During welding three distinct areas are formed, i.e.: the weld metal to be melted, the heat affected area that undergoes the thermal cycle of the welding process, and the parent metal that is unaffected during the operation. By applying post-weld heat treatment, the material properties can be adjusted to the desired, but the temperature changes induced by the welding result in undesirable effects on the original microstructure, meaning a deviation of the mechanical properties in and around the weld from the material not affected by welding (Fig. 2). This represents a number of possibilities in terms of the generation of undesirable residual stress [4].

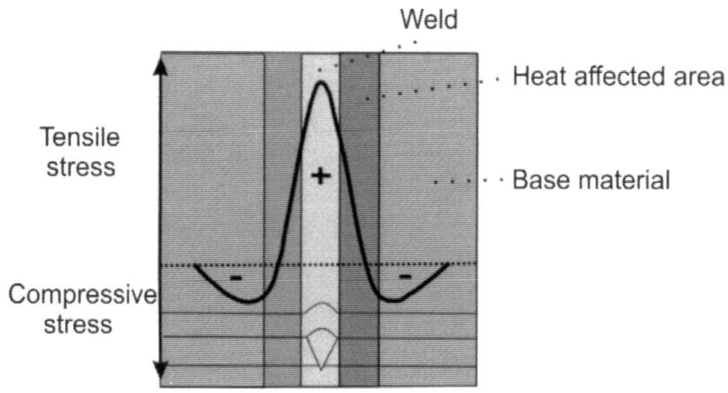

Fig. 2: Distribution of residual stresses created by welding

Welding is a process that is based on local heat generation by a moving heat source. The material to be welded reaches the melting point very quickly and then cools down rapidly causing microstructural and mechanical changes in properties and creating residual stresses [5].

The residual stresses present in the workpiece are the result of uneven expansion and compression of the weld and the base material due to the uneven heat distribution during the welding process. Usually, the residual stresses recorded in the weld joints are tensile and have a negative effect on the welded components. On the other hand, the residual compressive stresses can have beneficial effects on welded material. Tensile residual stresses can cause cracks, while compressive stresses can improve component quality. Tensile stresses are generated as a result of material shrinkage, while compressive stresses are generated by hardening and phase transformation. Both types of stresses exist in the weld joint, but their distribution depends on the position of the weld. The properties of a welded joint structure can be improved by applying specific processing to a particular location depending on residual stress distribution data. Fort that reason it is important to identify the weld residual stresses and their distribution to improve the quality and reduce the negative effects in the welded components [5].

Evaluation of Residual Stresses

Detecting and measuring residual stresses is important in predicting the life of a component or structure, analysing deformations and detecting failure reasons. All measurement techniques are divided into non-destructive, semi-destructive and destructive (see Table 1). The choice of a measurement method depends largely on the information we need to obtain about the specimen. An important aspect of such measurements is the knowledge of each measurement technique, their possibilities and limitations [6].

In practice, the determination of residual stresses in structures and machine components is usually based on the hole-drilling method. This method has the advantages of high accuracy and reliability, a standardized measurement process and appropriate practical application. This is a semi-destructive method because in most cases the drilled hole is very small in comparison to the specimen size. The drilled hole is often also repairable or negligible [7]. Strain gauge rosettes are used for recording relaxed deformations by hole drilling from which residual stresses are calculated [8].

Table 1: Distribution of methods by the destruction

Non-destructive method	Semi-destructive method	Destructive method
Barghausen noise	Hole drilling method	Partitioning method
RTG diffraction	Ring-Core method	Contour method
Neutron diffraction		
Ultrasonic method		

The strain gauge rosette differs from the classical strain gauge in that it has three measuring grids. Their position and rotation relative to each other on the rosette are precisely defined. Using a conventional drilling method, a hole is drilled into the centre of the rosette and deformations are measured in the material around. By the Ring-Core method, deformations are measured at the centre of the strain gauge while drilling is carried out around the rosette, where the annular groove is drilled. In Fig. 3, these methods are schematically illustrated, where D_0 is the diameter of the drilled hole and D is the mean diameter of the rosette [5]. H expresses the maximum hole depth if the measurement is still quantifiable, and ε_1, ε_2, ε_3 represent the relaxed deformations on the strain gauge sensors [8].

The Ring-Core method has the advantage of greater relaxed deformations. Nevertheless, in most cases, the conventional hole drilling method is used because of ease use and less damage to the specimen during drilling [7].

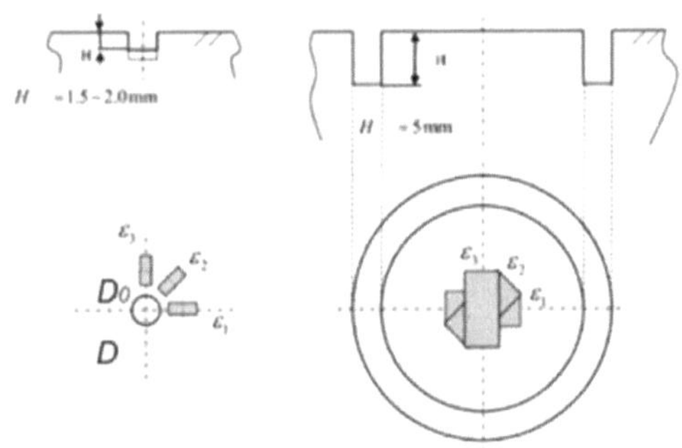

Fig. 3: Methods for measuring residual stresses with strain gauges, a) conventional hole drilling method, b) Ring-Core method

The scheme for investigating the released strains at any point of plane stress is shown in Fig. 4 [9].

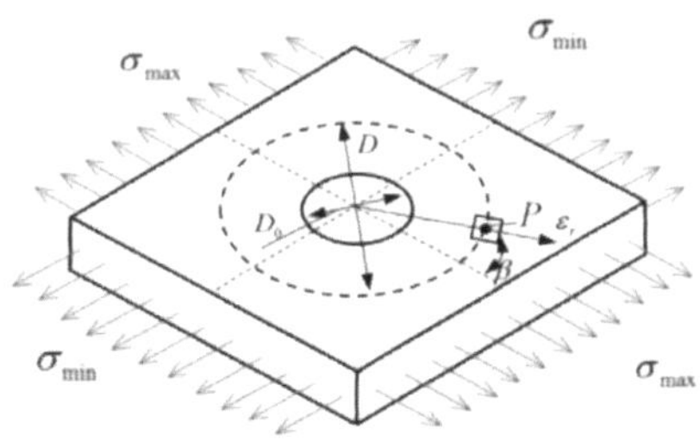

Fig. 4: Investigation of relaxed deformations at any point of plane stress

The general expression for the released relative radial deformation at the point, in the case of biaxial residual stress, can be expressed by

$$\varepsilon_r = (\bar{A} + \bar{B}\cos 2\beta)\sigma_{\max} + (\bar{A} - \bar{B}\cos 2\beta)\sigma_{\min} \quad (1)$$

where ε_r is the represents the released deformation in the radial direction registered by the strain gauge centered at point P, $\bar{A}, \bar{B}$ are the calibration coefficients, β is the angle measured counterclockwise from the direction of the strain gauge to the direction $\sigma_{\max}$, $\sigma_{\max}$ is the maximum normal stress at the hole before drilling, $\sigma_{\min}$ is the minimal principal normal stress at the hole before drilling, D is the mean diameter of the strain gauge rosette and D_0 is the hole diameter [10].

The coefficients $\bar{A}, \bar{B}$ are obtained empirically, it is by experimental calibration or numerical modelling using FEM. When introducing another independent variable - dimensionless hole depth, the generalized form of functions for coefficients can be expressed by

$$\bar{A} = f_A(E, \mu, r, Z/D)$$
$$\bar{B} = f_B(E, \mu, r, Z/D).$$

The procedure for measuring and evaluating the residual stresses have been described in standard ASTM E837-13a [11]. In the drilling method, it is necessary to recognise what type of specimen we use. We distinguish between "thin" and "thick" specimens. A specimen can be considered "thick" if its thickness is at least 1.2 D. When drilling such a specimen, the blind hole principle is applied, the drilling is performed in-depth increments. After each measurement step the strain gauge sensor records the relaxed deformations ε_1, ε_2, ε_3. According to standard ASTM E 837-13a, the measurement should be performed to a maximum depth corresponding to a ratio Z/D = 0.4, where the constant Z represents the depth of the hole and the constant D represents the mean diameter of the strain gauge rosette. A specimen of less than 0.4 D thick is considered to be 'thin'. In this case, the measurement is performed simultaneously, without increments. This rule for the through-hole applies, which means that the specimen is drilled in one step and only one strain relief value is obtained from all strain gauge sensors ε_1, ε_2, ε_3 [3].

Experimental determination of residual stresses near welded joints

The experimental part was focused on the verification of residual stresses near the weld joint by using optical methods as well as the hole drilling method. The aim was to compare residual stresses at individual sites of the test specimen, close and outside of the weld joint. This comparison determined the distribution of residual stresses at individual locations near and outside the weld joint.

First of all, the chemical composition of the specimen was determined, and based on the obtained data it was determined that the material was S235J2 with a thickness of 13 mm. The material characteristics required for the experiment are determined by the elastic modulus E = 200 GPa, Poisson's ratio μ = 0.3, ultimate strength R_m = 360-510 MPa and upper yield point R_{eH} = 235 MPa. The weld joint was made by electric arc welding with a melting electrode E-R 117 (Fig. 5).

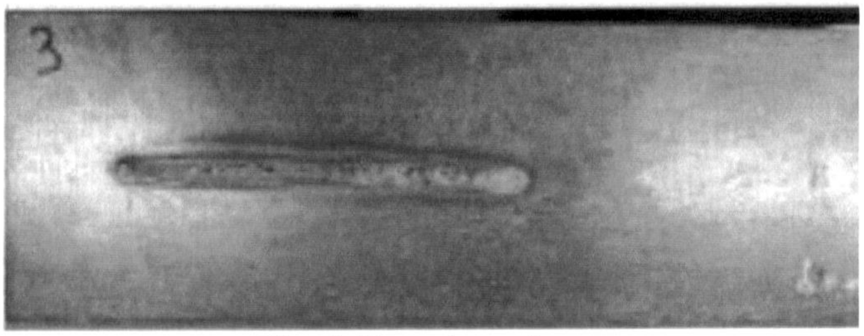

Fig. 5: Tested specimen

Fig. 6 shows the locations of the measurement points on the evaluated specimen.

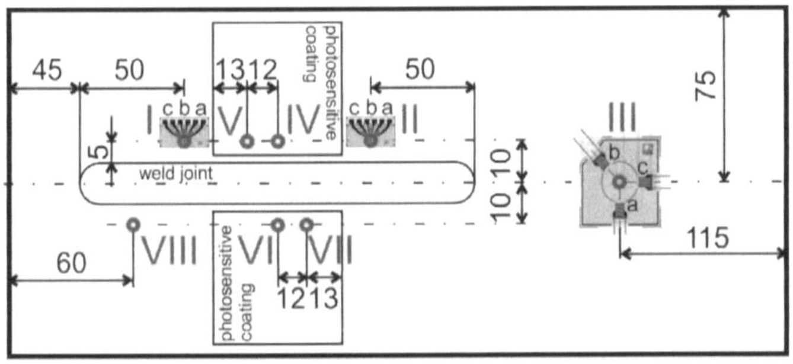

Fig. 6: Locations of the measurement points

As can be seen in Fig. 1, high tensile stresses are located 10 to 20 mm from the center of the weld joint. For this reason, the distance from the center of the weld joint was 10 mm (approximately 5 mm from the weld edge). Eight measurement points were selected for comparison (Fig. 6). Two of them (I and II) using strain gauges near the weld edge, each 50 mm from the ends of the weld joint and two photosensitive coatings between them with drilled holes (IV to VII), which were evaluated by Photostress. The hole VIII was drilled near the weld joint 60 mm from the edge of the specimen and evaluated by the DIC method. Location III was selected outside of the affected area to compare drilling data without interfering residual stresses generated by welding to the results. Three drilling devices were used in the experiment. It was a high-speed drilling device SINT MTS 3000 (Fig. 7), a high-speed and low-speed drilling device RS-200 (Fig. 8). Finally, a prototype of a device enabling the evaluation of released strains in individual steps of drilling by optical methods was also tested (DIC, Photostress) (Fig. 9). The P3 data bus was used for RS-200 measurements. The P3 reads the strain values during the RS-200 measurement at low speed, high-speed drilling.

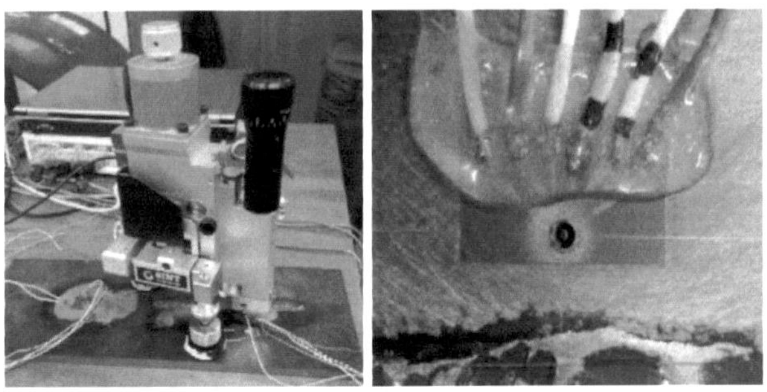

a) b)

Fig. 7: Drilling device SINT MTS 3000 a) whole view b) detail of the measured point

a) b)

Fig. 8: RS-200 drilling device and P3 data bus a) whole view b) detail of the measured point

After drilling the specimen at eight measuring points, we obtained the values of released deformations, from which were evaluated the residual stresses in selected locations. The residual stress values obtained by drilling with SINT MTS 3000 and RS-200 are given in (Table 2). The angle α was measured from the grid *a* of strain gauge rosettes (see Fig. 6).

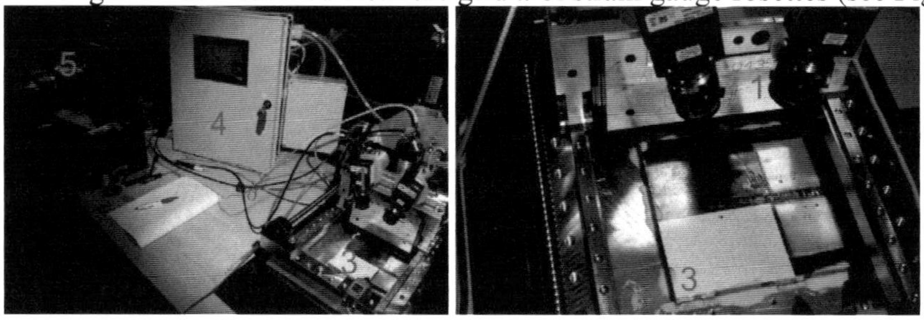

a) b)

Fig. 9: Drilling device using Photostress and DIC methods to quantify residual stresses by standard ASTM E 837-13a, where 1 couple of DIC cameras with high definition, 2 drilling device, 3 specimen, 4 control unit, 5 measuring and evaluating device a) whole view b) detail of the measured point

Table 2: The values of residual stresses reached by the hole drilling method

Hole Drilling Method			
Location	σ_{min} [MPa]	σ_{max} [MPa]	α [°]
I	145,67	563,37	-45,33
II	169	388	-51
III	15	10	-79

The SINT MTS 3000, RS 200 and the newly designed drilling device were used to compare the measurements near the weld joint. For measurements near the weld joint was used the strain gauges type RY61 from the manufacturer HBM. From the measured deformations, it is possible to assume tensile residual stress at the places near the weld joint, because the values of released strains were negative. For reference measurements outside the weld area, the measurement was performed with the RS-200 hand drilling device and the strain gauge type RY21. The measured deformations at this point were positive, indicating that there were compressive residual stresses at the drilling point.

Because the workplace has extensive experience using experimental methods in practical tasks, the results of experimental measurements of residual stress determination by drilling method were verified by optical methods Photostress and DIC. It should be noted that the use of optical methods (DIC, ESPI) in combination with drilling is solved by several authors [12-16]. The advantage of these non-contact methods is, among other things, the possibility of full-field analysis in the research area.

The experimental measurement procedure for the Photostress method was used to evaluate the specimen with applied optically sensitive coating Vishay PS-1D. In Fig. 10 there is a view of a specimen with an applied coating. For a "rough" comparison of the results, was drilled 3.2 mm blind hole to a depth 0.5 mm and 1.5 mm on both sides of the specimen. A newly designed drilling device was used for drilling (Fig. 11). Fig. 12 shows the locations where residual stress values have been read. Measured points have been chosen on the edge of drilled holes close to the weld joint because residual stresses were measured as close as possible to the weld joint.

Fig. 10: Specimen with applied photosensitive coatings

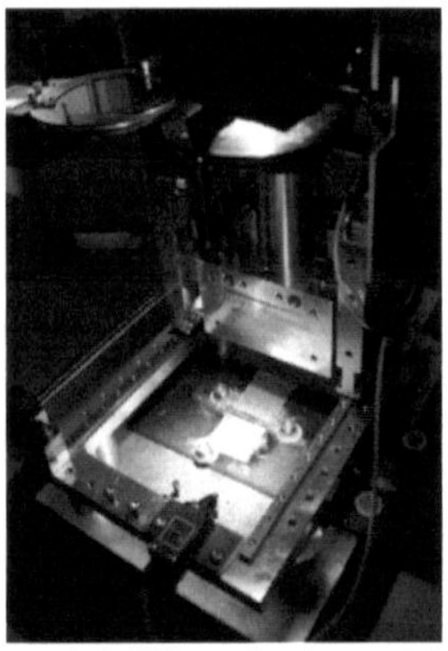

Fig. 11: Newly designed drilling device with Photostress module

a) b)

Fig. 12: Location of the examined points on the specimen a) position of drilled blind hole b) detail of examined points at the edge of the drilled blind hole

After drilling the blind hole, one point was selected as close as possible to the weld joint on each drilled hole (see Fig. 12b). To determine the stress levels at the examined points, it was necessary to measure the corresponding fringe order from the compensator. The process of measuring the fringe order using a compensator is shown in Fig. 13. The comparison of the initial and final state during measuring the fringe order is shown in Fig. 14. According to the orientation of the compensator at the investigated point on the edge of the blind hole, it is possible to determine whether it is tensile resp. compressive stresses. On the edge of the hole, one of the main stresses is equal to zero and therefore the non-zero stress is determined from the relation

$$\sigma = \frac{E}{1+\mu} \cdot N \cdot f \qquad (2)$$

where N is the line order, f is the photosensitive constant, μ is the Poisson's ratio and E is Young's modulus of elasticity.

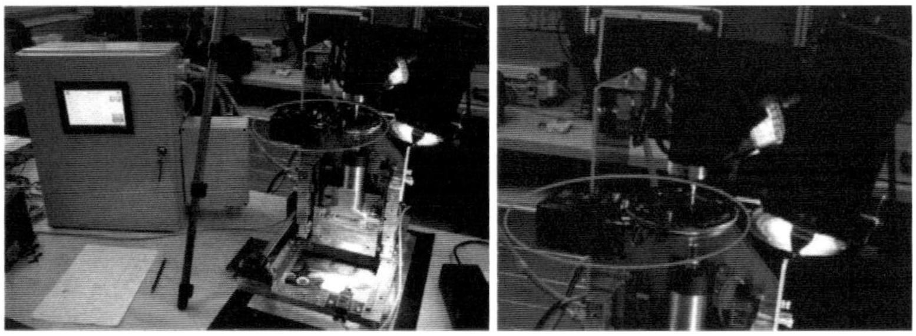

a) b)
Fig. 13: Measuring the fringe order using a compensator a) whole view, b) detail view

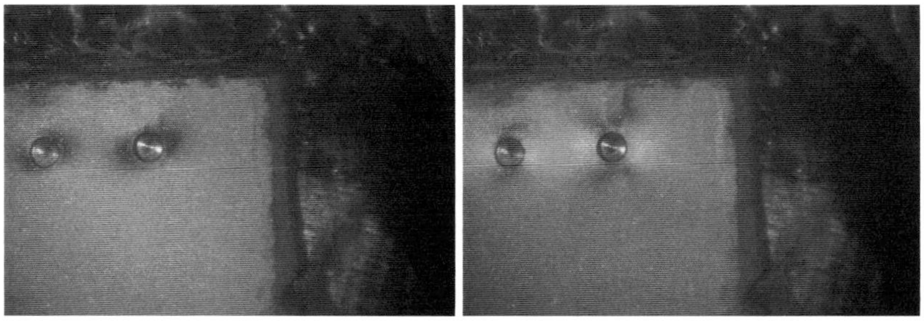

a) b)
Fig. 14: The process of measuring a fringe order a) initial state b) final state

In Table 3 shows the measured fringe orders using the compensator as well as the depths of the individual holes, the directions and the resulting values of the stresses at the individual locations.

Table 3: Values of residual stresses measured by Photostress

Location	Depth [mm]	Orientation of compensator	Fringe order N [-]	Stress value σ Photostress [MPa]
IV	0,5	=	0,5	291
V	1,5	=	0,59	344
VI	0,5	=	0,36	210
VII	1,5	=	0,56	326

Based on the results obtained by the Photostress method, it can be stated that in measured locations were tensile stresses thus negative residual stresses.

The experimental measurement process for the DIC method is based on applying a stochastic pattern to a specimen near the weld joint. In Fig. 15 there is a view of a specimen with a created pattern and a drilled hole near the weld joint. To compare the results was drilled a 3.2 mm blind hole to a depth of 2 mm and 10 mm from the edge of the weld joint. Drilling was performed using a newly designed drilling device.

a) b)

Fig. 15: Specimen with the applied stochastic pattern a) whole view, b) hole drilled near the weld joint

The reference images with 2452x2056px resolution (pixel density ca. 53px/mm) captured by the 3D DIC system Q-400 Dantec Dynamics can be seen in Fig. 16. As the results of strain analysis performed by the DIC method, which are used to calculate the residual stresses due to the standardized methodology, mentioned in ASTM E 837-13a, were markedly affected by the noise and did not correspond to the results obtained by strain gage as well as Photostress method. The successive change of displacement X and Y, respectively, during the hole-drilling obtained by 3D DIC system Q-400 Dantec Dynamics is depicted in Fig. 17.

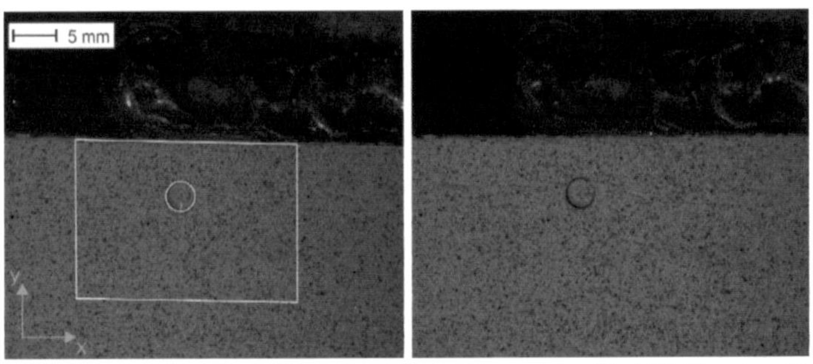

Fig.16: Reference images captured by the 3D DIC system Q-400 Dantec Dynamics with depicted evaluation mask

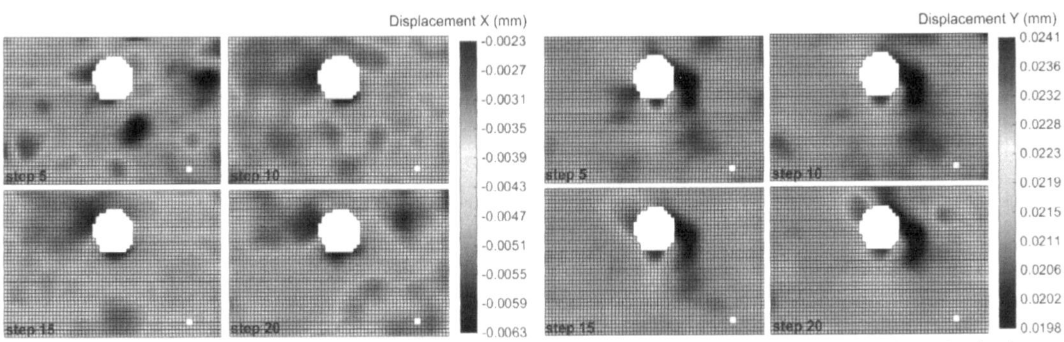

Fig.17: Successive change of displacement X and Y, respective, obtained during the hole-drilling by DIC

Due to obtained results from DIC affected by the noise, the authors' plan is to propose a methodology to compute the residual stresses from the obtained displacements.

Conclusions

The paper presents the procedure of experimental measurement on specimen with welded joint using three experimental methods. However, as already mentioned at the beginning of the paper, welded joints in structures present a certain risk because they can act as stress concentrators. Due to the negative effect of the welds on the structure, it is necessary to consider the possibilities of reducing these stresses. For this reason, identifying the weld residual stresses and their distribution is important to improve the quality and reduce the negative effects in the welded components. From the results obtained from drilling and the Photostress method can be found tensile residual stresses in the investigated area, which is consistent with theoretical assumptions. Due to results from DIC affected by noise, the authors plan methodology for computing the residual stresses from displacement.

Acknowledgments

This paper was supported by project VEGA 1/0290/18, APVV 15-0435 and ITMS 26220220182.

References

[1] Residual Stress., Integration de la notion des. Online:

[2] R. Nedin, V. Dudarev, Some aspects of modeling and identification of inhomogeneous residual stress, Engineering Structures, 151, (2017), 391-405

[3] Micro-Measurement Vishay Precision Group, Measurement of Residual Stresses by the Hole-Drilling* Strain Gage Method, Tech Note TN-503, (2010), 19-34

[4] W. Rae, Z. Lomas, Measurement of residual stress and microstructural evolution in electron beam welded Ti-6Al-4V using multiple techniques, Materials Characterisation, 132, (2017), 10-19

[5] N. Nadhira, M. Sarizam, Review on Welding Residual Stress, ARPN Journal of Engineering and Applied Sciences, 9, (2016), 6166-6175

[6] D. Walker, Residual Stress Measurement Techniques, Advanced Materials & Processes, (2001), 30-33

[7] G. S. Schajer, Hole drilling Residual Stress measurement at 75: Origins, Advancer, Opportunities, Experimental Mechanics, 50, (2010), 245-253.

[8] M. Steinzig, D. Upshaw, J. Rasty, Influence of Drilling Parameters on the Accuracy of Hole-Drilling Residual Stress Measurements, Experimental Mechanics, 54, (2014), 1537-1543

[9] A. S. Kobayashi, Handbook on Experimental Mechanics, Society for Experimental Mechanics, Vch Pub, 1993.

[10] G. S. Schajer, Practical residual stress measurement methods, John Wiley & Sons Ltd, 2013.

[11] ASTM E 837-13a: Standard Test Method for Determining Residual Stresses by the Hole-Drilling Strain-Gage Method.

[12] Y. Min M. Hong Z. Xi L. Jian, Determination of residual stress by use of phase-shifting moiré interferometry and hole-drilling method, 44, (2006), 68-79

[13] G. S. Schajer, Application of Finite Element Calculations to Residual Stress Measurements, Journal of Engineering Materials and Technology, 103, (1981), 157-163

[14] M. Hagara, F. Trebuňa, M. Pástor, R. Huňady, P. Lengvarský, Analysis of the aspects of residual stresses quantification performed by 3D DIC combined with standardized hole-drilling method, Measurement, 137, (2019), 238-256

[15] E. Ponslet, M. Steinzig, Residual Stress Measurement Using the Hole Drilling Method and Laser Speckle Interferometry. Part II: Analysis Technique, 27, (2003), 17-21

[16] G. S. Schajer, M. Steinzig, Full-field calculation of hole drilling residual stresses from electronic speckle pattern interferometry data, Experimental Mechanics, 45, (2005), 526-532

EAN 2020
58th conference on experimental stress analysis

Testing Carbon Fibers for 3D Printing

PAVLÍČEK P.[1,a], FOJTÍK F.[1,b], FUSEK M.[1,c]

[1]Deparment of Applied Mechanics, Faculty of Mechanical Engineering, VŠB – Technical University of Ostrava; 17. listopadu 2172/15, 708 00 Ostrava, Czech Republic

[a]pavel.pavlicek@vsb.cz, [b]frantisek.fojtík@vsb.cz, [c]martin.fusek@vsb.cz

Keywords: Composites, 3D printing, Tensile test, Carbon fiber

Abstract. This paper is focused on innovative method of the tensile testing of the carbon fibers in 3D printing. Carbon fibers have good mechanical properties especially high strength and Young modulus. 3D printing of carbon composites is progressive method which give new design and production possibilities. This material is used in automotive and aerospace industry. Composites in the design have good properties mainly in the mass reduction which follow trend decreasing of emissions.

Introduction

Carbon fibers belong to progressive materials, which are used in design of modern mechanical parts. Main advantages of carbon fibers are high stiffness and strength while maintaining a low weight. This material has some disadvantages mainly low ductility and fracture resistance. In this work, the carbon fiber used in 3D printing of composite in 3D printer type Markforged X7 is examined [1].

3D printing of composites is an innovative method of design of mechanical parts which cannot be produced using conventional technologies. Testing of mechanical properties of carbon fiber is important for the design of mechanical parts and for the prediction of material behaviour under loading.

Mechanical properties of the fiber is an important parameter for the prediction of mechanical properties of composites under different mixing ratio. This paper describes the method of fibers testing inspirited by standard ASTM D4018-17 [1].

Description of Test

The testing of fibers has some difficulties, especially when the small cross-sections are tested where the specimen can be damaged by the compression force and high stress concentration in clamping area.

This problem is solved in ASTM D4018-17 standard [2] where both ends of fiber are filled by resin which creates the clamping area. Another standard for fiber testing is ASTM D3379. However, this method is not suitable for large fiber diameters and high forces.

For clamping of fibers the new type of clamping zone has been designed. This can minimize the concentration of stress in the transition area between the fiber and resin. The form is made of aluminium.

Design of Aluminium Form

The form is made of two parts - form to be filled and base plate. Both parts are connected using bolts. Material of the form is aluminium plate 3 mm thick, machined by electro-spark machining technology or by laser cutting for the comparison of machining methods used. The middle part of the form is milled so that the carbon fiber is in the center of the form.

The removal force depends on the surface roughness of the specimen. Thus, it is necessary to minimize the grain size on vertical surfaces of the form. The optimal surface roughness is between Ra 0.4 and Ra 0.8. Another possibility is to incline vertical surfaces of the form by 2-3°. Base plate is only milled on its outer sides, the frontal surfaces did not need to be machined due to the sufficient roughness produced by the production technology of the plate. In the corners of the form and the base plate threaded holes for their mutual connection are drilled.

The form and the base plate must not be distorted and surfaces must abut without clearances. In the Fig. 1 the technical drawing of the form is presented.

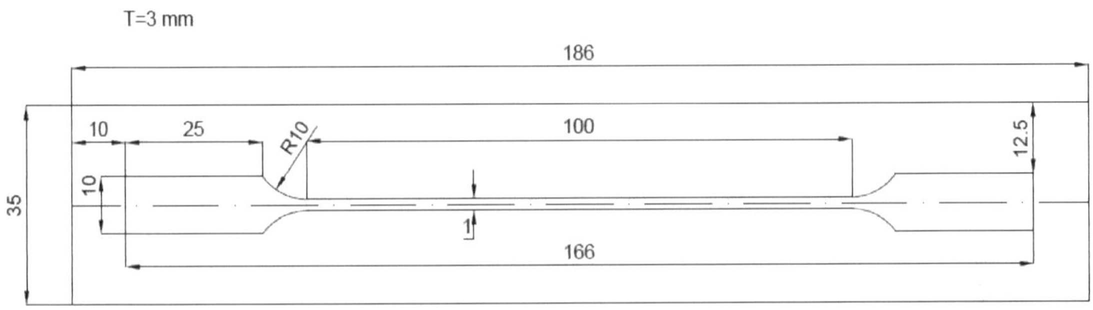

Fig 1: Drawing of form

Preparation of Specimen

The form and the fiber must be cleaned with isopropyl alcohol before the bonding, then it is necessary to connect the form and the base plate by bolted joints. After the connection, surfaces of the form are covered by the separator which prevents the bonding of the resin with the form. For this purpose, hot paraffin or wax is used. ¨

The separator must be removed from the middle part of the form. For the correct alignment of the fibre into the form, ends of the middle part of the fibre need to pressed using toothpick or wax. Polyurethane resin is mixed in the recommended mixing ratio. After the mixing, it is important to wait some time until the chemical reaction starts. After this step both ends are filled by resin.

The curing acceleration using higher temperature is not recommended because higher temperature can cause the change of mechanical properties of the fiber and resin is more fragile. This property was observed during specimen preparation in authors´ laboratory.

Specimen Clamping

Clamping specimen into the test machine must be done parallel with the axis of the deformation. Clamping the specimen is realized using jagged jaws with minimal clamping force. Too high force causes the failure of the bonded joint and the pramature failure of the specimen. This property was observed during test realization in authors´ laboratory.

Carbon Fiber Testing

The fiber has been measured by pasameter and digital-optical microscope before the testing. The fiber has the eliptical cross-section with the difference of halfaxis length of 0.03 mm. Due to this small value, the elipsis can be considered as the circle with diameter of 0.437 mm and cross-sectional area of 0.150 mm^2.

Carbon fiber is tested using the Testometric M500-50CT machine with 50000 N force sensor. For testing, ten specimens have been prepared. The loading velocity was set from 0.5 - 10 mm/min with the minimal initial preload. Values 0.5, 1 and 10 mm/min are infromative values only.

Deformation of the specimen is overtaken from the displacement of the crossbar sincethe usa of the extensometer was not possible due to the missaligment of jaws causing the initial deformation of the specimen.. The deformation of the specimen could be measured by the optical extensometer. However, this device was not availabe at the time of the experiment. After the exceeding the strength limit, the fiber broke at multiple locations into the fragments around 3-4 cm long.

For the calculation of mechanical properties equations from Table 1 are used [3]. In Table 2, the nomenclature for Table 1 is described. In Fig 2., the secant modulus is decribed.

Table 1: Used Equation

Stress	$\sigma_1 = \frac{F}{A}$	[MPa]	(1.1)
Strain	$\epsilon_1 = \frac{\Delta L}{L_o}$	[-]	(1.2)
Secant modulus	$E_t = \frac{\sigma_{\epsilon_{0.25}} - \sigma_{\epsilon_{0.005}}}{\epsilon_{0.25} - \epsilon_{0.005}}$	[GPa]	(1.3)

Table 2: Nomenclature used

σ_1	First principal stress	[MPa]
F	Force	[N]
A	Area	[mm^2]
ΔL	Change of length	[mm]
L_o	Initial length	[mm]
E_t	Secant modulus	[MPa]
ϵ_1	First principal strain	[-]
$\sigma_{\epsilon_{0.25}}$	Stress at 0.25 strain	[MPa]
$\sigma_{\epsilon_{0.005}}$	Stress at 0.005 strain	[MPa]
$\epsilon_{0.25}$	0.25 strain	[-]
$\epsilon_{0.005}$	0.005 strain	[-]

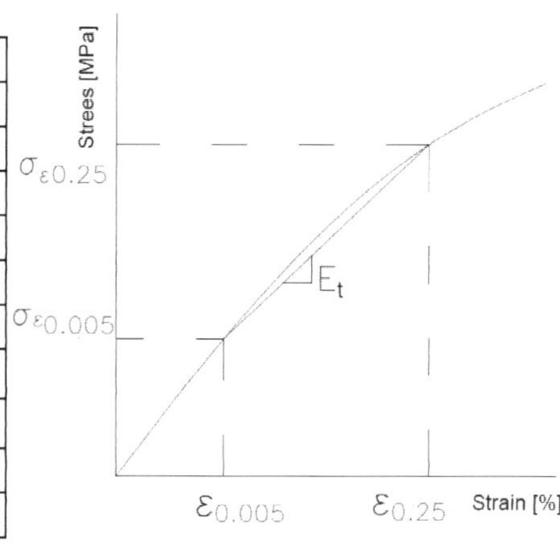

Fig. 2: Secant Modulus

For calculation of the Secant modulus the equation (1.3) is used. Optionally, the least squares method can be used.

Analysis Results

Results of tensile testing were evaluated acording to ISO-2602 [4] where monitored prameters are Young modulus, ultimate strength and ultimate strain.

Data from tensile test has been statisticaly evaluated for loading velocity 5 mm/min, so results can be considered asaverage values with 95 percent confidence interval.

Table 3: Mechanical properties

Velocity [mm/min]	Secant modulus E_t [GPa]	Strength σ_l [MPa]	Fracture Strain ϵ_l[-]
0.5	72938	1666	0.021
1	82925	1717	0.020
5	75918±933	1479±81	0.182±0.002
10	75342	1737	0.021

In the Table 3, mechanical properties for four loading velocities from 0.5 -10 mm/min are presented. Results show the small difference between individual strain rates. Maximum strength measured is 1479±81 MPa where this value exceeds **1.85** times the value guaranteed by producer of the 3D printer but the producer uses the different tensile test method D3039 [4].

In the Graph 1, stress-strain curves from testing are shown.

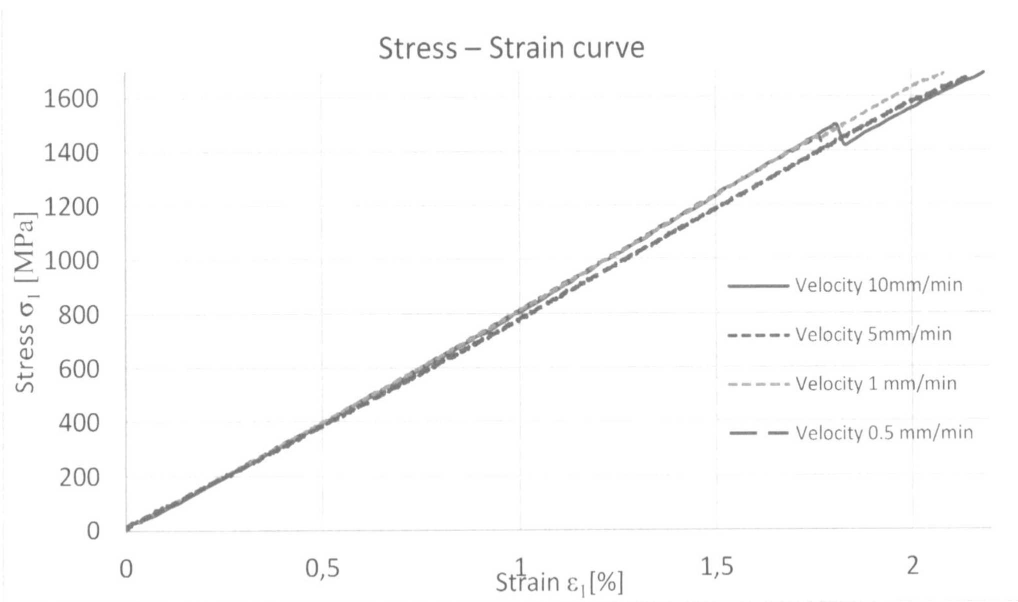

Graph 1: Stress – Strain curve σ_1 vs ϵ_1

In the Graph 1, the linear character of the material response on loading is shown. This corresponds to the behaviour of the brittle material with negligible plasticity effect. This fact is confirmed by the microscope image when the fracture inclination is approximately 45°.

The image of the fracture and the single diameter of the fiber are shown in Fig.3,4.

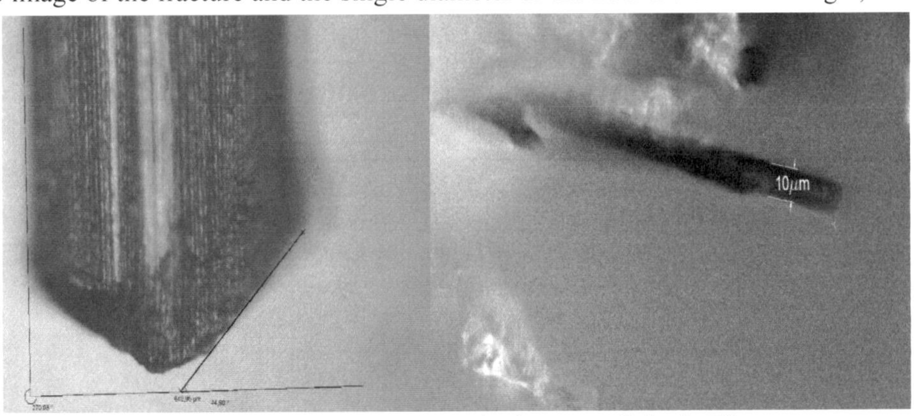

Fig 3,4: Fracture angle and Diameter of single Fiber

The fibre with the diameter of 10 μm is commonly used in industry and others applications According to ASTM D4018-99, the type of fracture is XGU where: X stands for eXplosive failure, G stands for location of failure near to the Grip and U stands for Unknown reason of the fracture.

Fracture into small fragments can be explained with the stress wave propagating along the length of the fiber. This could cause the cause fracture of the fiber on multiple locations.

Conclusion

This paper describes the tensile test of the carbon fibre for 3D printing. Resarch has shown that the method is apliccabe and gives good results with low scatering of values. This method describes newest trends in testing of materials.

Big advantage of this method is the simlipicity of specimen producution specimen. The form is applicable for other materials as well.

Microscopy images gave interesting values of fracture of carbon fibers when 45° correspond to a brittle fracture.

Next research is focused onto the prediction of fibers fracture using the acoustic emmisions and further on theeffect of fibre pre-heating on mechanical properties.

Additionaly, the effect of different clamping shape on mechanical properties of the fiber and the type of failure will be evaluated. This parameters are imporant for determination of mechanical properties of 3D printed composites.

Acknowledgement

Supported by grant contest – EF17_049/0008407 "*Innovative and additive technology of production – new technological solution of 3D print metals and composites*" (provided by Ministry of Education Youth and Sports of the Czech Republic).

References

[1] 3D printer Markforged X7, available from https://3dwiser.com/3d-tiskarny/3d-tiskarny-markforged/markforged-x3-x5-x7/

[2] ASTM D4018-17, Standard Test Methods for Properties of Continuous Filament Carbon and Graphite Fiber Tows, ASTM International, West Conshohocken, PA, 2017, www.astm.org

[3] K. Frydrýšek, Basic strength and elasticity of materials, Ostrava, 2016. ISBN 978-80-248-3870-0.

[4] ČSN ISO 2602 (010231) Statistic interpretation of test results, Estimation of average, Confidence interval, Prague, Czech institute of normalization, 1980.

[5] ASTM D3039 / D3039M-17, Standard Test Method for Tensile Properties of Matrix Composite Materials, ASTM International, West Conshohocken, PA,2017, www.astm.org

EAN 2020
58th conference on experimental stress analysis

Evaluation of the Acoustic Emission Measurement performed during Freeze-Thaw Cycles Loading

PAZDERA L.[a], TOPOLAR L.[b], KOCAB D.[c], VYMAZAL T.[d], MIKULASEK K.[e]

Faculty of Civil Engineering, Brno University of Technology, Veveri 331/95, 60200 Brno, Czech Republic

[a]pazdera.l@fce.vutbr.cz, [b]libor.topolar@vutbr.cz, [c]dalibor.kocab@vutbr.cz, [d]tomas.vymazal@vutbr.cz, [e]mikulasek.k@fce.vutbr.cz

Keywords: acoustic emission, freeze-thaw, cycles, concrete, evaluation

Abstract. The paper is concerned with the use of the acoustic emission method for testing the frost resistance of air-entrained concrete specimens. The drilling cores obtained in different places of a concrete pillar that was cast along its longer side (on high), that is edgewise, indicate a negligible influence of the casting direction. The acoustic emission method then shows that even if the mechanical properties remain constant, the changes inside the structure can be observed during the freeze-thaw cycles.

Introduction

Concrete blocks have been in use in the civil engineering since the very discovery of cement. During their entire service life, such blocks are exposed to various degradation factors (mechanical and chemical influences, temperature changes, etc.). After several winter and summer seasons, the influence of temperatures is essential. Knowing the impact of temperature fluctuations over time is necessary for the research and, even more, for the civil engineering practice. Apart from the degradation impact of high temperatures, that of low ones is among the most destructive operational factors for many concrete products [1,2]. Freeze-thaw cycles may have a very negative impact on the durability of concrete structures. In terms of research, it is suitable to observe the behaviour of concrete as early as during the actual cyclic freeze-thaw loading [3-5].

Due to its capillarity, concrete is soaked with water. Under freezing then, the temperature fluctuations inside the structure create stress that may damage its structure.

The assessment of the influence of freeze-thaw cycles is usually based on simply monitoring different mechanical characteristics such as moduli of elasticity, etc. [6] Such quantities are obtained after a certain number of cycles. Thus, a concrete is not described continually. In this way, the behaviour is estimated based on different specimens by statistical methods. To observe the status of a specimens, it is suitable to use the non-destructive testing methods [7-10]. Based on acoustic phenomena, the acoustic emission method makes it possible to observe the behaviour of each specimen continually during the entire freeze-thaw cycling process. The acoustic emission signals recorded provided a basis for a more detailed evaluation of materials [11,12].

Acoustic emission events are related to acoustic ultrasound wave that occur in the event of the material being exposed to cracking, that is, plastic deformation. The acoustic waves can be initiated releasing the related energy by micro cracks generated by internal loading. Acoustic emission can cover a wide scale of inaudible and audible frequencies. The acoustic waves captured are usually transformed into electric signals by piezoelectric sensors. The volume of acoustic activity generally depends on the amount of energy released, the distance and orientation of the source in relation to the sensor. The signals are then amplified and recorded in a data collection system. The signals of acoustic waves obtained in this way are then analyzed to determine the level of the resulting damage [13,14].

Experimental set up

To test the frost resistance of air-entrained concrete, drilling cores were used taken from a concrete pillar with parameters shown in Tab. 1. The prescribed water/cement ratio was 0.46. The concrete block was 2.4 m high, 1.8 m wide, and 0.45 m deep. Drilling cores were taken from this air-entrained concrete pillar of different diameters. For the acoustic emission method application, specimens were used with diameters of 150 mm and a length of 400 mm. Such specimens were exposed to a hundred freeze-thaw cycles to be subsequently tested for tensile splitting strength [15].

The acoustic emission sensors were placed on the surface of the concrete specimens. Applied sensors were with diameter 22 mm, height 52 mm, and with internal impedance converter & preamplifier 35 dB. The declared frequency range was up to 600 kHz. The XEDO acoustic emission system made by Dakel (Czech) recorded acoustic emission hits.

Based on a test of fresh concrete, the following data were determined: density 2288 kg/m^3, flow 460/450 mm, slump 180 mm, fresh concrete temperature 28 °C, and air content 5.0 %. The concreting was done in the vertical direction. The concrete was compacted using an immersion vibrator and treated to prevent moisture leakage.

Note that concrete frost resistance tests are done by alternately freezing and thawing water-saturated beams. One freeze-thaw cycle consists of four-hour freezing and two-hour thawing lasting six hours. The specimens were frozen at -18 °C and thawed in water warmed at +20 °C.

To carry out the acoustic emission measurements, a single acoustic emission sensor was attached to the upper base of each specimen as shown in Fig. 1.

Fig. 1: The layout of the test specimens in a freezing device during the frost resistance test.

Table 1: Composition of fresh air-entrained concrete

Material	Content per 1 m^3 of fresh concrete
Cement CEM I 42.5 R	390
Sand 0-4 mm	810
Aggregate 4-8 mm	160
Aggregate 8-16 mm	760
Superplasticising admixture	1.0
Air-entraining admixture	0.6
Workability enhancing admixture	1.6
Water	185

Results

The acoustic emission events were evaluated based on the ring down, i.e., the overshoot counts exceeding a preset threshold (N_{AE}). The overshoot count was evaluated for each loading cycle. Thus, the measurement of 100 freeze-thaw cycles took 600 hours or 25 days. Two specimens were selected to present the evaluation. The first specimen with a wider diameter of 150 mm, that is, a drilling core from the upper part of the concrete pillar, denoted by pt9. The second specimen with a diameter of 100 mm, that is, a drilling core from the lower part of the concrete pillar denoted by pt8.

Depending on the number of cycles N, the frost-resistance coefficient is calculated from the ratio of tensile splitting strengths f_{ct} with N cycles to the reference value (0 cycles) using the equation

$$I_{ft,N} = \frac{f_{ct,N}}{f_{ct,0}} \quad (1)$$

Table 2: Averages mechanical properties

Specimen	Diameter 100 mm			Diameter 150 mm		
Properties / Cycles	0	50	100	0	50	100
Transverse tensile strength f_{ct} [MPa]	3.55	3.45	3.60	3.20	3.25	3.25
Frost-resistance ratio I_{ft}		0.97	1.01		1.02	1.02

In view of the mechanical properties by Tab. 2, measured for reference (without cycles), specimens after 50 and 100 cycles, it can be concluded that the damage of the specimen during the freeze-thaw test were negligible since the tensile splitting strength values measured remain virtually constant with the number of cycles changing, thus, the frost-resistance ratio is close to one.

As compared to the mechanical properties obtained by a destructive method from different specimens with subsequent statistical processing, the acoustic emission method shows the current state of a single specimen. A statistical approach can of course be applied as well, but it is substantially more complicated.

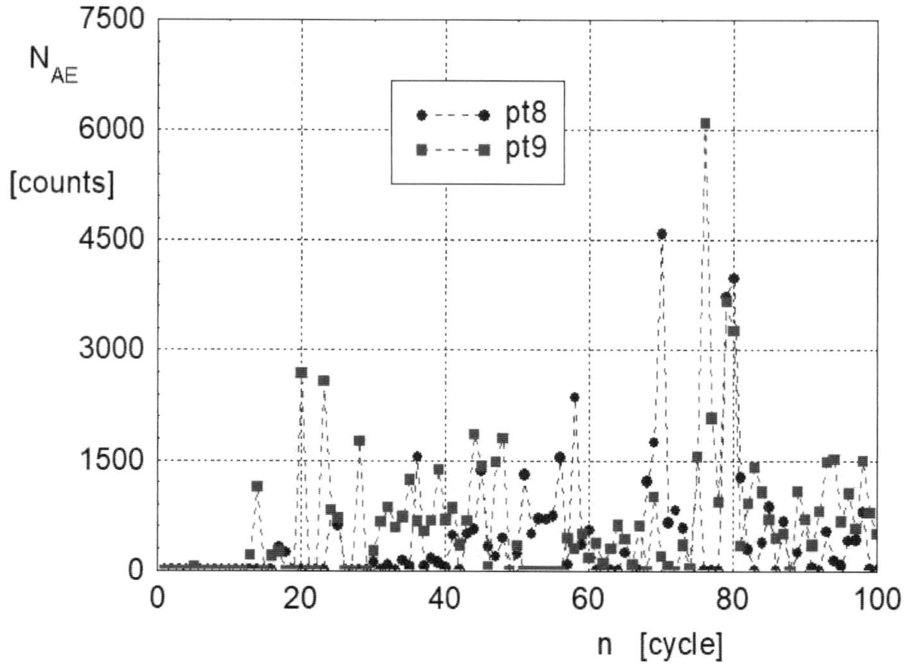

Fig. 2: Acoustic emission overshoot counts N_{AE} versus the number of cycles n for two specimens

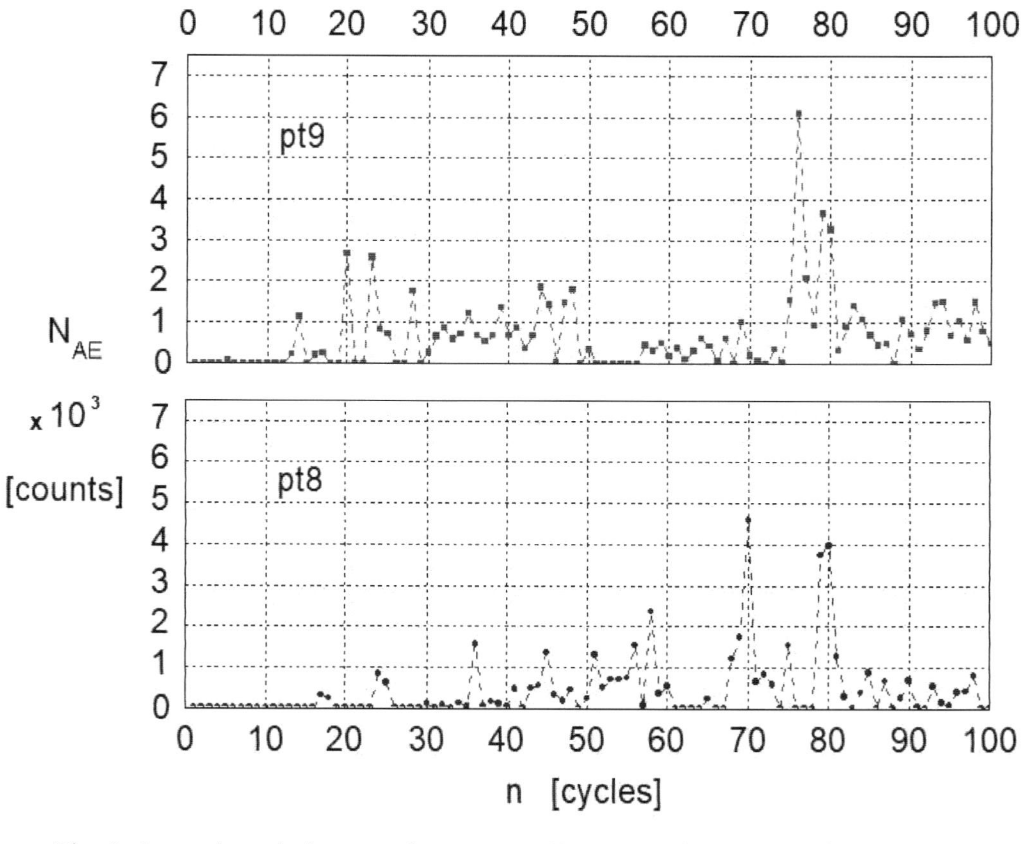

Fig. 3: Acoustic emission overshoot counts N_{AE} versus the number of cycles n for two specimens in separate graphs

Fig. 2 shows the acoustic emission ring-downs for both specimens. Fig. 3 is then used for an easier description of each signal, with one graph corresponding to each specimen. In the event of specimens pt8 and pt9 having the same mechanical properties and as specimen pt8 has a smaller size, its acoustic emission activity can be expected to be less. For the smaller pt8 specimen, the total ring down equals 43160 counts while, for the larger pt9 specimen, it equals 67046 counts. This, however, may not be viewed as a general rule as the acoustic emission activity determined may be influenced by the size of the active cracks as well as by its position within the specimen.

By Fig. 2, the first acoustic emission appears after ten cycles. With the smaller specimen, it occurs somewhat later. The first about twenty cycles show great concrete resistance to cyclic temperature loading. After about fifty cycles, the structure of the material adapts to the external factors with the acoustic emission activity dropping significantly. Next, between 70 and 80 cycles, the acoustic emission activity rises steeply to drop down again.

Conclusions

Measurement and assessment by the acoustic emission method are instrumental in observing the inner structure of air-entrained concrete. It can be concluded that, at present, there is hardly any other method more appropriate than the acoustic emission method for monitoring the state of the inner structure of concrete between the freeze-thaw cycles.

The results obtained make it clear that, despite insignificant changes in the mechanical properties, the acoustic emission activity was recorded in the structure of the specimens. Their size, however, is negligible in terms of the overall structure of the specimens. Determining the total service life would necessitate carrying out experiments until the structure was actually destroyed, which would require a substantial increase in the number of freeze-thaw cycles.

Acknowledgement

This article has been worked out and supported under the project Czech Science Foundation GACR No. GA 19-22708S "New approaches to predicting air-entrained concrete durability by means of determination of pore size distribution".

References

[1] H. Mastori, P. Piluso, J.F. Haquet, R. Denoyel, M. Antoni, Limestone-siliceous and siliceous concretes thermal damaging at high temperature, Construction and Building Materials, 228 (2019) UNSP 116671

[2] V. Marcantonio, D. Monarca, A. Colantoni, M. Cecchini, Ultrasonic waves for materials evaluation in fatigue, thermal and corrosion damage: A review, Mechanical Systems and Signal Processing, 120 (2019) 32-42

[3] Y.X. Zou, X.D. Shen, X.B. Zuo, H.J. Xue, G.F. Li, Experimental study on microstructure evolution of aeolian sand concrete under the coupling freeze-thaw cycles and carbonation, European Journal of Environmental and Civil Engineering (2020)

[4] L. Pazdera, Topolar L., Application acoustic emission method during concrete frost resistance, Russian Journal of Nondestructive Testing, 50 2 (2014) 127-131

[5] Y Farnam, D. Bentz, A. Sakulich, D. Flynn,J. Weiss, Measuring Freeze and Thaw Damage in Mortars Containing Deicing Salt Using a Low Temperature Longitudinal

Guarded Comparative Calorimeter and Acoustic Emission (AE-LGCC), Advances in Civil Engineering Materials, 3,1 (2014)

[6] D. Kocab, L. Topolar, B. Kucharczykova, R. Halamova, The analysis of acoustic emission signals detected during the loading of cement-based materials, Engineering Failure Analysis, 99 (2019) 18-25

[7] M. Korenska, L. Pazdera, L. Ritickova, Resonant inspection - Interesting non-destructive testing tools for determine quality of tested specimens, Previous Experience and Current Innovations in Non-Destructive Testing, in 6th International Conference of the Slovenian-Society-for-Non-Destructive-Testing, Portoroz, Slovenia, (2001) 45-48

[8] I. Plskova, Z. Chobola, M. Matysik, Assessment of Ceramic Tile Frost Resistance by Means of the Frequency Inspection Method, Ceramics-Silikaty, 55,2 (2011) 176-182

[9] J. Smutny, V. Nohal, Vibration Analysis in the Gravel Ballast by Measuring Stone Method, Akustika, 25 {2016) 22-28

[10] V. Petraskova, Correctness Testing and Equality Testing in Evaluation of Acoustic Laboratory Measurements, Akustika, 26 (2016) 20-28

[11] Z.L. Zhou, Y.C. Rui, J. Zhou, L.J. Dong, L.J. Chen, X. Cai, R.S. Cheng, A New Closed-Form Solution for Acoustic Emission Source Location in the Presence of Outliers, Applied Sciences-Basel, 8,6 (2018) 949

[12] L. Topolar, L. Pazdera, B. Kucharczykova, J. Smutny, K. Mikulasek, Using Acoustic Emission Methods to Monitor Cement Composites during Setting and Hardening, Applied Sciences-Basel, 7,5 (2017) 451

[13] L. Pazdera, R. Cajka, L. Topolar, P. Mateckova, V. Bilek, O. Sucharda, Measurement and Utilization of Acoustic Emission for the Analysis and Monitoring of Concrete Slabs on the Subsoil, Periodica Polytechnica-Civil Engineering, 63,2 (2019) 608-620

[14] Y. Wang, S.J. Chen, S.J. Liu, H.X. Hu, Best Wavelet Basis for Wavelet Transforms in Acoustic Emission Signals of Concrete Damage Process, Russian Journal of Nondestructive Testing, 52, 3 (2016) 125-133

[15] CSN 73 1322 Determination of frost resistance of concrete (1968)

EAN 2020
58th conference on experimental stress analysis

Experimental Investigation of Vibration Parameters of Sandwich Beam with Viscoelastic Core

PETRÍKOVÁ I[1,a], MARVALOVÁ B.[1,b], SIVČÁK M.[1,c], TRAN H.N.[1,d]

[1]Department of Applied Mechanics; Faculty of Mechanical Engineering; Technical University of Liberec; Studentská 2, 461 17 Liberec, Czech Republic

[a]iva.petrikova@tul.cz, [b]bohda.marvalova@tul.cz, [c]michal.sivcak@tul.cz, [d]nam.tran.huu@tul.cz

Keywords: Sandwich beam vibration, Experimental, Magnetorheological elastomer core, Frequency dependency, Amplitude dependency, Magnetic field

Abstract. This study deals with the vibrational measurement of a sandwich beam with carbon-epoxy composite skins and a viscoelastic core made of magnetorheological elastomer (MRE). The forced vibration measurements at different frequencies were performed under the action of gradually increasing magnetic field intensity. The stiffness and damping properties of the active constrained MRE core layer and the dynamic properties of the sandwich beam were evaluated. The experimental results show a reduction of mechanical vibrations together with the shift of the natural frequencies and vibration amplitude caused by the change of magnetic field intensity.

Introduction

Magnetorheological elastomers (MREs) are smart composites that are made of micro-sized magnetic particles dispersed in a non-magnetic elastomeric matrix. The mechanical and rheological properties of MREs can be controlled rapidly and reversibly by the application of an external magnetic field which leads to the so-called magnetorheological (MR) effect [1]. The control of MREs enabling the immediate change of their properties in the real time has made them very promising for a wide range of applications, such as damping elements in the vibration absorbers, vibration isolators, sensing devices, vehicle seat suspension, engine mounts, actuators to control the flow, and adaptive stiffness devices [2].

Matrices of MREs are made usually of soft viscoelastic materials such as silicone rubber, natural rubber, polyurethane, and thermoplastic elastomers [3]. The common type of fillers for MREs are carbonyl iron particles of micrometer size with high saturation magnetization. The properties of MREs strongly depend on the distribution of iron particles which can be either random or aligned in chains. The distribution structure of particles is determined by whether or not the magnetic field is applied during the cross-linking. The isotropic MREs have randomly distributed particles and the anisotropic MREs have the chain-like particle structure caused by the applied magnetic field during the cross-linking. The MR effect is far more pronounced in the anisotropic MREs [4].

The dynamic properties of isotropic and anisotropic MREs are usually measured using DMA tests in order to determine their suitability for potential use in vibration damping [4]. The dynamic properties as the storage and loss moduli and the loss factor depend on different frequencies and amplitudes and on the magnitude of the applied external magnetic field. A number of constitutive models have been developed to describe the rheological properties of MREs and the dependence of their dynamic properties on the frequency and amplitude of vibration and on the intensity of the magnetic field. Recently, the constitutive models based on fractional calculus have been used for viscoelastic modelling the behavior of MREs. The model parameters are determined by fitting the experimental data [4].

Beam-type structures are widely used in the fields of mechanical engineering. An optimum structural design requires an accurate analysis of their dynamic characteristics [5]. Sandwich beams with the core made of smart material such as MRE enable the continuous change of stiffness and damping properties with varying external magnetic field [6,7]. This study investigates the dynamical response of an exciting sandwich beam with an MRE core layer in the presence of an externally applied magnetic field.

Experiment

A sandwich beam 145 × 20 × 7 mm^3 consists of a 5 mm thick MRE core glued between the outer layers with 1 mm thickness and of an aluminum insert serving to attach the beam to a shaker. The outer layers are of carbon-epoxy composite and were made by hand-laying of 4 layers of KC 200g / m^2 plain carbon fabric and LH 289 epoxy resin with H289 hardener supplied by Havel Composites.

Components of MRE core are carbonyl iron particles of spherical shapes with 2–5 μm in diameter supplied by Sigma-Aldrich and liquid silicone ZA13 and its catalyst produced by Zhermack S.P.A. The core layer is of isotropic MRE with 27% volume fraction of carbonyl iron particles. The detailed procedure for the fabrication of the MRE core layer and its dynamic properties are given in previous articles by the authors [3,4]. The core was glued between the two composite layers by Loctite Super Attak Activator Adhesive.

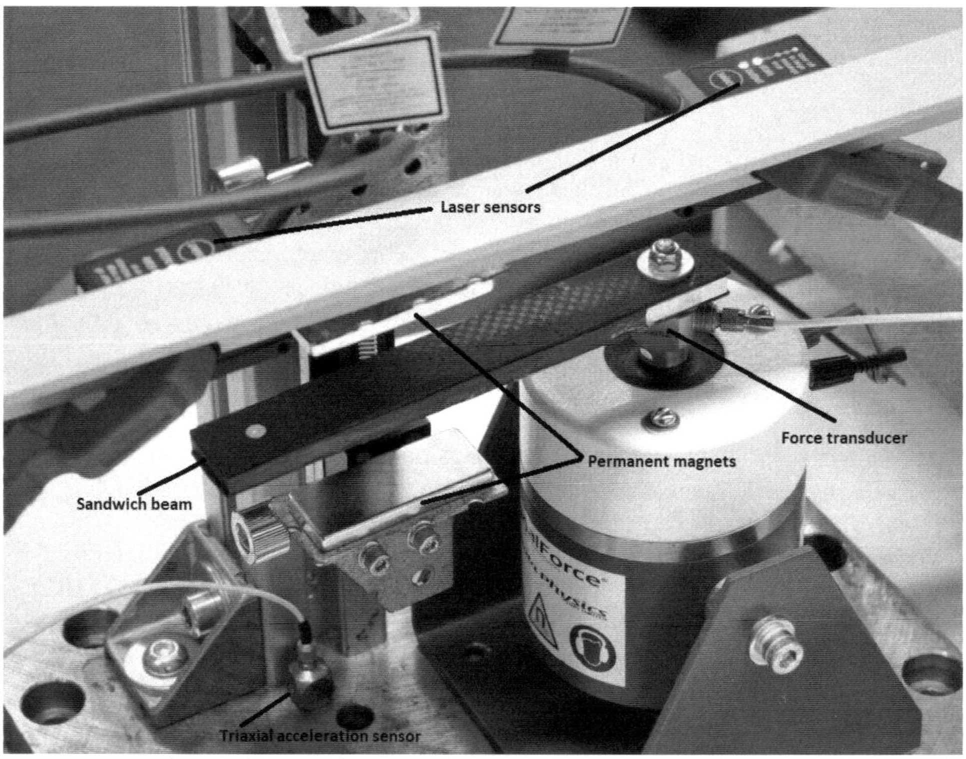

Fig. 1: Experimental setup: the sandwich beam is attached to the shaker at one end - two permanent magnets creating a magnetic field perpendicular to the beam

Experimental setup. Tested sandwich composite beam in Fig. 1 is fixed at its end to Permanent Magnet Shaker SignalForce GW-V4 with PA30E Power Amplifier supplied by

DataPhysics. Kistler force transducer 220 N is placed between the end of the beam and the shaker to monitor the excitation force and to prevent damage of the shaker whose maximum allowed sine force is 17.8 N. The shaker is firmly attached to a heavy steel plate, which rests on a cork board that dampens vibrations and this base lies on a solid table. Kistler Triaxial acceleration sensor is fixed on the steel base plate. The real-time signal from this sensor is used to verify that the transmission of unwanted vibrations is eliminated.

The excitation signal for the SignalForce GW-V4 shaker is generated using the Spider-80X dynamic signal analyzer with Spider-DSA software (Crystal Instruments) and amplified with PA30E Power Amplifier. The amplification is tuned manually using 50 Hz sine excitation signal until the shape of the force signal is sinusoidal with amplitude 1 N.

The external magnetic field is generated by two permanent magnets NdFeB 50x25x12 mm type N35 supplied by WAMAG spol. s r.o. The magnetic field acts in the vertical direction perpendicularly to the skin layers of the beam. The magnets are fixed on high-precision dovetail slides which move synchronously on the vertical rack by a pinion mechanism. This device allows to precisely adjust the distance between the magnets and thus the gradually change of the intensity of the magnetic field. The magnetic flux density is measured with portable Bell-5180 Gauss/Tesla meter with Standard Transverse Probe STD18-0404 supplied by Magnetic Sciences.

Two laser triangulation displacement sensors ILD1420-10 provided by Micro-Epsilon are placed above the tested beam. The first sensor reads the displacement of the target placed at the point of beam fixation to the shaker and the second sensor measures the displacement of the target at the free end of the beam.

Experimental procedure. Forced vibration tests were performed using sinusoidal harmonic loading over a range of frequencies in order to determine the resonant frequencies at different magnitudes of the external magnetic field. The input signal was linear sine sweep in Fig. 2 between the frequencies f_1=1 Hz and f_2=100 Hz and of time duration T=200 s

$$f(t) = f_1 + \frac{f_2 - f_1}{T} t \tag{1}$$

The intensity of the magnetic field was gradually increased between tests. The magnetic field depends on the distance of two permanent magnets and also on the magnetic properties of the beam that is inserted between them. The measured magnetic field for the different distances of magnets is presented in Fig. 3.

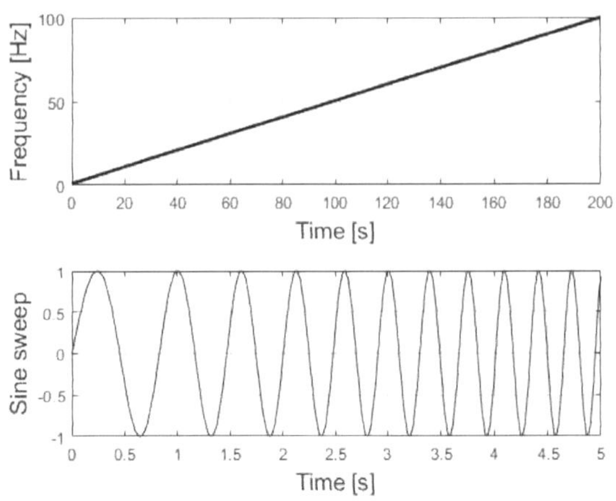

Fig. 2: Sine sweep of forced vibration signal

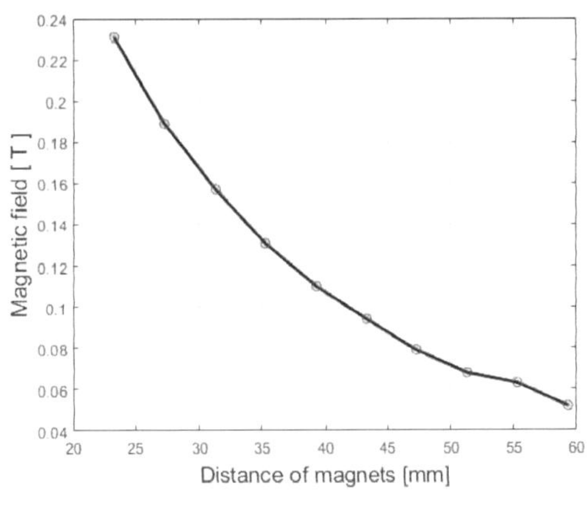

Fig. 3: Magnetic field vs magnets distance

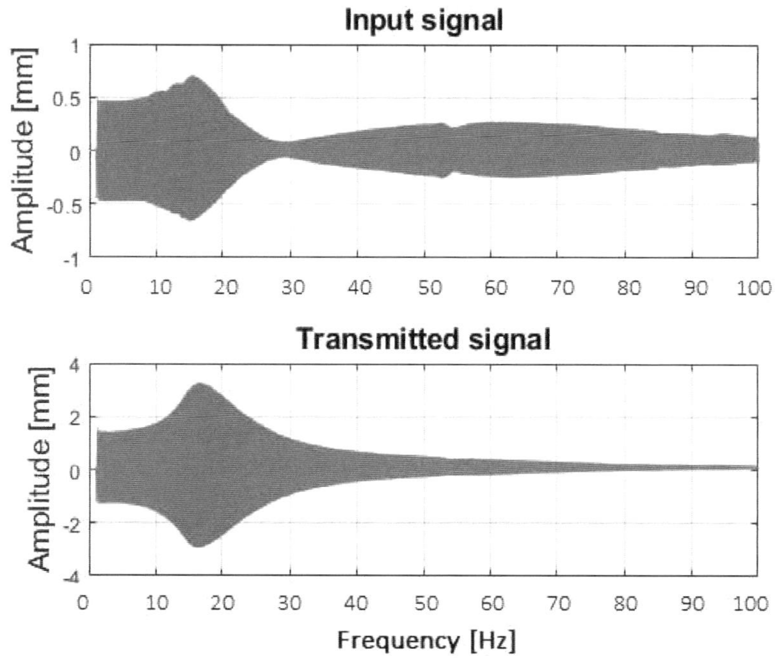

Fig. 4. Displacements of points at the shaker head (above) and at the end of beam (below) measured by laser sensors with sampling frequency 4 kHz

A total of 20 frequency sweep tests were performed with the values of magnetic flux density increasing gradually from 0.0 T to 0.236 T. The displacements of two points of the beam were measured by the laser sensors with the sampling frequency of 4 kHz. Point 1 is on the head of the shaker and point 2 is on the free end of the beam.

Evaluation of experimental data. Amplitude transmissibility was determined for each test based on the recorded data of displacements (deflection) of two points see Fig.4. The amplitude transmissibility is the ratio of the amplitude $a2(f_i)$ of the deflection at the free end of the beam and the amplitude $a1(f_i)$ of the displacement of the shaker head. Both amplitudes are evaluated in Matlab at the same time which corresponds to the selected excitation frequency as can be seen in Fig. 5. The transmissibility was evaluated for frequencies f_i=1,2,3,...,100 Hz.

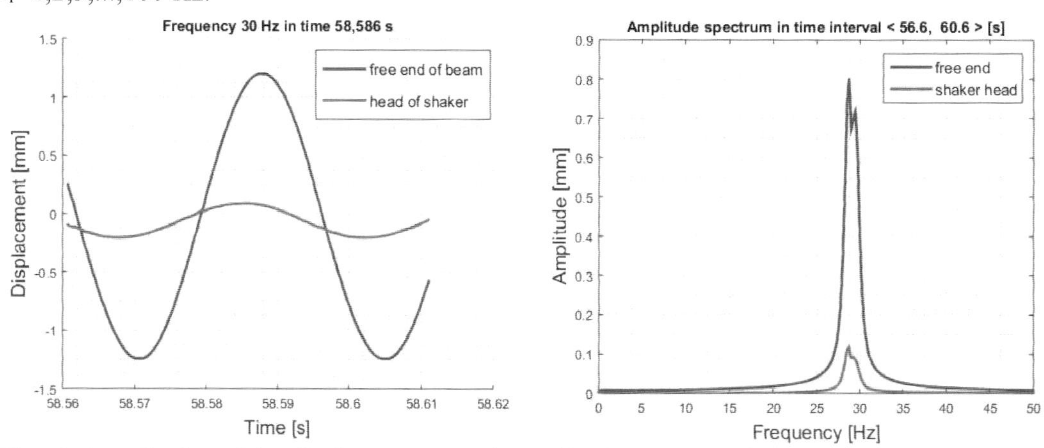

Fig. 5. Amplitudes of the displacements and amplitude spectrum for frequency around 30 Hz

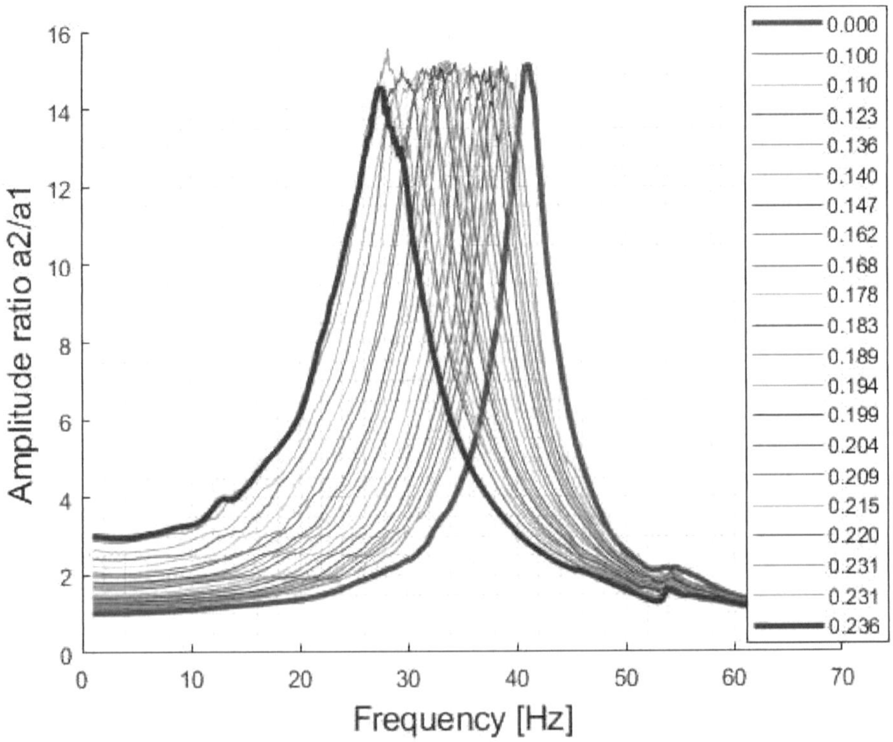

Fig. 6. Amplitude transmissibility a2/a1 in the frequency range 1-70 Hz for the magnetic field increasing gradually from 0 to 0.236 T.

The peaks of amplitude transmissibility curves shift to the lower frequencies as the magnetic field increases as is depicted in Fig. 7.

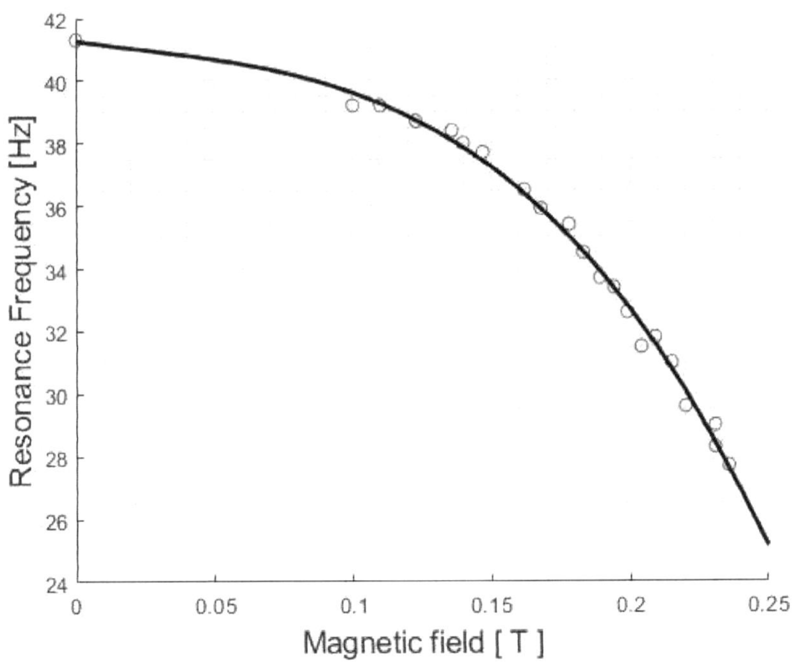

Fig. 7. Decrease of first resonant frequency due to increasing magnetic field

Conclusions

The forced vibration of the sandwich beam with carbon-epoxy composite skins and a viscoelastic MRE core has been examined. The results of experimental research of the forced vibration of the sandwich beam with the constrained layer show that the amplitude and velocity of the beam vibration have changed with varying the external magnetic field. The stiffness and damping properties of the beam vary due to a change in the rheological properties of the MRE in the magnetic field. The stiffness of the beam and the damping of the active constrained layer increased.

We can conclude that the first resonant frequencies of the beam decreased with increasing intensity of the magnetic field as can be seen from the graph in Fig.7. The decrease in frequency is quite significant and is around 35%.

The forced harmonic motion of the three-layer beam in which the damping constrained core layer core is sandwiched between two elastic layers can be described by differential wave equation of sixth-order [8,9]. In this equation the material of viscoelastic core is characterised by complex shear modulus $G^*=G(1+i\beta)$. The dynamic parameters of MRE core material are already determined by authors [3,4]. The numerical modeling of vibrations of sandwich beams with MRE core will be the subject of further research.

The experimental research and numerical modelling of the vibrations of sandwich beam with MRE constrained layer is the promising direction in the search of application of this smart magnetorheological material.

Acknowledgement

This work was supported by the Ministry of Education, Youth and Sports of the Czech Republic and the European Union - European Structural and Investment Funds in the frames of Operational Programme Research, Development and Education - project Hybrid Materials for Hierarchical Structures (HyHi, Reg. No. CZ.02.1.01/0.0/0.0/16_019/0000843).

References

[1] J.D. Carlson, M.R. Jolly, MR fluid, foam and elastomer devices, Mechatronics (2000) 555-569.

[2] Y. Li, J. Li, W. Li, H. Du, A state-of-the-art review on magnetorheological elastomer devices. Smart materials and structures 23 (2014) 123001.

[3] T.H. Nam, I. Petríková, B. Marvalová, Experimental characterization and viscoelastic modeling of isotropic and anisotropic magnetorheological elastomers, Polymer Testing 81 (2020) 106272.

[4] T.H. Nam, I. Petríková, B. Marvalová, Experimental and numerical research of dynamic mechanical properties of magneto-sensitive elastomeric composites, in: Proceedings of the 11th European Conference on Constitutive Models for Rubber, Nantes, France, Taylor & Francis, 2019, pp.138 – 143.

[5] M.A. Cantera, M. Behrooz, R.F. Gibson, Modeling of magneto-mechanical response of magnetorheological elastomers (MRE) and MRE-based systems: a review. Smart Materials and Structures 26 (2017) 023001.

[6] F. de Souza Eloy, et al. A numerical-experimental dynamic analysis of composite sandwich beam with magnetorheological elastomer honeycomb core. Composite Structures 209, (2019) 242-257.

[7] T. Szmidt, D. Pisarski, C. Bajer, B. Dyniewicz, Double-beam cantilever structure with embedded intelligent damping block. Dynamics and control. Journal of Sound and Vibration 401 (2017) 127-138.

[8] D.J. Mead, The measurement of the loss factors of beams and plates with constrained and unconstrained damping layers: A critical assessment. Journal of sound and vibration 300 (2007) 744-762.

[9] D.J. Mead, S. Markus, The forced vibration of a three-layer, damped sandwich beam with arbitrary boundary conditions. Journal of sound and vibration 10 (1969) 163-175.

Experimental Analysis of Degradation of Bending Properties of Composite Materials After Accelerating Aging

PETRŮ M.[1,a], WANG X.[1,b]

[1]Institute for Nanomaterials, Advanced Technologies and Innovation, Technical University of Liberec, Studentska 2, Liberec, 461 17, Czech Republic

[a]michal.petru@tul.cz, [b]xiaomeng.wang@tul.cz

Keywords: Experimental analysis, bending properties, flexural strength, flexural modulus composite materials, degradation, accelerating aging

Abstract. Eco-friendly fiber reinforced polymer (FRP) has gained much attention recently because of its low density, low cost and abundant resources. Flax FRP (FFRP) is a commonly used eco-friendly FRP in various fields, such as automobile, packaging, and building construction. In this paper was prepared experimental analysis of degradation of bending properties of FFRP after accelerating aging. Accelerated aging tests are useful to predict the performance of FFRP only if such tests can be shown to correlate with natural aging. The experimental study carried out in this paper aimed at developing a better understanding of the effect of accelerated aging on the mechanical behavior of FFRP. The flexural strength and flexural modulus of FFRP decrease significantly with aging time.

Introduction

In recent years, the use of flax fiber as a potential alternative to traditional fiber such as glass fiber and carbon fiber in FRP has gained attention among researchers [1-4, 10-11]. Compared to glass fiber and carbon fiber, flax fiber presents low density, low cost and easy recyclability. However, the porous structure of flax fiber leads to high water absorbability [5]. The interaction between hydrophilic fiber and hydrophobic matrix results in weakening of fiber/matrix interface, dimensional instability, matrix cracking and degradation of mechanical properties of the FFRP [6]. Some experimental studies have been done to study the effect of humidity and temperature on mechanical behavior of FFRP. The weakening of matrix-fiber interface is the main damage mechanism caused by water aging of the FFRP. In [7] found that the residual tensile strength of FFRP subjected to salt solution is higher than GFRP. Also authors reported that after 38 days of hygrothermal aging with a relative humidity of 90%, the elastic modulus and tensile strength of FFRP decrease by 58% and 52% [8].

Approach of study of bending properties of samples after accelerating aging

The FFRP specimen is made of approximately 30% fiber in volume. The dimension of FFRP specimen in rectangular form is 50mm×10mm×1.35mm. Accelerated aging tests are useful to predict the performance of FFRP only if such tests can be shown to correlate with natural aging. In 2018 was carried out natural aging test in Nanjing (China) [5]. Results of bending properties of FFRP after natural aging are show in Fig.1. And the decreasing rate of the flexural properties slows down with the increase of aging time. Compared with reference group, the flexural strength decreases by 11.17%, 14.89%, 15.49%, and the flexural modulus decreases by 21.33%, 32.28%, 35.77% after experimental days.

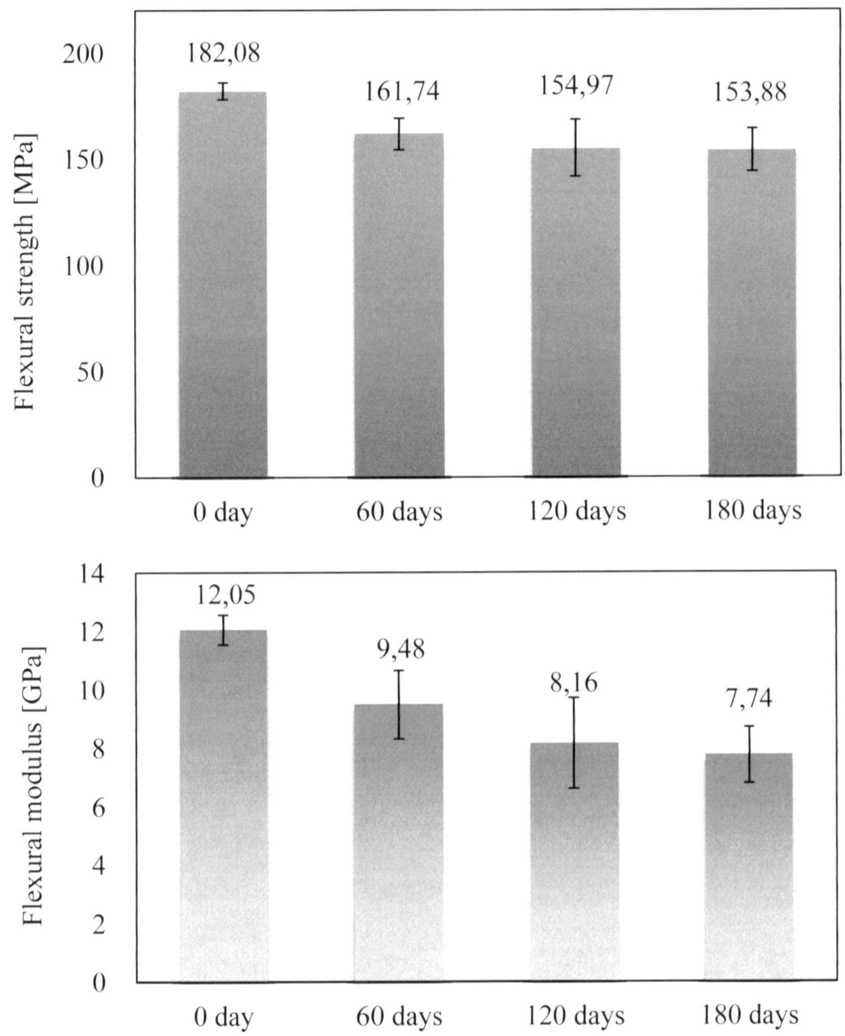

Fig. 1: Bending properties of FFRP after natural aging: (above) Flexural strength, (below) Flexural modulus

Therefore, it is important to find the conversion relationship between accelerated aging and natural aging. According to equation as described in [5] is the acceleration factor k_t can be obtained as (1).

$$k_t = \frac{t_1}{t_2} = \frac{e^{-C/(T_2\phi_2)}}{e^{-C/(T_1\phi_1)}}, \tag{1}$$

where t_1 is the natural aging time, t_2 is the accelerated aging time, T_1 and ϕ_1 are the temperature and relative humidity under natural aging conditions, T_2 and ϕ_2 are the temperature and relative humidity under accelerated aging environment, C is experimental coefficient.

Another method of accelerated moisture absorbed estimation mentioned in [9] is given by equation (2).

$$1/k_t = \frac{t_2}{t_1} = \frac{D_1}{D_2}, D = D_0 \exp(-C/T) \quad (2)$$

In this mathematical equation, the quotient of two diffusion coefficients is taken as the acceleration coefficient. Where D_0 and C is the diffusion-related constants at room temperature.

Experimental analysis of bending properties of samples after accelerating aging

The specimens were periodically taken out and tested under three-point loading to assess the flexural properties (Fig. 2). All experimental values were obtained by an average value of five specimens. The loading rate was 2 mm/min. Load and deflection were recorded by the test device.

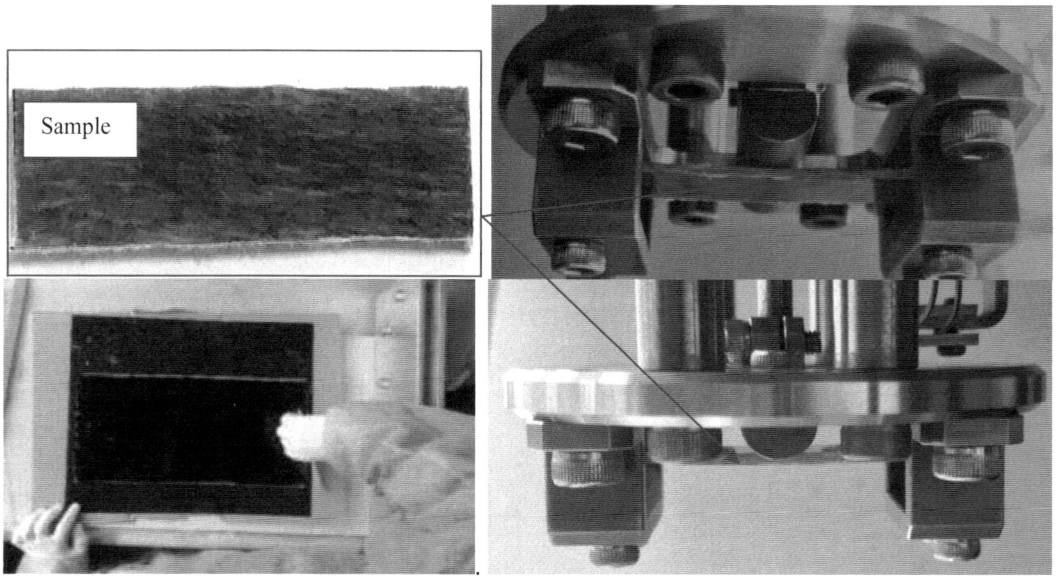

Fig. 2: Preparing of sample 50mm×10mm×1.35mm (left), device for experiment (right)

Results and discussion

To determine the correlation of accelerated aging test to natural aging test, the daily temperature and humidity during natural aging test were recorded. Corresponding accelerated aging time is calculated according to acceleration factor k_t in Eq. (1). Results are in Tab.1. According to Eq. (1), the accelerated aging time corresponding to each natural aging period can be obtained as $t_2 = \frac{t_1}{k_t}$. Where t_2 is the accelerated aging time, t_1 is the natural aging time.

The accelerated aging time corresponding to 60 days, 120 days, and 180 days of natural aging is 1.08 h, 37.2 h, and 167.8 h is in Fig.3. Accelerated aging tests were carried out according to the calculated time. For example, the bending strength of FFRP under accelerated aging is 20.28%, 16.71%, and 13.51% higher than the natural aging. It mainly because only two factors (temperature and humidity) are considered in the accelerated test, and the specimens will be affected not only by temperature and humidity, but also by light, rainwater, chemical erosion, et al. The bending modulus of specimens under accelerated aging corresponding to 180 days of natural aging is lower, this phenomenon may be caused by higher water uptake.

As shown in **Chyba! Nenalezen zdroj odkazů.**, the color of the specimens has changed obviously after aging. And with the increase of aging time, the boundary between matrix and fiber is more and more obvious. The comparison between the flexural strength and flexural modulus of FFRP under natural aging and corresponding accelerated aging is show in Fig. 1. The results show that the flexural strength and flexural modulus under both aging conditions decrease with time. The flexural strength obtained from the corresponding accelerated aging is slightly larger than those obtained from the natural aging test.

Tab. 1: Results of k_t factor

Period	1	3	5	6	7	8	9
k_t	5.60×10^{76}	2.51×10^{5}	2.18×10^{3}	2.49×10^{2}	3.90×10^{2}	3.26×10^{2}	39.70
Period	10	12	15	18			
k_t	58.57	40.68	13.7658	35.098			

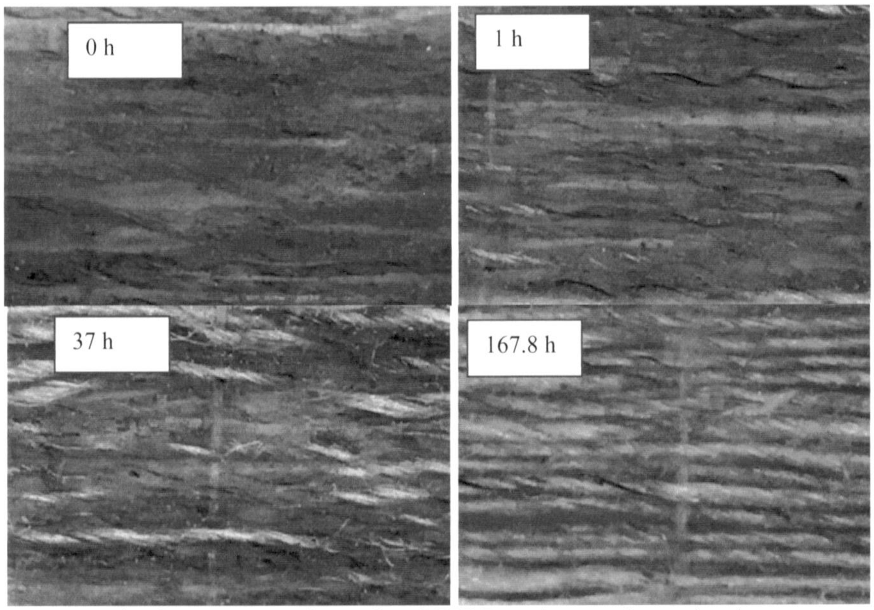

Fig. 3: Damage of sample of FFRP under accelerated aging

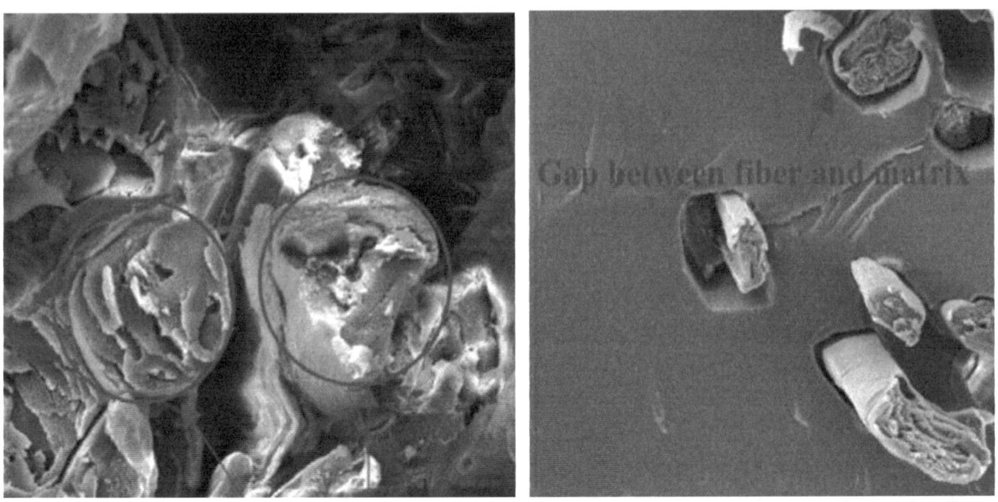

Fig. 4: Detail of morphology structure of FFRP sample: before under accelerated aging (left), after 167.8 hour under accelerated aging (right)

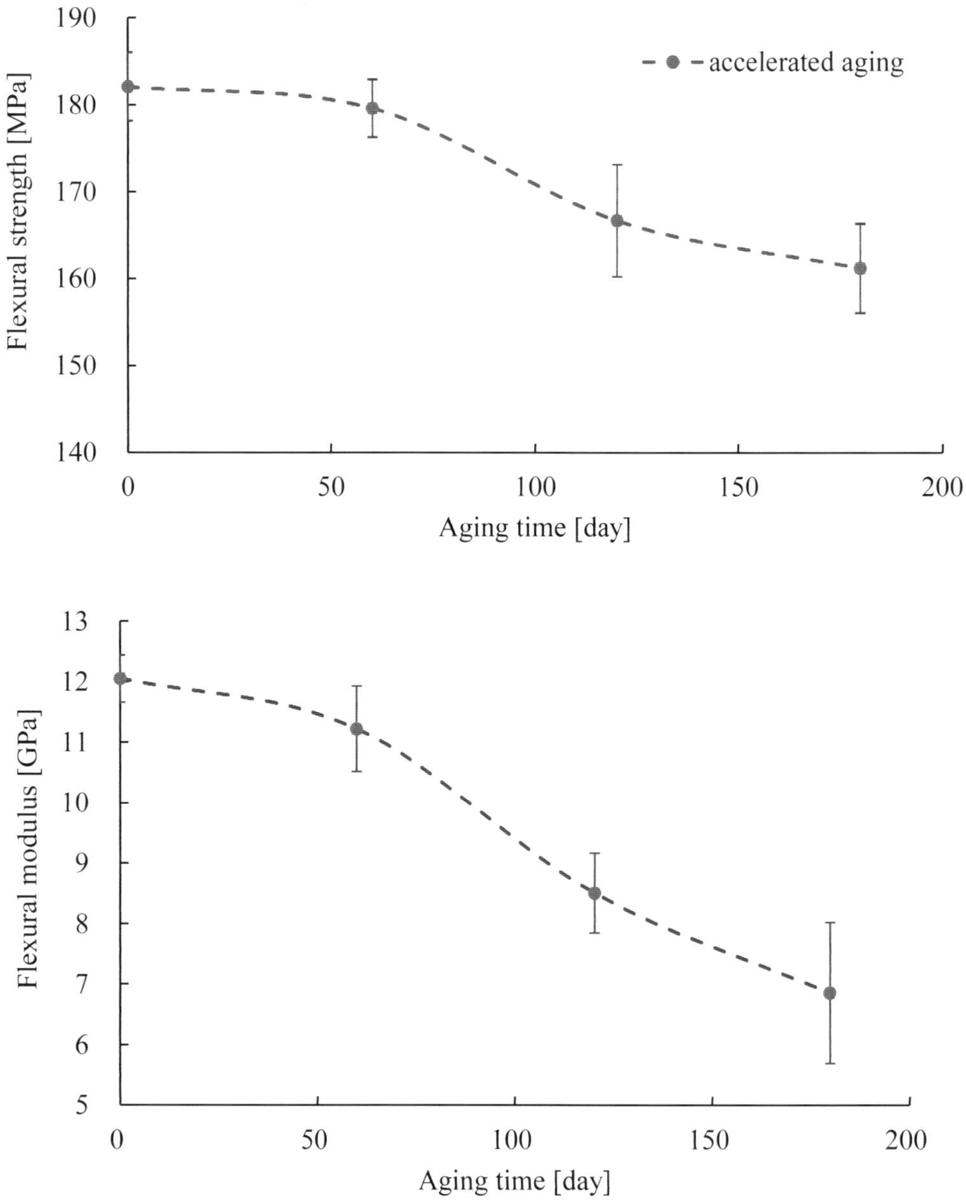

Fig. 1 Results of bending properties of FFRP under accelerated aging: (above) flexural strength, (below) flexural modulus

Conclusions

This article focused on an experimental and model-based study aimed at developing a better understanding of the effect of natural ageing and accelerated ageing on the mechanical behaviour of a flax fibre reinforced composite. The bending force and the FFRP bending module decreases significantly with age. The decreasing rate of bending properties slows with increasing age. The accuracy of the model results was verified by comparing the results of the rapid ageing and natural ageing test.

The results of the model may have applications in the durability analysis of other composite materials reinforced with natural fibres, and influencing factors such as ultraviolet light, chemical erosion, stress and so on may be introduced by adjusting parameters.

Acknowledgement

This work is partially supported by the Ministry of Education, Youth and Sports of the Czech Republic and the European Union (European Structural and Investment Funds - Operational Programme Research, Development and Education) in the frames of the project "Modular platform for autonomous chassis of specialized electric vehicles for freight and equipment transportation", Reg. No. CZ.02.1.01/0.0/0.0/16_025/0007293.

References

[1] P. Hana, A. Inneman, V. Daniel, L. Sieger, M. Petru, Mechanical properties of carbon fiber composites for applications in space, Proceedings of SPIE - The International Society for Optical Engineering, Vol. 9442, 2015

[2] X. Wang, M. Petrů, H. Yu, The effect of surface treatment on the creep behavior of flax fiber reinforced composites under hygrothermal aging conditions, Construction and Building Materials. 208 (2019) 220–227.

[3] M. Petrů, T. Martinec, J. Mlýnek, Numerical model description of fibres winding process for new technology of winding fibres on the frames, Manufacturing Technology. 16 (2016) 778–785.

[4] M. Petrů, O. Novak, P. Lepsik, Analysis and measurement of the charge intensity of the selected electrosp. Electrodes, Applied Mechanics and Materials, 486 (2014) 217-222.

[5] X. Wang, M. Petrů, Degradation of bending properties of flax fiber reinforced polymer after natural aging and accelerated aging, Construction and Building Materials. 208 (2019) 220–227.

[6] E.H. Saidane, D. Scida, M. Assarar, R. Ayad, Assessment of 3D moisture diffusion parameters on flax/epoxy composites, Compos. A Appl. Sci. Manuf. 80 (2016) 53-60.

[7] A. Regazzi, R. Léger, S. Corn, P. Ienny, Modelling of hydrothermal aging of short flax fiber reinforced composites, Compos. A Appl. Sci. Manuf. 90 (2016) 559-566.

[8] L. Yan, N. Chouw, K. Jayaraman, Effect of UV and water spraying on the mechanical properties of flax fabric reinforced polymer composites used for civil engineering applications, Mater. Design, 71 (2015) 17-25.

[9] S. Jia, Q. Ji, J. Qin, T. Zou, Prediction of influence of hygrothermal environment on mech. properties of CFRP laminates, Modern Plastics Process. Appl., 30 (2018) 23-27.

[10] X. Wang, M. Petrů, Mode I fracture evaluation of CFRP-to-concrete interfaces subject to aggressive environments agents: Freeze-thaw cycles, acid and alkaline solution, Composites Part B: Engineering,168, 581-588

[11] J. Mlynek, M. Petrů, T. Martinec, S. Koloor. Fabrication of High-Quality Polymer Composite Frame by a New Method of Fiber Winding Process, Polymers,12(5), 1-30.

EAN 2020
58th conference on experimental stress analysis

Delamination Strength of an Arterial Wall

PETŘIVÝ Z.[1,a], HORNÝ L.[1,b], ADÁMEK T.[2,c]

[1]Faculty of Mechanical Engineering, Czech Technical University in Prague, Technická 4, 160 00 Praha 6, Czech Republic

[2]Department of Forensic Medicine and Toxicology, Regional Hospital Liberec, Husova 357/10, 460 63 Liberec, Czech Republic

[a]zdenek.petrivy@fs.cvut.cz, [b]lukas.horny@fs.cvut.cz, [c]tomas.adamek@nemlib.cz

Keywords: Age-related changes, Aorta, Delamination, Dissection, Peeling test

Abstract. Aortic dissection is a life-threatening disease. It is manifested by a separation of the layers of an artery wall, and may end with total rupture and internal haemorrhaging. The exact conditions under which the dissection is initiated remain largely unexplored. In this study, we would like to contribute to the elucidation of these conditions. It is generally accepted that the dissection propagates by a delamination mechanism. The main objective is to collect a set of experimental data to be able to describe age-related changes in the delamination strength of the human aorta. Preliminary results of the peeling tests show that a strong correlation between age and delamination strength exists. This is in accordance with clinical observations which reveal that elderly people are more susceptible to the dissection than young individuals.

Introduction

Aortic dissection is a life-threatening disease manifested by a separation of the layers of an artery wall [1–3]. Although one could consider aortic dissection to be a relatively rare disease, the incidence is typically reported ranging from 3 to 6 cases per 100 000 per year, however the lethality of the dissection is rather high. According to [4], 37% of patients who reach the hospital alive die in next 30 days, and approximately 20% patients die before they receive medical intervention [5].

During the dissection, blood enters the wall and causes the delamination of its layers and consequently flows in a false lumen. A rupture of the weakened aorta may occur. The exact tear initiation mechanism is unknown and mechanical events underlying dissection have not been definitively established so far. In this study, we would like to contribute to elucidation of conditions under which the dissection appears by means of experimental research, focused on delamination properties of the human aorta. Our research is planned as a three-year study and this is only a preliminary report. The main objective is to collect a set of experimental data to be able to describe age-related changes in the delamination strength of the human aorta in detail.

Methods

A method adopted to characterize delamination properties of the aorta is the so-called peeling test. This experimental protocol was, in the context of biomechanics of the aortic dissection, introduced by Sommer in 2008 [1] and adopted in further studies focused on delamination properties of arteries [6–9]. It resembles the mode I crack opening that is widely used in fracture mechanics, see Fig. 1. The main advantage of this experimental technique consists in the controllable crack propagation which allows for the quantification of fracture energy.

As indicated in Fig. 1, when the clamps move apart, forces induced by them open the crack front and delamination takes place. The intact portion of the sample shortens as the clamps continue in their movement. In the final stage of the experiment, the tested sample falls apart into two separate parts.

The experiments were carried out with the help of the multipurpose tensile testing machine Zwick/Roell (Messphysik). Both the delamination force F (the force that is necessary to increase a tear length), and the tear lengths were recorded on PC. Subsequently, the delamination strength was computed as a ratio of the force to a width of the sample. During the experiment, clamps moved with the velocity 0.1 $mm \cdot s^{-1}$, and HBM U9C +/- 25N force transducers were used.

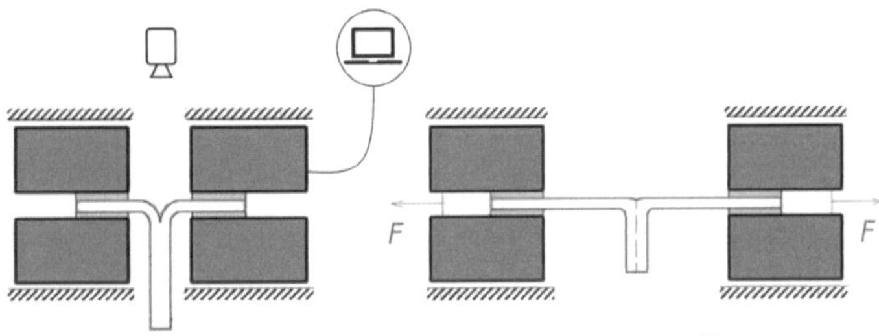

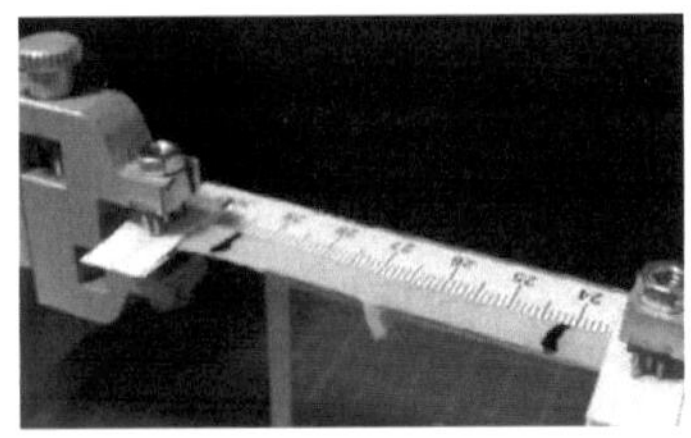

Fig. 1: Peeling test.

Materials

Samples of the human descendent thoracic aorta were obtained in regular autopsies carried out in the Department of Forensic Medicine and Toxicology at the Regional Hospital Liberec. The usage of the post-mortem tissue has been approved by the Ethical Committee of the Regional Hospital Liberec.

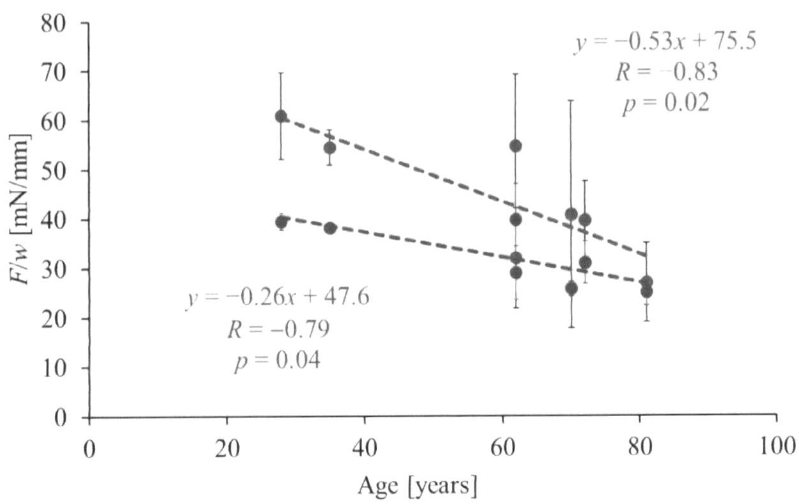

Fig. 2: Age-related changes in the delamination strength (mean delamination force per mean width of the sample). The brown line corresponds to circumferentially oriented samples, and the blue line was obtained in the linear regression of observations collected in experiments with samples aligned in the longitudinal direction.

The total number of donors was 7. Donors' age ranged from 28 to 81 years (an average ± sample standard deviation was 59 ± 20 years). Post mortem interval between death and experiment ranged from 2 to 6.5 days (an average ± sample standard deviation was 3.6 ± 1.7

day). Until the experiment, the samples were stored in refrigerators at 5°C. No putrefaction changes were observed. Three strips of the aorta aligned in the circumferential direction and three strips in the longitudinal direction were cut from each donor.

Results and Conclusions

Results of preliminary experiments are shown in Fig 2. It is a graph of the force per width of the sample necessary to propagate a crack front along the strip obtained at a given donor's age. Each point expresses the mean computed from three strips. Error bars indicate sample standard deviation. The regression lines, which were created separately for longitudinal and circumferential samples, clearly show that the delamination strength correlates negatively with age. Pearson correlation coefficient is $R = -0.83$, and $R = -0.79$, for longitudinal and circumferential direction, respectively, and in both cases reached statistical significance. It is in accordance with our working hypothesis that gradual deterioration of the crack resistance accompanies ageing of our arteries.

However, further experiments will be necessary to confirm it definitively, because the hypothesis as such may not be accepted at an initial reading. One could argue that we rather should expect age-related stiffening because of overall crosslinking and calcification which are well-known consequences of the ageing of human arteries. On the other hand, this contradiction between crack propagation properties and elastic properties of bulk material, is for example well-known for brittle materials which typically are stiff but at the same time very fragile.

Collected data also indicates that the delamination properties of the human thoracic aorta are anisotropic. It is because the delamination strength of the longitudinally oriented samples was always higher than the strength observed for the samples aligned in the circumferential direction. This is in accordance with results available in the literature [1,6,9] which also suggest that delamination resistance is a direction-dependent property.

Acknowledgement

This study has been supported by Czech Science Foundation in the project GA20-11186 entitled "Mechanics of Arterial Delamination and Crack Propagation".

References

[1] G. Sommer, T.C. Gasser, P. Regitnig, M. Auer, G.A. Holzapfel, Dissection properties of the human aortic media: An experimental study, J. Biomech. Eng., 130 (2008) art. no. 021007.

[2] C.A. Nienaber, R.E. Clough, Management of acute aortic dissection, The Lancet, 385 (2015) 800–811.

[3] T.-Y. Yeh, C.-Y. Chen, J.-W. Huang, C.-C. Chiu, W.-T. Lai, Y.-B. Huang, Epidemiology and medication utilization pattern of aortic dissection in Taiwan: A population-based study, Medicine, 94 (2015) art. no. e1522.

[4] C. Olsson, S. Thelin, E. Ståhle, A. Ekbom, F. Granath, Thoracic aortic aneurysm and dissection: Increasing prevalence and improved outcomes reported in a nationwide population-based study of more than 14 000 cases from 1987 to 2002, Circulation, 114 (2006) 2611–2618.

[5] I. Mészáros, J. Mórocz, J. Szlávi, J. Schmidt, L. Tornóci, L. Nagy, L. Szép, Epidemiology and clinicopathology of aortic dissection: A population- based longitudinal study over 27 years, Chest, 117 (2000) 1271–1278.

[6] J. Tong, T. Cohnert, P. Regitnig, J. Kohlbacher, R. Birner-Gruenberger, A.J. Schriefl, G. Sommer, G.A. Holzapfel, Variations of dissection properties and mass fractions with thrombus age in human abdominal aortic aneurysms, J. Biomech., 47 (2014) 14–23.

[7] Y. Wang, J. A. Johnson, F. G. Spinale, M. A. Sutton & S. M. Lessner, Quantitative Measurement of Dissection Resistance in Intimal and Medial Layers of Human Coronary Arteries, Exp. Mech., 54 (2014) 677–683.

[8] M. Kozuń, M. Kobielarz, A. Chwiłkowska, C. Pezowicz, The impact of development of atherosclerosis on delamination resistance of the thoracic aortic wall, J. Mechan. Behav. Biomed. Mater., 79 (2018) 292–300.

[9] M. Kozuń, Delamination properties of the human thoracic arterial wall with early stage of atherosclerosis lesions, J. Theor. App. Mech., 54 (2016) 229–238.

EAN 2020
58th conference on experimental stress analysis

Methodology of Methylene Blue Test for Earth Analysis

PLAČEK OTCOVSKÁ T.[1,a], MUŽÍKOVÁ B.[1,b], PADEVĚT P.[1,c]

[1]Deparment of Mechanics K132; Faculty of Civil Engineering; Czech Technical University in Prague; Thákurova 7/2077, 166 29 Praha 6 Dejvice, Czech Republic

[a]tereza.otcovska@fsv.cvut.cz, [b]barbora.muzikova@fsv.cvut.cz, [c]pavel.padevet@fsv.cvut.cz

Keywords: Methylene blue dye, Clay, Adsorption, Stirring time,

Abstract. Unfired earth is historical building material which is getting to forefront of interest. Final properties of unfired earth are influenced by composition of earth mixtures. Designing unfired earth building structures is complicated because composition of natural earth is various. The essential component of earth mixtures is clay because clay fulfils a function of binder. Methylene blue test could be a good method for analysis of unknown earth and it could help determine the type and amount of clay in the earth mixtures. A necessary condition for the relevant results from the experiment is to follow the procedure. Experiences from experimental measurements are described in this paper. Based on the results from the first experimental measurement, a modification of the procedure was proposed. Part of the experiment is stirring an earth sample with a dye solution. Stirring time is an important factor for a successful measurement.

Introduction

Unfired earth is a building material which had become minority in 2nd half of 20th century because of modern building materials such as steel or concrete have have became readily available. Nowadays, the unfired earth is getting back to forefront of interest for its properties that fit into principles of a sustainable building and improving microclimate in interiors [1–4].

Earth mixture is a basic material for production of unfired earth constructions. Uncertain composition of natural earth is the main reason why using of unfired earth in construction is complicated. It is because of final properties of unfired earth constructions are greatly influenced by the earth composition [4-7].

Methylene blue test is described in norm ČSN EN 933-9. Methylene blue test could be a good method for analysis of the earth composition. Basic idea of this research is that every kind of clay (main component of earth mixture) has different adsorption capacity of methylene blue dye. If adsorption capacity of individual clay is known, this test could be a good method for analyzing earth composition [7-9].

The goal of the research is to determine the adsorption rate of individual kinds of clays. Relevant results could be obtained only when procedure of experimental measurement is carefully followed. The methylene blue test procedure was modified upon gained experience. This experience and modification of methodology are described in this paper.

Methylene Blue Test

For methylene blue test a solution of methylene blue die (methylthioninium chloride - $C_{16}H_{18}ClN_3S$) by concentration 10 g/l is used. Methylene blue solution is inserted into earth sample (suspension of earth and distilled water) by a burette - Fig. 1). Earth sample with the solution is stirred and after that a drop of the mixture is placed on a filter paper.

This process is repeated and each sample drop with increasing amount of methylene blue solution is marked (Fig. 2). Methylene blue test is finished when a blue ring spreads around a drop sample (mark 3 in Fig. 2) because clay has already adsorbed the maximum amount of methylene blue.

Fig. 1: Dye dosed into the earth sample

Dye solution consumption increases with increasing amount of clay in earth samples. Amount of dye adsorbed also depends on a kind of clay. Amount of dye adsorbed is defined as the weight amount of dye in grams that is adsorbed by 1 kilogram of test material. It is a multiple of amount of amount of solution (m_s) used and amount of methylene blue in 1 g of solution ($m_{MB,1ml}$) divided by amount of tested material (m_{ES}) (Eq. 1).

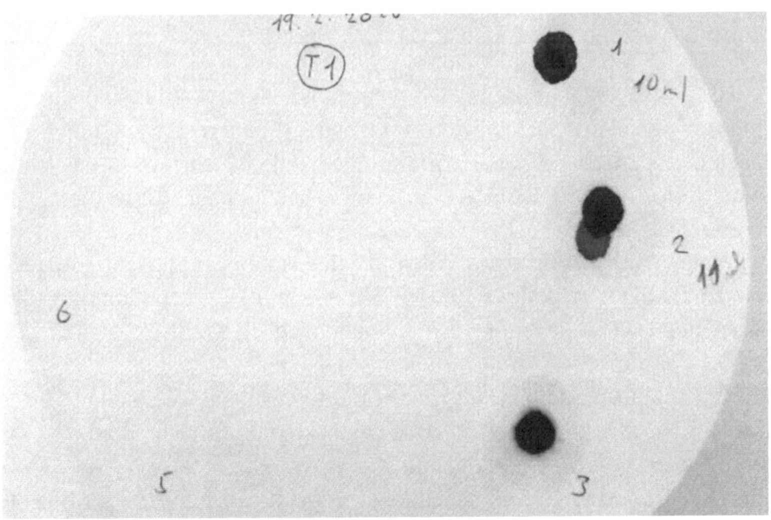

Fig. 2: A result of methylene blue test

$$M_{MB} = \frac{m_s . m_{MB,1ml} . 1000}{m_{ES}} \quad (1)$$

$M_{MB} -$ *amount of dye absorbed* $\left[\frac{g}{kg}\right]$

$m_s -$ *amount of methylene blue solution* $[g]$

$m_{MB,1ml} -$ *amount of methylene blue in 1 g of methylene blue solution* $[g]$

$m_{ES} -$ *amounto of tested material* $[g]$

First Results from Methylene Blue Test and Reason for Modification of Methodology

At beginning of our research was stirring time 1 minute for each dose of methylene blue solution. In some cases were stirring time longer but there was not clear rule for it.

Results of experiment were not repeatable. Three adsorption measures were made for sample B4-75 (11.25 g kaolinite clay, 3.75 g sand). Obtained values of adsorption are rather different. Amount of adsorbed dye, amount and number of doses of solution and stirring time are shown in Table 1. It can be seen that adsorption is smaller when stirring time is shorter. This is clearly visible in Figure 3.

Table 1: Amount of adsorbed dye by kaolinite clay before amendment of stirring time (sample B4-75: 11.25 g kaolinite clay, 3.75 g sand)

Measure	Number of dose	Amount of solution [ml]	Stirring time [min]	Total stirring time [min]	Total amount of solution [ml]	Dye adsorbed [g/kg]
1	1	30	3	10	47	31
	2	5	1			
	3	2	1			
	4	2	1			
	5	2	1			
	6	2	1			
	7	2	1			
	8	2	1			
2	1	40	3	4	42	27.7
	2	2	1			
3	1	25	3	3	35	23.1

Another experimental measurement was made to verify this finding. Based on the experience gained, it was stated that the amount of adsorbed dye is dependent on stirring time. Therefore, was defined stirring time 1 minute for each 1 millilitre of methylene blue dye solution of concentration 10 g/kg inserted into each earth sample. Repeatability of results was satisfying after this modification.

Table 1 and 2 show results and procedure of the same experiment measurement with a different procedure. Table 1 describes an experiment with shorter stirring time and Table 2 describes an experiment with stirring time 1 minute for each millilitre of dye solution in earth sample.

Table 1 and 2 show that the first dose of solution to earth sample was the largest. This first part of experiment most affected result in case of short mixing time. The first dose makes up more than half of total volume of dye solution in an earth sample but stirring time was lengthened by only 2 minutes. Therefore, there was probably not enough time for methylene blue dye adsorption. As in this case, other experimental measurements were performed before modification of the methodology.

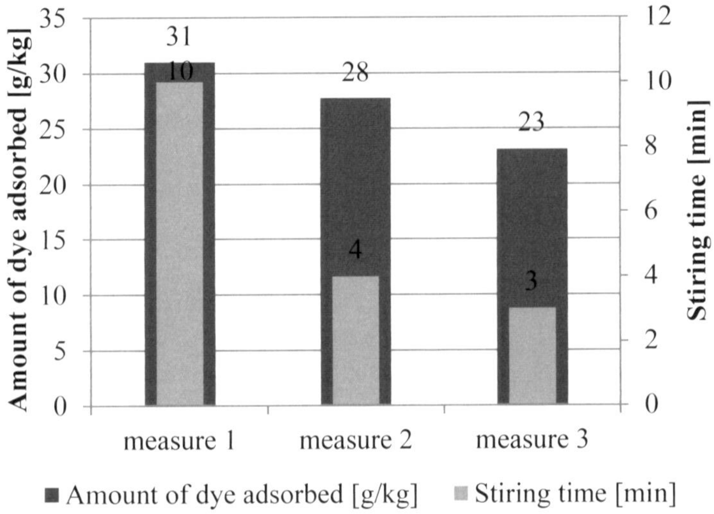

Fig. 3: Amount of adsorbed dye by kaolinite clay (sample B4-75: 11.25 g kaolinite clay, 3.75 g sand)

Table 2: Amount of adsorbed dye by kaolinite clay with stirring time 1 minute for each millilitre (sample B4-75: 11.25 g kaolinite clay, 3.75 g sand)

Measure	Number of dose	Amount of solution [ml]	Stirring time [min]	Total stirring time [min]	Total amount of solution [ml]	Dye adsorbed [g/kg]
1	1	16	16	70	70	46.2
	2	14	14			
	3	10	10			
	4	10	10			
	5	10	10			
	6	10	10			
2	1	30	30	67	67.2	44.4
	2	15	15			
	3	15	15			
	4	3	3			
	5	4.2	4			
3	1	25	25	65	65	42.9
	2	17	17			
	3	10	10			
	4	5	5			
	5	3	3			
	6	5	5			

Results after Modification of Methodology

To get an idea of stirring time influence an adsorption capacity of kaolinitic (B4) and illite-kaolinitic (KR) clay was compared. Composition of results of earth samples before and after modification of methodology is shown in Table 3 and Figure 4, 5.

Table 3: Composition of results of earth samples before and after modification of methodology (earth samples with stirring time of 1 minute for each millilitre of dye solution are emphasizing by *italics)*

Earth sample /clay	Clay/ sand ratio	Amount of clay [g]	Average stirring time [min]	Average adsorbed dye [g/kg]	Relative standard deviation [%]
B4-100	100/0	15	18.3	42.7	17.0
/kaolinite	*100/0*	*15*	*87.7*	*57.8*	*3.5*
B4-75	75/25	11.25	5	27.3	4.6
/kaolinite	*75/25*	*11.25*	*67.4*	*44.5*	*3.7*
B4-50	50/50	7.5	4.7	18.3	8.3
/kaolinite	*50/50*	*7.5*	*44.3*	*29.3*	*11.6*
B4-25	25/75	3.75	3.7	9.0	8.4
/kaolinite	*25/75*	*3.75*	*22*	*14.5*	*4.5*
KR-100	100/0	15	21.3	27.1	11.2
/Illite-kaolinitic	*100/0*	*15*	*48.7*	*32.1*	*6.6*
KR-75	75/25	11.25	8.7	25.1	13.2
/Illite-kaolinitic	*75/25*	*11.25*	*38.7*	*25.5*	*14.2*
KR-50	50/50	7.5	5.7	18.5	7.1
/Illite-kaolinitic	*50/50*	*7.5*	*27.3*	*18.0*	*7.6*
KR-25	25/75	3.75	2.0	7.5	20.4
/Illite-kaolinitic	*25/75*	*3.75*	*13.2*	*8.7*	*12.4*

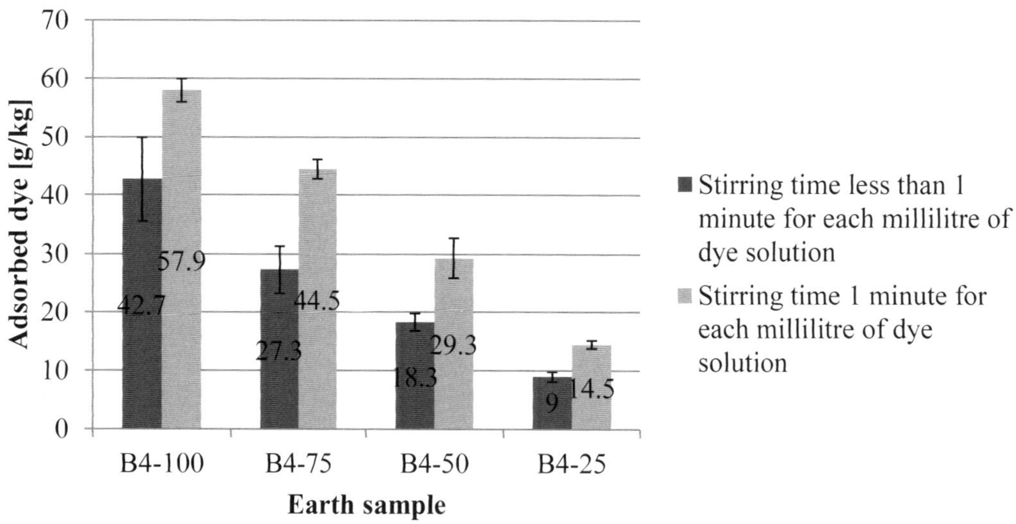

Fig. 4: Amount of adsorbed dye by kaolinite clay (B4) before and after modification of methodology

The results show that stirring time is significant differently. Stirring time for kaolinitic clay is crucial (Fig. 4). Stirring time for illite-kaolinitic clay is less significant (Fig. 5). Comparison of the stirring time influence on amount of adsorbed dye is expressed as a percentage. Amount of adsorbed dye by earth sample which stirring time was 1 minute for each millilitre of dye solution is 100 % (it is green column in Figure 4, 5).

Difference of adsorption capacity of kaolinitic clay is at least 26 % (Fig. 4, earth sample B4-100) but in one case is this difference more than 38 % (Fig. 4, earth sample B4-75).

Influence of stirring time is significantly smaller for illite-kaolinite clay. In two cases are values of adsorption dye more than 13 % (Fig. 5, earth sample KR-100, KR-25), in one case is this difference only 1.6 % (Fig. 5, earth sample KR-75) and in one case adsorption capacity is 2.8 % larger for experimental measure with shorter stirring time (Fig. 5, earth sample KR-50).

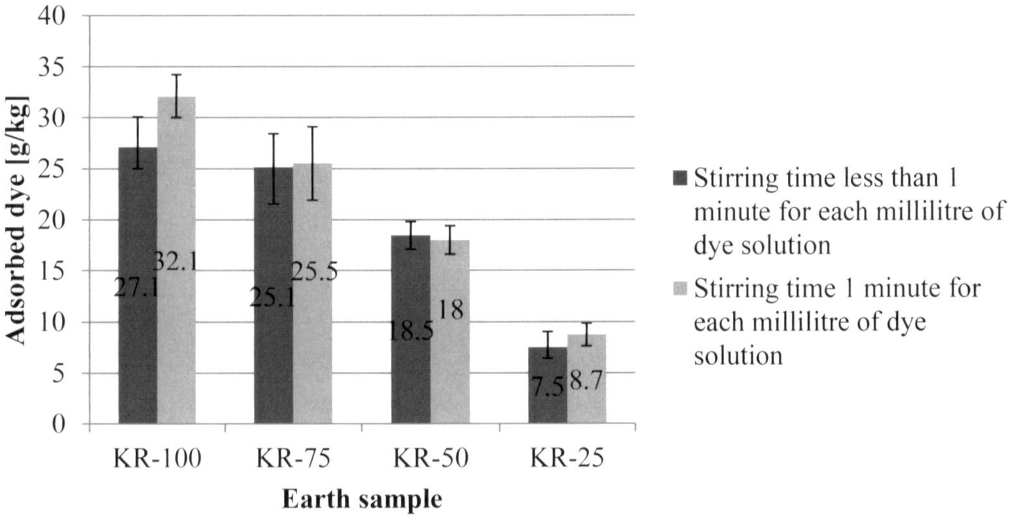

Fig. 5: Amount of adsorbed dye by illite-kaolinite clay (KR) before and after modification of methodology

Conclusion

Methylene blue test can be a good method for analyzing of earth composition. In order to obtain relevant results it is necessary to keep with the methodology. Insufficient stirring time has a different effect on results of experimental measurements depending on kind of clay. Stirring time is essential for kaolinite clay. On the contrary, stirring time for illite-kaolinite clay is less significant. Because methylene blue test should be used for analysis of composition of unknown earth, it is necessary to adhere to the prescribed stirring time.

Acknowledgement

The financial support of this experiment by the Czech Science Foundation (GAČR project NO. 18-10884S) and Faculty of Civil Engineering, Czech Technical University in Prague (SGS project No. SGS19/148/OHK1/3T/11) is gratefully acknowledged.

References

[1] P. Jaquin, Ch. Augarde. Earth building : history, science and conservation [online]. Report. B.m.: Bracknell : IHS BRE Press. 2012 [vid. 2017-02-16]. http://www.envia.bl.uk//handle/123456789/4131

[2] A. Jing. A History of the Great Wall of China. B.m.: World Scientific Publishing Company, 2015. ISBN ISBN: 978-1-938368-32-5.

[3] Agenda 21 - United Nations Environment Programme (UNEP) [online]. [vid. 2017-02-14]. http://www.unep.org/Documents.Multilingual/Default.asp?documentid=52

[4] ČSN 1168-1939 - PODMÍNKY PRO ZEDNICKÉ A PŘIDRUŽENÉ PRÁCE POZEMNÍCH STAVEB. 1951. bibtex: csn_csn_1951

[5] G. Minke. *Building With Earth* [online]. B.m.: Ökobuch Verlag, Staufen, 2006 [vid. 2015-08-18]. ISBN 978-3-0346-0822-0. http://archive.org/details/Gernot_Minke-Building_With_Earth

[6] I. Žabičková. *Hliněné stavby* [online]. Brno: Era 21, 2002 [vid. 2015-08-18]. ISBN 80-86517-21-7. http://www.kosmas.cz/knihy/107760/hlinene-stavby/

[7] P. Walker. *Rammed Earth: Design and Construction Guidelines*. Watford: IHS BRE Press, 2010. ISBN 978-1-86081-734-2.

[8] Z. Weiss. Jílové Minerály: Jejich Nanostruktura a Využití. Karolinum,,vyd. 1. edn., 2005.

[9] EVROPSKÝ VÝBOR PRO NORMALIZACI. ČSN EN 933-9, Zkoušení geometrických vlastností kameniva - Část 9: Posouzení jemných částic - Zkouška methylenovou modří. 2013.

EAN 2020
58th conference on experimental stress analysis

Experimental Dynamic Analysis of the Railway Bridge Near Žatec

PLACHÝ T.[1,a], POLÁK M.[1,b], RYJÁČEK P.[2,c]

[1]Department of Mechanics, Faculty of Civil Engineering, Czech Technical university in Prague, Thákurova 7, 166 29 Praha 6, Czech Republic

[2] Department of Steel and Wooden Structures, Faculty of Civil Engineering, Czech Technical university in Prague, Thákurova 7, 166 29 Praha 6, Czech Republic

[a]plachy@fsv.cvut.cz, [b]polak@fsv.cvut.cz, [c]pavel.ryjacek@fsv.cvut.cz

Keywords: Dynamic analysis, Railway bridge, Modal analysis, Natural frequency, Mode shape

Abstract. The paper describes the experimental dynamic analysis of the railway bridge near Žatec in the Czech Republic. The main aim of this analysis was to verify the compliance between measured quantities and calculated ones determined on the finite element model (FEM) for improvement of the FE model. The analysis was divided into two parts, modal analysis and dynamic load test. The used measurement system with dynamic range 160 dB enabled to measure the response of the bridge to train passages and also ambient vibration of the bridge between the passages without changing the setting of the measurement system.

Introduction

Steel bridges are an important part of bridge structures in the Czech Republic, especially on the railway network. Many of them are older than one hundred years and they are still in operation. Despite a certain degree of their degradation, the quick replacement of large number of bridges is unrealistic. Thus it is necessary to prove their building state and bearing capacity. For that assessment many factors must be considered, e.g. the stiffness of joints, spatial behaviour of the structure, the influence of corrosion [1] and fatigue life.

The current computational methods are highly advanced and allow to perform intensive computational analysis. Nevertheless, the input parameters for these models and their aptness can be obtained in many cases only by properly and efficiently designed experiments [2], [3] and [4]. The use of the test results for validation of the computational model can help to refine it enough that unsatisfactory bridge structure becomes at least limited satisfactory.

Description of the bridge

The studied railway bridge consists of two independent structures K01 and K02 (Fig. 1) and its total length is 116 m. Both bridge independent structures are identical, they are 58 meters long and consist of a riveted steel structure with truss main girders with an intermediate bridge deck (Fig. 2). The height of the main girders is 3.53 m The pillars and abutments are made of stone masonry (Fig. 1 and Fig. 2).

The railway track is transferred across the bridge in a straight and partially in a transition curve. The width of the clearance profile 2.5 m is kept only at niches on the pillar. On the rest of the bridge, the width of the clearance profile does not even reach 2.2 m.

Fig. 1: The view on the whole bridge

Fig. 2: The view on the first investigated half of the bridge

Measurement system

The sensors and measurement stations were prepared to positions in the pauses between regular train passages, which were approximately two hours.

The vibrations of the bridge were measured by uniaxial piezoelectric acceleration transducers 8344 Brüel&Kjær, which were placed in five cross sections, one position on the left upper chord and one position on the right upper chord in each cross section (Fig. 3 – red dots). The measurement of vibration was realized in vertical direction (Fig. 3 – from Az 041 to Az162) and in horizontal direction perpendicular to the longitudinal axes of the bridge (Fig. 3 – from Ay041 to Ay162). The transducers were mounted to the steel structure using magnets. They were put to all points on the upper chords at once, thus the response was measured in these points simultaneously.

The modal analysis was supplemented by measurement at next 10 points at contacts of the cross members of the deck with the main girders (Fig. 3). For the modal analysis, the Ambient Vibration Testing (AVT) system was used. It means, the measurements were realized between train passages by measuring only responses to ambient forces (wind, microseismicity, etc.).

The response of the bridge to regular train passages were also measured. The passages of trains with known characteristics were used for the dynamic load test.

Two interconnected measurement stations Front-end 3560-B-120 and 3050-B-040 Brüel&Kjær were used for data acquisition and the data analysis was done using the software PULSE and MS Excel. The evaluation of the natural frequencies and modes of vibration of the footbridge were done in the software MEScopeVES of the company Vibrant Technology.

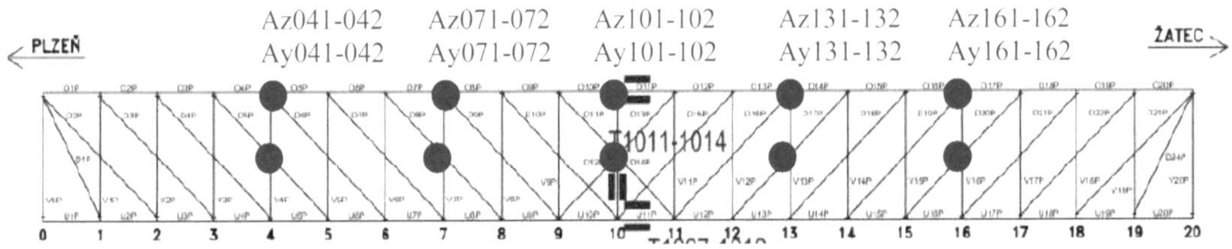

Fig. 3: The measured points marked in the longitudinal section of the railway bridge

Results of the dynamic analysis

The nine natural frequencies were evaluated in the frequency range 1.0 to 11.0 Hz during the experimental modal analysis of the investigated bridge [5]. The natural frequencies, which were compared with the theoretical results, are mentioned in Table 1.

Table 1: Comparison of the measured and calculated natural frequencies

Natural frequencies calculated		Natural frequencies measured		Frequency difference	Limit deviation	Character of the mode shape
No.: (j)	$f_{(j)}$ [Hz]	No.: (j)	$f_{(j)}$ [Hz]	$\Delta_{(j)}$ [%]	$\Delta_{(j)}$ [%]	
(1)	1.58	(1)	1.67	-5.7	+10 ; -15	1st horizontal bending mode shape
(2)	3.75	(2)	3.84	-2.4	+10 ; -15	2nd horizontal bending mode shape
(3)	3.87	(3)	3.91	-1.0	+10 ; -15	1st vertical bending mode shape
(4)	4.46	(4)	4.45	0.2	+/-15.2	1st torsional mode shape
(5)	6.10		x		+/-15.6	1st longitudinal mode shape
(6)	6.31	(5)	6.26	0.8	+/-15.6	3rd horizontal bending mode shape
(7)	8.66	(6)	8.58	0.9	+/-16.2	4th horizontal bending mode shape
(8)	9.73	(7)	9.80	-0.7	+/-16.5	2nd torsional mode shape
(9)	10.28	(8)	10.09	1.8	+/-16.7	2nd vertical bending mode shape
(10)	10.66	(9)	10.66	0.0	+/-16.8	mode shape of the upper flanges of the main girders

Notice: x – According to the theoretical calculation, the bridge vibrates in its longitudinal direction (X axes of the bridge) in the 5th natural mode shape. In this direction, the vibration was not measured.

The mode shapes corresponding to the natural frequencies were evaluated too and they were visually compared with the theoretical ones calculated on the FE model. The bridge was modelled in 3D with the use of beam elements. The real section properties were used. The truss joints and bearings were modelled as semi-rigid, according to the real stiffness. The material data for the assessment were used from the coupon tests. The FE model is described in more detail in [6]. The comparison is shown in Figures 4 - 12. In the figures, there are plan views (upper one) and elevations (lower one) of the theoretical mode shapes (on the left side) and experimental mode shapes (on the right side) of natural vibration.

The frequency differences meet all limits prescribed in the standard ČSN 736209 (see Table 1 – Limit deviations), the corresponding theoretical and experimental mode shapes have the same number of node lines at the same positions.

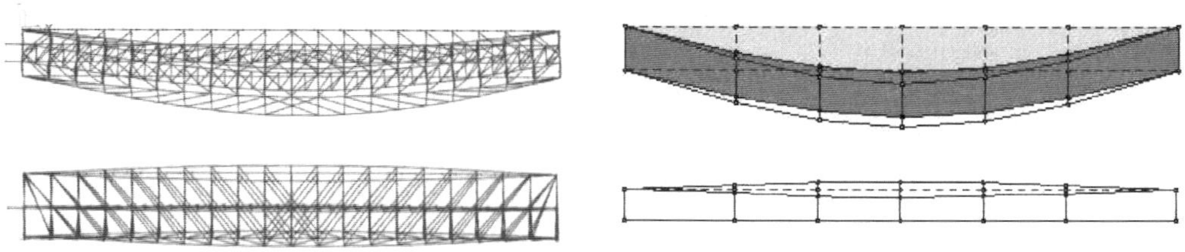

Fig. 4: Comparison of the 1st theoretical mode shape (left) of natural vibration with the 1st experimental one (right).

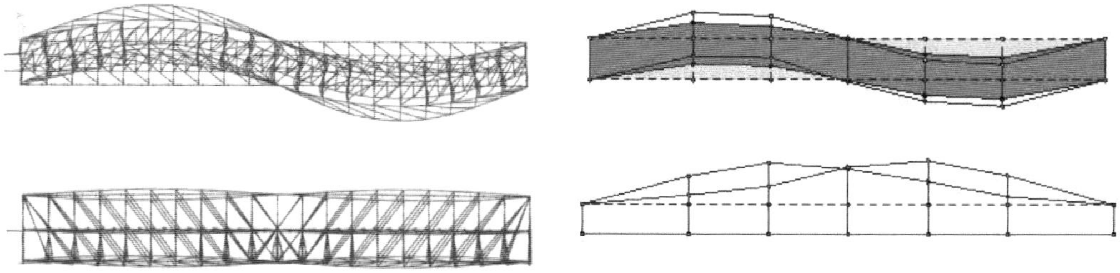

Fig. 5: Comparison of the 2nd theoretical mode shape (left) of natural vibration with the 2nd experimental one (right)

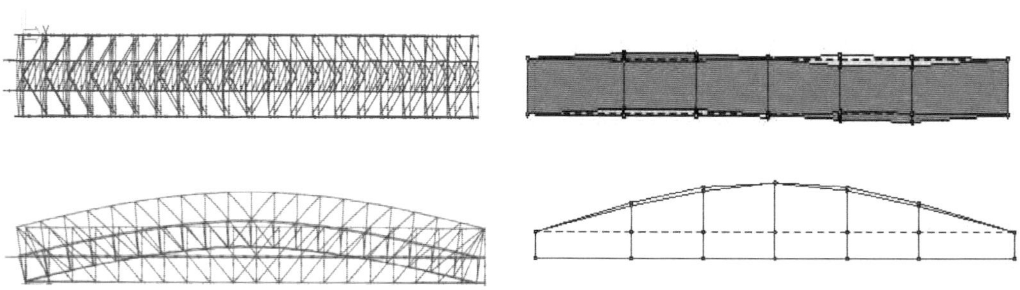

Fig. 6: Comparison of the 3rd theoretical mode shape (left) of natural vibration with the 3rd experimental one (right)

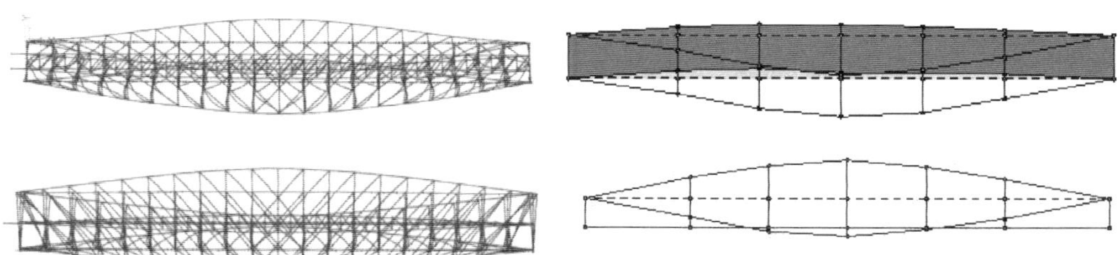

Fig. 7: Comparison of the 4th theoretical mode shape (left) of natural vibration with the 4th experimental one (right)

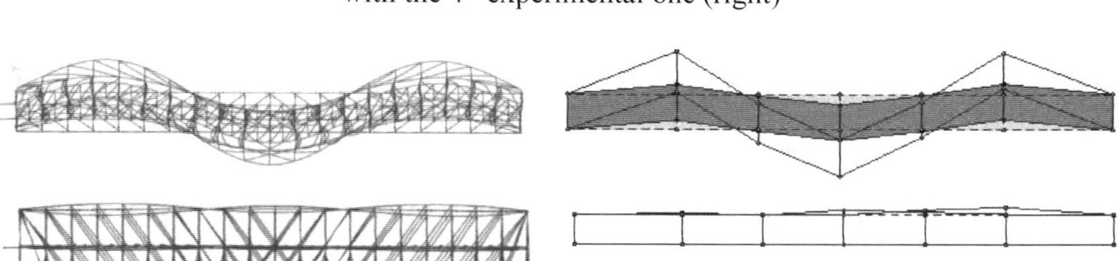

Fig. 8: Comparison of the 6th theoretical mode shape (left) of natural vibration with the 5th experimental one (right)

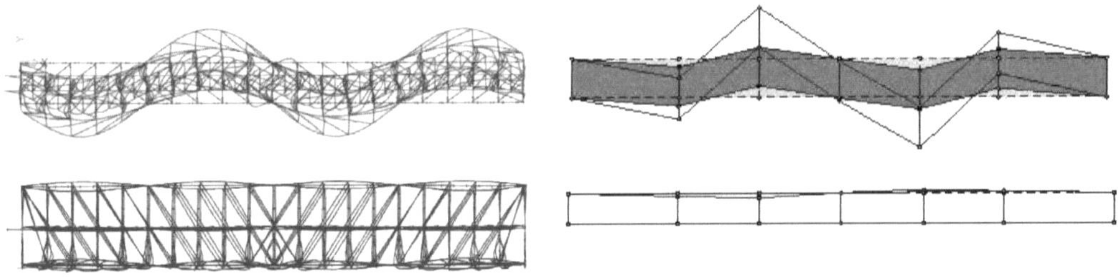

Fig. 9: Comparison of the 7th theoretical mode shape (left) of natural vibration with the 6th experimental one (right)

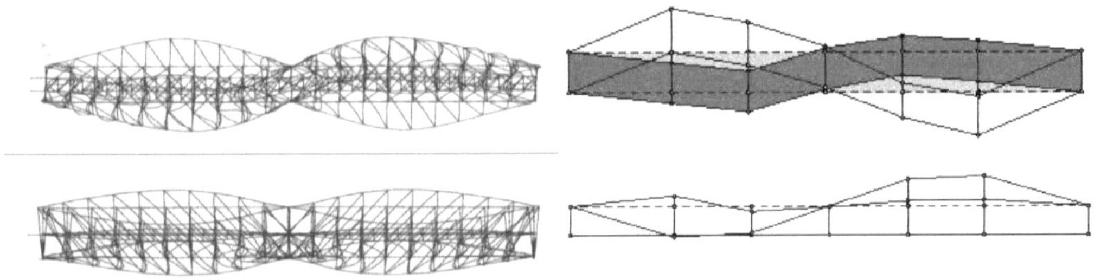

Fig. 10: Comparison of the 8th theoretical mode shape (left) of natural vibration with the 7th experimental one (right)

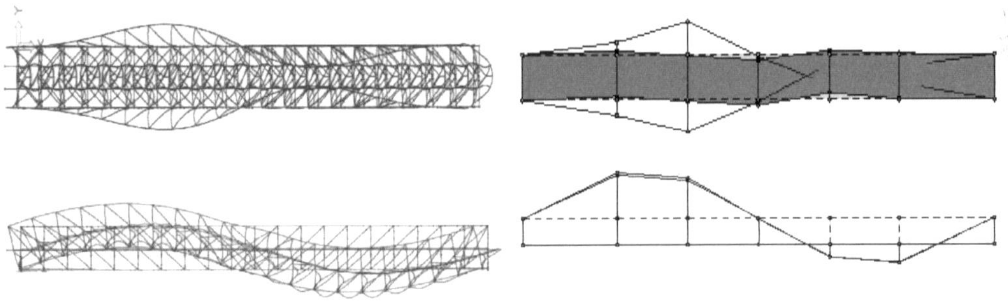

Fig. 11: Comparison of the 9th theoretical mode shape (left) of natural vibration with the 8th experimental one (right)

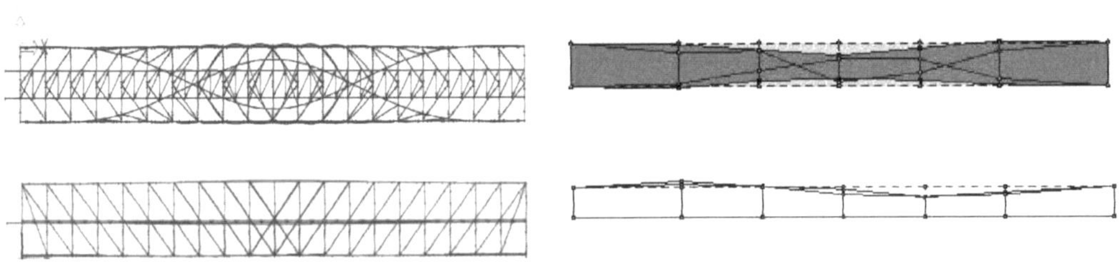

Fig. 12: Comparison of the 10th theoretical mode shape (left) of natural vibration with the 9th experimental one (right)

From all the recorded vibrations of the monitored railway bridge caused by the train passages, the extreme amplitudes of the measured accelerations were evaluated, which are clearly summarized in the Table 2.

All acquired records of acceleration were additionally adjusted by filtration. A low-pass Butterworth filter with cut-off frequency set to 30 Hz was used.

Table 2: Extreme amplitudes of vibration in the middle of the monitored span of the bridge

Train crossing the bridge	Extreme amplitudes of vibration					
	Az - 101		Az - 102		Ay - 102	
	Min.	Max.	Min.	Max.	Min.	Max.
	[m/s^2]	[m/s^2]	[m/s^2]	[m/s^2]	[m/s^2]	[m/s^2]
Passenger train	-0.57	0.59	-0.55	0.55	-0.82	0.88
Passenger train	-0.43	0.41	-0.36	0.51	-0.52	0.67
Passenger train	-0.49	0.38	-0.38	0.50	-0.71	0.73
Passenger train	-0.60	0.51	-0.49	0.45	-0.69	0.70
Goods train	-1.01	0.78	-0.76	0.82	-1.05	1.42
Passenger train	-0.46	0.46	-0.63	0.44	-0.66	0.71
Passenger train	-0.70	0.68	-0.46	0.50	-0.68	0.84

Conclusions

The experimental dynamic analysis of the railway bridge near Žatec in the Czech Republic is presented in the paper. The simplified modal analysis using AVT system which was carried out on this bridge only in five cross-sections was good enough to measure and reliably identify nine natural frequencies and natural modes of vibration of the bridge. Based on comparison between calculated and measured characteristics the FE model was modified. The agreement between finally calculated and measured modal characteristics was good and met all limits prescribed in the standard ČSN 736209 (Table 1). The validated model of the bridge can be used for apt static and dynamic calculation of the bridge in its real immediate state.

The paper was supported by Czech Technical University in Prague as the project No. SGS19/148/OHK1/3T/11

References

[1] J. Appuhamy, M. Ohga, T. Kaita, R. Dissanayake, Reduction of Ultimate Strength due to Corrosion – A Finite Element Computational Method, International Journal of Engineering, 5 (2011) 2

[2] T. Plachý, M. Polák, P. Ryjáček, Assessment of an Old Steel Railway Bridge Using Dynamic Tests, Procedia Engineering, 199 (2017) 3053–3058

[3] M. Macho, P. Ryjáček, The impact of the severe corrosion on the structural behavior of steel bridge members, in: Advances and Trends in Engineering Sciences and Technologies - Proceedings of the International Conference on Engineering Sciences and Technologies, ESaT 2015, pp. 123-128

[4] P. Ryjáček, M. Macho, V. Stančík, M. Polák, Deterioration and assessment of steel bridges, in: Maintenance, Monitoring, Safety, Risk and Resilience of Bridges and Bridge Networks - Proceedings of the 8th International Conference on Bridge Maintenance, Safety and Management, IABMAS 2016 , pp. 1188-1195

[5] P. Ryjáček, M. Polák, T. Plachý, J. Litoš, L. Vodehnal, M. Glöckner, Static and Dynamic Verification Test of the Bridge in TU 0502, km 200,916 – Žatec, Report (in Czech), 2018, pp. 50.

[6] P. Ryjáček, P. Očadlík. Recalculation of the bridge load capacity in TU 502, km 200,916 - Žatec, Report (in Czech), 2018, pp. 191.

EAN 2020
58th conference on experimental stress analysis

Effects of Additives Increasing Biological Resistance of Gypsum on Mechanical Properties of Resulting Composite

PROŠEK Z.[1,2,a], TESÁREK P.[1,b]

[1]Deparment of Mechanics, Faculty of Civil Engineering, CTU – Czech Technical University in Prague, Thákurova 7, 166 29 Prague, Czech Republic

[2]University Centre for Energy Efficient Buildings, CTU – Czech Technical University in Prague, Třinecká 1024, 273 43 Buštěhrad, Czech Republic

[a]zdenek.prosek@fsv.cvut.cz, [b]pavel.tesarek@fsv.cvut.cz

Keywords: Gypsum, Compressive Strength, Flexural Strength, Silver Nitrate, Additives

Abstract. The article focuses on the effects of additives increasing biological resistance of gypsum on mechanical properties of resulting gypsum composite. In this study we use Silver Nitrate as an additive that improves biological resistance of gypsum composites. We investigate influence of four different amount of Silver Nitrate ($AgNO_3$), namely, 0.5 wt. %, 1 wt. %, 1.5 wt. % and 2 wt. %. The determined mechanical properties were flexural strength and compressive strength. The mechanical properties were investigated on the 7 days old samples with dimensions equal to 40 × 40 × 160 mm. The results obtained from these samples were compared with reference material. The results show a negative effect of $AgNO_3$ on flexural strength.

Introduction

The gypsum is widely used material in civil engineering especially for production of plaster-board, gypsum plaster, gypsum masonry blocks [1]. Gypsum composites are more environmentally friendly and production is faster than cement composites. Mainly due to lower greenhouse gas production and sustainability [2]. Use of gypsum composites in civil engineering is significantly restricted because they have low moisture resistance. The moisture content of the material adversely affects the formation of molds and the reproduction of microorganisms. [3]. Molds belong to microorganisms, that commonly appear on inner and outer surfaces of building materials and can cause sick building syndrome which limits or completely restrict the use of the building [4]. From these reason, it is very important created biological resistance materials that resist mold formation on their surface and also in the inner of material.

Biological resistance of gypsum composites can be improved by adding antibacterial components (such as silver [5], TiO_2[6], cooper [7]) or hydrophobizing additives (such as waxes [8], saponifying agents [9], emulsifier [10]). This paper deals directly with using silver as additives increasing biological resistance. The silver was almost abandoned when penicillin were discovered. But today, with the emergence of antibiotic-resistant strains, it has gained new interest [11]. Silver was reported to be an efficient bactericidal antibacterial agent in the several works [12-14]. Also gypsum composite with antibacterial components is being developed and researched [15], but some of its properties are still unknown, for example mechanical properties which is important properties for using materials in civil engineering.

Materials and Samples

Testing samples were composed of gypsum from Počerady (Knauf, Praha) and different amount of Silver Nitrate ($AgNO_3$), which increases the biological resistance of the resulting composites. Four concentrations of Silver Nitrate were tested: 0.5 wt. %, 1.0 wt. %, 1.5 wt. % and 2.0 wt. %. The samples were compared with reference sample composed of gypsum only. In all cases, the water coefficient was equal to 0.6. The samples composition is shown in Table 1. Each set consisted of 6 samples with dimension equal to 40 × 40 × 160 mm.

Table 1: Composition of the testing samples

Set	Gypsum [g]	TiO2 [g]	Water [g]	Bulk density (kg/m^3)
ref.	1500	0	900	1071 ± 10
0.5 %	1492.5	7.5	895.5	1062 ± 8
1.0 %	1485	15	891	1062 ± 1
1.5 %	1477.5	22.5	886.5	1062 ± 4
2.0 %	1470	30	882	1065 ± 17

Experimental Methods

The flexural and the compressive strength were determined using the Web Tiv Ravestein devices on 7 days old the hardened gypsum samples. The resulting value was determined as the average of values excluding the lowest and highest values, and the standard deviation was determined.

The flexural strength was determined by three-point bending test. The test is performed on a beam with dimensions of 40 × 40 × 160 mm supported on two supports, which is loaded by force in the centre of the beam. The flexural strength was determined from the relationship between achieved the maximum force and the dimensions of beam at the fracture point as

$$f_b = \frac{3 \cdot F_{b,max} \cdot L}{2 \cdot h \cdot b} \tag{1}$$

where $F_{b,max}$ is the maximum force reached during the test, L is a distance between supports equal to 100 mm, h is height of the sample and b is width of the sample [16]. Six samples were tested from each set. The testing was displacement controlled at a constant rate of 1 mm/min.

The compressive strength was determined by uniaxial compressive test on the half samples with dimensions of 40 × 40 × about 80 mm (broken parts of samples after bending test). The loading area was 40 × 40 mm. Twelve samples were tested from each set. The testing was displacement controlled at a constant rate of 3 mm/min. The compressive strength calculation was performed according to the equation [16]:

$$f_c = \frac{F_{c,max}}{h \cdot b} \tag{2}$$

where $F_{c,max}a$ is the maximum force reached during the test.

Results and Discussion

Fig.1 shows investigated mechanical properties. The results show the effect of the Silver Nitrate on the flexural strength and the compressive strength of gypsum composites. The results show that the Silver Nitrate has a negative effect on the resulting flexural strength, with a difference of approximately 0.5 MPa for 0.5 % concentration of $AgNO_3$ and 1 MPa for higher $AgNO_3$ concentrations (1.0 %, 1.5 %, 2.0 %). It can be concluded that the introduction of inert $AgNO_3$ result in a less compact structure of hardened gypsum samples, leading to a deterioration of flexural strength.

The compressive strength was approximately the same for all tested samples (differences were up to the standard deviation) except for the 1% concentration of $AgNO_3$. The compressive strength samples with 1% concentration of $AgNO_3$ was about 3 MPa lower than reference samples.

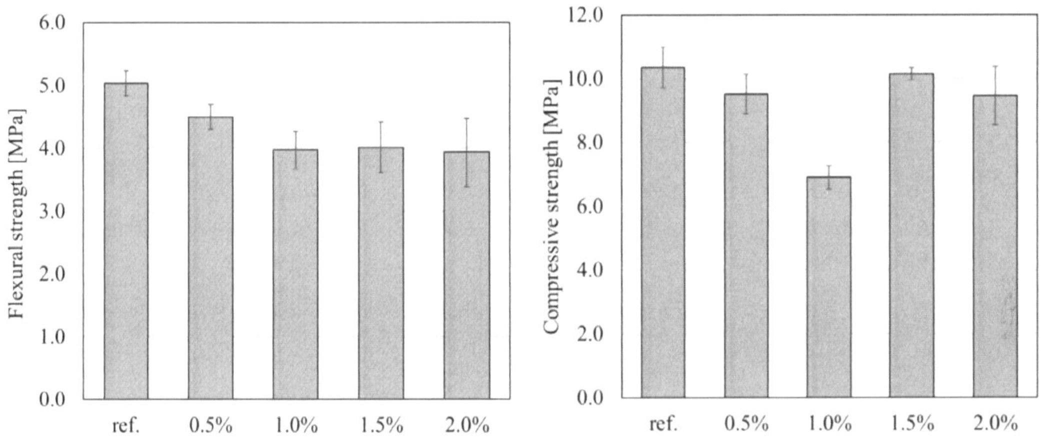

Fig. 1: Comparison of flexural strength – left and compressive strength – right (with standard deviations)

Conclusions

This work was focused on the effects of Silver Nitrate increasing biological resistance of gypsum on mechanical properties of resulting composites. The tested samples were composed of Gypsum from Počerady and Silver Nitrate. The difference between individual samples was in the concentration of Silver Nitrate. Based on the results, it can be concluded that results the introduction of inert Silver Nitrate result in a less compact structure of hardened gypsum samples, leading to a deterioration of flexural strength and the compressive strength was approximately the same for all tested samples.

In the future, the research will focus on confirming effects of Silver Nitrate on mechanical properties gypsum composites by using scanning electron microscope with X-ray microanalysis by confirms the inert effect of Silver Nitrate on the phase composition of the composite.

Acknowledgement

This paper was financially supported by the TA CR research a project under the number TJ02000235 and by Czech Technical University in Prague under No. SGS project SGS19/148/OHK1/3T/11. The authors also thank company Knauf, Praha Ltd.

References

[1] M. Doležalová, A. Vimmrová, Moisture influence on compressive strength of ternary gypsum-based binders, IP Conference proceedings. 1863 (2017) 23–31.

[2] N. Lushnikova, L. Dvorkin. Sustainability of gypsum products as a construction material, Sustainability of Constr. Materials. Woodhead Publishing, 2016 643-681.

[3] J. Krejsová, M. Doležalová, Resistance of Mortars with Gypsum, Lime and Composite Binders against Molds, Acta Polytechnica CTU Proceedings 21 (2019) 16–20.

[4] A. Ghaffarianhoseini, H. AlWaer, H. Omrany, A. Ghaffarianhoseini, C. Alalouch, D. Clements-Croome, J. Tookey, Sick building syndrome: are we doing enough?. Architectural Science Review 61 (2018), 99-121.

[5] M. Rai, A. Yadav, A. Gade, Silver nanoparticles as a new generation of antimicrobials, Biotechnology Advances 27 (2009) 76–83.

[6] A. M. Allahverdiyev, E. S. Abamor, M. Bagirova, M. Rafailovich, Antimicrobial effects of TiO_2 and Ag_2O nanoparticles against drug-resistant bacteria and leishmania parasites, Future Microbiology 6 (2011) 933–940.

[7] G. Ren, D. Hu, E. W. C. Cheng, M. A. Vargas-Reus, P. Reip, R. P. Allaker, Characterisation of copper oxide nanoparticles for antimicrobial applications, International Journal of Antimicrobial Agents 33 (2009) 587–590.

[8] G. Mumford, R. W. Phillips, Dimensional change in wax patterns during setting of gypsum investments. Journal of dental research, 37 (1958) 351-358.

[9] M. N. Razali, M. A. A. Aziz, W. N. A. W. M. Hamdan, N. A. M. Salehan, M. Y. Rosli, Synthesis of Grease from Waste Oils And Red Gypsum. Aust. J. Basic Appl. Sci. 11 (2017) 154-159.

[10] Y. Zhang, K. Wang, W. Tao, D. Li, Preparation of microencapsulated phase change materials used graphene oxide to improve thermal stability and its incorporation in gypsum materials. Construction and Building Materials 224 (2019) 48-56.

[11] I. Chopra, The increasing use of silver-based products as antimicrobial agents: a useful development or a cause for concern?, Jour. of Antimicrobial Chemotherapy 59 (2007) 587–590.

[12] G. D. da Silva, E. J. Guidelli, G.M. de Queiroz-Fernandes, M. R. M. Chaves, O. Baffa, A. Kinoshita, Silver nanoparticles in building materials for environment protection against microorganisms. Inter. jour. of environ. scie. and tech., 16 (2019), 1239-1248.

[13] M. Banach, R. Szczygłowska, J. Pulit, M. Bryk, Building materials with antifungal efficacy enriched with silver nanoparticles. Chem Sci J, 5 (2014.

[14] Q. Qiang, N. Hongwei, X. Wei, X. Zhong, W. Yong, Antibacterial mechanism of silver. Journal of Wuhan University of Science and Technology 2 (2007).

[15] R. Casado, Construction technology of internal gypsum block partitioning, Brick and Block Masonry Conference (2012).

[16] ČSN EN 13279-2 - Sádrová pojiva a sádrové malty pro vnitřní omítky - Část 2: Zkušební metody. Praha: Úřad pro technickou normalizaci, metrologii a státní zkušebnictví, 2014, p. 24, Třídící znak 72 2486.

EAN 2020
58th conference on experimental stress analysis

Dynamic Response Mesurement during the High-Speed Impact

RAŠKA J.[1,a], OBERTHOR M.[1,b]

[1]Czech Aerospace Research Centre (VZLÚ), Beranových 130, 199 05 Praha – Letňany, Czech Republic

[a]raska@vzlu.cz, [b]oberthor@vzlu.cz

Keywords: High-speed impact, Bird-strike, Hail-strike, Impact projectile, Parasitic resonances, Virtual load cells record

Abstract. The material of the aircraft structure must resist to the high-speed impact – it means the impact of the bird or of the hail at the flight speed (bird-strike or hail-strike). Proof of this resistance can be made by experimental or by the numerical way. The recent development of the numerical methods takes advantage of the numerical way. However, the necessity to verify used numerical models creates new challenges on the experiments. The goal is to measure and to record a great deal of data during the very short time of the high-speed impact.

Use of the high-speed load cells makes possible to record the time-dependent force response of the impact event. However, the major difficulty of this approach is the existence of the parasitic resonances. The objective of the stand design is therefore to minimize these parasitic resonances, due to the stand structure dynamics.

In this way, the special equipment for the experimental research of the dynamic response during the high-speed impact was designed and realized. The stand is destined to support the specimen within impact of projectile (bird or hail) ejected from the pneumatic gun. The dynamic response measurement is realized by means of load cells.

Introduction

Airworthiness requirements claim the resistance of the aircraft structures to the high-speed impact – it means the impact of the bird or of the hail at the flight speed (bird-strike or hail-strike). Proof of this resistance can be made by experimental or by numerical way. The recent development of the numerical methods, especially the explicit solution of the finite element method, takes advantage of the numerical way. However, the necessity to verify used numerical models – particularly the model of the material damage and the constitutive model of the impact body (projectile) – creates the new challenges on the experiments. The goal is to measure and to record the great deal of data during the very short time of the high-speed impact. This is technically difficult, but necessary for the numerical models verification.

Use of the high-speed load cells makes possible to record the time-dependent force response of the impact event. However, the major difficulty of this approach is the existence of the parasitic resonances. The experimental stand, the support of the studied specimen, has the eigen dynamic response and in consequence, the final measured dynamic response is coupled with the stand structure one. The objective of the stand design was therefore to minimize these parasitic resonances, due to the stand structure dynamics. For this reason, at the first draught, the equipment was optimized by means of FEM numerical simulations.

Specimen

The experimental stand was designed for the large flat panels, convenient for the advanced developmental experiments. Disadvantage of the standard square (or rectangle) specimen form is the risk of the non-regular flaws (see fig.1) [2]. The specimen is typically over-loaded at the centers of the edges – risk of the cracks and the disruption –, the specific form damages can appear at the corners. To avoid these types of flaws, the optimal specimen form is the circle. For the technological reasons, the octahedron was adopted (see fig. 2).

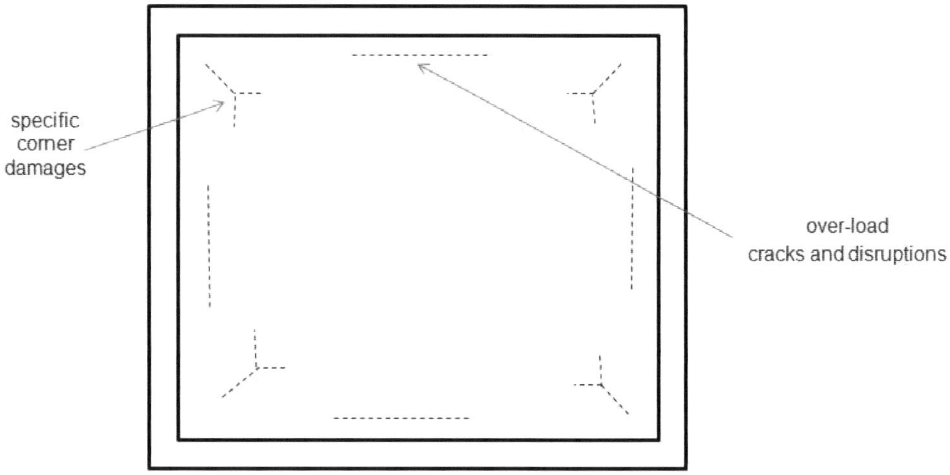

Fig. 1: Typical non-regulars flaws of the standard square form specimen

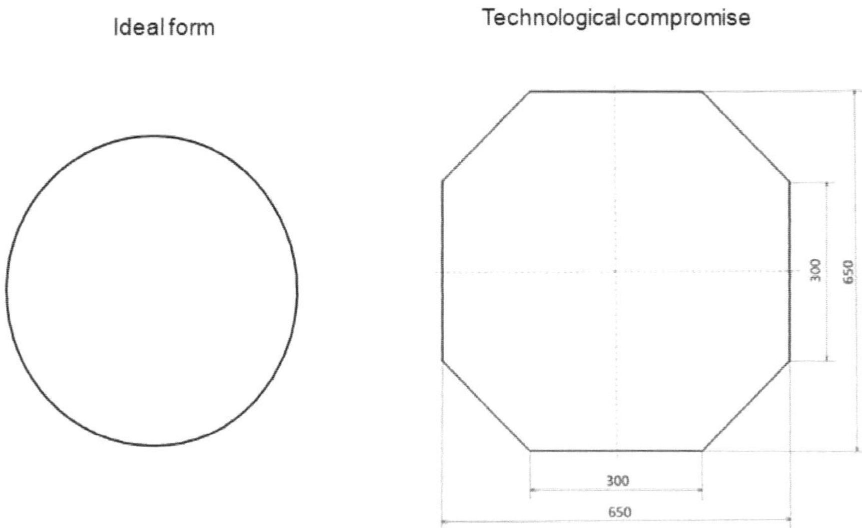

Fig. 2: Optimised specimen form

Experimental stand conception

Three basic objectives were applied for the equipment design. At first, the stand is destined to support the specimen within impact of projectile (bird or hail) ejected from the pneumatic gun. Secondly, the equipment is adapted to the dynamic response measurement and recording by means of load cells.

At last, the stand design enables the impact in any angle (from 0° to 90°), without change of the first contact point. It means, that in all cases, the gun axis points at the centre of the specimen. For any impact angle, the first contact of the bird or of the hail with the specimen occurs in his centre.

For this reason, the stand structure is divided in two parts. The movable part is connecting with the fixed one by means of pivots placed in the plane of the specimen (see fig. 3). In consequence, the gun axis points always at the same point, at the centre of the specimen [4].

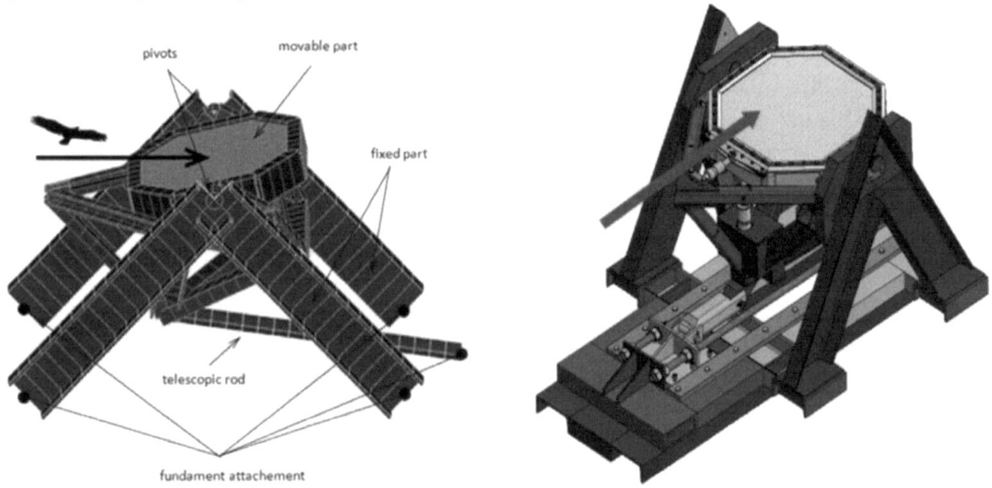

Fig. 3: Experimental stand configuration for low-angel impact [3], [4]

Fig.4: Experimental stand configuration for perpendicular impact [3], [4]

The impact angle is controlled on principle by the telescopic rod. As is demonstrated on the fig. 3, if the telescopic rod is prolonged, the impact angle decreases. By contrast, the short telescopic rod gives the large impact angle. On the fig. 4, the case of the perpendicular impact is demonstrated. The real, detailed design is a little different (see the right parts of the fig. 3 and of the fig. 4) [3]

Numerical simulation and optimisation

The major difficulty of the time-dependent force response measurement during impact event, by means of high-speed load cells, is the existence of the parasitic resonances. [3] The experimental stand has the eigen dynamic response and in consequence, the final measured dynamic response is coupled with the stand structure one. For this reason, the minimization of these parasitic resonances was a priority objective.

At the first draught, to detailed design, the FEM numerical simulations were applied. The finite element model was loaded by the time-dependent force function, simulating the contact interaction between the projectile (bird or hail) and the specimen. The number of appropriate force functions was deduced by an analytical way (with some arbitrary assumptions), by an empirical way (experimental results) and by a numerical simulations (explicit FEM solution, liquid like bird model). [5]

The time-dependent force function, applied on the complete structure model (specimen, stand structure and load cells model) cause a dynamic response. The dynamic behavior is analyzed by means of numerical FE simulation. The essential result is the time-dependent virtual load cells record – it means the computed internal force of the FE elements, simulating the load cells. In the ideal case, the stand structure parasitic resonances are completely negligeable and in consequence, the virtual load cells record corresponds to the applied force via equilibrium conditions. In the real case, the deviations from the equilibrium condition are incurred by the parasitic resonances influence. The goal of the optimization is to minimize, as much as possible, the parasitic resonances for all applied force functions. [1]

The example, demonstrated on the fig. 5, shows the case of the force applied in the specimen plane (the force due to the friction between the projectile and the specimen in case of small angle impact). Theoretically, only one load cell is loaded. According to the equilibrium condition, the load of this load cell is equal to the applied force.

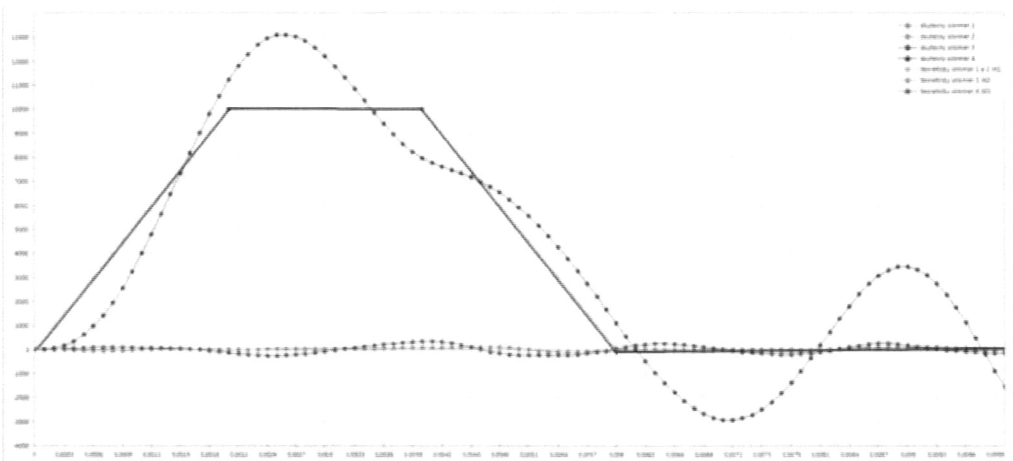

Fig.5: Example of the virtual load cells records – in specimen plane applied force [3]

As is observed on the fig. 5, the applied force (black solid line) is not exactly equal to the virtual load cell record. The applied force function has a trapezoid form, after the short impact pulse, the applied force is zero. The virtual load cell record oscillates arround the applied force function. This deviations are relatively important, however, the oscillations at only one dominant frequency are easy to eliminate by filtration. The virtual records of other load cells are very close to the theoretical value, very close to zero.

Conclusions

The special experimental equipment for the research of the dynamic response during the high-speed impact, was designed, developed and realized. The challenge was to measure and to record the great deal of data during the very short time of the high-speed impact, useful for the complex numerical model verification.

The stand was destined to support the specimen during impact of projectile (bird or hail) ejected from the pneumatic gun. The specimen is defined by the large flat panel with the special octahedron form, to avoid the non-regular flaws. The stand design enables the impact in any angle (from 0° to 90°), without change of the contact point. Independently on the impact angle setting, the gun axis points always at the same point, at the centre of the specimen.

The dynamic response measurement is realized by means of load cells. At the first draught, to detailed design, the equipment was designed and optimized by means of FEM numerical simulations in objective to minimize the parasitic resonances of the stand structure.

References

[1] R. Doubrava, M. Oberthor, P. Bělský, M. Dvořák, K. Doubrava, Experimental Verification of Jet Engine Composite Inlet for Bird and Hail Impact Resistance, Conference proceeding of 56th konference on experimental stress analysis (EAN 2018), ISBN 978-80-270-4062-9

[2] R. Doubrava, M. Oberthor, P. Bělský, J. Raška, Bird and hail impact resistance analysis on jet engine composite air inlet, MATEC Web Conf. Volume 188, 2018, Art. No. 04006, DOI: https://doi.org/10.1051/matecconf/201818804006

[3] R. Doubrava, M. Oberthor, P. Bělský, O. Vích, An Improvement of measurement technique for high speed impact tests analysis, EAN 2019 Conference Proceedings ISBN 978-80-214-5753-9

[4] J. Raška, Numerické simulace měřící soustavy pro vysokorychlostní impakty a výpočtové ověření návrhu zkušebního zařízení pro výzkum dynamické odezvy, VZLÚ report R-6656, 2016

[5] R. Hedayati, M. Sadighi, Bird Strike An Experimental, Theoretical, and Numerical Investigation, ISBN: 978-0-08-100093-9, Elsevir Ltd 2016

[6] K. Hyonny, Impact Damage Formation on Composite Aircraft Structures, UCSD Federal Aviation Administration JAMS Paper,2014

EAN 2020
58th conference on experimental stress analysis

Certification Requirements to Additive Manufactured Parts Used in Airframe Primary Structures

RŮŽEK R.[1,a], HOMOLA P.[1,b]

[1] VZLÚ - Czech Aerospace Research Centre, Aviation Division, Beranových 130, 199 05 Praha 9, Czech Republic

[a]ruzek@vzlu.cz, [b]homola@vzlu.cz

Keywords: Additive manufacturing, Metal, Aerospace, Certification, Regulations

Abstract. Paper introduces review and defines the main requirements of regulations and standards relating to certification of airframe primary structural parts made by additive manufacturing (AM), i.e. 3D printing. The aim of paper was to classify requirements to certification tests in compliance with airworthiness regulations. Paper gives a base for design of an experimental program supporting certification of the selected part of a structure. The paper is focused mainly on application in a military service. Requirements are specified in compliance with EASA (European Union Aviation Safety Agency) and EMACC (European Military Airworthiness Certification Criteria) regulations.

Introduction

Additive manufacturing (AM) is a form of manufacturing, which is subsequently derived from a rapid prototyping technology since the 1990's. Application of AM is still not extensively used technology in the Czech aerospace industry for principle structural elements of airframes. On the other side it can be stated that the certification practises are not fully proceeded in the world until now. In this sense, the main aim of this paper is to provide a review of airworthiness regulations and standards with the scope for definition of an experimental programme for certification of selected primary aircraft structure part in compliance with airworthiness regulations.

Airworthiness regulations base

CS (Certification Specification) issued by EASA (European Union Aviation Safety Agency), FAR (Federal Aviation Regulations) issued by FAA (Federal Aviation Administration), and MASAAG (Military Aircraft Structures Airworthiness Advisory Group) regulations for airworthiness of structures are fundamental.

Requirements of each regulation differ according to type and size of structure and according to scope of service. In case of a primary structural part of the airframe certification, it is relevant to apply the requirements of CS-23 (Certification Specifications for Normal-Category Aeroplanes) [1], or CS-25 (Large Airplanes) [2] valid in EU (European Union), or their equivalents FAR-23 [3] a FAR 25 [4] issued by FAA. Requirements in detail and ways for their complying are defined in a row of supporting and mandatory ruling, such as FAA a EASA Advisory Circulars (AC), Acceptable Means of Compliance (AMC), Certification memo (CM), Guidance Material (GM) a others standards, recommendations and handbooks – for example MMPDS (Metallic materials properties Development and Standardization), CMH (Composite Material Handbook), Joint service Specification Guides (JSSG), Defence Standards (DEF-

STAN), NATO STANAG (NATO Standardization Agreement), European Military Airworthiness Certification Criteria (EMACC), American Society for Testing and Materials (ASTM standards), etc.

From the material and manufacturing process application and certification point of view, the following paragraphs are critical CS-25 (Amendment 23, 2019) paragraphs:

- CS 25.603 Materials and fabrication (for composite materials AMC 20-29):
- CS 25.605 Manufacturing processes:
- CS 25.613 Material strength properties and material design values:

The regulation defines statistical base of characteristics/material properties A, or B. Properties „A“ a „B“ are defined in „The Metallic Materials Properties Development and Standardization (MMPDS) handbook“ or ESDU 00932 „Metallic Materials Data Handbook.“
Formulations with similarly meaning can be find in CS-23 (or FAR23 and FAR25) regulation although the new version of C-23 regulation reformulates and renumbers individual paragraphs. The new version formulates discussed requirements in paragraph CS 23.2260.
Test specimens used in certification process must be manufactured from materials manufactured based on certified manufacturing procedures. Specimens form and testing methods must:

(i) comply to worldwide recognized standards, or
(ii) comply to requirements described in detail in appropriate handbooks, or
(iii) be realized in compliance with approved testing plan which contains test specimen definitions and definition of testing.

Additionally, CS 25.571 Damage Tolerance and fatigue properties (see AMC 25.571) must be taking into account for principle elements and structures. In compliance with CS25.571 paragraph the Damage Tolerance philosophy application is impractical, it is allowed to apply Safe Life philosophy. This structure must be proved by analyses supported by experiments. Experiments must prove that the structure sustains repeated loads expected in the service without detection of any cracks. Appropriate safety factor must be applied in this case. Additional NDI (non-destructive inspection) procedures must be defined.

Number of tests for experimental certification

Material, structural parts and full-scale structures strength values have to be defined based on knowledge of “Engineering characteristics”, which are influenced by material, technology, manufacturing procedure and assembly. From these reasons the certification of structure must be complex and have to consist of tests pyramid from small coupons up to full scale structures – so called „Building block Diagram (BBD)” – see Fig. 1.
The three static statistic values for material characterisation are recognized by MMPDS standard [5]. A-, B- and S-basis of material characteristics are defined. Relationship between them depends on used data analysis (Normal, Weibull, Pearson or Non-parametric distribution), size of specimens, asymmetry, etc. In general, MMPDS defines number of 100 specimens for A- and B-basis design values as the smallest allowable number of specimens, whereas these values have to be evaluated from 10 heats and 10 lots. Minimum number of 299 specimens must be tested in case of the nonparametric data evaluation is required.

The A-basis means 95% confidence of 99% exceedance. The B-basis values mean 95% confidence of 90% exceedance. An S-basis value is taken from a specification (issued by industrial institutions or standardization groups as for example SAE Aerospace Material, ASTM, etc.) or federal or military standards (MIL-STD-970 [6]). According to MMPDS minimum number of 30 specimens is required for S-basis value definition.

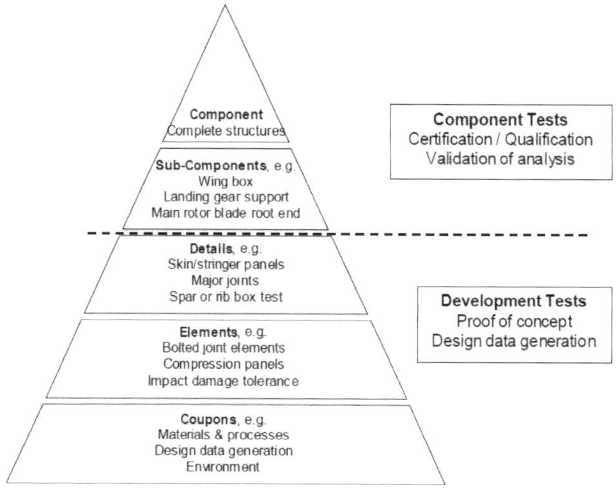

Fig. 1: Schematic BBD for certification purposes

Fundamental requirements for AM in the military field are specified namely in DEF-STAN 00-970 [7], MASAAG - Guidance Note on the Qualification and Certification of Additive Manufactured Parts for Military Aviation [8], EMACC Handbook [9] and ISO and ASTM standards. Regulations DEF STAN 00-970 Part 3, Part 4 or Part 5 are very similar to CS-23 or CS-25. MASAAG [8] deals with qualification and certification of AM metal parts labelled as Grade A. Any of part manufactured by AM have to go through certification procedure as a significant change of the original structure. This is caused by the fact that AM represents both manufacturing process and material change and often also design change. A-, B- and S-basis is defined similarly to MMPDS, only the basis is named as Grade A or Grade B structure. Moreover, US JSSG 2006 [10] guide defines that A-basis is strictly required for design of critical parts of the structure.

Generally, the B-basis for each failure mode should be evaluated in compliance with DEF STAN 00-932 [11]. This requirement could be severe and impractical. For this case, material characteristic could be evaluated based on at least 15 specimens. Then, the values must be evaluated under the worst environment in case of these conditions significantly influenced strength. In case of 15 specimens, 3 batch of materials must be used with minimum number of 5 specimens from each supplier. Exceptionally, it is acceptable to classify allowable values for particular material batches in case of mean value is changing in each batch and together coefficient of variation remains the same. Then B-value for individual batch must be decreased as compared to reference value. The B-value increase is not allowed.

Allowable stress or strength must be defined for each Grade A structure part. The strength is usually defined as B-basis [7]. Allowable of strength has not to be higher than 1.5-fold of design limit load. At the same time, no undesirable deformations are allowed under load level at least of 75% design limit load [7].

Variability of AM process have to be analysed also from fatigue behaviour point of view. As well definition of part resistance to environmental degradation (corrosion, stress corrosion, etc.) and part protection against these effects.

Specific requirements for AM

Special standards focused especially on AM technology were published by US National Institute of Standards and Technology (NIST), namely:

- NISTIR 7847 [12] and NISTIR 7873 [13] – review of ASTM and ISO test methods
- NISTIR 8005 [14] - summary of ASTM standards and general.

ASTM society published several standards focused especially for AM. These standards can be sorted to following basic groups:

- Terminology (ISO/ASTM52900-15, ISO/ASTM 52921).
- Design (ISO/ASTM 52915-16 a 52910-18).
- Materials and procedures (ASTM standards: F2924-14, F3001-14, F3049-14, F3055-14a, F3056-14e1, F3091/F3091M-14, F3184-16, F3187-16, F3213-17, F3301-18a, F3302-18, F3303-18 and F3318-18, ISO/ASTM52901-16).
- Experimental methods (ASTM standards F2971-13 and F3122-14, ISO/ASTM52921-13).

Mechanical properties of AM parts depend on manufacturing plan, build platform, build directions, origin point, part location, part orientation, etc. in generally. Based on this coordinate system, building plan and type of bounding box with positioning of test specimens shall be designed and manufactured. Fig. 2 shows an example of various specimen orientation.

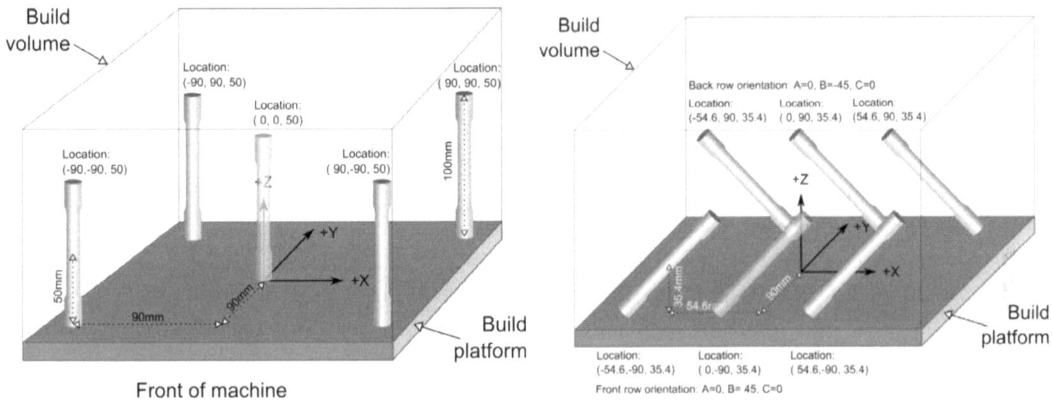

Fig. 2: Part location and orientation: round bar specimens with Z Orientation (left) and round bar specimens oriented at B+45 from Z and B–45 from Z (right)

It can be stated that up to now no complex standard for valuation of AM exists as compared to ordinary manufacturing processes or normalization of AM characteristics with respect to optimized selection of manufacturing parameters (print parameters, laser moving, surface treatment, etc.). Also, no manual for specification of design processes exists.

No complex qualification or certification procedure for AM in aviation industry is available. Some examples of qualification systems designed for selected manufacturing processes are listed in NASA MSFC-STD-3716, „Standard for Additively Manufactured Spaceflight Hardware by Laser Powder Bed Fusion in Metals“ [15] and in the new AWS D20.1 standard [16] (requirements for qualification of machines, manufacturing processes, operators, planning and inspection personnel).

For Class (grade) A components, a minimum of four witness specimens shall be distributed through the model. Three of the witness specimens are to be tested [20]. The console can be classified as Grade B structure. Then, at least of two parts must be put into the build model [20] (i.e. process including part design, orientation/area/build arrangement, including surrounding structure, resources for manufacturing and demonstrative specimens).

The tension test specimens in each standard qualification build shall include the following orientations: the tensile axis within the X-Y plane, the tensile axis along the Z axis, and the tensile axis at angle 45° (±15) from the Z axis (see Fig. 3). For Class (Grade) A and Class (Grade) B components, 54 tension test specimens are required at minimum. These 54 specimens shall be comprised of tension test specimens from each of the eight corners (or at four, equally-spaced extremes at the edges of the build platform and four at the top of the build volume) and in the centre of the build volume to be qualified [16]. The specimens are designed in compliance with ASTM E8 (Standard Test Method for Tension Testing of Metallic Materials) standard. All post-manufacturing processes, with exception of machining and surface polishing, have to satisfy to actual procedures.

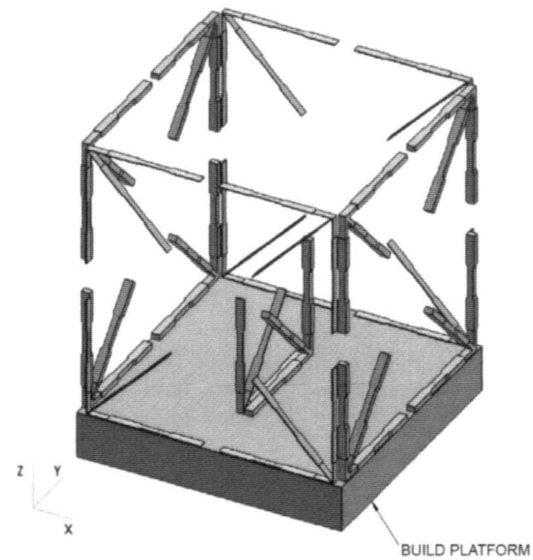

Fig. 3: Design example of a standard qualification build for „Powder Bed Fusion" [16]

Conclusions

Paper reviews and summarizes airworthiness regulation and standard requirements related with certification of parts and structures made by AM technology (3D printing) for aerospace applications. The aim of paper was to classify and define requirements for certification tests in compliance with EASA and EMACC regulations. Paper gives a base for design of experimental program supporting certification of a structure part made by AM. In the field of static characterisation, it will be necessary to prove the Young modulus of elasticity, tensile strength, yield strength and ductility for various orientations in the building block. Minimum number of 300 specimens should be manufactured and tested. Fatigue behaviour of AM part have to be investigated for the orientation and placement in building block with worst expected characteristics. Number of tested fatigue test specimens has to be in compliance with requirements of ASTM standard. Both as build and machined specimens are designed.

Acknowledgement

Work was performed in the framework of TA ČR project nr. TN01000029/05 – Design, optimization and evaluation of the primary structural part made by additive manufacturing.

References

[1] CS-23. Normal, Utility, Aerobatic and Commuter Aeroplanes. Certification specification. EU: EASA.

[2] CS-25. Large Aeroplanes. Certification specification. EU: EASA.

[3] FAR, Part 23. Small Airplane Certification Process Study. Federal Aviation Regulations. U.S.: FAA.

[4] FAR, Part-25. Airworthiness Standards: Transport Category Airplanes. Federal Aviation Regulations. U.S.: FAA.

[5] Metallic Materials Properties Development and Standardization (MMPDS) Handbook - 12. U.S.: Battelle Memorial Institute, 2017.

[6] MIL-STD-970 Standards and Specifications, Order of Preference for Selection of. U.S.: Department of Defence USA, 1987.

[7] DEF STAN 00970. Design and Airworthiness Requirements for Service Aircraft. U.K.: Ministry of Defence, Issue 13, 2015.

[8] Lunt, M. Guidance Note on the Qualification and Certification of Additive. U.S.: Naval Air Squadron, 2018. MASAAG Paper 124, Issue1.

[9] EMACC Handbook. European Military Airworthiness Certification Criteria. EU: European Defence Agency, 2018. Sv. Issue 3.0.

[10] JSSG-2006 Joint Service Specification Guide. Aircraft Structures. U.S.: Department of Defence, 1998.

[11] DEF STAN 00-932 Metallic Materials Data Handbook. U.K.: MODUK - British Defense Standards (MODUK), 2000.

[12] Slotwinski J., Cooke, A., Moylan, Sh. Mechanical Properties Testing for Metal Parts Made via Additive Manufacturing: A Review of the State of the Art of Mechanical Property Testing. U.S.: National Institute of Standard and Technology, U.S. Department of Commerce, 2012. NISTIR 7847.

[13] Cooke, A., Slotwinski. Properties of Metal Powders for Additive Manufacturing: A Review of the State of the Art of Metal Powder Property Testing. U.S. : National Institute of Standarda and Technology, U.S. Department of Commerce, 2012. NISTIR 7873.

[14] Slotwinski, J., Moylan, S. Applicability of Existing Materials Testing Standards for Additive Manufacturing Materials. U.S.: National Institute of Standard and Technology, U.S. Department of Commerce, 2014. NISTIR 8005.

[15] NASA MSFC-STD-3716 Standard for Additively Manufactured Spaceflight Hardware by Laser Powder Bed Fusion in Metals. Alabama, George C. Marshall Space Flight Center, 2017.

[16] AWS D20.1/D20.1M. Specification for Fabrication of Metal Components using Additive Manufacturing. U.S.: American Welding Society, 2019.

EAN 2020
58th conference on experimental stress analysis

Study of Degradation of Carbon and Basalt Fibres in Alkaline Environment

RYVOLOVÁ M.[1,a], BAKALOVA T.[1,b], SVOBODOVÁ L. [1,c]

[1]Department of Material Science, Faculty of Mechanical Engineering, Technical University of Liberec, Studentská 2, 460 17 Liberec, Czech Republic

[a]martina.ryvolova@tul.cz, [b]totka.bakalova@tul.cz, [c]lucie.svobodova@tul.cz

Keywords: basalt and carbon fibres, tensile strength, alkali resistance, degradation, image analysis

Abstract. Basalt and carbon grids are used for building application as reinforcement for cementitious mortars and geopolymers composites. The environment in which these materials are incorporated alkaline and aggressive with pH ≥ 12. Influence and differences of long-term exposure of fibres in two liquids was observed; demineralized water and potassium water glass. Defects or cracks due to the effect of potassium water glass were expected to appear on the surface of the basalt fibres. It was stated that basalt fibres (BFs) treated with demi water and carbon fibres (CFs) treated with both liquids remained without significant damage within three months. The change in mechanical properties and the occurrence of degradation processes in composites were confirmed by using tensile strength test and using image analysis based on scanning electron microscopy (SEM).

Introduction

Grid (warp knitted fabric) based on the network structure made of basalt, carbon or glass fibres is a new option for reinforcing concrete, cement mortar or geopolymer composite. A typical property of building materials is a strongly alkaline environment (concrete pH 12, geopolymer composite pH 12-14), which causes degradation of mainly mineral reinforcements (basalt and glass fibres). Carbon fiber is corrosion resistant and chemically stable; it is fire resistant and non-flammable, with good thermal and electrically conductivity. Composite materials with carbon reinforcements are commonly available in automotive, aerospace and shipbuilding industry, machinery construction, energy production and intervene to medicine, sport and lifestyle [1]. Carbon fibres have a great chemical resistance [7, 9]. More interesting is investigation of thermal stability, thermal stability in aggressive environment and flammability of CFs and geopolymer composite with carbon fillers [13, 14, 15].

Basalt fibre (BF) is classified as technical, inorganic, mineral fibre. The fibre is made of basalt rock. The technology of basalt fibre production is similar to that of glass fibres. Production is based on melt spinning at a temperature of 1500-1700 °C. Basalt fabric usually reinforces plastics constructions (BFRP – basalt fibres reinforced plastics). Production of basalt fibres is cheaper than production of carbon fibres. Basalt fibre is a natural product. Recycling BFRP and other products reinforced basalt fibres has more environmentally friendly impact than disposal CFRP (carbon fibres reinforced plastics). They have outstanding temperature stability, large thermal range of applicability -200 to +700 °C, they are incombustible, and they have a good UV-stability, high corrosion resistance, low humidity absorption. Carbon fibers are resistant in aggressive chemical environments [2].

Literature sources [3-7] more often describe the degradation of glass fibres due to an acidic or alkaline environment. The degradation of glass fibres in an alkaline environment is characterized by the occurrence of mineral incrustations and their peeling out from the fibre surface. The process is quantified by the changing of the fibre diameter and by the tensile strength value or Young's modulus [3]. Basalt fibers degrade in a similar way as glass fibers, both types are mineral origin. Studies [4, 5, 6] describe the evaluation of changes in the mechanical strength of fibre grids influenced by an alkaline environment for different temperature and concentration. Accelerating aging test was carried out for warp bundles from carbon, basalt and anti-corrosion glass fibres. Tests of durability provided following experimental results: carbon fibres have very good resistance to alkali even at higher temperature. Degradation of the mineral fibres was demonstrated by a tensile strength test, which showed a decrease in strength after immersion in an alkaline solution. Higher temperature and concentration accelerate their degradation process. Authors [7] use analysis of chemical composition (EDX), weight loss and SEM analysis for description and quantification the degradation process of carbon, basalt and glass fibres in deionized water, acid, alkaline, and saline solutions. Carbon fibres proved very good chemical resistance against different types of liquids. Weight loss and decrease of fibers diameter (degradation) was typical phenomena of mineral fibers in alkaline environment. Authors [8] observed chemical resistance of basalt fibres in water, salt, acid, and alkaline environments. The degradation of basalt fibres in alkaline media and in water was described, where the fiber diameter and fibre strength decreased significantly. The cause of these phenomena is a chemical reaction between the fibre and the solution. The diffusion rate is related to OH- concentration and is influenced by the environment temperature. SEM microscopy is very suitable to show and detect corrosion shell morphology and texture changes. However, no calculation based on image analysis is available until today. Only verbal description of fiber surface or diameter changes is used in a wide range of papers [9, 11]. Image analysis proceedings are able to show and detailly calculate the shape, size, amount, texture of objects on the surface of the fibres; unfortunately, a few researchers use this method in their research [12].

In this paper, the aim is to find the relationship between the change in mechanical properties and the results of the evaluation of image analysis. By conducting a series of experiments, we monitored the evolving degradation processes in basalt fibers in comparison with durable carbon fibers. Further study will be directed to the development of a proper methodology for the evaluation of degradation processes taking place on the surface of fibers in the volume of composite materials.

Materials and Methods

For experiment was used fibre rovings of witch grids are made. Carbon roving STS40 F13 24K from producer TohoTenax Europe GmbH has density 1.67 g/cm^3, linear density 1700 tex, tensile modulus 240 GPa, elongation 1.7 %, and filament diameter 7 μm. Basalt roving from producer Deutsche Basalt Faser GmbH has density 2.7 g/cm^3, linear density 2400 tex, tensile modulus 84 GPa, elongation 3 %, filament diameter 13 μm. Mineral composition of basalt roving present Table 1 [2].

Table 1: Mineral composition (weight ratio %)

	SiO_2	Al_2O_3	Fe_2O_3	CaO	MgO	TiO_2	Na_2O	others
Min	45	12	5	6	3	0.9	2.5	2.0
Max	50	19	15	12	7	2.0	6.0	3.5

Experiment was based on comparison progress of degradation; chemical resistant carbon fibre roving (CR) and inorganic basalt fibre roving (BR) in alkaline environment was compared.

The test was realized according to Standard ČSN ISO 695 - Resistance to alkali [16]. Carbon (CRs) and basalt rovings (BRs) were treated with potassium water glass (KWG, concentrated solution of $K_2(SiO)_4$) and demineralized water (DW) at specified time intervals: two and six weeks (W2; W6), three, six and twelve months (M3, M6, M12). Experiment was carried at ambient temperature.

Samples for tensile strength test were prepared according to Standard ISO 11566 [17] with small differences, which respected rovings properties and dimensions. Rovings were sticked to paper frames. Samples prepared in this way minimize the possibility of roving break or other damage. The degree of mechanical properties degradation was observed with the help of tensile strength test according to EN ISO 13934 [18] (device Testometric). Clamping length was 100 mm and speed of test was 10 mms^{-1}. Each samples group contained ten samples. Scanning Electron Microscopy (UHR FE-SEM Carl Zeiss ULTRA Plus) was used for fibre surface changing investigation and results was processed using image analysis methods.

Image analysis of SEM pictures was performed within Matlab software (The MathWorks, Inc.); image was first corrected through the '*adapthisteq*' function (adaptive histogram equalization enhances the image contrast) and '*imadjust*' function (corrects image intensity values), then the image was converted to a grayscale form (8 bit image). The fibre diameter was obtain through the distance transform. Euclidean distance (straight-line distance) was calculated using the function 'bwdist, euclidean'. This computes the sum of pixels between each pixel and the nearest non-zero pixel; the maximum value of distance transform corresponds to fibre diameter determined at each point along the length of a fibre. The fibre surface roughness was determined along the fibre boundary, based to relative contrast of fibre to background, see green straight line in Figure 1. Low fibre roughness corresponds to a straight fibre with one-level gradient step to background along the whole border (see course of surface roughness on graph in Figure 1). The roughness is calculated as an average deviation of shadow level gradient to background.

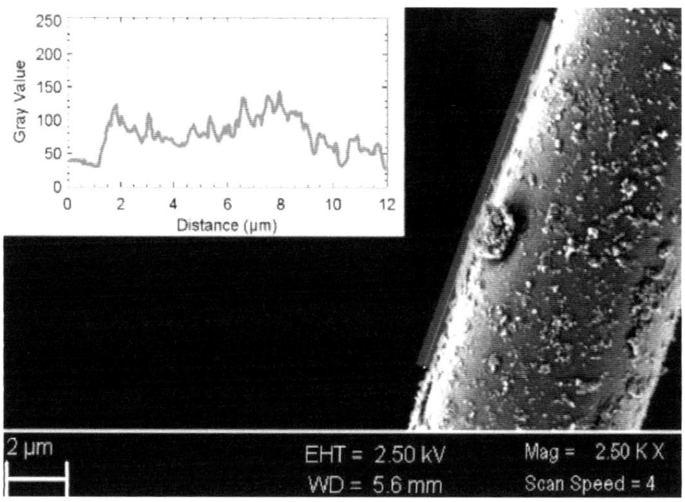

Fig. 1: Image analysis, representation of the roughness calculation along the fibre.

Results

Table 2 demonstrates changes of average strength values of basalt and carbon fibre rovings. Comparison between reference sample (REF), samples treated with demi water (DW) and samples treated with potassium water glass (WGK) shows long-term influence on strength of rovings.

Table 2: Result of tensile strength test – F_{max} [N]

	REF	W2	W6	M3	M6	M12
BR/WGK	758 ± 51	239 ± 32	151 ± 31	97 ± 2	-*	-*
BR/DW	758 ± 51	478 ± 53	404 ± 40	413 ±5 4	408 ± 54	440 ± 82
CR/WGK	700 ± 44	821 ± 78	774 ± 64	791 ± 51	794 ± 89	726 ± 109
CR/DW	700 ± 45	736 ± 92	681 ± 103	672 ± 30	648 ± 108	730 ± 179

**Basalt roving fibers were unusable for mechanical testing after about 4 months.*

Table 2 presents significant changes of BRs strength treated with water glass; strength of rovings decrease significantly depending on time, fibres dissolved in liquid after 3 months. Dissolved sizing on surface of fibres – the result is a reduction of strength values between sample REF and strength values of other basalt rovings. The demi water decreased mechanical properties of basalt rovings by 42 % after long-term effects. The tensile strength of carbon roving treated with H_2O oscillates around references value; water dissolved sizing and it violated cohesion of carbon fibres bundle and simultaneously strength of carbon roving. Water glass influenced strength of carbon roving; chemical reaction between components of WGK and components of sizing can be probably the reason of this change.

Figure 3 presents tensile strength loss of basalt fibres and of carbon fibres. Tensile strength loss between reference value of BR and condition after 2W is 42%. Generally, strength loss after long-term influence DW moves from 42 % to 34 %.

Demi water first acts on the sizing, which is dissolved over time. Subsequently, the surface of the fibers changes, with small amounts of oxides beginning to be precipitated from the surface.. The reason for the strength decrease was slow creation of high-silica layer, which protect fibre surface simultaneously. The trend of strength loss after influence of potassium water glass is clearly declining. Decrease after W2 is 64 %, after three months was decrease almost 90 %, after six months are basalt roving completely dissolved. Tensile strength loss between reference strength of CRs and condition after 2W is -5%. For CF was observed the opposite trend. Following strength increase is from 3% to 7.5%. The reason is probably the disruption and slow dissolution of sizing of fibres. SEM analysis supports the assumption – surface of carbon fibres was smooth and without sizing residues gradually. In case of interaction between KWG and carbon rovings strength grew up on average about 13 %. For determination of source, it is necessary use chemical analysis

For comparison in study [8] changes in chemical composition was described together with degradation process. Ions type of Na, K, and Al are missing in the chemical composition of the treated fibre compared to the reference sample. When evaluating the surface of fiber treatments, it must be considered that the fibre consists of a core and an outer layer. The content of insoluble oxides of Mg, Fe, Ca, and Ti increases in this outer layer. BF retains 99 % of its original weight in water at 25 °C, the change of strength is small. The fibre treated with boiling water forms a high-silica layer on the surface, which causes a 20 % increase in strength; it has a protective function (covers cracks on fiber surface) and prolongs the degradation process. Images of EDS and SEM analysis in experiments confirmed the achieved results of mechanical tests. Authors [9] examined different corrosion mechanism of basalt fibres in NaOH solution and cement solutions. Solutions represented alkaline environment, pH value 12. Corrosion process is dependent on calcium ions concentration. NaOH solution causes creation and development of corrosion shell; it causes dissolution of fiber surfaces layer.

Aging of basalt fibres in study [10, 11] was realized according to Standards (ASTM C 1203-91, ISO 695:1991). EDX-analysis proved changes in weight ratio of elements between reference samples of BFs and fibres treated alkaline solution. Article describes creation of corrosion shell on surface of BFs in detail in NaOH solution. Cement mixture allows only local attack on surface – holes and cracks, creates small holes on surface of fibres, which are connected together over time. Process ends with fibre damage. Authors in [12] used these methods, they confirmed the fundamental effect of temperature and concentration on the rate of degradation. A detailed description of the degradation processes presented as the corrosion shell formation phases was related to the changes in the tensile strength of the fibre. The strength of the fibres does not trend to decrease only; it usually varies according to the current phase of degradation. The strength of the fibre decreases depending on the gradual formation and increase of mineral incrustations amount. When the corrosion shell is formed, it is peeled from the fibre and the fibre strength increases again (not to its original value). The process is repeated over time until the fibre is completely dissolved.

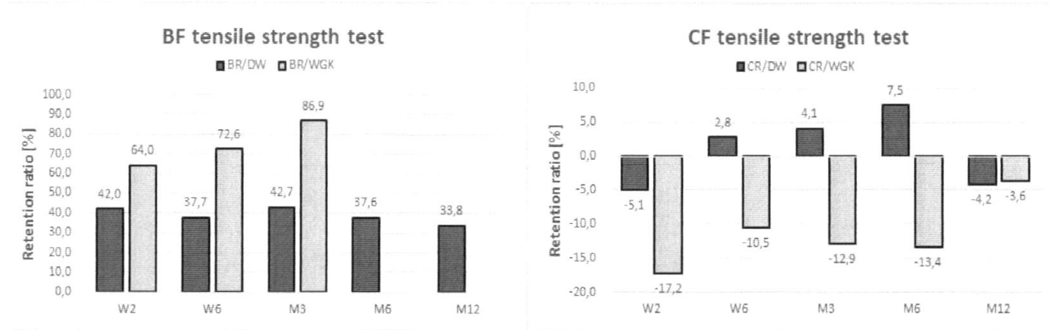

Fig. 3: Tensile strength loss in basalt and carbon rovings treated DW and KWG.

Figure 4 shows differences between basalt fibres (BFs) surface. Picture in the left side presents sizing on surface of reference sample (without other influence). Picture in the middle presents remaining of sizing on surface after treatment with demi water and picture in the ride presents corrosion coating - result of alkaline liquid influence after 3 months. Figure 5 shows carbon fibres surface; in the left - sizing on reference sample surface and clear carbon fibres surface after treated of demi water (12M) and potassium water glass (12M). Carbon fibres (CFs) surface is clean and smooth in both cases, without residues of sizing and without damage.

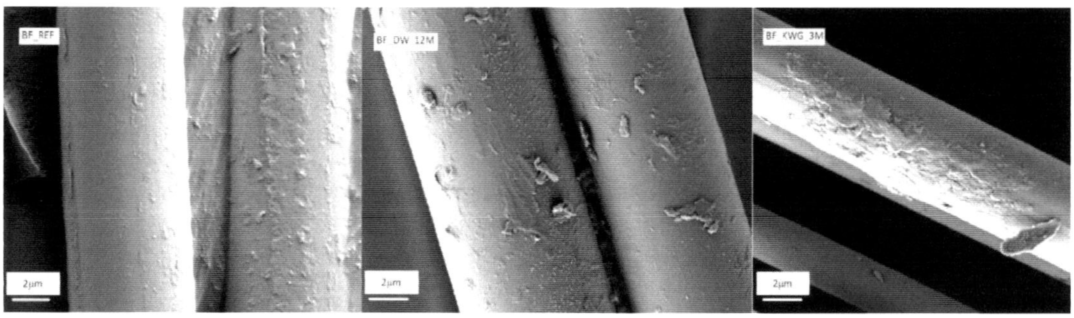

Fig. 4: Surface of BFs – reference sample, 12M in demi water, 3M in potassium water glass.

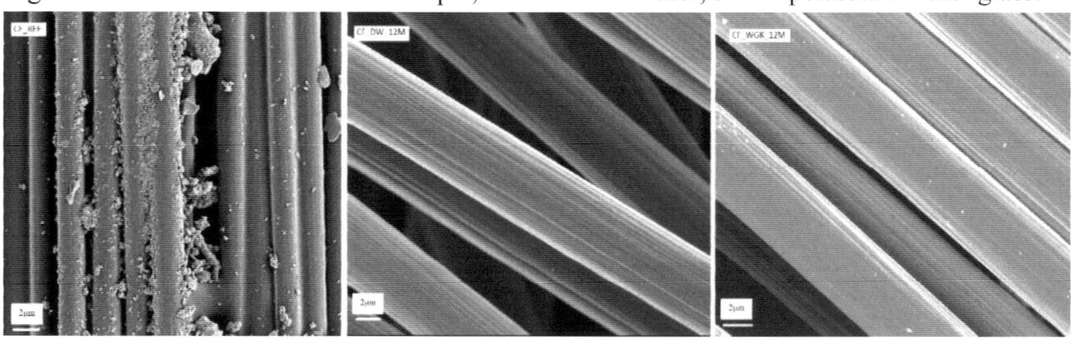

Fig. 5: Surface of CFs – reference sample, 12M in demi water, 12M in potassium water glass.

Results of image analysis (diameter and roughness determination) are depicted in Figure 6. Carbon fibres test show very low change in fibre diameter within 6 weeks (only the deviation of the diameter measurement is different for some samples). Fibre surface roughness (expressed in percentage in relation to the reference sample in zero-time, which is equal to 100 %) decrease in demi water, and increase in water glass (KWG). Basalt fibres test show almost 50 % decreasing in fibre diameter in KWG; and of up to 16 % changing of fibre diameter in demi water. Fibre surface roughness increase in demi water; and has no significant changes in water glass.

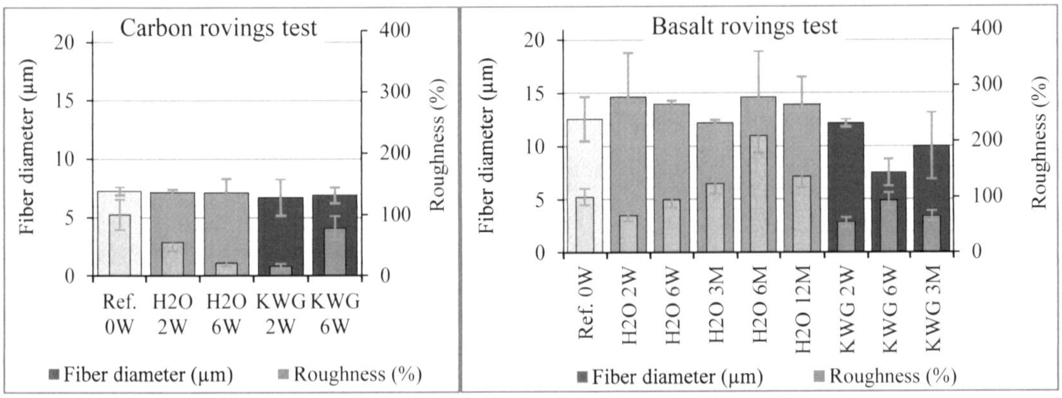

Fig. 6: Image analysis result showing fibres diameter (as blue) and fibres surface roughness (as orange).

A combination of the following methods was used to determine the degree of degradation of basalt and carbon fibers: tensile strength test, SEM and image analysis. SEM analysis and image analysis provided information about corrosion processes and development of corrosion aggregates on surface of basalt fibres. Chemical stability was assumed for carbon fibres, which was confirmed by laboratory experiments. Tensile strength test was supported by image analysis results. Both methods (SEM and image analysis) provide reliable information on fibre degradation due to the solutions used. Research shows that using image analysis, when using multiple SEM images changes in thickness and roughness of the examined fibres can be well monitored.

Conclusion

Experiment based on long-term influence of demineralized water and aggressive alkaline solution brought following results.

1. The presence of water affected the strength of BRs (decrease compared to the reference sample by approximately 40 %). Image analysis showed fibre swelling, the fibre diameter increased and gradually decreased. The contact of CRs with water showed only the dissolution of the sizing of fibres, which caused a slight decrease in strength.
2. Alkaline environment (in this case - potassium water glass) caused dissolution of BRs. Result confirmed tensile strength test and results of image analysis (decrease of diameter and increase of roughness).
3. Results for determination of CRs degradation is not possible conclusively express. Strength of carbon rovings has grown after treatment by alkaline environment. Image analysis did not reveal the cause. There is a need to use chemical analysis for exact determination of reason.

Acknowledgement

This work was supported by the Student Grant Competition of the Technical University of Liberec under the project No. SGS-2020-5019.

The work was supported by the Ministry of Industry and Trade in the framework of the targeted support of the Application, Call VII, the Operational Program Enterprise and Innovations for Competitiveness – "Research and development of textile products focusing on knitting technology and using aqueous chemicals", reg. number CZ.01.1.02/0.0/0.0/19_262/0020121.

References

[1] DataSheet TohoTenax, www.tohotenax.com

[2] DataSheet Basaltfaser Roving 2400tex. Spezification, https://www.deutsche-basalt-faser.de

[3] G. W. Ehrenstein, Polymers composite materials. Prague: Scientia, 2009, ISBN 9788086960296

[4] N. W. Portal, N. Silva, K. Malaga, U. Mueller, P. Billberg, Durability Study of Textile Fibre Reinforcement, in: Proceedings of the 27th Biennial National Conference of the Concrete Institute of Australia in conjunction with the 69th RILEM Week "Construction Innovations. Research into Practice", Institute of Australia, 2015, pp. 408-413.

[5] N. W. Portal and col, Tensile behaviour of textile reinforcement under accelerated ageing conditions, Journal of Building Engineering 5, 2016, pp. 57–66.

[6] B. Lia, H. Xionga, J. Jianga, X. Dou, Tensile behavior of basalt textile grid reinforced Engineering Cementitious Composite, Composites Part B 156 (2019) 185–200

[7] C. Scheffler, S.L.Gao, R. Plonka, E. Mäder, S. Hempel, M. Buttler, V. Mechtcherine, Interphase modification of alkali-resistant glass fibres and carbon fibres for textiles reinforced concrete I: Fibre properties and durability, Composites Science and Technology 69, 2009, pp.531-538

[8] S. Ying, X. Zhou, Chemical and Thermal Resistance of Basalt Fiber in Inclement Environments, Journal of Wuhan University of Technology-Mater. Sci. Ed. June 2013, DOI 10.1007/s11595-013-0731-4

[9] P. Cousin, M. Hassan, P.V. Vijay, M. Robert and B. Benmokrane, Chemical resistance of carbon, basalt and glass fibers used in FRP reinforcing bars, Journal of Composite Materials, 2019, Vol. 53(26–27) pp. 3651–3670.

[10] G. Wu, X. Wang Z. Wu, Z. Dong and G. Zhang, Durability of basalt fibers and composites in corrosive environments, Journal of Composite Materials 2015, Vol. 49(7) 873–887

[11] C. Scheffler and col , Aging of alkali-resistant glass and basalt fibers in alkaline solutions: Evaluation of the failure stress by Weibull distribution function;; Journal of Non-Crystalline Solids 355 (2009) 2588–2595

[12] Ch.Tang, H. Jiang, X. Zhang, G. Li and J. Cui, Corrosion Behavior and Mechanism of Basalt Fibers in Sodium Hydroxide Solution, Materials 2018, 11, 1381; doi:10.3390/ma11081381

[13] P. Behera, V. Baheti, J. Militky, S. Naeem, Microstructure and mechanical properties of carbon microfiber reinforced geopolymers at elevated temperatures; Construction and Building Materials 160, 2018, pp. 733–743

[14] D.H. Tran, D. Kroisová, P. Louda, O. Bortnovsky, P. Bezucha, Effect of curing temperature on flexural properties of silica-based geopolymer carbon reinforced composite, JAMME, Dec 2009

[15] P. He, D. Jia, T. Lin, M. Wang, Y. Zhou, Effects of high-temperature heat treatment on the mechanical properties of unidirectional carbon fiber reinforced geopolymer composites, Ceramics International 36, 2010, pp. 1447–1453

[16] ČSN ISO 695 (ISO 3585:1998) Borosilicate glass 3,3 – Properties, http://csnonlinefirmy.agentura-cas.cz

[17] ISO 11566:1996, Carbon fibre — Determination of the tensile properties of single-filament specimens, https://www.iso.org

[18] EN ISO 13934-1:201, Textiles – Tensile properties of fabrics – Part 1: Determination of maximum force and elongation at maximum force using the strip method, http://csnonlinefirmy.agentura-cas.cz

Comparison of Micromotion of Different Femoral Stems During Gait: A Quantitative CT-Based Finite Element Analysis

ŘEHOUNEK L.[1,a], MIMATA H.[2,b]

[1]Department of Mechanics, Faculty of Civil Engineering, CTU in Prague, Thákurova 7, 166 29 Prague 6 - Dejvice, Czechia

[2]Research Center of Computational Mechanics, Inc., Tokyu Togoshi Building 8F, 1-7-1 Togoshi, Shinagawa-ku, Tokyo, Japan

[a]lubos.rehounek@fsv.cvut.cz, [b]mimata@rccm.co.jp

Keywords: Micromotion, Femoral, CT analysis, FEA, Gait

Abstract. With the increasing availability of CT scans, it is now possible to use patient CT data for tailoring a patient-specific treatment method. The CT-based FEM software MECHANICAL FINDER was used to evaluate the values of micromotion of 4 different femoral stems in simulation of THA (total hip arthroplasty) with real CT patient femoral data and STL files of femoral implants. The analysis was done using literature-obtained muscle forces and boundary conditions of the human femur at an arbitrary gait cycle. As expected, it was found that the shortest stem exhibits the highest values of micromotion (L = 80 mm, total micromotion up to 39 μm) compared to a longer stem (L = 100 mm, total micromotion up to 28 μm). Even longer stems (L = 120 mm and L = 140 mm) show values ranging between the first two. This approach shows potential viability of using simulations to predict osseointegration in patient-specific conditions available for medical practitioners. Experimental verification is still necessary.

Introduction

Aseptic loosening of the acetabular and femoral component as a result of loss of primary stability is the main reason of early implant failure in THA patients. As a measure of osseointegration, micromotion monitoring is a viable tool that can determine the movement of the femoral component relative to the bone. While some small movement (generally above 40 μm) can be beneficial for bone ingrowth, displacements larger than 150 μm completely inhibit bone ingrowth [1]. Large micromotion is generally bad for osseointegration, especially in cemented implants where motion can lead to cracks and residue of methylmethacrylate [2]. Micromotion, just like osseointegration, is also directly linked with bone quality of the individual, which depends mainly on their diet, physical activity, age, sex and disease [3].

Objectives

The main goal of the research is to determine the influence of femoral stem length on the micromotion along different axes of motion. As excessive motion is undesirable, numerical methods, such as FEM can serve as a precursor to the design of novel implants or to analyze an individual patient's susceptibility to micromotion of the implant stem in the femur. Therefore, the presented method aims to provide a tool that medical practitioners can potentially use in practice to determine the optimal implant and/or optimal implant position for a specific patient based on calibrated CT data.

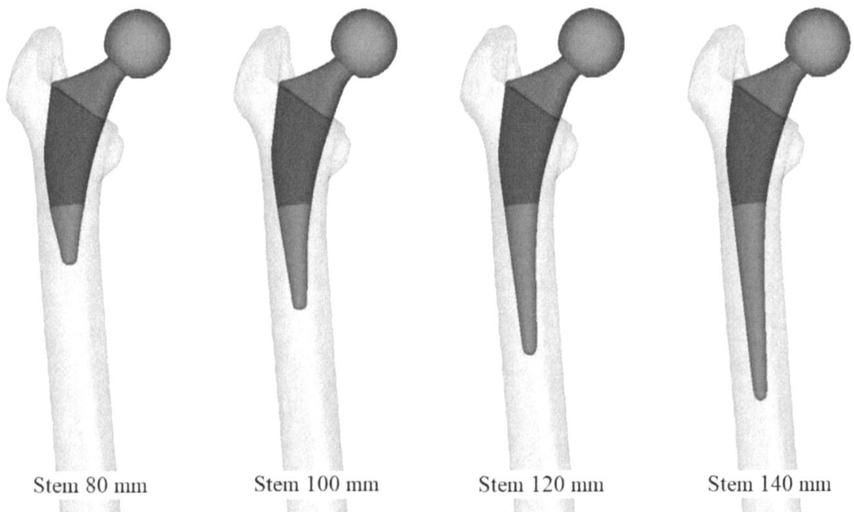

Fig. 1: The length, shape and placement of all analysed femoral stems

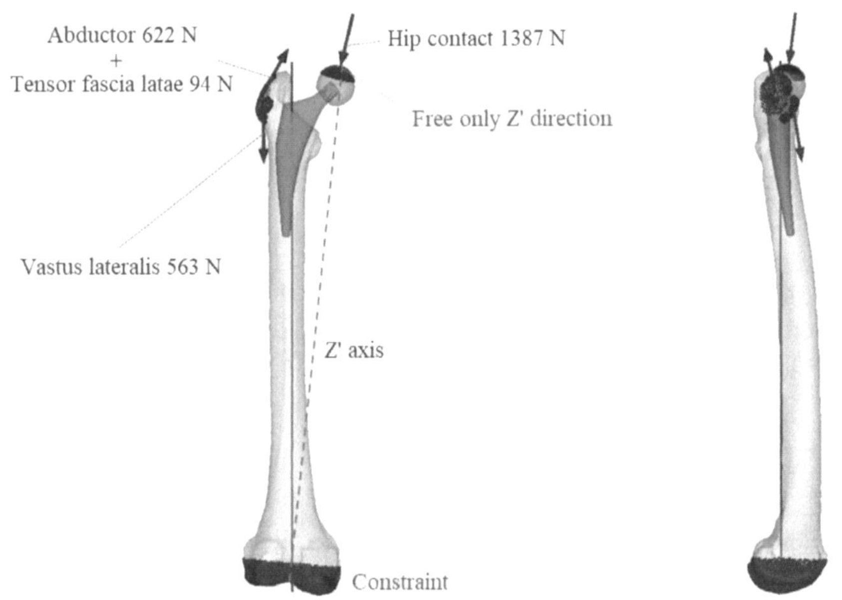

Fig. 2: The boundary conditions applied to the model of human femur. BW = 600 N. Derived from [4]

Study Design & Methods

The presented study focuses on FEM analyses of different variants of stems used in femoral implants with regard to micromotion between the implant and bone. Four geometrical solutions of the implant stem are presented. All implant variants have conventional straight stems and they vary in length (Fig. 1). The used software is MECHANICAL FINDER, a FEM-based practical tool utilizing patient-specific CT data to obtain a realistic, inhomogeneous material properties from calibrated CT scans. An anonymized CT scan, which is the subject of the research, has been calibrated by a phantom and served to create a 3D model of the femur. Real STL implant data have then been imported into the software to simulate a realistic post-operation environment. The global matrix of stiffness is then derived based on specified bone material model, which reads calibrated CT values (Fig. 3). Therefore, the software is able to create a fully inhomogeneous model of bone of individual patients. Boundary conditions are applied from each muscle forces and hip contact force in an arbitral cycle of gait (Fig. 2 [4]). The measure of osseointegration used in this study is stem micromotion relative to the human femur in the lateral-medial, posterior-anterior and proximal-distal axes.

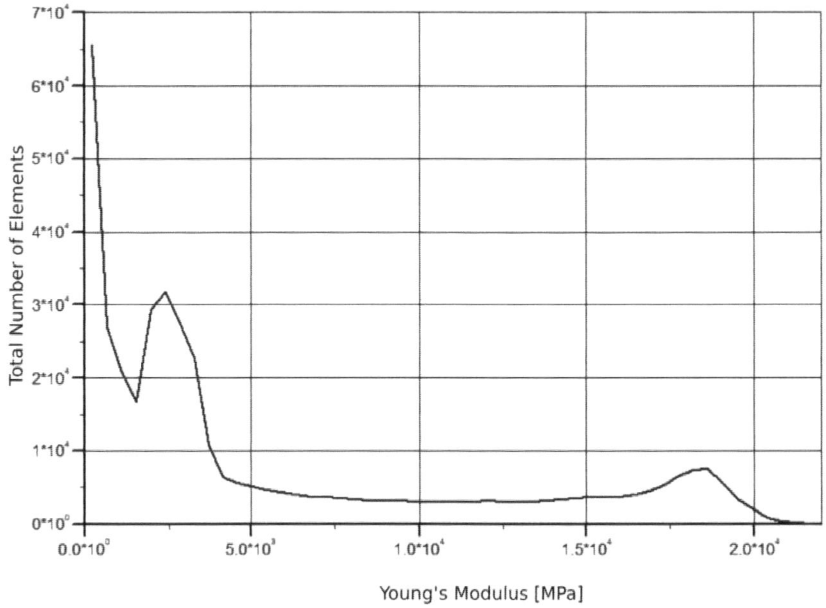

Fig. 3: The Young's Modulus histogram of the analyzed patient-specific CT scan

The FE mesh composed of linear (1st order) tetrahedral elements with element-unique values of Young's Modulus derived from the calibrated CT scan (Fig. 3). The number of elements for each analyzed variant, as well as the computation times, are shown in Table 1. The values of Young's Modulus were in the range of (minimum threshold) 0.001-21928 MPa. The minimum value is set so as to represent the softest materials in the bone while still maintaining some stiffness.

Table 1: Numbers of FE elements and length of computation for different stem variants

Stem variant	Number of elements	Length of computation [s]
L = 80 mm	502979	3510
L = 100 mm	517041	4577
L = 120 mm	526886	7471
L = 140 mm	543475	8936

Results

Micromotion between the stem and the femur was below 150 μm in stem types during gait, although their distribution was different, as shown in Fig. 3-6. The commonly recognized criterion for osseointegration (micromotion < 150 μm) was satisfied. The shortest stem (L = 80 mm) has shown the greatest values of micromotion (39 μm) and the least amount of micromotion was recorded for the second smallest (L = 100 mm) stem (28 μm). Variants L = 120 mm and L = 140 mm have shown values of 31 μm and 34 μm, respectively.

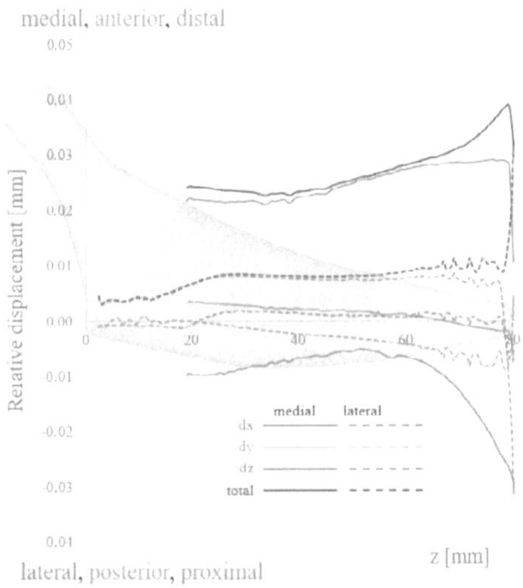

Fig. 4: The micromotion of the shortest (L = 80 mm) stem variant

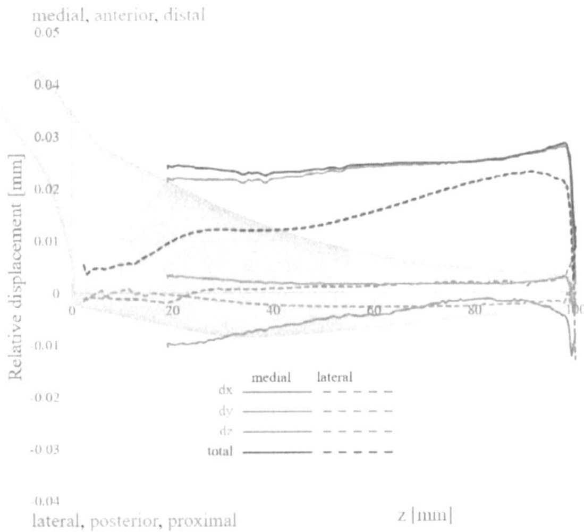

Fig. 5: The micromotion of the L = 100 mm stem variant

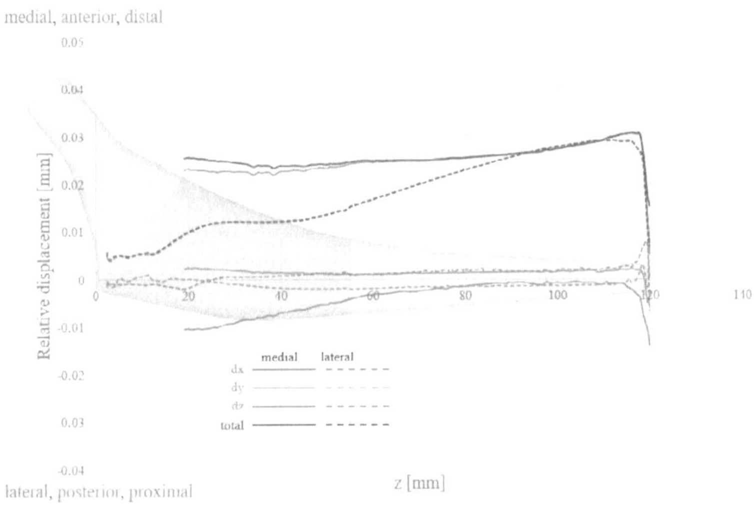

Fig. 6: The micromotion of the L = 120 mm stem variant

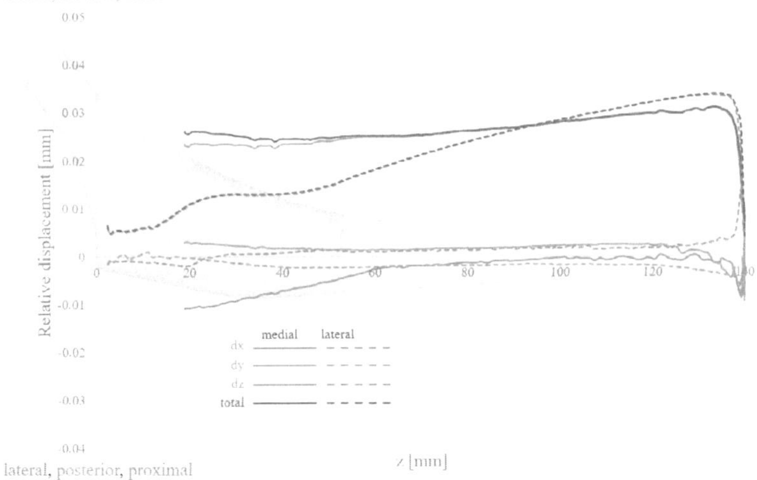

Fig. 7: The micromotion of the longest (L = 140 mm) stem variant

Conclusions

In this analysis, we only analyzed the gait cycle on four implants which differ in length. If the load does not exceed extreme values, micromotion is well within the tolerance required for bone osseointegration (< 150 μm) for all implant stem variants. The maximum values of micromotion seem to stabilize among the 3 longest implant variants (length above 80 mm). However, stress shielding might be more important when considering long-term bone atrophy in THA patients. As choosing the right implant (length and shape) for a patient poses the question of bone quality, a simulation showing the stress distributions of different implant variants for a specific CT scan of a patient might also be desirable.

Acknowledgment

The financial support by the Faculty of Civil Engineering, Czech Technical University in Prague (SGS project No. SGS20/155/OHK1/3T/11) is gratefully acknowledged.

References

[1] R.M. Pilliar, J.M. Lee and C.D.D.S. Maniatopoulos, Observations on the effect of movement on bone ingrowth into porous-surfaced implants, Clin. Orthop. Relat. Res. 208 (1986) 108–113.

[2] D.W. Burke et al., Micromotion of cemented and uncemented femoral components, Bone Joint J. 73-B. 1 (1991) 33–37.

[3] L. Řehounek, F. Denk and A. Jíra. Numerical Simulations of Static Strength of Novel Dental Implants, Acta Polytechnica 58. 4 (2018) 259–263.

[4] B-A. Behrens et al., Numerical investigations on the strain-adaptive bone remodelling in the periprosthetic femur: influence of the boundary conditions, Biomed Eng 8. 7 (2009).

EAN 2020
58th conference on experimental stress analysis

Modelling and Testing of 3D Printed Flexible Structures with Three-pointed Star Pattern Used in Biomedical Applications

ŘEPA V.[1,a], MARŠÁLEK P.[1,2,b], PROKOP J.[2,3,4,c], ŠOTOLA M.[1,d], STARECZEK D.[1,e], RYBANSKÝ D.[1,f], HALAMA R.[1,g]

[1]Deparment of Applied Mechanics, Faculty of Mechanical Engineering, VŠB – Technical University of Ostrava, 17. listopadu 2172/15, 708 00 Ostrava, Czech Republic

[2]Marschal Innovative Technologies s.r.o., Masarykova 409/26, 602 00 Brno, Czech Republic

[3]Department of Surgical Studies, Faculty of Medicine, University of Ostrava, Dvorakova 7, 701 03 Ostrava, Czech Republic

[4]Department of Surgery, University Hospital Ostrava, 17. listopadu 1790/5, 708 00 Ostrava, Czech Republic

[a]vojtech.repa.st@vsb.cz, [b]pavel.marsalek@vsb.cz, [c]jiri.prokop@fno.cz, [d]martin.sotola@vsb.cz, [e]david.stareczek@vsb.cz, [f]david.rybansky@vsb.cz, [g]radim.halama@vsb.cz

Keywords: Flexible, 3D print, Stiffness, Structure, Simulation, Biomedical, Material, PA12

Abstract. The objective of this work is to determine, using laboratory testing and computer modelling, the stiffness of flexible structures with three-pointed star pattern made by 3D printing method. In this paper, we present the design of the experiment and the development of the computer model as well as evaluating the structure's stiffness. At the end of the work, a validated computer model is used for a parametric study of the three-pointed star pattern. The obtained results can be used for designing of the biomedical applications.

Introduction

The flexible structure (FST) represents a thin shell with openings pattern that decreases the structure's stiffness in the normal direction. With such a structure, parts with desired deformation behaviour or aesthetically modified parts with preserved structural integrity can be designed [1, 2, 3]. Due to the enormous advantages of 3D printing methods such as Selective Laser Sintering (SLS) or Multi-Jet Fusion (MJF), the complicated shape of FST can be printed without supports and the final shape is created without additional machining. By using Polyamide 12 (PA12), also known as nylon, the resulting smooth structure can be used for biomedical applications. These applications for their purposes must meet strictly stiffness requirements which are established using the medical examination because FST must adapt to the user body.

In this paper, the determination of the stiffness of designed FST with three-pointed star pattern printed by SLS method using material PA12 is presented. Firstly, the stiffness of FST is determined by experiment. In the second part, the computer model using the Finite Element (FE) method simulating the experiment is created. The results of the FST stiffness obtained from the computer model are validated by the experiment results. Parametric study of the stiffness of FST with three-pointed star pattern is done with this validated computer model.

Experiment and stiffness determination

The experiment is inspired by the laboratory testing of Cranial Orthoses published in [4]. The specimen representing FST with a half sphere-shape with inner radius R_1 = 65.0 mm is clamped into testing device and gradually loaded by plunger with dimension R_2 = 65.0 mm and h = 18.0 mm, see Fig. 1. The load is generated and measured using TESTOMETRIC 500-50CT universal testing machine.

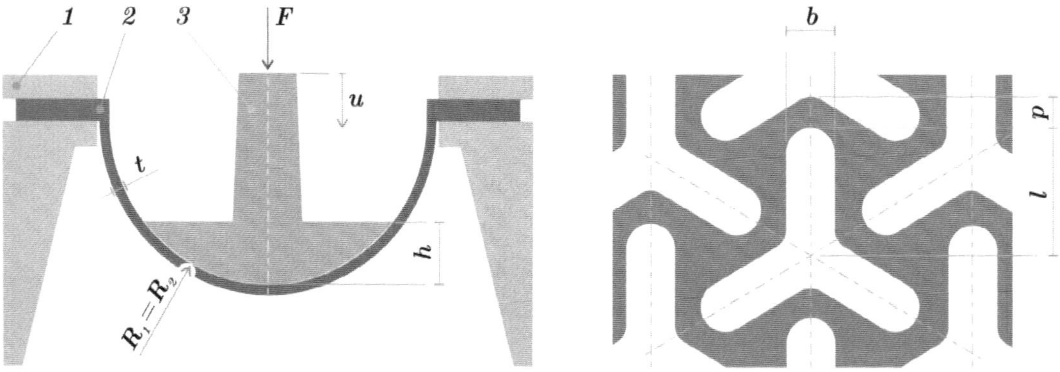

Fig. 1: Three-pointed star pattern (right) and scheme of testing device (left):
1 – testing device, 2 – specimen, 3 – plunger

For experiment purposes, two specimens marked by letter A with three-pointed star pattern employing parameters b = 2.00 mm, l = 5.00 mm and p = 2.35 mm were printed, see Table 2. The designed thickness of the first specimen A is t = 1.00 mm, the thickness of the second specimen A is t = 1.25 mm. The measured relationship between applied force F and deflection u of specimens A is depicted on Fig. 2. The stiffness of specimen A is determined by the linear approximation of the measured data with excluded nonlinear part in the beginning of loading.

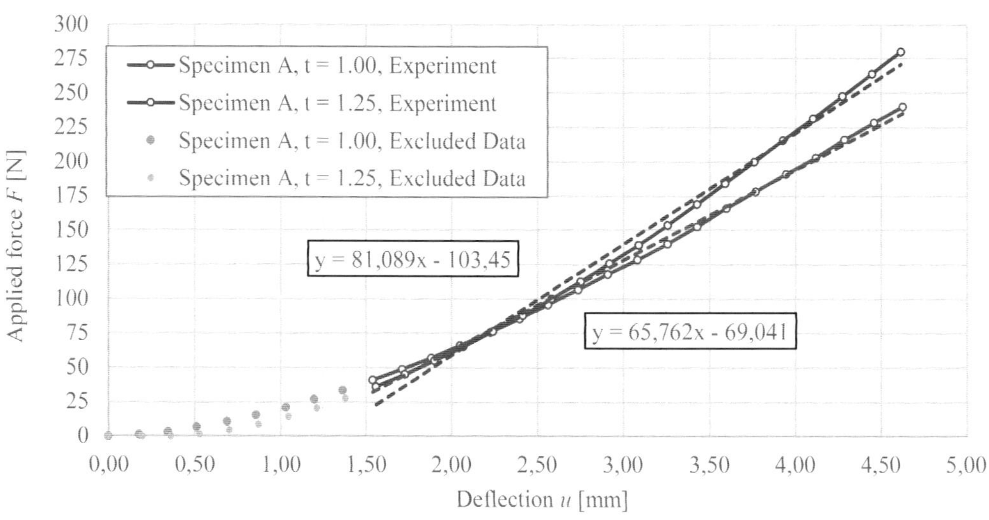

Fig. 2 The relationship between applied force F and measured deflection u with stiffness determination

Simulations

To be able to replace FST experiment measuring with a virtual one, a computer model is developed. This model uses the Finite Element (FE) method for mathematical description and is created in SW ANSYS Workbench 2019 R3 and it consists of two parts. The first part represents the parametric geometrical model of the specimen with subjected three-point star pattern, second one is used for prescribing boundary conditions corresponding to the specimen loading by plunger. Since plunger is much stiffer than tested structure, static loading is gradually increased and applied on a contact area by using ANSYS remote force function with deformable behaviour. The friction between plunger and specimen is neglected. Loading of the structure is gradually increased until loading force value $F = 200$ N is reached. Measured thicknesses of printed Specimen A are lower than required, which was caused by manufacturing tolerances. As a result of SLS printing process, real thicknesses t_r are decreased by 0.20 mm.

To capture specimen physical behaviour together with large deformations and deflection, nonlinear static structural analysis using finite elements for shell discretization (SHELL181) are used in computer model [5]. Linear isotropic elastic material model of material is used with Young's modulus $E = 1\ 224$ MPa and Poisson ratio $\mu = 0.39$ [6, 7]. FE model consisted of 100 000 elements and 90 000 nodes.

The calculated result from a computer model is relationship between applied force F and deflection u, from which stiffness is determined by linear approximation again. The high-quality calibration between experiment and simulation is achieved (Table 1). The vector displacement field correspond to loading force $F = 200$ N is shown in Fig. 3.

Table 1: Comparison of determined stiffness k from experiment and simulation

Designed thickness t [mm]	Real thickness t_r [mm]	Experiment k [N/mm]	Simulation k [N/mm]	Difference [%]
1.00	0.80	65.76	68.97	4.90
1.25	1.05	81.09	88.48	9.10

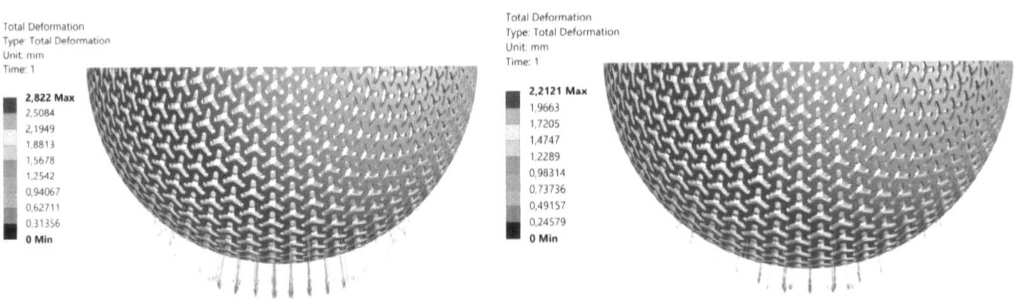

Fig. 3 The vector displacement fields corresponding to the applied force $F = 200$ N for specimen A, real thickness $t_r = 0.80$ mm (left) and $t_r = 1.05$ mm (right)

Parametric study

Parametric study of the FST stiffness employing the presented computer model was done. The stiffness is evaluated all together for fifteen variants of the pattern and real thicknesses. Three geometries of specimens with different tree-point star pattern were designed. The pattern geometry of first specimen A is described above. The pattern of the second specimen B has increased width between pattern openings p by 0.25 mm corresponding to parameters $b = 2.00$ mm, $l = 5.00$ mm, $p = 2.60$ mm. In case of the pattern of third specimen C, pattern openings parameters b and l were increased by 0.25 mm, i.e. $b = 2.25$ mm, $l = 5.25$ mm, $p = 2.35$ mm. Each specimen was thought in five real thicknesses ranging from 0.80 mm to 1.80 mm with the step size 0.25 mm. The pattern parameters are presented in Table 2.

Table 2: Tree-point star pattern parameters

Pattern parameters	Opening parameter b [mm]	Opening parameter l [mm]	Width between pattern openings p [mm]
Specimen A	2.00	5.00	2.35
Specimen B	2.00	5.00	2.60
Specimen C	2.25	5.25	2.35

Obtained results are listed in Table 3. It could be seen, that the influence of the determined stiffness k grows with increasing the real thickness t_r. Also, it is apparent from Fig. 4, the changing of the opening parameters b, l and the width between pattern openings p has a significant effect on the stiffness.

Table 3: Evaluation of determined stiffness from parametric study

Real thickness [mm]	0.80	1.05	1.30	1.55	1.80
Flexible structure	Determined stiffness k [N/mm]				
Specimen A	68.97	88.48	107.64	126.41	144.85
Specimen B	83.68	107.88	131.68	155.02	177.91
Specimen C	62.02	79.37	96.40	113.10	129.51

Conclusions

In this paper, two ways of determining the stiffness of the 3D printed FST are presented. A new experimental jig was constructed replacing traditional tensile and flexural testing techniques used for internal structure properties testing [8, 9]. The second one is a numerical analysis based on FE method. It was demonstrated that developed computer model is able to describe behaviour of FST and as such could be used to evaluate the FST stiffness. The high-quality calibration between experiment and simulation is achieved. In the end of the work, the computer model has been used for parametric study including three-point star pattern parameters and real thickness variation (Fig. 4). The proposed experimental technique has been found as an important complementary tool in the design of the 3D printed FST with tree-point star pattern by non-linear computer modelling, employing FE method.

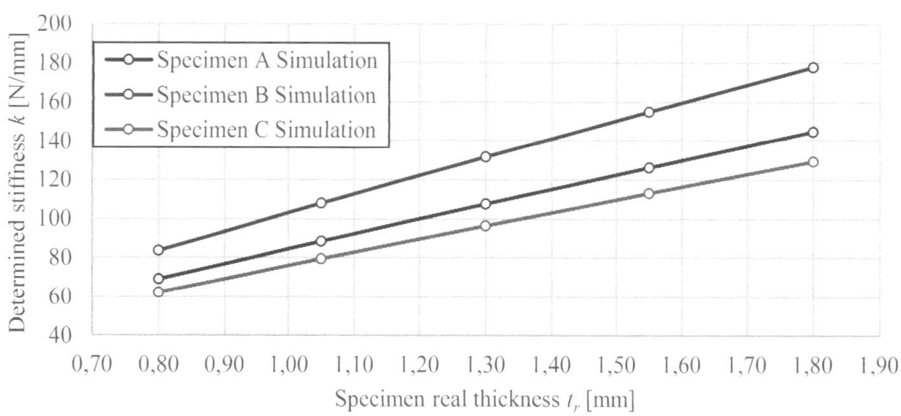

Fig. 4 Relationship between determined stiffness k and specimen real thickness t_r

Acknowledgements

This work was supported by The Ministry of Education, Youth and Sports from the Specific Research Project SP2020/23, by The Technology Agency of the Czech Republic in the frame of the project TN01000024 National Competence Center-Cybernetics and Artificial Intelligence and by Structural Funds of the European Union within the project Innovative and additive manufacturing technology – new technological solutions for 3D printing of metals and composite materials, reg. no. CZ.02.1.01/0.0/0.0/17_049/0008407.

References

[1] Ch. Schumacher, S. Marschner, M. Cross, B. Thomaszewski. Mechanical characterization of structured sheet materials. *ACM Transactions on Graphics*. 2018, 37(4), 1-15. DOI: 10.1145/3197517.3201278. ISSN 07300301.

[2] Ch. Schumacher, B. Thomaszewski, M. Gross. Stenciling: Designing Structurally-Sound Surfaces with Decorative Patterns. *Computer Graphics Forum*. 2016, 35(5), 101-110. DOI: 10.1111/cgf.12967. ISSN 01677055.

[3] B. Bickel, M. Bächer, M. A. Otaduy, H. R. Lee, H. Pfister, M. Gross, W. Matusik. Design and fabrication of materials with desired deformation behavior. *ACM SIGGRAPH 2010 papers on – SIGGRAPH '10*. New York, New York, USA: ACM Press, 2010, 2010, 1-. DOI: 10.1145/1833349.1778800. ISBN 9781450302104.

[4] P. Maršálek, A. Grygar, T. Karásek, T. Brzobohatý. *Virtual prototyping of 3D printed cranial orthoses by finite element analysis*. 2019, DOI: 10.1063/1.5114332.

[5] T. Kmínek, P. Maršálek, T. Karásek. *Analysis of Steel Tanks for Water Storage Using Shell Elements*. 2019, DOI: 10.1063/1.5114329.

[6] N. Lammens, I. De Baere, W. Van Paepegem. On the orthotropic elasto-plastic material response of additively manufactured polyamide 12. *7th bi-annual international conference of Polymers & moulds innovations*. Ghent, Belgium, 2016.

[7] D. Stoia, E. Linul, L. Marsavina. Influence of Manufacturing Parameters on Mechanical Properties of Porous Materials by Selective Laser Sintering. *Materials*. 2019, 12(6). DOI: 10.3390/ma12060871. ISSN 1996-1944.

[8] F. Dorčiak, M. Vaško, F. Bárnik, J. Majko. Comparison of experimental flexural test with FE analysis for specimen with different size and shape of internal structure created by 3D printing's. *IOP Conference Series: Materials Science and Engineering*. 2020, 776(1), Article number 012078. DOI: 10.1088/1757-899x/776/1/012078. ISSN 1757-899X.

[9] F. Dorčiak, M. Vaško, M. Handrik, F. Bárnik, J. Majko. Tensile test for specimen with different size and shape of inner structures created by 3D printing. *Transportation Research Procedia*. 2019, 40, 671-677. DOI: 10.1016/j.trpro.2019.07.095. ISSN 23521457.

EAN 2020
58th conference on experimental stress analysis

Properties of Respiratory Sound and Their Use

SKALICKÝ D.[1,a], KOUCKÝ V.[2,b], VÍTĚZNÍK M.[3,d], LOPOT F.[1,c]

[1]Department of Anatomy and Biomechanics, Faculty of Physical Education and Sport, Charles University in Prague, José Martího 31, 162 52 Prague 6, Czech Republic

[2]Department of Pneumology, 2nd Faculty of Medicine, Charles University in Prague and Motol University Hospital, V Úvalu 84, 150 06 Prague 5, Czech Republic

[3]Faculty of Biomedical Engineering, Czech Technical University in Prague, nám. Sítná 3105, 272 01 Kladno 2, Czech Republic

[a]skalickydavid@seznam.cz, [b]Vaclav.Koucky@seznam.cz, [c]flopot@seznam.cz, [d]martin.viteznik@gmail.com

Keywords: Respiratory sound, Signal processing, Filtration, Analysis, Respiratory phases

Abstract. Sound as an oscillation of acoustic pressure is a manifestation that accompanies every mechanical movement in the Earth's atmosphere. Every change of mechanical system movement causes change in the progress of acoustic pressure and therefore recorded sound contains lot of information about movement of the specified system. If the sound recording is processed and analyzed correctly, it is possible to detect changes in the system, analyse a source of these changes and thus find reason of change of movement. Method of system's disorder detection using analysis of generated sound is used also in medicine. Auscultation - examination of the patient by listening – is often used method for lung function examination and the first used method for respiratory disease diagnosis (e.g. asthma). This method is performed using phonendoscope. Unfortunately this method has its limitations, such as the low sensitivity of human ear and also congestion of recording due to unwanted noise. Therefore, both the detection of small changes in the respiratory sound and diseases diagnosis in their initial phase is quite difficult. But early disease diagnosis is important because it means more efficiently treatment [1]. To obtain better and more reliable data on the patient's condition and for easier disease diagnosis using breath sound analysis, it is necessary to use sufficiently sensitive sensor to sound record, to process the record well and precisely.

Respiratory sound

The respiratory sound is sound in frequency range from 20 Hz to 2000 Hz [2]. This sound arises due to flowing air in the airways.

The amplitude of each sound frequency is different depending on the area of sound listening. The sound with frequencies up to 600 Hz go through lung parenchyma better than the sound with frequencies over 600 Hz. These frequencies can be detected on jugulum in better way, while the convenient area for lower frequencies detection is lung lobe. The respiratory sound can be affected also by size of patient's body and his health condition [1].

Breath system of healthy man produces a normal respiratory sound – sound without strange artifacts. If the man suffers from some respiratory disease, abnormal breath sounds and foreign artifacts (called also like bronchial sounds [3]) can arise in his breath sound. These artifacts are produced by turbulent flow that arises in airways due to bronchial obstructions [2]. Bronchial obstruction arises like a consequence of airway inflammation [4]. Bronchial sounds - sounds with abnormal high amplitude of specific frequencies are typical in many diseases with airway obstruction, like in asthma and in chronic bronchitis [3]. The bronchial sounds are defined sometime like wheezing, crackles and snores [3]. Frequency range and duration time of these artifacts is different - the literature states the frequency range from 100 Hz to 2000 Hz [2, 3]. Discovering of these artifacts in the breath sound is important to correct diagnosis.

Respiratory sound recording

Respiratory sound recording is necessary for its detailed analysis and correct diagnosis. It is important to use quality record equipment for obtain recording suitable for analysis – sensitive sensor and quality sound card. Unfortunately is impossible to obtain recording only with respiratory sound that is described in previous part, because the sound is always affected by unwanted noise. This noise can be produced by patient's body or by its surrounding.

Patient's body noise. Movement of muscles and working of organs is a large source of noise. For elimination of noise of muscles movement is necessary to limit body movement as much as possible. The noise produced by organs (mainly cardiovascular sound) is a low-frequency sound that occurs in frequency range up to 50Hz. For elimination of this noise in a sound recording is possible to use high-pass filter [5].

Ambient noise. It is clear that for quality sound recording is necessary to make recording in quiet environment. Unfortunately quiet environment doesn't ensure quality sound recording because due to manipulation with record equipment arise defects in sensor that generate a lot of noise. Elimination of this noise depends on doctor's skill and experience.

Signal processing

Breath sound includes many properties and artefacts that could be suppressed by unwanted noise. Correct signal processing is important step to discover suppressed artefacts and to obtain relevant results from sound analysis. To discover all significant frequencies in the sound, it is necessary to use sampling frequency twice higher as the highest frequency in the sound for sound recording (Nyquist–Shannon sampling theorem). In this research was used a phonendoscope Littmann 3200 for sound record. This stethoscope records sound with sampling frequency 4000 Hz. Now they are not expected higher frequencies than 2000 Hz in the recorded sound. For further recorded signal processing was used Matlab software.

Basic filtration. As the basic operation of signal processing, the filtration using high-pass and low-pass filter is applied. This signal modification is convenient for elimination of consistent disturbing elements with high amplitude. Such an unwanted sound with defined frequency level and high amplitude is primarily cardiovascular sound. Heart due to its position near lung affects the complete respiratory sound significantly. Such sound is much disorganized, searched properties and artefacts are drowned out and sound that is plotted as a progress of acoustic pressure is very chaotic - Fig. 1.

When this recording is modified using suitable filter, the sound is much clear and the progress of acoustic pressure is more visible. The definition of the correct filter depends on the application and the expected result. If the high-pass filter 500 Hz and low-pass filter 1600 Hz are used for recording filtration, some significant frequencies are removed from the sound, but in the new frequency range other sound properties like is alternation of inhalation and exhalation phases will appear - Fig. 2.

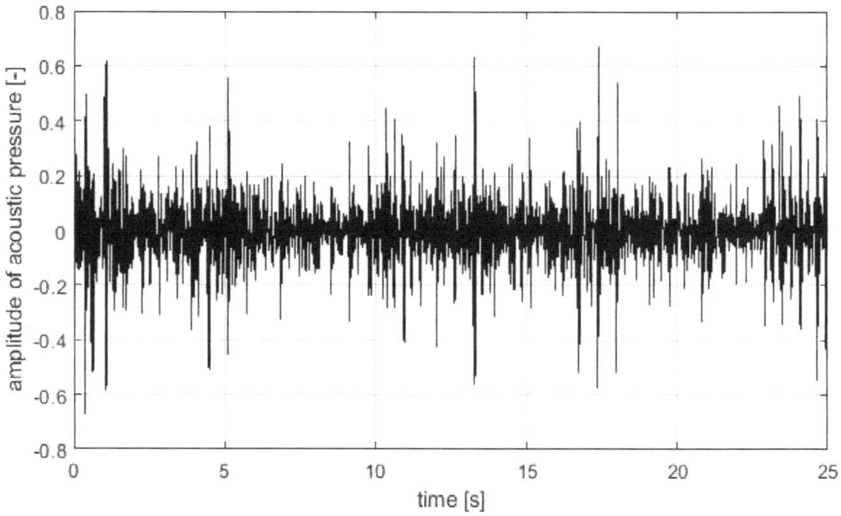

Fig. 1: Unfiltered respiratory sound of healthy man – chaotic progress of acoustic pressure due to noise in the sound

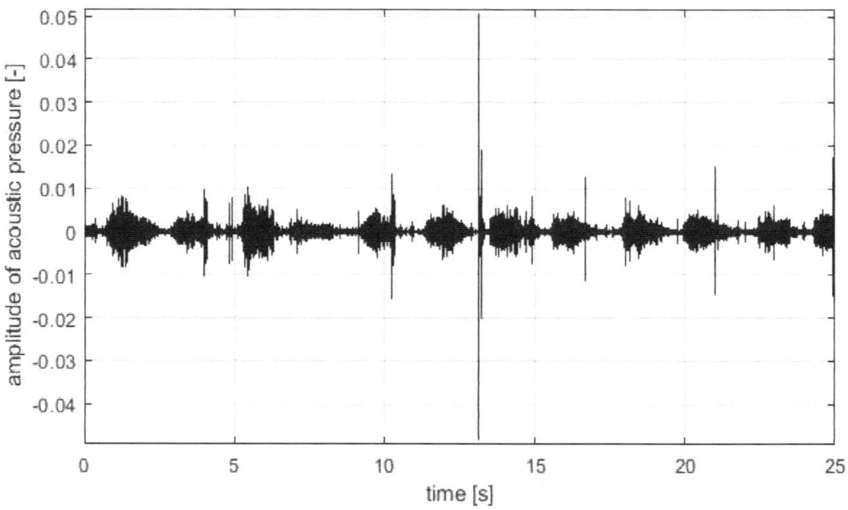

Fig. 2: Respiratory sound filtered using high-pass (500Hz) and low-pass filter (1600Hz) – the same healthy man. The filter eliminates high amplitude part of sound and recording contains now breath sound mainly. In the pressure progress, the alternation of inhalation and exhalation phases is obvious. Now, this part of the signal is located on low amplitude level

Ability of these phases detection is important for easy orientation in frequency spectrum and for creation of average curves of inhalation and exhalation. With these phases also sound defects generated due to manipulation with record equipment are obvious in pressure progress - e.g.: hitting the sensor or rubbing the sensor against the skin. These sound defects arise aperiodically in sound recording and they are very short (time of duration around 0,02s) with high amplitudes and wide frequency ranges. Therefore, the standard filtration can't be use for their elimination – Fig. 3. Unfortunately, these artefacts make above mentioned detection more difficult.

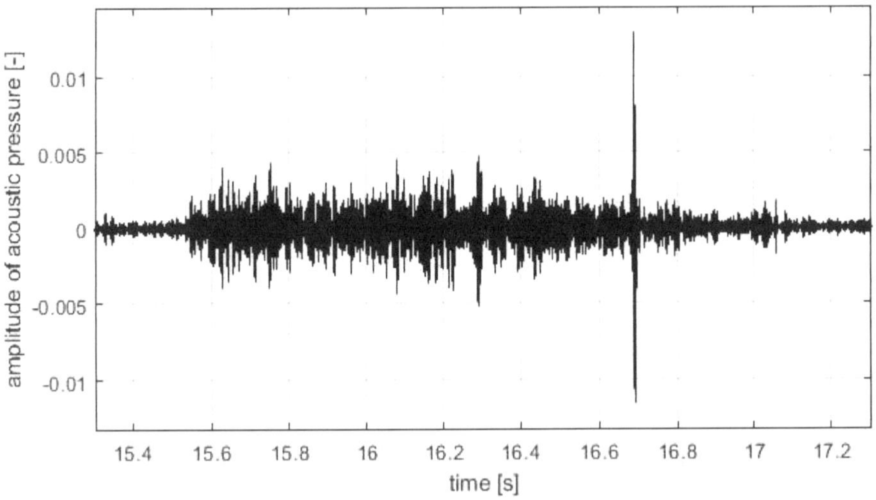

Fig. 3: The short and amplitude distinct sound defect (in time 16,7s) caused by manipulation with record equipment

Own filtration and respiratory phases detection. The alternation of inspiratory and expiratory phases is closely related to the movement of air in the airways. The air flows through the airways during inspiration and expiration, while at the time of the respiratory phases change, air movement is minimal. Because the air movement in the airways generates sound, hold true that the larger (faster) movement of air generates a louder sound. This louder sound has higher amplitude deviation in the pressure progress than quietly sound during time of phases change – Fig. 4. This feature is basic for respiratory phases identification.

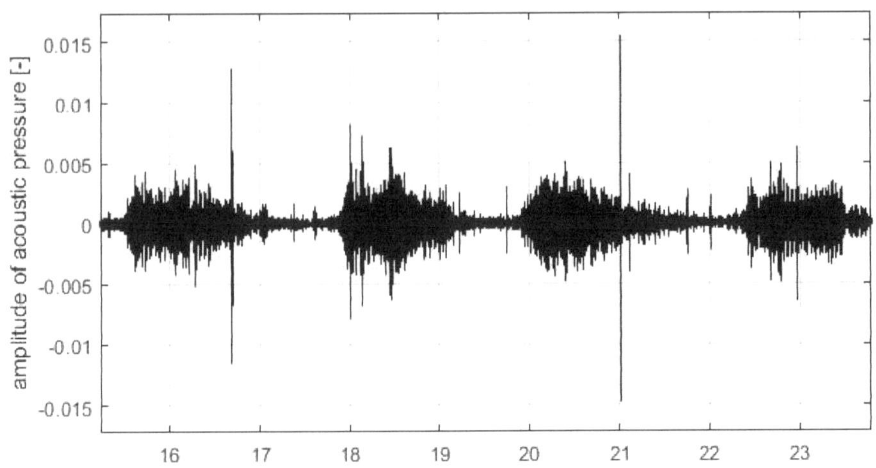

Fig. 4: Inspiratory and expiratory phases alternation - at the time of the respiratory phases change is amplitude of acoustic pressure minimal

For easier identification maximal and minimal deviation of pressure amplitude, it is appropriate to create envelope of the signal. The upper part of the envelope includes positive values of amplitudes only, while the whole signal oscillates between positive and negative values, and therefore it can be effectively smoothed. Subsequently, the boundaries of the respiratory phases as the minima of the smoothed curve can be defined and found.

Unfortunately, this curve is greatly affected by defects caused by sensor manipulation. Therefore, it is necessary in suitable way to adjust the curve and eliminate effect of the sound defects – Fig. 5.

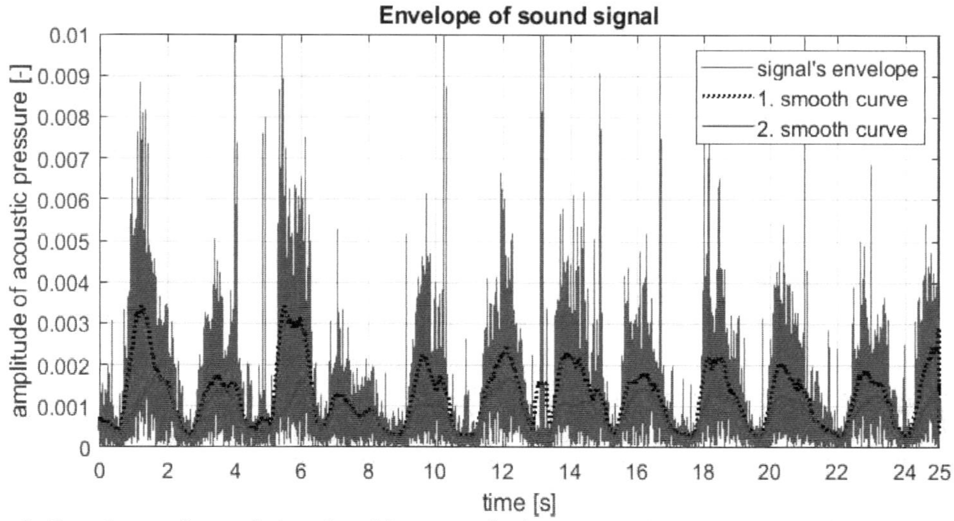

Fig. 5: Envelope of sound signal and its smoothed curves – the boundaries of the respiratory phases as the minima of the 2nd smooth curve

Because the breath sound is different for each person, it is necessary to specify more features of sound for correct identification of respiratory phases. A suitable and easily definable feature is the approximate rate of change of respiratory phases (duration of one phase). This feature varies according to patient's age and its healthy condition – Table 1. Therefore, it is necessary to define specific duration of one phase individually for each patient.

Table 1: Respiratory rate according to patient's age [6]

Age of patient	Respiratory rate
neonate	40 – 60 / min
1 month	40 – 45 / min
6 month	35 – 40 / min
1 year	30 – 35 / min
5 – 6 years	25 / min
10 – 12 years	20 – 22 / min
14 – 15 years	18 – 20 / min

If the envelope is properly adjusted and the respiratory rate is correctly defined, the boundaries of the respiratory phases can be determined – Fig. 6. The accuracy of boundaries definition depends on more factors mainly on respiratory rate. If this rate is low, a transition part between phases with minimal amplitude is not exactly defined. Nevertheless, this definition of the boundaries of the respiratory phases is well useful.

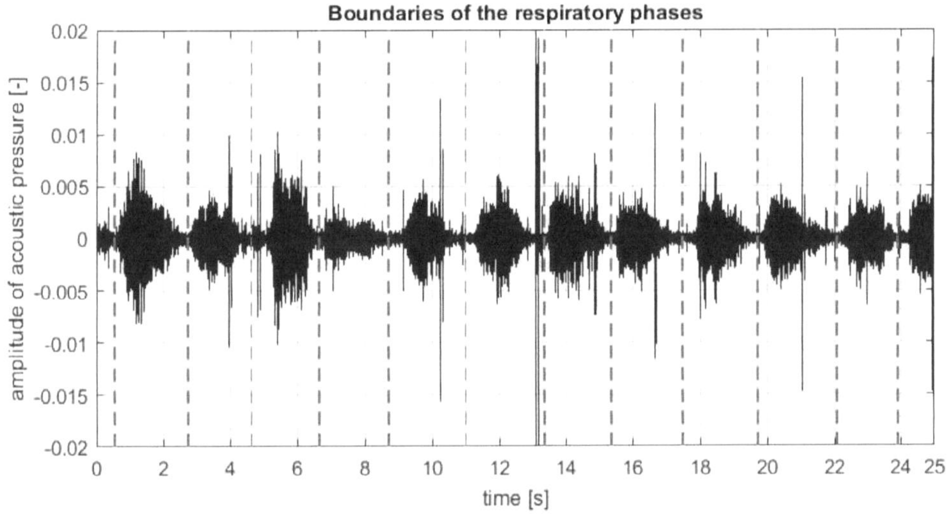

Fig. 6: Sound signal with border lines – dashed vertical lines

Table 2: Basic data of the measured patient

Sex	female
Age	13 years
Health condition	without bronchial obstruction
Recording location	jugulum

Results – use of the boundaries definition

These defined border lines are very useful for orientation in frequency spectrum of the respiratory sound – Fig. 7.

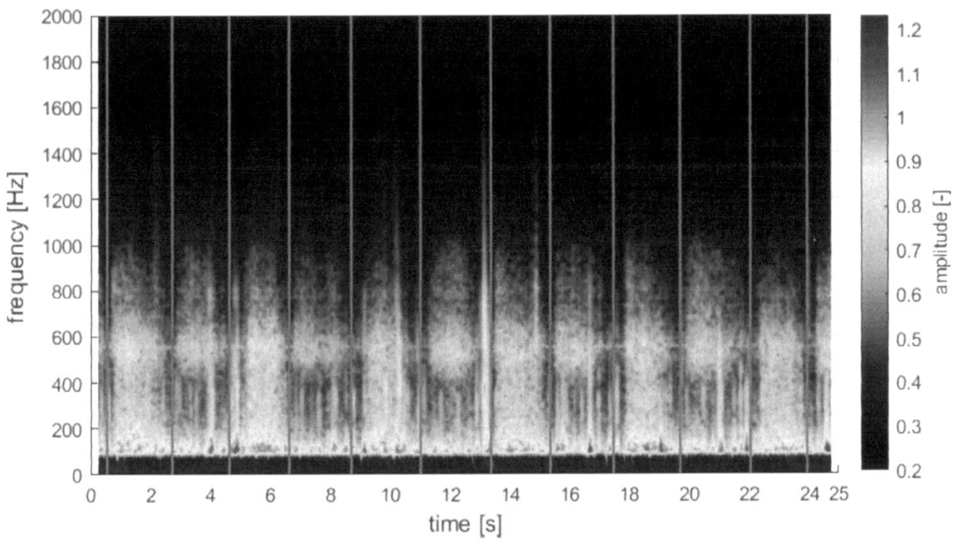

Fig. 7: The frequency spectrum of respiratory sound with vertical border lines. In the spectrum are distinct sound defects caused by manipulation with record equipment (amplitude distinct areas between 200 Hz – 600 Hz with time duration roughly 0,02s). In the spectrum are obvious also two distinct horizontal lines – 1110 Hz and 1330 Hz. These are with the highest probability an effect of background noise

If the boundaries of the respiratory phases are known, the average curves of inhalation and exhalation phases can be plotted – Fig. 8.

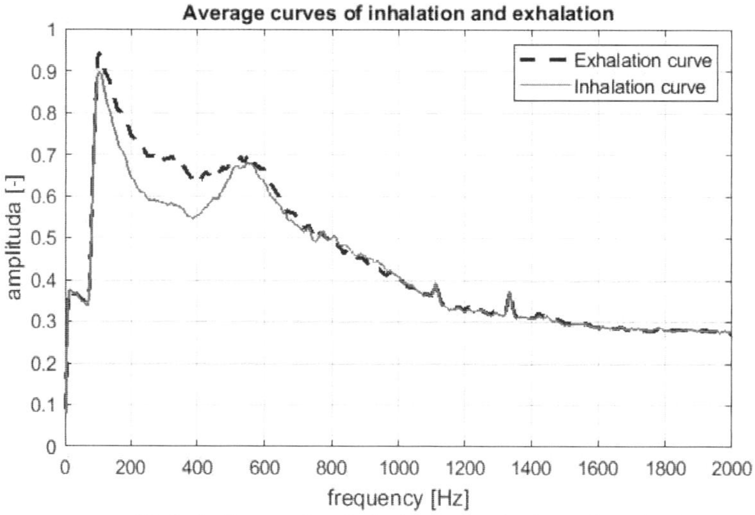

Fig. 8: Average curves of inhalation and exhalation – is obvious that the important frequency area is between 100 Hz – 600 Hz

Using these curves, it is possible to eliminate sounds defects in the frequency spectrum, at least partially and thus make the spectrum more transparent. Because of this elimination, it is possible to discover the searched artefacts in the frequency spectrum mentioned above.

Conclusions

It is obvious that the respiratory sound contains many features and artefacts which can be effectively used in medicine practise and research. For the best use, it is necessary to be able to discover and define these properties. This could outline a way to further process and use the signal efficiently.

The application of such audio signal processing in medicine is wide – from online monitoring of patient lung function to deep analysis of the sound and detection of artefacts in the breath sound recording, that are not detectable using auscultation. This method of the artefacts detection and so early discovering of bronchial obstructions in the airways would be very helpful for respiratory diseases diagnose in infants to 5 years old. In this case, the traditional methods – e.g. spirometry – are not effective. Moreover this diagnosis method is non-invasive and its using in practice is easy because for obtain of sound signal can be used device based on classic stethoscope that is able to make sound record. Therefore, no demands are placed on the cooperation of a small patient, the possibility of patient stress is reduced, and so it is possible to obtain more accurate data for analysis.

The discovering of the searched artefacts in the sound frequency spectrum is not easy, it depends on properties of sound and quality of the sound recording. Therefore, the semiautomatic or automatic differentiation of respiratory phases in the spectrum is important.

Acknowledgements

This text is supported by GAUK grant No. 9022/2018

References

[1] D. Skalický, F. Lopot, V. Koucký, Respiratory sounds as a source of information in asthma diagnosis, Clinician and Technology 2017, vol. 47(2), ISSN 2336-5552.

[2] H. Pasterkamp, S. S. Kraman, G. R. Wodicka, Respiratory Sounds, American Journal of Respiratory and Critical Care Medicine. 1997, ISSN 1073-449x.

[3] A.R.A. Sovijärvi, L.P. Malmberg, G. Charbonneau, J. Vanderschoot, F. Dalmasso, C. Sacco, M. Rossi, J.E. Earis, Characteristics of breath sounds and adventitious respiratory sounds, ©ERS Journals Ltd 2000, ISSN 0905-9180.

[4] J. G. Ayres, Astma, 1. vyd. Praha: Grada 2001, ISBN 80-247-0091-3.

[5] J.E. Earis, B.M.G. Cheetham, Current methods used for computerized respiratory sound analysis, ©ERS Journals Ltd 2000, ISSN 0905-9180.

[6] D. Skalický, Využití elektronického fonendoskopu v diagnostice astmatu, ČVUT Praha 2016.

EAN 2020
58th conference on experimental stress analysis

The Analysis of Dynamic Effects in the Exchange Part of Turnouts

SMUTNÝ J.[1,a], JANOŠTÍK D.[1,b], NOHÁL V.[1,c] , PAZDERA L.[2,a]

[1]Institute of Railway Structures and Constructions, Faculty of Civil Engineering; Brno University of Technology; Veveri 331/95, 602 69 Brno, Czech Republic

[2]Institute of Physics, Faculty of Civil Engineering; Brno University of Technology; Zizkova 331/17, 602 69 Brno, Czech Republic

[a]smutny.j@fce.vutbr.cz, [b]janostik.d@fce.vutbr.cz, [c]nohal.v@fce.vutbr.cz, [d]pazdera.l@fce.vutbr.cz

Keywords: Railway turnout, dynamic effects, Vibration analysis in time and frequency domain, Non-linear time-frequency transformation, Wigner-Ville transformation

Abstract. This paper deals with description and application of the modified Wigner-Villa's transformation for dynamic analysis. The use in the field of railway constructions testing represents a quite interesting application area of the transformation. This paper contains mathematical description of the technique, a case study from area of railway turnout's dynamic analysis and practical experience obtained and recommendations for its practical use. The described analysis is focused on dynamic processes in the exchange area of turnouts and their comparison with dynamic processes in the frog part of the turnouts.

Introduction

The basic claim for the particular parts of the railway track is their functional reliability and the safety of the interconnected railway operation. Although the superstructure has been developed to almost perfection during more than 150 years of the railway history, it is still possible to find new technical solutions. This especially applies to various types of turnout structures, various rail fastenings, various types of rail pads, sleepers etc.

Permanent pressure to increase the transport speed and the operational load on the railway resulted in huge development of new technologies. The decision on modernization of the corridors became the impulse for development of all areas of the railway transport, either in the area of vehicles and infrastructure. In conformity with the above-mentioned trend, application of new experimental processes should continue for evaluation of the quality and suitability of individual constructional solutions [1].

The level of dynamic effects in the railway superstructure are influenced by its quality, operational and technical conditions, and climatic events and, most importantly by dynamic load on the railway vehicle axle. The fact also relates to construction of the railway turnouts representing the key points of the railway roads [2].

The railway turnout structures represent one of the key points of the railway transport. Therefore it is necessary to pay a special attention to these structures. The turnout structure consists of a number of structural components of different properties which in summary must fulfil the required functions. Among the most important properties is especially their reliability and safety within the framework of the railway operation. In accordance to this trend, the development and application of new experimental procedures for assessing the quality and effectiveness of individual design solutions must also proceed.

A classic simple turnout consists of three main parts - a frog part, a middle part and an exchange (Fig. 1). The frog part consists of a frog, a retainer and an outer rail. There is a crossing of rails and separation of rails in two different directions. The frog forms the place where the wheel passes a gap (depression) for the wheel rim of the vehicle, which moves in another direction. In order to avoid an impact or climbing of the flange on the tip of the frog, the second wheel of the same wheelset is guided by a retaining device, which does not allow it to move away from the outer rail.

Fig. 1 The picture of a turnout with a description of the most important parts

The middle part of the turnout actually forms a tangle of tracks. It connects the exchange and the frog part. The interchange is a moving part of the turnout, on which the track branches in two directions. The main parts of the exchange consist of tongues, supports and a turnout. The tongue forms the only moving structure of the carriageway. The position of the tongues determines the direction of further movement of the train set while driving. The tongues are usually stored on stool bases. Under normal circumstances, one tongue always rests on the support. A motorized converter is used to adjust the tongues. To secure them in the appropriate position there is an exchange lock. From the above, it is clear that especially the frog and exchange part belong to the key parts of turnouts in terms of dynamic effects. Let us now briefly consider the dynamic processes in both areas.

High dynamic forces can be expected in the area of the transition of the train set wheel from the wing rail to the tip of the heart, where, depending on the quality of the geometry of the transition, a dynamic impact occurs [3].

After the impact of the wheel, vibrational energy propagates downwards from the surface of the heart to the relatively rigid base and sleepers on which the heart is mounted, then the vibrational energy is transferred from the sleeper to the gravel bed. A small part of the energy is captured by the internal damping of the frog itself, another small part is captured by the base and the pad under the tip of the heart, the pad under the base and also the sleeper. Due to the fact that the whole system is relatively rigid, a large part is transferred to the gravel bed and subsoil. Here, the sharp edges of the gravel bed are stressed and they begin to abrade. The resulting fine-grained parts adversely affect the cohesion of the grains and free spaces begin to form under the sleeper. If the turnout is not sufficiently supported, the geometry of the transition from the wing rail to the tip of the heart will collapse and the whole process of degradation will be greatly accelerated.

From theoretical assumptions and computer simulations [4], it is concluded that another area requiring increased attention is the exchange area. The mechanism of vibration transmission in the exchange part is similar to that in the frog area, assuming lower vibration values. This is due to the uninterrupted running surface of the rail, when the ankle gradually approaches and rests on the tongue rail. The intensity of vibrations is damped by the rail grate, when a small part is damped already in the rail and its rubber pad. Further vibration damping occurs in the sleeper. The undamped component of the vibration energy is also transferred to the gravel bed. The question is how much of this energy is compared to the energy propagating in the frog area.

Wigner-Villa's transformation

Wigner´s distribution was proposed in 1932 by Professor Wigner in the field of quantum physics and about 15 years later, it was adapted for the area of signal analysis by the French scientist Ville. Wigner-Villa's transformation is defined for the time-frequency domain by relation [5]

$$WVT_x(t,f) = \int_{-\infty}^{\infty} x\left(t+\frac{\tau}{2}\right)\cdot x^*\left(t-\frac{\tau}{2}\right)\cdot e^{(-j\cdot 2\cdot\pi\cdot f\cdot\tau)}\cdot d\tau, \tag{1}$$

where '*' represents a complex conjunction, t is time, f is frequency, τ is shift along the time axis, x is time representation of the signal $x(t)$ and $WVT_x(t,f)$ is a time-frequency representation of the input signal. The equation (1) shows that it is essentially the Fourier transformation of relation $x(t+\tau/2)\cdot x^*(t-\tau/2)$, so the functions $x(\tau/2)$ and the complex conjugate of $x^*(-\tau/2)$ at some point in time t. From the equation (1) it is also apparent that the Wigner-Villa's transformation is a complex function in the time-frequency space. Another interpretation of equation (1) shows that this transformation can be interpreted as a combination of the autocorrelation function [6]

$$C(t,\tau) = x\left(t+\frac{\tau}{2}\right)\cdot x^*\left(t-\frac{\tau}{2}\right) \tag{2}$$

and Fourier transformation

$$WVT_x(t,f) = \int_{-\infty}^{\infty} C(\tau)\cdot e^{(-j\cdot 2\cdot\pi\cdot f\cdot\tau)}\cdot d\tau. \tag{3}$$

Unlike for example the Short time Fourier method, in which the resolution is limited by window function, Wigner´s-Villa's spectrum provides good resolution both in frequency and in time domain. The important feature here is that the calculation is not limited by Heisenberg uncertainty principle. Although the calculation of the coefficients of Wigner-Villa's transformation is not limited by Heisenberg's uncertainty principle, certain problems may arise in calculating multicomponent signal, which is generated by the sum of two or more signals [7, 8].

In principle, there are two reasons to modify the basic properties of Wigner-Villa's transformation. The first reason is that in practice it is not possible to integrate from -∞ to +-∞, but the calculation can be carried out only for a limited signals. The second reason is to try to eliminate the effect of cross-components, which often oscillate heavily. Restrictions of both problems can be achieved by calculating the modified of Wigner-Villa's transformation (often called pseudo or smoothed) by equation [9]

$$PWVT_x(t,f)=\int_{-\infty}^{\infty} h(\tau)\cdot x\left(t+\frac{\tau}{2}\right)\cdot x^*\left(t-\frac{\tau}{2}\right)\cdot e^{(-j\cdot 2\cdot \pi\cdot f\cdot \tau)}\cdot d\tau \qquad (4)$$

where the function $h(\tau)$ is a window function with maximum at $\tau = 0$. A Gaussian curve is often used as the function $h(\tau)$. Such a solution quite effectively suppresses interference (smoothes), but deteriorates resolution. Smoothed Wigner-Villa's transformation is therefore a certain compromise between smoothing and resolution [10].

Case study

To compare the dynamic effects in the exchange and the heart section, turnout No. 59 (Fig. 2) was selected in the so-called Pardubice head of the Choceň railway station with a left turn direction (turnout designed for a speed of 80 km·h^{-1}). The selected turnout is of type J60-1:14-760-zlp,L,p,ČZP,b,KS,ZPT. Trainsets may pass through this turnout in a straight line at a line speed of 160 km·h^{-1}. The turnout is equipped with a system of railway superstructure UIC 60 with Vossloh Skl 24 fastening and concrete sleepers.

A measurement methodology certified by the Ministry of Transport was used to measure the dynamic effects acting on the turnout [11]. This methodology consists of three parts. The first involves measuring the motion behaviour of a structure. The second area includes the propagation of vibrations around the frog and the exchange of turnouts and especially the effects of vibrations on the gravel bed. In the third one, the measurement of force is performed.

In total, two basic measuring areas with 3 measuring points were set in the measured turnout, to which acceleration sensors were attached (A4 - rail, A7 - sleeper and A0 - gravel ballast). The vibrations in the gravel bed were measured by an acceleration sensor placed on the rod (it served as a waveguide) implemented in the gravel bed. In the exchange part, the sensors were placed on the heel of the support, then on the sleeper in the area of the passage of the ankle from the support to the tongue rail and measuring rod, which was embedded in the gravel bed near the passage of the wheel. Simultaneously with the measurement of vibration characteristics, a large number of other parameters were measured. Their comparison and analysis is not part of this paper.

Fig. 2: The view of frog (left) and exchange (on the right) part of turnout

Autonomous S-Box dataloggers developed at the authors' workplace were used for the measurements. As part of the measurement campaign, the passage of a total of 30 trains was recorded, for which, in addition to the time course of vibration acceleration, their other parameters were also recorded - travel time, running speed, number of cars, and type of locomotive and direction of travel. For their own evaluation, these trains were divided into suitable categories. Note that these characteristics were determined by the measuring system itself. From the measured records, the basic vibration characteristics, ie the effective value of the acceleration a_{ef} and the amplitude spectra for the individual train categories, were calculated for each measured point and each position of the acceleration sensor. A table and basic graphs were created from the calculated data of the effective value of acceleration and these were compared.

Tab. 1: The comparison of the maximum values of vibration acceleration

Train set	Turnout	Effective values of vibration acceleration $[m.s^{-2}]$					Travel speed[km/h]
		A0Z	A7Z	A4Z	A5X	A6Y	
Leo Express	59, frog part	20	43	108	25	32	163
	59 exchange part	2	38	40	10	17	160
RegioJet	59, frog part	10	26	95	17	26	156
	59 exchange part	3	24	49	11	20	154
150	59, frog part	2	10	33	6	10	83
	59 exchange part	2	40	65	15	25	67
151	59, frog part	7	24	46	8	13	120
	59 exchange part	2	26	42	10	18	101

Legend
- in the vertical direction
- in the longitudinal direction
- in the transverse direction

It is clear from Table 1 that higher values of effective acceleration were achieved in the frog part of the turnout compared to the exchange part. This is true for most trainsets except the trains with the 150 locomotive type. Similar findings also show the analysis and comparison of time records and their amplitude frequency spectra.

To present a comparison of the dynamic stress in the exchange and the frog part of the turnout, the passage of the RegioJet train set at a travel speed of 156 $km \cdot h^{-1}$ was chosen (Fig. 3, Fig 4). Due to the limited space of the paper, the authors focused on a more detailed analysis of the measured data from the sensor, which was placed on the measuring rod in the gravel bed. This provides vertical values of the vibration acceleration at the attachment point. The analysis consists of images composed of three graphs.

The upper graph shows the time course of the oscillation acceleration. The left graph shows the amplitude spectrum of the vibration response calculated using the Fourier transform. The middle graph then shows a 3D representation of the time-frequency course of the amplitude spectrum of the vibration response. The smoothed form of the Wigner-Ville transform in discrete form was used to calculate the time-frequency spectrum. Note that a reference value of 10 $m \cdot s^{-2}$ was used in the calculation of the frequency and time-frequency spectrum.

From the comparison of time records, it is clear that higher values were measured during the transition of the wheels of the set from the wing rail to the tip of the frog.

Frequency spectra show that significant frequency components obtained from signals occur in the frog part in a lower frequency range, from about 120 Hz to 1 kHz.

In the case of the exchange part, the significant frequency components are distributed higher, in the interval from about 500 Hz to 1.5 kHz. Note that frequency components above 160 dB are considered significant. It is also evident that the significant frequencies are significantly higher in the frog area compared to the exchange part. The same conclusions are drawn with time-frequency analysis using the Wigner-Ville transformation method.

Even in the case of this analysis, it is clear that higher peaks in the time-frequency plane were achieved in the case of passing the trainsets over the frog part. It is also clear that the highest values of the spectrum were found at relatively low frequencies up to 500 Hz. It is clear from the exchange part of the turnout that lower values of the time-frequency spectrum have been achieved. However, this is spread over a wider frequency and time interval.

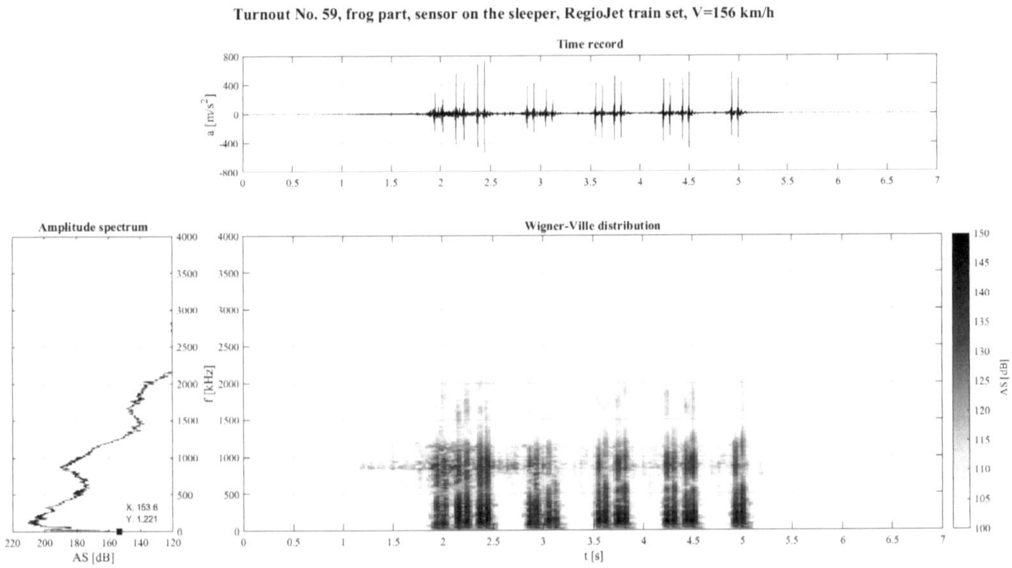

Fig. 3: Time, frequency and time-frequency characteristics of the vibrations, sensor on the sleeper, turnout no. 59 frog part, vertically direction, RegioJet, speed 156 km·h^{-1}

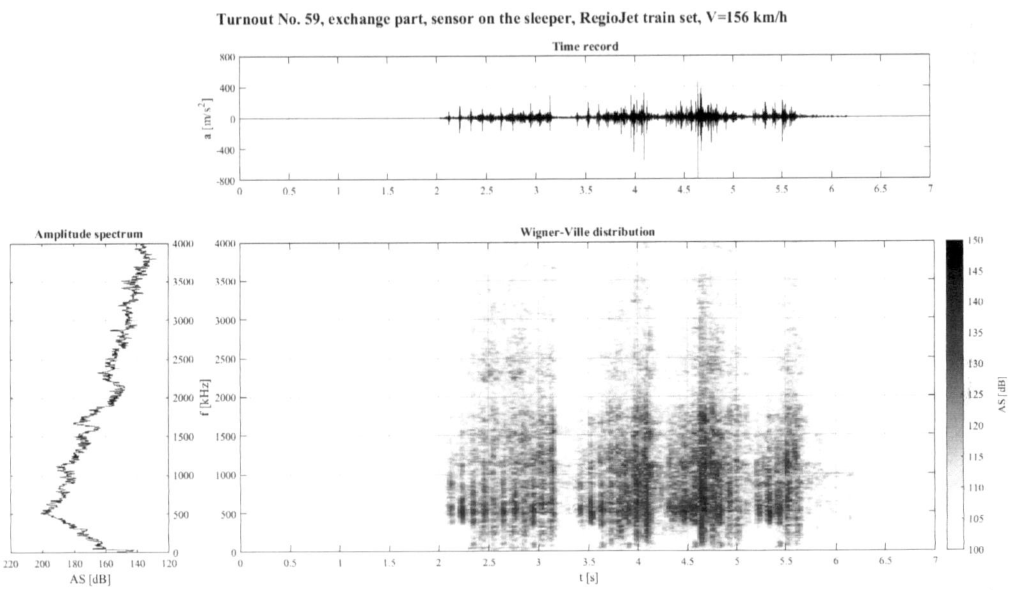

Fig. 4: Time, frequency and time-frequency characteristics of the vibrations, sensor on the sleeper, turnout no. 59 exchange part, vertically direction, RegioJet, speed 156 km·h^{-1}

Conclusions

The railway turnout No. 59, was selected for presentation of the modified Wigner-Villa's transformation in the analysis process of dynamic load from the train units. The selected railway turnout is located in the Choceň station. The results from the exchange section of the turnout were compared with those measured in the frog part of the same turnout.

The modified Wigner-Villa's transformation offers a comprehensive tool for analysing non-stationary signals in particular. Characteristic feature of presented transformation is the fact that its resulting distinguishment in time and frequency in not limited by Heisenberg principle of indefiniteness. This fact includes the high distinguishing ability in the time frequency plane that gives rise to a "precise" localisation of important frequency components in time. This method gives a fast and accurate localisation of frequency components included in the measured signal.

Acknowledgments

This paper has been worked out under the research project FAST-S-20-6408, The Analysis of dynamic response of railway structures by the method of higher order spectra.

References

[1] M. Moravcik, Analysis of vehicle bogie effects on track structure-nonstationary analysis of dynamic response, Communications, Volume 13, 3/2011, EDIS Zilina, ISSN 1335-4205.

[2] E. Kassa, C. Andersson, J. C. O. Nielsen, Simulation of dynamic interaction between train and railway turnout, Vehicle System Dynamics, pp 247 -258, Volume 44, 2006 - Issue 3, ISSN 0042-3114, 2007.

[3] S. Kaewunruen: Monitoring structural deterioration of railway turnout systems via dynamic wheel/rail interaction, Case Studies in Nondestructive Testing and Evaluation, Elsevier, Volume 1, 2014, Pages 19-24, ISSN 2214-6571.

[4] W. Kollment, P. O'Leary, M. Harker, U. Oßberger, S. Eck, Towards condition monitoring of railway points: Instrumentation, measurement and signal processing, 2016 IEEE International Instrumentation and Measurement Technology Conference Proceedings, Taipei, 2016, pp. 1-6, ISBN 978-1-4673-9221-1.

[5] J. Smutny, Measurement and Analysis of Dynamic and Acoustic Parameters of Rail Fastening, NDT & E International - Independent Nondestructive Testing and Evaluation, 2004, Volume 37, Issues 8, ELSEVIER, pp. 119-129, ISSN 0963-8695.

[6] J. Smutny, L. Pazdera, New Techniques in Analysis of Dynamic Parameters Rail Fastening, InSight, The Journal of The British Institute of Non-Destructive Testing. Vol 46. No 10. October. 2004. pp. 612-615. ISSN 13542575.

[7] D. Kakofengitis, O. Steuernagel, (2017). "Wigner's quantum phase space current in weakly an harmonic weakly excited two-state systems" European Physical Journal Plus, Springer Science + Business Media, ISSN 21905444, 2017

[8] R. R. Sharma, R. B. Pachori, Improved eigenvalue decomposition-based approach for reducing cross-terms in Wigner-Ville distribution, Circuits, Systems, and Signal Processing, 2018, ISSN 0278-081X.

[9] R. B. Pachori, A. Nishad, Cross-terms reduction in the Wigner–Ville distribution using tunable-Q wavelet transform, Signal Processing, Volume 120, Elsevier, March 2016, Pages 288-304, ISSN: 0165-1684.

[10] R.R. Sharma and R.B. Pachori, Improved eigenvalue decomposition-based approach for reducing cross-terms in Wigner-Ville distribution, Circuits, Systems, and Signal Processing, 2018, Springer nature, ISBN 978-981-10-6918-5

[11] J. Smutny, L. Pazdera, I. Vukusic, Dynamic effects on switches evaluation, The certified methodology, Ministry of Transportation, 124/2014-710-VV/1.

EAN 2020
58th conference on experimental stress analysis

Dynamic Analysis of Carbon Woven Reinforced Epoxy Matrix Composites Used for Parts of Electric Vehicles

SRB P.[1,a], RYVOLOVÁ M.[1,b]

[1]Institute for Nanomaterials, Advanced Technology and Innovation, Bendlova 1407/7, Liberec, Czech Republic

[a]pavel.srb@tul.cz, [b]martina.ryvolova@tul.cz

Keywords: Composites, Carbon fibres, DMA, Three-point bending

Abstract. This paper deals with evaluation of mechanical properties of fibre composites using an experimental technique based on dynamic mechanical analysis (DMA). Fibre composite material are increasingly being used in various industries due to their properties. Composites excel in high strength at low weight compared to conventional materials. Carbon woven reinforced epoxy matrix composites is a very promising type of composite material. Electric vehicles are a desirable replacement for vehicles with combustion engines, and the weight is a key parameter of their usability. The low density of composite materials is very advantageous for part design of electric vehicles. Fibre composites can be used not only for visual parts but also for structural and safety elements. Three types of carbon composites, with different configuration, were used for this study. The chosen test method was three-point bending at frequency 10 Hz and variable temperature. The performed measurement shows different storage modulus of each type of sample.

Introduction

Transportation sector is using mainly fossil fuels. This sector alone consumes nearly two-thirds of global oil consumption. Transportation is responsible for 23 % of CO_2 emissions connected to energy [1]. The current trend in vehicle development puts great emphasis on reducing harmful emissions and energy consumption. Legislative regulations lead manufacturers to reduce vehicle weight and search for new ways of propulsion.

A possible solution is the development of electric vehicles. According to study of material composition of conventional passenger vehicles [2] there is 72.3 % metal of the total weight. For electric vehicles, weight is one of the key parameters, so manufacturers are forced to look for new low-density construction materials. Continuous fiber composites are becoming very attractive materials for application in many industries. Their high strength and stiffness to weight ratios are their main advantages over metallic structures. However, their low impact tolerance, easy delamination and poor damage tolerance behaviors are the main disadvantages [3]. Fiber composite materials are increasingly used also in the automotive industry. Usage of carbon fiber composites can reduce weight up to 65 % [4]. At present time, composites are used not only for visual parts but also for structural and safety elements. When composite part is properly designed, it can absorb a large amount of deformation energy. Compared to conventional metal deformation parts based on plastic deformation, composite deformation elements can show not only lower weight but also significantly higher specific absorbed energy [5]. With the gradual improvement of the production process, the properties are improved and also price of composite products is decreasing. These are other reasons for increased use of composite materials in vehicle production.

Fiber reinforced polymer is a composite material made of a polymer matrix reinforced with fibers. The polymeric matrix is most often epoxy, polyester or phenolic. The fibers are usually carbon, glass, or aramid, plant fibers are becoming increasingly popular due to ecology. Carbon fiber has been described as a fiber containing at least 90 % carbon obtained by the controlled pyrolysis of appropriate fibers. Conventional carbon fibers typically have diameter around 7 μm. The carbon fiber blanks for the composite are most often prepared by textile processing techniques of weaving, knitting and braiding. Weaving can be done in a conventional manner to produce two-dimensional fabric as well in a multilayer weaving that can create three-dimensional fabric. Braiding is suited to the manufacture of narrow width flat or tubular fabric. 3D braiding machines produce spatially shaped braiding fabrics with different shape of cross section. Fabric produced by knitting is regarded as two-dimensional, but multiaxial warp knitting machines are capable of producing multilayered fabrics with layers oriented from 0°to 90° - so called non-crimp fabrics (NCF) or multiaxial fabrics. All layers are connected using threads, which is insert using needles to form a 3D structure [6].

Dynamic mechanical analysis (DMA) is very effective method for determining the morphology and viscoelastic properties of polymers and composite material. The storage modulus (E') is typically related to the Young's modulus, determines material ability to store energy applicable for future purpose. Loss modulus (E'') is a viscous response of the material and regarded as material tendency to dissipate energy applied to it. Loss modulus is also associated with internal friction. Tan δ is mechanical damping factor defined as the ratio of loss and storage modulus (tan δ = E''/E'). A low value of tan δ shows high elasticity of material. Increasing bonding interface fiber/matrix results reduction of tan δ [7].

Materials

For experiment three type of composite with fibre reinforcement were used. Sample 1 – carbon woven fabric, standard 6K, plain weave, sample 2 - carbon woven fabric, 12K - WD, plain weave, sample 3 - carbon woven fabric, spread tow 12K, plain weave. Table 1 contains detailed specification of fabric.

Table 1: Parameters of carbon fabrics

Fabric	Weave	Weight [gm^{-2}]	Thickness [mm]	Sett 0°/0°
1	Plain 6K	200	0.55	520/520
2	Plain 12K	193	0.35	120/120
3	Plain 12K	160	0.19	67/67

Composite from one layer of carbon woven fabrics with plain and epoxy resin CHS-EPOXY 531 was made using hand lay-up moulding process (weight ratio of hardener 1:27, curing temperature +20°C and curing time 24 hours). CFRP (carbon fiber reinforced polymer) with reinforcements 1-3 you can see in Fig 1.

For this experiment, the volume fraction was not the main parameter of the composite samples. The aim was to produce composite samples with the same thickness. Samples for the experiment were prepared by the same method. Defining the differences in mechanical properties between individual samples is based on the different processing of the input roving from which the reinforcements are made (reinforcement - fabrics with plains weave, carbon roving 6K and 12K). In the first case, it is a classic roving processing - the result is a bundle of parallel-arranged fibrils, the cross section is circular. In the case of the second sample, it is a WD - flat tow adjustment; the result is an arrangement of fibrils in a tape with a rectangular cross-section. In the third case, it is a type of spread tow; the fibrils are arranged in parallel in an ultra-thin tape with a minimum achievable thickness.

Fig. 1: Carbon composite samples

The standard prescribes a minimum sample thickness of 1 mm that is a fact. In this case, an effort was made to achieve the same minimum thickness for all samples, although the difference between the highest and lowest value of the reinforcement thickness is 0.35 mm. The same volume fraction could only be maintained by adding extra layers. In such a case, it would not be possible to monitor directly how the reinforcement parameters affect the properties of the composite. Single-layer samples were chosen to make the difference in mechanical properties depending on the processing of the input roving stand out.

Experiment

The aim of carried experiment was compare elastic behavior of different types of long fiber carbon composites under dynamic loading. For this test was use dynamic mechanical analyzer TA DMA Q800. Used method was the 3-point bending test Fig.2, that is together with the 4-points bend probably the most appropriate method useable for testing layered composites [8]. The carried experiment consists single frequency with constant strain amplitude. Because the properties of the composite material are temperature dependent, the measurements were carried out for different temperatures.

Fig. 2: DMA Q800 device with 3-point bending clamps

Before the main testing, it is appropriate to set the value of the geometry factor. For the 3-point bending clamps with rectangular samples it is possible to determine the optimal sample size according to (1):

$$GF = \frac{L^3}{4t^3W}\left[1 + \frac{12}{5}(1+\nu)\left(\frac{t}{L}\right)^2\right] \quad (1)$$

Where: L is sample length, W is width, t is sample thickness, ν is Poisson's ratio. Parameters are shown in table 2, ν is estimated value.

Table 2: Parameters of tested samples

Parameter:	Sample 1	Sample 2	Sample 3
L [mm]	20	20	20
t [mm]	0.66	0.65	0.64
W [mm]	13.42	12.64	12.29
ν [-]	0.3	0.3	0.3
GF [1/mm]	520	954	622

Obtained values of geometric factor correspond with recommended values for device DMA Q800. The recommended thickness of samples is 1/10 to 1/32 of the span (length) of the chosen 3-point bending sample supports. Used samples thickness is about 0.6 mm, which is at the limit of the recommended value for 20 mm span supports. Unfortunately supports with smaller span were not available. The thickness should be as uniform as possible across the sample because, cube of the thickness is used in the modulus calculation a 3% error in thickness becomes a 10% error in the calculated modulus. For composite samples the amplitude of 10 to 50 μm should yield good results [9]. In this measurement amplitude was set to 15 μm, preload force to 0.01 N, frequency 10 Hz and temperature in the range 30 - 50 °C.

Results

In the fig. 3 are shown results of dynamic 3-point bending test. It is possible to see dependency of storage modulus on temperature (range 30°C – 50°C). The results are shown for three different samples of carbon composites. Sample number 3 shows the highest value of storage modulus, and at the its decreasing with temperature is the most significant.

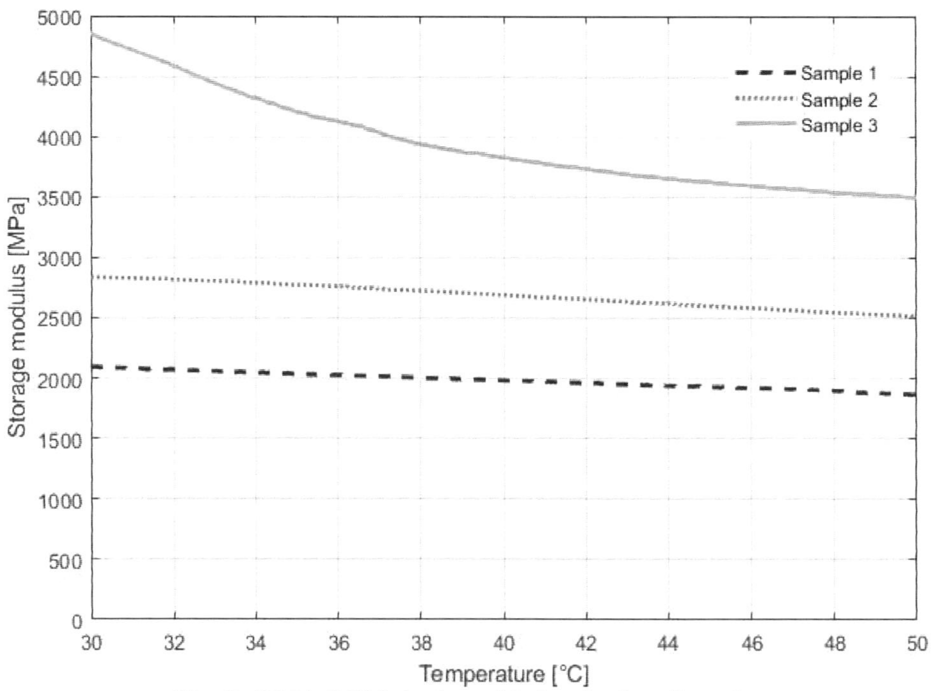

Fig. 3: DMA Q800 device with 3-point bending clamps

Conclusion

Mechanical properties of three types of carbon composite were compared using a dynamic mechanical analyzer. The measurement shown different characteristics at different temperatures. Storage modulus values were obtained at a frequency of 10 Hz for a temperature range of 30°C – 50°C. The parameters found can help for selection suitable composite material for a particular application.

Acknowledgment:

The result was obtained through the financial support of the Ministry of Education, Youth and Sports of the Czech Republic and the European Union (European Structural and Investment Funds - Operational Programme Research, Development and Education) in the frames of the project "Modular platform for autonomous chassis of specialized electric vehicles for freight and equipment transportation", Reg. No. CZ.02.1.01/0.0/0.0/16_025/0007293.

References

[1] K. Habib, S. T. Hansdóttir, H. Habib, Critical metals for electromobility: Global demand scenarios for passenger vehicles 2015–2050, Resources, Conservation and Recycling, Volume 154, 2020, ISSN 0921-3449, https://doi.org/10.1016/j.resconrec.2019.104603.

[2] J. Sullivan, J. Kelly, A., Elgowainy, Vehicle Materials: Material Composition of Powertrain Systems. Argonne National Laboratory, Argonne, 2015

[3] E. Selver, Tensile and flexural properties of glass and carbon fibre composites reinforced with silica nanoparticles and polyethylene glycol. Journal of Industrial Textiles, 49(6), 809–832. https://doi.org/10.1177/1528083719827368, 2020

[4] S. Balakrishnan, H. Seidlitz, Potential repair techniques for automotive composites: A review. Composites Part B: Engineering, 145, 28-38. doi:10.1016/j.compositesb.2018.03.016, 2018

[5] JM. Hormann, M. Wacker. Simulation of the Crash Performance of Crash Boxes based on Advanced Thermoplastic Composite. 5th European LS-DYNA Conference :DYNAmore, 2005

[6] M. A. Masuelli Introduction of Fibre-Reinforced Polymers − Polymers and Composites: Concepts, Properties and Processes, Fiber Reinforced Polymers - The Technology Applied for Concrete Repair, Martin Alberto Masuelli, IntechOpen, DOI: 10.5772/54629. January 23rd 2013. Available from: https://www.intechopen.com/books/fiber-reinforced-polymers-the-technology-applied-for-concrete-repair/introduction-of-fibre-reinforced-polymers-polymers-and-composites-concepts-properties-and-processes

[7] N. Saba, M. Jawaid, Othman Y. Alothman, M.T. Paridah, A review on dynamic mechanical properties of natural fibre reinforced polymer composites, Construction and Building Materials, Volume 106, 2016, Pages 149-159, ISSN 0950-0618, https://doi.org/10.1016/j.conbuildmat.2015.12.075.

[8] P. Kulhavý, V. Fliegel, P. Lepšík, P. Srb, Dynamic analysis of 3D printed and moulded composites with carbon microfibers, Experimental Stress Analysis - 56th International

[9] QDMA Manual, TA Instruments - Waters LLC, New Castle, DE 19720, 2006

Influence of Annealing Time on Tensile Strength and Change of Dimensions of PLA Samples Produced by Fused Filament Fabrication

SUDER J.[1,a], MACHALLA V.[2,b], ŠAFÁŘ M.[3,c], HECZKO D.[1,d], ZEMAN Z.[1,e]

[1]Deparment of Robotics, Faculty of Mechanical Engineering, VSB – Technical University of Ostrava; 17. listopadu 2172/15, 708 00 Ostrava, Czech Republic

[2]Deparment of Applied Mechanics, Faculty of Mechanical Engineering, VSB – Technical University of Ostrava, 17. Listopadu 2172/15, 708 00 Ostrava, Czech Republic

[3]Deparment of Power Engineering, Faculty of Mechanical Engineering, VSB – Technical University of Ostrava, 17. Listopadu 2172/15, 708 00 Ostrava, Czech Republic

[a]jiri.suder@vsb.cz, [b]vojtech.machalla@vsb.cz, [c]michal.safar@vsb.cz, [d]dominik.heczko@vsb.cz, [e]zdenek.zeman@vsb.cz

Keywords: 3D print, Annealing, PLA, FFF, Strength test

Abstract

Annealing is a well-known technology of heat treatment, which is commonly used for metallic materials. Nowadays, with the growth of 3D printing technologies, the first experiments with annealing of plastic prints are starting to appear. One of the most important parameters of plastic annealing is the annealing time. This article deals with the experimental determination of the influence of annealing time on tensile strength and change of dimensions of printed samples from PLA. The samples were annealed in the industrial oven for 15, 30, 45, 60 and 240 minutes at 110 °C. After gradual cooling in air, they were measured and tested on a tensometer, which produces stress-elongation curves. The measurement results showed that all annealed samples changed in size depending on the position at which they were printed. For all different annealing times, the deformations of the annealed samples were approximately the same. The tensile strength of all annealed samples increased compared to the non-annealed ones. Also, no dependence between tensile strength and annealing time was found.

Introduction

Additive Manufacturing currently plays one of the key roles in design [1]. The most commonly used Additive Manufacturing technologies are Fused Filament Fabrication (FFF) or Fused Deposition Modeling (FDM)[2]. Both of these technologies are identical, but FDM is a registered trademark of Stratasys, so other printer manufacturers cannot use it and use the FFF designation. This technology is also currently the most widespread in the area of hobby 3D printers, whose prices are decreasing rapidly. This is due to the expiration of protection patents for this technology, and so the production of these printers has increased [3].

Although production by FFF technology is relatively simple, it is necessary to have good knowledge of the parameters and capabilities of both the technology and the materials itself to achieve the desired outcome [4]. There are many materials available for 3D printing on the market today. Each of them has different properties and therefore is suitable for different purposes. They also behave differently before, during and after printing [1]. In some cases, the prints are further processed after printing (post-processing). By doing so, for example, improved surface quality, required dimensions, and geometric accuracy, or desired material properties can be achieved [5].

Polylactic acid (PLA) is one of the most widely used materials for FFF [6]. It is a non-toxic (suitable for use in the human body) bioabsorbable polymer. According to the United States Food and Drug Administration, it is classified as a safe material and can be used in the food industry as well as in medical industry [7]. In FFF technology, it is mainly used because it is easy to print, has good mechanical properties, and low cost. After the object is printed, it can be heat-treated, which changes its properties. One of the heat treatment processes is annealing.

Annealing is a process whereby a printed plastic part is heated to a specific temperature, then maintained at that temperature for some time and then allowed to cool to room temperature in air. This process reduces the internal stress in the plastic part, that has been generated by thermal expansion during printing [8] and improves the resulting strength of the part [9].

Several researches about the PLA annealing process were already conducted, which has shown that the annealed samples are reaching greater strength than the non-annealed ones. [1,9,10]. This article adds to those studies [1,9,10] with an experimental determination of the influence of the time of annealing on the resulting tensile strength of the printed samples and the change in their dimensions. The samples were printed by FFF method from PLA material.

Among other things, the results of the experiment will help in the application of 3D printing in several robotics systems such as snake robotic devices [11], walking robotic devices [12] and modular systems [13] which are being developed in our department, where annealing of printed parts could help to improve the parameters of these devices.

Description of Experiments

All the samples were made by FFF technology on ORIGINAL PRUSA I3 MK3S printer from Prusament PLA [14]. The used basic printing parameters are shown in Table 1.

Table 1: The used basic printing parameters

Nozzle width [mm]	Layer height [mm]	Infill [%]	Number of perimeters [-]	Nozzle temperature [°C]	Bed temperature [°C]
0.4	0.2	100	2	215	60

All samples were printed in the same position as shown in Fig.1 on the left. The dimensions of the test samples are shown in Fig. 1 on the right.

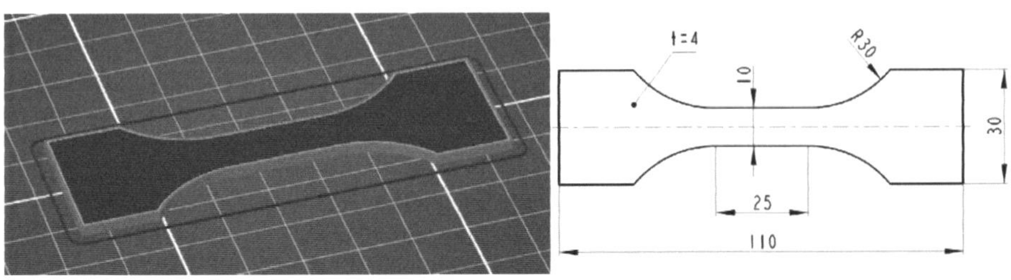

Fig. 1: Left – sample position on the bed, Right – sample dimensions

A total of 60 samples were printed. The basic dimensions of each sample were measured after printing and cooling it to room temperature. Then, they were annealed in the sets of 10 pieces in a Memmert UN 30 dryer over a period of 15, 30, 45, 60 and 240 minutes to 110 ° C, which is the recommended annealing temperature for PLA [10]. After they cooled to room temperature, the basic dimensions were re-measured to determine the formed deformation.

The samples were tested for tensile strength on a Testometric M500 / 50CT at room temperature. In addition to the annealed samples, 10 pieces of non-annealed samples were tested for subsequent comparison. Fig. 2 shows a test machine with tearing test sample.

Fig. 2: Test machine Testometric M500/50CT with tearing test sample

Results

The experiment focused on detecting the change in sample's dimensions. The change in sample's geometric shape was not measured. The dimensions changed for all annealed samples. Dimensions were measured at the locations shown in Fig. 3. After annealing, the height (H1) of all samples has increased and all other dimensions decreased.

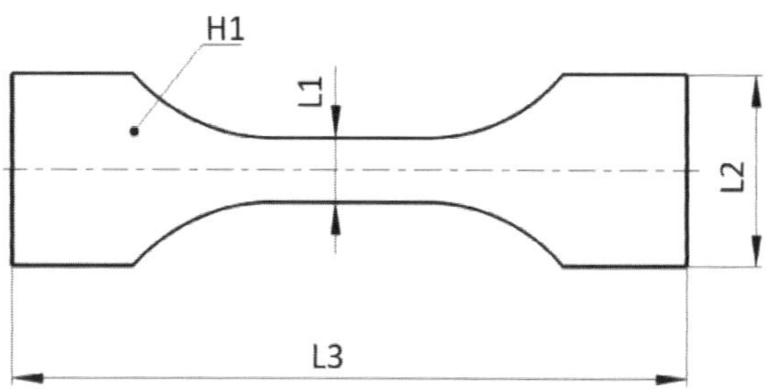

Fig. 3: Measured dimensions

Table 2 shows the average percentage change in dimension by annealing time.
From the table values, it can be seen that the average difference in dimension change between different annealing times for H1 is 1.05% and that for all other dimensions it does not exceed 0.5%. Thus, the annealing time has almost no effect on the resulting dimension change. On average, at all different annealing times, H1 increased by 12.14%, L1 decreased by 3.73%, L2 decreased by 5.80% and L3 decreased by 6.47%. Fig. 4 shows the influence of annealing time on dimension change.

Table 2: Average percentage change in dimension by annealing time

Annealing time [min]	ΔH1 [%]	ΔL1 [%]	ΔL2 [%]	ΔL3 [%]
15	12.10	-3.85	-5.84	-6.69
30	11.69	-3.78	-5.83	-6.46
45	11.93	-3.75	-6.03	-6.51
60	12.74	-3.39	-5.57	-6.26
240	12.25	-3.88	-5.72	-6.44

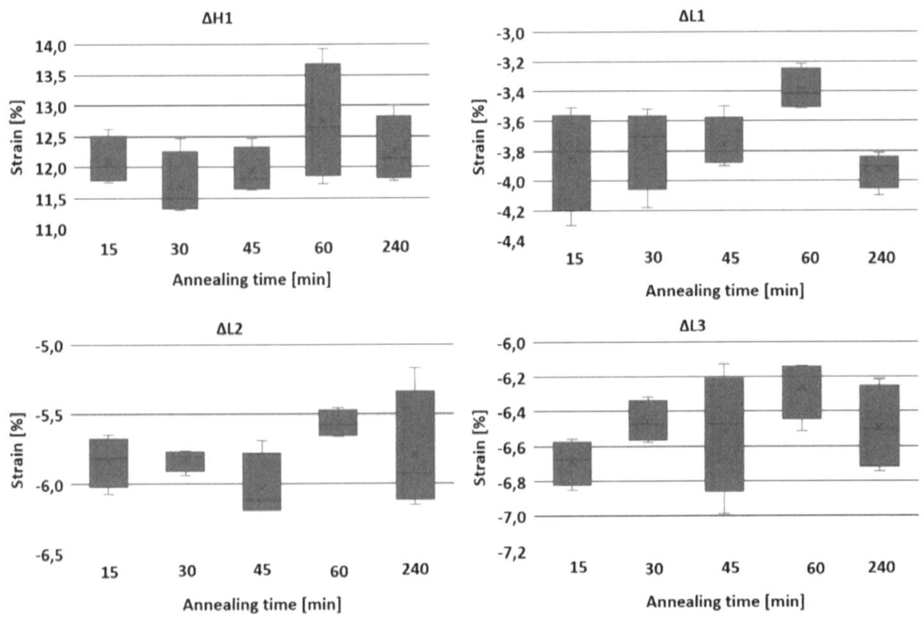

Fig. 4: Boxplots of the influence of annealing time on dimension change

Individual samples were then tested for tensile strength on a test machine. Fig. 1. 5 shows selected stress-elongation curves for individual annealing times. The curves show an increase in the strength value of the annealed samples.

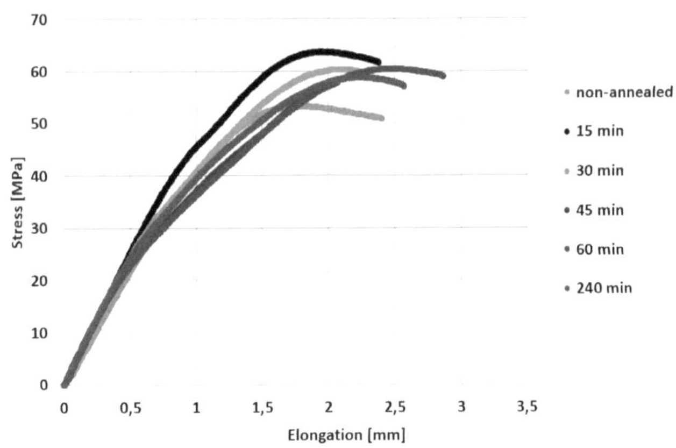

Fig. 5: Stress – elongation curve

Table 3 shows the average values of the tensile strength and the percentage increase in strength (compared to non-annealed samples). The experiment showed that annealing increases the tensile strength by approximately 10% and that the change of annealing time had minimal impact on it.

Table 3: Average values of the tensile strength and the percentage increase in strength

Annealing time [min]	Tensile strength [MPa]	Percentage increase in strength [%]
Non-annealed	53.86	-
15	61.15	13.54
30	59.03	9.59
45	59.89	11.20
60	62.28	15.63
240	60.51	12.34

Fig. 6 shows the influence of annealing time on tensile strength.

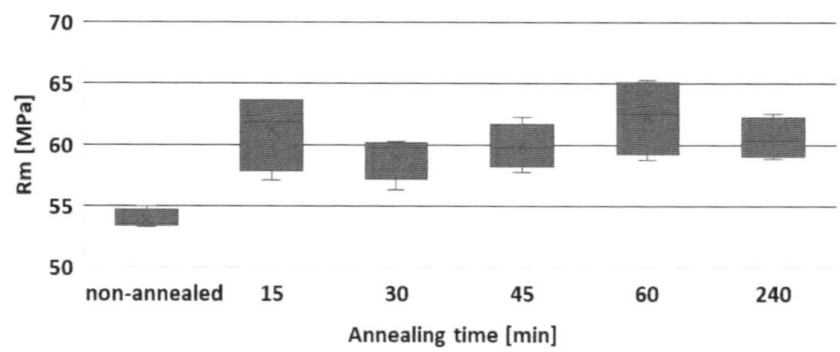

Fig. 6: Boxplots of tensile strength

Conclusions

All annealed samples have changed their dimensions as shown in Tab. 2 and FIG. 4. These dimension changes are dependent on the position at which the samples were printed. In the direction perpendicular to the printing surface (dimension H1 in Fig. 3), the dimensions of the test samples increased by an average of 12.14%. In other directions, the dimensions were reduced. The experiment did not prove the influence of the annealing time on the amount of deformation. It should be noted that the annealing time must be long enough for the entire material to be heated to a given temperature. Thereafter, the annealing time has no effect on the dimension change.

All annealed samples showed higher values of tensile strength as shown in Tab. 3, FIG. 5 and FIG. 6. On average, the tensile strength was increased by 10%. The experiment did not prove any influence of the annealing time on the tensile strength, once the entire material was heated to a given temperature.

The results of this work will be used for further research in the field of annealing of plastic materials for FFF technology.

Acknowledgment

This article has been elaborated under support of the project Research Centre of Advanced Mechatronic Systems, reg. no. CZ.02.1.01/0.0/0.0/16_019/0000867 in the frame of the Operational Program Research, Development and Education. This article has been also supported by specific research project SP2020/141 and financed by the state budget of the Czech Republic.

References

[1] J. Beniak, M. Holdy, P. Križan, and M. Matúš, "Research on parameters optimization for the Additive Manufacturing process," in *Transportation Research Procedia*, 2019, vol. 40, pp. 144–149, doi: 10.1016/j.trpro.2019.07.024.

[2] M. Jin, C. Neuber, and H. W. Schmidt, "Tailoring polypropylene for extrusion-based additive manufacturing," *Addit. Manuf.*, vol. 33, p. 101101, 2020, doi: 10.1016/j.addma.2020.101101.

[3] J. Beniak, P. Križan, and M. Matúš, "A comparison of the tensile strength of plastic parts produced by a fused deposition modeling device," *Acta Polytech.*, vol. 55, no. 6, pp. 359–365, 2015, doi: 10.14311/AP.2015.55.0359.

[4] J. Lipina, V. Krys, and F. Fojtík, "Tensile Test on Samples Produced by Rapid Prototyping Technology with a Higher Number of Contours," *INES 2018 - IEEE 22nd Int. Conf. Intell. Eng. Syst. Proc.*, pp. 000431–000436, 2018, doi: 10.1109/INES.2018.8523963.

[5] N. N. Kumbhar and A. V. Mulay, "Post Processing Methods used to Improve Surface Finish of Products which are Manufactured by Additive Manufacturing Technologies: A Review," *J. Inst. Eng. Ser. C*, vol. 99, no. 4, pp. 481–487, 2018, doi: 10.1007/s40032-016-0340-z.

[6] M. Van der Walt, T. Crabtree, and C. Albantow, "PLA as a suitable 3D printing thermoplastic for use in external beam radiotherapy," *Australas. Phys. Eng. Sci. Med.*, vol. 42, no. 4, pp. 1165–1176, 2019, doi: 10.1007/s13246-019-00818-6.

[7] S. Farah, D. G. Anderson, and R. Langer, "Physical and mechanical properties of PLA, and their functions in widespread applications — A comprehensive review," *Advanced Drug Delivery Reviews*. 2016, doi: 10.1016/j.addr.2016.06.012.

[8] I. Gajdos, J. Slota, E. Spisak, T. Jachowicz, and A. Tor-Swiatek, "Structure and tensile properties evaluation of samples produced by Fused Deposition Modeling," *Open Eng.*, vol. 6, no. 1, pp. 86–89, 2016, doi: 10.1515/eng-2016-0011.

[9] S. Bhandari, R. A. Lopez-Anido, and D. J. Gardner, "Enhancing the interlayer tensile strength of 3D printed short carbon fiber reinforced PETG and PLA composites via annealing," *Addit. Manuf.*, vol. 30, p. 100922, 2019, doi: 10.1016/j.addma.2019.100922.

[10] Jakub Kočí, "How to improve your 3D prints with annealing - Prusa Printers," 05-Dec-2019. [Online]. Available: https://blog.prusaprinters.org/how-to-improve-your-3d-prints-with-annealing/. [Accessed: 01-Mar-2020].

[11] I. Virgala, M. Dovica, M. Kelemen, E. Prada, and Z. Bobovský, "Snake robot movement in the pipe using concertina locomotion," *Appl. Mech. Mater.*, vol. 611, pp. 121–129, 2014, doi: 10.4028/www.scientific.net/AMM.611.121.

[12] Z. Bobovský, "Design of geometrical parameters for walking mechanism leg with use of matlab algorithm and simmechanics," *Appl. Mech. Mater.*, vol. 656, pp. 164–170, 2014, doi: 10.4028/www.scientific.net/AMM.656.164.

[13] Z. Bobovský, P. Novák, V. Krys, and T. Kot, "The module for a self-reconfigurable robotic system," *SAMI 2014 - IEEE 12th Int. Symp. Appl. Mach. Intell. Informatics, Proc.*, pp. 275–280, 2014, doi: 10.1109/SAMI.2014.6822421.

[14] Prusament PLA Prusa Galaxy Black 1kg - Prusa Research. [Online]. Available: https://shop.prusa3d.com/en/prusament/711-prusament-pla-prusa-galaxy-black-1kg.html. [Accessed: 03-Mar-2020].

EAN 2020
58th conference on experimental stress analysis

Effect of Ply-Drop on Stress and Strain Fields in Tapered Thermoplastic Matrix Composites

ŠEDEK J.[1,a], KADLEC M.[1,b]

[1]VZLU – Czech Aerospace Research Centre, Beranových 130, 199 05 Prague - Letňany, Czech Republic

[a]sedek@vzlu.cz, [b]kadlec@vzlu.cz

Keywords: Composite, FEM, Ply-drop, PPS, Thermoplastic

Abstract. The analysis of ply-drop in carbon fibre reinforced thermoplastic composite with PPS matrix was carried out using numerical modelling. The results of fatigue tests yielded the ply-drop as a critical point when cyclically loaded in tension. Several delaminations were found in phractography in the area of ply-drop. The finite element model was build using ply-by-ply technique, that enables to evaluate the stress and strain fields in each lamina. The concentrations at the interface of terminated plies and resin rich region were identified. The terminating ply with orientation of 0/90° causes high stress concentrations of shear stress in connection with adjacent plies and high strains in resin rich region. The proper choice of terminated ply location can extend the fatigue behaviour considerably.

Introduction

Woven composites are used in aerospace due to its high strength and stiffness-to-weight ratio. In order to reach this goal, the thickness is reduced in locations of lower loading resulting in terminating of plies (ply-drops), that are generally stress and strain concentrators. Its effect on strength, fatigue characteristics and fracture mechanism has not yet been widely investigated for the modern aerospace-grade composites with thermoplastic matrix.

Several types of thickness variations by ply-drops can be classified according to Ref. [1] as external, internal (longitudinal or transverse), and mid-plane ply-drop constructions. Basically, a ply-drop can be embedded between plies or at the surface. The bonded top plies located at the continued ones can suffer by separation owing to interlaminar cracks emanating from free surface as studied in Ref. [2]. Inserting semipregs to proper structure areas can achieve more uniform stress distribution and lead to the desired thickness change and the inclination to separation of adjacent plies can be suppressed as well.

Modelling the stress and strain field at ply-drop can show the issues of concentrations and help to optimize the design of ply stacking. In present work, the shear stress concentration on ply drops and longitudinal stress on the tapering transition were analysed by modelling. In fatigue loading, these concentrations cause delamination initiated by loading in mode I or II as observed in Ref [3].

Fatigue test. Presented analysis was carried out investigating the delamination occurring during fatigue tests in ply-drops of tailored blank C/PPS thermoplastic composite as shown by fractography in Fig. 1. The delamination raised by fatigue loading typically initiated at the termination of the ply oriented 0/90°. The fabric had a five-harness satin weave, so the crossing of yarns causes the wavy yarn laying. The thickness transition was made from 16 to 11 layers, i.e. the layup before transition was $[0/90, \pm45]_{4s}$ and after that $[(0/90, \pm45)_2, \overline{0/90}]_s$. The nominal thicknesses of the sections were 4.96 mm for 16 layers, 3.41 mm for 11 layers and the tapered angle was 6°. During manufacturing of thermoplastic composite, the layers of tailored blanks are positioned, stacked together and cured in the die. Nevertheless, some slip can occur and finally the layers can be displaced in the range of mm. Such slip can be seen in Fig. 1, where layers ±45° and 0/90° should terminate together.

The fatigue test was performed under tension loading with frequencies in the range of 0.5–15 Hz and the load ratio of 0.05. The symmetric tension specimens with thinner part with 11 layers in the centre were tested. The thicker part was clamped on both sides. Mainly two fracture modes were present due to fatigue: Fracture at a ply-drop and fracture away from the ply-drops. The former one was analysed in more detail as described in following section.

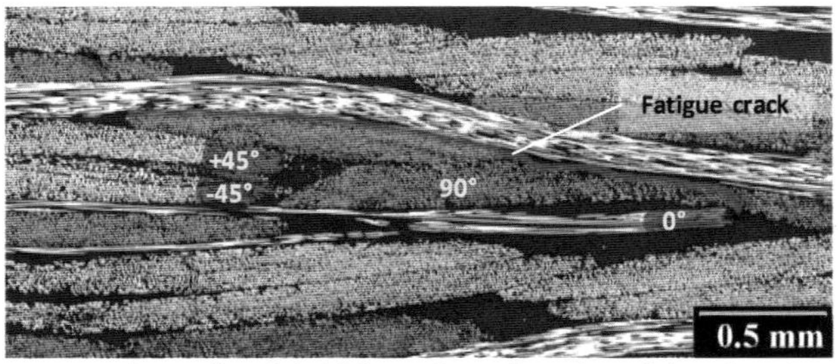

Fig. 1: Laminate cross-section after 34 000 cycles, showing delamination (fatigue crack) initiated from 0/90° ply drop.

Modelling. Woven composites are heterogeneous materials that are not easy to characterize computationally. There are several difficulties owing to hierarchical structure and microstructure. The modelling is being done in the macro (component), meso (fabrics) and micro level (yarn, tows) according to structure detail. The meso-scale analysis requires at least the characterisation of plies and the detailed 2D or 3D model shall be used. Such model can evaluate internal stress, interlaminar stress and can suggest stress risers and delamination.

Investigating fatigue cracks the 3D finite-element model of ply-drop was created using ply-by-ply technique, that enables to evaluate the stress and strain fields in each lamina. Some idealisation in the model was necessary to apply. The void volume at the ply transition was filled by resin assuming ends of two adjacent plies ±45° and 0/90° at the same location, but it was not generally true (note remark above). That volume was filled by PPS resin assumed as an isotropic material with Young's modulus of 3800 MPa and Poisson's ratio 0.36. Each ply was represented as 3D continuum with orthotropic material applied. The parameters were used according to Ref. [4] as follows: $E_{11} = E_{22} = 85$ GPa, $E_{33} = 10.5$ GPa, $G_{12} = 4.1$ GPa, $G_{13} = G_{23} = 3.5$ GPa, $\nu_{12} = 0.05$ and $\nu_{13} = \nu_{23} = 0.41$. Only an ideally linearly elastic material model was used in the analysis.

Analysis results. The FE model was loaded by 40 kN in tension. The results show some concentrations in ply-drop region. The highest longitudinal stress peak is located at the interface of the 0/90° ply and the resin rich region. The lower position (straight side of the specimen) is worse. The in-plane stress plot indicates the symmetry of the layup with stress peaks at continuous ply oriented ±45° caused by the termination of the ply oriented 0/90°, but the out-of-plane shear stress plot does not show clear stress symmetry (see Fig. 2) owing to tapered part in the top of the detail. Nevertheless, the peaks at the interface of continuous ±45° ply and the terminated one oriented 0/90° are clear. No severe stress peak is present in the middle at terminated ply oriented 0/90°. The stress plot is shown in Fig. 2 together with direct notation of the stress values in details of ply transitions. The red ellipses mark peak values.

In strain plot shown in Fig. 2, it is obvious that high concentrations are caused by termination of plies with orientation of 0/90° next to resin rich regions.

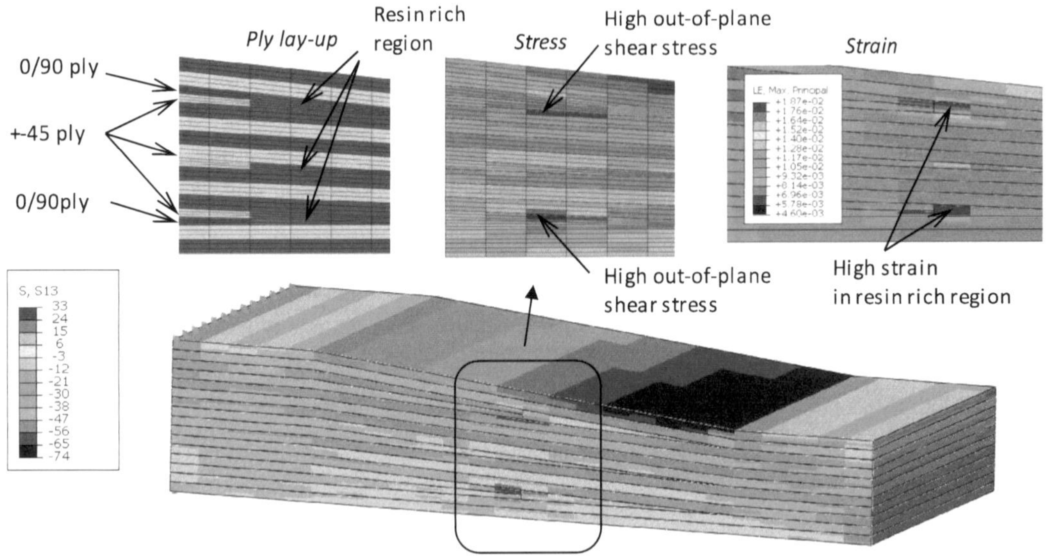

Fig. 2: FE half-model of the ply-drop; results at the load of 40 kN

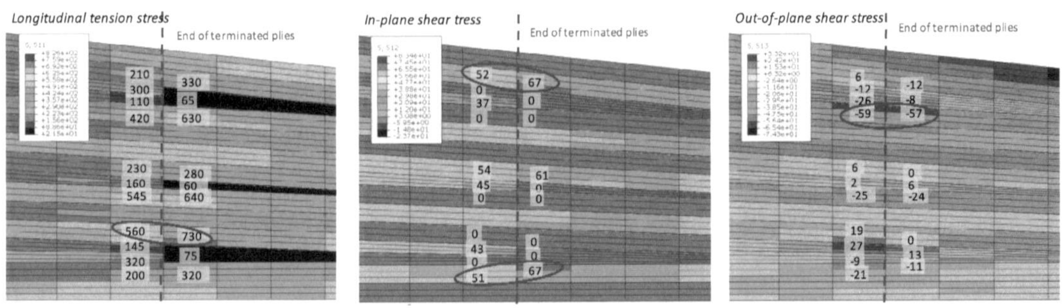

Fig. 3: Stress results in detail at the load of 40 kN; red ellipses marks peak values.

Discussion

From the point of view of crack initiation, the surface phenomenon should be considered. The fatigue crack initiates at the surface [5], therefore the interface of the resin and the fibre is critical. The difference in longitudinal stress between the 0/90° play and the resin region is 330 MPa, that is considerably high compared to 93 MPa of strength of pure PPS resin according to Ref. [6]. It suggests, that during loading, plastic deformation occurs. The results show linear strains in resin of 1.9 %, but due to high interface tension stress, the resin will be apparently plastically deformed and decrease the stress peak.

The FE results are consistent with experimental observations. Some fatigue cracks appeared at the termination point of the ply oriented 0/90° adjacent to the continuous one oriented ±45° (see Fig.1) and their growth continued along the ply oriented 0°. The growth is evidently driven by high shear stress. The other interesting finding was the indispensable heating during the fatigue test described in Ref. [3]. The heating of the specimen was greater together with increasing frequency of the cyclic loading. In addition to heating due to material damping, it could be caused by dissipation of the energy during plastic deformation and after the delamination, by friction between plies.

Conclusions

The FE analysis of the ply-drop in woven composite with thermoplastic matrix indicated several difficulties with the terminating plies. Less stress peak was determined in the middle of the specimen.

Without doubts, it can be said that the ply-drop is critical point with impact on fatigue behaviour. In analysed design, the ply-drop configuration is not the favourable solution owing to termination of two layers together. This approach enlarges the void filled by the resin and during fatigue loading the delamination occurs primarily in this location. The proper choice of terminated ply location can extend the fatigue behaviour considerably, e.g. to end the 0/90° plies in the thicker section, and the +-45° plies in the thinner section where the stress is higher.

Acknowledgements

This work was funded by an institutional support from the Ministry of Industry and Trade of the Czech Republic

References

[1] K. He, S., Hoa, R. Ganesan, The study of tapered laminated composite structures: a review, Compos. Sci. Technol. 60 (2000) 2643–2657. doi:10.1016/S0266-3538(00)00138-X.

[2] J. Šedek, R. Hron, M. Kadlec, Bond Joint Analysis of Thermoplastic Composite Made from Stacked Tailored Blanks. Applied Mechanics and Materials, Vol. 827, 2015. pp.161-168, ISSN 1662-7482. DOI: 10.4028/www.scientific.net/AMM.827.161.

[3] R. Růžek, M. Kadlec, L. Petrusová, Effect of fatigue loading rate on lifespan and temperature of tailored blank C/PPS thermoplastic composite, International Journal of Fatigue 113 (2018) 253–263

[4] B. Vieille, V. M. Casado, C. Bouvet, 2014. Influence of matrix toughness and ductility on the compression-after-impact behavior of woven-ply thermoplastic- and thermosetting-composites: A comparative study. Compos. Struct. 110, 207–218.

[5] J. Schijve, Fatigue of Structures and Materials, 2. issue, Springer, 2010, 621 pages, ISBN-13: 978-1-4020-6807-2

[6] Mitsubishi Chemical Advanced Materials, Techtron 1000, product data sheet, available at: <http://qepp.matweb.com/search/DataSheet.aspx?Bassnum=P1SM37>

EAN 2020
58th conference on experimental stress analysis

Defect Detectability in Composite Plates with Variable Thickness Using Lamb Waves

ŠEDKOVÁ L.[1,a], ŠEDEK J.[1,b]

[1]VZLU – Czech Aerospace Research Centre, Beranových 130, 199 05 Prague - Letňany, Czech Republic

[a]sedkova@vzlu.cz, [b]sedek@vzlu.cz

Keywords: Impact damage, structural health monitoring, ultrasonic guided waves, time delay

Abstract. In this work thermoplastic composite plates with constant and variable thickness were subjected to multiple impact damages. The network of PZT transducers was utilized to excite Lamb waves at different frequencies. Influence of the variable thickness on the guided wave propagation was evaluated and compared to the constant thickness plate. Optimal frequency was chosen from the point of a clear mode shape. Signals before and after impacts were evaluated using two parameters: correlation coefficient and time delay of the particular wave mode. Finally, RAPID (Reconstruction Algorithm for the Probabilistic Imaging of Damage) algorithm was used to visualize impact damages. Damage induced time delay parameter showed correct impact localization on both plates. However, time delays measured on the variable thickness plate were about four times smaller.

Introduction

Defect detectability plays an essential role in the area of non-destructive testing (NDT) and structural health monitoring (SHM) and ultrasonic guided wave testing is considered as one of the promising methods [1]. Aircraft composite parts suffer from impact damages which can consequently cause failure threatening the structural integrity. Therefore, new composite materials and structural parts are tested for impact damages that should be reliably detected using different NDT and SHM systems. In the area of NDT ultrasonic phased array methods and laser shearography belong to the reliable detection and visualization techniques of impact damages [2, 3]. SHM research area is focused on the guided wave testing, measurements using FBG (Fiber Bragg Grating) sensors, and also a promising work regarding electrical resistance tomography (ERT) have been done lately [4].
Algorithms for impact damage visualization by means of guided waves are described for example in [1, 5]. RAPID algorithm is probably the most exploited one. It enables to use any signal parameter evaluating signal before and after the damage [6, 7].

Materials and methods

Test specimens were manufactured from carbon composite with thermoplastic matrix ABS5045. The first plate (P1) had dimensions of 280×530 mm and thickness of 2.2 mm and the second plate (P2) had dimensions of 240×615 mm and thickness of 3.5-4.3 mm. The second plate with variable thickness was stacked from 4 or 5 glass layers and 10 or 15 carbon layers according to the thickness. PZT disc transducers (Stem, Inc.) used for guided wave measurements were placed along the edges and served as both actuators and sensors. Measurements were performed before and after each impact. Standard 5-cycle sine wave modulated by Hanning window was used as the actuation signal for the frequencies of 40, 50, 60, 70, 80, 100, 150, 200, 250 and 300 kHz. The frequency of 50 kHz was chosen based on the clear A0 mode shape (no distorsion of the wave packet).

Signal processing

Impact damage evaluation was performed using 3 methods. All of them are based on the signal comparison of all the actuation–sensor pairs before and after the damage. The first one exploited SW Acess provided with the Acellent Scan Genie II measuring device. The correct determination of the impact location was unconsistent. It was not possible to choose one frequency and methodology to correctly locate all the impacts. Therefore, the RAPID algorithm was programmed in order to use arbitrary signal parameters. Correlation coefficient and time delay was used as a damage index (DI). Evaluation of the wave time delay was performed according to the following steps:

1. Group velocity determination of the A0 mode in all actuator–sensor directions
2. Extraction of the particular wave modes
3. Time delay determination of the maximum amplitude of the relevant waves

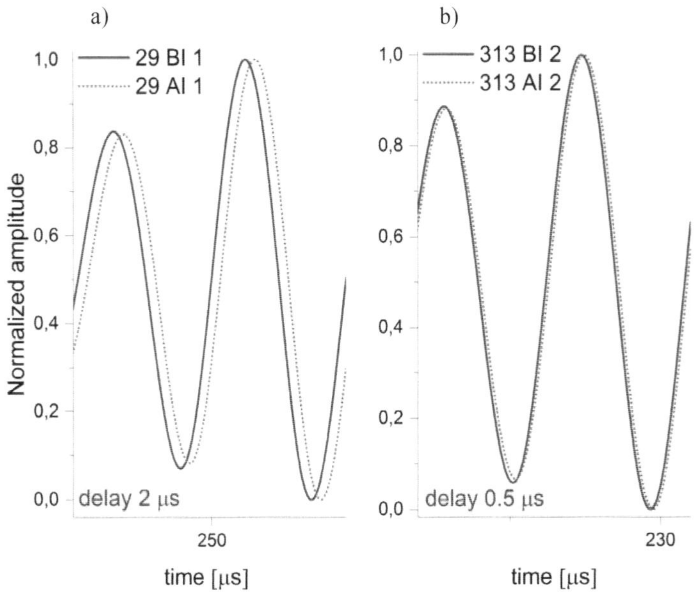

Fig. 1: Comparison of the wave delay for a) constant thickness plate and b) variable thickness plate

Results

The comparison of the three methods is shown in Fig. 2 and Fig. 3 for the constant thickness plate and variable thickness plate, respectively. Designations a), b), c) denote evaluation using SW Acess, RAPID - Correlation coefficient and RAPID - Wave time delay, respectively. The time delay parameter showed correct localization of all impacts. The real impact location is marked with the red cross. Delays for the variable thickness plate were about 4 times smaller than for the constant thickness plate (Fig. 1). However, normalized values are used in the RAPID algorithm, so the impact is clearly visualized.

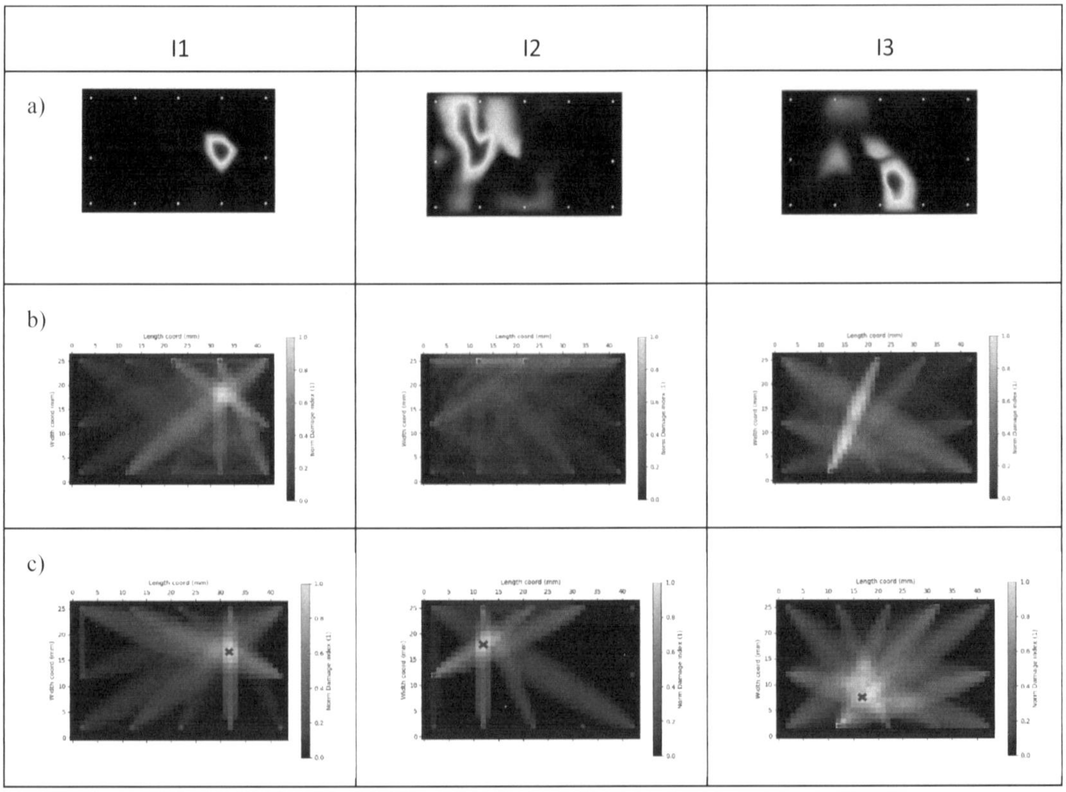

Fig. 2: Comparison of impact evaluation for the constant thickness plate using a) SW Acess, b) RAPID - correlation coefficient, c) RAPID - time delay

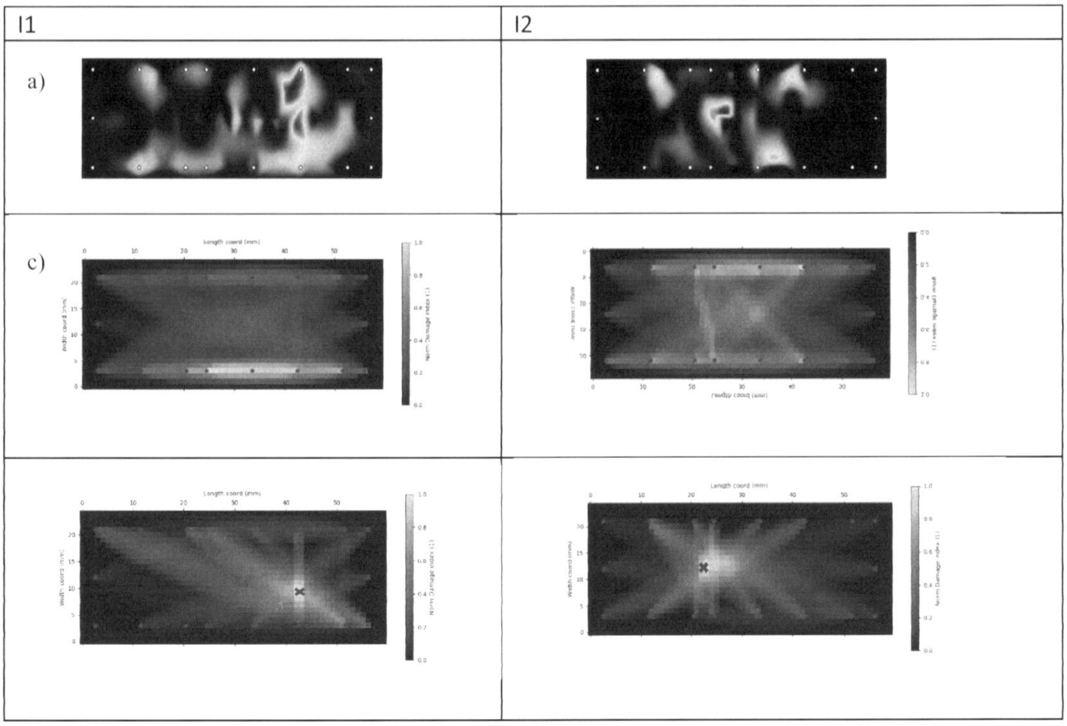

Fig. 3: Comparison of impact evaluation for the variable thickness plate using a) SW Acess, b) RAPID - correlation coefficient, c) RAPID - time delay

Conclusion

Two thermoplastic composite plates were subjected to multiple impact damages. Three methods were used to damage evaluation. SW Acess with unknown algorithm and RAPID algorithm with two different DI (correlation coefficient and time delay) were used to evaluate impacts. Results using SW Acess did not locate correctly all impact damages at the given frequency so this method was not effective. Correlation coefficient used in RAPID algorithm did not show correct results either. Time delay extraction of the relevant waves detected correctly location of all 5 impacts. However, time delays measured on the variable thickness plate had four times smaller values than on the constant thickness plate. Normalized values in the RAPID algorithm enabled clear visualization. Significantly smaller time delays are probably caused by the material itself. The variable thickness plate contains multiple glass layers which influence the guided wave propagation.

It can be concluded that extraction of the particular wave is essential, especially for the variable thickness plate. Using the same time window length for different actuator-sensor distances include multiple wave reflections which can cause false damage indications.

References

[1] J.L. Rose, Ultrasonic Guided Waves in Solid Media, Cambridge University Press, Cambridge, 2014.

[2] M. Kadlec, R. Růžek, A Comparison of Laser Shearography and C-Scan for Assessing a Glass/Epoxy Laminate Impact Damage. Applied Composite Materials 19 (2012) 393-407.

[3] P. Bělský, M. Kadlec, Capability of non-destructive techniques in evaluating damage to composite sandwich structures, International Journal of Structural Integrity 10 (2019) 56-370.

[4] J. Cagáň, L.Michalcová, Impact Damage Detection in CFRP Composite via Electrical Resistance Tomography by Means of Statistical Processing, Journal of Nondestructive Evaluation, 2020

[5] M. Tabatabaeipour, J. Hettler, S. Delrue, K. van den Abeele, Reconstruction Algorithm for Probabilistic Inspection of Damage (RAPID) in Composites, 11th European Conference on Non-Destructive Testing (ECNDT 2014)

[6] Z. Sharif-Khodaei, M. H. Aliabadi, Assessment of delay-and-sum algorithms for damage detection in aluminium and composite plates, Smart Materials and Structures, 23(7), (2014) 075007.

[7] M. Bonet et al., Identification of Barely Visible Impact Damages on a Stiffened Composite Panel with a Probability-based Approach. Structural Health Monitoring 2015 (2015)

EAN 2020
58th conference on experimental stress analysis

Design of the Composite Frame of the Zero-Generation Commercial Electric Vehicle

ŠEVČÍK L.[1,a]

[1]Deparment of Machine Design, Faculty of Mechanical Engineering, TUL – Technical University in Liberec, Studentská 1402/2, 46117 Liberec, Czech Republic

[a]ladislav.sevcik@tul.cz

Keywords: Special Electro Vehicle, Composite Frame, Topological optimisation, Frame Design

Abstract. The project "Modular Platform for Autonomous Chassis of Specialized Electric Vehicles for Freight and Equipment Transportation" solves commercial electric vehicle designed for field work. The aim is to move the car autonomously in the nature. Car performed the assigned tasks, and use augmented realities - call Microsoft Windows Holographic HoloLens - for orientation. Originally, the chassis had eight engines, each wheel was individually driven and steered. However, this was not managed in real time. In this way, the electric vehicle was adapted to only one driven axle and two steering wheels. This arrangement is already managed in real time. On paved terrain, an electric car should indicate 800kg of cargo including its own weight. The content of the paper is a design of composite frame. The starting point for optimization is the existing duralumin frame. This has been topologically optimized in terms of dimensional possibilities and mounting of vehicle units and equipment. The frame is glued from composite boards. The boards are made by pressing of composite carbon fibres with acrylic resin. Bonding will be done with Cyberbond adhesive. Cybercryl acrylic adhesives are methyl - methacrylate based and designed for structural bonding of engineered thermoplastics, thermosets, nylons, composites, fibre coat and metal in any combination. Ideal for applications that require high tensile and peel strength. A commercial electric vehicle with a duralumin frame is currently being tested and a frame glued with composite boards is being prepared.

Introduction

Within the project "Modular Platform for Autonomous Chassis of Specialized Electric Vehicles for Freight and Equipment Transportation" the frame for the zero generation vehicle was solved. This vehicle will be tested for basic functions, especially autonomous driving and terrain recognition. Furthermore, the project solves the composite frame of this commercial electric vehicle. The frame load capacity is up to 300 kg. The aim is to create a frame by topological optimization to meet the requirements for geometry and fastening of elements, torsional yield and load capacity and deformation. The frame type is a self-supporting composite glued from planar shaped plates. The calculation was solved in CAD program CREO by topological optimization and ANSYS WORKBENCH solvers. The starting point for optimizing was a duralumin machined frame.

Default aluminium alloy frame

The basic shape of the zero-generation vehicle is shown in Fig. 1. The battery and control electronics are located inside of the frame. But battery and electronic was removed for frame, as be better its visibility. The car drives is 1200 W stepper motor and the front wheels drive two stepper motors with encoders. The car has no brakes and in case of sudden need of braking turns the wheels against each other.

The starting point of the optimization was a torsion-yielding assembled frame made of machined duralumin parts, see Fig. 2.

Fig. 1: Frame with wheels and drives

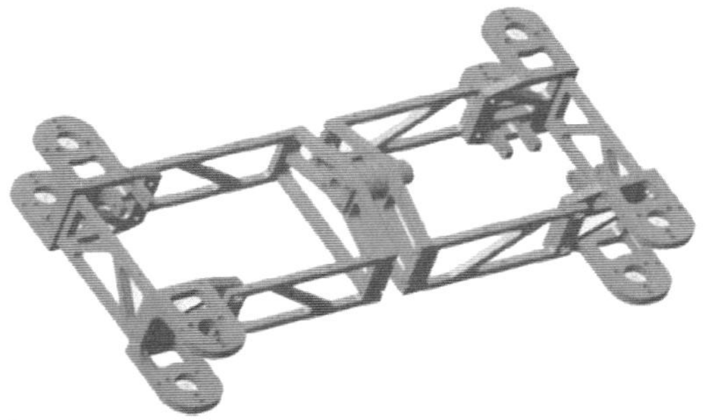

Fig.2: Dural frame ALMgSi alloy

The calculation was carried out by attaching the frame to the wheels. The points were supported, but allowed axial displacements due to frame deformation. The force 8000 N was entered at the hinge in the center of the frame. This is the worst case scenario that can occur with uneven load distribution. Except for the hinge, all the parts of the frame parts were taken together as unmountable - glued together. Physical values for the calculation were as follows: frame material AlMgSi6, Young's modulus E = 7.1e10 Pa, poison number $\nu = 0.33$, specific gravity $\varrho = 2770$ kg/m^3, frame volume V = 5.59e6 mm^3, frame weight m = 15.5 kg, $R_m = 380$ MPa, $R_{0,2} = 210$ MPa.

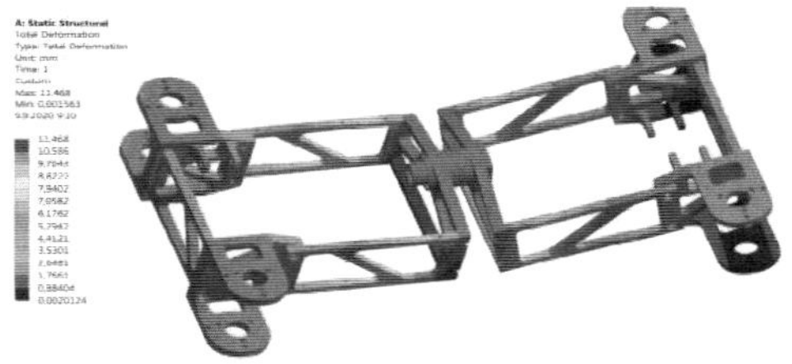

Fig. 3: Total frame deformation

Fig.4: Reduced frame tension

FEM calculation solver was used as static structural analysis with the elastic material model and small deflections. The calculation resulted in a maximum deformation value of 11,5 mm for a given load of 3000 N. Furthermore, the approximate value of equivalent stress in the carrier parts 380 MPa.

Glue by methacrylate

The frame is designed from five glued boards with topologically optimized ribs. These boards must be glued together in the largest possible area and in places with low tensile stress. The joints should be particularly stressed by shear stresses. For the calculations it was necessary to determine the values of the glued joint, especially in shear. A glue with a sales designation was used Cybercryl - methyl-methacrylate based and designed for structural bonding of engineered thermoplastics, thermosets, nylons, composites, fibre coat and metal in any combination. The shear strength of the joint of the steel and composite plate is stated by the manufacturer as 28 MPa. Since this value may be vary, we have tried to create an optimum bonded composite pipe joint and verify the shear stress value. The composite pipe of Fig. 6 was inserted into a steel pipe and sealed with glue. The joint was loaded on the shredder, Fig. 5. Several measurements were made. The average value was determined from five measurements.

The measurement results depend mainly on the surface roughness, a mean value of 1.6 μm was chosen, further on the quality of degreasing of the surface before gluing and especially on the glue thickness. There was a gap to make thickness of glue less than 0.1 mm. The glue was applied to both parts of the joint and the parts slid together. In a few hours the glue had hardened.

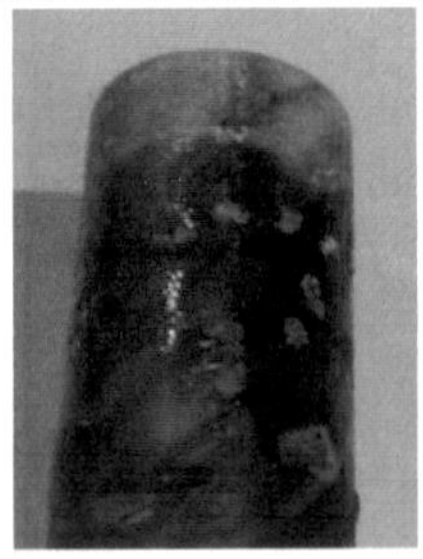

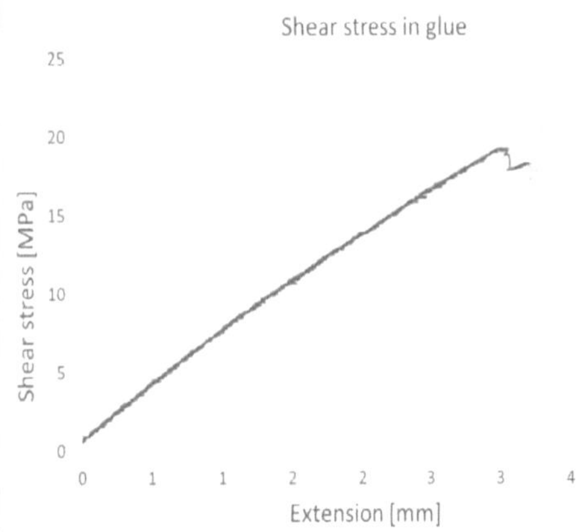

Fig. 5: Loaded joint test   Fig.6: Composite tube   Fig 7. Shear stress

The shear stress in the adhesive surface was calculated from the measured tensile force and the adhesive bonded area. The resulting value of the measured shear stress is shown in Fig. 7.

The result of tests was an average shear stress of 19 MPa. The manufacturer's reported values could not be achieved. This is probably because our joint did not have the required adhesive thickness over the entire joint area.

Design of carbon composite frame

Composite boards are produced by pressing a special semi-finished product - prepreg [1,2]. The ribbed elements are an optimized rib system. Values for calculating the glued joint were determined by measurement [3,4]. In calculations, the x-axis is the longitudinal axis of the car. Mechanical values of ACRYL AND CARBON fiber used with values $E_x = 1.2e11$ Pa, $E_y = 8.6e9$ Pa, $E_z = 8.6e9$ Pa, $\nu_{xy} = 0.27$, $\nu_{yz} = 0.4$, $\nu_{xz} = 0.27$, further specific density $\varrho = 1490$ kg/m^3, frame volume V = 2.31e6 mm^3, frame weight m = 3.44 kg and Rm = 850MPa. The results of the calculation are shown in Fig. 8 and Fig. 9.

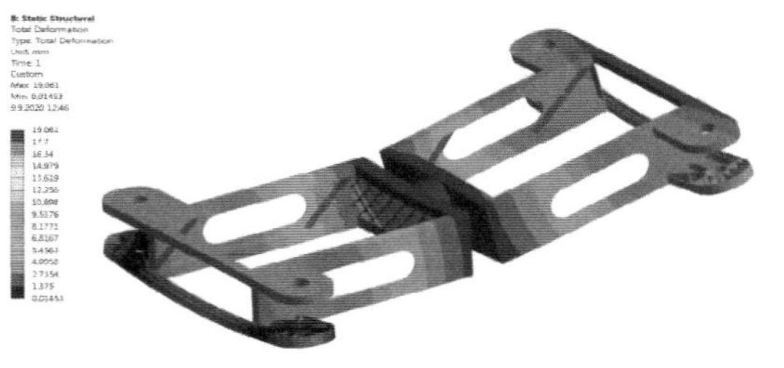

Fig. 8: Total deformation of the composite frame

Fig.9: Reduced stress in the composite frame

Fig.10: Maximum shear stress in the composite frame

The result of the calculation was a maximum deformation value of 19 mm for a given load of 3000 N. Furthermore, the approximate value of equivalent stress in the carrier parts was 450 MPa. The maximal shear stress, see Fig.10, reaches a maximum value of 230 MPa. At the bonding surfaces of the individual composite parts, the shear stress is lower and does not exceed 30 MPa.

Conclusion

The proposed composite frame for a given contract load shows 8 mm more deformation, same maximum of stress and 4.5 less mass. Since the frame was made by machining of duralumin blanks, the price difference between the frames will not be large.

The frame was glued from several composite boards. The boards were glued by Cybercryl - methyl methacrylate. Shear stress measurements in the adhesive were made. An average shear stress of 30 MPa was achieved. In fact, the shear stress value may be higher, since in reality the composite sheets will stick together. However, the measurement was done by gluing the steel and the composite. The steel was chosen for attachment to the blasting machine.

ACKNOWLEDGMENT

The result was obtained through the financial support of the Ministry of Education, Youth and Sports of the Czech Republic and the European Union (European Structural and Investment Funds - Operational Programme Research, Development and Education) in the frames of the project "Modular platform for autonomous chassis of specialized electric vehicles for freight and equipment transportation", Reg. No. CZ.02.1.01/0.0/0.0/16_025/0007293.

References

[1] O. Grimes, Design and Finite Element Analysis of a Composite Monocoque Chassis for the Shell, Eco Marathon Car Project, Coventry University 2012

[2] C. Kirby, Formula SAE Monocoque Design and Validation Using a Foam Core and Carbon Fibre Skins, South Dakota School of Mines and Technology 2007

[3] M. Biron, The Plastics Industry Economic Overview, Thermosets and Composites - Material Selection, Applications, Manufacturing and Cost Analysis, 2014, pp. 25-103.

[4] A. Elmarakbi, A. Wheatley, D. Warren and S. Das, Advanced Composite Materials for Automotive Applications, Chichester, UK: John Wiley & Sons, 2014.

EAN 2020
58th conference on experimental stress analysis

A Rheological Model of Fabric Elongation According to Lethersich

ŠIMIĆ PENAVA D.[1,a], PENAVA Ž.[2,b] and SKENDER, A.[1,c]

[1]Faculty of Civil Engineering, University of Zagreb, Kačićeva 26, 10000 Zagreb, Croatia

[2]Faculty of Textile Technology, University of Zagreb, Pr. b. Filipovića 28a, 10000 Zagreb, Croatia

[a]diana.simic.penava@grad.unizg.hr, [b]zpenava@ttf.unizg.hr, [c]ana.skender@grad.unizg.hr

Keywords: Rheology, Cotton woven fabric, Twill weave, Lethersich's model, Differential equations of elongation

Abstract. Under tensile loading, tearing of fabric or unwanted viscoelastic or plastic deformations occur. In order to avoid the appearance of plastic deformations in the fabric, it is necessary to know in advance which tensile force will cause such deformations. An appropriate rheological model that describes the behaviour of fabric, depending on its geometrical and structural parameters under tensile stress, curve (F-ε), should be set. For the purpose of testing the rheological properties of fabric, various types of raw cotton fabrics have been deposited in the twill weave with same warp density and different weft densities. The samples were made on the OMNIplus 800 tt air-jet loom machine Picanol. Based on the experimental results obtained to stretch the fabric samples, a rheological Lethersich's model was introduced. Appropriate differential equations for the tested sample patterns were made and resolved, resulting in the dependence between tensile force and relative strain. The rheological model according to Lethersich describes well the process of elongation of cotton woven fabric in twill weave under static test conditions.

Introduction

The use of textile materials in various industries is increasing, and knowledge of their physical and mechanical properties is very important. When measuring, the mechanical properties of fabric are affected by nonlinear viscoelasticity, friction between fibres, yarns, fabric density, geometrical changes during connection to external forces, and changes in temperature and humidity. Tensile forces cause tearing of the woven fabric or the appearance of viscoelastic or plastic deformations [1]. Such deformations are undesirable because undesirable effects, poor quality of the finished product, e.g. woven fabric, can only be observed in the final stages of processing. Therefore, the good quality of the finished product is achieved if during the use of woven fabric such conditions are ensured in which the deformation will be within the elastic area.

In order to describe the behaviour of the woven fabric subjected to load, it is generally accepted that the woven fabric is a continuous medium, so the continuum mechanics is discussed [2, 3]. A link between stress and deformation around the arbitrary point of the fabric should be established. These connections are called the mechanical characteristics of the material. The science involved in this is called rheology [4, 5]. In order to avoid the appearance of plastic deformation in the fabric, one should know in advance at which tensile force such deformations will occur. An appropriate rheological model that describes well the fabric behaviour should be fitted. Rheology is a science that deals with the physics of deformation, and its primary goal is to establish a connection between forces, that is, stresses and their derivations over time, with the resulting deformations due to the action of these forces and their derivations over time [6]. The basic rheological properties are elasticity, viscosity and plasticity. Due to the testing of the mechanical properties of fabrics, the setting of appropriate rheological models and a good analysis of the change in load in different fabric production processes, the allowable values of the stress force of the fabrics should move within the limits of the occurrence of elastic deformations, and on the basis of this force value optimization and complete automation of the corresponding processes is performed. The aim of this paper is to set rheological model that describes relatively well the behaviour of a cotton fabric in twill weave depending on its geometrical and structural parameters at tensile load, i.e. curve (F-ε).

Relationship between stress and strain

The functional relationship between force (stress) and strain cannot be determined theoretically, but only experimentally by testing samples of fabrics. The experiment determines the relationship between forces (stresses) and strain in the form of a characteristic diagram under certain conditions. Tensile tests are carried out on woven fabric, and the experimentally measured data of force F and elongation ε are presented in the form of a characteristic force-elongation diagram (F-ε) for the woven fabric, Fig. 1. The diagram shown in Fig. 1 has several characteristic points:

F_{max} - the maximum force acting on the fabric sample
F_{pr} - ultimate force immediately before the interruption of the fabric sample
F_{inf} - force at the inflection point
F_e - force at the limit of elasticity (Hook's law applies)
ε_{max} - maximum elongation of the fabric sample at maximum force
ε_{pr} - elongation of the fabric sample at break
ε_{inf} - elongation at the inflection point
ε_e - elastic elongation (deformation) that disappears after unloading
W_{max} - area (work) below the curve to maximum force
W_{pr} - area (work) below the curve to the breaking force
W_{inf} - area (work) below the curve to the force at the inflection point

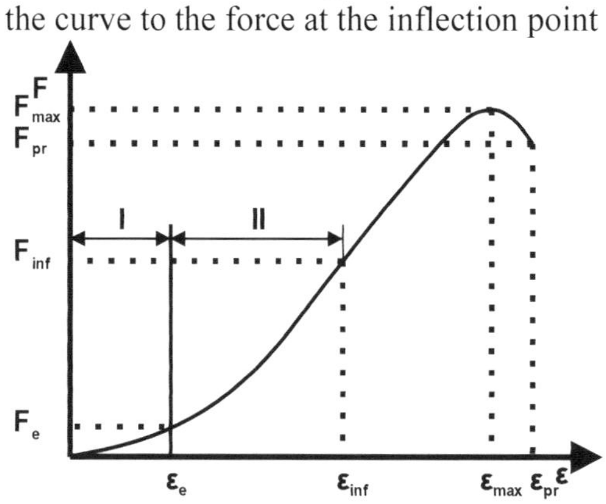

Fig. 1: The characteristic diagram of the force - elongation (F-ε) of the woven fabric

Area I is up to the elasticity limit (F_e, ε_e) and is called the elastic region that can be represented by the rheological Maxwell model. Area II is from the elasticity limit (F_e, ε_e) to the inflection point (F_{inf}, ε_{inf}) and is a viscoelastic region that can be described by the Burger's model formed by the serial connection of the Maxwell and Kelvin models.

The elasticity limits of woven fabrics are determined on the basis of the force-elongation diagram F(ε) and on the basis of F'(ε) and F"(ε). Fig. 2 represents the force-elongation function as well as its first derivation and second derivation. The maximum of the first derivation indicates the permissible load up to the limit of which the fabric shows elastic properties.

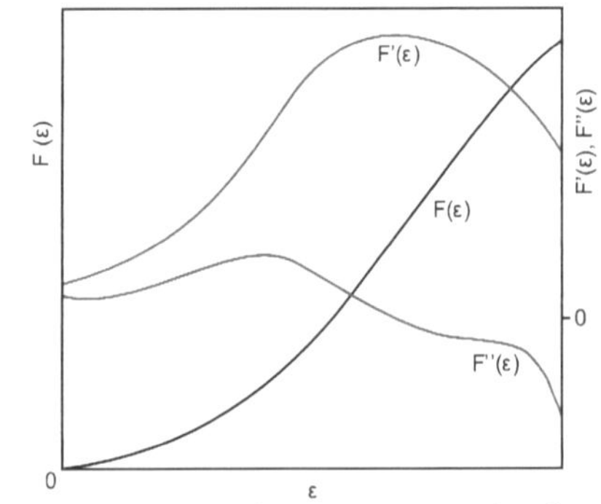

Fig. 2: Force-elongation diagram of the function F(ε), the first F'(ε) and the second F''(ε) derivation of the function

Up to this point, the woven fabric shows greater resistance to the tensile forces (F'(ε) function increases). When the function of the first derivation reaches its maximum, at that point the second derivation equals 0. Then a faster deformation of the woven fabric occurs until the destruction of the material (the function F'(ε) decreases).

Basic rheology settings

The rheological behaviour of a material is determined by some relation containing stresses, deformations, and their derivations over time:

$$f(\sigma_{ij}, \dot{\sigma}_{ij}, \varepsilon_{kl}, \dot{\varepsilon}_{kl}) = 0 \tag{1}$$

This relationship is called rheological equation of the state of the material, Eq. (1). Rheological equations contain some scalar values that characterize the rheological properties of materials and are called rheological constants, e.g. modulus of elasticity or coefficient of viscosity. Stresses and deformations are called rheological variables. Rheological equations contain variable values: stresses, strains, strain rate and rate of change of stress [7, 8]. The basic rheological models are Hooke's, Newton's, St. Venant's model. Real bodies have at the same time properties of elasticity, viscosity and plasticity in different form and relationship. By combining simple elements, a body model can be formed to describe the behaviour of real materials [9].

The Hooke's model, i.e. the linear elastic body "H", represents the relation between the stress deviator S_{ij} and the strain deviator e_{ij}, and this relationship is linear. An ideal elastic Hooke's material is shown by an elastic spring, Fig. 3a. For this model relation $S_{ij} = 2 \cdot G \cdot e_{ij}$ is valid.

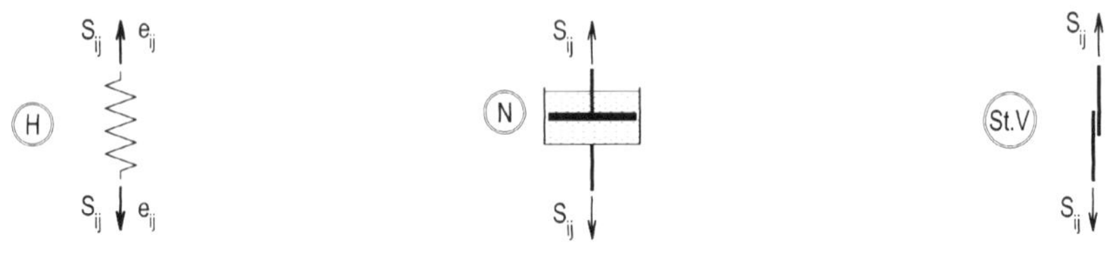

Fig. 3: Basic rheological models: a) Hooke's model, b) Newton's model, c) St. Venant's model

Newton's model „N", Fig. 3b, is a viscous silencer and represents a model of a viscous fluid. For a uniaxial stress state, it is defined as $\sigma = \eta\dot{\varepsilon}$ where η is Newton's coefficient of viscosity and $\dot{\varepsilon}$ is the strain rate. Saint Venant's model „St.V", Fig. 3c, describes the plastic deformation of the material. For a uniaxial stress state, it is defined as $\sigma = \sigma_T$.

The complex rheological models (materials)

Models of base materials H, N and St.V can be interconnected in different ways, thus obtaining models of some complex materials that have complex rheological properties.

The Maxwell's model is a serial connection between Hooke's and Newton's element, also called an elastoviscous fluid having the rheological formula M = H — N, Fig. 4a. In the case of uniaxial stress state, the following expression applies:

$$\dot{\varepsilon} = \frac{1}{E}\dot{\sigma} + \frac{1}{\eta}\sigma \tag{2}$$

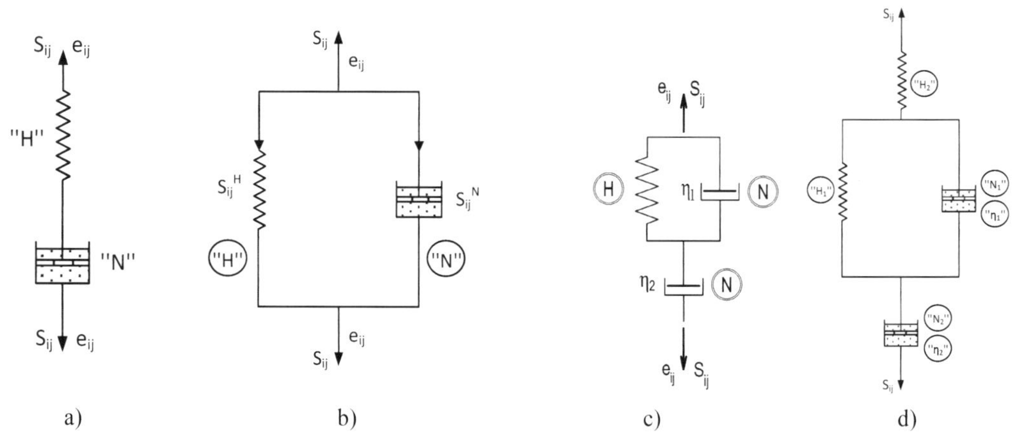

Fig. 4: Complex rheological models: a) Maxwell's model, b) Kelvin's model, c) Lethersich's model, d) Burger's model

The Kelvin / Voight model of material is formed by the parallel connecting of Hooke's and Newton's material, that is symbolically K = H | N, Fig. 4b. It is a viscoelastic material that slowly reaches the final deformation, retains it for a long time without further noticeable increase, and when unloaded, this deformation is slowly lost, and the body returns to its original form. For uniaxial stress states the following expression is valid:

$$\sigma = E\varepsilon + \eta\dot{\varepsilon} \tag{3}$$

Lethersich's model is suitable for testing the rheological properties of elongation cotton yarns, cotton woven fabrics of different types of weaves and densities, describing the behaviour of geotextiles, Fig. 4c. This material is represented by the model as a serial connection between Kelvin's and Newton's elements, so its rheological formula is L = K — N. The expression for the rate of deformation of the body according to the Lethersich model can also be written in the following form:

$$\dot{\varepsilon} = \frac{\sigma}{\eta_N} + \frac{\sigma}{\eta_K} - \frac{\sigma}{\eta_K} \cdot \exp\left(-\frac{E_K}{\eta_K}t\right) \cdot \left[\varepsilon_0 + \frac{1}{\eta_K}\int \sigma \cdot \exp\left(\frac{E_K}{\eta_K}t\right)dt\right] \tag{4}$$

where η_N is the viscosity coefficient of the Newton body, η_K is the viscosity coefficient of the Kelvin model, E_K is the coefficient of elasticity of the Kelvin model and ε_0 is the initial elongation. By derivation by time and arranging of the Eq. (4), the differential equation of the rheological model of cotton woven fabric is obtained in the form:

$$\eta_K\ddot{\varepsilon} + E_K \cdot \eta_K \cdot \dot{\varepsilon} = \dot{\sigma}(\eta_N + \eta_K) + \sigma \cdot E_K \tag{5}$$

Since in this case $\dot{\varepsilon} = const.$ and $\ddot{\varepsilon} = 0$, the Eq. (5) takes the following form:

$$\dot{\sigma}(\eta_N + \eta_K) + \sigma \cdot E_K = E_K \cdot \eta_K \cdot \dot{\varepsilon} \tag{6}$$

The solution of the differential Eq. (6) can be represented by the Eq. (7):

$$\sigma = -C \cdot \exp\left(-\frac{E_K}{\eta_N + \eta_K} \cdot t\right) + \eta_N \cdot \dot{\varepsilon} \tag{7}$$

The integration constant C is determined from the initial conditions, for $t = 0$, $\sigma = 0$. The dependence of stress on the time after the determination of the integration constant can be represented by the Eq. (8):

$$\sigma = \eta_N \cdot \dot{\varepsilon}\left[1 - \exp\left(\frac{l_0}{100 \cdot v \cdot \tau_r} \cdot \varepsilon\right)\right] \tag{8}$$

where $\tau_r = (\eta_N + \eta_K)/E_K$ is relaxation time, l_0 is the initial sample length, v is the test speed and ε elongation.

The Burger's model is applied to textile materials which have significant plastic deformation, Fig. 4d. This model is composed as a serial connection between Maxwell's and Kelvin's material, for its properties somewhat correspond to viscoelastic behaviour of materials. Its rheological formula is: B = M — K = (H — N) — (H | N). For the uniaxial state of stress, the following equation is obtained:

$$\left(\eta_1 \frac{\partial^2}{\partial t^2} + E_1 \frac{\partial}{\partial t}\right)\varepsilon = \left[\frac{\eta_1}{E_2}\frac{\partial^2}{\partial t^2} + \left(1 + \frac{\eta_1}{\eta_2} + \frac{E_1}{E_2}\right)\frac{\partial}{\partial t} + \frac{E_1}{\eta_2}\right]\sigma \tag{9}$$

Experimental part

In the experimental part of the paper, tests were carried out on elongation cotton woven fabric specimens in twill weave with the same warp density and different weft densities. Tensile properties of all specimens were tested according to standard ISO 13934-1:2008 using the strip method for measuring fabric strength and its elongation on a tensile strength tester Textechno Statimat M. The measurement results were collected and stored on the hard disk by a computer program of tensile strength tester. In this way, force-elongation (F-ε) curves were obtained. For the purposes of this testing standard specimens with dimensions 350 x 50 mm were cut, clamped in clamps of the tensile tester at a distance of l_0 = 200 mm with the pulling speed of 100 mm/min and subjected to uniaxial tensile load till rupture. It can be stated that the deformation rate is constant. Sample break time is 20 s. The specimens were cut in warp direction and weft direction. The direction of action of the tensile force during the test is always the same. Five tests were done for each mentioned direction of force action on the fabric specimen. Before testing all specimens were conditioned under the conditions of standard atmosphere (relative air humidity 65 ± 2%, at a temperature of 20 ± 2°C). In order to achieve balanced humidity, the yarn stood at standard conditions for 24 h before testing.

Table 1: Test results for basic fabric parameters

		Warp direction			Weft direction				
Fabric structure	Fabric tag	Yarn fibres	Yarn count (tex)	Density (cm^{-1})	Yarn fibres	Yarn count (tex)	Density (cm^{-1})	Weight (g/m^2)	Thickness fabric (mm)
	K12	Cotton	30.3	24.0	Cotton	30.3	12.0	114.89	0.362
Twill	K18	Cotton	30.3	24.1	Cotton	30.3	18.1	134.84	0.363
1/3	K24	Cotton	30.3	24.1	Cotton	30.3	24.2	156.91	0.364

Raw cotton fabrics were weaved for the purpose of testing the rheological properties of the woven fabrics. Fabrics specimens of the stated structural characteristics were made on an OMNIplus 800 tt air-jet loom Picanol. In Table 1 are the actual (measured) values of structural parameters of the raw woven fabrics. Yarn linear density was determined by the gravimetric method according to standard ISO 2060:1994. Number of threads per unit length was determined according to standard ISO 7211-2:1984. Standard ISO 5084:1996 describes a method for the determination of the thickness of fabric. The same yarn was used for the weft and the warp. Determination of the density of warp and weft threads was carried out using computer-controlled (stereo) microscope DinoLite.

Results of testing

Specimens are tested when the tensile force acts in the warp and weft direction. As a result, another letter is introduced into the fabric labels. The first letter in the fabric label indicates the type of weave, the second letter indicates the direction of force action, and the number indicates the weft density, e.g. KP18 - twill weave, the force in the weft direction, the weft density 18.1 cm^{-1}. Diagrams of the experimental mean values of tensile force F and its longitudinal deformation (elongation) ε on fabric samples, the first derivation of the function F'(ε) and the second derivation of the function F"(ε) when the force acts in the warp and weft direction are shown in Figures 5-7.

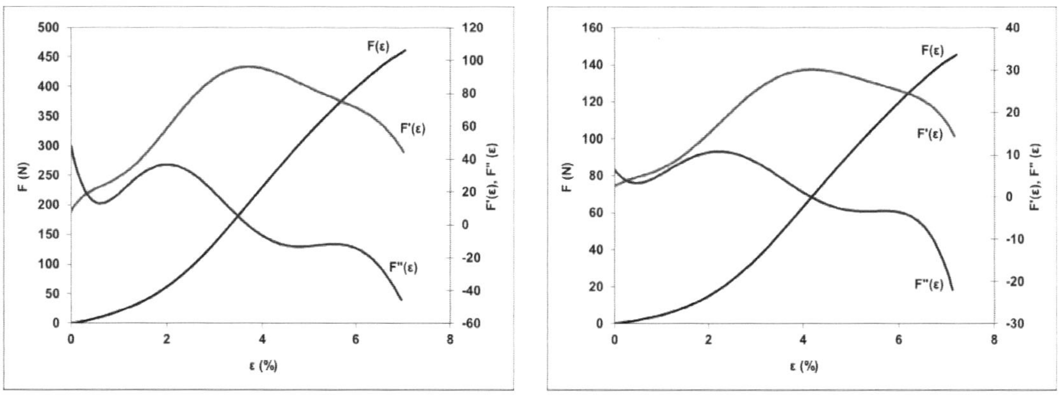

a) b)

Fig. 5: Diagram force-elongation (F-ε), the first derivation of a function F'(ε), the second derivation of a function F"(ε): a) specimen KO12, b) specimen KP12

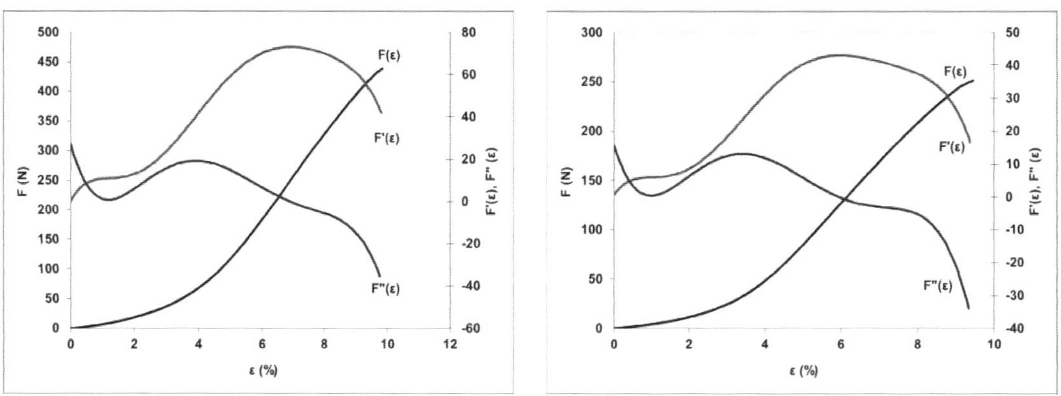

a) b)

Fig. 6: Diagram force-elongation (F-ε), the first derivation of a function F'(ε), the second derivation of a function F"(ε): a) specimen KO18, b) specimen KP18

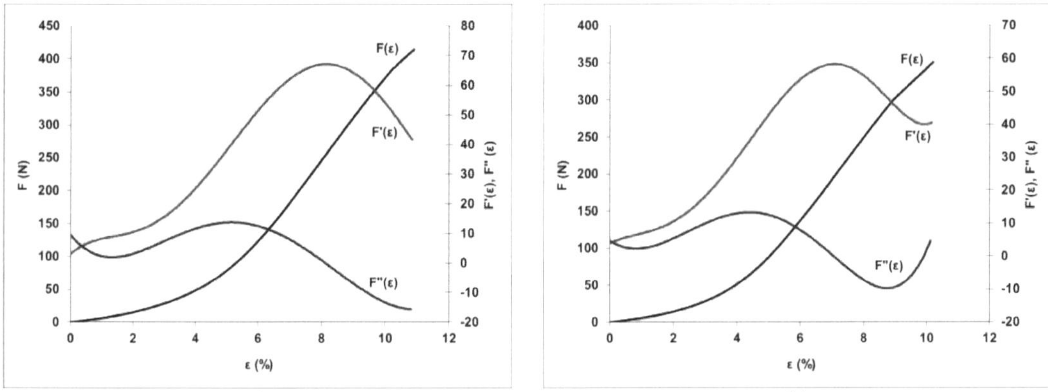

a) b)

Fig. 7: Diagram force-elongation (F-ε), the first derivation of a function F'(ε), the second derivation of a function F''(ε): a) specimen KO24, b) specimen KP24

The mean values of the test results of maximum force F_{max}, maximum elongation ε_{max}, work to maximum force W_{max}, ultimate force F_{pr}, elongation at break ε_{pr}, work up to break W_{pr}, force at the inflection point F_{inf}, elongation at the inflection point ε_{inf} and work up to the inflection are shown in Table 2.

Table 2: Mean values of F_{max}, ε_{max}, W_{max}, F_{pr}, ε_{pr}, W_{pr}, F_{inf}, ε_{inf}, W_{inf}

Fabric tag	F_{max} (N)	ε_{max} (%)	W_{max} (Ncm)	F_{pr} (N)	ε_{pr} (%)	W_{pr} (Ncm)	F_{inf} (N)	ε_{inf} (%)	W_{inf} (Ncm)
KO12	461.7	7.04	282.9	461.7	7.04	282.9	96.6	3.70	69.1
KP12	145.5	7.20	85.7	145.5	7.20	85.7	30.1	4.17	26.1
KO18	438.9	9.84	311.9	438.9	9.84	311.9	73.2	6.92	97.0
KP18	252.2	9.48	182.8	252.2	9.48	182.8	43.1	5.96	40.6
KO24	414.4	10.88	313.6	414.4	10.88	313.6	67.2	8.15	117.8
KP24	350.6	10.20	260.7	350.6	10.20	260.7	58.7	7.04	94.3

Setting rheological model of woven fabric to the inflection point

Predicting the behaviour of cotton woven fabrics to the inflection point when a tensile force is acting can be represented by rheological material according to the Lethersich model. When extending on a tensile tester, the deformation rate has a constant value. The initial length of the test tube is 0.2 m. The pulling speed of the clamps of the tensile tester is 100 mm/min. By introducing these data into equation (8) and assuming that the viscosity coefficients are equal ($\eta_N = \eta_K$), Eq. (10) is obtained:

$$\sigma = a \cdot [1 - e^{-b \cdot \varepsilon}] \tag{10}$$

By fitting the experimentally obtained data in the form of Eq. (10), the values of the coefficients "a" and "b" were determined [10, 11]. In Table 3 are the values of the coefficients "a" and "b" for models which are obtained on the basis of a rheological model based on experimental data for fabrics obtained by extending woven fabric specimens.

Table 3: The coefficients of the model to the inflection point

Fabric tag	a	b	Determination coefficient R^2	Root Mean Square Error RMSE (N)
KO12	-3842.4	-0.49443	0.99671	4.150
KP12	-1032.3	-0.48431	0.99480	1.813
KO18	-1649.1	-0.41312	0.99776	3.274
KP18	-772.1	-0.49015	0.99767	1.641
KO24	-1662.5	-0.35088	0.99833	2.890
KP24	-1559.3	-0.37338	0.99656	3.624

Diagrams of the mean values of the F-ε results obtained experimentally and according to the Lethersich's model are shown in Fig. 8-10.

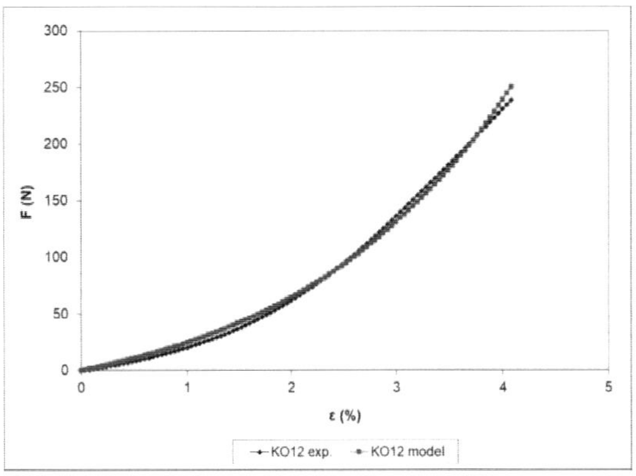

a)

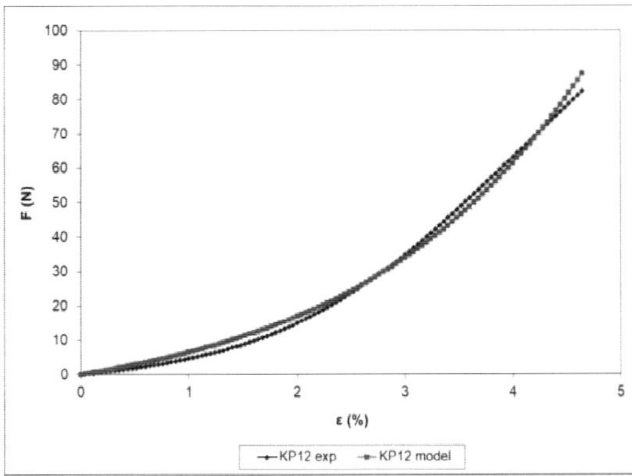

b)

Fig. 8: Diagram force-elongation (F-ε) obtained experimentally and by model: a) for specimen KO12, b) for specimen KP12

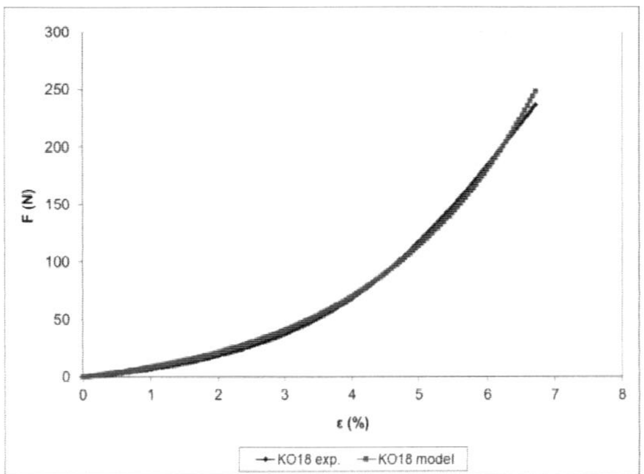

a)

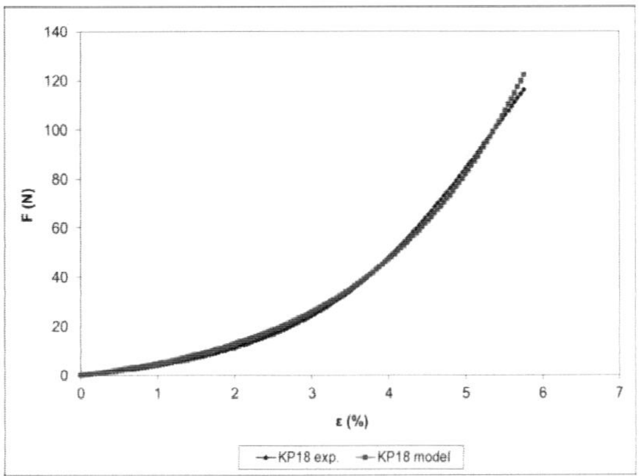

b)

Fig. 9: Diagram force-elongation (F-ε) obtained experimentally and by model: a) for specimen KO18, b) for specimen KP18

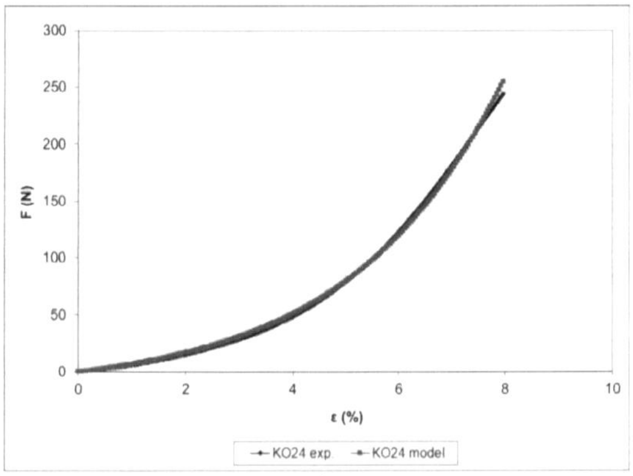

a)

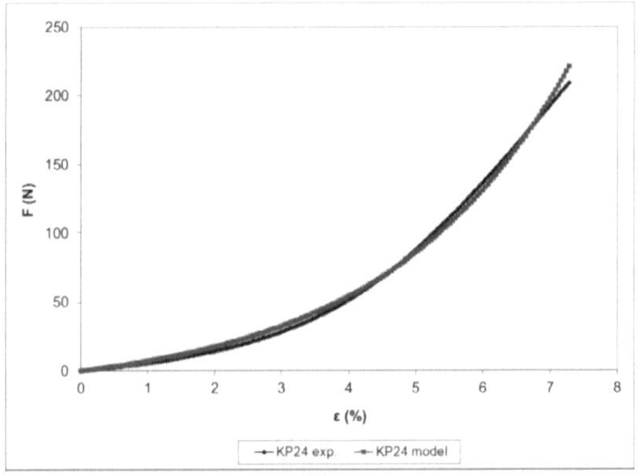

b)

Fig. 10: Diagram force-elongation (F-ε) obtained experimentally and by model: a) for specimen KO24, b) for specimen KP24

By reviewing the diagrams from Fig. 8 to Fig. 10, formed on the basis of experimental results (black) and the results obtained using the model (red) it is noted that they do not have a linear form, which confirms that the woven fabrics do not behave ideally elastic even at low stresses. Based on the experimental results and the shape of the force-elongation curves, it can be concluded that viscoelastic behaviour of the fabrics is observed in the analysed area. The elastic limit is essentially the limit to which elastic deformations in the material dominate. After the elastic limit, a higher rate of material deformation occurs, and the fabric structure is disrupted. Therefore, the fabric elasticity limit represents the boundary load at which the deformations occurring in the fabric will not significantly affect the stability of the structure and the durability of the fabric. By analysing the diagram, it can be concluded that the Lethersich's model, Fig. 4c, correctly describes the behaviour of the cotton woven fabric in twill weave in the zone of elastic deformation.

Conclusions

The mechanical properties of the woven fabrics are starting parameters for the design and adjusting the parameters in the textile industry. By defining the limit of elasticity of fabrics, one comes to the knowledge of the boundary intensities of the forces to which the fabrics can be subjected, without compromising their quality. The anisotropic woven fabric structure contributes to different, sometimes difficult to explain, fabric behaviours during elongation. The term "elastic limit" defines the limit to which elastic deformations in the fabric dominate. This is the limit when the material begins to deform faster under stress and should therefore be considered as the limit of permissible loads.

The mechanical properties of fabrics can be predicted using rheological models. Each type of deformation of real materials is described by a basic model, or the fabric behaviour is represented by complex models created by a combination of basic models.

Based on the experimental results obtained for elongation fabric specimens under standard test conditions, a combination of basic rheological models, set rheological models of fabric elongation, the corresponding differential equations for the tested fabric specimens were derived and solved, thus obtaining the dependence between the tensile force (stress) and the elongation. Rheological models describe the process of elongation cotton woven fabric relatively well under static test conditions.

Lethersich model was used to describe the behaviour of cotton woven fabrics in twill weave. This model can describe the behaviour of cotton fabrics subjected to the action of the tensile force in the zone dominated by elastic deformation. In the zone of elastic deformation, that is, up to the inflection point, the force-elongation relation for cotton woven fabrics is defined. The estimation error between the model and the experimentally obtained values is very low, so it can be concluded that the developed method can serve to predict the behaviour of cotton woven fabrics in the zone of elastic deformation. The obtained determination coefficients (R^2) show a high correlation ($R^2 > 0.99$) between relation of model and the experimentally obtained results, and they are slightly different (only in the third decimal place).

The force-elongation curve estimation error expressed through the root mean square error (RMSE) is the smallest for the model of woven fabric in twill weave (KP18) and is (164.1 cN) and the highest is for model of woven fabric in twill weave (KP24) whit highest density and is (362.4 cN). Mechanical properties of fabrics depend on their structural solutions, as well as on technological conditions of manufacture. The structural and physical-mechanical characteristics of the fabrics, the surface mass of the fabrics and the parameters of the weaving process play the most important role. Knowledge of the interconnectedness of the mechanical characteristics of fabrics provides the possibility of their proper design depending on the future purpose, which can contribute to the savings of raw materials and energy. Developed Lethersich's model can be used to predict the limit load of cotton woven fabric after which irreversible deformations of these materials occur.

References

[1] Z. Kostrenčić, Teorija elastičnosti, Školska knjiga, Zagreb, 1982.

[2] L. Šuklje, Rheological aspects of soil mechanics, John Wiley, London, 1969.

[3] O.U. Dmitriev, N.A. Osmin, Koefficient uskorenia ekspressnogo metoda ocenki obryvnosti prazi, Tehnologia tekstilnoj promyslennosti, 33 (1990) 2, 28-30.

[4] O.U. Dmitriev, Opredelenie parametrov cetyrehelementnoj modeli mehaniceskih svojstv v tekstilnyh materialah, Tehnologia tekstilnoj promyslennosti, 45 (2002) 2, 17-19.

[5] A. Koblyakov, Laboratory Practice in the Study of Textile Materials, Mir Publishers, Moscow, 1989.

[6] R. Postle, S. de Jong, The Rheology of Woven and Knitted Fabrics, Part 1: Fabric Geometry and Force Methods of Analysis Applied to Fabric Mechanics, Sen'i Kikai Gakkaishi (Journal of the Textile Machinery Society of Japan), 34 (1981) 5, P264-P273.

[7] R. Postle, S. de Jong, The Rheology of Woven and Knitted Fabrics, Part 2: Energy Minimization Techniques Applied to Elastic Mechanism of Fabric Deformation, Sen'i Kikai Gakkaishi (Journal of the Textile Machinery Society of Japan), 34 (1981) 7, P344-P353.

[8] R. Postle, S. de Jong, The Rheology of Woven and Knitted Fabrics, Part 3: Objective Specification of Fabric Mechanical Behaviour and Inelastic Mechanisms of Deformation and Recovery, Sen'i Kikai Gakkaishi (Journal of the Textile Machinery Society of Japan), 34 (1981) 10, P442-P460.

[9] V. Lasić, M. Srdjak, I. Stupica, Reološki modeli dvoosnovinih pletiva, Tekstil 39 (1990) 11, 656-660.

[10] X. Zhang, Bagging of Woven Fabrics: The Rheological Mechanism and Predictions, Journal of the Textile Institute, 92 (2001) 3, 235-255.

[11] M. Tarfaoui, J. Drean, S. Akesbi, Predicting the Stress-Strain Behavior of Woven Fabrics Using the Finite Element Method, Textile Research Journal, 71 (2001) 9, 790-795.

EAN 2020
58th conference on experimental stress analysis

Experimental Measurement of Twin-Disc Wear Resistance Test by Laser Profilometer

ŠMACH J.[1,a], MAREK M.[1,2,b], ŠOFER M.[1,c], FUSEK M.[1,d]

[1]Deparment of Applied Mechanics, Faculty of Mechanical Engineering, VŠB – Technical University of Ostrava, 17. listopadu 2172/15, 708 00 Ostrava, Czech Republic

[2]Deparment of Technical Studies, College of Polytechnics Jihlava, Tolstého 16, 586 01 Jihlava, Czech Republic

[a]jiri.smach@vsb.cz, [b]martin.marek@vsb.cz, [c]michal.sofer@vsb.cz, [d]martin.fusek@vsb.cz

Keywords: twin-disc test, wear resistance, profilometer measuring.

Abstract. In this work, two laser profilometers are used for measuring the specimen dimensions in real time during twin-disc test. The proposed technique allows to characterize the changes in the specimen dimensions during the test in itself and doesn´t require to interrupt the test at the same time. Using the real-time values of the specimen dimensional contours, it is possible to calculate instantaneous value of slip ratio or the contact path width.

Introduction

Thanks to progressive development of rails and raising travel speed of trains, the requirements for material quality are constantly growing. Nowadays, materials are tested using different dedicated experimental testing rigs, like SUROS, etc [1]. In our case, twin-disc test has been used as a substitute of the real in-service material loading process [2, 3]. In the past, several authors have to interrupt the test for measurement of the specimen dimensions, especially their diameter and the contact path width [4, 5]. This interrupting, however, is not favourable for automated testing. Also, if there is a requirement on the surface ratcheting measurement, a dedicated destructive method has to be used, which will not allow further use of the tested specimen in following tests [6]. One of modern contactless methods of research is chromatic confocal microscopy which is non-contact probe [7]. Or technology based on floating giant-magnetoresistance for search microcrack under surface [8]. Proposed paper deals with measurement procedure of specimen dimensional contour in real time, which is especially convenient for the calculation of the slip ratio or estimation of the contact area width.

Experimental procedure

The experimental rig was constructed by the INOVA company. The measurement of the specimen contours has been performed using Keyence LV-7200 laser profilers, which were mounted on self-developed holders, manufactured using 3D printing technology (Fig. 1). Controller for these profilers was used Keyence LJ-V7001P. Each laser profilers divided the signal into 800 points parallel to the axis of rotation and the sample frequency was set to 500 Hz. The material of specimens was Class C (AAR M107 Standart) [9], which were used for subsequent twin-disc tests, with additional hardening of the rail samples. Initial diameter of the wheel rim sample was 70 mm, while the diameter of rail was 220 mm. The twin-disc test was scheduled for 50 000 cycles. The initial Hertzian contact pressure was set to 1200 MPa with the initial slip ratio equal to 2 %. The appearance of the measured contour is shown on Figure 2.

Fig. 1: INOVA twin-disc experimental rig with Keyence LV-7200 laser profiler

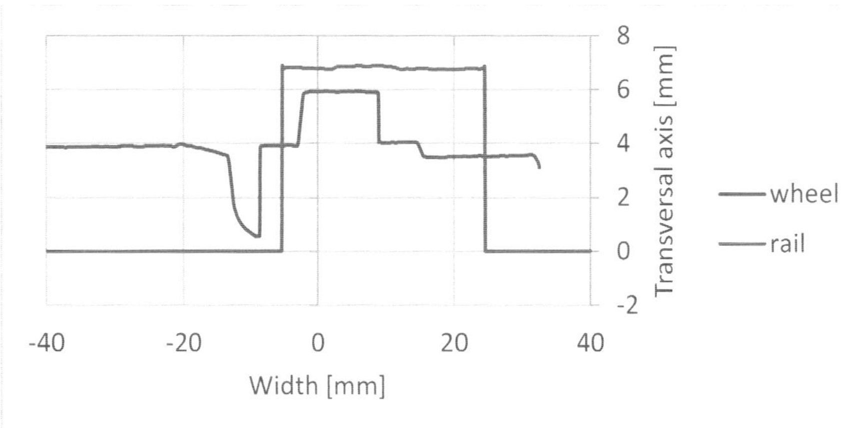

Fig. 2: Measured contour line by Keyence LV-7200 laser profiler

Result

Figure 3 respectively Figure 4 show change of contact radius of wheel and rail respectively as a function of number of cycles. Figure 5 shows the variation of slip ratio over the number of cycles. First 10,000 cycles approximately carried out a stabilization of test in itself, the results during this period are only tentative. From the next part of the evaluation it is possible to determine the trend of the course.

At the end of the test, the diameter of wheel sample was smaller about 0.5mm, while the wheel specimen diameter decreases at about 0.2mm. The final change of diameter was verified using callipers. The actual slip ratio was calculated on the proportion of peripheral speeds based on the measured diameter and the stable specimen´s revolutions per minute refer to formula (1).

$$s = \frac{v_1 - v_2}{\frac{v_1 + v_2}{2}} = \frac{\frac{d_1}{2} \cdot \omega_1 - \frac{d_2}{2} \cdot \omega_2}{\frac{\frac{d_1}{2} \cdot \omega_1 + \frac{d_2}{2} \cdot \omega_2}{2}} = \frac{2(d_1 \cdot \omega_1 - d_2 \cdot \omega_2)}{d_1 \cdot \omega_1 + d_2 \cdot \omega_2} \qquad (1)$$

In this particular experiment, the slip ratio decreased from 2 % to approximately 1.2 % after only 50000 cycles. Presented approach is relatively simple for measuring specimen's diameter in real time with additional calculation of the slip ratio.

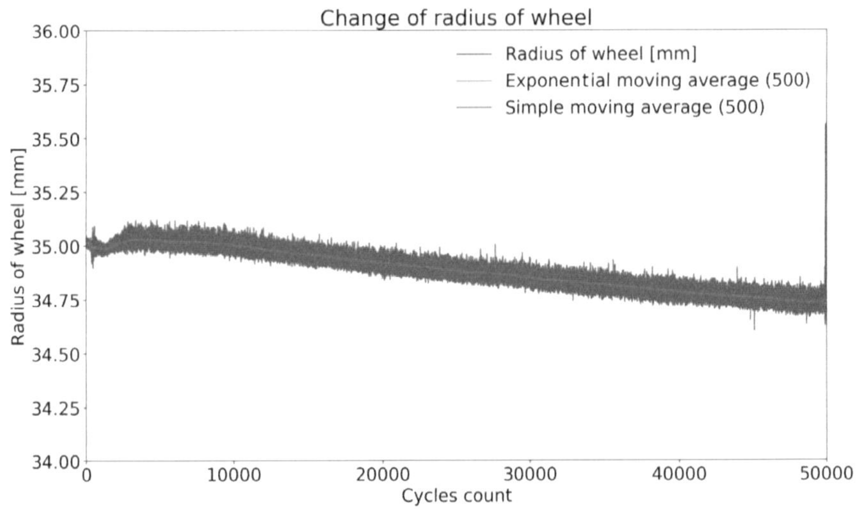

Fig. 3: Change of radius of wheel at number of cycles

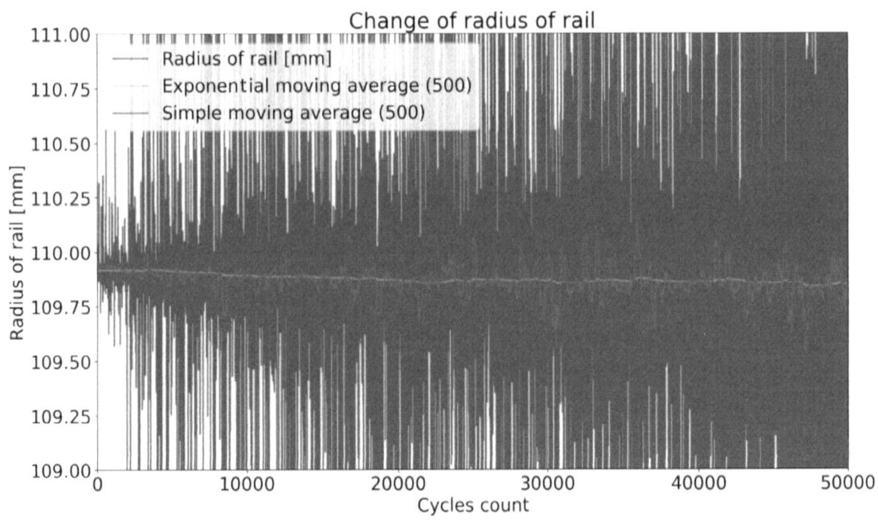

Fig. 4: Change of radius of rail at number of cycles

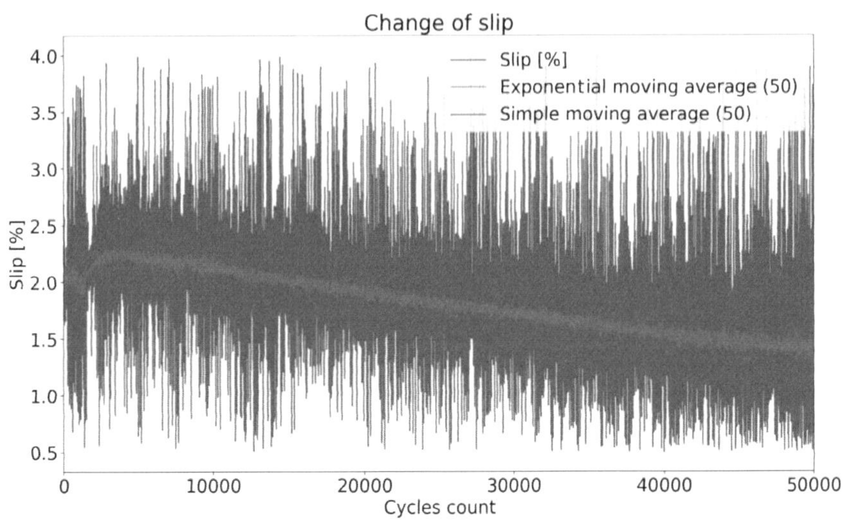

Fig. 5: Change of slip ratio at number of cycles

The oscillation of the raw data was caused by the oscillation of the piston cylinder, which was controlled by PLC. Small changes of pressure in hydraulic circuit in short time caused noticeable vibrations of the lodge, which is a reason why exponential and simple moving average for the results smoothing has been used and refer to formula (2) respectively formula (3).

$$EMA_i = x_i \cdot k + EMA_{i-1} \cdot (1 - k) \tag{2}$$

$$SMA_i = \frac{1}{n}\sum_{i=0}^{n-1} x_{n-i} \tag{3}$$

Conclusions

Presented approach has proven to be a suitable method for estimating the specimen´s diameter, which are subjected to twin-disc test. Using the already known value of the diameter and together with the known revolutions per minute, it is possible to estimate the slip ratio in real time and controlling this value at the same time as well. This method can be useful in comparison of the new materials for railways wheelset under the same initial conditions.

The verification of the method requires many additional tests to confirm its suitability, since the estimation of the slip ratio is mainly fraught with a random effect in terms of the measurement error.

Acknowledgement

This work was supported by the Specific Research Project (SP2020/23). The support of BONATRANS Group a.s. company is also acknowledgements.

References

[1] W.J.,Wang, S.R. Lewis, R.Lewis, A.Beagles, C.G.He, Q.Y.Liu, The role of slip ratio in rolling contact fatigue of rail materials under wet conditions, Wear, p. 376-377, 2016. 10.1016/j.wear.2016.12.049.

[2] J. Šmach, M. Šofer, L. Čepová, R. Halama, Problematics of twin-disc wear resistance tests for selected wheel rim materials. In: Experimental Stress Analysis 2017: 55th international scientific conference. May 30th - June 1st, 2017, Nový Smokovec. Košice: Technical University of Košice, 2017. p. 50-56. ISBN 978-80-553-3167-6.

[3] A. Ramalho, M. Esteves, P. Marta, Friction and wear behavior of rolling-sliding steel contacts, Wear, 302, p. 1468-1480, 2013. https://doi.org/10.1016/j.wear.2012.12.008

[4] M. Sofer, R. Fajkoš, R. Halama, Influence of induction hardening on wear resistance in case of rolling contact, Strojnicky Casopis, 66, p. 17-26, 2016. https://doi.org/10.1515/scjme-2016-0007

[5] J. Šmach, R. Halama, M. Marek. Advanced testing in rolling contact fatigue for wheel material investigation. In: Applied Mechanics 2018 - Conference Proceedings. Plzeň: Západočeská univerzita, 2018. p. 139-141. ISBN 978-80-261-0766-8.

[6] R. Halama, M. Šofer, J. Rojíček, F. Fojtík, K. Kolařík, A method for predicting ratcheting and wear in rolling contact fatigue, taking technological residual stresses into consideration, Tehnicki Vjesnik, 24, p. 7-14, 2017. https://doi.org/10.17559/TV-20140224213525

[7] P. Andersson, B. Hemming, Determination of wear volumes by chromatic confocal measurements during twin-disc tests with cast iron and steel, Wear, 338-339, p. 95-104, 2015. https://doi.org/10.1016/j.wear.2015.05.011

[8] K. Nagato, K. Shintani, T. Hamaguchi, M. Nakao, Real-time detection of microcracks with floating giant-magnetoresistance sensor in twin-disk sliding tests, CIRP Annals, 66(1), p. 539-542, 2017. https://doi.org/10.1016/j.cirp.2017.04.078

[9] AAR M-107/M-208 Standard: AAR Manual of Standards and Recommended Practices Wheels and Axles Wheels, Carbon Steel Specification M-107/M-208, Pueblo, CO, USA, 2011.

EAN 2020
58th conference on experimental stress analysis

Damage Analysis of Composite CFRP Tubes Using Acoustic Emission Monitoring

ŠOFER M.[1,a], CIENCIALA J.[1,b], FUSEK M.[1,c], PAVLÍČEK P.[1,d], MORAVEC R.[2,e]

[1]Deparment of Applied Mechanics, Faculty of Mechanical Engineering, VŠB – Technical University of Ostrava, 17. listopadu 2172/15, 708 00 Ostrava, Czech Republic

[2] Havel Composites CZ s.r.o., Svésedlice 67, 783 54 Přáslavice, Czech Republic

[a]michal.sofer@vsb.cz, [b]jakub.cienciala@vsb.cz, [c]martin.fusek@vsb.cz, [d]pavel.pavlicek@vsb.cz, [e]moravecr@havel-composites.com

Keywords: Acoustic emission; CFRP composites; Unsupervised learning; Paddle

Abstract. The utilization of acoustic emission monitoring across different sectors with various applications is evident especially in the field of composites. In this work, three carbon fiber reinforced polymer composite tube types, intended for paddle production, are subjected to mechanical bending test up to their ultimate limit. With the use of acoustic emission monitoring and subsequent postprocessing of the data by unsupervised pattern recognition approach, it was possible to characterize the damage evolution for given composite tube type.

Introduction

Over the past decades, carbon fiber reinforced polymer (CFRP) composites have shown a constant increase in a variety of applications such as car or aircraft components, sports and medical equipment [1] and recently also in additive manufacturing [2-4]. Their main benefit lies primarily in relatively large strength/weight ratio or the ability to customize the material properties for dedicated purposes by changing the stacking sequence and related fiber orientation. One of the specific sectors, where the CFRP composites are widely applied, is canoeing with related manufacturing of the paddles. This sport equipment is during its lifecycle exposed to cycling bending, which can, besides the material aging process, to a certain extend affect the paddle stiffness. The basic study of the mechanical response of given structure, in our case the CFRP composite tube forming the paddle, can be realized using universal testing machine with dedicated equipment. To gain more detailed insight into damage monitoring process on micro-scale, we have to incorporate any additional method such as acoustic emission (AE) testing [5]. AE monitoring during mechanical loading of given structure is an effective tool for studying the evolution of damage processes in the material [6]. Generally, the method is even capable of detecting the onset of plastic deformation [7], which has the character of white noise with low energy [8]. The advantages of the AE method are apparent in the case of analysis of more complex materials such as composites. The main objective of the presented study is to experimentally simulate the operation of the three types of CFRP paddle tubes to its ultimate limit state including monitoring the damage evolution process using AE method. The measured AE data are subsequently analyzed with use of unsupervised pattern recognition approach. Based on the reported signal properties of different failure modes in composites from various studies, the authors we were able to classify the measured AE signals accordingly.

Test sample characterization

The experiments were carried out on three types of CFRP tubes with different number of layers, their orientation and woven fiber density of used material (see Figure 1), where each type of CFRP tube has been represented by three test samples. Samples labeled as „A“ were manufactured using four layers of unidirectional carbon woven fabric with density 200 g/m^2 and one layer of aramid/carbon woven fabric [0°-90°] with density 175 g/m^2 and average wall thickness of 1,45 mm. The production of samples labeled as „B“ included the use of two layers of unidirectional carbon woven fabric with density 300 g/m^2 and one layer of carbon woven fabric [0°-90°] with density 280 g/m^2 with average wall thickness of 0,9 mm, while samples labeled as „C“ were manufactured using solely four layers of unidirectional carbon woven fabric with density 300 g/m^2 with average wall thickness of 1,42 mm.

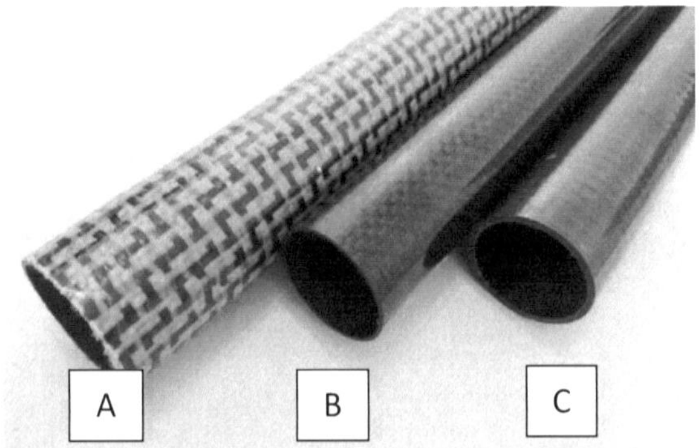

Fig. 1: CFRP composite tubes under test

Experimental procedure

Three point bending test procedure has been selected in order to simulate as much as possible the real nature of the loading process during the use of the paddle. The realization of the experiments was carried out on universal testing machine Testometric M500-50CT with dedicated weldment, which enables with its moving parts to modify its geometry for a wide variety of such experiments. The distance between the supports was equal to 1040 millimeters, while the force has been acting 440 millimeters from AE sensor #1 (see Fig. 2).

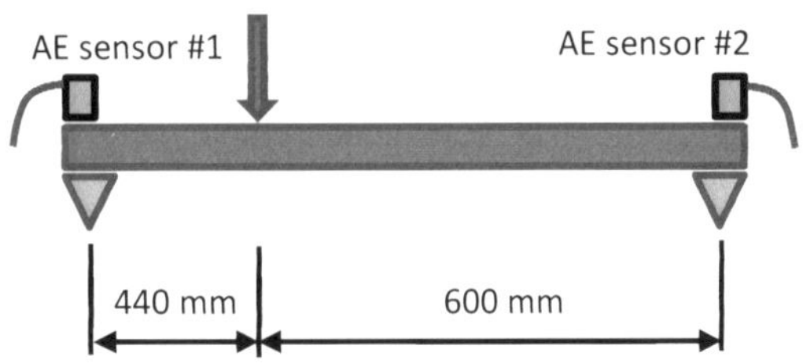

Fig. 2: In-Situ photograph of the test rig including specimen equipped with AE sensors (left); schematic representation of the three point bending test (right)

The tested tube with attached AE sensors was placed in a plastic pipe to prevent damage to the AE sensors and other equipment due to sudden integrity violation. The supports were covered by thin felt to allow free movement of the tube during its bending. The test has been deformation-controlled with the the upper anvil speed equal to 10 mm/min.

Acoustic emission monitoring

The acoustic emission activity was monitored using Vallen AMSY-6 AE system with two utilized measuring channels (ASIP-2A dual channel signal processor card), equipped with AEP5H 34 dB preamplifiers and broad-band Vallen VS-900 AE sensors, which have been attached onto the tube with use of oil-based plasticine. The sampling frequency of the AE data was set to 10 MHz while the transient data (wave transients) were sampled with 20 MHz and frequency range between 50 and 1100 kHz. Since the AE sources, which will occur in the close vicinity of the force acting point, will be located at a relatively greater distance and, at the same time, the composite exhibits higher rate of signal attenuation, it was necessary to set a lower value of threshold, in our case equal to 32 dB. The measured data were then filtered using the linear localization condition, for which has been known the approximate AE source location, i.e. force acting point at the distance of 440 mm from AE sensor #1 (see Fig. 2).

The filtered data were followingly analyzed with Vallen VisualClass software package, which uses the pattern recognition method [9] to associate similar waveform types into separate groups. Due to the nature of the task, an unsupervised learning approach [10] was chosen. The procedure starts with loading the selected database into Vallen VisualClass software, where in the beginning, the user has to specify the number of time windows including their span and the starting point in the time domain related to the origin defined by detection threshold (see Figure 3 - left).

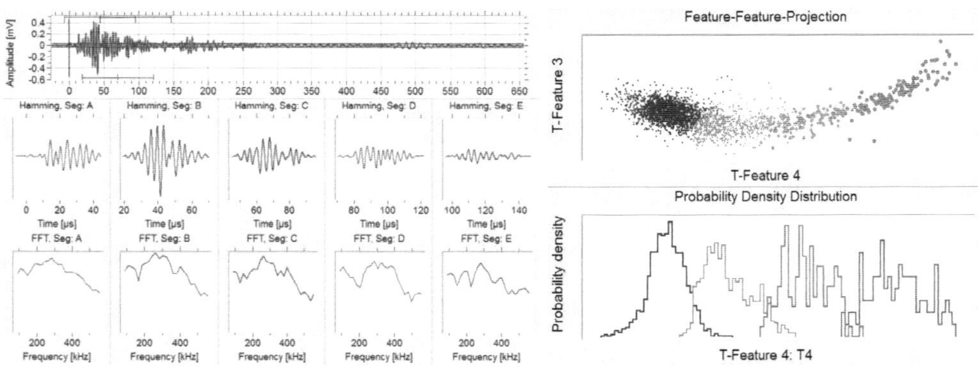

Fig. 3: Setting up the Hamming windowed time segments with corresponding results in frequency domain (left); Transformed features projection – results for four selected clusters (right)

The current analysis incorporates following settings: Number of time segments: 5; Size of single time segment in terms of points: 1024; Rel. trigger offset: -128 points; min/max frequency limit: 50/700 kHz. The software then performs the assembly of multidimensional feature vector, which size depends on the chosen number of time segments including their size. After the calculation process will be obtained the basic feature space, which is then linearly transformed for maximizing inter-class distance and minimizing the intra-class extension, at the same time (see Figure 3 - right). The results are then transferred into VisualAE software for further postprocessing. Four clusters were chosen for subsequent analysis, since the additional increase of the number of clusters did not lead to better differentiation of individual transients. One sample from each series has been subjected to the attenuation measurement of the AE signal in order to properly display the real AE signal amplitude in subsequent data analysis. The results of the attenuation measurement are shown in table 1.

Table 1: Attenuation measurement

Sample series	Near field attenuation [dB/m]	Far field attenuation [dB/m]
A	90	33,3
B	66,6	33,2
C	222,2	36,2

Results and discussion

Figure 4 shows relation between force and displacement of the anvil for individual A/B/C production run samples. The highest stiffness is being reached by C series samples, followed by A and B series samples. All three manufacturing modifications show within their group very similar trend in terms of the force-displacement course except A series sample A2, which exhibits marginally lower stiffness.

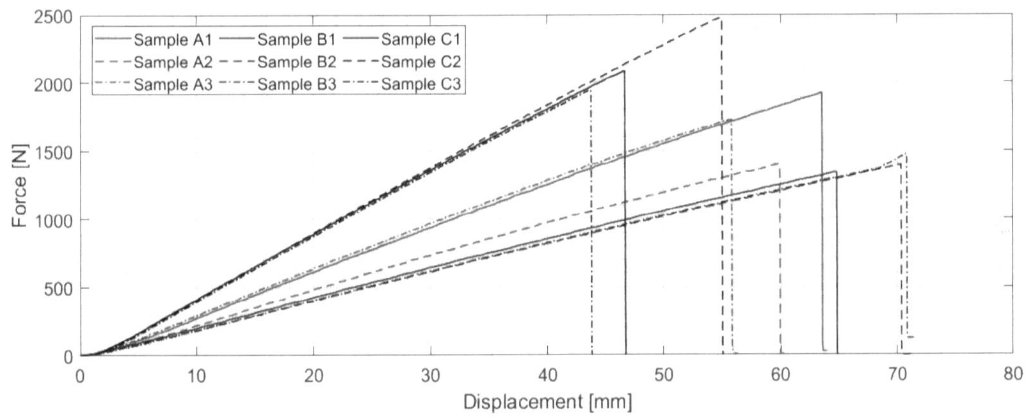

Fig. 4: Force as the function of displacement for individual A/B/C production run samples

The above conclusion however is not valid for the maximum force across individual series, where can be registered differences from 10 to 28 percent related to the maximal achieved force in each production run. The reason for the results variation can be found in the production itself. A somewhat similar trend can be registered in case of number of located AE events across the 0÷F_{max} range for individual samples (see Figure 5), where a relatively large variation is registered.

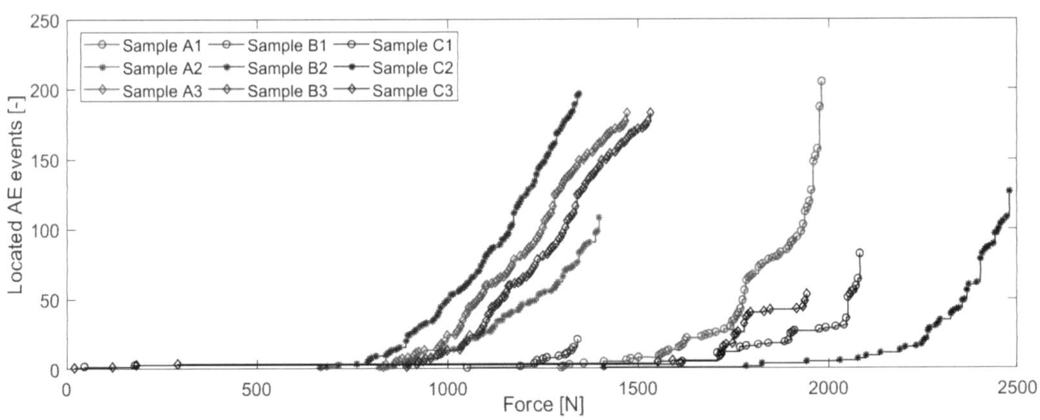

Fig. 5: Located event as the function of force for individual A/B/C production run samples

Following figure display the energy-force relation for individual samples from A/B/C series. The maximum value of released AE energy is for all samples between 10^8 and 10^9 aJ. The difference, however, lies in the character how the energy is being released during the loading process. A and B series specimen exhibit almost gradual AE energy release, with the difference in the final loading stage. While A series specimen tend to gradually continue with the cumulation of the AE events and gradual release of the AE energy, B series samples tend to suddenly lose integrity without any warning phase. A completely different behavior can be found in case of C series samples, which have considerably larger energy per event ratio with a lack of any warning phase before the integrity lose.

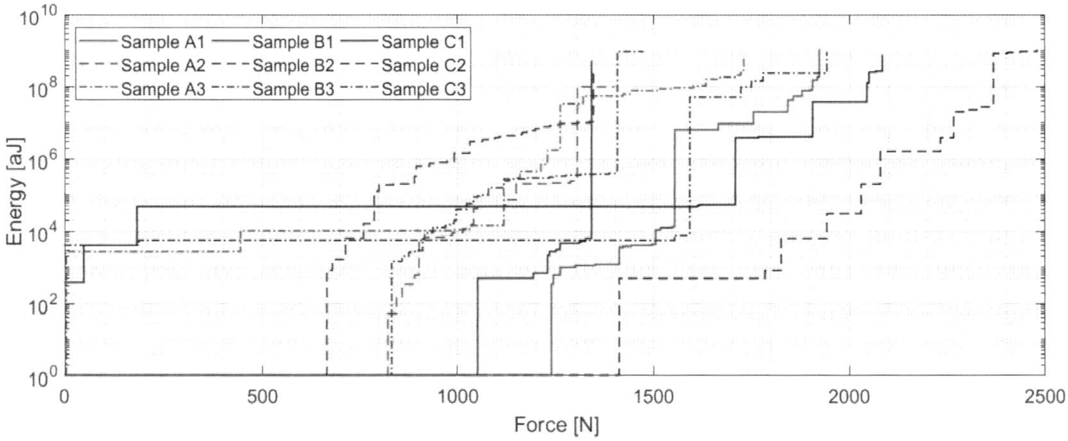

Fig. 6: Energy as the function of force for individual A/B/C production run samples

The issue of identification of failure mechanisms in composites is relatively complicated, as evidenced by a number of publications, which surprisingly do not give a clear answer in the area of amplitude and frequency spectra [11-15]. The performed unsupervised pattern recognition analysis revealed four groups of signals with characteristic properties. The signals in the first class are characterized by high amplitude in most cases exceeding 95 dB_{AE} with energy value usually above $1 \cdot 10^6$ aJ and frequencies in the span from 50 kHz to 150 kHz (see Fig. 7), whereas the class also contains signals with frequency content above 300 kHz. The second class is characterized by amplitudes usually below 80 dB_{AE} with AE energy in the order of thousands to tens of thousands of aJ and the frequency in the 50÷450 kHz range (see Fig. 8). The third class is represented by signals with amplitude in the 70÷90 dB_{AE} range with AE energy in the order of ten thousand aJ and the frequency in the 50÷300 kHz range (see Fig. 9). The fourth class is characterized by signals in the 90÷110 dB_{AE} amplitude range, AE energy in $1 \cdot 10^5 \div 1 \cdot 10^6$ aJ and with frequency in the 50÷200 kHz range (see Fig. 10).

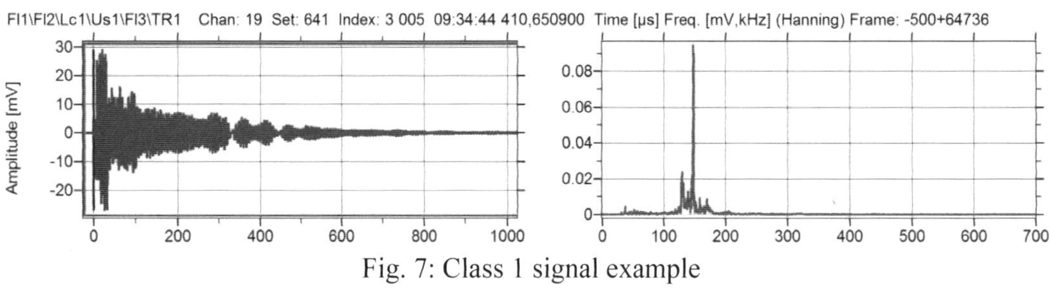

Fig. 7: Class 1 signal example

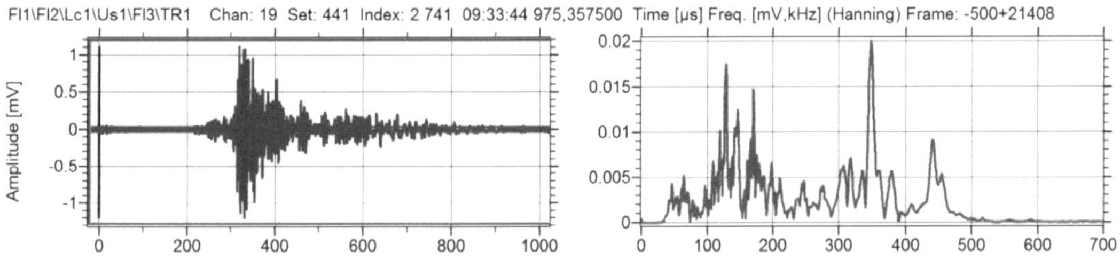

Fig. 8: Class 2 signal example

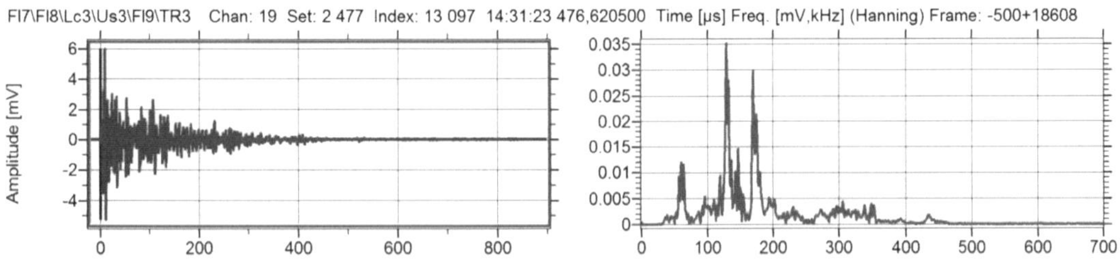

Fig. 9: Class 3 signal example

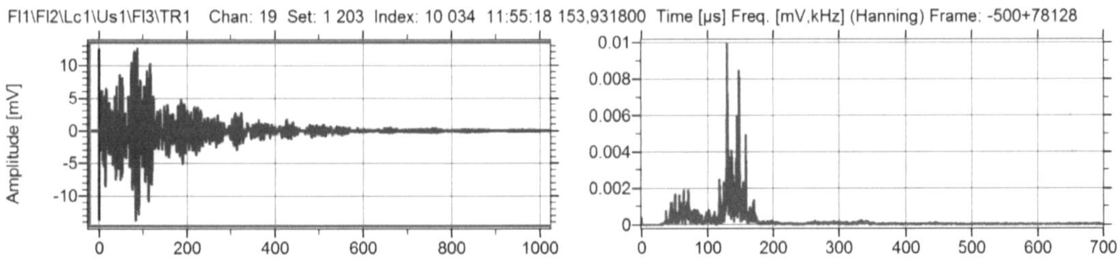

Fig. 10: Class 4 signal example

Figures 11-13 show the cluster classification of localized AE events as the function of time and distance corrected amplitude for individual A/B/C production run samples. Almost in all cases starts the damage with located events assigned to cluster 2, i.e. characterized by lower amplitudes and 50÷150 kHz frequency range, which can be most likely assigned to the matrix cracking, followed by signals with high frequency content, usually above 300 kHz, coming from fiber breakage. A similar categorization can be found in case of AE signals belonging to class 3, whose are characterized by higher amplitudes and AE energies, but slightly lower frequency ranges.

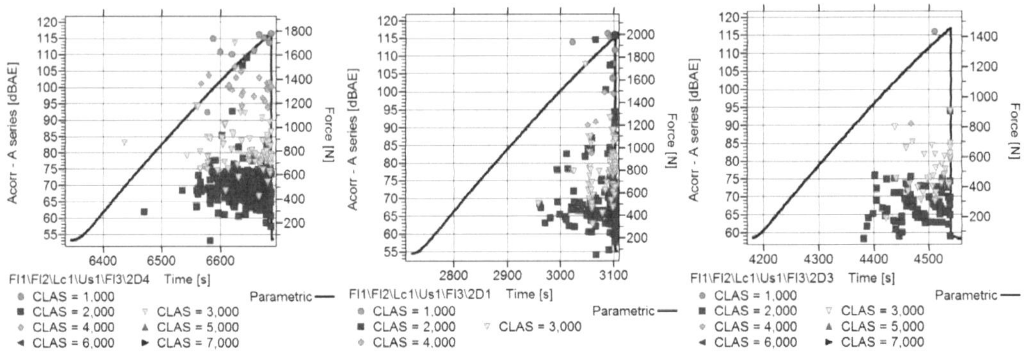

Fig. 11: A series sample results

Classes 1 and 4 on the other hand share the highest energy as well as amplitude values, while having a lower frequency range, however even in this case can be found a small group of signals with a frequency content reaching up to 450 kHz. The reason for occurrence of both low and high frequency signals in the mentioned classes lies in the overlapping of individual clusters between each other (see Fig. 3). Based on the given parameters and, at the same time, using available literature can we conclude, that classes 1 and 4 most likely represent the interface decohesion and in some cases also the fiber breakage (signals with higher frequency content). Figure 14 additionally shows the typical frequency spectrum for fiber breakage in case of C1 sample. Generally speaking, figures 11-13 show different character of the damage evolution process particularly in the percentage of the presence of class 1 and class 4 signals, which should represent the interface decohesion and higher volume fiber breakage events due to different number of unidirectional woven fabric layers with different density at the same time. Especially in case of C series samples can be the fiber breakage clearly identified by high frequency class 1 signals over 110 dB_{AE}, which start to occur near the ultimate limit of the loading force.

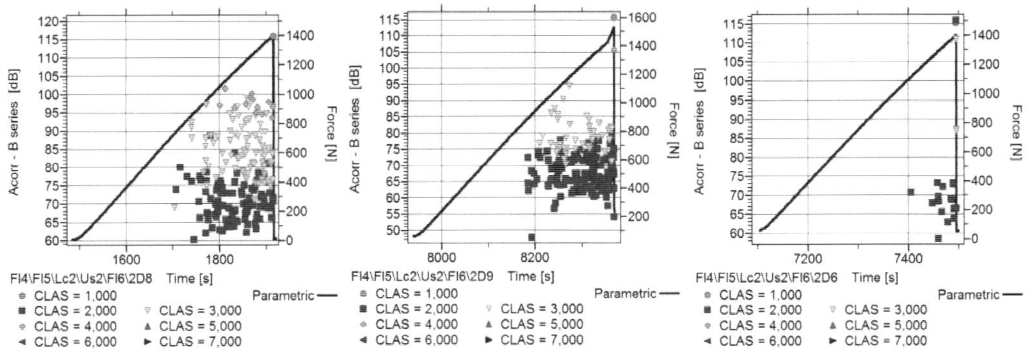

Fig. 12: B series sample results

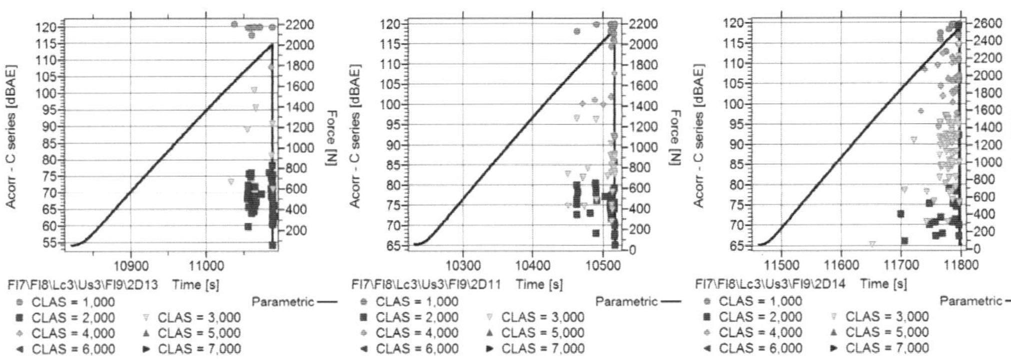

Fig. 13: C series sample results

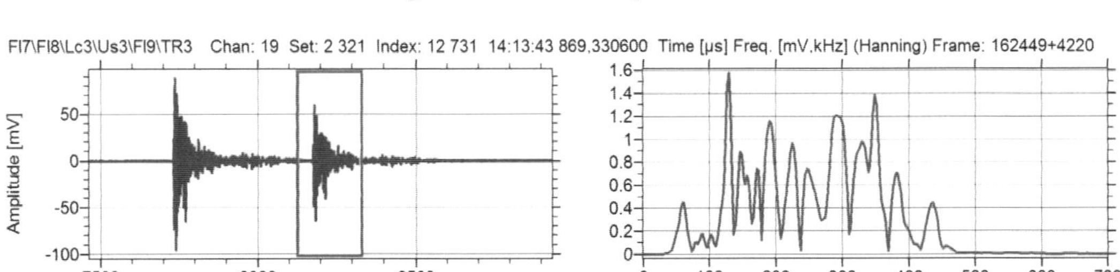

Fig. 14: Fiber breakage at $0,98*F_{max}$ force value with amplitude exceeding 120 dB_{AE}, Class 1 (C1 sample)

On the other hand, A series samples, especially sample A1, report a relatively large number of high frequency class 2 signals, which also belong to fiber breakage. However, as mentioned above, such signals will begin do occur after their low frequency counterparts, whose report the matrix cracking process. In conclusion we can say that the fiber breakage failure mechanism occurs in a wide spectrum of amplitudes, while the matrix cracking process occupies amplitudes up to 60÷70 dB_{AE}. AE events with higher amplitude and AE energy mainly belong to interface decohesion and/or higher volume of fiber breakages.

Conclusion

AE monitoring technique together with an unsupervised pattern recognition technique has been used to analyze the damage evolution of the three types of CFRP composite samples in order to identify the different damage mechanisms of considered materials during three point bending test. The analysis resulted in cluster classification, which to a certain extend represented matrix cracking, fiber breakage or interface decohesion. Due to a partial overlap of individual clusters between each other can be the mentioned phenomena registered in multiple clusters, which is particularly evident in the case of fiber breakage or matrix cracking. In spite of the work done, there are a number of questions, which have to be answered in a future research activities, such as more specific properties of the AE signal belonging to the fiber breakage or interface decohesion.

Acknowledgment

This work was supported by Specific Research (SP2020/23) and by project Innovative and additive manufacturing technology—new technological solutions for 3D printing of metals and composite materials - CZ.02.1.01/0.0/0.0/17_049/0008407. The authors would also like to thank the Havel Composites Ltd. company for providing the test samples and consulting during the elaboration of the paper.

References

[1] R. Unnthorsson, T. P. Runarsson, M.T. Jonsson, Acoustic emission based fatigue failure criterion for CFRP, Int. J. of Fat. 30 (2008) 11-20.

[2] J. Majko, M. Handrik, M. Vaško, M. Sága, Influence of Fiber Deposition and Orientation on Stress Distribution in Specimens Produced Using 3D Printing, Strojnícky časopis - Journal of Mechanical Engineering, 69 (2019), 81-88.

[3] F. Dorčiak, M. Vaško, M. Handrik, F. Bárnik, J. Majko, Tensile test for specimen with different size and shape of inner structures created by 3D printing, Transportation Research Procedia 40 (2019), 671-677.

[4] J. Majko, M. Sága, M. Vaško, M. Handrik, F. Barnik, F. Dorčiak, FEM analysis of long-fibre composite structures created by 3D printing, Transportation Research Procedia 40 (2019), 792-799.

[5] C. U. Grosse, Acoustic Emission Testing, Berlin Heidelberg: Springer-Verlag, 2018.

[6] M. Šofer, P. Kučera, E. Mazancová, L. Krejčí, Acoustic emission and fractographic analysis of seamless steel pressure cylinders with artificial flaws under hydrostatic burst testing, J. of Non. Ev. 38 (2019) 84-96.

[7] R. Čech, P. Horyl, P. Maršálek, Modelling of Two-Seat Connection to the Frame of Rail Wagon in Terms of Resistance at Impact Test, Strojnícky časopis - Journal of Mechanical Engineering, 66 (2016), 101-106.

[8] R. Kocich, M. Cagala, J. Crha, P. Kozelsky, Character of acoustic emission signal generated during plastic deformation, 30th European Conference on Acoustic Emission Testing & 7th International Conference on Acoustic Emission, University of Granada, 12-15 September 2012.

[9] J. Tang, S. Soua, C. Mares, T.-H. Gan, Pattern Recognition Approach to acoustic emission data originating from fatigue of wind turbine blades, Sensors 17 (2017) 2507-2519.

[10] J.H. Vallen, G. Ruske, Acoustic emission source discrimination by analyzing short-term frequency spectra.

[11] N. Godin, S. Huguet, R. Gaertner, Integration of the Kohonen's self-organising map and k-means algorithm for the segmentation of the AE data collected during tensile tests on cross-ply composites. NDT & E International 38 (2005) 299–309.

[12] J.R. Wadim, Acoustic Emission Applications; D. Endevco: San Juan Capistrano, CA, USA, 1978.

[13] K. Komai, K. Minoshima, T. Shibutani, Investigations of the fracture mechanism of carbon/epoxy composites by AE Signal Analyses. JSME international journal 34 (1991) 381–388.

[14] H.-J. Jong, Transverse cracking in a cross-ply composite laminate - detection in acoustic emission and source characterization. Journal of Composite Materials 40 (2006) 37–69.

[15] M. Gresil, M.N. Saleh, C. Soutis, Transverse crack detection in 3D angle interlock glass fibre composites using acoustic emission, Materials 9 (2016) 699-719.

Simulation of Tram-Pedestrian Collision with Validated Windshield Material Model

ŠPIRK S.[1,a], ŠPIČKA J.[2,b], VYCHYTIL J.[2,c]

[1]Regional Technological Institute, University of West Bohemia in Pilsen, Univerzitni 8, 306 14 Pilsen, Czech Republic

[2]New Technologies - Research Centre, University of West Bohemia in Pilsen, Univerzitni 8

301 00 Pilsen, Czech Republic

[a]spirks@rti.zcu.cz, [b]spicka@ntc.zcu.cz, [c]jvychyti@ntc.zcu.cz

Keywords: Tram, Pedestrian, Crash, Windshield model

Abstract. The rail industry has been significantly affected by the passive safety technology in the last few years. The tram front-end design must fulfil the new requirements for pedestrian passive safety performance in the near future. The requirements are connected with a newly prepared technical guide "Tramway front end design" prepared by Technical Agency for ropeways and Guided Transport Systems [1]. This paper describes research connected with new tram front-end design safe for pedestrian. The brief description of collision scenario and used human-body model "VIRTHUMAN" is provided. The numerical simulations (from field of passive safety) are supported by experiments. The interesting part is the numerical model of the tram windshield which is experimentally validated and completely described (even by table with material parameters). Some major simulations results are discussed at the end of paper.

Introduction

This paper is focused on the simulation of the pedestrian and vehicle collision. The legislation in this area is still under preparation, although there exists a significant pressure on the tram front-end design crashworthiness. The research is supported by the cooperation with the worldwide tram developer acting in Czech Republic. The main attention here is paid to windshield of tram as significant safety feature. The material model of windshield layers (glass, PVB foil, glass) is described with all material parameters. The behaviour of tram windshield model is compared with experiment to prove the model fidelity. The full-scale tram front-end collision simulations with pedestrian are also presented. The results of the simulations are significantly influenced by the windshield model, therefore it is necessary to have a precise material model, validated on an experimental data. The results can predict the crash performance of new tram front-end and consequently the injury risk of the pedestrian.

Collision scenario

The collision scenario defined in the technical guide is based on the statistical data of the tram to pedestrian collisions and also follows the automotive safety scenarios, defined in EuroNCAP [7]. The collision scenario is defined as a moving tram hitting the pedestrian (moving or standing) from his side. The pedestrian is moving perpendicular to the tram trajectory, in front of its front end. The technical report divides the impact into three phases, where the first phase is considered as an impact of the vehicle to the pedestrian. Second phase is an impact of the pedestrian onto the ground and the third impact phase deals with the scenario, where the pedestrian lays on the railway (ground) and can be overrun with the vehicle. The scope of the technical report is focused on the first and third phases. The second phases is connected mainly with urban engineering and material of the surroundings (grass, concrete, pavement, asphalt etc.)

For the first collision scenario (type A) the pedestrians involved in the collision are specified to be mid-size male (175 cm, 78 kg – 50th percentile) and 6 years old (YO) child (110 cm, 24 kg). The report also defines possible impact area and impact zones, with respect to the shape of the vehicle, for more specification, see [1].

The collision scenario evaluating of the first impact considers the tram moving with the initial velocity equals to 20 km/h and pedestrian standing still, left side to the vehicle, one step forward (not specified which leg to be forward) and the lateral position relative to the vehicle has two specifications (H-point with respect to the tram):

- 15 % value of half of the tram width
- 50 % value of half of the tram width

The vehicle does not stop (not loaded with any deceleration pulse) only energy lost due to the impact. The pedestrian injury risk is monitored only with the Head Injury Criteria (HIC), which should not exceed threshold 1000 [8].

Evaluation of the third impact, collision scenario type B, (overrun of the pedestrian) is tested via 4 scenarios (each of them with the adult and child dummy). For this particular test, the dummies are specified to be "adult rescue dummy" (183 cm, 75 kg) and "child rescue dummy" (122 cm, 17 kg). The testing scenarios are defined as follows:

- Test 1: transverse to the rail, centred
- Test 2: transverse to the rail, off centre (hip on the rail)
- Test 3: lengthwise on the rail, centred (feet pointing towards the tram)
- Test 4: lengthwise on the rail, off centre (hip on the rail, feet pointing towards the tram)

This technical report also describes the protective technology to be used and how to be used, the gaps between dummy and vehicle etc. The initial velocity of the tram in this scenario is 25 km/h and after reaching specified position it starts to break within emergency breaking until it stops. The objective of this test is to verify capabilities of the vehicle during crash, with the following parameters [1]:

- to stop any part of the rescue mannequin before the first wheel set
- not to jam the rescue mannequin at its thighs, chest, or head
- not to sever one of the rescue mannequin's limbs so that the rescue mannequin should remain intact
- to push the rescue mannequin away so that it does not come into contact with the wheels
- not to trigger any debris or fracture on impact with the rescue mannequin (risk of aggravating injuries)

The full test protocol is available in the technical report, where all settings of the test are specified. The conclusion of the test indicates whether it meets the objective or not. There is no threshold value specified to pass the tests. Only the position of the pedestrian with respect to the tram is monitored.

Human body model VIRTHUMAN

To represent a pedestrian in the collision scenario, Virthuman (Fig. 1) model is considered. It is a virtual human body model which skeleton is formed of multi-body structure (MBS). Outer surface of the model is formed of deformable segments that are connected to the skeleton via nonlinear springs and dampers to account for deformability of soft tissues. Individual rigid bodies of the MBS structure are interconnected via kinematics joints. Moreover, additional "breakable" joints are considered in lower extremities to account for possible fractures of both femur and tibia of the pedestrian in the collision scenario. The model has been validated extensively to ensure its biofidelity [6]. The basic referential model (50th percentile male) can be scaled using the parameters of height, weight, age and gender. In this case, the 50th percentile male was used corresponding to the Hybrid III dummy (male, 172 cm, 78 kg). Due to the MBS structure, the model is easy to position. In this case, the positions as defined in the chapter "Collision scenario" were considered for the model. There is an embedded algorithm in the model to evaluate standard injury criteria for individual body parts as defined by EuroNCAP testing procedures [7]. In particular, Head Injury Criterion is used in this study to predict injury sustained by the pedestrian in the collision with the tram.

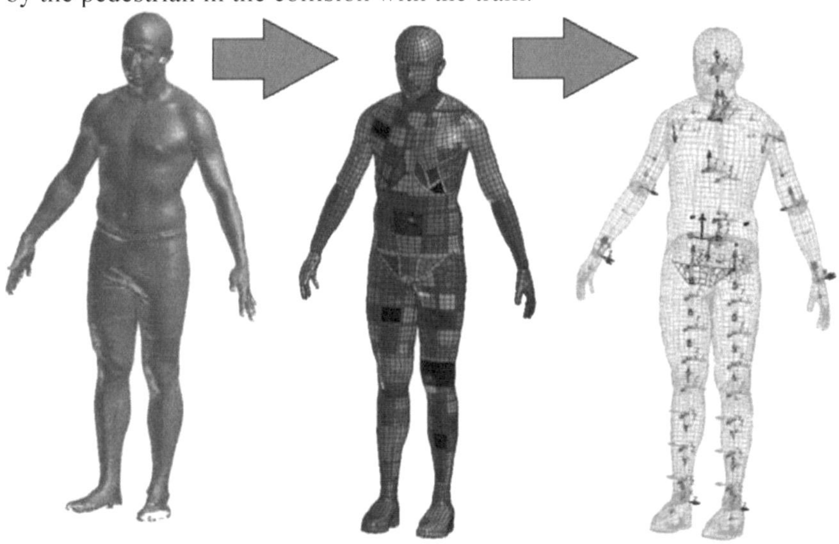

Fig. 1: VIRTHUMAN model CAD data, FE model and Multibody structure

Numerical model of collision

The tram-pedestrian collision is modelled with numerical software (Visual Performance Solution) and human body model described above (Fig. 2). For the dynamic structural analysis (with significant nonlinearities) the explicit integration method is used. The model of the vehicle is created mainly with quad and brick elements with one gauss integration point. The contacts and links (node to segment connection) are realized by penalty algorithm. The model was discretised into 1.6 million elements with the smallest element characteristic length 2 mm (leading to a time step 5e-6 ms). The simulation time of the defined scenario is 390 ms. Mechanical properties of steel S235 and 1.4301 are used from literature. The top shell cover is made from polymer (acrylonitrile butadiene styrene) with acceptable fire protection and recycling possibilities. Unfortunately mechanical properties of this material used in simulation are confidential (courtesy of the company).

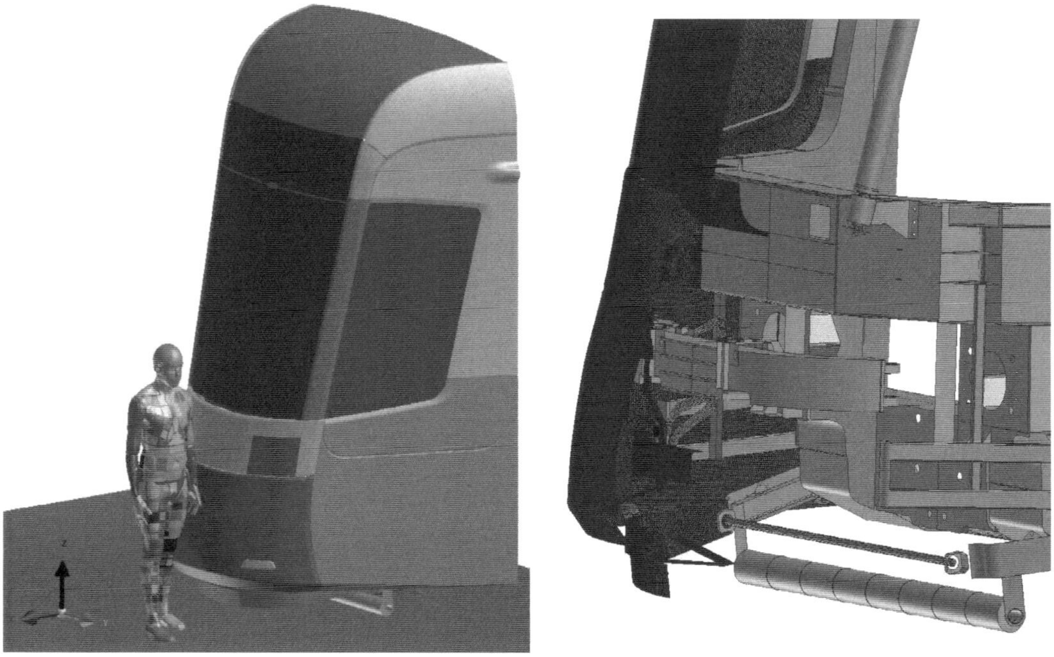

Fig. 2: Numerical model of the collision (left) and detail of tram FE structure (right)

Windshield material model

The approach for windshield modelling is based on the studies connected with automotive industry. The FE model of laminated glass is composed of two outer layers of glass and one inner layer of polyvinyl butyral (PVB) [3]. All layers are modelled with shell elements connected by *tide* node-segment link. The linear acceleration of the head-form impactor were determined by the critical fracture stress [4]. The aim of this work is to build a standard tram windshield model and to experimentally validate this model. The main difference (between road and rail vehicle) is mainly in the thickness of glass and PVB layers. The initial prediction is, that the material behaviour of layers is very similar for the tram and road vehicle. The PVB foil is modelled as an isotropic nonlinear viscoelastic shell element of Maxwell type:

$$\sigma = k(1 - e^{-w\varepsilon})(1 + h_1\varepsilon + h_2\varepsilon^2)\left(\frac{\dot{\varepsilon}}{\dot{\varepsilon}_{ref}}\right)^m \tag{1}$$

Where $\dot{\varepsilon}$ is the plastic strain rate, $\dot{\varepsilon}_{ref}$ is the reference strain rate, and k, w, m, h_1, h_2 are material constants. The glass is modelled as linear elastic material with a brittle failure criterion. For the fracture definition the Rankine criterion is used and the fracture occurs when the maximum principal stress exceeds the critical value.

$$\sigma = \begin{bmatrix} (1-d_1)\sigma_{11} & (1-d_{max})\sigma_{12} & (1-d_{max})\sigma_{13} \\ (1-d_{max})\sigma_{12} & (1-d_2)\sigma_{22} & (1-d_{max})\sigma_{23} \\ (1-d_{max})\sigma_{13} & (1-d_{max})\sigma_{23} & 0 \end{bmatrix} \tag{1}$$

where σ is the damaged stress tensor, $\sigma_{11}, \sigma_{12}, .., \sigma_{23}$ are components of undamaged tensor, d_1 and d_2 are damage values in two directions and d_{max} is maximum of d_1 and d_2.

The detailed description of material model parameters is presented below (units mm kg ms). These material parameters are used for tram windshield in simulation with three layers (3 mm outside glass, 0.9 mm PVB foil, 3 mm inside glass) connected with TIED links.

Table 1: Parameters of material model GLASS

MAT TYP	RHO	BELYTS.-TSAY REDUCED INTEGRATION	STIFFNES ELASTIC HOURGLASS	QUADRATIC VISCOSITY MULTIPLY	
126	2.5e-6	0	0	1	
E	NU	MEMBRANE HOURGLASS	OUT OF PLANE HOURGLASS	ROTATION HOURGLASS	TRANS SHARE
70	0.2	0.01	0.01	0.01	0.8333
SIGMAC	TIME FILTER	STIF DAMPING			
0.031	0.01	0.1			

Table 2: Parameters of material model PVB-foil (nonlinear viscoelastic model)

MAT TYP	RHO	BELYTS.-TSAY REDUCED INTEGRATION	STIFFNES ELASTIC HOURGLASS	QUADRATIC VISCOSITY MULTIPLY	
121	1e-6	0	0	1	
E	NU	MEMBRANE HOURGLASS	OUT OF PLANE HOURGLASS	ROTATION HOURGLASS	TRANS SHARE
9	0.39	0.01	0.01	0.01	0.8333
G-SHELL PARAM	k	m	h1	h2	w
	0.007	1.33	1.35	0	3

Experimental validation of windshield numerical model

In order to validate the material model, the simple pendulum test and its numerical simulation (Fig. 3) is used. The pendulum is made of steel profile (50x30 mm, thickness 2 mm, length 2000 mm and 4.41 kg of mass) and the ball (150 mm of diameter and 4.7 kg of mass). The part of windshield is placed on the extruded polystyrene (with known properties) with circular opening with diameter 300 mm. More than 10 tests was performed to get statistically significant results. In most of the tests, the ball was falling from full height (2000 mm). Few tests were executed from a smaller height and also with initial crack on glass or different shape and size of glass. These experimental results are not described here, but significant difference was observed only in the condition of smaller initial height, that theoretically allow the validation for range of initial velocities. The windshield model was validated only for full height with the impact velocity 6 m/s. The acceleration was recorded with accelerometer (PCB 352C33 S/N 120471) and displacement was recorded with high-speed camera (Photron fastcam SA X2 RV). This kinematics conditions are similar to pedestrian head impact during collision with tram (initial speed 20 km/h). The impactor is considered as a rigid to validate the glass model (not for head injury prediction where biofidelic impactor is necessary).

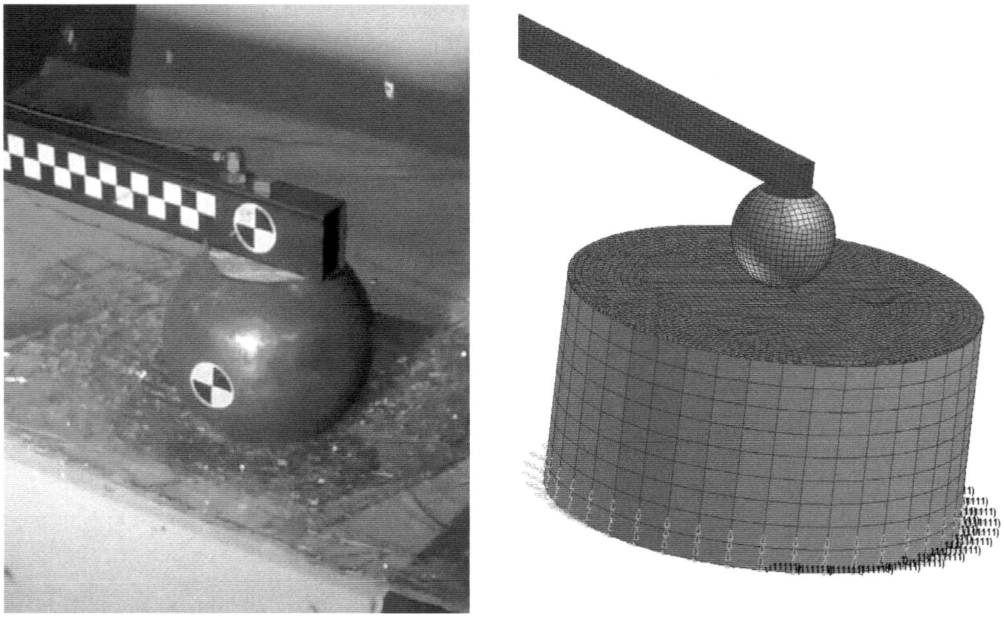

Fig. 3: Experimental test of windshield (left) and simulation for material validation (right)

It is clear that this model cannot exactly predict the crack shape in detail, but it has been discovered that in repeated experiments and simulations the influence of crack shape difference is insignificant. In this phenomena the crack shape has some general similarities in radial and circular direction (Fig 4). Moreover, the acceleration results are influenced by the steel rod oscillations. This is one of the possible improvements for the further experiments.

Fig. 4: The windshield glass with damage tensor directions

The comparison of the experimental data and validated numerical model (described above) shows very good coincidence. This coincidence is adequate and good enough for safety simulations and for head impact injury predictions in case of windshield-head impact. The acceleration-time curves are very close to each other (Fig. 5) what indicates similar result of HIC criterion. The deformation characteristics result in a good agreement of an experimental and simulation curves (Fig. 6).

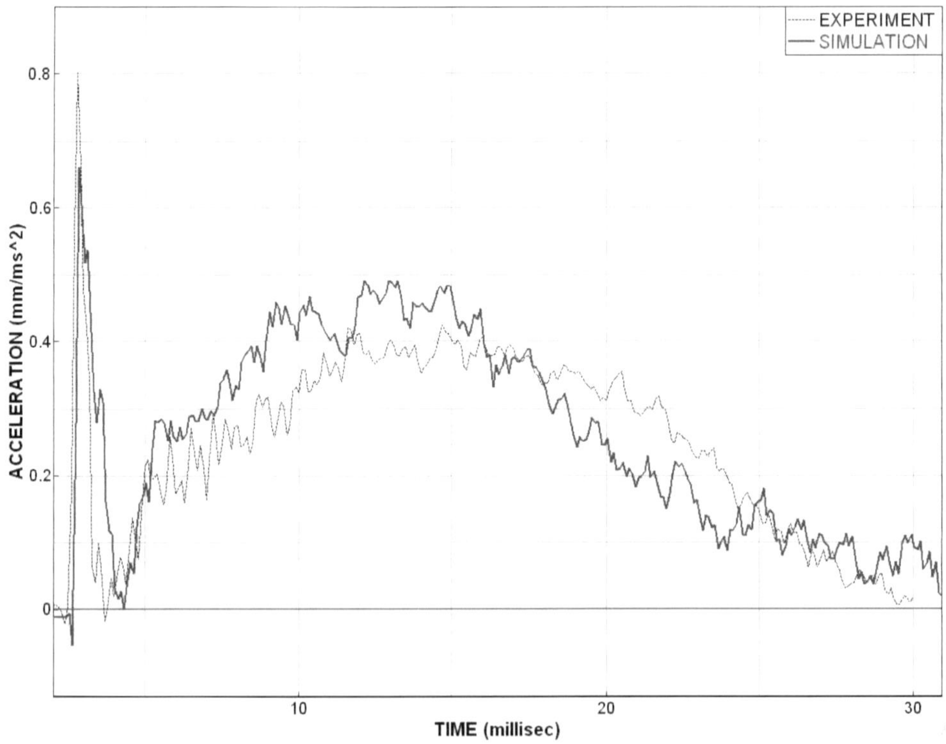

Fig. 5: Plot of the acceleration vs time

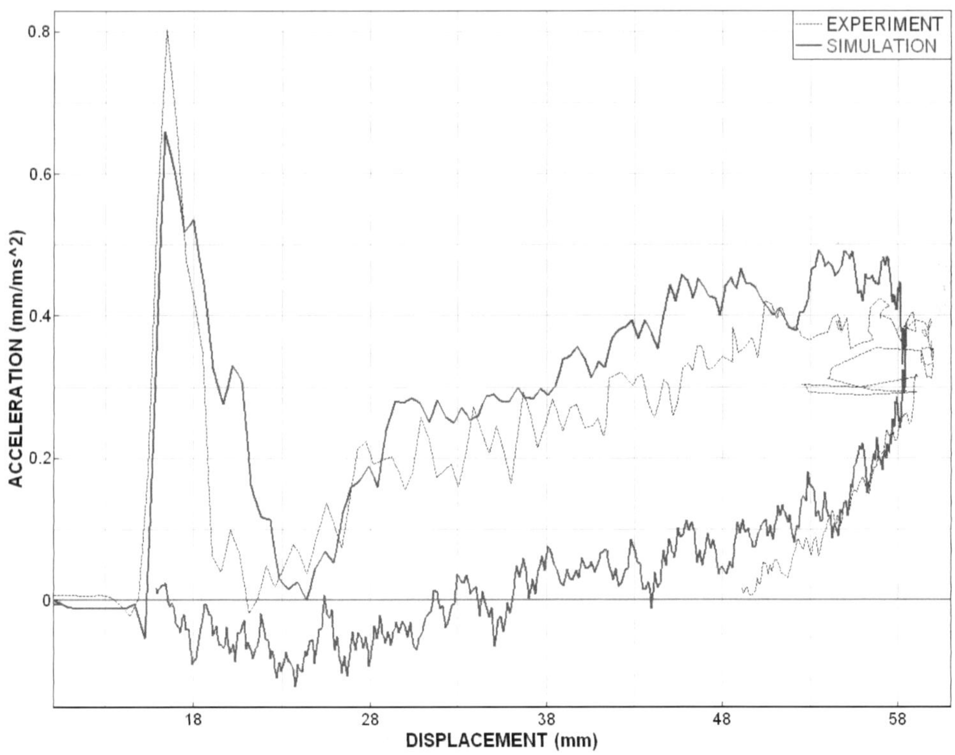

Fig. 6: Plot of the acceleration vs displacement

Simulations results

There are two different set-ups of the simulations. The first type is a collision with standing pedestrian. The second one is a collision with laying pedestrian. All simulations are performed with the same model. The main reason for material model validation is the simulation from field of passive safety. The sequence figures in the time (Fig. 7) shows the detail of the head impact to the tram front-end. This part of collision is the most important, since the head injury connected with significant severity occurs here. It is clearly visible, that the head impact occurs directly to the windshield. However, the head acceleration does not exceed limit of 0.8 mm/ms^2 and the HIC criterion is only 234 (bellow the threshold 1000). This indicates low injury risk of the head.

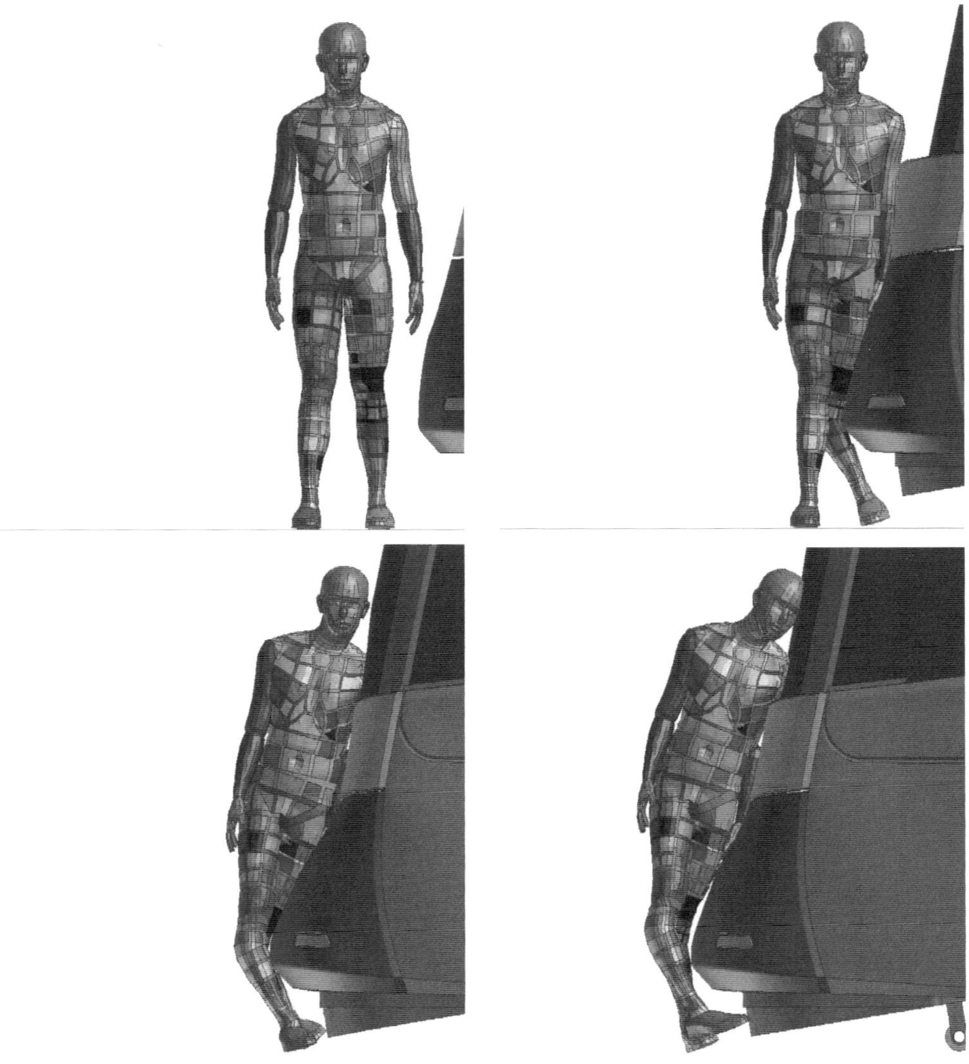

Fig. 7: The results of simulation (in time 0, 25, 50, 75 ms), where the head impacts the windshield

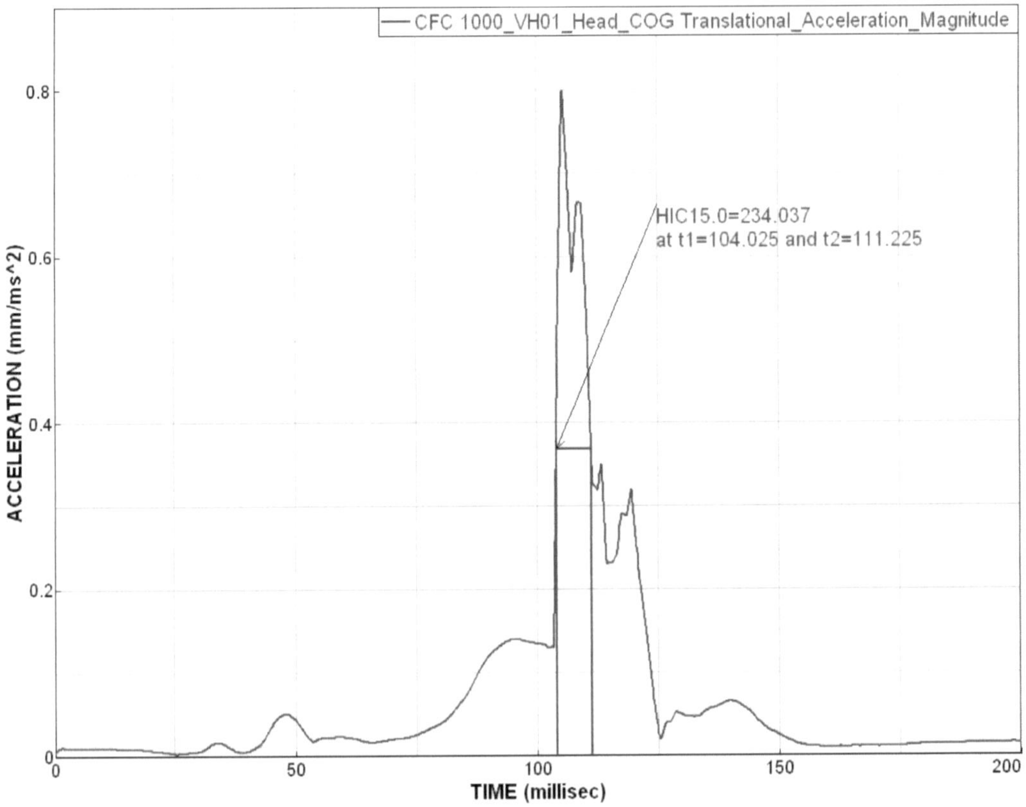

Fig. 8: The acceleration of the COG filtered by CFC1000 filter with HIC15 injury criterion assessment

The collision of tram and laying pedestrian (Fig. 9) shows how the requirements for the pedestrian anti-crush mechanism are met. With the advantages of the simulations the effect of mechanism with correct clearance (100 mm) is visible. It can be said that the injury of laying pedestrian during collision with tram without pedestrian anti-crush mechanism are absolutely fatal (the most of trams has no pedestrian anti-crush mechanism). The design of new tram can save a lot of lives and significantly reduce number and severity of injuries.

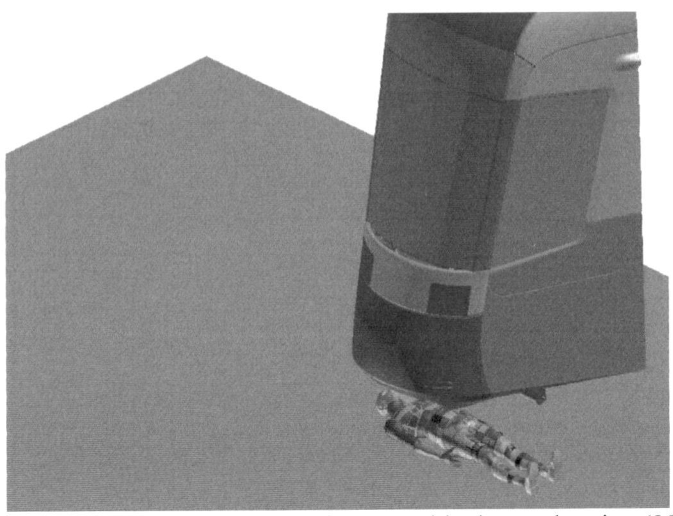

Fig. 9: The result of simulation of tram and laying pedestrian (90 ms)

Conclusions

The collision of tram-pedestrian described in paper was simulated with the advantage of the human body model "Virthuman". The injury of the pedestrian head (HIC) is highly influenced by the windshield behaviour. The simulation indicates that the defined tram to pedestrian crash scenario results in the HIC value (234) significantly smaller than the threshold limit of 1000. The first results of the experiments suggest very similar material behaviour of tram and road vehicle windshield. Therefore it is possible to use the material model of glass and PVB foil from an automotive industry with modified thickness. The paper contains description of experimental material validation and also material model of windshield (with detail material parameters). The simple pendulum test has some inaccuracies (see above), which can be further improved, but the experimental observations are acceptable for material validation. The results of the simulation (with experimentally validated material) indicate that the windshield is feature with good crashworthiness. The design of tram with low height of bottom windshield is feasible.

Acknowledgment

This work has been supported by the project TRIO FV20441 "Research and development of safe tram face" provided by the Ministry of Industry and Trade of Czech Republic.

References

[1] A. Guesset, Tramway front end design: technical guide, STRMTG, 2016.

[2] J. Vychytil, L. Hyncik, J. Manas, P. Pavlata, R. Striegler, T. Moser, R. Valasek, Prediction of Injury Risk in Pedestrian Accidents Using Virtual Human Model VIRTHUMAN: Real Case and Parametric Study, 2016, SAE Technical Papers.

[3] Kosiński P., Osiński J., Laminated windshield breakage modelling in the context of headform impact homologation tests, Int. J. of Applied Mechanics and Engineering, 2015, vol.20, No.1, pp.87-96.

[4] Y. Peng, J. Yang, C. Deck, R Willinger, Finite element modelling of crash test behaviour for windshield laminated glass, International Journal of Impact Engineering 57, 2013.

[5] EuroNCAP, Assessment protocol – Adult occupant protection, version 6.0, July 2013.

[6] J. Vychytil, J. Manas, H. Cechova, S. Spirk, L. Hyncik, L. Kovar: Scalable Multi-Purpose Virtual Human Model for Future Safety Assessment, SAE Technical Paper 2014-01-0534, doi:10.4271/2014-01-0534.

[7] EuroNCAP: Assessment Protocol – Vulnerable Road User Protection, Version 10.0.2, July 2019.

[8] SCHMITT, Kai-Uwe, et al. Trauma biomechanics. Berlin: Springer, 2010.

EAN 2020
58th conference on experimental stress analysis

An Experimental Investigation of Yield Surfaces Anisotropies

ŠTEFAN J. [1,a], PARMA S. [1,b], MAREK R. [1,c], PLEŠEK J. [1,d], FEIGENBAUM H. [2,e] CIOCANEL C. [2,f]

[1]Deparment of Impact and Waves in Solids, Institute of Thermomechanics of the Czech Academy of Sciences, Dolejškova 1402/5, 182 00 Praha, Czech Republic

[2]College of Engineering, Informatics, and Applied Sciences, Northern Arizona University, 2112 S Huffer Lane, Flagstaff, AZ 86 011, US.

[a]stefan@it.cas.cz, [b]parma@it.cas.cz, [c]marek@it.cas.cz, [d]plesek@it.cas.cz, [e]heidi.feigenbaum@nau.edu, [f]constantin.ciocanel@nau.edu

Keywords: Experimental mechanics, Yield surface, Directional distortional hardening, Metallic materials

Abstract. This contribution is dedicated to a rigorous experimental procedure for evaluation of yield surfaces of common metallic materials. This yield surface tracing procedure employs thin-walled tubular specimens to identify individual yield points of the material under arbitrary combinations of axial load and torque. A yield point is identified on the basis of a prescribed threshold for the effective plastic strain that is being continuously and fully automatically evaluated throughout the test. The experimental results generated with this tracing method are promising, leading to shapes of the yield surfaces conform with the von Mises criterion. The proposed methodology can effectively capture the YS shape.

Introduction

This work aims to establish a rigorous experimental procedure for evaluation of yield surfaces of common metallic materials. The concept of a yield surface is embedded in the phenomenological plasticity theory, and careful identification of yield surface shape is critical to the development and calibration of phenomenological models of plasticity [1–3]. In accordance with the theory, the yield surface (YS) can be defined as a region within a particular stress space bounding the elastic domain. When the stress state is elevated to a state on the YS, plastic flow occurs and the yield surface evolves. This is referred to as strain hardening. While evolving, the yield surface undergoes changes of size, position, and shape in a stress space that correspond to isotropic hardening, kinematic hardening, and distortional hardening, respectively.

Knowledge of YS shapes and their evolution under complicated modes of loading is useful for a number of applications including, but not limited to, the processes of forming of material products, and predicting the behaviour under cyclic loading (e.g., during earthquakes, in service conditions, repetitive wind or wave loading) leading to the accumulation of plastic strain. This is known as ratcheting. In particular, recent work [4] uses YS shape change to help develop models aimed at improving predictions of multiaxial ratcheting behaviour, since conventional models that neglect the YS distortion do not generally provide satisfactory predictions in this complex loading case. Experimental evaluation of strength anisotropy is also very important for additively manufactured materials [5].

Experiments

The experimental method in this work is based on mechanical testing of thin-walled tubular specimens under combined axial and torsional loading. Individual yield points constituting a yield surface within the axial stress – shear stress space ($\sigma - \sqrt{3}\tau$) are detected by performing a sequence of loads ("probes") characterized by particular levels of intensities of both loading components (i.e., the axial force, and the torque). Each probe indicates a yield point on the yield surface of the material under investigation. The experiments are conducted with an Instron-type biaxial testing machine and with an Epsilon-Tech biaxial contact extensometer enabling simultaneous measurement of axial and shear strains.

For the detection of a yield point, the approach presented in reference [6] was adopted. It is a type of proof strain method defining a yield point as a point for which a predetermined amount of von Mises effective plastic strain is developed. Under the combination of axial load and torque, the von Mises effective plastic strain reads

$$\Delta\varepsilon_{eff} = \sqrt{\varepsilon_{pl.}{}^2 + \frac{1}{3}\gamma_{pl.}{}^2}, \tag{1}$$

where the axial plastic strain $\varepsilon_{pl.}$and the shear plastic strain $\gamma_{pl.}$ are given, respectively, by

$$\varepsilon_{pl.} = \varepsilon - \varepsilon_{el.} - \varepsilon_R, \quad \gamma_{pl.} = \gamma - \gamma_{el.} - \gamma_R. \tag{2}$$

Here, the total strain components ε and γ are directly measured by the extensometer, the elastic strain components $\varepsilon_{el.}$ and $\gamma_{el.}$ are determined from load by making use of the Hooke's law, and ε_R and γ_R are residual strains as illustrated in Fig. 1.

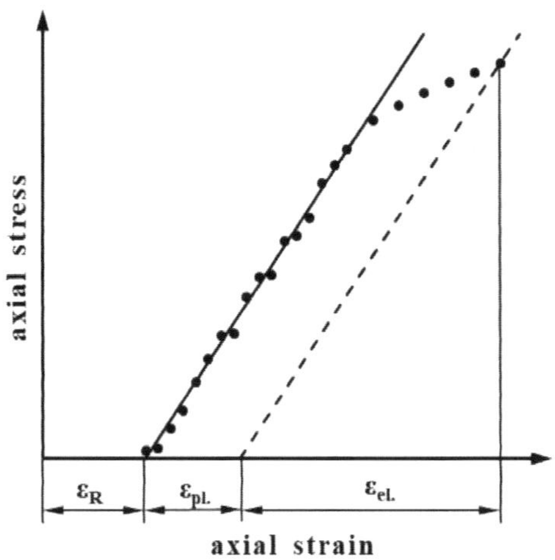

Fig. 1: Schematic representation of the axial stress – strain curve

In this work, the effective plastic strain threshold $\Delta\varepsilon_{eff} = 200\ \mu\varepsilon$ was used to define yielding. A vital part of the procedure is an algorithm that continuously and automatically performs a least squares linear regression analysis to fit the elastic part of the stress – strain curve during loading of the specimen. This algorithm has two primary outputs: the elastic moduli (E, G) as slopes, and the residual strains (ε_R, γ_R) as offsets. Actual values of elastic parts of strain ($\varepsilon_{el.}$, $\gamma_{el.}$) are calculated from the elastic moduli and subsequently used to evaluate the total effective plastic strain accumulated up to the current strain state of the specimen, via Eq. (2). Hence, the procedure is fully automatized for evaluating unique yield points. Once a yield point is detected, the loading ceases, the specimen is unloaded to a state inside the YS (generally a zero stress), and a new probing path is started; thus a single specimen is used to fully characterize a yield surface.

The experimental procedures presented herein were validated by performing evaluation of the initial yield surfaces of common metallic materials, i.e., evaluation of yield surfaces of materials in the as-manufacture state with no additional plastic pre-load. At first, a non-alloyed cold-drawn ferritic steel was used. In the second experiment, a hot rolled carbon steel of C60R grade (W. Nr. 1.1223) was investigated. The geometry of the test specimen used in these experiments is depicted in Fig. 2.

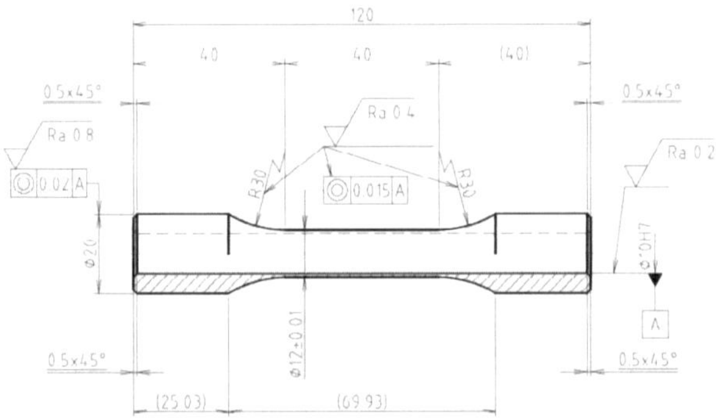

Fig. 2: Design and dimensions of the test specimen manufactured from C60R steel (the dimensions are stated in millimeters)

Results and discussion

Fig. 3 shows the yield surfaces evaluated on thin-walled tubular specimens manufactured from a non-alloyed ferritic steel (A), and on a carbon steel of C60R grade (B). The YS shapes reported in Fig. 3 are relatively well predicted by the von Mises yield criterion, as expected, because the material/specimen has not been pre-loaded. These preliminary results suggest that the proposed methodology proposed can effectively capture the YS shape.

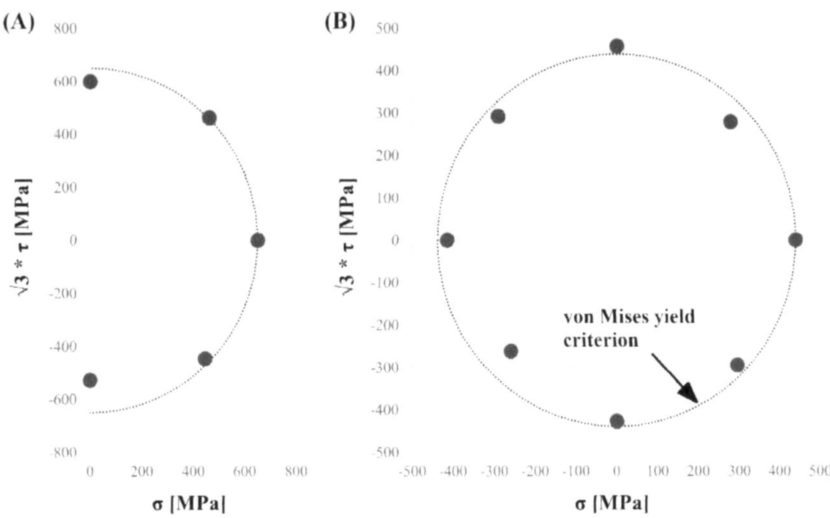

Fig. 3: A tensile part of the initial yield surface of a non-alloyed cold-drawn ferritic steel (A), and a full initial yield surface of a C60R steel (B)

Tab. 1: Results of yield surfaces tracing experiments.

Probe	Non-alloyed steel		C60R steel	
No.	σ_{YP} [MPa]	τ_{YP} [MPa]	σ_{YP} [MPa]	τ_{YP} [MPa]
1	649.5	0.0	438.8	0.0
2	0.0	346.1	–415.4	0.0
3	0.0	–304.9	0.0	264.0
4	462.2	266.8	0.0	–246.7
5	446.5	–257.7	277.8	160.6
6	—	—	295.3	–169.7
7	—	—	–291.0	168.3
8	—	—	–259.3	–150.7

Conclusions

In this paper, a rigorous experimental method for the evaluation of yield surfaces is presented. The method uses algorithms enabling continuous and fully automatized evaluation of elastic moduli and the actual amount of effective plastic strain, which are being used to calculate a threshold plastic strain which defines the trigger of a yield point. It was shown that the first couple of experiments performed on specimens of conventional metallic materials provided promising results as the yield surfaces shapes conformed with the von Mises yield criterion. Future work will focus on improving the method, characterizing other steels and capturing the YS distortion due to strain hardening under various pre-loads.

Acknowledgements

This work was supported by US Army Research Laboratory and the US Army Research Office under Grant No. W911NF-19-1-0040, the Czech Science Foundation under Grant No. GA19-03282S, and by the Ministry of Education, Youth and Sports of the Czech Republic under Grant No. LTAUSA18199.

References

[1] S. Parma, J. Plešek, R. Marek, Z. Hrubý, H. P. Feigenbaum, Y. F. Dafalias, Calibration of a simple directional distortional hardening model for metal plasticity, Int. J. Solids Struct. 143 (2018) 113–124.

[2] R. Marek, J. Plešek, Z. Hrubý, S. Parma, H. P. Feigenbaum, Y. F. Dafalias, Numerical Implementation of A Model With Directional Distortional Hardening, J. Eng. Mech. 141 (2015).

[3] H. P. Feigenbaum, Y. F. Dafalias, Directional distortional hardening at large plastic deformations, Int. J. Solids Struct. 51 (2014) 3904–3918.

[4] Ch. A. Welling, R. Marek, H. P. Feigenbaum, Y. P. Dafalias, J. Plesek, Z. Hruby, S. Parma, Numerical convergence in simulations of multiaxial ratcheting with directional distortional hardening, Int. J. Solids Struct. 126–127 (2017) 105–121.

[5] B. Mooney, K. I. Kourousis, R. Raghavendra, D. Agius, Process phenomena influencing the tensile and anisotropic characteristics of additively manufactured maraging steel, Mat. Sci. Eng. A 745 (2019), 115-125.

[6] H. C. Wu, W. C. Yeh, On the Experimental Determination of Yield Surfaces and Some Results of Annealed 304 Stainless Steel, Int. J. Plast. 7 (1991) 803–826.

EAN 2020
58th conference on experimental stress analysis

Changing the Stress State of the Track Superstructure while Strengthening the Subgrade

TIUTKIN O.[1,a], NEDUZHA L.[2,b], KALIVODA J.[3,c]

[1]Faculty of Bridges, Department of Bridges and Tunnels, Dnipro National University of Railway Transport named after Academician V. Lazaryan, Lazaryan St. 2, 49010, Dnipro, Ukraine

[2]Faculty of Transport Engineering, Department of Theoretical and Structural Mechanics, Dnipro National University of Railway Transport named after Academician V. Lazaryan, Lazaryan St. 2, 49010, Dnipro, Ukraine

[3]Faculty of Mechanical Engineering, Department of Automotive, Combustion Engine and Railway Engineering, Czech Technical University in Prague, Technická 4, 166 07, Prague 6, Czech Republic

[a]alexeytutkin@gmail.com, [b]nlorhen@i.ua, [c]jan.kalivoda@fs.cvut.cz

Keywords: finite-element model, stress state, rolling stock, track superstructure, subgrade, strengthening

Abstract. Strengthening the subgrade with the help of various technologies significantly affects the stress state of the track superstructure. The strengthening of the subgrade with piles, leads to an abrupt stiffening of the track superstructure, which is problematic for its normal operation. Two finite element models were created to determine the change in the stress state of the track superstructure when strengthening the subgrade. They fully reflect the geometric, deformation and power characteristics of a real subgrade, which is strengthened by piles. Individual finite elements of the models are provided with the deformation characteristics of the steel rails, reinforced concrete sleepers, soil subgrade, ballast and soil-cement material of piles. The authors carried out the calculation of stress state of the track superstructure for two finite element models. Results are obtained and analyzed to help in choosing the most effective option for strengthening the subgrade.

Introduction

At the current stage, among the many issues in transport improving traffic safety is relevant. It depends on many factors including the interaction of parts in the "rolling stock – track superstructure – subgrade" system. A comprehensive analysis of these parts involves the research of the stress state when the rolling stock operations, taking into account the features of the track superstructure (TS) [1]. For example, geometrical irregularities and rails unevenness [1,2], their geometric position in the plane (circular and transition curves) [1,3], parameters of TS [4], etc. Since the rolling stock operation is characterized by significant dynamic loads [5,6], they, in turn, move in the form of impacts on the "track superstructure – subgrade" system.

Physical modelling is an important part of the researches that provides possibilities to correct theoretical foundations [7–10]. A number of methods are existing for solving the problem of straightening of the subgrade in transport [11–15]. One of which there is strengthening the subgrade by various technologies, which significantly affects the stress state of the track superstructure.

Researches of the stress state of the track superstructure

In order to determine the influence of the subgrade straightening on the stress state of the track superstructure two finite-element models were built using a professional complex SCAD. They correspond to the strengthening of the subgrade by soil-cement piles. Option 1 – piles are located near a rail (Fig. 1, a); Option 2 – piles are located near the ballast section (Fig. 1, b).

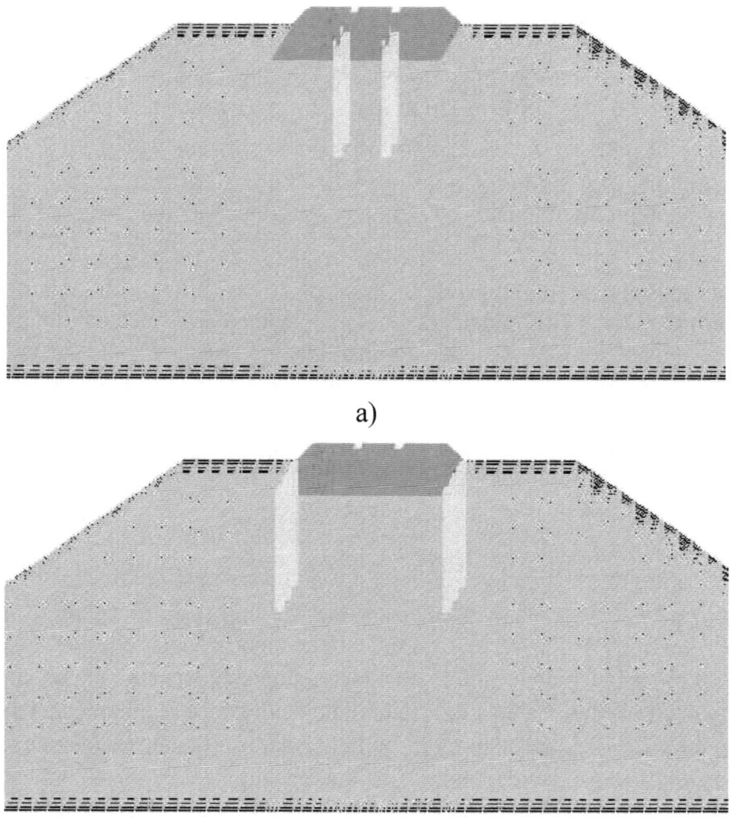

a)

b)

Fig. 1: Finite-element models of the subgrade: a) Option 1; b) Option 2

The models fully reflect the geometric, deformation and power characteristics of a real subgrade which is strengthened by piles [13,16,17]. At the same time, the individual finite elements are given the deformation characteristics of steel rails, reinforced concrete of sleepers, soil of the subgrade, crushed stone of ballast and soil-cement material of piles.

The total number of model nodes is 28 203 pcs. (about 85 thousand degrees of freedom, the task is considered to be large-sized), the number of finite elements is 31 572 pcs. The finite elements in the model are accepted as compatible ones, that is, all nodes of neighboring elements coincide, which positively affects the accuracy of the solution. The finite elements of the "prism" and "tetrahedron" types were taken from the SCAD complex library (license number is F755B84 (KMBKB RA 4810)). These are isoparametric elements with a maximum size of 0.2×0.2×0.2 m (there are elements with a size of 0.25×0.25×0.25 m, not more than 5... 7% of the total number of finite elements in the model). Model dimensions: length (base) is 40.6 m, width is 1.64 m, height is 11.1 m (6 m of which is the height of the subgrade).

The finite elements of the model were provided with the corresponding stress-strain properties: 1) the basis of the subgrade is loam, specific gravity γ=1.9 t/m^3, modulus of elasticity

E=25 MPa, Poisson's ratio ν=0.3; 2) ballast section is crushed stone, specific gravity γ=2.2 t/m^3, modulus of elasticity E=150 MPa, Poisson's ratio ν=0.2; 3) a rail is steel, specific gravity γ=7.7 t/m^3, modulus of elasticity E=2.1 GPa, Poisson's ratio ν=0.2; 4) a pile – specific gravity γ=2.3 t/m^3, modulus E=200 MPa, Poisson's ratio ν=0.2; 5) a sleeper is reinforced concrete, specific gravity γ=2.5 t/m^3, modulus of elasticity E=0.4 GPa, Poisson's ratio ν=0.2.

Boundary conditions are imposed on the model: below is the ban in displacement along all three axes x, y and z, on the sides of the base is the ban on x and y axes, on the transverse sides of the model is the ban on y axis (flat deformation condition). The top and slopes of the model are free from boundary conditions.

The train was accepted as a model load, the axle pressure was assumed to be equal to the normative pressure from the train P=23.5 t per axle. The load was applied to the finite element node that simulates the rail, the nature of the action is the point contact.

For the two finite-element models the stress state of the track superstructure was calculated along the x (horizontal) and z (vertical) axes in the place of a sleeper and in the place between sleepers. As an example, we present the characteristic results of the performed calculations (Fig. 2–5).

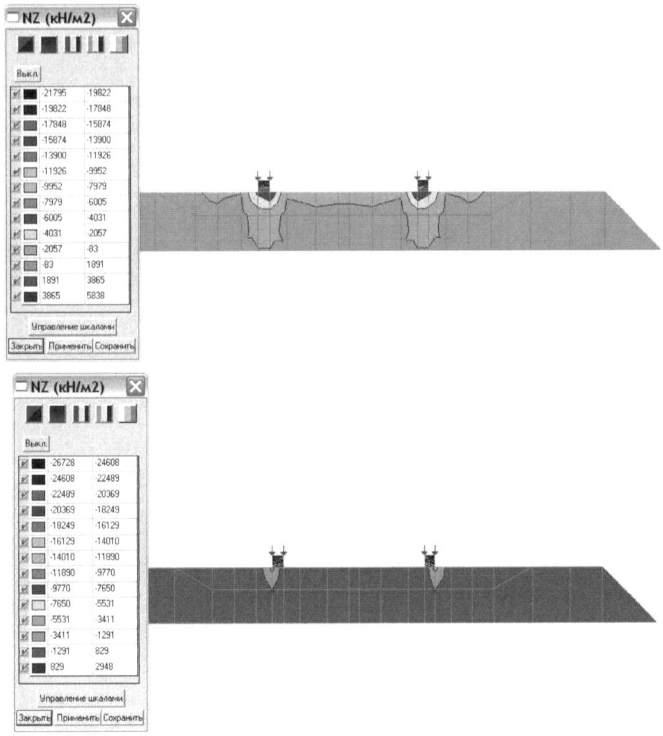

Fig. 2: Isolines and isofields of vertical normal stresses in the fragment of models (Option 1)

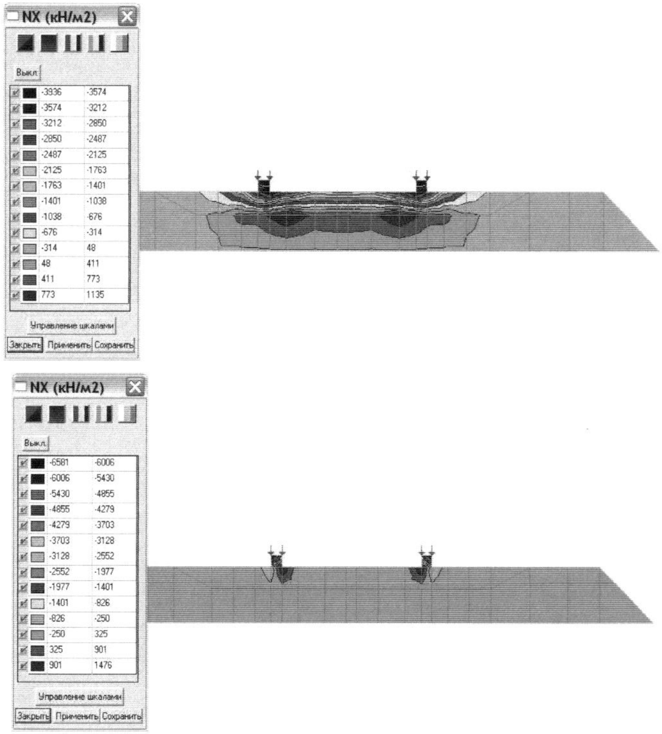

Fig. 3: Isolines and isofields of horizontal normal stresses in the fragment of models (Option 1)

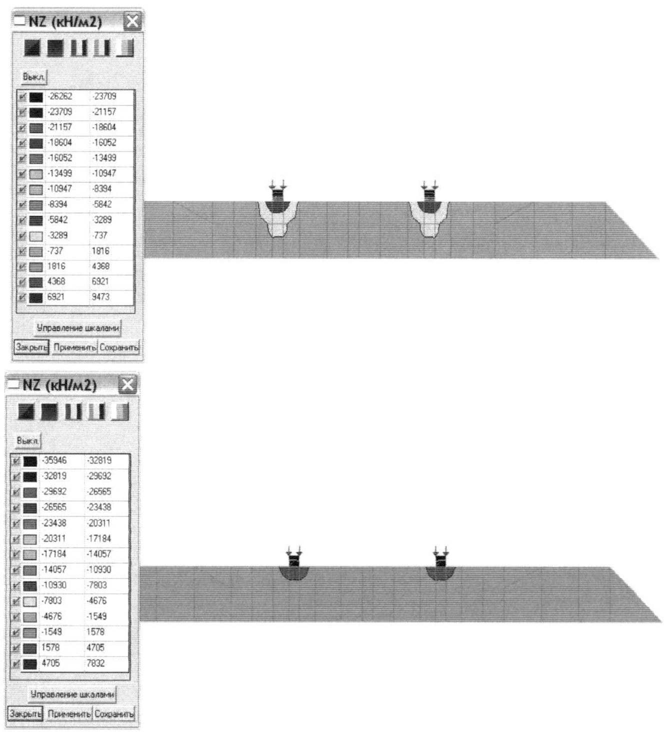

Fig. 4: Isolines and isofields of vertical normal stresses in the fragment of models (Option 2)

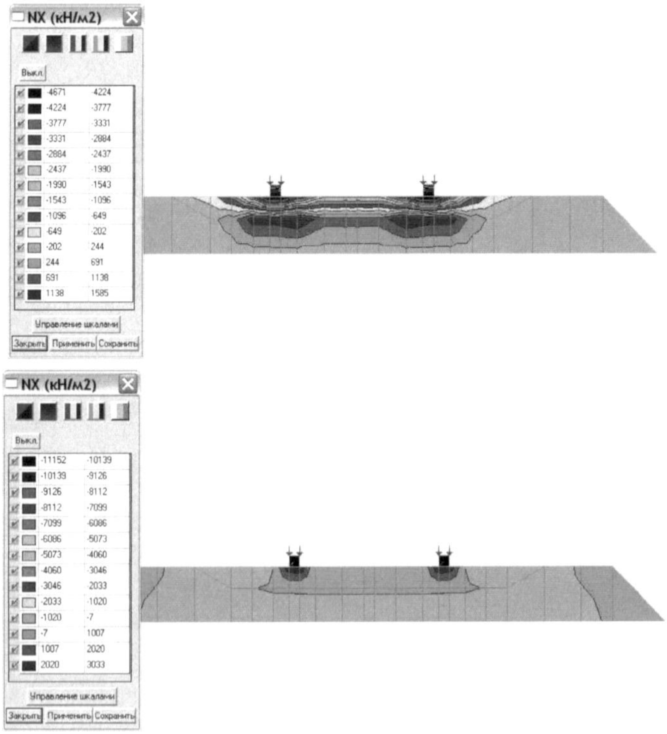

Fig. 5: Isolines and isofields of horizontal normal stresses in the fragment of models (Option 2)

The distribution patterns of normal vertical stresses (Fig. 2 – 3) in the ballast section reach -3400 … -4000 kPa (mean value is 1700 kPa) and -4600 … -5800 kPa (mean value is 1800 kPa) according to Options 1 and 2, which is significantly less than the strength of the ballast material.

The obtained results showed that the horizontal stresses in the TS have the following maximum values (Fig. 4–5): Option 1 is -3936 kPa for finding the load in the place of a sleeper and – -6581 kPa in the place between sleepers; Option 2 is -4671 kPa for finding the load in the place of a sleeper and – -11152 kPa in the place between sleepers, which are on the surface of rails and sleepers, they do not cause any overstresses.

Conclusions

The system of soil-cement piles, which is analyzed in the finite element analysis, introduces a significant positive fluctuation into the "TS – the subgrade" system. This is expressed in the significant heterogeneity of the stress state around the piles, which take on a significant part of the impact of the rolling stock.

Both of the proposed options for strengthening the subgrade are such that have a positive effect on the subsystems of the overall "TS – subgrade" system. In each of the options of strengthening there is a clear decrease in a stress state, which indicates an increase in the load capacity of the subgrade.

After analyzing the obtained results, the most effective (from the stress state viewpoint) location of the subgrade strengthening piles was chosen. This is the Option 1 (piles are located near the rail). As a following step the tuning of the FEA model of "TS – subgrade" system by measurement of the properties of crushed stone, subgrade soil, and soil-cement piles is planned. In the future work the influence of subgrade strengthening also on the dynamic characteristics of the rolling stock will be investigated.

References

[1] A. Kuzyshyn, S. Kostritsa, L. Ursulyak, A. Batig, J. Sobolevska, O. Voznyak, Research of the impact of geometric unevenness of the railway track on the dynamic parameters of the railway rolling stock with two-stage spring suspension, in: IOP Conference Series: Materials Science and Engineering 664(1), 012024, 2019.

[2] I. Bondarenko, O. Lunys, L. Neduzha, R. Keršys, Dynamic track irregularities modeling when studying rolling stock dynamics, in: Proceedings of 23rd International Scientific Conference Transport Means 2019, pt. III, Trakai, Lithuania, pp. 1014–1019.

[3] V. Hauser, O.S. Nozhenko, K.O. Kravchenko, M. Loulová, J. Gerlici, T. Lack, Impact of wheelset steering and wheel profile geometry to the vehicle behavior when passing curved track, Manufacturing Technology 17(3) (2017) 306–312.

[4] D.M. Kurhan, Features of perception of loading elements of the railway track at high speeds of the movement, Science and Transport Progress 56(2) (2015) 136–145.

[5] I. Klimenko, J. Kalivoda, L. Neduzha, Parameter optimization of the locomotive running gear, in: Proceedings of 22nd International Scientific Conference Transport Means 2018, pt. III, Trakai, Lithuania, pp. 1095–1098.

[6] O. Lunys, L. Neduzha, V. Tatarinova, Stability research of the main-line locomotive movement, in: Proceedings of 23rd International Scientific Conference Transport Means 2019, pt. III, Trakai, Lithuania, pp. 1341–1345.

[7] P. Bauer, J. Kalivoda, System of axle-box force measurement for experimental railway bogie, in: Experimental Stress Analysis – 56th International Scientific Conference, EAN 2018 – Conference Proceedings 2018, Harrachov, Czech Republic, pp. 9–16.

[8] S. Myamlin, L. Neduzha, Ž. Urbutis, Research of Innovations of Diesel Locomotives and Bogies, in: Procedia Engineering 9th International Scientific Conference Transbaltica 2015, 134, Vilnius, Lithuania, pp. 469–474.

[9] Y. Bezin, D. Farrington, C. Penny, B. Temple, S. Iwnicki, The dynamic response of slab track constructions and their benefit with respect to conventional ballasted track, Vehicle System Dynamics 48(SUPPL. 1) (2010) 175–193.

[10] D. Ignatenko, O. Tiutkin, V. Petrenko, A. Alkhdour, Application of centrifugal modeling for the study of landscape structure stability, International Journal of Civil Engineering and Technology (IJCIET) 10(01) (2019), 2179–2187.

[11] J. Kalivoda, P. Bauer, Mechatronic Bogie for Roller Rig Tests, in: The Dynamics of Vehicles on Roads and Tracks – Proceedings of the 24th Symposium of the International Association for Vehicle System Dynamics, IAVSD 2015, Graz, Austria, pp. 899–908.

[12] I. Klimenko, L. Černiauskaite, L. Neduzha, O. Ochkasov, Mathematical Simulation of Spatial Oscillations of the "Underframe-Track" System Interaction, in 12th International Conference Intelligent Technologies in Logistics and Mechatronics Systems – ITELMS'2018, Panevėžys, Lithuania, pp. 105–114.

[13] O. Pshinko, V. Petrenko, A. Tiutkin, V. Andrieiev, A. Gubar, D. Ihnatenko, R. Markui, Comparative analysis of calculation results of supporting structure of soil-cement piles, in: Proceedings of 23rd International Scientific Conference Transport Means 2019, pt. II, Palanga, Lithuania, pp. 820–828.

[14] S.V. Raksha, P.G. Anofriev, V.M. Bohomaz, O.S. Kuropiatnyk, Mathematical and S-models of cargo oscillations during movement of bridge crane, Naukovyi Visnyk Natsionalnoho Hirnychoho Universytetu 2 (2019) 108–115.

[15] D.O. Bannikov, Analysis of the causes of accidents of steel capacitive structures for bulk materials, Metallurgical and Mining Industry 5 (2011) 91–96.

[16] V.I. Krysan, V.V. Krysan, V. Petrenko, O. Tiutkin, V. Andreev, Improving the safety of soil foundations when they are restored using soil-cement elements, in: 2nd International Scientific and Practical Conference "Energy-Optimal Technologies, Logistic and Safety on Transport" (EOT-2019).

[17] A.V. Radkevych, V.D. Petrenko, O.L. Tiutkin, V.S. Andrieiev, N.A. Mukhina, Comparative analysis of the parameters of the strength of the subgrade at the transition to the higher axial loading up to 25 t, in: IOP Conference Series: Materials Science and Engineering, 708(1), 012024, 2019.

Pilot Study of Linear Localization Using the Acoustic Emission Method During a Semi-Circular Bending Test

TOPOLÁŘ L.[1, a]

[1]Brno University of Technology, Faculty of Civil Engineering, Veveří 331/95, 602 00 Brno, Czech Republic

[a]Libor.Topolar@vut.cz

Keywords: Linear localization, Acoustic emission method, Semi-circular bending test, Alkali-activated material, Crack formation

Abstract. The paper deals with the topic of linear localization of acoustic emission sources in fine-grained alkali-activated composites. The accuracy of the localization of acoustic emission events depends on the homogeneity of the environment with regard to the propagating ultrasonic waves. The pilot experiments were conducted on three placements of the acoustic emission sensors relative to the axis of the test specimen in a semi-circular bending configuration. The aim of these pilot experiments was to assess the influence of the acoustic emission sensor placement relative to the stress concentration on the localization of the acoustic emission events.

Introduction

The acoustic emission method is a very effective method for the detection of various levels of structural damage in different types of materials [1]. Accurate localization of acoustic emission sources can be used to derive crack initiation and propagation [2-4]. Substantial effort has been devoted to the study of accurate localization of acoustic emission sources and a number of algorithms have been devised to determine the source of acoustic emissions. These algorithms can be divided into two groups - non-iterative (they have the same velocity for all stations, so they are inflexible in dealing with variable velocity models) and iterative (the derivative method, the sequential search method, the genetic algorithm and the simplex method) [5-8].

The given algorithms are based on the assumption that acoustic waves propagate directly from the source to the sensor along a straight line. That is to say, the environment in which the waves propagate is sufficiently homogeneous for the generated acoustic waves. The described algorithms lose accuracy in the case of two distinct environments or in the case of refraction of the generated waves [9,10].

The pilot experiments involved linear localization of the acoustic emission sources. We focused on the influence of the sensor placement on the recorded localized acoustic emission events. At the same time, the experiments aimed to find the ideal placement for acoustic emission sensors relative to the stress concentration for a planned series of experiments.

Experiment description

The experiments were conducted in a semi-circular bending configuration [11]. The advantage of this configuration is the initiation of a tensile crack even under compressive load see Fig. 1 and photographs in Fig. 2. The experiment parameters were as follows:

- disk radius R = 50 mm,
- span between the supports S = 80 mm,
- notch length a = 25 mm,
- notch angle of inclination β = 50°.

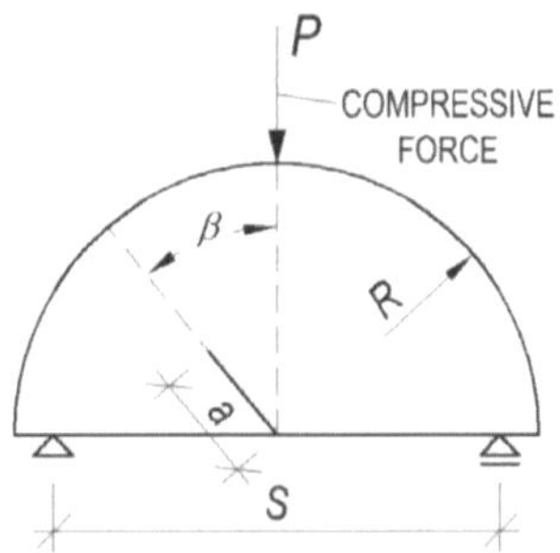

Fig. 1: Diagram of the semi-circular bending configuration used in the experiments [11]

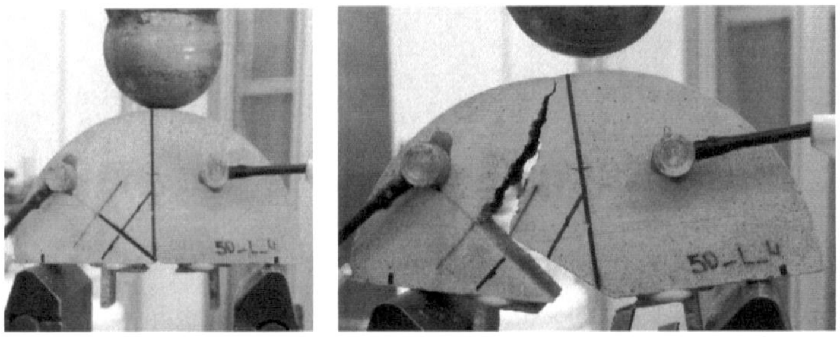

Fig 2: Photographs of the experiment before and after the fracture

The AE activity was observed by DAKEL-ZEDO system. DAKEL-ZEDO is a modular system for measuring AE that can be used throughout the entire spectrum of the application of this diagnostic method, for example, the detection and localization of the formation and development of the failure of materials. To measure this, the four-channel unit ZEDO-AE were used. The sensors were attached on the specimen surface with beeswax. The measuring AE parameters were set as follows frequency range 80–400 kHz; pre-amplifier 35 dB; software amplifier 20dB. The one sensor (guard-sensor) was mounted on the test equipment to catch the noise of background.

Results and discussion

The pilot tests were conducted on a fine-grained alkali-activated material. The dose of activator was adjusted to 6% Na_2O concerning the slag weight. Water to slag ratio was 0.45 (including water from the activator). Sand to slag ratio was 3:1 (by weight). Siliceous sand was with a grain size of up to 2 mm (consisting of three fractions according to EN 196-1). This fine-grained composite material is expected to contain minimal structural inhomogeneities that would prevent the passage of the generated mechanical waves. Three placements for the acoustic emission sensors were selected, see Figs. 3–5. Each sensor placement was subjected to determination of the localized position (vertical axis – the zero value is the position of the sensor that is in the notch section of the test specimen, left sensor) and amplitude of the acoustic emission (circle diameter) in relation to time. A graph of the course of the loading force in time is also presented.

In configuration A (Fig. 3), the sensors were placed symmetrically 30.0 mm from the axis of the test specimen. The left sensor, to which the distance of the localized events is related, was at a direct distance of 17.7 mm from the end of the notch. The graph shows a cluster of localized events at the distance of 18.0 mm at the moment of reaching the maximum force. This cluster is probably connected to the formation of a crack emerging from the end of the notch. Later, more significant events are evident near the test specimen axis. This is related to the shift of the crack front towards the test specimen axis.

In configuration B (Fig. 4), the left sensor was placed 20.0 mm from the test specimen axis and the right sensor 30.0 mm from the test specimen axis. At the same time, the left sensor was at a direct distance of 14.0 mm from the end of the notch. In this configuration, it can be observed that no events near the end of the notch were localized at the moment of reaching the maximum force. This was probably caused by the small distance of one of the sensors from the end of the notch.

In configuration C (Fig. 5), the left sensor was placed 30.0 mm from the test specimen axis while the right sensor was placed 20.0 mm from the test specimen axis. The left sensor was therefore at the direct distance of 17.7 mm from the end of the notch, as in configuration A. An event was localised at the distance of 17.3 mm from the left sensor at the moment of reaching the maximum force and probably corresponds to the formation of a crack front. Other localized events again indicate the shift of the crack front towards the axis of the test specimen.

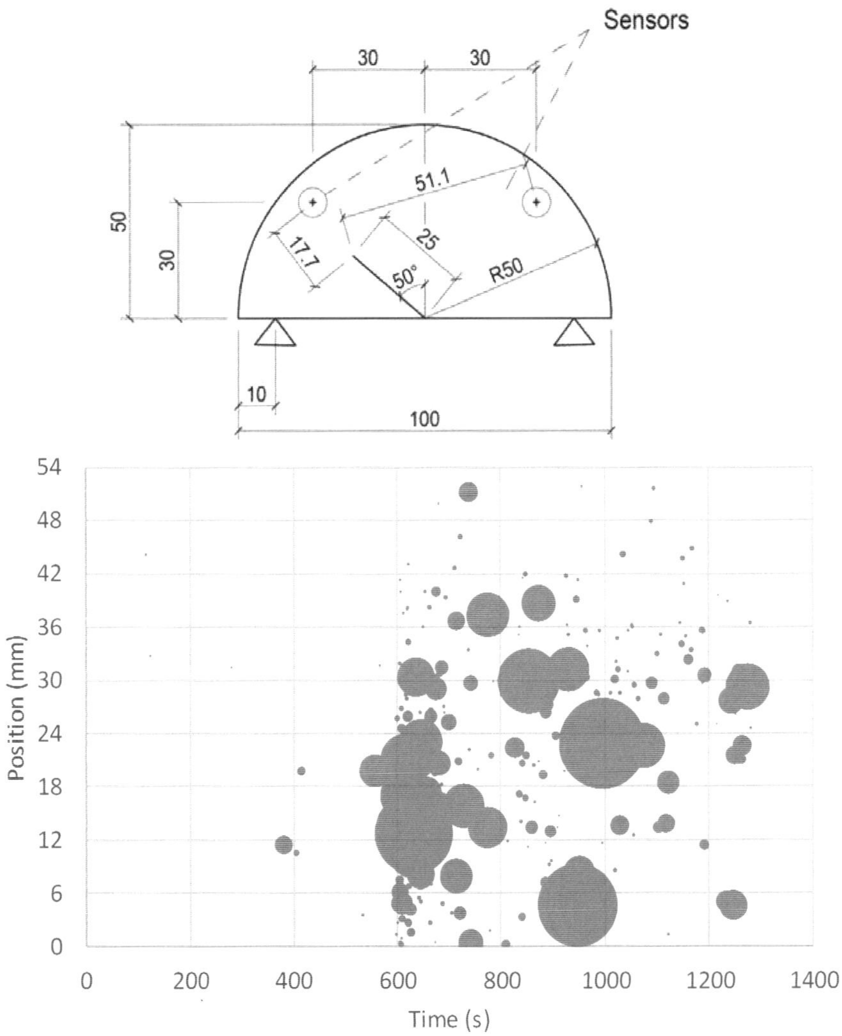

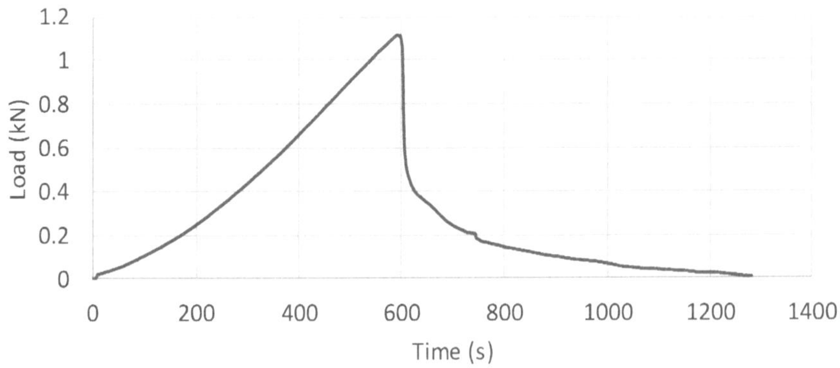

Fig. 3: Diagram of configuration A with results
(middle graph - vertical axis – the zero value is the position of the sensor that is in the notch section of the test specimen, left sensor)

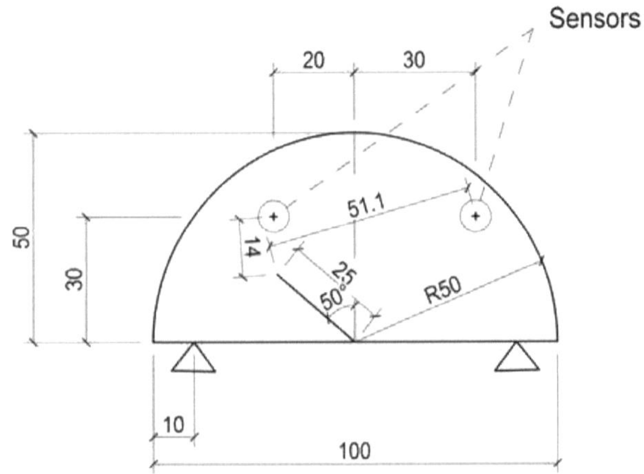

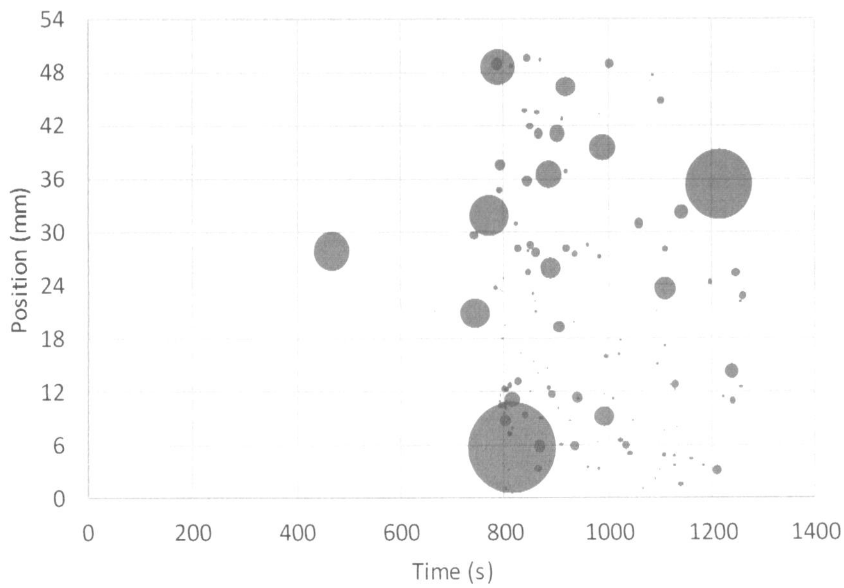

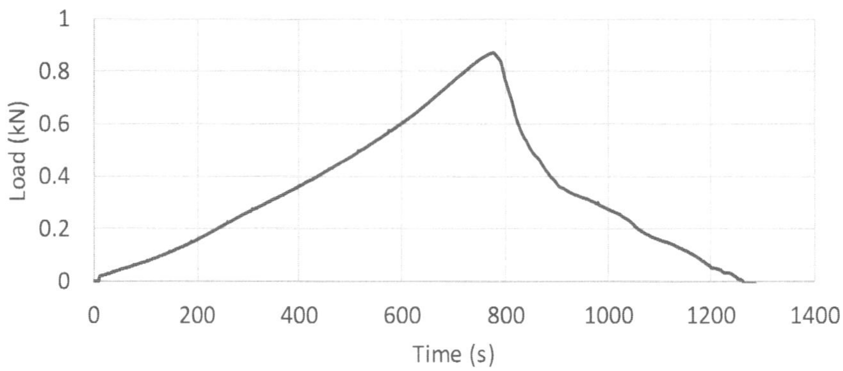

Fig. 4: Diagram of configuration B with results
(middle graph - vertical axis – the zero value is the position of the sensor that is in the notch section of the test specimen, left sensor)

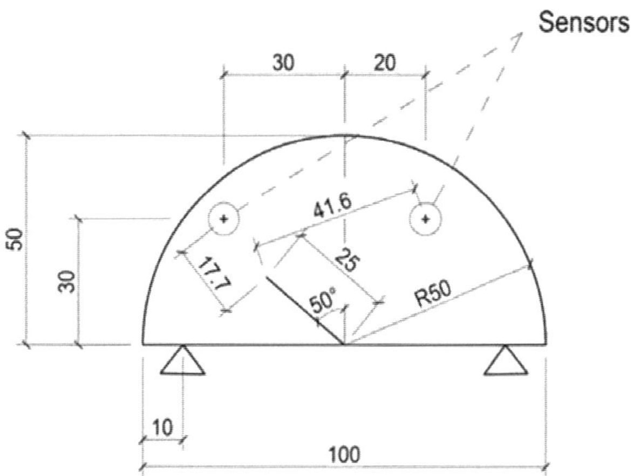

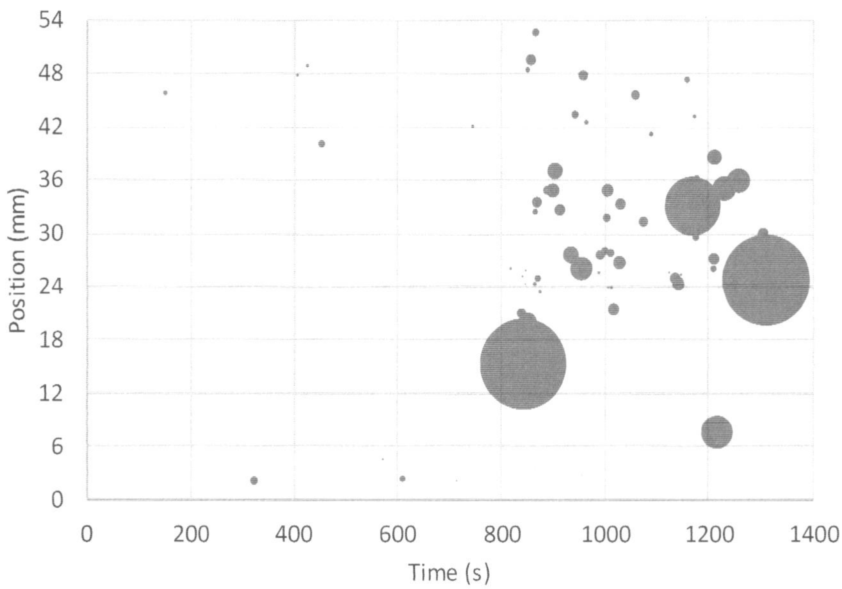

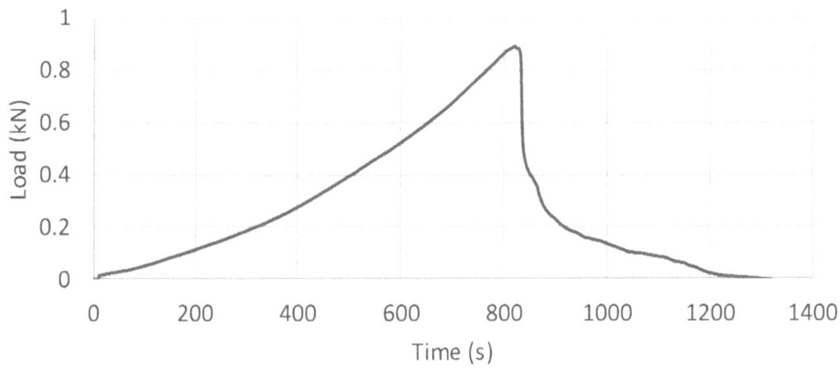

Fig. 5: Diagram of configuration C with results
(middle graph - vertical axis – the zero value is the position of the sensor that is in the notch section of the test specimen, left sensor)

Conclusion

The pilot experiments sought to identify a suitable position for the placement of the sensors in relation to the axis of the test specimen. Based on the results presented above, configuration B was discarded since it failed to successfully locate the acoustic emission events that would correspond to the theoretical assumption. The localized events also failed to match the physical outcome of the experiment. Configuration A was also discarded although its results were more consistent with both the theoretical assumption and the physical result. It was discarded because of a large number of false acoustic emission signals.

Configuration C was therefore selected for further experiments since the localized acoustic emission signals corresponded to both the theory and the physical result. At the same time, this configuration managed to filter out a large number of false signals. This is probably due to suitable distances of the individual sensors from the notch and from the axis of the test specimen.

Acknowledgement

This outcome has been achieved with the financial support of the Czech Science Foundation, project No. 18-12289Y "Advanced characterization of crack propagation in composites based on alkali activated matrix".

References

[1] Ch.U. Grosse, M. Ohtsu (ed.), Acoustic emission testing, Springer Science & Business Media, 2008.

[2] M. Ohtsu, Prospective applications of AE measurements to infra-dock of concrete structures, Constr. Build. Mater. 158 (2018) 1134–1142.

[3] M.N. Noorsuhada, An overview on fatigue damage assessment of reinforced concrete structures with the aid of acoustic emission technique, Constr. Build. Mater. 112 (2016) 424–439.

[4] S.C. Paul, et al., Acoustic emission for characterising the crack propagation in strain-hardening cement-based composites (SHCC), Cem. Concr. Res. 69 (2015) 19–24.

[5] B. Debecker, A. Vervoort, Localization by acoustic emission in transversely isotropic slate, Adv. Acoust. Vib., 2011 (2011).

[6] W. Huang, W. Zhang, F. Li, Acoustic emission source location using a distributed feedback fiber laser rosette, Sensors 13.10 (2013) 14041–14054.

[7] Z. Zhou, et al., A new closed-form solution for acoustic emission source location in the presence of outliers, Appl. Sci. 8.6 (2018) 949.

[8] A.G. Beattie, Acoustic emission non-destructive testing of structures using source location techniques, Albuquerque and Livermore (2013).

[9] S. Sengupta, A.K. Datta, P. Topdar, Structural damage localization by acoustic emission technique: A state of the art review, Lat. Am. J. Solids. Stru. 12.8 (2015) 1565–1582.

[10] Z. Zhou, et al., Experimental study on the location of an acoustic emission source considering refraction in different media, Sci. Rep. 7.1 (2017) 1-13.

[11] L. Malikova, et al., Multi-parameter fracture mechanics, Frattura Integr. Strutt. 13.49 (2019) 65–73.

Residual Stresses of Laser-Hardened Railway Axles

TROJAN K.[1,a], ČAPEK J.[1,b], GANEV N.[1,c], ČERNÝ I.[2,d], KEC J.[2,e], NĚMEČEK S.[3,f]

[1]Department of Solid State Engineering, Faculty of Nuclear Sciences and Physical Engineering, Czech Technical University in Prague, Trojanova 13, 120 00 Prague, Czech Republic

[2]SVÚM a.s., Tovární 2053, 250 88 Čelákovice, Czech Republic

[3]RAPTECH s.r.o., U Vodárny 473, 330 08 Zruč-Senec, Czech Republic

[a]karel.trojan@fjfi.cvut.cz, [b]jiri.capek@fjfi.cvut.cz, [c]nikolaj.ganev@fjfi.cvut.cz, [d]ivo.cerny@seznam.cz, [e]kec@svum.cz, [f]nemecek@raptech.cz

Keywords: Railway axles, Laser hardening, Residual stresses, Fatigue testing

Abstract. The paper describes the influence of laser hardening on the state of residual stresses and high-cycle fatigue resistance of railway axles made of the EA1N steel. It has been found that the compressive residual stresses along the surface of the hardened track in axial direction, which is more important from the technological point of view, significantly increase fatigue strength. With the optimized position of the laser tracks on the inner edge of the axle seats, these compressive stresses can play a crucial role in significantly increasing the fatigue resistance of the axles under operation.

Introduction

This research is aimed at the development of the technology of surface treatment of axles using laser hardening to increase resistance to fatigue failure under press fits. Railway non-powered axles made of standard EA1N steel were examined. These axles are connected with wheels by press fitting, performed usually at room temperature. Failures with fatal consequences can occur on the axle seats due to ongoing mechanical processes during operation [1]. The stress on the axle seats has the character of both high-cycle to very high-cycle fatigue and fretting fatigue. Both mechanisms, in particular, the fretting fatigue mechanism under compression, develop over a large number of load cycles – considering the service life of railway vehicles even more than 10^8 cycles [2]. By researching optimum combinations of laser surface treatment parameters, it is anticipated to achieve optimum structure and surface properties of the axle at critical points, including particularly fatigue resistance being positively affected by compressive residual stresses. Therefore, it is necessary to demonstrate a sufficient state of residual stresses on the axle seats and the higher fatigue life of the axle.

Experiment

For experimental tests, a forged turned semi-finished product made of the EA1N steel was used, see Fig. 1. The hardening was carried out circumferentially, following the expected hardening method for the axles to be operated, on axle seats with a controlled hardening temperature of 1150 °C, and a beam speed of 3 mm/s. Part of the sample was air-quenched, as is common for laser hardening, the second set of experimental samples were cooled with a water spray.

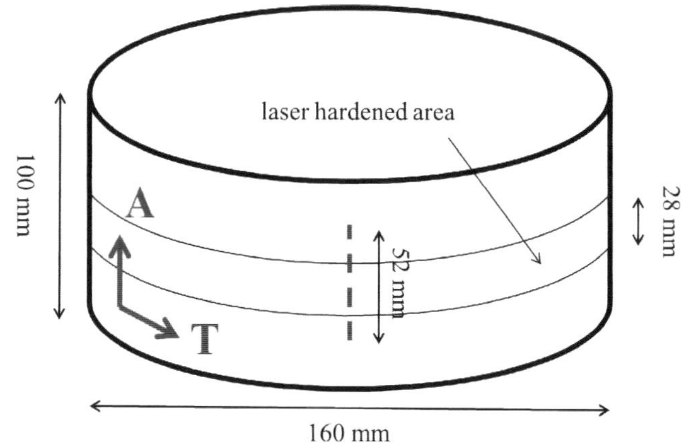

Fig. 1: Diagram of analysed samples

Metallographic specimens were hot embedded into conductive bakelite resin with the carbon filler. After the embedding, the specimens were prepared by machine grinding and polishing. Structure was revealed by etching in 4 % Nital (98 ml ethanol + 2 ml HNO_3). Microstructure was investigated by optical microscope Zeiss Axio Observer Z1M.

Residual stresses were determined by X-ray diffraction, {*211*} diffraction line of α-Fe was measured using Cr*Kα* radiation. The residual stresses across the hardened track, see Fig. 2 and the dashed line in Fig. 1, and consequently, the depth profiles of residual stresses were determined in the middle of the hardened track. The values of residual stresses were calculated from lattice deformations determined on the basis of experimental dependencies of $2\theta(\sin^2\psi)$ assuming uniaxial and resp. biaxial state of residual stress (θ is the diffraction angle, ψ the angle between the sample surface and the diffracting lattice planes). The diffraction angle was determined as the centre of gravity of the Cr$K\alpha_1\alpha_2$ doublet diffracted by the lattice planes {*211*} of the α-Fe phase. The X-ray elastic constants $\frac{1}{2}s_2 = 5.76\ \text{TPa}^{-1}$, $s_1 = -1.25\ \text{TPa}^{-1}$ were used for the stress calculation. The experimental error given for each value is the standard deviation according to the "$\sin^2\psi$" residual stress calculation algorithm [3]. The residual stresses were obtained in the axial A and tangential T directions. Nevertheless, the axle axial direction is more important from the technological point of view and, in this direction, the main stresses occur. The FWHM value (Full Width at Half Maximum) was determined from the analysed diffraction line {*211*} at a tilt of $\psi = 0°$, error of determination does not exceed 0.05° 2θ.

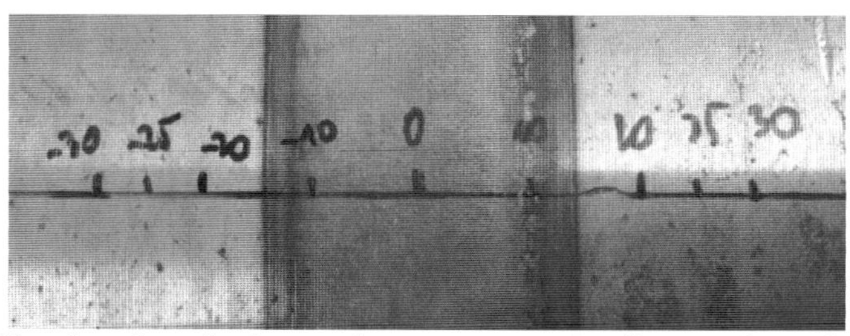

Fig. 2: Photo of hardened track, the numbers represent the distance from the track centre in millimetres

For the determination of the depth profiles of the macroscopic residual stresses, electrolytic etching of the surface layers of the material was used using a PROTO Electrolytic Polisher with the electrolyte A. The etched area was defined by a mask of Ø 14 mm. The depth of the removed layer was measured using a micrometre gauge.

Although results and analyses of residual stresses are the main subject of this paper, it is worthy to mention also fatigue resistance, which may be by residual stresses affected. Fatigue loading under press fits is very complicate and complex. Moreover, its simulation and experimental modelling are not easy and so it cannot be fully described and analysed within this paper, but some limited information about the experimental approach can be provided. As mentioned in the Introduction, two quite different fatigue mechanisms act on the surface of the axle seat under the press fit – "pure" high cycle fatigue and fretting fatigue. In this part of the investigations, only the first mechanism was studied. The problem was, how to simulate actual surface cyclic stresses, which are axial due to the bending loading of the axle in service. Eventually, after detailed analyses and considerations, it was decided that the best and likely also the only approach is to prepare three-point-bend specimens perpendicular to the surface laser track, with the laser track in the central part.

High-cycle fatigue tests were performed at different stress ranges to obtain the whole S-N curve including endurance limit. Load asymmetry was R = 0.1 (preloaded for security reasons), frequency 68 Hz. Test span was 80 mm. The central loading point was on the opposite side to the laser track. Cross section of the specimens was approximately 22 mm (height) x 20 mm (width).

Results and discussion

Cross-section of laser hardened track is shown in Fig. 3. The depth of track is 1.5–2 mm, which is sufficiently thick to affect the initiation and decelerate fatigue crack growth and friction corrosion cracks. Microstructure in laser-hardened track consists of martensite and retained austenite, see Fig. 4a. Under this full-transformed laser-hardened track, heat-affected zone (HAZ) formed by bainite and granular pearlite was observed, see Fig. 4b. The microstructure of the EA1N steel is ferritic-perlitic, see Fig. 4c - ferrite is equiaxial and fine with grain mean size 5.9 μm, colonies of pearlite has lamellar micromorphology.

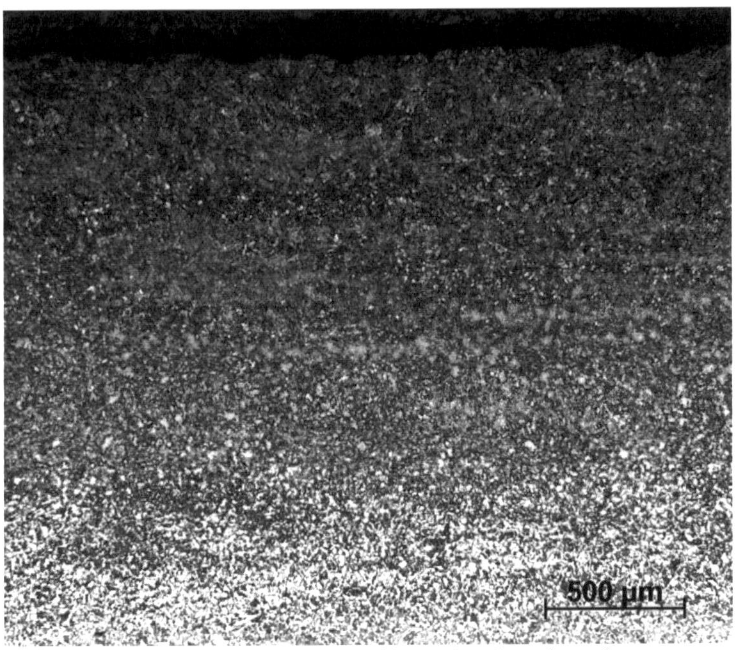

Fig. 3: Cross-section of laser hardened track

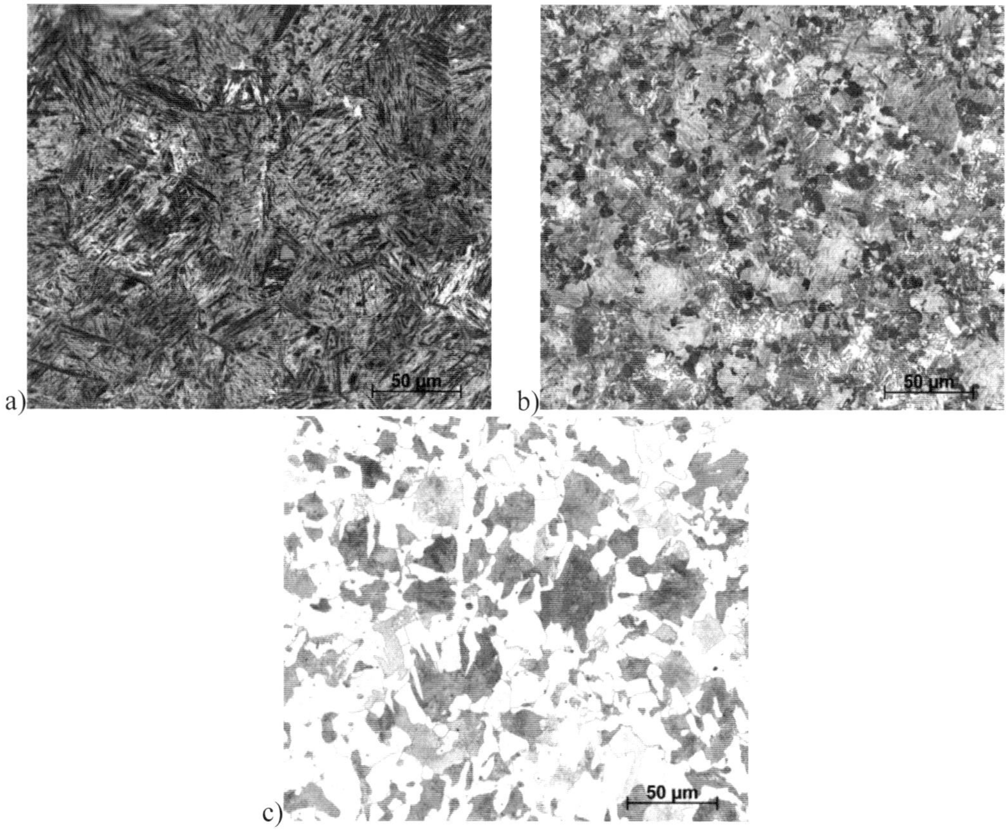

Fig. 4: Microstructure in: a) laser hardened track, b) HAZ and c) base metal

Analysed residual stresses on the surface of hardened track for the air-quenched (AQ) and also for the water-sprayed (WS) sample are compressive, on the contrary, residual stresses of bulk material (machined surface not affected by hardening, approx. 14 mm from the centre of hardened track) are tensile, see Fig. 5. The AQ sample shows slightly lower compressive residual stresses in the track, but this difference is not significant to compensate for the technological problems of quenching with a water spray. In addition, in the technologically significant direction A, the residual stresses are almost identical within the measurement inaccuracy as for the WS sample. Residual stresses of the bulk material vary by up to 400 MPa around the circumference, which is most likely caused by the production of the semi-finished product. However, the average value of 72 residual stress measurements around the circumference of the AQ sample in the middle of the hardened track is -101 ± 45 MPa. Therefore, the state of the residual stress after laser hardening is homogeneous around the circumference.

The FWHM values of the bulk material are homogeneous within the error, see Fig. 6. The FWHM values are higher in the hardened track compared to the bulk material. This could be a result of higher microdeformation and/or dislocation density, and/or smaller crystallite size. Additional experiments, especially diffraction phase analysis, are needed to identify the cause. Sharp maxima of the FWHM values approx. 10 mm from the hardened track axis can be caused by a step change of the diffraction angle of the $\{211\}$ planes. Due to the size of the irradiated area and the radiation beam divergence, the irradiated areas diffract with both the "smaller" and "larger" diffraction angles, resulting in an increase in the FWHM parameter. The step change in diffraction angle is caused by a change in the material microstructure between the bulk material (ferrite-pearlite) and the hardened area (martensite and retained austenite). The FWHM values of the bulk material of the samples are identical. Thus, the microstructure of the bulk material is homogeneous, however, the residual stresses vary.

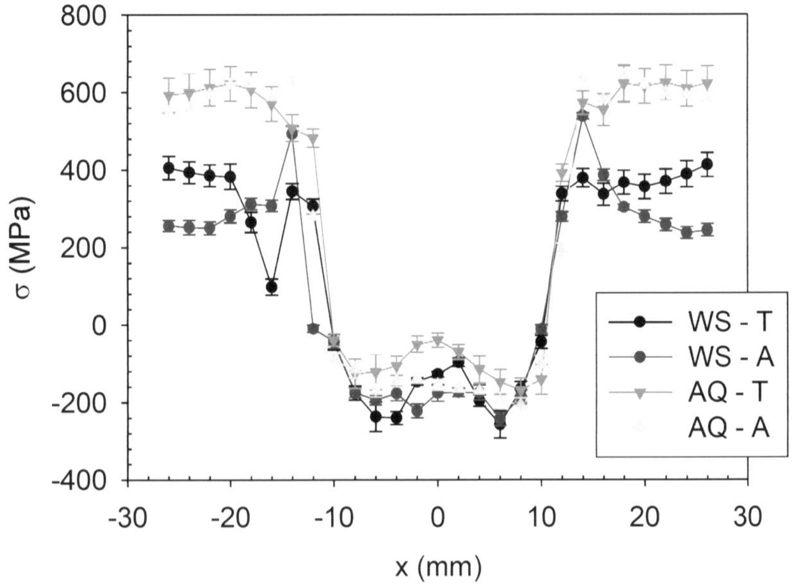

Fig. 5: Distribution of surface macroscopic residual stresses across the hardened track for air-quenched (AQ) and water-sprayed (WS) sample in the axial (A) and transversal (T) direction

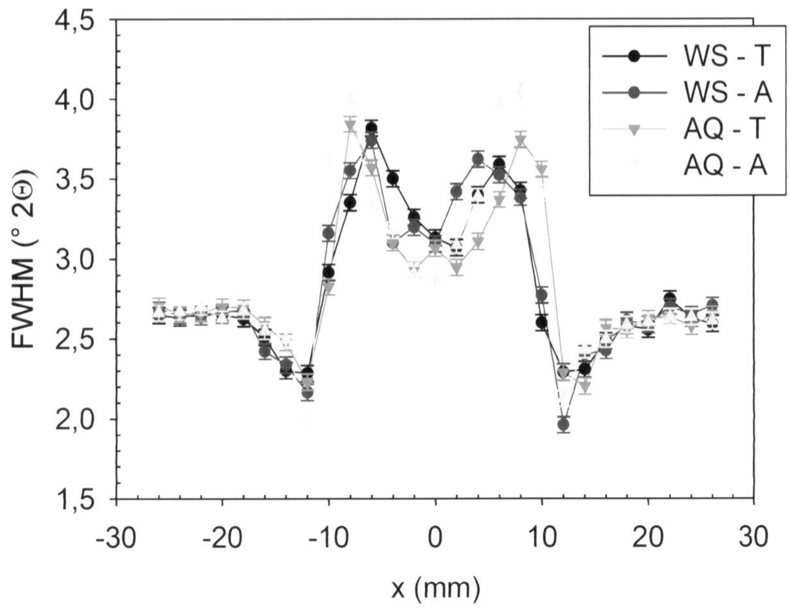

Fig. 6: Distribution of FWHM values across the hardened track for air-quenched (AQ) and water-sprayed (WS) sample in the axial (A) and transversal (T) direction

The depth profiles of the macroscopic residual stresses and FWHM values in axial direction are almost identical for both investigated samples, see Fig. 7. Compressive residual stresses up to a depth of ca 200–300 μm are an important positive factor for the fatigue resistance of the axles under operation. Compressive residual stresses of the WS sample are approx. 50 MPa higher than in the case of the AQ one. The greatest difference can be seen in the 300 μm thick surface layer. FWHM values are approx. 0.1° 2θ higher compared to AQ sample. These differences are not significant in terms of measurement accuracy, and again, slither higher compressive residual stress values do not justify the use of technologically more complicated quenching with a water spray. From the depth approx. 1000 μm the values of residual stresses and 1500 μm for FWHM are stable. The decrease in FWHM values indicates a decreasing microdeformation, dislocation density and/or increasing crystallite size. Low FWHM values refer to coarser grained bulk material, see Fig. 4.

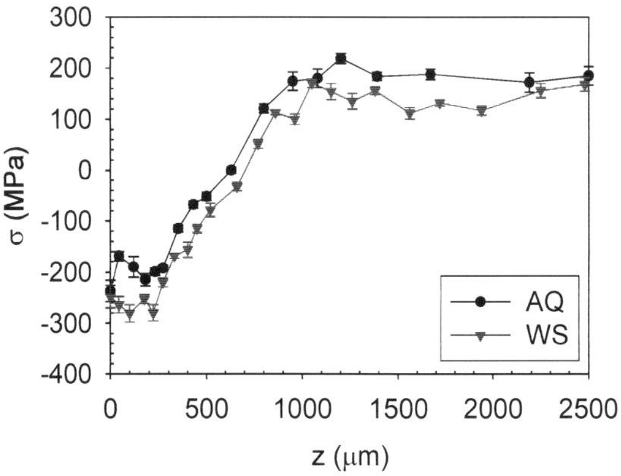

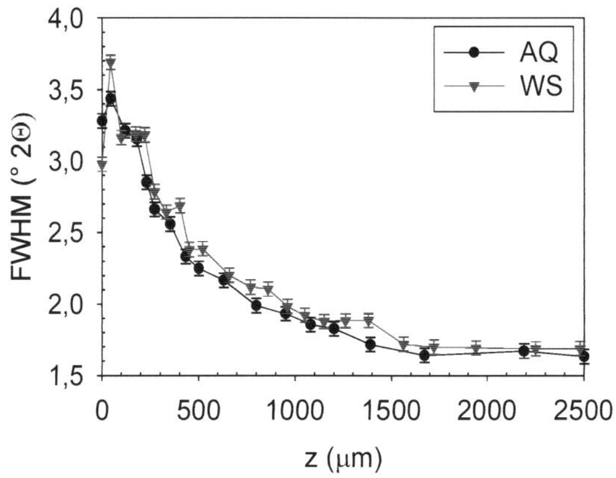

Fig. 7: Depth profiles of the macroscopic residual stresses and FWHM values for air-quenched (AQ) and water-sprayed (WS) sample in axial (A) direction

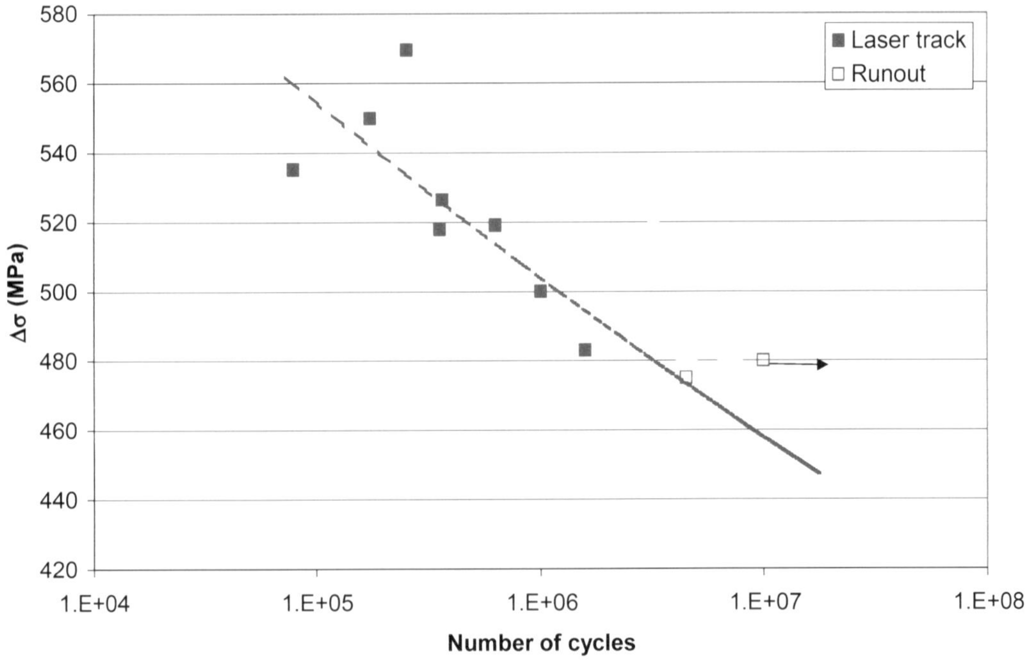

Fig. 8: Results of the high-cycle fatigue tests

Results of the first stage of high-cycle fatigue tests are in Fig. 8. During these tests, with the central load point exactly opposite the centre of the laser track, failure occurred always at the track centre. So, the diagram represents fatigue strength of the laser treated surface itself, not the boundary between the track and untreated material.

Though it is not easy to recalculate fatigue strength for different load asymmetries, the results are encouraging. The first estimations indicate that fatigue strength of the full axle will be higher than requested by railway standards and higher than for laser untreated axle. This is consistent with the measurement of residual stresses, where in the centre of the track, compressive stresses favourable for fatigue resistance occurred, see Fig. 5. Note, however, that the critical area of premature failure likely will be the boundary between the track and untreated material.

Conclusions

In terms of fatigue failure of the axle seats, compressive residual stresses in the surface layers of the laser track in the axial direction can be evaluated as the most valuable. With the optimized position of the laser tracks on the inner edge of the axle seats, these compressive stresses can play a crucial role in the process of retardation or even complete arrest of physically short cracks, thereby significantly increasing the fatigue resistance of the axles under operation. The paper is completed by results of high-cycle fatigue tests indicating a significant increase of fatigue strength inside laser traces on one hand and partial reduction of fatigue strength in the area of boundaries between the laser trace and untreated surface.

Acknowledgements

Measurements were supported by the project TH04010475 of the Technology Agency of the Czech Republic. This work was supported by the Grant Agency of the Czech Technical University in Prague, grant No. SGS19/190/OHK4/3T/14.

References

[1] G. Gürer, C. H. Gür, Failure Analysis of Fretting Fatigue Initiation and Growth on Railway Axle Press-Fits, Eng. Fail. Anal. 84 (2018) 151–166.

[2] V. Linhart, I. Černý, An Effect of Strength of Railway Axle Steels on Fatigue Resistance Under Press Fit, Eng. Fract. Mech. 78 (2011) 731–741.

[3] I. Kraus, N. Ganev, Residual Stress and Stress Gradients, in: Industrial Applications of X-Ray Diffraction, New York, USA, 2000, pp. 793–811.

EAN 2020
58th conference on experimental stress analysis

Residual Stress Measurement Uncertainty of the Hole Drilling Method

VÁCLAVÍK J.[1,a], WEINBERG O.[1,b], NÉMET J.[2,c], BOHDAN P.[2,d]

[1]Výzkumný a zkušební ústav Plzeň s.r.o., Tylova 1581/46, 301 00 Plzeň, Czech Republic

[2]Regional Technological Institute, Faculty of Mechanical Engineering, University of West Bohemia, Univerzitní 8, 306 14 Plzeň, Czech Republic

[a]vaclavik@vzuplzen.cz, [b]weinberg@vzuplzen.cz, [c]nemet@vzuplzen.cz, [d]pbohdan@rti.zcu.cz

Keywords: Residual stresses, Hole drilling method, Uncertainty

Abstract. The goal of presented work is discussion about residual stress uncertainty evaluation using hole-drilling method based on results of software SINT EVAL 7.2. The uncertainty for some input parameters are presented on model example for homogenous tensile stress field just using integral evaluation method. The results of comparative inter-laboratory test including uncertainties are shown including their evaluation in probability domain. It has been shown that the uncertainty is highest at the surface and at the final drilling depth and for ideal conditions was estimated to be 33 % at the surface.

Introduction

The hole drilling method [1] is standardised for long years [2], [3]. At the same time the test for presence of residual stresses is an objective of many testing laboratories. The estimation of uncertainty must be added to test results also for this type of measurement. The demand for accredited tests goes even further with necessity of comparison of the distributions of obtained results and allowable limits based on probability approach.

During evolution of evaluation methods from the procedure based on one step measurement [4] followed with multiple power series method [5] ending with the integral through profile method [6], [7] the uncertainty estimation became rather bit complicated.

According [3] the uncertainty of determination of the uniform stresses by means of the power-series method is ±10%, no estimation is made for integral method. In [5] an extensive analysis of a measuring error was carried out based on the uncertainties of the most of the input parameters which results in a similar error stated in the standard [4]. The main source of the uncertainties specified herein is the eccentricity of the hole being drilled (5%), stress induced by drilling (5.5%), a diameter of the hole being drilled and the material constants. These considerations are valid if all the requirements of the standard [3] for the measuring accuracy have been met (both the hole misalignment and depth accuracy less than ±0.004·gauge circle D, strain reading accuracy and stability ±1 $\mu\varepsilon$).

According [1] for incremental hole drilling the maximum uncertainty for given test was about ±40 MPa (15 %) at the surface and ±24 MPa (20 %) on the last drilled layer. The smallest one was estimated in the middle of whole drilled depth.

Calculation of uncertainties of the hole drilling method

Recently introduced software SINT Eval 7.2 [7] includes very good estimation of the uncertainty for several hole-drilling methods used for non-homogenous residual stress profile evaluation. Operator can in detail set the uncertainties and their distributions of all variables, coming into the calculation: Young's modulus E, Poisson's ratio μ, strain measurement ε, gage factor k, hole diameter D_0, zero depth offset z_0, and depth measurements z_i. Some systematic errors can be eliminated e.g. correction of end mill fillet, form of strain gauge rosette, hole eccentricity and elastic-plastic stress state. As a result, the combined stress uncertainty U expanded by a coverage factor k for each evaluated layer is computed, tabled and presented in a chart as upper and bottom tolerance limits around the computed residual stress mean value.

Model example. For examination of the influence of basic input parameters to the resulting uncertainty, the set of 20 released strains was generated at equal steps from 0.05 mm to 1 mm for in depth homogenous tensile stress field $\sigma = 100$ MPa. This was made solving directly the basic equation for integral method and using the relaxation constants given in [3] for 1/16 in. rosette; calculation was made inside MS Excel. The resulting set of released strains ε_{ai}, ε_{bi}, ε_{ci} was loaded to the Eval 7.2 software and the residual stresses were evaluated. The obtained residual stress profile using the ASTM E837-13a method for non-homogenous stress field is presented in Fig. 1, curve σ_{min}, σ_{max}. The residual principal stresses perfectly fit the input value of 100 MPa which indicates, that both Eval and our calculations were made correctly.

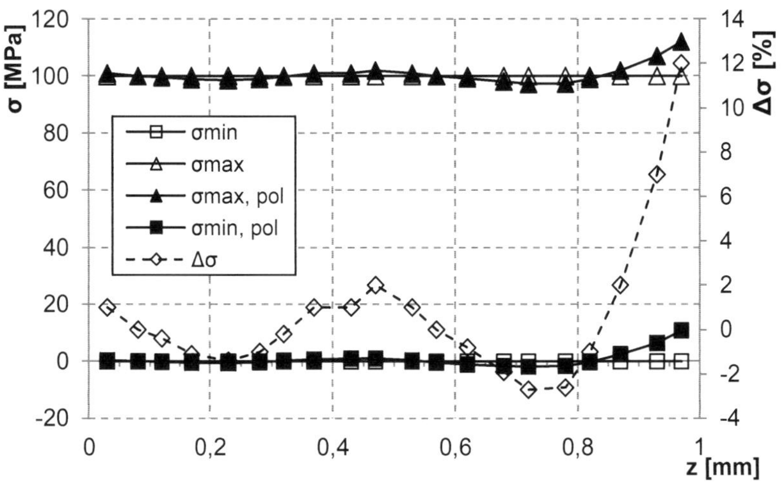

Fig. 1: Ideal stress profile for pure tension and distortion using polynomial interpolation

One of the possible smoothing of measured set of strains before the same calculation (and often used), is to interpolate them with the polynomial. You can see in Fig. 1, curves marked as "pol", what happens, if you use this EVAL option. The evaluated stress field near the whole depth changes to unrealistic value. This result, obtained for ideal calculated strains, may be much worse for real data from the measurement. The reason is that the polynomial may change the slopes of relaxation curves for all gages in another way. The difference in the whole depth causes the lack of the data for the interpolation from the curves right sides. It is therefore recommended to drill the whole some steps (e.g. four) above the required depth.

Using uncertainty computation with the software Eval 7.2 is presented in Fig. 2 again for the above mentioned model example for the pure tension. The tolerance limits for 95.4 % coverage probability were evaluated for two cases of input uncertainties. The first case are pre-set "Eval" values labelled as $\pm U(\sigma)$, the second case are uncertainty values specified by ASTM standard as the worst accuracy of measured strains, hole dimensions and the depth (labelled as $\pm U(\sigma)$, ASTM). In Table 1, these expanded uncertainties are evaluated for three depths including input values for standard deviations of selected parameters. Here also the uncertainty calculation for pre-set input value of measuring amplifier SPIDER8 (third example) and the calculation for the lower accuracy of depth measurement (when the manually controlled device as Vishay RS 200 is used) are presented.

It is obvious, that the highest uncertainties are at the surface and at the whole depth. The uncertainty at the whole depth increases with the increase uncertainty of the measured input values. For pre-set optimistic parameters the uncertainty is about 33 % at the surface and 23.3 % at the whole depth; for lower input uncertainties the stress uncertainty may reach about 100 %. On the other hand, the uncertainty in the middle depth does not change significantly from depth 0.3 mm to 0.7 mm (uncertainty is here lower than 10 %).

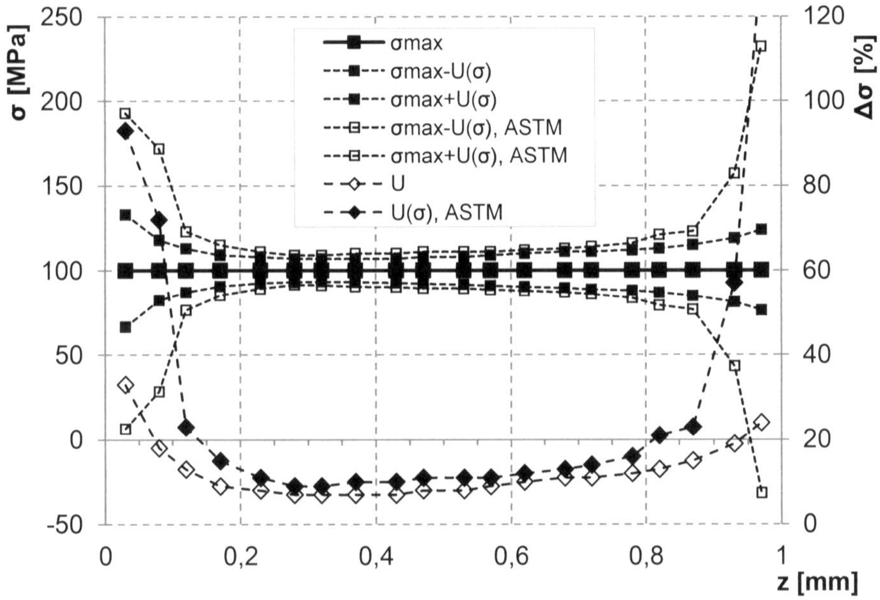

Fig. 2: Uncertainty tolerance limits for pre-set and critical ASTM input uncertainties

Table 1: Uncertainty at three depths for selected input parameters

Uncertainty estimation	Parameters		Pre-set	Depth with lower accuracy	SPIDER8	ASTM
Material	E	[%]	3	3	3	3
	μ	[%]	2	2	2	2
Gauge factor	k	[%]	1	1	1	1
Amplifier			pre-set	pre-set	SPIDER8	pre-set
	ε	[μm·m^{-1}]	0.32	0.32	1.83	0.32
	class	[%]	0.05	0.05	0.1	0.05
Hole (turbine)	D_0	[mm]	0.01	0.01	0.01	0.02
Depth	zero depth	[mm]	0.01	0.01	0.02	0.01
	accuracy	[mm]	0.001	0.01	0.02	0.001
	step by step	[mm]	0.002	0.01	0.02	0.002
	repeatability	[mm]	0.002	0.01	0.01	0.002
Uncertainties	$U(\sigma)_{z=0.05}$	[%]	33.3	52	75.5	93.4
	$U(\sigma)_{z=0.35}$	[%]	6.8	7.4	7.6	8.6
	$U(\sigma)_{z=0.95}$	[%]	23.6	62.5	86.2	128.6

Using the uncertainties of the hole drilling method in praxis

Comparative tests. The comparison of evaluated profiles of residual stresses measured on a bisected part of the railway axle made from annealed A1N steel (Fig. 3) using the software [6] was performed. Three different drilling techniques were used: high speed air turbine with SINT MTS3000 (D_0=1.8 mm), low speed end mill with Vishay RS200 (D_0=4 mm and 2 mm) and end mill with SINT MTS3000 (D_0=4 mm) (both 30000 rpm). Here only the minimum principle stresses of two drilling techniques made with air turbine (RTI) and end mill (VZÚ) are presented (Fig. 4). The upper and lower tolerance limits (stress uncertainty $U(\sigma)$ for $k = 2$) for both laboratories are drawn with dashed lines. The most of the input values of uncertainties were chosen from EVAL 7.2 database and the rest was estimated. The standard [3] requirements for the input uncertainty were not exceeded. For the evaluation the EVAL ASTM E837-13 advanced method was used.

Fig. 3: Investigated half section of the axle and used drilling device Vishay RS-200

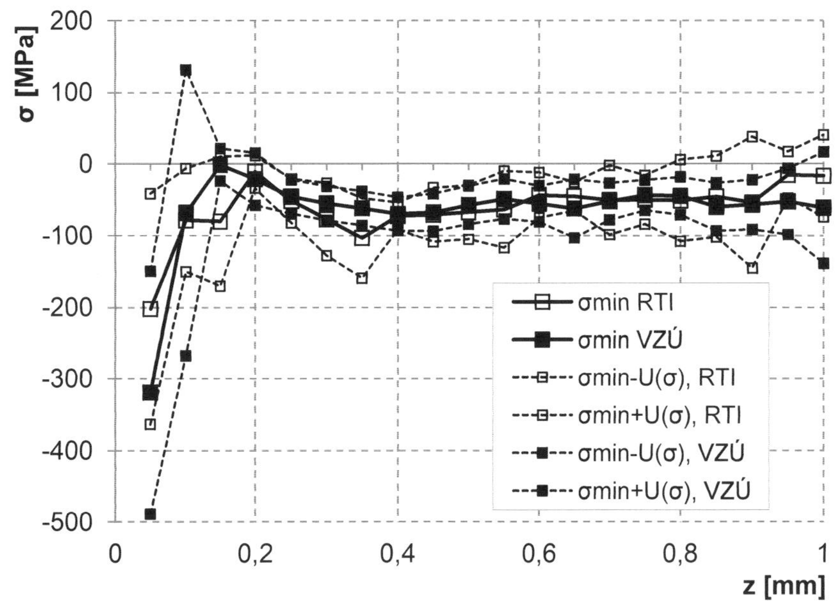

Fig. 4: Compared residual stress profiles including tolerance limits for 2 drilling techniques

The largest differences between evaluated residual stress profiles are at the surface and in the 1 mm depth, which corresponds to the uncertainty tolerance limits.

In Fig. 5 (left) only the upper tolerance limits for the results from the laboratory VZÚ are presented. It is interesting, that depth profile for real measurement is not as smooth as in model example and is a function of the stress values at individual steps. This observation is valid also for relative uncertainties. In real measurement the evaluated stress profile is also not ideally smooth. The large changes between steps may probably be caused also due to drilling errors. Good praxis may be to compare these changes with the uncertainty envelope – the changes should be inside the tolerance limits (Fig. 5, right). In other way it is necessary to correct the uncertainties of input parameters.

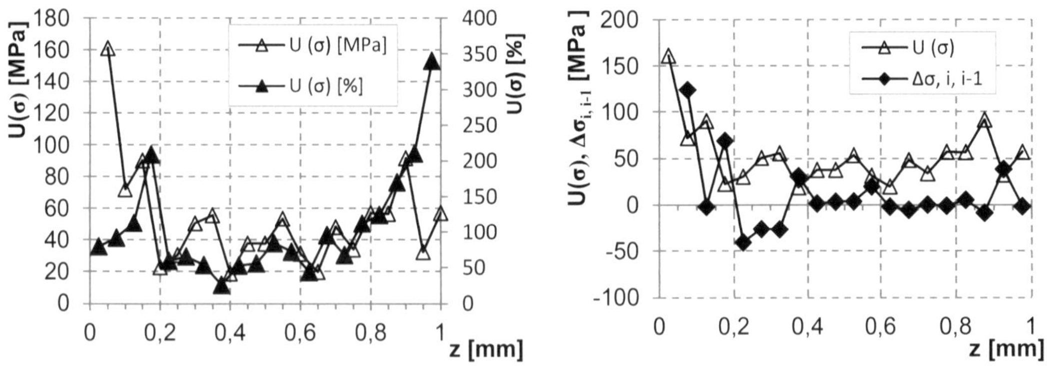

Fig. 5 Error analysis of real measurement (left absolute and relative values, right comparison of the uncertainty with the stress difference between two adjacent steps)

Uncertainty assessment. For obtained uncertainties it is necessary to decide, if the results are good or bad. For here presented inter laboratory test is good praxis to compare the results using the E_n – score (1), calculated from combined expanded uncertainty for each step. For satisfactory results this factor should be within the interval ±1.

$$E_n = \frac{\sigma_1 - \sigma_2}{\sqrt{U_{\sigma_1}^2 - U_{\sigma_2}^2}} \qquad (1)$$

Obtained profile of this comparative characteristic for presented inter-laboratory test is presented in Fig. 6, left. From the point of view of this assessment the most problematic are the middle depth, where the theoretical uncertainties are low, which for the higher differences of mean values can cause unsatisfactory results.

For laboratory tests it is necessary to make assessment of coincidence of the results with the customer specification. This was made using mean values in the past. However the probability assessment has to be made at present. E.g. for our presented test the measured residual stress near the surface has to be lower than the limit value of 100 MPa. An example of this access is made in Fig. 6, right. The distribution of uncertainties at the surface, in the middle and the whole depth are presented here with Gaussian approximation. It is obvious, that the evaluated stress on all depth is lower than 100 MPa with probability approaching level 1. Should the mean measured value be more closely to 100 MPa and the probability density function wold intersect the limit value, the fulfilling the specification would be with lower probability, which might not be accepted from the customer. This is the driving force for the increasing of the accuracy measurement of the testing laboratories.

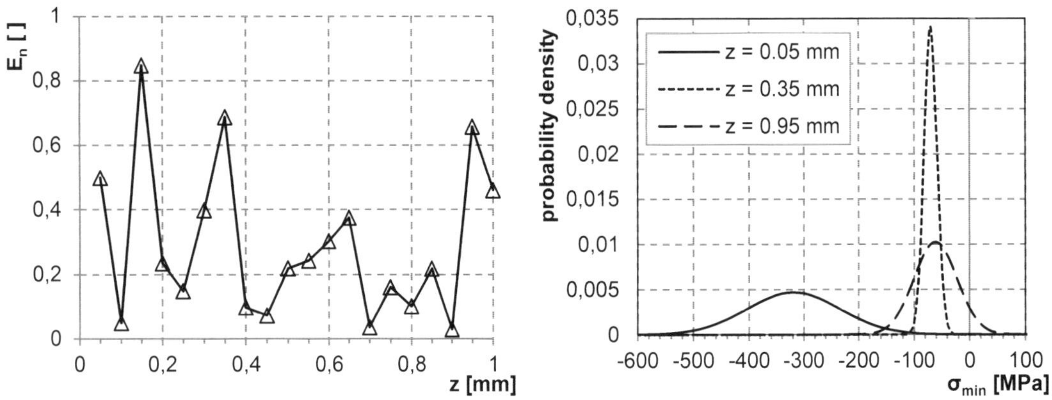

Fig. 6: Uncertainty assessment – inter-laboratory test (left), uncertainty probability (right)

Conclusions

It was shown, that the uncertainty of the hole drilling method is for high accuracy measurement with stepping motor about 33 % at the surface, 8 % in the middle depth and about 24 % in the whole depth. With decreasing the measurement accuracy the uncertainty increases especially at the whole depth. Using ASTM recommended limits of the measurement accuracy the uncertainty increases at the whole depth about one order.

The uncertainties of input values to the calculation of resulting uncertainty have to be selected very carefully, otherwise unrealistic low or high values can be obtained.

The resulting uncertainties have to be included to the measurement results because it is common praxis that the results are given for 95.4 % probability. If the limit stress is e.g. 100 MPa at the surface, the specification is fulfilled, when the measured mean value is max. 67 MPa.

The article has originated in the framework of the institutional support for the long-term conceptual development of the research organization.

References

[1] Gary S. Schajer, Practical Residual Stress Measurement Methods, Willey & Sons Ltd., 2013.

[2] J. Václavík, P. Bohdan, S. Holý, O. Weinberg Residual Stress Evaluation According ASTM E837-08 Revision, Proceedings of 47th International Conference on Experimental Stress Analysis 2009, Sychrov, June 2009.

[3] ASTM E837-13a. Standard Test Method for Determining Residual Stresses by the Hole-Drilling Strain-Gage Method.

[4] R. Oettel, The Determination of Uncertainties in Residual Stress Measurement (Using the hole drilling technique), Code of practice, No.15, Issue 1, September 2000.

[5] M. Scafidi, E. Valentini, B.Zuccarello, Error and Uncertainty Analysis of the Residual Stresses Computed by Using the Hole Drilling Method, Strain 47 (2011) 301-312.

[6] M.Vlk, M. Peč, Some Recognitions Obtained by Software EVAL-7, Proceedings of 53th International Conference on Experimental Stress Analysis 2015, Český Krumlov, June 2015, pp. 477-478.

[7] MTS3000-Restan EVAL 7.2 Software, Operating and Maintenance Manual, SINT Technology s.r.I., 2019.

EAN 2020
58th conference on experimental stress analysis

Experimental Tests of Braided Composite Tubes with Kapton Tape

ZEMČÍK R.[1,a], ZEMČÍK H.[1,b], KRYSTEK J.[1,c], KROUPA T.[1,d], PIHERA J.[2,e], PROSR P.[2,f], KOMÁREK J.[3,g], FRIEDL J.[3,h]

[1]NTIS – New Technologies for the Information Society, Faculty of Applied Sciences, University of West Bohemia, Technická 8, 301 00 Plzeň, Czech Republic

[2]University of West Bohemia, Faculty of Electrical Engineering, Regional Innovation Centre for Electrical Engineering (RICE), Univerzitní 8, 301 00 Plzeň, Czech Republic

[3]Technofiber, s.r.o., Bulharská 2172/122, 612 00 Brno, Czech Republic

[a]zemcik@kme.zcu.cz, [b]hzemcik@kme.zcu.cz, [c]krystek@kme.zcu.cz, [d]kroupa@kme.zcu.cz, [e]pihera@ket.zcu.cz, [f]prosr@ket.zcu.cz, [g]j.komarek@technofiber.cz, [h]ninkt@yahoo.com

Keywords: Composite material, Braided textile, Polyimide film, Four-point bending test, CT scan

Abstract. This paper focuses on experimental investigation of mechanical properties of composite tubes with separated layers and Kapton tape inclusions. The tubes are made by braiding of glass fiber bundles on a cylindrical mandrel and application of epoxy resin. The standard configuration is compared with two other configurations. First, the layers are simply separated along the circumference, thus the reinforcing fibers are no longer continuous. Second, the Kapton tape is inserted between separated layers. This configuration should have special electrical insulating properties. The tubes are subjected to four-point bending test. The variation of stiffness and load-carrying capacity of the tested configurations are analyzed.

Introduction

Kapton is a material developed by DuPont company. It is commonly used for insulation of electric and electronic devices. It offers excellent, physical, electrical, thermal and chemical resistant characteristics. It is a polyimide and it is usually available in a form of a homogeneous film wound onto a drum. In this work the Kapton tape is used within the layup of glass-epoxy (Letoxit LH198 resin, Letoxit EM231 hardener) braided composite tubes in order to increase the electrical insulation capability in case of high-voltage applications.

The goal is to investigate the mechanical properties of three configurations of the braided tubes using four-point bending test and to mutually compare their stiffness and strength. The stiffness is defined here as the ratio of force vs. deflection and the strength means the load-carrying capacity (i.e. the maximum force reached) during the bending test.

Material and Apparatus

The first configuration (V1) of the tested structures is represented by purely composite tube having 6 layers of glass-epoxy braided textile reinforcement (approx. ±45°). The second type (V2) has layer which were separated along the whole circumference (two separation per layer). The third type (V3) has layers that are separated in the same was as V2 but there is a 20 mm wide ring of Kapton tape inserted along the circumference between each separated textile parts. The layer separation is shown in Fig. 1 and the scheme of its placement along the tubes's length is shown in Fig. 2. The analysis of mechanical properties of various polyimide films (Kapton) from the anisotropy point of view was performed previously [1]. The tubes were manufactured by Technofiber s.r.o. company. The dimensions were as follows: outer diameter 100 mm, inner diameter 89 mm, length approx. 1000 mm. The bending tests were performed using Zwick/Roell Z050 machine with custom 3D-printed supports and cork layer (see Fig. 3). The internal structure of the tubes was investigated using CT scans and subsequent segmentation analysis (see Fig. 6).

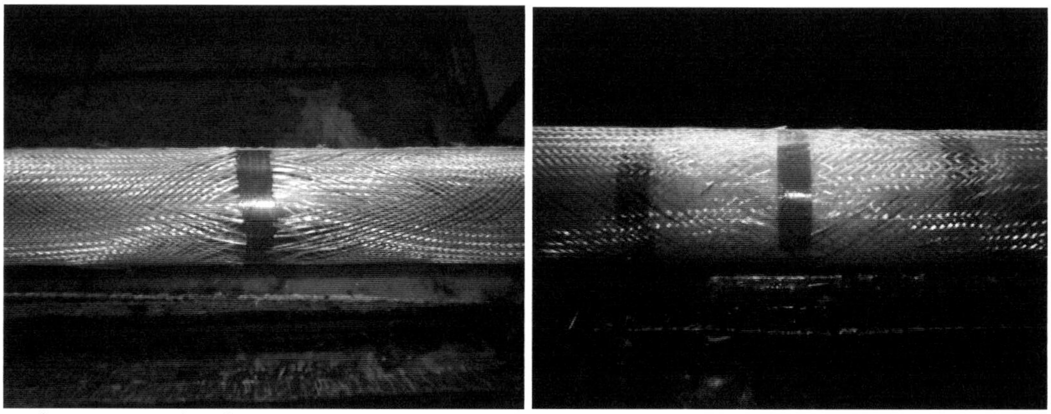

Fig. 1 Example of Kapton tape applied in the first layer (left) and in three different layers (right).

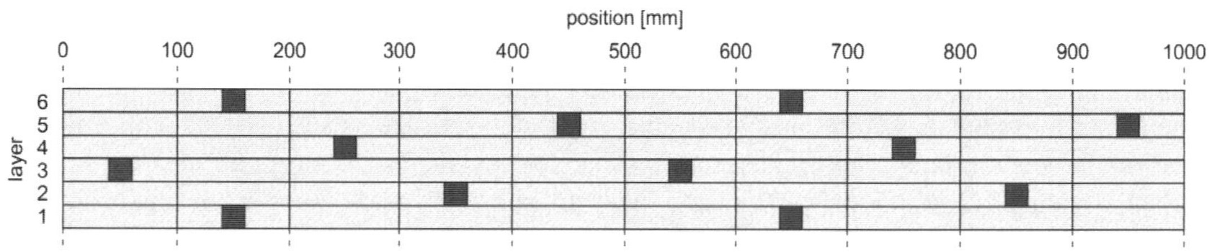

Fig. 2 Scheme of locations of layer separations indicated by red marks. Layer 1 is at the inner diameter and layer 6 is at the outer diameter.

Results and Conclusions

Five tubes for each configuration V1, V2 and V3 were tested. The dependencies between the loading force F and the displacement of the upper support u were measured. The obtained curves are shown in Fig. 4. Examples of ruptured tubes are shown in Fig. 5. The values of the maximum measured force at tube rupture (the load capacity) are summarized in Tab. 1 for the three configurations. The tube stiffness calculated as dF/du on the interval $u \in (5;\ 15)$ mm (linear regression on the dominantly linear part of the curves) is summarized in Tab. 2.

It can be concluded that both the layer separation and the presence of the inserted Kapton tape have certain influence on both the bending stiffness (curve slope) and the load-carrying capacity (maximum force), as could be expected. The mean load capacity of tubes with continuous reinforcement was notably larger than that of the tubes with separated layers. However, the differences between configurations V2 and V3 were smaller than those within these configuration. Therefore, it cannot be concluded which type of separation has which influence on the load capacity. It will be necessary to test larger set of tubes for each configuration and to avoid undesirable geometrical imperfections of the tubes (such as variable thickness or the reinforcement orientation).

Similar conclusion can be drawn from the stiffness point of view. The tubes with continuous reinforcement have the largest mean stiffness and the smallest deviation. The separation causes decrease in the stiffness and increase in the deviation. The effect is even more significant in the case of tubes with Kapton tape.

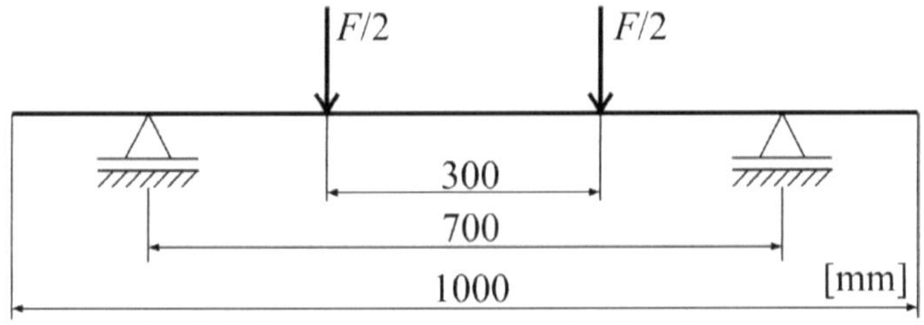

Fig. 3 Scheme of the four-point bending test.

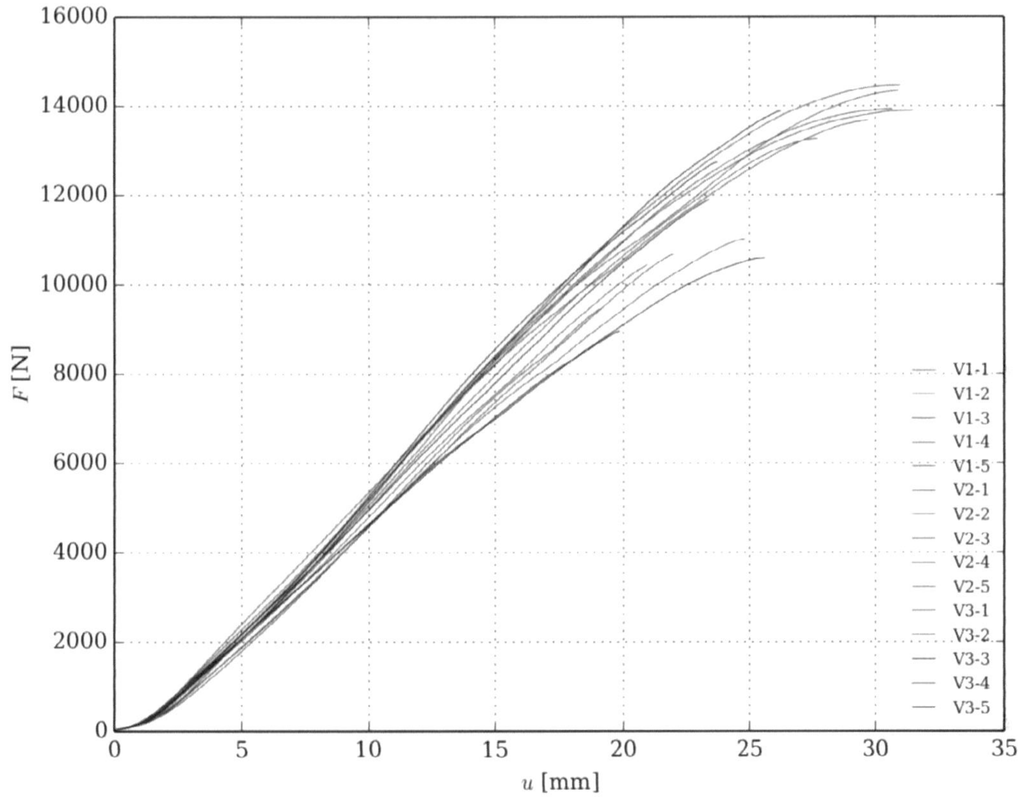

Fig. 4 Measured force–displacement dependencies for all tested tubes.

Fig. 5 Photographs of used apparatus and examples of ruptured tubes (top – V1, middle – V2, bottom – V3).

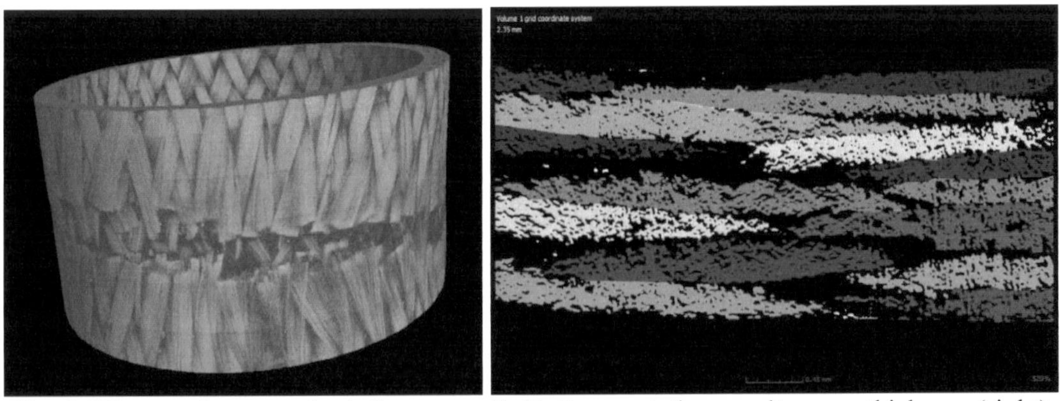

Fig. 6 CT scan of separated reinforcement (left). Segmentation result across thickness (right).

Table 1: Numerical comparison of measured load-carrying capacity values.

Maximum force F [kN]	Configuration V1	Configuration V2	Configuration V3
Tube 1	13.678	13.274	10.595
Tube 2	13.897	10.677	11.891
Tube 3	14.461	9.459	8.957
Tube 4	13.933	11.017	12.739
Tube 5	14.333	10.437	13.888
Mean	14.061	10.973	11.615
Median	13.934	10.677	11.892
Standard deviation	0.325	1.411	1.911
Minimum	13.678	9.460	8.958
Maximum	14.461	13.274	13.888

Table 2: Numerical comparison of measured stiffness values.

Tube stiffness [kN/m]	Configuration V1	Configuration V2	Configuration V3
Tube 1	590.09	587.71	507.30
Tube 2	607.40	538.16	562.28
Tube 3	608.43	572.50	511.32
Tube 4	599.57	541.07	629.06
Tube 5	622.66	553.44	623.94
Mean	605.63	558.57	566.78
Median	607.40	553.44	562.28
Standard deviation	12.03	21.17	58.69
Minimum	590.09	538.16	507.30
Maximum	622.66	587.71	629.06

Acknowledgments

This publication was supported by project TRIO FV30382 "Braided composite construction in electrical engineering – NAPREKON" of Czech Ministry of Industry and Trade.

References

[1] T. Kroupa, H. Zemčík, R. Zemčík, P. Kadlec, J. Pihera. Mechanical testing of Kapton films, In: 36th Danubia-Adria Symposium on Advances in Experimental Mechanics, Extended Abstracts, pp. 143–144, Plzeň, 2019.

Experimental Verification of the Bending Stiffness of the Wall for the Program for Design of High Walls in Dry Construction

ZOBAL J.[1,a], BITTNAR Z.[1,b], FAJMAN P.[1,c], NYČ M. [1,d], JÁRA R.[2,e], POŠTA J.[2,f], HATAJ M.[2,g]

[1]Czech Technical University in Prague, Faculty of Civil Engineering, Department of Mechanics, Thákurova 7, 166 29 Prague 6, Czech Republic

[2]University Centre for Energy Efficient Buildings of Czech Technical University in Prague, Třinecká 1024, 273 43 Buštěhrad, Czech Republic Republic

[a]ondrej.zobal@fsv.cvut.cz, [b]bittnar@fsv.cvut.cz, [c]fajman@fsv.cvut.cz, [d]nyc.m@atlas.cz, [e]robert.jara@cvut.cz, [f]jan.posta@cvut.cz, [g]martin.hataj@cvut.cz

Keywords: High wall, Dry construction, Bending stiffness, Calculation program

Abstract. This paper presents the results of experimental determination of the bending stiffness of a high wall made of plasterboard and newly developed calculation tool for the design of high walls. The experiments and the calculation followed the relevant standards and the design of the calculation program is based on the optimization of different approaches.

Introduction

The great boom of the dry construction in the Czech Republic is in progress on since the turn of the 20th and 21st centuries, when this technology has significantly improved here. The dry construction has found application not only in the realization of temporary buildings, but also in reconstruction and new buildings. The plasterboard is used for vertical and horizontal constructions. The main advantages are quick construction, elimination of wet process, good acoustic properties and fire resistance. The disadvantage is higher sensitivity to moisture [1].

The plasterboard is increasingly used for the realization of high walls and partitions. In practice, the limit height is 4 m, a more complex static assessment is needed for the design of higher structures. The design and assessment of the high wall in the Czech Republic are governed by the Eurocode, specifically ČSN EN 1991-1-1 a ČSN EN 1991-1-4 [2,3]. The problem is that this methodology does not take into account some important parameters such as the effect of vibration and pressure (only the line load is considered in the Eurocode). However, these parameters are taken into account by the German standard DIN 4103-1 [4]. Another difference between these methodologies is that the Eurocode does not prescribe an individual assessment for walls higher than 12 m. It is, therefore, necessary to complete the theoretical basis in the Czech Republic and the lack of basic research on the parameters necessary for the relevant assessment. One such parameter is the bending stiffness of the wall. The aim of this project is to compare both methodologies with regard to maximum safety of constructions and to create an optimized experimentally verified calculation tool for design and assessment.

Methodology of experiments

Material and mechanical properties of the individual supporting elements of the panel and fastening devices are normally verified experimentally. But the coupling is verified due to technologically and economically demanding experiments minimally, mostly necessary parameters are calculated. It was, therefore, necessary to carry out several selected experiments to verify our calculation tool. This paper describes a full-scale experiment for determination of the bending stiffness of a high wall.

The experiments were first performed on a test panels with dimensions 1250 × 7000 mm (presented in this paper), then on larger panels, a total of 5 full scale experiments were performed. The static scheme of the experiment corresponded to a four-point bend as shown in Fig. 1. The experiment was prepared according to standards ČSN 73 2030 a ČSN 73 2035 [5,6]. The test setup can be seen in Fig. 2.

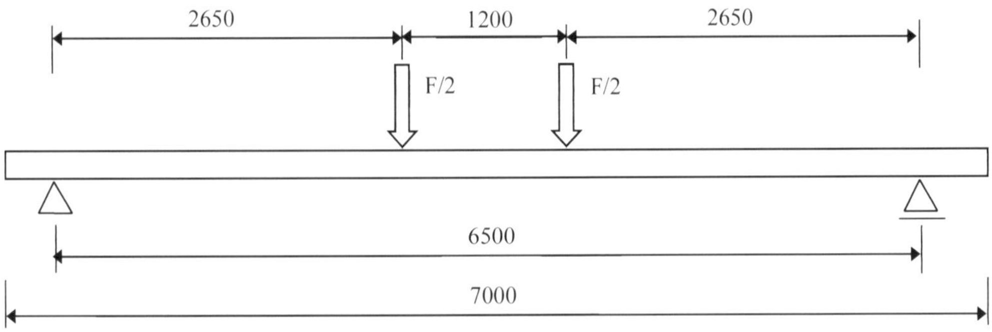

Fig. 1: Static schema of of the bending stiffness of a high wall.

Fig. 2: Set up of experimental verification of the bending stiffness of the wall panel.

Calculation program

The calculation program was created in MS Excel and allows to evaluate the construction of a high wall after entering the basic parameters. The procedure can be divided into 3 phases:

1) input of panel geometry and selection of plasterboard and profile,
2) input load,
3) static calculation and assessment (ultimate and serviceability limit state).

The selection of plasterboard and profile automatically completes the necessary cross-sectional characteristics and other properties from the prepared database. The considered load is a combination of methodologies EN a DIN [2-4].

Evaluation and discussion

The calculation model takes into account the coupling of plasterboard with the profil and the nonlinear behavior of the structure. To correctly determine the effect of coupling, it is necessary to perform experimental tests from which the stiffness of the panel is obtained and then the coefficient of the effect of coupling. Other unknowns are obtained from the working diagram of the load and ultimate strength of the tested element. A nonlinear behavior is derived from these quantities and it is defined in the model as trilinear depending on the ultimate strength of the panel. It should be noted that for high panels, the serviceability limit state, which is located in the first linear part of the diagram, is decisive (Fig. 3).

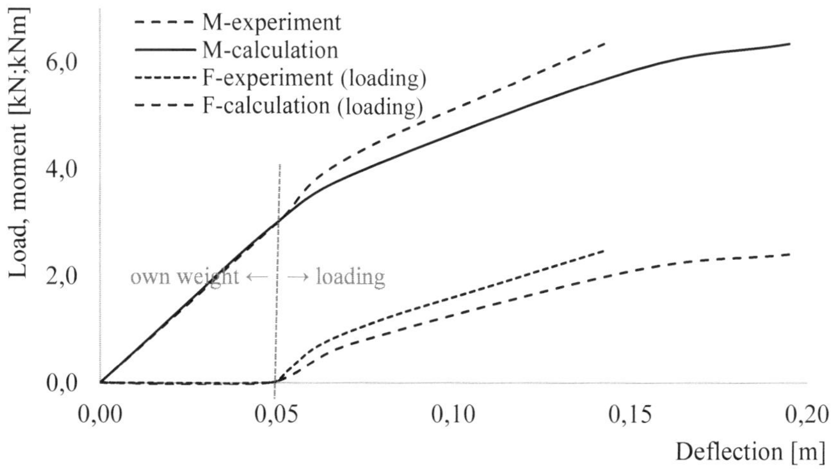

Fig. 3: Comparison of the courses of moments and loads depending on the deflection (average value of experiment vs calculation)

Conclusions

The design and assessment of high walls in the field of dry construction have been optimized based on an analysis of standard used methodologies used in the Czech Republic and Europe (e.g. Germany). Subsequently, a computational program was created, which is verified by selected experiments, such as the presented determination of the bending stiffness of the high wall. Furthermore, these are experiments to verify the frame stiffness of the wall according to ČSN EN 594 [7] and several types of load-bearing capacity of fasteners - ČSN EN 383 a ČSN EN 14 566 [8,9].

Acknowledgement

This paper has been supported by the Technology Agency of the Czech Republic TA ČR, project No. TN01000056 "Centre for Advanced Materials and Efficient Buildings".

References

[1] M. Nyč, Plasterboard, Grada, ISBN 80-247-0986-4, 2005.

[2] ČSN EN 1991-1-1 Eurocode 1: Actions on structures - Part 1-1: General actions - Densities, self-weight, imposed loads for buildings, CNI, Prague, 2004.

[3] ČSN EN 1991-1-4 Eurocode 1: Actions on structures - Part 1-4: General actions - Wind loads, CNI, Prague, 2007.

[4] DIN 4103-1 Internal non-loadbearing partitions - Part 1: Requirements and verification, Berlin, 2015.

[5] ČSN 73 2030 Static loading tests of building structures, CNI, Prague, 2019.

[6] ČSN 73 2035 Impact load testing of building elements, CNI, Prague, 1986.

[7] ČSN 594 Timber structures – Test methods – Racking strength and stiffness of timber frame wall panels, CNI, Prague, 2011.

[8] ČSN 383 Timber structures - Test methods - Determination of embedment strength and foundation values for dowel type fasteners, CNI, Prague, 2007.

[9] ČSN 14566 Mechanical fasteners for gypsum plasterboard systems – Definitions, requirements and test methods, CNI, Prague, 2010.

Index of Authors

EXPERIMENTÁLNÍ
MECHANIKA

GOAL: Design the World's Finest High Force DMA and Fatigue Testing System
NEW ElectroForce DMA3200
Combines decades of cutting-edge fatigue and world-leading Dynamic Mechanical Analysis technologies into a highly versatile testing platform.
• Patented frictionless ElectroForce® Motor
• 500N Force, high frequencies, billions of cycles
• Industry's only 10-year motor warranty
• Wide range of clamping systems and environmental control accessories
• TRIOS data analysis software
TA
tainstruments.com

XSIGHT

STATE OF THE ART STRAIN ANALYSIS SYSTEM FOR EXPERIMENT

Welcome X-Sight strain analysis system in your experiment and promote your lab to the next level.

This Digital Image Correlation based system brings the opportunity to employ a single system for a wide range of application like:

Axial & transversal strain measurement during common tests (tensile, multiaxial, compression, bending and shear)

Strain measurement of micro-samples with telecentric lenses or microscopes (gauge lengths <1mm)

High/Low temperature measurement with climatic chambers and furnaces

Strain monitoring of fatigue tests (free run or force synchronized up to 1kHz)

Angular twist measurement during torsional tests

Post-process strain and displacement distribution maps

www.xsight.eu
info@xsight.eu

Czech Society for Mechanics (CSM)
Dolejskova 5
Praha 8
182 00
Czech Republic

ISBN 978-1-7138-2783-2

9 781713 827832

CURRAN ASSOCIATES INC.
proceedings.com

ICM INTERNATIONAL CONGRESS OF MATHEMATICIANS 2022 JULY 6—14

PRIZE LECTURES

EDITED BY D. BELIAEV AND S. SMIRNOV